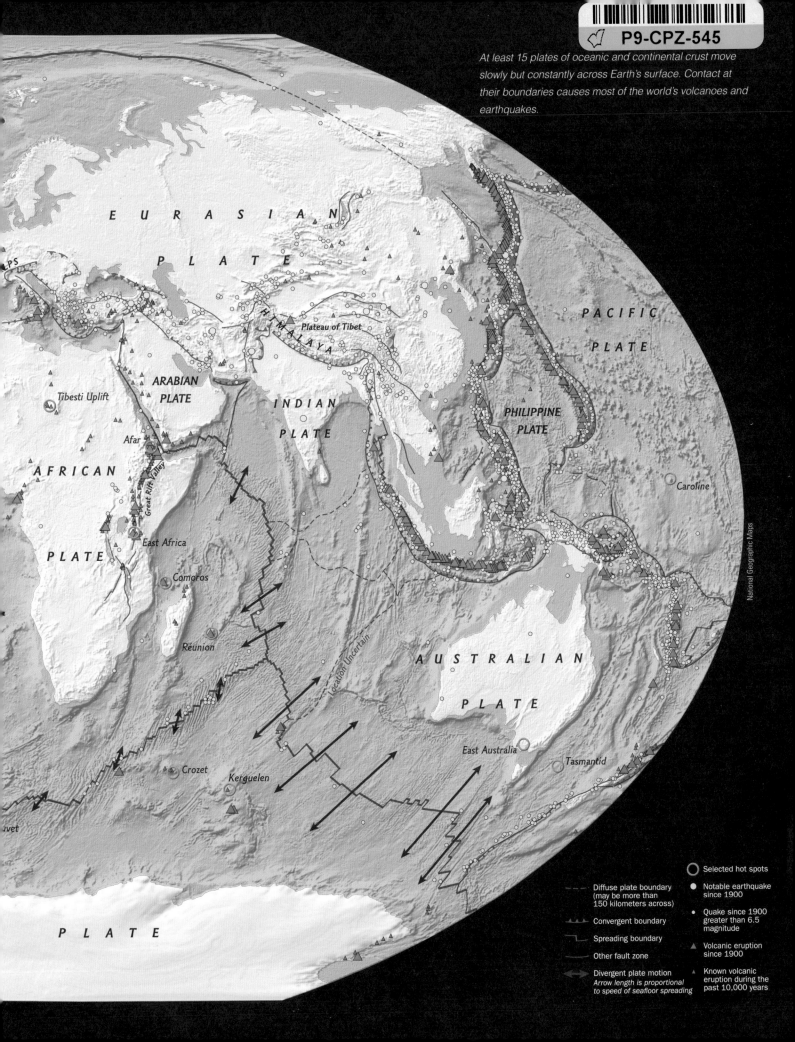

At least 15 plates of oceanic and continental crust move slowly but constantly across Earth's surface. Contact at their boundaries causes most of the world's volcanoes and earthquakes.

EURASIAN PLATE

PACIFIC PLATE

ARABIAN PLATE

Tibesti Uplift

Afar

INDIAN PLATE

HIMALAYA

Plateau of Tibet

PHILIPPINE PLATE

AFRICAN PLATE

Great Rift Valley

East Africa

Comoros

Réunion

Caroline

AUSTRALIAN PLATE

East Australia

Tasmantid

Crozet

Kerguelen

PLATE

National Geographic Maps

⬭ Selected hot spots

○ Notable earthquake since 1900

· Quake since 1900 greater than 6.5 magnitude

▲ Volcanic eruption since 1900

▲ Known volcanic eruption during the past 10,000 years

--- Diffuse plate boundary (may be more than 150 kilometers across)

Convergent boundary

Spreading boundary

Other fault zone

⟷ Divergent plate motion Arrow length is proportional to speed of seafloor spreading

Location Uncertain

Oceanography

An Invitation to Marine Science

9e

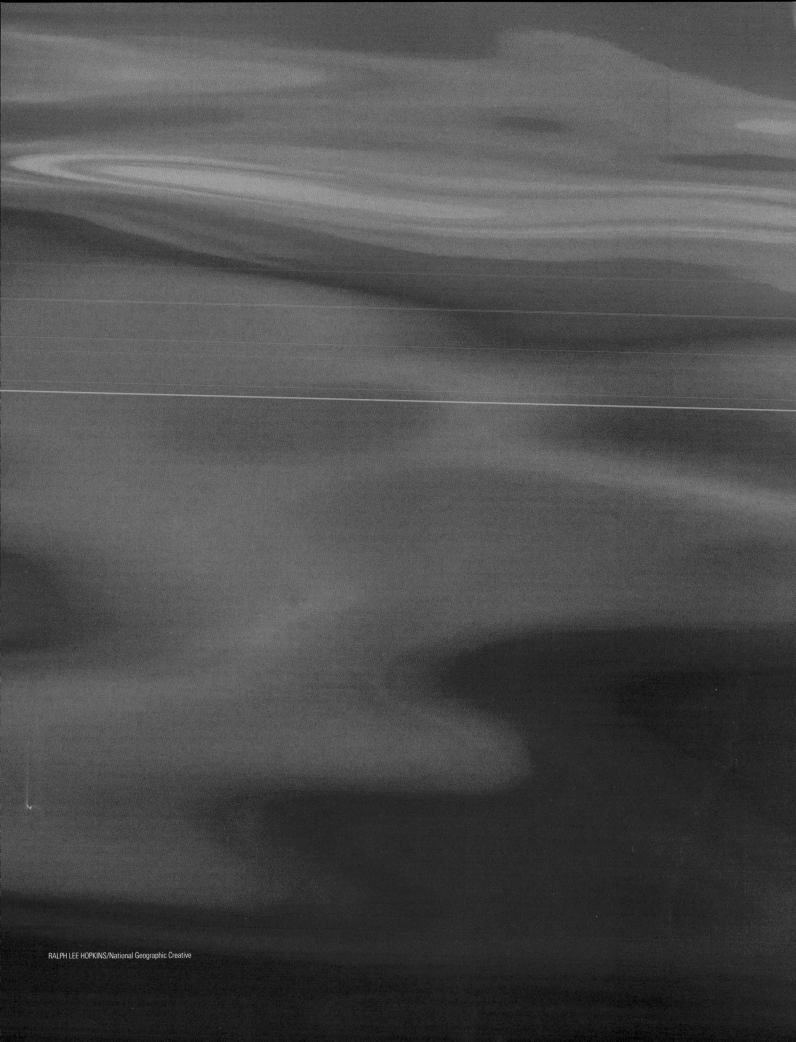

Oceanography

An Invitation to Marine Science 9e

Tom Garrison
Orange Coast College
University of Southern California

Robert Ellis
Orange Coast College

Australia • Brazil • Mexico • Singapore • United Kingdom • United States

Oceanography: An Invitation to Marine Science,
Ninth Edition
Tom Garrison and Robert Ellis

Product Director: Yolanda Cossio

Product Manager: Aileen Berg

Content Developer: Jake Warde

Product Assistant: Victor Luu

Media Developer: Stefanie Chase

Marketing Manager: Julie Schuster

Content Project Manager: Carol Samet

Art Director: Pam Galbreath

Manufacturing Planner: Becky Cross

Production Service: Graphic World Inc.

Photo Researcher: Carly Bergey, PreMedia Global

Text Researcher: Kristine Janssens,
    PreMedia Global

Copy Editor: Graphic World Inc.

Text and Cover Designer: Liz Harasymczuk

Cover Image: ©Willyam Bradberry/
    Shutterstock.com

Compositor: Graphic World Inc.

For product information and technology assistance, contact us at
Cengage Learning Customer & Sales Support, 1-800-354-9706.

For permission to use material from this text or product,
submit all requests online at **www.cengage.com/permissions.**
Further permissions questions can be e-mailed to
permissionrequest@cengage.com.

Library of Congress Control Number: 2014946235

Student Edition:

ISBN-13: 978-1-305-10516-4

ISBN-10: 1-305-10516-8

Loose-leaf Edition:

ISBN-13: 978-1-305-25428-2

ISBN-10: 1-305-25428-7

Cengage Learning
20 Channel Center Street
Boston, MA 02210
USA

Cengage Learning is a leading provider of customized learning solutions with office locations around the globe, including Singapore, the United Kingdom, Australia, Mexico, Brazil, and Japan. Locate your local office at **www.cengage.com/global.**

Cengage Learning products are represented in Canada by Nelson Education, Ltd.

To learn more about Cengage Learning Solutions, visit **www.cengage.com.**

Purchase any of our products at your local college store or at our preferred online store **www.cengagebrain.com.**

Printed in the United States of America
Print Number: 08   Print Year: 2019

To our families and our students,
our hope for the future.

Tom Garrison (Ph.D., University of Southern California) is emeritus professor of Marine Science at Orange Coast College (OCC) in Costa Mesa, California, one of the largest undergraduate marine science departments in the United States. Dr. Garrison also holds an adjunct professorship at the University of Southern California. He has been named the country's Outstanding Marine Educator by the National Marine Technology Society, is a founding member of COSEE, writes a regular column for the journal *Oceanography*, and has written for *National Geographic* magazine. He was a winner of the prestigious Salgo-Noren Foundation Award for Excellence in College Teaching. Dr. Garrison was an Emmy Award team participant as writer and science advisor for the PBS syndicated *Oceanus* television series and writer and science advisor for *The Endless Voyage*, a set of television programs in oceanography completed in 2003. His widely used textbooks in oceanography and marine science are the college market's best sellers. In 2009, the faculty of OCC selected Dr. Garrison as the institution's first Distinguished Professor, and in 2010, he was honored by the Association of Community College Trustees as the outstanding community college professor in western North America.

His interest in the ocean dates from his earliest memories. As he grew up with a U.S. Navy admiral as a dad, the subject was hard to avoid! He had the good fortune to meet great teachers who supported and encouraged this interest. Years as a midshipman and commissioned naval officer continued the marine emphasis; graduate school and 42+ years of teaching have allowed him to pass his oceanic enthusiasm to more than 65,000 students. Although he retired from full-time professoring in 2011, he continues to bother OCC staff and students on a regular basis.

Dr. Garrison travels extensively and most recently served as a guest lecturer at the University of Hong Kong, the University of Tasmania (Australia), and the National University of Singapore. He has been married to an astonishingly patient lady for more than 47 years, has a daughter who teaches in a local public school, a diligent son-in-law, two astonishingly cute granddaughters and a fresh new grandson, and a son who, along with his fashionista wife, works in international trade. He and his family live in and around Newport Beach, California, USA.

Hank Schellengerhaut, Orange Coast College PR department

Robert Ellis (M.E.S.M., University of California, Santa Barbara) has been teaching marine, earth, and environmental science courses in both the classroom and in the field since 2000. He currently serves as Assistant Professor in the Marine Science Department at Orange Coast College in southern California. When not on campus, Robert often helps to develop and teach international field courses in marine science and management in various parts of the Caribbean, Central America, and the South Pacific. His graduate work focused on Marine Resource Management at UC Santa Barbara, and he has participated in and managed research projects and educational programs in many parts of the world. He hopes to have the good fortune to continue to travel and explore the world with his wife, Katie, and son, Kalen.

Robert Ellis

Top photo: Brian J. Skerry/National Geographic Creative

Brief Contents

Contents

16 Marine Communities 454

This book was written to provide an *interesting*, clear, current, and reasonably comprehensive overview of the ocean sciences. It was designed for students who are curious about Earth's largest feature, but who may have little formal background in science.

Students bring a natural enthusiasm to their study of this subject, an enthusiasm that will be greatly enhanced by our partnership with the National Geographic Society. Access to more than 125 years of archival resources makes this National Geographic Learning text uniquely appealing. Even the most indifferent reader will perk up when presented with stories of encounters with huge waves, photos of giant squid, tales of exploration under the best and worst of circumstances, evidence that vast chunks of Earth's surface move slowly, news of Earth's past battering by asteroids, micrographs of glistening diatoms, and data showing the growing economic importance of seafood and marine materials. If pure spectacle is required to generate an initial interest in the study of science, oceanography wins hands down!

In the end, however, it is subtlety that triumphs. Studying the ocean re-instills in us the sense of wonder we all felt as children when we first encountered the natural world. There is much to tell. The story of the ocean is a story of change and chance—its history is written in the rocks, the water, and the genes of the millions of organisms that have evolved here.

The Ninth Edition

Our aim in writing this book was to produce a text that would enhance students' natural enthusiasm for the ocean. Our students have been involved in this book from the very beginning—indeed, it was their request for a readable, engaging, and thorough text that initiated the project a long time ago. Through the many years we have been writing textbooks, our enthusiasm for oceanic knowledge has increased (if that is possible), forcing our patient reviewers and editors to weed out an excessive number of exclamation points. But enthusiasm does shine through. One student reading the final manuscript of an earlier edition commented, "At last, a textbook that does not read like stereo instructions." Good!

This new edition builds on its predecessors. National Geographic resources have been instrumental in the book's focus on the *processes* of science and exploration. Decades of original art, charts and maps, explorers' diaries, data compilations, artifact collections, and historic photographs have been winnowed and included when appropriate. The experience has been exhilarating. Indeed, the National Geographic staff in Washington, D.C., has been *very* patient in tolerating authors whose every other word seemed to be "Wow!"

As before, a great many students have participated alongside professional marine scientists in the writing and reviewing process. In response to their recommendations, as well as those of instructors who have adopted the book and the many specialists and reviewers who contributed suggestions for strengthening the earlier editions, we have:

- **Modified every chapter to reflect current thought and recent research.** New discoveries concerning the establishment of Earth's age, the evolution of its atmosphere, the details of subduction, the sources of ocean water, the maintenance and measurement of salinity, and thermohaline circulation have been incorporated in the text. Recent developments in remote sensing are discussed. Material on the origin, evolution, and extent of life have been updated, as have recent developments in our understanding of oceanic food webs. Recent events are covered: IPCC data on global climate change, the 2010 Gulf of Mexico oil spill, the 2011 earthquake and tsunami in northern Japan, invasive species, coastal development in Dubai, and the ongoing collapse of fisheries.

- **Modified the illustration program** to incorporate National Geographic Society assets. The maps, charts, paintings, and photographs drawn from more than 125 years of Society archives have greatly enhanced the visual program for increased clarity and accuracy.

- **Emphasized the process of science throughout.** The first chapter's discussion of the nature of science has been expanded, and underlying assumptions and limitations are discussed throughout the book. Additional "How Do We Know?" boxes expand on this theme by describing how oceanographers know what they know about the ocean.

- **Added "Insights from an Explorer."** These text boxes highlight the experiences of National Geographic Explorers, men and women whose research has been supported by the National Geographic Society. They are among the top scientists in their respective fields, and their discoveries have significantly expanded our understanding of the ocean sciences.

- **Added new features to encourage active learning and develop critical thinking skills.** Selected figures are accompanied by "Thinking Beyond the Figure." These queries at the ends of the captions guide the readers to investigations of topics related to what they are learning in the chapter. Global Geo-Watch activities have been added to the end-of-chapter material. Each chapter ends with two sets of review questions. The first, "Thinking Critically," invites students to recall specific information covered in the chapter; the second, "Thinking Analytically," challenges students to apply what they have learned to novel situations.

- **Added relevant quotes** in highlighted windows from sources within each chapter and from famous individuals. The popu-

Tom Garrison

lar "Questions from Students" feature has been retained and expanded—these brief discussions address topics of immediate or controversial interest immediately after a chapter.

- **Developed Oceanography MindTap.** MindTap is well beyond an eBook, a homework solution or digital supplement, a resource center Web site, a course delivery platform, or a Learning Management System. MindTap is a new personal learning experience that combines all the digital assets—readings, multimedia, activities, and assessments—into a singular learning path to improve student outcomes.

Ocean Literacy and the Plan of the Book

Ocean literacy is the awareness and understanding of fundamental concepts about the history, functioning, contents, and utilization of the ocean. An ocean-literate person recognizes the influence of the ocean on his or her daily life, can communicate about the ocean in a meaningful way, and is able to make informed and responsible decisions regarding the ocean and its resources. **This book has been designed with ocean literacy guidelines firmly in mind**.

The book's plan is straightforward: Because all matter on Earth except hydrogen and some helium was generated in stars, our story of the ocean starts with stars. Have oceans evolved elsewhere? The history of marine science follows (with additional historical information sprinkled through later chapters). The theories of Earth structure and plate tectonics are presented next, as a base on which to build the explanation of bottom features that follows. A survey of ocean physics and chemistry prepares us for discussions of atmospheric circulation, classical physical oceanography, and coastal processes. Our look at marine biology begins with an overview of the problems and benefits of living in seawater, continues with a discussion of the production and consumption of food, and ends with taxonomic and ecological surveys of marine organisms. The last chapters treat marine resources and environmental concerns.

This icon 🌀 appears when our discussion turns toward the topic of global climate change. Oceanography is central to an understanding of this interesting and controversial set of ideas, so those areas have been expanded, emphasized, and clearly marked in this edition.

Running on the beach, watching waves. If you're 2, could anything be more fun?

Tom Garrison

Tom Garrison

A discerning shopper in a Hong Kong wet market selects the freshest seafood for her family's dinner.

As always in our books, *connections between disciplines* are emphasized throughout. Marine science draws on several fields of study, integrating the work of specialists into a unified whole. For example, a geologist studying the composition of marine sediments on the deep seabed must be aware of the biology and life histories of the organisms in the water above, the chemistry that affects the shells and skeletons of the creatures as they fall to the ocean floor, the physics of particle settling and water density and ocean currents, and the age and underlying geology of the study area. This book is organized to make those connections from the first.

Organization and Pedagogy

A broad view of marine science is presented in 18 chapters, each freestanding (or nearly so) to allow instructors to assign chapters in any order they find appropriate. Each chapter begins with a list of the **five or six most important concepts** to be covered. An engaging chapter opener photo and caption whets the appetite for the material to come.

The chapters are written in an **engaging style**. Terms are defined and principles developed in a straightforward manner. Some of the more complex ideas are initially outlined in broad brushstrokes, and then the same concepts are discussed again in greater depth after the reader has a clear view of the overall situation (a "spiral approach"). When appropriate to their meanings, the derivations of words are shown. **Measurements** are given in both metric (S.I.) and U.S. systems. At the request of a great many students, the units are written out (that is, we write *kilometer* rather than *km*) to avoid ambiguity and for ease of reading.

The photos, charts, graphs, and paintings in the **extensive illustration program** have been chosen for their utility, clarity, and beauty. **Heads and subheads** are usually written as complete sentences for clarity, with the main heads sequentially numbered. A set of **Concept Checks** concludes each chapter's major sections. The answers are provided in the book's dedicated Web site.

Also concluding each chapter is a **Questions from Students** section. These questions are ones that students have asked us over the years. This material is an important extension of the chapters and occasionally contains key words and illustrations. Each chapter ends with an array of study materials for students, beginning with **Chapter in Perspective**, a narrative review of the chapter just concluded. Important **terms and concepts to remember** are listed next; these are also defined in an extensive **glossary** in the back of the book. **Study Questions** are also included in each chapter.

Appendixes explain measurements and conversions, geological time, absolute and relative dating, latitude and longi-

tude, chart projections, taxonomy, and the Law of the Sea. For students interested in joining us in our life's work, the second to last appendix discusses **jobs in marine science**.

The book has been thoroughly **student tested**. You need not feel intimidated by the concepts—this material has been mastered by students just like you. Read slowly and go step by step through any parts that give you trouble. Your predecessors have found the ideas presented here to be useful, inspiring, and applicable to their lives. Best of all, they have found the subject to be *interesting*!

Suggestions for Using This Book

1. **Begin with a preview.** Scout the territory ahead—note the photo and caption that begin each chapter; flip through the assigned pages, reading only the headings and subheadings; look at the figures and read any captions that catch your attention.

2. **Keep a pen and paper handy.** Jot down a few questions—any questions—that this quick glance stimulates. *Why* is the deep ocean cold if the inside of Earth is so hot? *What* makes storm conditions like those seen in the eastern United States in 2014? *Where* did sea salt come from? *Will* climate change actually be a problem? *Does* anybody still hunt whales? *How* do we know how old Earth is? Writing questions will help you focus when you start studying.

3. **Now read in small but concentrated doses.** Each chapter is written in a sequence and tells a story. The logical progression of ideas is going somewhere. Find and follow the organization of the chapter. Stop to read the "Brief Review" sections. Flip back and forth to review and preview.

4. **Strive to be actively engaged!** Write marginal notes, underline occasional passages (underlining whole sections is seldom useful), write more questions, draw on the diagrams, check off subjects as you master them, make flashcards while you read (if you find them helpful), *use your book*!

5. **Monitor your understanding.** If you start at the beginning of the chapter, you will have little trouble understanding the concepts as they unfold. But if you find yourself at the bottom of the page having only scanned (rather than understood) the material, stop there and start that part again. Look ahead to see where we're going. Remember, students just like you have been here before, and we have listened to their comments to make the material as clear as we can. This book was written for you.

6. **Enjoy the journey.** Your instructor and teaching assistants would be glad to share their understanding and appreciation of marine science with you—you have only to ask. Students, instructors, and authors all work together

toward a common goal: an appreciation of the beauty and interrelationships a growing understanding of the ocean can provide.

Identification of intertidal organisms is a pleasant summer morning challenge.

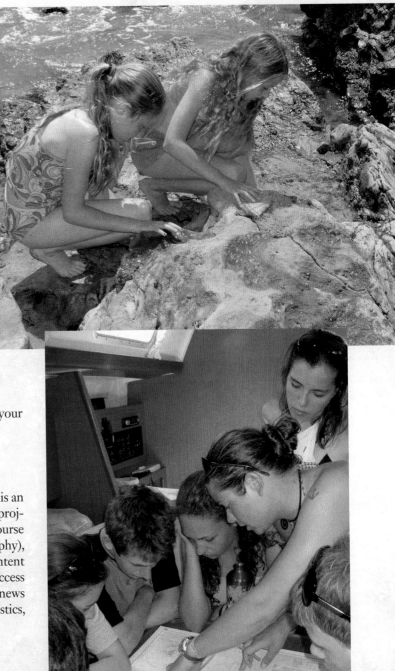
Tom Garrison

Instructor Resources

Instructor Companion Site

Everything you need for your course in one place! This collection of book-specific lecture and class tools is available online via www.cengage.com/login. Access and download PowerPoint presentations, images, instructor's manual, videos, and more.

Cognero Test Bank

Cengage Learning Testing Powered by Cognero is a flexible, online system that allows you to:

- author, edit, and manage test-bank content from multiple Cengage Learning solutions
- create multiple test versions in an instant
- deliver tests from your Learning Management System, your classroom, or wherever you want

Global Geoscience Watch

Updated several times a day, the Global Geoscience Watch is an ideal one-stop site for classroom discussion and research projects for all things geoscience! Broken into the four key course areas (Geography, Geology, Meteorology, and Oceanography), this site makes it easy for you to get to the most relevant content available for your course. You and your students will have access to the latest information from trusted academic sources, news outlets, and magazines. You will also receive access to statistics, primary sources, case studies, podcasts, and much more!

Robert Ellis

A group of students learns navigational techniques before setting sail.

Student Resources

Oceanography MindTap

MindTap is well beyond an eBook, a homework solution or digital supplement, a resource center Web site, a course delivery platform, or a Learning Management System. MindTap is a new personal learning experience that combines all the digital assets—readings, multimedia, activities, and assessments—into a singular learning path to improve learning outcomes.

Acknowledgments

Many years ago, Jack Carey, the grand master of college textbook publishing, willed the first edition of this book into being. His suggestions have been combined with those of more than 1,400 undergraduate students and 190 reviewers to contribute to our continuously growing understanding of marine science. Donald Lovejoy, Stanley Ulanski, Richard Yuretich, Ronald Johnson, John Mylroie, and Steve Lund at the senior author's alma mater, the University of Southern California, deserve special recognition for many years of patient direction. For this edition, we have especially depended on the expert advice of Allen J. Costa, Tidewater Community College–Norfolk Campus; Brent Lewis, Coastal Carolina University; Carrie E. Schweitzer, Kent State University at Stark; Charles Greene, Cornell University; Len Pietrafesa, North Carolina State University & Coastal Carolina University; Ryan P. Mulligan, East Carolina University; Stephanie Schwabe, University of Kentucky; Heather Miller, Grand Valley State University; Randall J. Adsit, East Los Angeles College; Calvin Prothro, Onondaga Community College; Michelle Hardee, Asnuntuck Community College; Mike Valentine, University of Puget Sound; David Gillikin, Union College; Joe Staton, University of South Carolina–Beaufort; Michele Hoffman, Columbia College; Joseph Gorga, Diablo Valley College–San Ramon.

Our long-suffering departmental colleagues Dennis Kelly, Karen Baker, Mary Blasius, Erik Bender, Robert Profeta, and Steven Hatosy again should be awarded medals for putting up with the two of us, answering hundreds of our questions, and being so forbearing through the book's lengthy gestation period. Thanks also to our dean, Robert Mendoza, and our college president, Dennis Harkins, for supporting this project and encouraging our faculty to teach, conduct research, and be involved in community service. Our past and present department teaching assistants deserve praise as well, especially Timothy Heuer, Peter Hernandez, Kevin-Dan Williams, Spencer Wonder, Lili Clark, Belen Cairo, Briana Trevino, Bristol Coon, David Krueger, and Velvet Park.

Yet another round of gold medals should go to our families for being patient (well, *relatively* patient) during those years of days and nights when we were holed up in our respective dark reference-littered caves, throwing chicken bones out the door and listening to *really* loud Telemann and Bach recordings, again working late on The Book. Thank you Marsha, Jeanne, Greg, Grace, Sarah, John, Dinara, Alem, Katie, and Kalen for your love and understanding. The many friends and colleagues that we have bounced ideas off deserve special recognition, including Jenell Schwab, Mary Arbogast, Joana Tavares-Reager, Chris Krajacic, Sarah Sikich, and Andy Balendy.

The people who provided pictures and drawings have worked miracles to obtain the remarkable images in these pages. To mention just a few: Gerald Können allowed us to use his extraordinary image of a broken rainbow to illustrate seawater's index of refraction; Gerald Kuhn sent classics taken by his late SIO colleague Francis Shepard; Vincent Courtillot of the University of Paris contributed the remarkable photo of the Aden Rift; Catherine Devine at Cornell provided time-lapse graphics of tsunami propagation; Robert Headland of the Scott Polar Research Institute in Cambridge searched out prints of polar subjects; Charles Hollister at Woods Hole kindly provided seafloor photos from his important books; Andreas Rechnitzer and Don Walsh recalled their exciting days with *Trieste;* and Bruce Hall, Pat Mason, Ron Romanoski, Ted Delaca, William Cochlan, Christopher Ralling, Mark McMahon, John Shelton, Alistair Black, Howard Spero, Eric Bender, Ken-ichi Inoue, and Norman Cole contributed beautiful slides. Seran Gibbard provided the highest-resolution images yet made of the surface of Titan, and Michael Malin forwarded truly beautiful images of erosion on Mars. Herbert Kawainui Kane again allowed us to reprint his magnificent paintings of Hawai'ian subjects. Deborah Day and Cindy Clark at Scripps Institution, Jutta Voss-Diestelkamp at the Alfred Wegener Institut in Bremerhaven, and David Taylor at the Centre for Maritime Research in Greenwich dug through their archives one more time. Don Dixon, William Hartmann, Ron Miller, and William Kaufmann provided paintings, Dan Burton sent photos, and Andrew Goodwillie printed customized charts. Bryndís Brandsdóttir of the Science Institute, University of Iceland, patiently showed me the jaw-slackening Thingvillir rift. Wim van Egmond contributed striking photomicrographs of diatoms, forams, and copepods. Kim Fulton-Bennett of MBARI found extraordinarily beautiful photos of delicate midwater animals. Peter Ramsay at Marine Geosolutions, Ltd., of South Africa, sent state-of-the-art side-scan sonar images. Michael Boss kindly contributed his images of Admiral Zheng He's astonishing *beochuan.* Bill Haxby at Lamont provided truly beautiful seabed scans. Karen Riedel helped with DSDP core images. James Ingle offered a desk and breathing room at Stanford whenever it was needed. NOAA, JOI, NASA, USGS, the Smithsonian Institution, the Royal Geographical Society, the U.S. Navy, and the U.S. Coast Guard came through time and again, as did private organizations like Alcoa Aluminum, Cunard, Shell Oil, The Maersk Line, Grumman Aviation, Breitling-SA, CNN, Associated Press, MobileEdge, and the *Los Angeles Times*. The Woods Hole team was also generous—especially Robin Hurst, Jack Cook, Larry Madin, and Ruth Curry. Thanks also to WHOI researchers Philip Richardson, William Schmitz, Susumu Honjo, Doug Webb, James Broda, Albert Bradley, John Waterbury, and Kathy Patterson, who all provided photographs, diagrams, and advice. Individuals with special expertise have also been willing to share: Hank Brandli processed satellite digital images of storms, Peter Sloss at the National Geophysical Data Center helped me sort through computer-generated seabed images, Steven Grand of the University of Texas provided a descending deep-slab image, Hans-Peter Bunge of Princeton patiently explained mantle-core dynamics, Michael Gentry again mined the archives of the Johnson Space Center for Earth images, Jurrie van der Woulde at JPL and Gene Feldman at NASA helped with images of oceans here and elsewhere, John Maxtone-Graham of New York's Seaport Museum found me a rogue wave picture, Ed Ricketts, Jr., contributed a portrait of his father, and professor Lynton Land of the University of Texas sent a rare photo of a turbidity current. Michael Latz at Scripps Institution taught me about bioluminescence. Thomas Maher, retired vice-provost and friend, led the senior author and his son on a personal inspection of the Gulf Stream and other fluid wonders. Dr. Wyss Yim of Hong Kong University offered suggestions and references (as well as unending hospitality and dim sum), and Dr. Shouye Yang of Tongji University in Shanghai graciously explained his research and shared plans for China's expansion into the field of oceanography. Tommi Lahtonen sent images of a Norwegian maelstrom. Kim Fulton-Bennett of MBARI shared some astonishing photos of midwater organisms. Neil Holbrook at Australia's University of Tasmania taught us about Sydney's Hawkesbury sandstone and bagpipes simultaneously. Dave Sandwell at Scripps shared his astonishing satellite-generated imagery of the seabed. Rick Grigg at the University of Hawai'i encouraged us to tackle some tricky bits of wave physics. Dr. Wilhelm Weinrebe of GEOMAR in Kiel, Germany, arranged for the use of bathymetric images of unprecedented resolution and clarity. Ulrke Schulte-Rahde at L-3Com sent images of the latest side-scan sonar installations. Dr. Steve Hatosy at UCI provided training in marine microbiology. Ruth Curry at WHOI added to the senior author's understanding of ocean circulation. The staff of The Viking Ship Museum and The *Fram*

Father, son, ocean—learning marine science is a joy at any age.

Tom Garrison

Despite a severe California drought, these supratidal plants are sustained by heavy morning fogs.

Tom Garrison

A tourist photographs the steerboard of a restored Viking longship.

The Cengage/Brooks-Cole team performed the customary miracles. The charge was led by Jake Warde, whose patience, understanding, and panic-proof demeanor was a constant model and inspiration. The text was polished by the late Mary Arbogast, a good friend and the best text editor in the Orion arm of the galaxy (and who will be *greatly* missed), and Marcie McGuire, the copy editor who saved us from many errors. Carley Bergey worked tirelessly to assist in photo research and permissions, and Dan Fitzgerald and Carol Samet were in charge of production. Stefanie Chase developed the content for Oceanography MindTap. The amazing Yolanda Cossio and our always-upbeat editor, Aileen Berg, kept us all running in the same direction. What skill!

Our unending thanks to all.

A Goal and a Gift

The goal of all this effort: *To allow you to gain an oceanic perspective.* "Perspective" means being able to view things in terms of their relative importance or relationship to one another. An oceanic perspective lets you see this misnamed planet in a new light and helps you plan for its future. You will see that water, continents, seafloors, sunlight, storms, seaweeds, and society are connected in subtle and beautiful ways.

The ocean's greatest gift to humanity is intellectual—the constant challenge its restless mass presents. Let yourself be swept into this book and the class it accompanies. Give yourself time to ponder: "Meditation and water are wedded forever," wrote Herman Melville in *Moby Dick*. Take pleasure in the natural world. Ask questions of your instructors and TAs, read some of the references, and try your hand at the questions at the ends of the chapters.

Though you will find discouraging news in the last chapters, be optimistic. Take pleasure in the natural world. Please write to us when you find errors or if you have comments. Above all, *enjoy yourself.* Learning something new is a permanent pleasure!

Tom Garrison
Orange Coast College
University of Southern California
tomgarrison@sbcglobal.net

Robert Ellis
Orange Coast College
rellis@occ.cccd.edu

Museum in Oslo were kind in allowing access to their magnificent ships. Liping Zhou made encouraging contributions from Peking University, Beijing. Adam Spitzer at Nanyang Technical University and Pavel Tkalich and Soo Chin Liew at the National University of Singapore were gracious and patient hosts. Gray Williams provided an occasional home base (and welcome cold beer) at Hong Kong University's magnificently sited Swire Institute of Marine Science. Without their inestimable goodwill, a project like this would not be possible.

The National Geographic team was understanding of our needs and deadlines. Erin West deserves a medal for her work with the Society's Explorers and Grantees mentioned in the text. The photo and image archive staff, led by Bill Bonner, opened the magic door and allowed these overwhelmed writer-oceanographers unfettered access. Maureen Flynn and James McClelland of the National Geographic Maps provided updates to many of the spectacular National Geographic maps and graphics in the text.

Tom Garrison

Foreword

by Dr. Enric Sala

Rebecca Hale/National Geographic Creative

Dr. Enric Sala
Marine Ecologist
National Geographic Explorer-in-Residence

Imagine that an alien were visiting Earth for the first time to explore its landscape and inhabitants. It would have a 72% chance to "land" the spaceship on the ocean, because this is how much of our planet is ocean. But if the alien asked us about this largest feature of our planet, there would be huge gaps in our description. We have more detailed maps of the surface of the moon than of the bottom of the ocean, and we have explored in detail less than 5% of the ocean. Unfortunately, the technology that allows humans to descend into the deep only became available long after we had begun to degrade the ocean. Our understanding of ocean life is a by-product of studying degraded ecosystems, like trying to understand how a car functions by studying a car wreck in a junkyard.

Despite these shortcomings, we have learned enough about the ocean to know that it regulates our climate, is vital for the water cycle that creates the rain that waters our forests and fields, produces more than half of the oxygen we breathe, absorbs more than a quarter of the carbon dioxide we put in the atmosphere, gives us almost 100 million metric tonnes of seafood every year, and provides jobs and livelihoods for hundreds of millions of people worldwide. Despite everything the ocean gives us for free, we seem determined to degrade marine life, by taking out of the ocean what we like (seafood) and throwing in what we don't want (pollution, carbon dioxide, excess heat). These insults reduce the capacity of the ocean to provide all these goods and services that are essential to our well-being.

To find solutions to these problems, we need to start by reviewing what we know about the ocean. And I cannot think of a better way than through Tom Garrison's *Oceanography*. Professor Garrison does a terrific job showing why there is so much water in the ocean, why it is salty, why there are ocean currents, and how currents influence the climate. Most importantly, this book frames oceanography in the context of global change and shows clearly how to distinguish between natural changes and human impacts. If I gave that alien explorer an instruction manual for the ocean to take to his planet, I would give him this book.

1 The Origin of the Ocean

KEY CONCEPTS

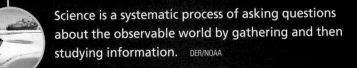

Science is a systematic process of asking questions about the observable world by gathering and then studying information. DER/NOAA

The universe's observable mass consists mostly of hydrogen atoms. The heavy elements we see around us were constructed in stars. Rogelio Bernal Andreo/NASA Images

Earth is density stratified—that is, as Earth formed, gravity pulled the heaviest materials (iron, nickel) to its center as lighter minerals rose to the surface. Earth's first solid surface formed about 4.6 billion years ago. © Cengage Learning

Life probably originated in the ocean shortly after it formed. © Cengage Learning

Water, even liquid water, appears to be present in other places in our solar system. Tom Garrison

An artist imagines an Earth-like planet orbiting a distant star. Water worlds may not be rare in the universe, but we know of only one: the beautiful blue sphere we call home.
NASA/Ames/JPL-Caltech

Media Connection

Start off this chapter by listening to a podcast featuring National Geographic Explorer Kevin Hand as he discusses Europa and what he hopes to find on Jupiter's fourth largest moon. Visit www.cengagebrain.com to access MindTap, a complete digital course that includes this podcast and other resources.

1.1 Earth Is an Ocean World

Think of oceanography as the story of the ocean. In this first chapter, the main character—the world ocean—is introduced in broad brushstrokes. We begin our investigation of the ocean with an overview of the process of science and then look at the long and often surprising story of how the ocean came to be.

Imagine, for a moment, that you had never seen this place—this ocean world—this poorly named Earth. As worlds go, you would surely find this one singularly beautiful and exceptionally rare. But the sun warming its surface is not rare—there are billions of similar stars in our home galaxy. The atoms that compose Earth are not rare—every kind of atom known here is found in endless quantity in the nearby universe. The water that makes our home planet shine a gleaming blue from a distance is not rare—there is much more water on our neighboring planets. The fact of the seasons, the free-flowing atmosphere, the daily sunrise and sunset, the rocky ground, the changes with the passage of time—none is rare.

What *is* extraordinary is a happy combination of circumstances. Our planet's orbit is roughly circular around a stable star. Earth is large enough to hold an atmosphere, but not so large that its gravity would overwhelm. Its neighborhood is tranquil—supernovae have not seared its surface with radiation. Our planet generates enough warmth to recycle its interior and generate the raw materials of atmosphere and ocean but is not so hot that lava fills vast lowlands or roasts complex molecules. Best of all, our distance from the sun allows Earth's abundant surface water to exist in the liquid state. Ours is an ocean world (**Figure 1.1 and Table 1.1**).

The **ocean**[1] may be defined as the vast body of saline water that occupies the depressions of Earth's surface. More than 97% of the water on or near Earth's surface is contained in the ocean;

---

[1]When an important new term is introduced and defined, it is printed in boldface type. These terms are listed at the end of the chapter and defined in the Glossary.

Table 1.1 Some Statistics for the World Ocean

- Total area: 331,441,932 square kilometers (127,970,445 square miles)
- Total volume: 1,303,155,354 cubic kilometers (312,643,596 cubic miles)
- Total mass: 1.41 billion billion metric tons (1.55 billion billion tons)
- Average depth: 3,682 meters (12,081 feet)
- Greatest depth: 10,994 meters (36,070 feet)
- Mean ocean crust thickness: 6.5 kilometers (4.04 miles)
- Average temperature: 3.9°Celsius (39.0°Fahrenheit)
- Average salinity: 34,482 grams per kilogram (0.56 ounces per pound); 3.4%
- Average elevation of land: 840 meters (2,772 feet)
- Age: 4.5 billion years
- Future: Uncertain

© Cengage Learning

Figure 1.1 Dominating Earth's surface is a single great ocean of liquid water. This ocean moderates temperature and dramatically influences weather. The dry land on which nearly all of human history has unfolded is hardly visible from space, for nearly three quarters of the planet is covered by water. In this 2011 photograph from NASA's *Aqua* satellite, North America's wrinkled western edge is seen beneath a thin veil of atmosphere. Strong easterly winds blow dust into the vast Pacific. *Oceanus* would surely be a better name for our watery home.

NASA Earth Observatory

about 2.5% is held in land ice, groundwater, and all the freshwater lakes and rivers. If all Earth's surface water were gathered into a sphere, its diameter would measure only 1,380 kilometers (860 miles) **(Figure 1.2)**.

Traditionally, we have divided the ocean into artificial compartments called *oceans* and *seas*, using the boundaries of continents and imaginary lines such as the equator. In fact, the ocean has few dependable natural divisions, only one great mass of water. The Pacific and Atlantic oceans, the Mediterranean and Baltic seas, so named for our convenience, are in reality only temporary features of a single **world ocean**. In this book we refer to the ocean *as a single entity*, with subtly different characteristics at different locations but with very few natural partitions. Such a view emphasizes the interdependence of ocean and land, life and water, atmospheric and oceanic circulation, and natural and human-made environments.

On a *human* scale, the ocean is impressively large—it covers 331 million square kilometers (128 million square miles) of Earth's surface.[2] The average depth of the ocean is about 3,682 meters (12,081 feet); the volume of seawater is 1.3 billion cubic kilometers (312 million cubic miles); the average temperature a cool 3.9°C (39°F). Its mass is a staggering 1.41 billion *billion* metric tons. If Earth's contours were leveled to a smooth ball, the ocean would cover it to a depth of 2,686 meters (8,810 feet). The average land elevation is only 840 meters (2,772 feet), but the average ocean depth is 4½ times

---

[2]Throughout this book, SI (metric) measurements precede American measurements. For a quick review of SI units and their abbreviations, please see Appendix 1.

The ocean has few dependable natural divisions, only one great mass of water. The Pacific and Atlantic oceans, the Mediterranean and Baltic seas, so named for our convenience, are in reality only temporary features of a single world ocean.

as great! The ocean borders most of Earth's largest cities—nearly half of the planet's 7 billion human inhabitants live within 240 kilometers (150 miles) of a coastline.

On a *planetary* scale, however, the ocean is insignificant. Its average depth is a tiny fraction of Earth's radius—the blue ink representing the ocean on an 8-inch paper globe is proportionally thicker. The ocean accounts for only slightly more than 0.02% of Earth's mass, or 0.13% of its volume. Much more water is trapped within Earth's hot interior than exists in its ocean and atmosphere.

CONCEPT CHECK

Before going on to the next section, check your understanding of some of the important ideas presented so far:

Why did we write that there is *one* world ocean? What about the Pacific and Atlantic oceans, the "Seven Seas"?

Which is greater: the average depth of the ocean or the average height of the continents above sea level?

Is most of Earth's water in the ocean?

Figure 1.2 The relative amount of water in various locations on or near Earth's surface. More than 97% of the water lies in the ocean. If all water at Earth's surface were gathered into a sphere, its diameter would measure only 1,380 kilometers (860 miles).

THINKING BEYOND THE FIGURE

Given the predominance of water at Earth's surface, why do you think there is any dry land at all?

a. Kevin Hand (JPL/Caltech), Jack Cook (Woods Hole Oceanographic Institution), Howard Perlman (USGS)/NASA Images

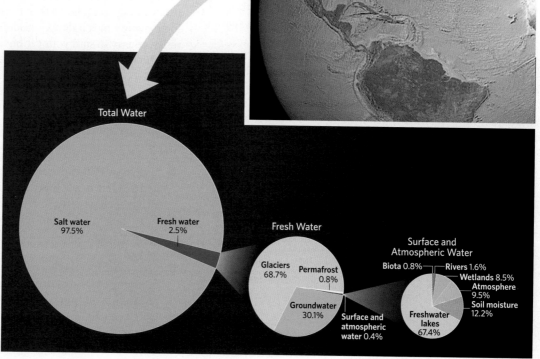
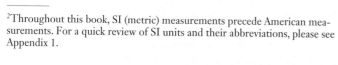

Total Water

Salt water 97.5% Fresh water 2.5%

Fresh Water

Glaciers 68.7% Permafrost 0.8% Groundwater 30.1% Surface and atmospheric water 0.4%

Surface and Atmospheric Water

Biota 0.8% Rivers 1.6% Wetlands 8.5% Atmosphere 9.5% Soil moisture 12.2% Freshwater lakes 67.4%

© Cengage Learning

1.2 Marine Scientists Use the Logic of Science to Study the Ocean

Marine science (or **oceanography**) is the process of discovering unifying principles in data obtained from the ocean, its associated life-forms, and its bordering lands. Marine science draws on several disciplines, integrating the fields of geology, physics, biology, chemistry, and engineering as they apply to the ocean and its surroundings. Nearly all marine scientists specialize in one area of research, but they also must be familiar with related specialties and appreciate the linkages between them.

- *Marine geologists* focus on questions such as the composition of inner Earth, the mobility of the crust, the characteristics of seafloor sediments, and the history of Earth's ocean, continents, and climate. Some of their work touches on areas of intense scientific and public concern, including earthquake prediction and the distribution of valuable resources.
- *Physical oceanographers* study and observe wave dynamics, currents, and ocean–atmosphere interaction.
- *Chemical oceanographers* study the ocean's dissolved solids and gases and their relationships to the geology and biology of the ocean as a whole.
- *Climate specialists* investigate the ocean's role in Earth's changing climate. Their predictions of long-term climate trends are becoming increasingly important as pollutants change Earth's atmosphere.
- *Marine biologists* work with the nature and distribution of marine organisms, the impact of oceanic and atmospheric pollutants on the organisms, the isolation of disease-fighting drugs from marine species, and the yields of fisheries.
- *Marine engineers* design and build oil platforms, ships, harbors, and other structures that enable us to use the ocean wisely.

Other marine specialists study the techniques of weather forecasting, ways to increase the safety of navigation, methods to generate electricity, and much more. **Figure 1.3** shows marine scientists in action.[3]

Marine scientists today are asking some critical questions about the origin of the ocean, the age of its basins, and the nature of the life-forms it has nurtured. We are fortunate to live at a time when scientific study may be able to answer some of those questions. **Science** is a systematic *process* of asking questions about the observable world by gathering and then studying information (data), but the information by itself is not science. Science *interprets* raw information by constructing a general explanation with which the information is compatible.

Scientists start with a question—a desire to understand something they have observed or measured. They then form a tentative explanation for the observation or measurement. This explanation is often called a working **hypothesis**, a speculation about the natural world that can be tested and verified or disproved by further observations and controlled experiments. (An **experiment** is a test that simplifies observation in nature or in

the laboratory by manipulating or controlling the conditions under which the observations are made.) Hypotheses consistently supported by observation, experiment, or historical exploration often evolve to become a **theory**, a statement that explains the observations.

Comprehensive constructs, known as **laws**, can also summarize experimental observations. Laws are principles explaining events in nature that have been observed to occur with unvarying uniformity under the same conditions. A law usually takes the form of a concise mathematical or verbal expression; a theory provides an *explanation* for the observations. *One is not "more true" than the other–both a law and a theory can be statements of facts.*

Theories and laws in science do not arise fully formed or all at once. Scientific thought progresses as a continuous chain of questioning, testing, and matching theories to observations. A theory is strengthened if new facts support it. If not, the theory is modified or a new explanation is sought (science is thus "self-correcting"). The power of science lies in its ability to operate *in reverse*; that is, in the use of a theory or law to predict and anticipate new facts to be observed.

This procedure, often called the **scientific method**, is an orderly process by which theories are verified or rejected. The scientific method rests upon the assumption that nature "plays fair"—that the rules governing natural phenomena do not change capriciously as our powers of questioning and observing improve. We believe that the answers to our questions about nature are *ultimately knowable*.

There is no one scientific method. Some researchers observe, describe, and report on some subject and leave it to others to hypothesize. Scientists don't have one single method in common—the general method they employ is a critical attitude about being *shown* rather than being *told*, and taking a logical approach to problem solving. The process is circular and collaborative—new theories and laws always suggest new questions. See **Figure 1.4**.

You've heard of the scientific method before but may have thought that scientific thinking was beyond your interest or ability. Nothing could be further from the truth—you use scientific logic many times a day. Consider your line of thinking if, later today, you try to start your car but are met only with silence. Your first thoughts (after the frustration subsided) would likely be these:

1. So! The car won't start!
2. *Why* won't the car start? (That second thought—*why*—is a very powerful bit of Western philosophy. Its implication: The car won't start for a *reason*, and that reason is *knowable*.)

You immediately begin to conduct a set of mental experiments:

3. You know that cars need electricity to start. You turn on the lights. They work. Electricity is present. The problem is not a lack of electricity.
4. Cars need air to combine with fuel in the engine. Is air present? You take a breath. Air? Yes. The problem is not lack of air.
5. Cars need fuel. Is there fuel? You turn on the ignition. The fuel gauge registers three-fourths full. (You also notice a fuel receipt in your pocket from yesterday.) Yes, there's fuel.
6. Cars need all of these things to be present *simultaneously* in order to start. You open the hood to look for loose wires or hoses interrupting flow. *AHA!* A wire is loose.

---

[3]Would you like to join us? Appendix 8 discusses careers in the marine sciences.

Robert Ellis

a A student research team attempts to identify a humpback whale by comparing its unique fluke pattern to previously cataloged individuals.

Figure 1.3 Doing marine science is sometimes anxious, sometimes routine, and always interesting.

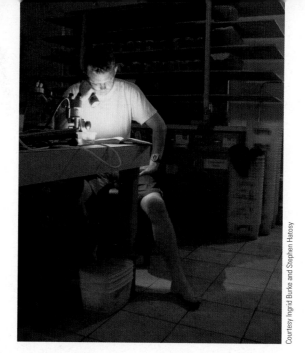

Courtesy Ingrid Burke and Stephen Hatosy

b Quiet, thoughtful study comes before an experiment is begun and after the data are obtained. A student works with a flashlight on a lab report during a power outage at the University of California's Moorea Research Station in the South Pacific.

c Sometimes marine scientists come up with an unanticipated surprise. Fortunately this marine worm is very, very small.

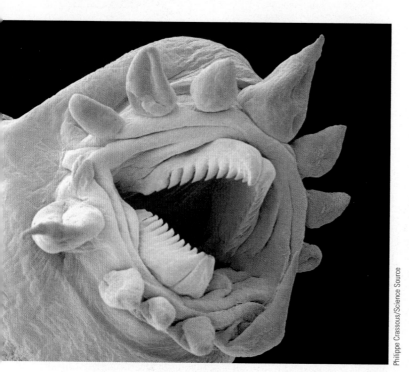

Philippe Crassous/Science Source

7 You put the wire back into place.

8 The car starts! Science wins! The question "why" is answered!

Or you could pursue an alternative line of thinking: You could decide that the spirits of car starting have somehow turned against you. Once you lose their confidence, your power over cars is greatly diminished, and you will almost certainly never be able to drive again. Maybe if you shake your keys over the hood of the car, the spirits will look favorably on you and the car might start, but you can't possibly fix anything yourself—these things are out of your hands. Your relationship with cars is over. (This line of reasoning is not very productive!)

Although clearly powerful in its implications and applications, nothing is ever shown to be *irrevocably* true by the scientific method. Still, the mechanism of science has provided durable, valuable conclusions that have withstood the test of time and immeasurably improved our lives. It is the best tool we have for exploring the natural world. Note that *science is neither a democratic process nor a popularity contest.* As we can sense from the current acrimonious debates over global climate change or even evolution, conclusions about the natural world that we reach by scientific process may not always be comfortable, easily understood, or immediately embraced. But if those conclusions consistently match observations, they may be considered true.

This textbook shows some of the results of the scientific process as they have been applied to the world ocean. It presents facts, interpretations of facts, examples, stories, and some of the crucial discoveries that have led to our present understanding of the ocean and the world on which it formed. As the results of science change, so will the ideas and interpretations presented in books like this one. As you read the chapters to come, you will see examples of scientific thought in boxes labelled: "How Do We Know. . .?"

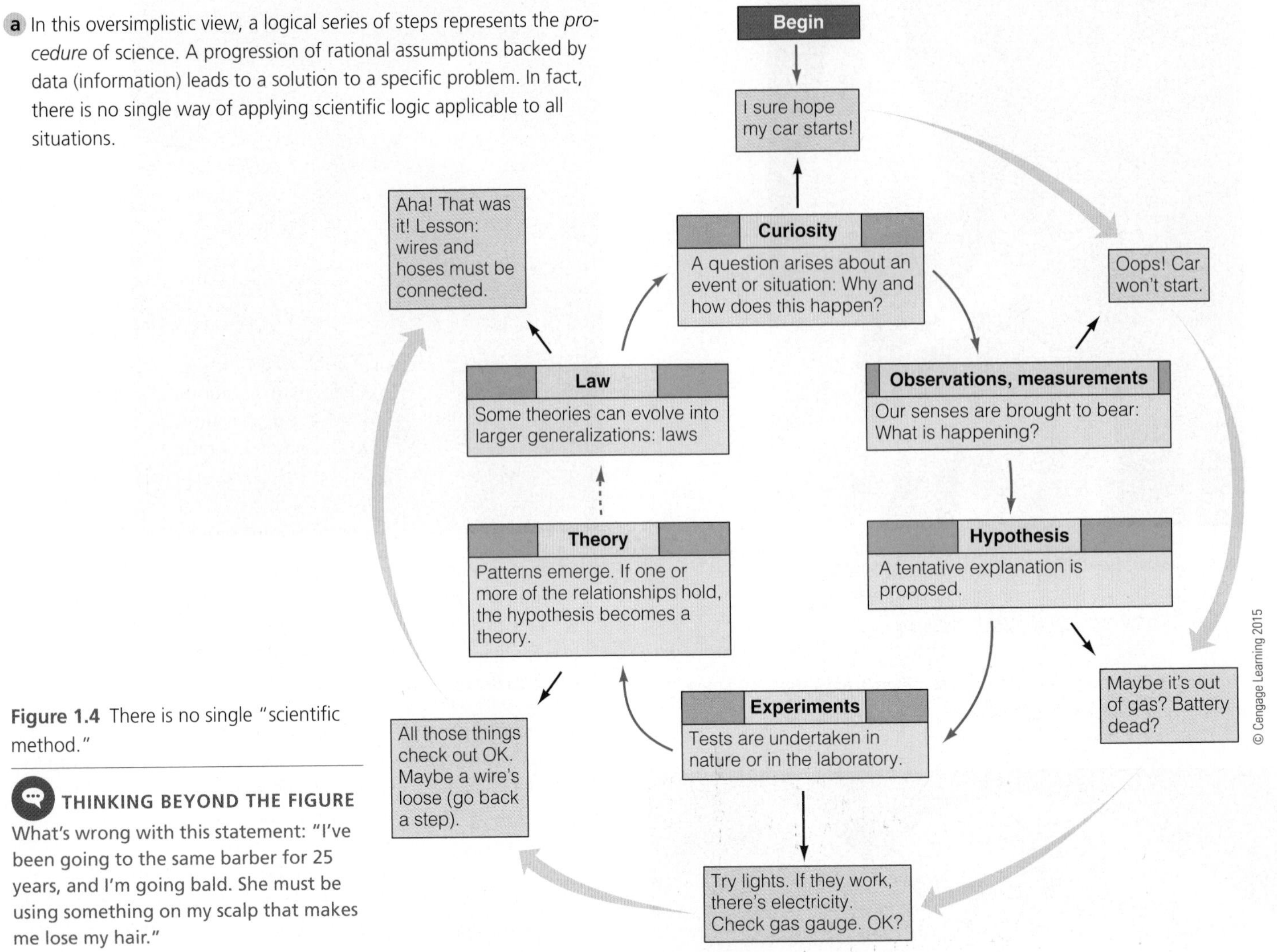

 a In this oversimplistic view, a logical series of steps represents the *procedure* of science. A progression of rational assumptions backed by data (information) leads to a solution to a specific problem. In fact, there is no single way of applying scientific logic applicable to all situations.

Begin

I sure hope my car starts!

Curiosity
A question arises about an event or situation: Why and how does this happen?

Oops! Car won't start.

Aha! That was it! Lesson: wires and hoses must be connected.

Law
Some theories can evolve into larger generalizations: laws

Observations, measurements
Our senses are brought to bear: What is happening?

Theory
Patterns emerge. If one or more of the relationships hold, the hypothesis becomes a theory.

Hypothesis
A tentative explanation is proposed.

Maybe it's out of gas? Battery dead?

All those things check out OK. Maybe a wire's loose (go back a step).

Experiments
Tests are undertaken in nature or in the laboratory.

Try lights. If they work, there's electricity. Check gas gauge. OK?

© Cengage Learning 2015

Figure 1.4 There is no single "scientific method."

💬 **THINKING BEYOND THE FIGURE**

What's wrong with this statement: "I've been going to the same barber for 25 years, and I'm going bald. She must be using something on my scalp that makes me lose my hair."

CONCEPT CHECK

Before going on to the next section, check your understanding of some of the important ideas presented so far:

Can the scientific method be applied to speculations about the natural world that are not subject to test or observation?

What is the nature of "truth" in science? Can anything be proven *absolutely* true?

What if, at the moment you shake the keys, the wires under the hood are jostled by a breeze and fall back into place? What if the car starts when you try it again? Can you see how superstition might arise?

1.3 Stars Form Seas

To understand the ocean, we need to understand how it formed and evolved through time. Since the world ocean is the largest feature of Earth's surface, it should not be surprising that we believe the origin of the ocean is linked to Earth's origin. The origin of Earth is linked to that of the solar system and the galaxies.

Theories may change as our knowledge and powers of observation change.

The formation of Earth and ocean is a long and wonderful story—one we've only recently begun to know. As you continue reading this chapter, you may be startled to discover that most of the atoms that make up Earth, its ocean, and its inhabitants were formed within stars billions of years ago. Stars spend their lives changing hydrogen and helium to heavier elements. As they die, some stars eject these elements into space during cataclysmic explosions. The sun and the planets, including Earth, condensed from a cloud of dust and gas enriched by the recycled remnants of exploded stars.

Our ocean did not come directly from that cloud, however. Most of the ocean formed later, as water vapor trapped in Earth's outer layers escaped to the surface through volcanic activity during the planet's youth. The vapor cooled and condensed to form an ocean. Comets may have delivered additional water to the new planet's surface. Life originated in the ocean soon after, developing and flourishing in the nurturing ocean for more than 3 billion years before venturing onto the unwelcoming continents.

Tom Garrison

b The underlying method of science describes an attitude. Scientists like to be shown why an idea is correct, rather than being told. All science is a work in progress, never completed. The external world, not internal conviction, must be the testing ground for scientific beliefs. Here, marine scientists are planning an experiment to better understand how small intertidal snails withstand the high temperatures of their tropical environment. They have a hypothesis and will design experimental steps to resolve it.

Stars Formed Early in the History of the Universe

The universe apparently had a beginning. The **big bang**, as that event is modestly named, occurred about 13.7 billion years ago. All of the mass and energy of the universe is thought to have been concentrated at a geometric point at the beginning of space and time, the moment when the expansion of the universe began. We don't know what initiated the expansion, but it continues today and will probably continue for billions of years, perhaps forever.

The very early universe was unimaginably hot, but as it expanded, it cooled. About a million years after the big bang, temperatures fell enough to permit the formation of atoms from the energy and particles that had predominated up to that time. Most of these atoms were hydrogen, then as now the most abundant form of matter in the universe. About a billion years after the big bang, this matter began to congeal into the first galaxies and stars.

Stars and Planets Are Contained within Galaxies

A **galaxy** is a huge, rotating aggregation of stars, dust, gas, and other debris held together by gravity. Our galaxy **(Figure 1.5)** is named the **Milky Way galaxy** (from the Greek *galaktos*, which means "milk").[4]

---

[4]Because they can be useful, as well as interesting, the derivations of words are sometimes included in the text.

Figure 1.5 A brilliant laser points toward the center of our Milky Way galaxy. (The beam is used to monitor conditions in Earth's upper atmosphere to provide a clearer image of the distant stars.) Analysis of these images suggests our home galaxy contains between 100 and 400 billion stars and is about 120,000 light-years in diameter. Our solar system lies about 27,000 light-years from the galactic center in a concentration of dust and gas called the Orion-Cygnus Arm. (A light-year is the distance light travels in one year: about 9.5 trillion kilometers or 6 trillion miles.) Our solar system orbits the galactic center at a speed of about 220 kilometers (138 miles) per second. This galaxy is one of the 54 galaxies comprising what astronomers have called the "local group."

 THINKING BEYOND THE FIGURE

Think for a moment: What does the term "local group" suggest?

ESO/Science Source

Figure 1.6 A filament of hot gas erupts from the face of our sun in September of 2012. Like all normal stars, the sun is powered by nuclear fusion—the welding together of small atoms to make larger ones. These violent reactions generate the heat, light, matter, and radiation that pour from stars into space. The entire Earth could easily fit beneath this filament's fiery arc. NASA Images

The **stars** that make up a galaxy are massive spheres of incandescent gases. They are usually intermingled with diffuse clouds of gas and debris. In spiral galaxies like the Milky Way, the stars are arrayed in curved arms radiating from the galactic center. Our part of the Milky Way is populated with many stars, but distances within a galaxy are so huge that the star nearest the sun is about 42 trillion kilometers (26 trillion miles) away. Astronomers tell us there are perhaps 100 billion galaxies in the universe and 100 billion stars in each galaxy. Imagine more stars in the Milky Way than grains of sand on a beach!

Our sun is a typical star **(Figure 1.6)**. The sun and its family of planets, called the **solar system**, are located about three fourths of the way out from the galaxy's center, in a spiral arm. We orbit the galaxy's brilliant core, taking about 230 million years to make one orbit—even though we are moving at about 280 kilometers per second (half a million miles an hour). Earth has made about 20 circuits of the galaxy since the ocean formed.

> **E**very chemical element heavier than hydrogen—most of the atoms that make up the planets, the ocean, and living things—was manufactured by the stars.

Stars Make Heavy Elements from Lighter Ones

As we will see, most of the Earth's substance and that of its ocean was formed by stars. Stars form in **nebulae**, large, diffuse clouds of dust and gas within galaxies. With the aid of telescopes and infrared-sensing satellites, astronomers have observed such clouds in our own and other galaxies. They have seen stars in different stages of development and have inferred a sequence in which these stages occur. The **condensation theory**, a theory based on this inference, explains how stars and planets are believed to form.

The life of a star begins when a diffuse area of a spinning nebula begins to shrink and heat up under the influence of its own weak gravity. Gradually, the cloudlike sphere flattens and condenses at the center into a knot of gases called a *protostar* (*protos*, "first"). The original diameter of the protostar may be many times the diameter of our solar system, but gravitational energy causes it to contract, and the compression raises its internal temperature. When the

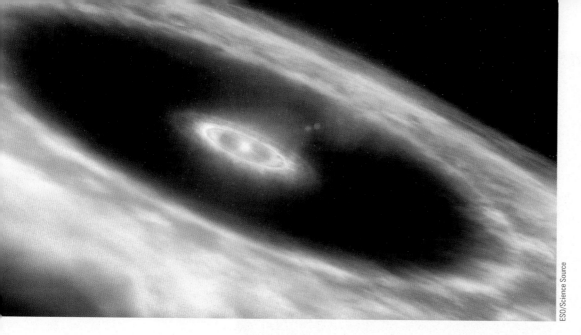

ESO/Science Source

Figure 1.7 An artist depicts the origin of our sun and its family of planets about 5 billion years ago. Near our star, the dust and gas has clumped together to form planetary embryos. Farther away in the distant cold, a halo of comets and debris is coalescing.

protostar reaches a temperature of about 10 million degrees Celsius (18 million degrees Fahrenheit), nuclear fusion begins. That is, hydrogen atoms begin to fuse to form helium, a process that liberates even more energy. This rapid release of energy, which marks the transition from *protostar* to *star*, stops the young star's shrinkage. (The process is shown in the top half of **Figure 1.7**.)

After fusion reactions begin, the star becomes stable—neither shrinking nor expanding, and burning its hydrogen fuel at a steady rate. Over a long and productive life, the star converts a large percentage of its hydrogen to atoms as heavy as carbon or oxygen.

This stable phase does not last forever, though. The life history and death of a star depend on its initial mass. When a medium-mass star (like our sun) begins to consume carbon and oxygen atoms, its energy output slowly rises and its body swells to a stage aptly named *red giant* by astronomers. The dying giant slowly pulsates, incinerating its planets and throwing off concentric shells of light gas enriched with these heavy elements. But most of the harvest of carbon and oxygen is forever trapped in the cooling ember at the star's heart.

Stars much more massive than the sun have shorter but more interesting lives. They, too, fuse hydrogen to form atoms as heavy as carbon and oxygen, but being larger and hotter, their internal nuclear reactions consume hydrogen at a much faster rate. In addition, higher core temperatures permit the formation of atoms—up to the mass of iron.

The dying phase of a massive star's life begins when its core—depleted of hydrogen—collapses in on itself. This rapid compression causes the star's internal temperature to soar. When the infalling material can no longer be compressed, the energy of the inward fall is converted to a cataclysmic expansion called a **supernova** (*nova*, "new" [star]). The explosive release of energy in a supernova is so sudden that the star is blown to bits, and its shattered mass accelerates outward at nearly the speed of light. The explosion lasts only about 30 seconds, but in that short time the nuclear forces holding apart individual atomic nuclei are overcome, and atoms heavier than iron are formed. The gold in your rings, the mercury in a thermometer, and the uranium in nuclear power plants were all created during such a brief and stupendous flash. The atoms produced by a star through millions of years of orderly fusion *and* the heavy atoms generated in a few moments of unimaginable chaos are sprayed into space (**Figure 1.8**). Every chemical element heavier than hydrogen—most of the atoms that make up the planets, the ocean, and living creatures—was manufactured by the stars.

Solar Systems Form by Accretion

Earth and its ocean formed as an indirect result of a supernova explosion. The thin cloud, or **solar nebula**, from which our sun and its planets formed was probably struck by the shock wave and some of the matter of an expanding supernova remnant. Indeed, the turbulence of the encounter may have caused the condensation of our solar system to begin. The solar nebula was affected in two important ways: First, the shock wave caused the condensing mass to spin; second, the nebula absorbed some of the heavy atoms from the passing supernova remnant. In other words, a massive star had to live its life (constructing elements in the process) and then undergo explosive disintegration in order to seed heavy elements back into the nebular nursery of dust and gas from which our solar system arose. The planets are made mostly of matter assembled in a star (or stars) that disappeared billions of years ago. We ourselves are also made of that stardust. Our bones and brains are composed of ancient atoms constructed by stellar fusion long before the solar system existed.

By about 5 billion years ago, the solar nebula was a rotating, disk-shaped mass of about 75% hydrogen, 23% helium, and 2% other material (including heavier elements, gases, dust, and ice). Like a spinning skater bringing in her arms, the nebula spun faster as it condensed. Material concentrated near its center became the protosun. Much of the outer material eventually became **planets**, the smaller bodies that orbit a star and do not shine by their own light.

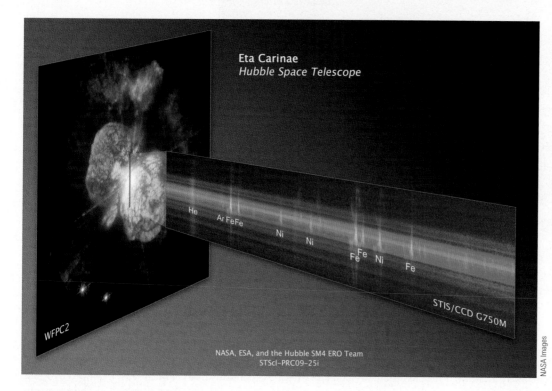

Eta Carinae
Hubble Space Telescope

He   Ar FeFe   Ni   Ni   Fe   Fe   Fe   Ni   Fe

WFPC2

STIS/CCD G750M

NASA, ESA, and the Hubble SM4 ERO Team
STScI–PRC09–25i

NASA Images

Figure 1.8 The dispersion of heavy elements. Light from this huge exploding star reached Earth about 160 years ago. A spectrum of its light (the ribbon extending to the right) shows evidence of iron (Fe) and nickel (Ni) in its shattered atmosphere. In the distant future it is possible that some of these elements might be swept into a new solar system.

💬 **THINKING BEYOND THE FIGURE**

Why do you suppose an oceanography textbook begins with a discussion of the origin of heavy elements?

New planets formed in the disk of dust and debris surrounding the young sun through a process known as **accretion**—the clumping of small particles into large masses. **Figure 1.9** shows a more detailed view of planet formation in the inner ring of Figure 1.7. Bigger clumps with stronger gravity pulled in most of the condensing matter. The planets of our outer solar system—Jupiter, Saturn, Uranus, and Neptune—were probably first to form. These giant planets are composed mostly of methane and ammonia ices because those gases can congeal only at cold temperatures. Near the protosun, where temperatures were higher, the first materials to solidify were substances with high boiling points, mainly metals and certain rocky minerals. The planet Mercury, closest to the sun, is mostly iron, because iron is a solid at high temperatures. Somewhat farther out, in the cooler re-

NASA/JPL-Caltech

Figure 1.9 Planet building in progress. Accretion of planets occurs when small particles clump into large masses. Heavy elements formed by exploding stars account for most of the planets' masses.

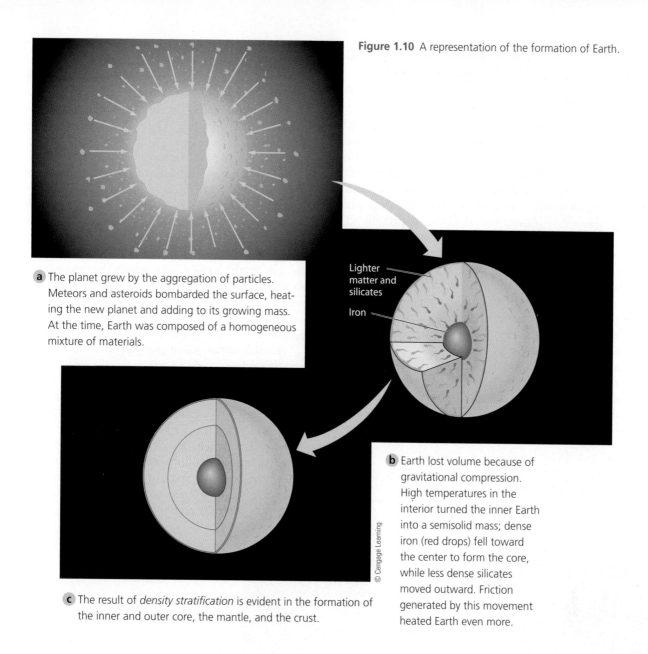

Figure 1.10 A representation of the formation of Earth.

a The planet grew by the aggregation of particles. Meteors and asteroids bombarded the surface, heating the new planet and adding to its growing mass. At the time, Earth was composed of a homogeneous mixture of materials.

Lighter matter and silicates

Iron

b Earth lost volume because of gravitational compression. High temperatures in the interior turned the inner Earth into a semisolid mass; dense iron (red drops) fell toward the center to form the core, while less dense silicates moved outward. Friction generated by this movement heated Earth even more.

© Cengage Learning

c The result of *density stratification* is evident in the formation of the inner and outer core, the mantle, and the crust.

gions, magnesium, silicon, water, and oxygen condensed. Methane and ammonia accumulated in the frigid outer zones. Earth's array of water, silicon-oxygen compounds, and metals results from its middle position within that accreting cloud.

The period of accretion lasted perhaps 30 million to 50 million years. The protosun became a star—our sun—when its internal temperature rose high enough to fuse atoms of hydrogen into helium. The violence of these nuclear reactions sent a solar wind of radiation sweeping past the inner planets, clearing the area of excess particles and ending the period of rapid accretion. Gases like those we now see on the giant outer planets may once have surrounded the inner planets, but this rush of solar energy and particles stripped them away.

This process is probably not rare. As we write this in December of 2013, 1,048 planets (in 794 planetary systems, including 175 multiple planetary systems) have been identified. Estimates of the frequency of systems strongly suggest that more than 50% of sun-like stars harbor at least one planet, and about 40 billion Earth-size planets are thought to orbit in the habitable zones of sun-like stars in the Milky Way galaxy. How many are ocean worlds?

CONCEPT CHECK

Before going on to the next section, check your understanding of some of the important ideas presented so far:

Can scientific inquiry probe further back in time than the big bang?

What element makes up most of the detectable mass in the universe?

Outline the main points in the condensation theory of star and planet formation.

Trace the life of a typical star.

How are the heaviest elements (uranium or gold) thought to be formed?

1.4 Earth, Ocean, and Atmosphere Accumulated in Layers Sorted by Density

The young Earth, formed by the accretion of cold particles, was probably chemically homogeneous throughout. Then, in the midst of the accretion phase, Earth's surface was heated by the impact of asteroids, comets, and other falling debris. This heat, combined with gravitational compression and heat from decaying radioactive elements accumulating deep within the newly assembled planet, caused Earth to partially melt. Gravity pulled most of the iron and nickel inward to form the planet's core. The sinking iron released huge amounts of gravitational energy, which, through friction, heated Earth even more. At the same time, a slush of lighter minerals—silicon, magnesium, aluminum, and oxygen-bonded compounds—rose toward the surface, forming Earth's crust **(Figure 1.10)**. This important process, called **density stratification**, lasted perhaps 100 million years.[5]

Then Earth began to cool. Its first surface is thought to have formed about 4.6 billion years ago. That surface did not remain undisturbed for long. Shortly after its formation, a planetary body somewhat larger than Mars smashed into the young Earth and broke apart. The metallic core fell into Earth's core

and joined with it, while most of the rocky mantle was ejected to form a ring of debris around Earth. The debris began condensing soon after and became our moon. The newly formed moon, still glowing from heat generated by the kinetic energy of infalling objects, is depicted in **Figure 1.11**. Could a similar cataclysm happen today? The issue is addressed on page 374 of Chapter 13.

Radiation from the energetic young sun had stripped away our planet's outermost layer of gases, its first atmosphere, but soon gases that had been trapped inside the forming planet burped to the surface to form a second atmosphere. This volcanic venting of volatile substances—including water vapor—is called **outgassing (Figure 1.12a)**. As the hot vapors rose, they condensed into clouds in the cool upper atmosphere. Though most of Earth's water was present in the solar nebula during the accretion phase, recent research suggests that a barrage of icy comets or asteroids from the outer reaches of the solar system colliding with Earth may also have contributed a portion of the accumulating mass of water, this ocean-to-be **(Figure 1.12b)**.

Earth's surface was so hot that no water could collect there, and no sunlight could penetrate the thick clouds. (A visitor approaching from space 4.4 billion years ago would have seen a vapor-shrouded sphere blanketed by lightning-stroked clouds.) After millions of years the upper clouds cooled enough for some of the outgassed water to form droplets. Hot rains fell toward Earth, only to boil back into the clouds again. As the surface became cooler, water collected in basins and began to dissolve minerals from the rocks. Some of the water evaporated, cooled, and fell again, but the minerals remained behind. The salty world ocean was gradually accumulating.

---

[5]**Density** is an expression of the relative heaviness of a substance; it is defined as the mass per unit volume, usually expressed in grams per cubic centimeter (g/cm^3). The density of pure water is $1\ g/cm^3$. Granite rock is about 2.7 times denser, at $2.7\ g/cm^3$.

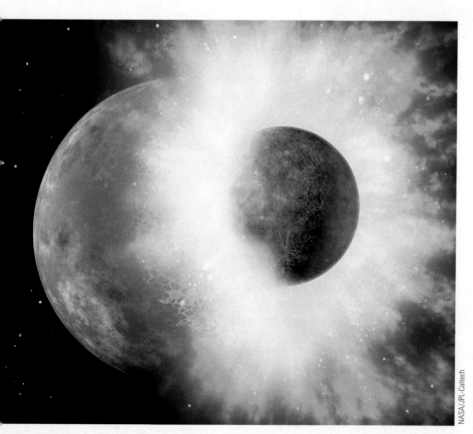

Figure 1.11 The first stage of the formation of the moon. A planetary body somewhat larger than Mars is thought to have smashed into the young Earth about 4.4 billion years ago. The rocky mantle of the impactor was ejected to form a ring of debris around Earth, and its metallic core fell into Earth's core and joined with it. Rocks brought from the lunar surface by Apollo astronauts suggest the ejected material condensed soon after to become our moon.

💬 **THINKING BEYOND THE FIGURE**
What do you think Earth would be like without its moon?

NASA/JPL-Caltech

a Outgassing. Volcanic gases add water vapor, carbon dioxide, nitrogen, and other gases to the atmosphere. Volcanism was a major factor in altering Earth's original atmosphere; later, the action of photosynthetic bacteria and plants was another.

Figure 1.12 Sources of the ocean.

💬 **THINKING BEYOND THE FIGURE**

Do you think the volume of the ocean is increasing, decreasing, or staying about the same over long periods of time?

b. Don Dixon/cosmographica.com

b Comets may have delivered some of Earth's surface water. Intense bombardment of the early Earth by large bodies—comets and asteroids—probably lasted until about 3.8 billion years ago. The inset shows a close-up of the nucleus of Comet Hartley-2 taken by the *EPOXI* spacecraft in November, 2010. Frozen carbon dioxide and water ice can be seen jetting from its surface.

NASA/JPL-Caltech/UMD

Tom Garrison

The Age of Earth and Ocean

The age estimates presented in this chapter (and the ones that follow) are derived from data obtained by many researchers using different sources and techniques. One source is meteorites, chunks of rock and metal formed at about the same time as the sun and planets and out of the same cloud. Many have fallen to Earth in recent times. We know from signs of radiation within these objects how long it has been since they were formed. That information, combined with the rate of radioactive decay of unstable atoms in meteorites, moon rocks, and in the oldest rocks on Earth, allows astronomers to make reasonably accurate estimates of how long ago these objects formed.[7]

Remnants of Earth's early surface are rare because (as you'll discover in Chapter 3) nearly all of that material has been recycled into our planet's interior. In 2008, geologists exploring the eastern shore of Hudson Bay in northern

Figure A Small grains of zircon found in rocks in Australia appear to be 4.35 billion years old.

Quebec encountered greenish-gray rocks that appeared to be of great age. Those rocks have been dated to about 3.9 billion years old. Small grains of zircon **(Figure A)** found within similar rocks in Australia are even older—some of them appear to be 4.35 billion years old (and so must have formed soon after the moon-

forming impact). Inclusions of quartz trapped within the zircons as they crystallized suggest that they formed within molten material rich in dissolved water and silica—evidence that early Earth cooled relatively quickly after the moon formed and supported substantial bodies of surface water. Other Canadian rocks contain 3.85-billion-year-old specks of carbon that bear a chemical fingerprint that many researchers feel could only have come from a living organism.

As for the age of the universe itself, by April 2002, astronomers had obtained very accurate measurements of its rate of expansion. By calculating backward, they found the universe would have begun its expansion 13.7 billion years ago.

By the way, regardless of surprisingly persistent opinion, essentially no evidence supports the contention that Earth is between 6,000 and 10,000 years old.

© Cengage Learning

These heavy rains may have lasted about 20 million years. Large amounts of water vapor and other gases continued to escape through volcanic vents during that time and for millions of years thereafter. The ocean grew deeper. Evidence suggests that Earth's crust grew thicker as well, perhaps in part from chemical reaction with oceanic compounds.

What was the temperature of the young ocean? Earth's surface temperature has fluctuated since the ocean's formation, but the extent of that fluctuation is another area of controversy. For the first chaotic quarter-billion years, the seas would have been hot and precipitation nearly constant, but that condition did not persist. Temperature variations have been common. Some scientists, for example, believe that the early sun's energy output was about 30% less than it is today. The ocean should have frozen. But differences in the quantity and composition (and even the shape) of particles and volcanic gases in Earth's atmosphere allowed heat to be retained and apparently permitted the ocean's surface to remain largely liquid for the next billion years. Colder periods followed, perhaps freezing the ocean to considerable depth (even at the equator) between 800 and 550 million years ago. Although we are presently unsure of the details, scientists are certain that climate change—often *drastic* climate change—has been a feature of Earth since the beginning.[6]

The composition of the early atmosphere was much different from today's. Geochemists believe it may have been rich in carbon dioxide, nitrogen, and water vapor, with traces of ammonia and methane. Beginning about 3.5 billion years ago, this mixture began a gradual alteration to its present composition, mostly nitrogen and oxygen. At first this change was brought about by carbon dioxide dissolving in seawater to form carbonic acid, then combining with crustal rocks. The chemical breakup of water vapor by sunlight high in the atmosphere also played a role. Then about 1.5 billion years later, the ancestors of today's green plants produced—by photosynthesis—enough oxygen to oxidize minerals dissolved in the ocean and surface sediments. Oxygen began to accumulate in the atmosphere. (This monumental event in Earth's history is called the *oxygen revolution*. You'll read about it in Chapter 15.)

---

[6]You'll find more on the specific topic of recent climate change in Chapters 8 and 18.).

[7]For information on radiometric dating, please see "How Do We Know? 3.1" on page 69.

1.5 Life Probably Originated in the Ocean

Life, at least as we know it, would be inconceivable without large quantities of water. Water can retain heat, moderate temperature, dissolve many chemicals, and suspend nutrients and wastes. These characteristics make it a mobile stage for the intricate biochemical reactions that allowed life to begin and prosper on Earth.

Life on Earth is formed of aggregations of a few basic kinds of carbon compounds. Where did the carbon compounds come from? There is growing consensus that most of the organic (that is, carbon-containing) materials in these compounds were transported to Earth by the comets, asteroids, meteors, and interplanetary dust particles that crashed into our planet during its birth. The young ocean was a thin broth of organic and inorganic compounds in solution.

In laboratory experiments, mixtures of dissolved compounds and gases thought to be similar to Earth's early atmosphere have been exposed to light, heat, and electrical sparks. These energized mixtures produce simple sugars and a few of the biologically important amino acids. They even produce small proteins and nucleotides (components of the molecules that transmit genetic information between generations). The main chemical requirement seems to be the absence (or near absence) of free oxygen, a compound that can disrupt any unprotected large molecule.

Did *life* form in these experiments? No. The compounds that formed are only building blocks of life. But the experiments do tell us something about the commonality and unity of life on Earth. The facts that these crucial compounds can be synthesized so easily and are present in virtually all living forms are probably not coincidental. Those compounds are "permitted" by physical laws and by the chemical composition of this planet. The experiment also underscores the special role of water in life processes. The fact that all life, from a jellyfish to a dusty desert weed, depends on saline water within its cells to dissolve and transport chemicals is certainly significant. It strongly suggests that simple, self-replicating—living—molecules arose somewhere in the early ocean. It also strongly suggests that all life on Earth is of common origin and ancestry.

The early steps in the evolution of living organisms from simple organic building blocks, a process known as **biosynthesis**, are still speculative. As noted earlier, planetary scientists suggest that the sun was faint in its youth. Perhaps Earth's atmospheric haze was opaque enough to block ultraviolet radiation that would have hindered the formation of complex molecules (and therefore life) at the ocean surface. The first living molecules might have arisen at great depths on clays or pyrite crystals at mineral-rich seeps on the ocean floor **(Figure 1.13)**.

A similar biosynthesis seems unlikely to occur today. Living things have changed the conditions in the ocean and atmosphere, and those changes are not consistent with any new origin of life. For one thing, green plants have filled the atmosphere

> **L**ife on Earth almost certainly evolved in the ocean; the cells of all life forms are still bathed in salty fluids.

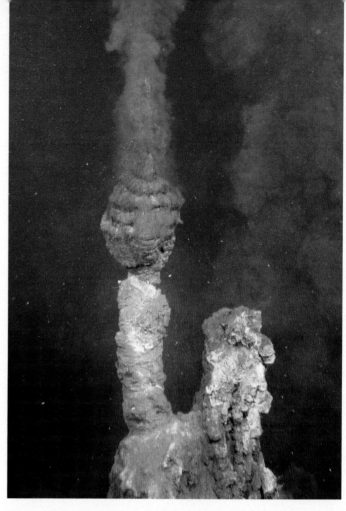

Figure 1.13 An environment for biosynthesis? Weak sunlight and unstable conditions on Earth's surface may have favored the origin of life on mineral surfaces near deep-ocean hydrothermal vents similar to the one shown here.

Image courtesy of New Zealand American Submarine Ring of Fire 2007 Exploration, NOAA Vents Program, the Institute of Geological & Nuclear Sciences and NOAA-OE

with oxygen. For another, some of this oxygen (as ozone) now blocks most of the dangerous wavelengths of light from reaching the surface of the ocean. And finally, the many tiny organisms present today would gladly scavenge any large organic molecules as food.

How long ago might life have begun? The oldest fossils yet found, from northwestern Australia, are between 3.4 billion and 3.5 billion years old **(Figure 1.14)**. They are remnants of fairly complex bacteria-like organisms, indicating that life must have originated even earlier, probably only a few hundred million years after a stable ocean formed. Evidence of an even more ancient beginning has been found in the form of carbon-based residues in some of the oldest rocks on Earth, from Akilia Island near Greenland ("How Do We Know? 1.1"). These 3.85-billion-year-old specks of carbon bear a chemical fingerprint that many researchers feel could only have come from a living organism. Life and Earth have grown old together; each has greatly influenced the other.

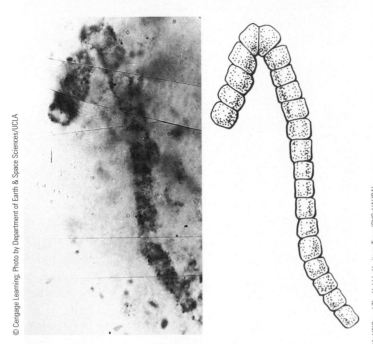

Figure 1.14 Fossil of a bacteria-like organism (with an artist's reconstruction) that photosynthesized and released oxygen into the atmosphere. Among the oldest fossils ever discovered, this microscopic filament from northwestern Australia is about 3.5 billion years old.

Figure 1.15 The end of a solar system? This glowing gas in this beautiful nebula once formed the outer layers of a sun-like star that exploded only 15,000 years ago. The inner loops are being ejected by a strong wind of particles from the remnant central star. If planets orbited this star, their shattered remnants are contained in the outward-rushing filaments at the periphery. Perhaps 5 billion years from now, observers 5,000 light-years away would see a similar sight as our sun passes the end of its life.

CONCEPT CHECK

Before going on to the next section, check your understanding of some of the important ideas presented so far:

Are the atoms and basic molecules that compose living things different from the molecules that make up nonliving things? Where were the atoms in living things formed?

How old is the oldest evidence for life on Earth? On what are those estimates based?

Was Earth's atmosphere rich in oxygen when life originated here?

💬 **THINKING BEYOND THE FIGURE**

Think of the long history of the atoms in your eyes that allow you to read and comprehend this sentence. Now think of their potential future. You might begin to sense something called "deep time."

1.6 What Will Be Earth's Future?

Our descendants may enjoy another 5 billion years of life on Earth as we know it today. But then our sun, like any other star, will begin to die. The sun is not massive enough to become a supernova, but after a billion-year cooling period, the re-energized sun's red-giant phase will engulf the inner planets. Its fiery atmosphere will expand to a radius greater than the orbit of Earth. The ocean and atmosphere, all evidence of life, the crust, and perhaps the whole planet, will be recycled into component atoms and hurled by shock waves into space (as in **Figure 1.15**). Our successors, if any, will have perished or fled to safer worlds. Its fuel exhausted and its energies spent, the sun will cool to a glowing ember and ultimately to a dark cinder. Perhaps a new system of star and planets will form someday from the debris of our remains.

A timeline that shows the history of past and future Earth appears in **Figure 1.16.**

CONCEPT CHECK

Before going on to the next section, check your understanding of some of the important ideas presented so far:

Have the particles that make up the atoms of your body existed for nearly all the age of the universe?

1.7 Are There Other Ocean Worlds?

As we have seen, planets with liquid on or near their surfaces may not be rare. Water itself is not scarce. In the solar system, for example, Jupiter has hundreds of times as much water as Earth does, nearly all of it in the form of ice. In 1998, ice was discovered in deep craters near our moon's south pole, and a few places on Mars are known to have a layer of ice just below a surface layer of blown sand. Astronomers have even located water molecules drifting free in space. But *liquid* water is unexpected.

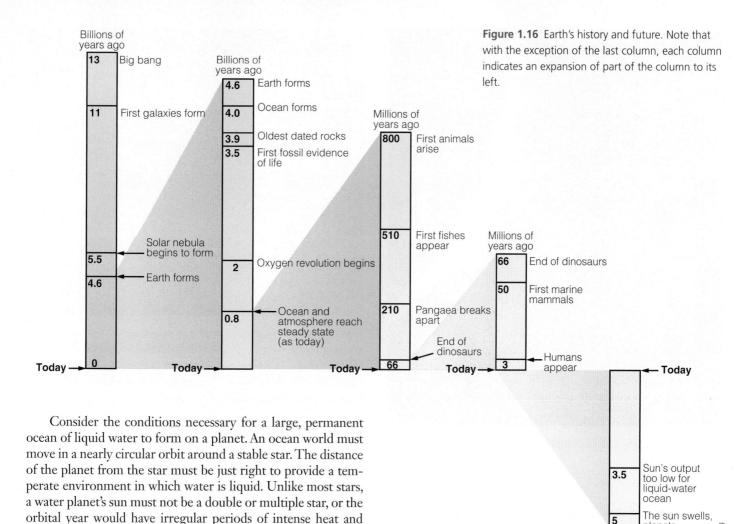

Figure 1.16 Earth's history and future. Note that with the exception of the last column, each column indicates an expansion of part of the column to its left.

Consider the conditions necessary for a large, permanent ocean of liquid water to form on a planet. An ocean world must move in a nearly circular orbit around a stable star. The distance of the planet from the star must be just right to provide a temperate environment in which water is liquid. Unlike most stars, a water planet's sun must not be a double or multiple star, or the orbital year would have irregular periods of intense heat and cold. The materials that accreted to form the planet must have included both water and substances capable of forming a solid crust. The planet must be large enough that its gravity will keep the atmosphere and ocean from drifting off into space.

Although it might be a bit premature to consider "comparative oceanography" as a career choice, researchers are increasingly certain that liquid water exists (or existed quite recently) on at least four other bodies in our solar system. We can begin to compare and contrast them.

Our Solar System's Outer Moons

The spacecraft *Galileo* passed close to Europa—a moon of Jupiter—in early 1997. Photos sent to Earth revealed a cracked, icy crust covering what appears to be a slushy mix of ice and water. *Galileo* also detected a distinctive magnetic field, the signature of a salty liquid-water ocean below the ice.

The volume of this ocean is astonishing. Though Europa is slightly smaller than our own moon, its ocean averages about 160 kilometers (100 miles) deep. The amount of water in its ocean is perhaps 40 times that of Earth's! Europa's ocean is probably kept liquid by heat escaping Europa's interior and by gravitational friction of tidal forces generated by Jupiter itself. Though the surface of ice is about 8 kilometers (5 miles) thick and as cold as the surface of Jupiter, the liquid interior of the ocean, cradled deep in rocky basins, may be warm enough to

sustain life. No continents emerge from this alien sea. A mission to the surface of Europa is being planned.

Ganymede, Jupiter's largest satellite, was surveyed by *Galileo* in May 2000; the photographs showed structures strikingly similar to those on Europa. Again, magnetometer data suggested a salty ocean beneath a moving, icy crust.

In November 2005, the *Cassini* spacecraft flew close behind Saturn's moon Enceladus. From this vantage point *Cassini's* cameras detected fountains of ice crystals shooting from gashes on the small moon's surface (**Figure 1.17**). The relative warmth of the plumes and the detection of accompanying molecules of methane and carbon dioxide suggest one more encrusted liquid-water ocean.

Mars

Europa and Ganymede—and perhaps Enceladus—may have icy oceans now, but Mars, a much nearer neighbor, may have had an ocean in the distant past. An ocean could have occupied the low places of the northern hemisphere of Mars between 3.2 billion and 1.2 billion years ago when conditions were warmer. Cur-

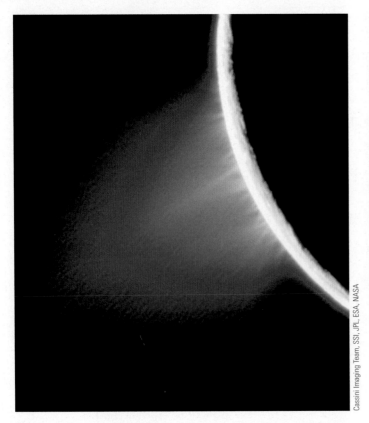

Figure 1.17 Fountains of ice shoot from the surface of Saturn's moon Enceladus.

Figure 1.18 In September of 2012, the Mars *Curiosity* rover photographed a fractured outcrop of smoothed streambed rock with surfaces eroded by water-driven pebbles. Some bits of gravel at the left of the frame show the characteristic rounded shapes that result from turbulence in stream flow, evidence of a once-wet Mars.

rent models suggest that early in its history, Mars had a thick atmosphere rich in carbon dioxide, much like the atmosphere of early Earth. Carbon dioxide is a "greenhouse" gas—it traps the sun's heat like the glass panels of a greenhouse. The atmosphere kept Mars warm and allowed water to flow freely. In September 2012, a camera aboard the Mars rover *Curiosity* sent photos from the surface showing clear evidence that water once flowed there. Figure 1.18 shows a fractured outcrop of smoothed streambed rock with surfaces eroded by water-driven pebbles. Some bits of gravel at the left of the frame show the characteristic rounded shapes that result from turbulence in stream flow.

Where is the water now? Over the eons, rocks on the Martian surface absorbed the carbon dioxide, and the atmosphere grew thin and cold. The ocean disappeared, its water binding to rocks or freezing beneath the planet's surface. Mars has become much colder in the past billion years, perhaps because of the loss of greenhouse gases in the atmosphere. If a large quantity of water is present today, most of it probably lies at the poles. In September 2008, the *Phoenix* lander excavated to permafrost beneath a thin layer of sediment in the northern Martian arctic plain.

Could wet pockets exist today? In August 2011, NASA released photographs from *Mars Reconnaissance Orbiter* that showed clear evidence of recently flowing water. These gullies darken in the springtime (as ice melts and water flows downhill?) and appear very young.

Titan

Must an ocean consist of liquid water? Hydrocarbons have been seen on the surface of Titan, Saturn's largest moon. In 1999, scientists using the huge W. M. Keck Telescope in Hawai'i detected cold, dark, infrared-absorbing organic matter surrounding a bright area about the size of Australia. By late 2004 the hardworking *Cassini* had photographed what appears to be a cold liquid ocean of methane, ethane, and other hydrocarbons, complete with islands, bays, and peninsulas (Figure 1.19). In early January 2005, *Cassini* detached a small probe (named *Huygens* in honor of the Dutch astronomer who discovered this moon) to travel to Titan. Its cameras photographed drainage channels and other continental details and then soft-landed on a solid surface.

Extrasolar Planets

As noted earlier, perhaps half of all sun-like stars have planets in orbit around them. Most of these planets were found by watching the wobbling path a star takes through space when influenced by the gravity of a massive companion planet. One of these was directly imaged in 2010, but its large size and distant orbit suggest it is a Jupiter-like body devoid of liquid water. But smaller and cooler planets with atmospheres containing water vapor and methane have been found around two stars nearer the sun (see "insight from a national geographic explorer 1.1").

As for the building blocks of life, in December 2005, researchers using the Spitzer Infrared Space Telescope detected the gaseous precursors of biochemicals common on Earth in the planet-forming region around a star about 375 light-years away.

Kevin Hand/National Geographic

DR. KEVIN HAND discusses life in extreme environments, some of which could mimic conditions on other planets.

We were headed there to study the microbes that live within rocks. Originally, no one thought life could exist in such a harsh, dry, cold environment, but back in the '70's Dr. Imre Friedmann, now at UW Seattle, inherited a rock sample from Antarctica that later led to him giving birth to the field of studying cryptoendolithic microbial communities (cryptoendolith, meaning within the pore spaces of rocks). Friedmann discovered complex communities of lichens (fungi 1 algae) and cyanobacteria thriving within the sandstone rocks of the Dry Valleys. Much remains to be understood about how these communities live, die, and ultimately modify the geological setting around them. We had all been working on specially designed instruments that would help us study these life-forms in the field. The idea was to bring the instruments to rocks, instead of bringing the rocks to the lab.

Studying microbes is tricky in that regard; if you change the environment they often change what they're doing. Thus, studying life in the lab often tells you one thing while in reality the little buggers may be doing something quite different in their home environs. Along with furthering our understanding of what makes these life-forms tick, NASA is also interested in seeing how our field instruments perform. The ultimate application would be for future robotic landers, a case where all of your instruments are sent off to a far and distant field with no one standing by to press the reset button if something goes wrong. At least two of our instruments had been proposed for the 2009 Mars Rover. Neither was selected, but we'll likely try again during the next round.

Source: http://www.spaceref.com/news/viewsr.html?pid515358
Newly discovered planet Gliese 581e.

Background photo: DANA BERRY/National Geographic

NASA/JPL-Caltech/Space Science Institute

NASA/JPL-Caltech/ASI/Cornell

Figure 1.19 The face of Saturn's moon, Titan, the only other body in the solar system known to possess liquid on its surface. Sunlight is shining on the edge of a huge circular storm just below the day/night line. Seas of liquid methane or ethane seen as dark patches cover a substantial portion of this large moon's surface. The inset shows a lake near Titan's north pole. It is a bit larger than Lake Superior, one of Earth's largest lakes.

Might organic gases be a common component in the accretion of solar systems?

Life and Oceans?

Could the presence of oxygen be a clue to the existence—or past existence—of life? Could it point to an ocean on a planet or moon? Since CO_2 is the "normal" composition for the atmosphere of a terrestrial planet, a large quantity of atmospheric oxygen would be unexpected—after all, oxygen is one of the most reactive of gases. Any oxygen in an atmosphere is likely to react with other materials. The red, rust-colored rocks typical of Mars almost certainly resulted from the oxidation (rusting) of iron-containing minerals. If a planet is found with lots of free oxygen, something is probably replenishing that oxygen.

That "something" is probably life. The action of photosynthetic organisms (including plants) produces excess oxygen. Without photosynthesis, Earth's atmosphere would be all but oxygen free. As noted in this chapter, life—at least on Earth—almost certainly originated in the ocean. If the atmosphere of distant planets contains significant quantities of oxygen, oceans and life might be possible. Scientists are close to technologies that will allow them to detect chemical signatures in the atmospheres of planets orbiting other stars.

Stay tuned!

Chapter in Perspective

J. Hester (ASU) et al., CXC, HST, NASA

In this chapter you learned Earth is a water planet, possibly one of few in the galaxy. An ocean covering 71% of its surface has greatly influenced its rocky crust and atmosphere. The ocean dominates Earth, and the average depth of the ocean is about 4½ times the average height of the continents above sea level. Life on Earth almost certainly evolved in the ocean; the cells of all life forms are still bathed in salty fluids.

We study our planet using the scientific method, a systematic *process* of asking and answering questions about the natural world. Marine science applies the scientific method to the ocean, the planet of which it is a part, and the living organisms dependent on the ocean.

Most of the atoms that make up Earth and its inhabitants were formed within stars. Stars form in the dusty spiral arms of galaxies and spend their lives changing hydrogen and helium to heavier elements. As they die, some stars eject these elements into space by cataclysmic explosions. The sun and the planets, including Earth, probably condensed from a cloud of dust and gas enriched by the recycled remnants of exploded stars. Earth formed by the accretion of cold particles about 4.6 billion years ago.

Heat from infalling debris and radioactive decay partially melted the planet, and density stratification occurred as heavy materials sank to its center and lighter materials migrated toward the surface. Our moon is thought to have been formed by debris ejected when a planetary body somewhat larger than Mars smashed into Earth.

The ocean formed later, as water vapor trapped in Earth's outer layers escaped to the surface through volcanic activity during the planet's youth. Comets may also have brought some water to Earth. Life originated in the ocean very soon after its formation—life and Earth have grown old together. We know of no other planet with a similar ocean, but water is abundant in interstellar clouds, and other water planets are not impossible to imagine.

In the next chapter you will learn that science and exploration have gone hand in hand. Voyaging for necessity evolved into voyaging for scientific and geographical discovery. The transition to scientific oceanography was complete when the *Challenger Report* was concluded in 1895. The rise of the great oceanographic institutions quickly followed, and those institutions and their funding agencies today mark our path into the future.

QUESTIONS FROM STUDENTS[8]

1. **You wrote that "Nothing is ever proven absolutely true by the scientific method." What good is it then? Can't we depend on the process of science?**

One philosopher of science has described truth as a liquid: it flows around ideas and is hard to grasp. The progressive improvement in our understanding of nature is subject to the limitations inherent in our observations. As our observations become more accurate, so do our conclusions about the natural world. But because observations (and interpretations of observations) are never perfect, truth can never be absolute. In the 1920s, for example, astronomers assumed that the universe was limited to our own Milky Way galaxy. Observations made with a large new telescope on Mt. Wilson in California by Harlow Shapley and Edwin Hubble allowed them to measure more distant objects. Galaxies were discovered in profusion, "like grains of sand on a beach," in Shapley's words.

This "fluidity" is not a disadvantage. Scientific thought is not bound by dogma. It is free to winnow good ideas from bad. It provides a durable framework on which to build a sustainable technological civilization. What we have accumulated so far is of inestimable practical and aesthetic value, and we have only scratched the surface.

2. **What's the difference between a law and a theory? People sometimes say, "It's just a theory. . ." to put down an idea.**

A *theory* is a synthesis of a large and important body of information about a related group of natural phenomena. A *law* refers to a body of observations that can be summarized in a short mathematical (or verbal) statement. One is not "more true" than the other—both can be statements of facts.

3. **Life appears to have arisen on Earth soon after the formation of a stable surface. Could life have formed on other planets?**

We have no evidence, direct or indirect, of life on other planets around our sun or elsewhere in the universe. Yet it seems provincial to assume that life could have arisen only here. The formation of organic molecules from simple chemicals receiving energy from lightning, heat, ultraviolet light, and other sources may be quite common, and increasing complexity in these compounds may be a universal phenomenon.

4. **Would life on other planets resemble life on Earth?**

Organisms elsewhere might be very different. Recall that life on this planet probably arose in the ocean, and all life-forms here carry an ocean of sorts within their bodies. On a planet without water, the organisms would surely be much different.

For example, on a hypothetical planet with an ammonia ocean, life would not have a structure of cells surrounded by lipid membranes. Lipid membranes are the sheets of fatty molecules that keep the inside of a cell separate from the environment, and ammonia prevents these membranes from forming. Without membranes, cells as we know them are not possible. Notwithstanding this argument, life need not be confined to planets with water. Other life-forms may exist, based on other "brews."

5. **Supernovas seem really important. Has anybody ever seen a supernova?**

They are, indeed—all the heavy elements that make up you and your surroundings were constructed in them. They are occasionally visible to the unaided eye. Light from an exploding star reached Earth in April or early May of the year 1054 C.E. Its position was recorded by Chinese and Arab astronomers; it was bright enough to be seen in daylight for 23 days. At its brightest it was far brighter than anything in the night sky and was said to cause blind spots in the eyes of those who gave it more than a passing glance. Astronomers have recently found an astonishingly dense remnant of the nova spinning at the center of the existing nebula (see **Figure 1.20**)—this star is 19 kilometers (12 miles) across and spins at a rate of 30.2 times per second!

Hundreds of distant novas are visible using large telescopes any time you care to look.

6. **How far away are exploding stars? What if a star became a nova in our neighborhood of the galaxy? Would we notice anything?**

Of course it depends on what you mean by "neighborhood," but the outcome of a nearby event could be ugly. The intense bursts of gamma rays and X-rays from a huge supernova (a hypernova) could sterilize everything in part of a galaxy's spiral arm—nothing alive based on water and proteins would survive. The radiation from the disintegration of a sun-like star would be less catastrophic (see again Figure 1.15). Astronomers have detected gamma ray bursts since the 1960s, but only in 2003 was a gamma burst directly associated with the first light from a hypernova. Fortunately the event happened in a distant galaxy.

Figure 1.20 This neutron star, a superdense remnant of the supernova of 1054 C.E., is called a "pulsar" because it emits rapid regular pulses of energy as it spins.

J. Hester (ASU) et al., CXC, HST, NASA

[8]Each chapter ends with a few questions students have asked us after a lecture or reading assignment. These questions and their answers may be interesting to you, too.

TERMS AND CONCEPTS TO REMEMBER

accretion	galaxy	ocean	solar nebula
big bang	hypothesis	oceanography	solar system
biosynthesis	laws	outgassing	stars
condensation theory	marine science	planets	supernova
density	Milky Way galaxy	science	theory
density stratification	nebula	scientific method	world ocean
experiment			

STUDY QUESTIONS

Thinking Critically

1. Why do we refer to only one world ocean? What about the Atlantic and Pacific oceans, or the Baltic and Mediterranean seas?
2. Which is greater—the average depth of the ocean or the average elevation of the continents?
3. Can the scientific method be applied to speculations about the natural world that are not subject to test or observation?
4. What are the major specialties within marine science?
5. Where did the Earth's heavy elements come from?
6. Where did Earth's surface water come from?
7. Considering what must happen to form them, do you think ocean worlds are relatively abundant in the galaxy? Why or why not?
8. Earth has had three distinct atmospheres. Where did each one come from, and what were the major constituents and causes of each?
9. How old is Earth? When did life arise? On what is that estimate based? How did the moon form?
10. What is biosynthesis? Where and when do researchers think it might have occurred on our planet? Could it happen again this afternoon?
11. Marine biologists sometimes say that all life-forms on Earth, even desert lizards and alpine plants, are marine. Can you think why?
12. How do we know what happened so long ago?
13. What is density stratification? What does it have to do with the present structure of Earth?
14. Do we know of the existence of other water planets?

Thinking Analytically

1. A light-year is the distance light can travel in 1 year. Light travels at 300,000 kilometers (186,000 miles) per second. Commercial television broadcasting began in 1939. Television signals travel at the speed of light. How far away would a space probe have to be before it could no longer detect those signals?
2. Density is mass per unit volume. Granite rock weighs about 2.7 g/cm^3, water weighs about 1.0 g/cm^3. Knowing their sizes, how might you determine whether Europa or Ganymede is hiding a large liquid water ocean beneath an icy crust?
3. Can you think of any way an astronomer could detect a large planet orbiting a star without actually seeing the planet? (Hint: How would the star move as the planet orbits it?)

GLOBAL GEOSCIENCE WATCH

Visit www.cengagebrain.com to access MindTap, a complete digital course which includes access to Global Geoscience Watch and more. Research the origins of life within the GREENR database by using the search term "origin of life." Use the information you find to write a short report in which you: identify some of the primary candidate locations where scientists believe life could have originated, describe how each of these sites could have contributed important requirements to early life, and explain which theory (if any) you believe is the most likely and why you came to this conclusion.

2 A History of Marine Science

KEY CONCEPTS

The ocean did not prevent the spread of humanity. By the time European explorers set out to "discover" the world, native peoples met them at nearly every landfall. Herbert Kane/National Geographic Image Collection

Any coastal culture skilled at raft building or small-boat navigation had economic and nutritional advantages over less skilled competitors.

Richard Schlecht/National Geographic Creative

The first global exploratory expeditions were undertaken by Chinese Admiral Zheng He beginning in 1405. Tom Garrison

The three expeditions of Captain James Cook, British Royal Navy, were perhaps the first to apply the principles of scientific investigation to the ocean.

SuperStock/SuperStock

The voyage of HMS *Challenger* (1872–1876) was the first extensive expedition dedicated exclusively to research. First page from the Journal of HMS Challenger, a personal diary by Pelham Aldrich, 1872 (pen & ink and w/c on paper), Aldrich, Admiral Pelham (1844–1930)/Royal Geographical Society, London, UK/The Bridgeman Art Library

Modern oceanography is guided by consortia of institutions and **governments.** Tom Garrison

---

A marine archaeologist surveys an 11th-century shipwreck off the Turkish coast. Recent research suggests seafaring began much earlier in human history. The Mediterranean island of Crete has been isolated from the mainland of Greece, Turkey, and Africa for 5 million years. The discovery of 130,000-year-old stone tools on the coast of Crete indicates that our hominid ancestors repeatedly crossed the open Mediterranean to settle the island.

Jonathan Blair/Corbis

25

Media Connection

Start off this chapter by listening to a podcast with Explorer Robert Ballard, as he revisits his explorations under the sea. Visit www.cengagebrain.com to access MindTap, a complete digital course that includes this podcast and other resources.

Figure 2.1 People able to use boats to move along a coastline to exploit rich fishing grounds had nutritional and strategic advantages over their landlocked competitors.

2.1 Understanding the Ocean Began with Voyaging for Trade and Exploration

It has taken a long time for humans to appreciate the nature of the world, but we're a restless and inquisitive lot, and despite the ocean's great size, we have populated nearly every inhabitable place. This fact was aptly illustrated when European explorers set out to "discover" the world, only to be met by native peoples at nearly every landfall! Clearly the ocean did not prevent the spread of humanity. The early history of marine science is closely associated with the history of voyaging.

Early Peoples Traveled the Ocean for Economic Reasons

Ocean transportation offers people the benefits of mobility and greater access to food supplies. Any coastal culture skilled at raft building or small-boat navigation would have economic and nutritional advantages over less skilled competitors **(Figure 2.1)**. Edible nearshore resources (fishes, shellfish, and so forth) have been hunted by hunter-gatherers for more than 150,000 years—the earliest evidence of marine foraging by our prehuman ancestors is found in South Africa.

The first direct evidence we have of **voyaging**, traveling on the ocean for a specific purpose, comes from records of trade in the Mediterranean Sea. The Egyptians organized shipborne commerce on the Nile River, but the first regular ocean traders were probably the Cretans or the Phoenicians, who inherited maritime supremacy in the Mediterranean after the Cretan civilizations were destroyed by earthquakes and political instability around 1200 B.C.E. Skilled sailors, the Phoenicians carried their wares through the Strait of Gibraltar to markets as distant as Britain and the west coast of Africa. Given the simple ships they used, this was quite an achievement.

The Greeks began to explore outside the Mediterranean into the Atlantic Ocean around 900–700 B.C.E. **(Figure 2.2)**. Early Greek seafarers noticed a current running from north to south beyond Gibraltar. Believing that only rivers had currents,

> The early history of marine science is closely associated with the history of voyaging—traveling on the ocean for a specific purpose.

they decided that this great mass of water, too wide to see across, was part of an immense flowing river. The Greek name for this river was *okeanos*. Our word "ocean" is derived from *oceanus*, a Latin variant of that root. Phoenician sailors were also very much at home in this "river," but like the Greeks, they rarely ventured out of sight of land.

As they went about their business, early mariners began to record information to make their voyages easier and safer—the location of rocks in a harbor, landmarks and the sailing times between them, the direction of currents. These first **cartographers** (chart makers) were probably Mediterranean traders who made routine journeys from producing areas to markets. Their first charts (from about 800 B.C.E.) were drawn to jog their memory for obvious features along the route. Today's **charts** are graphic representations that primarily depict water and water-related information. (*Maps* primarily represent land.) For more on maps and charts, please see Appendix 4.

In this early time, other cultures also traveled on the ocean. The Chinese began to engineer an extensive system of inland waterways, some of which connected with the Pacific Ocean, to make long-distance transport of goods more convenient. The Polynesian peoples had been moving easily among islands off the coasts of Southeast Asia and Indonesia since 3000 B.C.E. and were beginning to settle the

Figure 2.2 An artist reconstructs the oldest known Greek cargo ship. Used in trade around 390 B.C.E., the lead-sheathed wooden vessel is shown at port on the island of Rhodes. Sailors are loading amphorae of oil for transport to the mainland. A Greek bireme—a warship—is seen in the distance.

mid-Pacific islands. Though none of these civilizations had contact with the others, each developed methods of charting and navigation. All these early travelers were skilled at telling direction by the stars and by the position of the rising or setting sun.

Curiosity and commerce encouraged adventurous people to undertake ever more ambitious voyages. But these voyages were possible only with the coordination of astronomical direction finding (and knowledge of the shape and size of Earth), advanced shipbuilding technology, accurate graphic charts (not just written descriptions), and perhaps most important, a growing understanding of the ocean itself. **Marine science**, the organized study of the ocean, began with the technical studies of voyagers.

Systematic Study of the Ocean Began at the Library of Alexandria

Progress in applied marine science began at the **Library of Alexandria**, in Egypt. Founded in the third century B.C.E. at the behest of Alexander the Great, the library constituted history's greatest accumulation of ancient writings. The library and the adjacent museum could be considered the first university in the world. Scholars worked and researched there, and students came from around the Mediterranean to study. Written knowledge of all kinds—characteristics of nations, trade, natural wonders, artistic achievements, tourist sights, investment opportunities, and other items of interest to seafarers—was warehoused around its leafy courtyards. When any ship entered the harbor, the books (actually scrolls) it contained were by law removed and copied; the *copies* were returned to the owner and the originals kept for the library. Caravans arriving overland were also searched. Manuscripts describing the Mediterranean coast were of great interest. Traders quickly realized the competitive benefit of this information.

Marine science was only one of the library's many research areas. For 600 years, it was the greatest repository of wisdom of all kinds and the most influential institution of higher learning in the ancient world. Here, perhaps, was the first instance of cooperation between a university and the commercial community, a partnership that has paid dividends for both science and business ever since.

Eratosthenes Accurately Calculated the Size and Shape of Earth

The second librarian at Alexandria (from 235 B.C.E. until 192 B.C.E.) was the Greek astronomer, philosopher, and poet **Eratosthenes of Cyrene**. This remarkable man was the first to calculate the circumference of Earth. The Greek Pythagoreans had realized Earth was spherical by the sixth century B.C.E., but Eratosthenes was the first to estimate its true size.

Eratosthenes had heard from travelers returning from Syene (now Aswan, site of the great Nile dam) that at noon on the longest day of the year, the sun shone directly onto the waters of a deep, vertical well. In Alexandria, he noticed that a vertical pole cast a slight shadow on that day. He measured the shadow angle and found it to be a bit more than 7°, about 1/50 of a circle. He correctly assumed that the sun is a great distance from Earth, which means that the sun's rays would approach Syene and Alexandria in essentially parallel lines. If the sun were directly overhead at Syene but not directly overhead at Alexandria, then Earth's surface would have to be curved. But what was the *circumference* of Earth?

By studying the reports of camel caravan traders, he estimated the distance from Alexandria to Syene at about 785 kilometers (491 miles). Eratosthenes now had the two pieces of information needed to derive the circumference of Earth by geometry. **Figure 2.3** shows his method. The precise size of the units of length (stadia)

> **T**he size and shape of Earth was known by about 230 B.C.E., more than 1,700 years before Columbus's voyages.

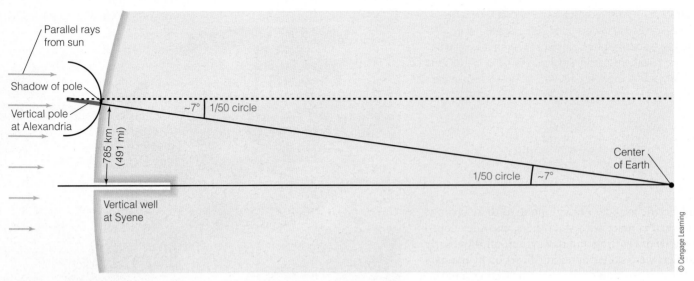

Figure 2.3 A diagram showing Eratosthenes's method for calculating the circumference of Earth. As described in the text, he used simple geometric reasoning based on the assumptions that Earth is spherical and that the sun is very far away. Using this method, he was able to discover the circumference to within about 8% of its true value. This knowledge was available more than 1,700 years before Columbus began his voyages. (The diagram is not drawn to scale.)

💬 **THINKING BEYOND THE FIGURE**

How would Eratosthenes's estimate have been different if the sun were actually closer to Earth?

Eratosthenes used is thought to have been 555 meters (607 yards), and historians estimate that his calculation, made in about 230 B.C.E., was accurate to within about 8% of the true value. Within a few hundred years most people in the West who had contact with the library or its scholars knew Earth's approximate size.

Cartography flourished. The first workable charts that represented a spherical surface on a flat sheet were developed by Alexandrian scholars. Latitude and longitude, systems of imaginary lines dividing the surface of Earth, were invented by Eratosthenes. **Latitude** lines were drawn parallel to the equator, and **longitude** lines ran from pole to pole **(Figure 2.4)**.

Figure 2.4 The world, according to a chart from the third century B.C.E., Eratosthenes drew latitude and longitude lines through important places rather than spacing them at regular intervals as we do today. The Alexandrian perception of the world is reflected in the size of the continents and the central position of Alexandria at the mouth of the Nile. This representation was published in the first volume of the *Challenger Report*. (More on the *Challenger* expedition will be found later in this chapter.)

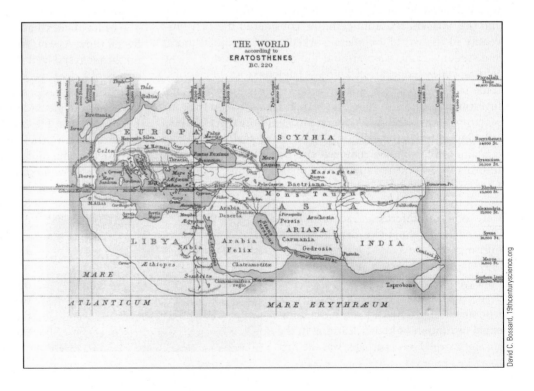

Mohsen Allam/Egypt Today

a The exact site of the Library of Alexandria had been lost to posterity until the early 1980s. By 2004, a theater and 13 classrooms had been unearthed. One of the classrooms is shown here.

TAYLOR S. KENNEDY/National Geographic Creative

b A modern Library of Alexandria opened in the spring of 2002. Sponsors of the new *Bibliotheca Alexandrina* hope it will become "a lighthouse of knowledge to the whole world." The goal of this conference center and storehouse is not to restore the past but to revive the ancient library's questing spirit.

Figure 2.5 The Alexandrian Library

Eratosthenes placed the lines through prominent landmarks and important places to create a convenient though irregular grid. Our present regular grid of latitude and longitude was invented by Hipparchus (c.165–c.127 B.C.E.), a librarian who divided the surface of Earth into 360 degrees. A later Egyptian–Greek, Claudius Ptolemy (90–168 C.E.), *oriented* charts by placing east to

the right and north at the top. Ptolemy's division of degrees into minutes and seconds of arc is still used by navigators. Latitude and longitude are explained in "How Do We Know? 2.1: Where We Are: Latitude and Longitude."

Ptolemy also introduced an "improvement" to Eratosthenes's surprisingly accurate estimate of Earth's circumference. Unfortunately, Ptolemy wrongly depended on flawed calculations of the effects of atmospheric refraction. He publicized an estimate of the size of Earth that was too small—about 70% of the true value. This error, coupled with his mistake of overestimating the size of Asia, greatly reduced the apparent width of the unknown part of the world between the Orient and Europe. More than 1,500 years later, these mistakes made it possible for Columbus to convince people he could reach Asia by sailing west.

Though it weathered the dissolution of Alexander's empire, the Library of Alexandria did not survive the subsequent period of Roman rule. The last librarian was Hypatia, the first notable woman mathematician, philosopher, and scientist. In Alexandria, she was a symbol of science and knowledge, concepts the early Christians identified with pagan practices. The mission of the library, as personified by the last librarian, antagonized the governors and citizens of the city of Alexandria. After years of rising tensions, in 415 C.E. a mob brutally murdered Hypatia and burned the library with all its contents. Most of the community of scholars dispersed, and Alexandria ceased to be a center of learning in the ancient world. The academic loss was incalculable, and trade suffered because shipowners no longer had a clearinghouse for updating the nautical charts and information they had come to depend on. All that remains of the library today is a remnant of an underground storage room and the floors of a few lecture halls (**Figure 2.5**). We will never know the true extent and influence of its collection of more than 700,000 irreplaceable scrolls.

Western intellectual development slackened during the aptly named Dark Ages that followed the fall of the Roman Empire in 476 C.E. For almost 1,000 years, until the European Renaissance, much of the progress in medicine, astronomy, philosophy, mathematics, and other vital fields of human endeavor was made by the Arabs or imported by them from Asia. For example, the Arabs used the Chinese-invented compass (shown later in Figure 2.11) for navigating caravans over seas of sand, and their understanding of the Indian Ocean's periodic winds—the monsoons—allowed an Arabian navigator to guide Vasco da Gama from East Africa to India in 1498.

But we're getting ahead of the story. Earlier, at the height of the Dark Ages, Vikings raided and explored to the south and west. Half a world away the Polynesians continued some of the most extraordinary voyages in history.

Seafaring Expanded Human Horizons

As one writer has noted, it's hard to differentiate true seafaring from a bit of boating gone horribly wrong! Accidental discovery was common in the early history of marine science, and its most dramatic practitioners were the Polynesians and the

Where We Are: Latitude and Longitude

A sphere has no edges, no beginnings or ends, so what should we use as a frame of reference for positioning and navigation? The question was first successfully addressed by geographers at the Library at Alexandria, in Egypt. In the third century B.C.E., Eratosthenes drew latitude and longitude lines through important places (see Figure 2.3). The Alexandrian perception of the world is reflected in the size of the continents and the central position of Alexandria.

A later Alexandrian scholar divided Earth into an orderly grid based on 360 increments, or "degrees" (*degre,* "step"). The equator was a natural dividing point for the north–south (latitude) positioning grid, but there was no natural dividing point for the east–west (longitude) grid. Not surprisingly, Alexandria was arbitrarily selected as the first "zero longitude" and a regular grid laid out east and west of that city.

The general scheme has withstood the test of time, but there has been controversy. Though use of the equator as "zero latitude" has never been in question, each seafaring country wanted the prestige of having the world's longitude centered on its capital. For centuries, maritime nations issued charts with their own longitude "zeros." After much political disagreement, nations agreed in 1884 that the Greenwich meridian near London would be the world's "zero longitude" **(Figures A–D)**. Given the accuracy of that meridian's known position and the long history and success of British navigation and timekeeping, Greenwich was an excellent choice.

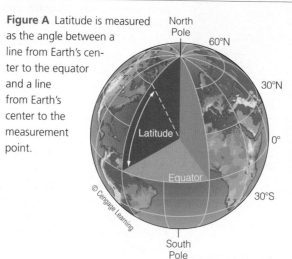

Figure A Latitude is measured as the angle between a line from Earth's center to the equator and a line from Earth's center to the measurement point.

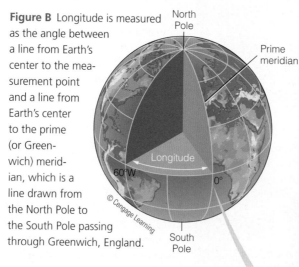

Figure B Longitude is measured as the angle between a line from Earth's center to the measurement point and a line from Earth's center to the prime (or Greenwich) meridian, which is a line drawn from the North Pole to the South Pole passing through Greenwich, England.

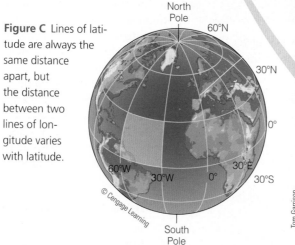

Figure C Lines of latitude are always the same distance apart, but the distance between two lines of longitude varies with latitude.

Figure D A tourist straddles the zero longitude line as it emerges from England's Greenwich Observatory.

Figure 2.6 The Polynesian triangle. Central Polynesia was settled between 500 and 1,000 B.C.E., but the explosive dispersion that led to the settlement of Hawai'i occurred about 800–1,000 C.E. Arrows show a possible direction and order of settlement. Note that South America was probably reached by Polynesian voyagers around the same time as their discovery of Hawai'i.

🗨 **THINKING BEYOND THE FIGURE**

Given their abilities, do you think Polynesians could have reached South America?

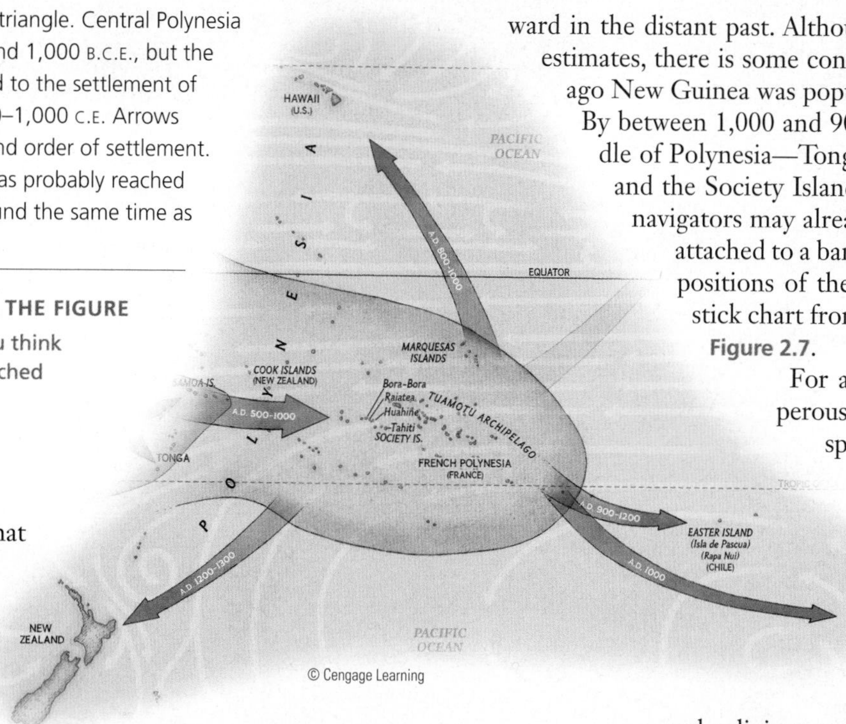

Vikings, civilizations that had no contact with each other (but were active at about the same time).

In the history of human migration, no voyaging saga is more inspiring than that of the colonization of **Polynesia**, the peopling of the central and eastern Pacific islands. A profound knowledge of the sea was required for these voyages, and the story of the Polynesians is a high point in our chronology of marine science applied to travel by sea.

The Polynesians are one of four cultures that inhabited some 10,000 islands scattered across nearly 26 million square kilometers (10 million square miles) of open Pacific Ocean **(Figure 2.6)**. The Southeast Asian ancestors of the Oceanian peoples, as these cultures are collectively called, spread east-

ward in the distant past. Although experts differ in their estimates, there is some consensus that by 6,000 years ago New Guinea was populated by these wanderers. By between 1,000 and 900 B.C.E. the so-called cradle of Polynesia—Tonga, Samoa, the Marquesas, and the Society Islands—was settled. Oceanian navigators may already have been using shells attached to a bamboo grid to represent the positions of their islands. A Micronesian stick chart from recent times is shown in **Figure 2.7**.

For a long and evidently prosperous period the Polynesians spread from island to island until the easily accessible islands had been colonized. Eventually, however, overpopulation and depletion of resources became a problem. Politics, intertribal tensions, and religious strife shook society. Groups of people scattered in all directions from some of the "cradle" islands during a period of explosive dispersion. Between 300 and 600 C.E., Polynesians successfully colonized nearly every inhabitable island within the vast triangular area shown in Figure 2.6. Easter Island was found against prevailing winds and currents, and the remote islands of Hawai'i were discovered and occupied. These were among the last places on Earth to be populated.

How did these risky voyages into unexplored territory come about? Religious warfare may have been the strongest stimulus to colonization. If the losers of a religious war were banished from the home islands under penalty of death, their only hope for survival was to reach a distant and hospitable new land.

Seafaring had been a long tradition in the home islands, but such trips called for radical new technology. Great dual-hulled sailing ships, some capable of transporting up to 100 people, were designed and built. New navigation techniques were perfected that depended on the positions of stars barely visible to the north. New ways of storing food, water, and seeds were devised. Whole populations left their home islands in fleets designed especially for long-distance discovery **(Figure 2.8)**. In some cases, fire was nurtured on board in case of landfall on an island that lacked volcanic flame. But a new island was only a possibility, a dream. How many fleets set out from the troubled homelands only to fall victim to storms, thirst, or other dangers?

Yet in that anxious time the Polynesians practiced and perfected their seafaring knowledge. To a skilled navigator, a change

Figure 2.7 A Micronesian stick chart. Islands (as shells) are not depicted in their fixed geographic relationship to one another; instead the shells show the islands in relation to the prevailing winds and ocean currents that would carry a canoe between them.

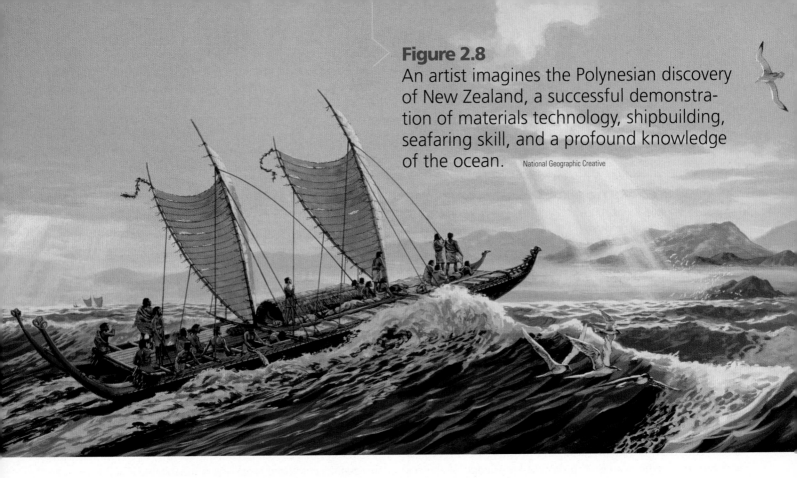

Figure 2.8
An artist imagines the Polynesian discovery of New Zealand, a successful demonstration of materials technology, shipbuilding, seafaring skill, and a profound knowledge of the ocean. National Geographic Creative

in the rhythmic set of waves against the hull could indicate an island out of sight over the horizon. The flight tracks of birds at dusk could suggest the direction of land. The positions of the stars told stories, as did the distant clouds over an unseen island. The smell of the water, or its temperature, or salinity, or color, conveyed information—as did the direction of the wind relative to the sun, and the type of marine life clustering near the boat. The sunrise colors, the sunset colors, hue of the moon—every nuance had meaning; every detail had been passed in ritual from father to son. The greatest Polynesian minds were navigators, and reaching Hawai'i was their greatest achievement.

Of all the islands colonized by the Polynesians, Hawai'i is farthest away, across an ocean whose guide stars were completely unknown to the southern navigators. The Hawai'ian Islands are isolated in the northern Pacific. There are no islands of any significance for more than 2,000 miles to the south. Moreover, Hawai'i lies beyond the equatorial doldrums, a hot and often windless stretch across which these pioneers must somehow have paddled. And yet some fortunate and knowledgeable people colonized Hawai'i sometime between 450 and 600 c.e. Try to imagine their feelings of relief and justification upon reaching a promised paradise under a new night sky. Think of that first approach to the high islands of Hawai'i, the first unlimited drink of fresh water, the first solid Earth after months of uncertainty!

Within a hundred years of their first arrival, Hawai'ian navigators were routinely piloting vessels on regular return trips to the Marquesas and the Society Islands (Tahiti and others). Some of the trips were undertaken to import needed food species to the newly found islands, but others were made to recruit new citizens and leaders to "green-clad Hawai'i."

At a time when seafarers of other civilizations sailed beside the comforting bulk of a charted coast, Polynesians looked to the open sea for sustenance, deliverance, and hope. Their great knowledge of the ocean protected them.

Viking Raiders Discovered North America

The Dark Ages were periodically punctuated by the raids of **Vikings**, bands of Scandinavian adventurers and treasure seekers whose remarkably fast, strong, and stable ships **(Figure 2.9)** enabled them to sail or row up rivers faster than a horse and rider could spread warning. Danish and Norwegian Vikings swept down the coast of Europe; they methodically pillaged Paris, robbed monasteries in Ireland, and looted Britain. The Swedish Vikings foraged as far away as Kiev and Constantinople! In 859 c.e., Vikings spent a week or so ashore in Morocco, rounding up prisoners for sale as slaves or to hold for ransom. Sixty-two Viking ships participated, a spectacular display of technology, sea power, seamanship, and navigation.

At first the Europeans were powerless against these marauders, but eventually the need for common defense overcame provincial hostility and xenophobia. One of the causes of the Renaissance in Europe may have been the experience of banding together for protection against these northern raiders.

When People First Arrived at Distant Places

The process of dating human migrations depends on a large number of seemingly unrelated variables, some of which may be surprising to you.

For example, the word used by the South American Incas for "sweet potato" is essentially identical to the word used by some Polynesians (*kūmara*). Does that mean Polynesians arrived in South America and settled there? The sweet potato has been radiocarbon dated in the Cook Islands of central Polynesia to 1000 C.E. Perhaps it was brought to Polynesia around 700 C.E., possibly by Polynesians who had traveled to South America and back. Then again, it is possible that South Americans themselves brought it into the Pacific on their travels. (You can look ahead to Chapter 9's illustrations of ocean currents to see if this makes sense.) As

for the seeds simply floating across the ocean from South America to Polynesia, that won't work—the seeds die quickly in saltwater. The plant needs people to spread by cuttings. So, was there contact between Polynesians and the Inca's ancient ancestors? The jury's still out on this one.

What about the Polynesian people themselves? Research published in 2011 suggests the ancestors of the Polynesians left the Asian mainland about 10,000 years ago and settled in Papua New Guinea about 6,000 years ago. These conclusions are based on analysis of DNA samples from 4,700 people in Southeast Asia (Vietnam, for example) and Polynesia. But the Polynesian language family has its roots in Taiwan, off the mainland coast of China, not in Southeast Asia. How to resolve

this dilemma? By bacteria! *Helicobacter pylori* is a bacterium that specifically colonizes the human stomach. Because it is unique to humans, it has spread and evolved genetically alongside its host since people emerged from Africa. The genetics of *Helicobacter* present in today's Asians and Polynesians indicate that humans spread from Taiwan in two waves. It was the second wave (carrying today's Maori strain) that traveled within humans to the Philippines and then on into the Polynesian home islands.

Combine the data from DNA, strains of bacteria, radiocarbon dating, language studies, and fossil remains, and scientists arrive at the dates and locations we've given in the text. But things change as scientific progress is made, and these dates and places will probably change, too.

The true genius of the Norwegian Vikings was revealed when they began to look westward. Iceland and Greenland had been discovered by ships blown off course during storms. Iceland was colonized by about 850 C.E., Greenland by 996. In an early voyage from Norway to Greenland in 986, a commuter named Bjarni Herjulfsson was blown past his goal by unfavorable winds. For about 5 days he sailed up and down the coast of a new land (which was, in fact, North America) without landing or making charts. His sketchy reports kindled a real-estate fever; Leif, son of Eric the Red, purchased Bjarni's ship and returned. His party found salmon-filled lakes, vines and grapes, and fodder for cattle in what was probably the northeastern tip of Newfoundland. In a bit of advertising overstatement, he called the place Vinland ("wine-land").

By 1000 C.E. the Norwegians had colonized Vinland. The settlements were modest, and at first relationships with the natives were encouraging. Unlike the Spanish, who landed in the New World 500 years later, the Norwegian colonists tried to cooperate with the locals, to learn from them, and to help them in a mutual pact of assistance. Unfortunately, misunderstandings arose, battles ensued, and north Atlantic weather turned colder. The colony had to be abandoned in 1020. The Norwegians lacked the numbers, the weapons, and the trading goods to make the colony a success.

> **T**he first global exploratory expeditions were undertaken by Chinese Admiral Zheng He in 1405.

The Vikings' oceanographic connection lies in their astonishing ships. Ships like the *Gokstad* (seen in this chapter's opener and Figure 2.9) were Europe's fastest, longest-ranging vessels and greatly ahead of their time. Ocean crossings in such ships were relatively dependable, and fleets could be constructed at reasonable expense. Their basic design was adaptable for long-range voyaging, yachting, cargo transfer, and even ceremonial use.

The Chinese Undertook Organized Voyages of Discovery

The extent of ancient Chinese contributions to oceanographic, geological, and geographic knowledge is only now becoming clear. By 1086, the Chinese philosopher Shen Kuo had deduced that Earth was of great age and that land had been shaped by sedimentary deposit, rock formation, uplift, and erosion over great spans of time (Figure 2.10). (It should be noted that until the mid-19th century, most western European scientists wrongly believed Earth to be between 6,000 and 10,000 years old.)

Later, shipbuilding and distant investigations began to occupy Chinese rulers. As the Dark Ages distracted Europeans, **Chinese navigators** became more skilled, and their vessels grew larger and more seaworthy. They then set out to explore the other side of the world. Between 1405 and 1433, Admiral Zheng He (pronounced "jung huh") commanded the greatest

Figure 2.9 Viking vessels like the *Gokstad* ship (seen here in an artist's conception) were the fastest, longest-ranging European vessels extant and greatly ahead of their time (the ninth century). Sturdy and intended for rapid travel over open water, the ship was 23.3 meters (76 feet) long and 5.25 meters (17 feet) wide. Thirty-two oarsmen sat on loose benches or chests, and holes for the oars could be covered by small disks so that water would not flow in when the ship heeled under sail. Cargo, weapons, and supplies were stored beneath movable deck plates, and a canvas tent could be rigged against the weather. Hervey Garret Smith/National Geographic Image Collection

fleet the world had ever known. At least 317 ships and 27,500 men undertook seven missions to explore the Indian Ocean, Indonesia, and around the tip of Africa into the Atlantic. Their aim: to display the wealth and power of the young Ming dynasty and to show kindness to people of distant places. The largest ship in the fleet, with nine masts and a length of 134 meters (440 feet) **(Figure 2.11)**, was a huge treasure ship carrying objects of the finest materials and craftsmanship. The mission of the fleet was not to accumulate such treasure but to give it away! Indeed, the primary purpose of these expeditions was to convince all nations with which the fleet had contact that China was the only truly civilized state and beyond any imaginable need for knowledge or assistance.

Many technical innovations had been required to make such an ambitious undertaking possible. In addition to inventing the compass, the Chinese invented the central rudder, watertight compartments, and sophisticated sails on multiple masts, all of which were critically important for the successful operation of large sailing vessels. Until Europeans adopted the rudder in about 1100, long-distance voyaging in a Western ship large enough to be stable in rough seas was usually difficult. Early Mediterranean traders and, later, the Polynesians and the Vikings had used specialized steering oars held against the right side (*steer-board* eventually became *starboard*) of their boats. Although this system worked well in protected waters, the small area of the steering oar (and the exposed position of the steersman) made it difficult to hold a course on long ocean passages. The centrally mounted, submerged rudder solved that problem. Also, dividing the ship into separate compartments below the waterline meant that flooding due to hull damage could be confined to a relatively small area of the ship, and the vessel could then be repaired and saved from sinking. Since sails provided the power to move, advances in sail design could drastically influence the success of any voyage. The Chinese fitted their trapezoidal or triangular sails with battens (pieces of bamboo inserted into stitched seams running the width of the sail) and placed the sails on multiple masts. The sails resembled venetian blinds covered with cloth. It was not necessary for Chinese sailors to climb the masts to unfurl the sails every time the wind changed; everything could be done from the deck with windlasses and

Figure 2.10 A painting by 11th-century Chinese artist Li Kung-Lin showing an anticlinal arch, an exposed cliff of layered and twisted rock strata. Philosophers in China realized that Earth was ancient and that land had been shaped by sedimentation, rock formation, uplift, and erosion over great spans of time. Their European counterparts didn't make this discovery until around 1800.

💬 **THINKING BEYOND THE FIGURE**
Why do you suppose the Chinese had a better grasp of Earth's age than the contemporary Europeans?

Painting by Li Kung-Lin showing the layers of the earth

lines. The shape of the sails made it easier to sail close to the wind in confined seaways.

Perhaps most astonishing of all, the Chinese fleet could stay at sea for nearly 4 months and cover at least 8,000 kilometers (5,000 miles) without re-provisioning. They distilled fresh water from seawater, grew fresh vegetables on board, provided luxurious staterooms for foreign ambassadors, and collected and cataloged large numbers of cultural artifacts and scientific specimens.

Despite enjoying these advances, the Chinese intentionally abandoned oceanic exploration in 1433. The political winds had changed, and the cost of the "reverse tribute" system was judged too great. Less than a century later, it was a crime to go to sea from China in a multimasted ship! In all, until late in the 20th century, the Chinese made very few contributions to our understanding of the ocean. Still, their voyaging technology filtered into the West and made subsequent discoveries possible.

CONCEPT CHECK

Before going on to the next section, check your understanding of some of the important ideas presented so far:

What advantages would a culture gain if it could use the ocean as a source of transport and resources?

How was the culture of the Library of Alexandria unique for its time? How was the size and shape of Earth calculated there?

What were the stimuli to Polynesian colonization? How were the long voyages accomplished?

What stimulated the Vikings to expand their exploration to the West? Were they able to exploit their discoveries?

What innovations did the Chinese bring to geology and ocean exploration? Why were their remarkable exploits abruptly discontinued?

2.2 The Age of European Discovery

Half a world away from their Polynesian and Chinese counterparts, Renaissance Europeans set out to explore the world by sea. They did not undertake exploration for its own sake, however; any voyage had to have a material goal. Trade between East and West had long been dependent on arduous and insecure desert caravan routes through the central Asian and Arabian deserts. This commerce was cut off in 1453 when the Turks captured Constantinople, and an alternative ocean route was needed.

Prince Henry Launched the European Age of Discovery

A European visionary who thought ocean exploration held the key to great wealth and successful trade was **Prince Henry the Navigator**, third son of the royal family of Portugal **(Figure 2.12)**. Prince Henry established a center at Sagres for the study of marine science and navigation "... through all the watery roads." Although he personally was not well traveled (he went to sea only twice in his life), captains under his patronage explored from 1451 to 1470, compiling detailed charts wherever they went. Henry's explorers pushed south into the unknown and opened the west coast of Africa to commerce. He sent out small, maneuverable ships designed for voyages of discovery and manned by well-trained crews. For navigation, his mariners used the **compass**—an instrument (invented in China in the fourth century B.C.E.) that points to a magnetic pole. Although Arab traders had brought the compass from China in the 12th century, navigators still considered it a magical tool. They concealed the compass in a special box (predecessor of today's binnacle) and consulted it out of view of the crew. Henry's students knew Earth was round, but because of the errors publicized by Claudius Ptolemy, they were wrong in their estimation of its size.

Hongnian Zhang/National Geographic Creative

A MING TREASURE SHIP
Perhaps 400 ft (122 m) long,
170 ft (52 m) wide

VASCO DA GAMA'S
SÃO GABRIEL
About 74 ft (23 m) long,
18 ft (5 m) wide

Gregory A. Harlin/National Geographic Stock

ASIA

IRAN

SAUDI
ARABIA

Persian Gulf

Hormuz

Jeddah   Mecca   *Arabian
Peninsula*

Red Sea

YEMEN   OMAN

Sanaa

Aden   Dhofar

SUDAN

Mukalla

*Arabian
Sea*

INDIA

BANGLADESH

Chittagong

CHINA

Great Wall   Beijing

MING EMPIRE

YUNNAN   JIANGSU

Kunyang

Origin of all
7 voyages   Nanjing

FUJIAN
Changle
Quanzhou
Xiamen

Dalian

*East
China
Sea*

Grand Canal

Yangtze

Kozhikode
(Calicut)

*Bay of
Bengal*

SOMALIA

KENYA

AFRICA

Nairobi
Malindi

Mogadishu
Baraawe

Pate I.
Lamu

*Swahili
coast*

Mombasa

TANZANIA

*Malabar
Coast*   Kochi
(Cochin)

(Quilon) Kollam

Jaffna
SRI LANKA
CEYLON

Colombo   Galle

MALDIVES

Dondra Head

*Andaman
Islands*
(India)

*Nicobar
Islands*
(India)

THAILAND
SIAM
Ayutthaya

CHAMPA

Hainan

CAMBODIA

VIETNAM

Qui Nhon

*South
China
Sea*

Banda Aceh
Semudera

Kelantan

MALAYSIA
Pahang
Malacca

EQUATOR

Sumatra

The Routes
—— Main route
- - - Subsidiary route
○ Major trading center

MOZ.

Present-day
boundaries shown
Scale varies in
this perspective.
Straight-line
distance between
Nanjing and
Mombasa is
5,730 miles
(9,221 kilometers).

ART BY GREG HARLIN
NG MAPS
©2005

INDIAN OCEAN

Palembang   INDONESIA

Java

Surabaya

National Geographic Maps/National Geographic Image Collection

Tom Garrison

Figure 2.11 The explorations of Chinese Admiral Zheng He.

a The purpose of the vast Chinese fleet was to show kindness to people of distant places and to warn that interference in Chinese affairs would be unwise. The fleet sailed the Pacific and Indian oceans between 1405 and 1433. At the end of his voyages, Zheng He (seen here) wrote: "We have traversed more than one hundred thousand li (64,000 kilometers, or 40,000 miles) of immense water spaces and have beheld in the ocean huge waves like mountains rising to the sky and we have set eyes on barbarian regions far away hidden in a blue transparency of light vapors, while our sails loftily unfurled like clouds day and night."[1]

b At least 10 ships of the types later used by Vasco da Gama or Christopher Columbus could fit on the treasure ship's 4,600-square-meter (50,000-square-foot) main deck. The rudder of one of these great ships stood 11 meters (36 feet) high—as long as Columbus's flagship Niña!

c The voyages of Zheng He, 1405–1433.

d A Chinese compass from the Ming era of exploration. The magnetized "spoon" rests on a bronze plate about 25 centimeters (10 inches) square. The handle of the "spoon" points south rather than north. The plate bears Chinese characters that denote the eight main directions.

💬 **THINKING BEYOND THE FIGURE**

After achieving such immense technological sophistication, what do you think prompted the Ming emperor to order these huge ships dismantled and exploration to cease?

A master mariner (and skilled salesman), **Christopher Columbus** "discovered" the New World quite by accident. Native Americans had been living on the continent for about 11,000 years, and the Norwegian Vikings had made about two dozen visits to a functioning colony on the continent 500 years before his noisy arrival; yet Columbus gets the credit. Why? Because his interesting souvenirs, exaggerated stories, inaccurate charts, and promises of vast wealth excited the imagination of royal courts. Columbus made North America a media event without ever sighting it!

Columbus wasn't trying to discover new lands. His intention was to pioneer a sea route to the rich and fabled lands of the East, made famous more than 200 years earlier in the overland travels of Marco Polo. As "Admiral of the Ocean Sea," Columbus was to have a financial interest in the trade routes he blazed. He was familiar with Prince Henry's work and, like all other competent contemporary navigators, knew Earth was spherical. He believed that by sailing west, he could come close to his eastern destination, whose latitude he thought he knew. Because of wishful thinking and dependence on Ptolemy's data, however, Columbus made the *smallest* estimate of Earth's size by any navigator in modern history; he assumed Earth to be only about half its actual size!

[1] F. Viviano, "China's Great Armada," *National Geographic*, vol. 208, no. 1, July 2005.

Not surprisingly, Columbus mistook the New World for his goal of India or Japan. He thought that the notable absence of wealthy cities and well-dressed inhabitants resulted from striking the coast too far north or south of his desired latitude. He made three more trips to the New World but went to his grave believing that he had found islands off the coast of Asia. He never saw the mainland of North America and never realized the size and configuration of the continents whose future he had so profoundly changed.

Other explorers quickly followed, and Columbus's error was soon corrected. Charts drawn as early as 1507 included the New World (**Figure 2.13**). Such charts perhaps inspired **Ferdinand Magellan** (**Figure 2.14**), a Portuguese navigator in the service of Spain, to believe that he could open a westerly trade route to the Orient. Unfortunately, the chart makers estimated the Americas and the Pacific Ocean to be much smaller than they actually are. Magellan was killed in the Philippines, and his men decided to continue sailing west around the world under the command of Juan Sebastián Elcano. Only 18 of the original crew of 260 survived, and they returned to Spain 3 years after they had set out. But they had proved it was possible to circumnavigate the globe.

The Magellan expedition's return to Spain in 1522 marks the end of the European Age of Discovery. An unpleasant era of exploitation of the human and natural resources of the Americas followed. Native empires were destroyed, and objects of

Pabkov/Shutterstock.com

Figure 2.12 Prince Henry of Portugal, the Navigator, looks westward from his monument in Portugal. In the mid-1400s, Henry established a center at Sagres for the study of marine science and navigation ". . . through all the watery roads."

Figure 2.13 The Waldseemüller Map, published in 1507—the first map to name America and to show the New World as separate from Asia. This is an image of the only known copy to survive of the 1,000 printed from 12 wood blocks. It was purchased in 2007 by the U.S. Library of Congress for US$10,000,000.

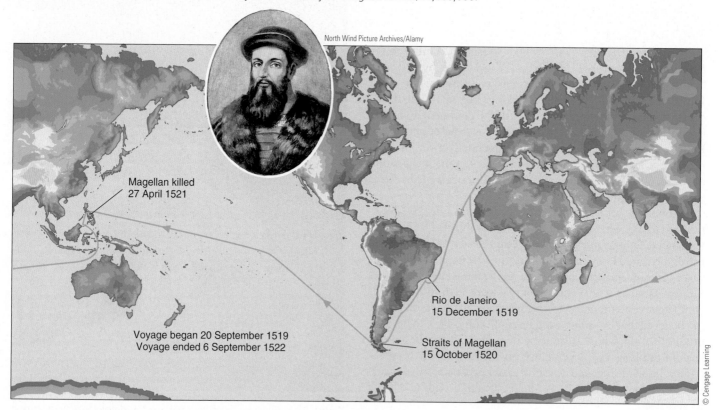

Magellan killed
27 April 1521

Rio de Janeiro
15 December 1519

Voyage began 20 September 1519
Voyage ended 6 September 1522

Straits of Magellan
15 October 1520

Figure 2.14 Ferdinand Magellan, a Portuguese explorer in service to Spain, whose expedition was first to circumnavigate the world, is shown with the track of his expedition. Magellan himself did not survive the voyage; only 18 out of 260 sailors managed to return after 3 years of dangerous travel.

💬 **THINKING BEYOND THE FIGURE**

Who was the *next* captain to circumnavigate Earth?

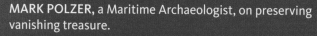

insight from
a national geographic explorer 2.1

MARK POLZER, a Maritime Archaeologist, on preserving vanishing treasure.

In terms of archeology, the ocean is the world's single greatest museum. The cultural resources it contains, as repository of so many shipwrecks, material objects, and submerged sites, preserve for us an astonishing amount of information about peoples and cultures from throughout our history. But just like much of the marine life in the ocean, our submerged cultural heritage is disappearing as well, and at an alarming rate. I hope that my work will help preserve some of that legacy and the information gleaned from it for future generations, and that it will inspire others to a greater awareness of this amazing asset and the dangers that threaten it.

Only once we appreciate the value of our cultural heritage in the ocean, and recognize that it belongs to all of us, and to posterity, will we be prompted to act to preserve it. —Mark Polzer

Source: http://www.nationalgeographic.com/explorers/bios/mark-polzer/
A diver rises from a Byzantine wreck with a basket full of artifacts.
Background photo: THOMAS J. ABERCROMBIE/National Geographic Creative

priceless cultural value were melted into coins to fund European warfare and greed.

CONCEPT CHECK

Before going on to the next section, check your understanding of some of the important ideas presented so far:

If he was not a voyager, why is Prince Henry of Portugal considered an important figure in marine exploration?

What were the main stimuli to European voyages of exploration during the Age of Discovery?

2.3 Voyaging Combined with Science to Advance Ocean Studies

British sea power arose after the Age of Discovery to compete with the colonial aspirations of France and Spain. Sailing ships required dependable supply and repair stations, especially in remote areas. The great powers sent out expeditions to claim appropriate locations, preferably inhabited by friendly peoples eager to help provision ships half a globe from home. The French sent Admiral Louis Antoine de Bougainville into the South Pacific in the mid-1760s. His 1768 claim for France of what is now called French Polynesia opened the area to the powerful European nations. The British followed immediately.

Captain James Cook: First Marine Scientist

Scientific oceanography begins with the departure from Plymouth Harbor in 1768 of HMS *Endeavour* under the capable command of **James Cook** of the British Royal Navy **(Figure 2.15)**. An intelligent and patient leader, Cook was also a skillful navigator, cartographer, writer, artist, diplomat, sailor, scientist, and dietitian. The primary reason for the voyage was to assert the British presence in the South Seas, but the expedition had numerous scientific goals as well. First, Cook conveyed several members of the Royal Society (a scientific research group) to Tahiti to observe the transit of Venus across the disk of the sun. Their measurements verified calculations of planetary orbits made earlier by Edmund Halley (later of comet fame) and others. Then, Cook turned south into unknown territory to search for a hypothetical southern continent, which some philosophers believed had to exist to balance the landmass of the Northern Hemisphere.

Cook and his men found and charted New Zealand, mapped Australia's Great Barrier Reef, marked the positions of numerous small islands, made notes on the natural history and human habitation of these distant places, and initiated friendly relations with many chiefs. Cook survived an epidemic of dysentery con-

tracted by the ship's company while ashore in Batavia (Jakarta) and sailed home to England, completing the voyage around the world in 1771. Because of his insistence on cleanliness and ventilation, and because his provisions included cress, sauerkraut, and citrus extracts, his sailors avoided scurvy, a vitamin C–deficiency disease that for centuries had decimated crews on long voyages.

The Admiralty was deeply impressed. Cook was promoted to the rank of commander and in 1772 was given command of the ships *Resolution* and *Adventure*, in which he embarked on one of the great voyages in scientific history. On this second voyage he charted Tonga and Easter Island and discovered New Caledonia in the Pacific and South Georgia in the Atlantic. He was first to circumnavigate the world at high latitudes. Though he sailed to 71° south latitude, he never sighted Antarctica. He returned home again in 1775.

Posted to the rank of captain, Cook set off in 1776 on his third, and last, expedition, in *Resolution* and *Discovery*.[2] His commission was to find a northwest passage around Canada and Alaska or a northeast passage above Siberia. He "discovered" the Hawai'ian Islands (Hawai'ians were there to greet him, of course) and charted the west coast of North America. After searching unsuccessfully for a passage across the top of the world, Cook retraced his route to Hawai'i to provision his ships for departure home. On 14 February 1779, after an elaborate farewell dinner with the chief of the island of Hawai'i, Cook and his officers

---

[2]The date 1776 rings a bell. Benjamin Franklin, himself an ocean scientist (as you'll soon learn), assured the British government that the young United States Navy would provide safe passage for Cook while the American Revolutionary war was in progress.

a In this 1776 painting by Nathaniel Dance, Cook is seen as a fully matured, self-confident captain who has twice circled the globe, penetrated into the Antarctic, and charted coastlines from Newfoundland to New Zealand.

SuperStock/SuperStock

Figure 2.15 Captain James Cook, Royal Navy.

b A cutaway of HMS *Endeavour.* The bluff but sturdy lines of the converted coal carrier are clearly evident, as are the cramped quarters endured by explorers and crew. A close look will reveal Captain Cook conferring with one of the naturalists in his cabin at the right.

Source: National Geographic Magazine Sept. 1971 pp. 308–309

prepared to return to *Resolution*, anchored in Kealakekua Bay. The Englishmen somehow angered the Hawai'ians and were beset by the crowd. Cook, among others, was killed in the fracas.

Cook deserves to be considered a scientist, as well as an explorer, because of the accuracy, thoroughness, and completeness in his descriptions. He and the scientists aboard took samples of marine life, land plants and animals, the ocean floor, and geological formations; they also reported the characteristics of these samples in their logbooks and journals. Cook's navigation was outstanding, and his charts of the Pacific were accurate enough to be used by the Allies in World War II invasions of the Pacific islands. He drew accurate conclusions, did not exaggerate his findings, and opened friendly diplomatic relations with many native populations. Cook recorded and successfully interpreted events in natural history, anthropology, and oceanography. Unlike many captains of his day, he cared for his men. He was a thoughtful and clear writer. This first marine scientist peacefully changed the map of the world more than any other explorer or scientist in history.

> **C**aptain James Cook deserves to be considered a scientist, as well as an explorer, because of the accuracy, thoroughness, and completeness in his descriptions.

a few days at sea. Indeed, they were governed by pendulums, which are useless in a rolling ship.

The key to the longitude problem was inventing a sturdy clock that ran at a constant rate under any circumstance, even the changeable conditions of a ship at sea. The problem was so pressing that in 1714 the British government formed The Board of Longitude to administer a set of prizes intended to encourage inventers—a £20,000 prize would be awarded to a method that could determine longitude within 30 nautical miles (56 kilometers). In 1728, **John Harrison**, a Yorkshire cabinetmaker, began working on a clock that would be accurate enough to determine longitude. His radical new timepiece, called a **chronometer**, was governed not by a pendulum but by a spring escapement. His first version was tested at sea in 1736, and Harrison was awarded £500 as encouragement to continue his efforts. Over the next 25 years he built three more clocks, culminating in 1760 in his Number Four **(Figure 2.16)**, perhaps the most famous timekeeper in the world.

Accurate Determination of Longitude Was the Key to Oceanic Exploration and Mapping

How did Cook (or Columbus, or any ocean explorer) know where he was? Unless explorers could record position accurately on a chart, exploration was essentially useless. They could not find their way home, nor could they or anyone else find the way back to the lands they had discovered.

At night, Columbus and his European predecessors used the stars to find latitude and, as a consequence, knew their position north or south of home. You can do this, too. In the Northern Hemisphere, take a simple protractor and measure the angle between the horizon, your eye, and the north polar star. The protractor reads approximately in degrees of latitude. To find the Indies, for example, Columbus dropped south to a line of latitude and followed it west. But to pinpoint a location, you need both latitude *and* the east–west position of longitude.

You can find longitude with a clock. First, determine local noon by observing the path of the shadow of a vertical shaft—it is shortest at noon—and set your clock accordingly. After traveling some distance to the west, you will notice that noon according to your *clock* no longer marks the time when the shadow of the *shaft* is shortest at your new location. If "clock" noon occurs 3 hours before "shaft" noon, you can do some simple math to see how far west of your starting point you have come. Earth turns toward the east, making one rotation of 360° in 24 hours, so its rotation rate is 15° per hour (360°/24 hours = 15°/hour). The 3-hour difference between "clock" noon and "shaft" noon puts you 45° west of your point of origin (3 × 15° = 45°). The more accurate the clock is (and the measurement of the shaft's shadow), the more accurate is your estimate of westward position.

The time method just described would work in theory, but in Columbus's time—and for many years afterward—no clocks were accurate enough to make this calculation practical after

Figure 2.16 The Number Four timekeeper, which won the £20,000 award offered by the British Board of Longitude. About the size of a modern wind-up alarm clock, it is functioning and on display at the National Maritime Museum in Greenwich, England.

National Maritime Museum, London/The Image Works

THINKING BEYOND THE FIGURE
Why was the Board of Longitude hesitant to pay Harrison the monetary reward for the chronometer?

Figure 2.17 A tourist peers through the zero longitude transit circle's northward extension in Greenwich, England. The longitude line may be seen on the pedestal extending toward the floor.

THINKING BEYOND THE FIGURE

Is there anything geographically special about the place where zero longitude is located? What about the zero latitude?

Tom Garrison

A sea trial of Number Four was begun in HMS *Deptford* in 1761. Harrison, too old and infirm to accompany the chronometer, sent his son and collaborator to tend the instrument. *Deptford* crossed the Atlantic from England to Jamaica and made a near-perfect landfall. After taking the clock's known error rate into account—its "rate of going" was 2⅔ seconds a day—the clock was found to be only 5 seconds slow. This would have meant an error in longitude of only 2.3 kilometers (1.4 miles), an astonishing achievement by the then-current standards of long-distance navigation.

Technically, Harrison had won the prize—Number Four had more than met the criteria set by the British Board of Longitude—but he was granted only part of the promised reward. Understandably, the officials would not hand over the money until it had been determined that the clock's secrets could be applied to quantity production. Just as understandably, Harrison did not wish his life's work compromised without compensation. He feared (correctly, as it turned out) that once the clockwork was examined by a competent watchmaker, his ideas would be copied. Finally, in 1769, a single copy was made; its success clinched Harrison's achievement. Captain Cook took a copy of Harrison's fourth chronometer on his last two voyages, but Harrison received the balance of the prize only in 1773 (when he was 80) and then only through the direct intervention of King George III.

All four of Harrison's chronometers are on view functioning in Britain's National Maritime Museum at Greenwich, in eastern London. Greenwich is an ideal site for the museum; in 1884 the Greenwich meridian, a longitude line at the naval observatory there, became "longitude zero" for the world **(Figure 2.17)**. Not since Eratosthenes's selection of Alexandria as the first "longitude zero" had Western nations recognized a common base for positioning.

CONCEPT CHECK

Before going on to the next section, check your understanding of some of the important ideas presented so far:

Captain James Cook has been called the first marine scientist. How might that description be justified?

Why was determining longitude so important? Why is it more difficult than determining latitude? How was the problem solved?

2.4 The First Scientific Expeditions Were Undertaken by Governments

Great as Cook's contributions undoubtedly were, his three voyages were not purely scientific expeditions. Cook was a serving British naval officer engaged in Crown business, mainly concerned with charting, foreign relations, and natural phenomena as they applied to Royal Navy matters. The first genuine *only-for-science* expedition may well have been the British *Challenger* expedition of 1872–1876, but the United States got into the act first with a hybrid expedition in 1838.

The United States Exploring Expedition Helped Establish Natural Science in America

After a 10-year argument over its potential merits, the **United States Exploring Expedition** was launched in 1838. It was also primarily a naval expedition, but its captain was somewhat freer in maneuvering orders than Cook had been. The work of the scientists aboard the flagship USS *Vincennes* and the expedition's five other vessels helped establish the natural sciences as reputable professions in America. Had it not been for the combative and disagreeable personality of its leader, Lieutenant Charles Wilkes **(Figure 2.18)**, this expedition might have become as famous as those of Cook or the later *Challenger* voyage.

The expedition departed on a 4-year circumnavigation. Its goals included showing the flag, whale scouting, mineral gathering, charting, observing, and pure exploration. One unusual goal was to disprove a peculiar theory that Earth was hollow and could be entered through huge holes at either pole.

Wilkes's team explored and charted a large sector of the east Antarctic coast and made observations that confirmed the landmass as a continent. A map of the Oregon Territory produced in 1841, one of 241 maps and charts drawn by members of the expedition, proved especially valuable when connected to the map of the Rocky Mountains prepared the following year by Captain John C. Fremont. Hawai'i was thoroughly explored, and Wilkes led an ascent of Mauna Loa, one of the two highest peaks of Hawai'i's largest island. James Dwight Dana, the Expedition's brilliant geologist, confirmed Charles Darwin's hypothesis of coral atoll formation (about which more will be

found in Chapter 12). The expedition returned with many scientific specimens and artifacts, which formed the nucleus of the collection of the newly established Smithsonian Institution in Washington, D.C. No evidence of polar holes was found.

Upon their return in 1842, Wilkes and his "scientifics" prepared a final report totaling 19 volumes of maps, text, and illustrations. The report is a landmark in the history of American scientific achievement.

Matthew Maury Discovered Worldwide Patterns of Winds and Ocean Currents

At about the time the Wilkes expedition returned, **Matthew Maury (Figure 2.19)**, a Virginian and fellow U.S. naval officer, became interested in exploiting winds and currents for commercial and naval purposes. After being crippled in a stagecoach accident, in 1842 Maury was given charge of the Navy's Depot of Charts and Instruments. There he studied a huge and neglected treasure trove of ships' logs, with their many regular readings of temperature and wind direction. By 1847 Maury had assembled much of this information into coherent wind and current charts. Maury began to issue these charts free to mariners in exchange for logs of their own new voyages.

Slowly a picture of planetary winds and currents began to emerge. Maury himself was a compiler, not a scientist, and he was vitally interested in the promotion of maritime commerce. His understanding of currents built on the work of **Benjamin Franklin**. Nearly a hundred years earlier, Franklin had noticed the peculiar fact that the fastest ships were not always the fastest ships; that is, hull speed did not always correlate with out-and-return time on the European run. Franklin's cousin, a Nantucket

Figure 2.18 Lieutenant Charles Wilkes soon after his return from the United States Exploring Expedition. Wilkes commanded the largest number of ships sent on such an expedition since the 15th century voyages of Chinese Admiral Zheng He.

Bettmann/CORBIS

Beverly Stautz/U.S. Naval Observatory Library

Figure 2.19 Matthew Fontaine Maury, compiler of winds and currents. Maury was perhaps the first person for whom oceanography was a full-time occupation.

Figure 2.20 Benjamin Franklin's 1769 chart of the Gulf Stream system. His cousin, Timothy Folger, discovered that Yankee whalers had learned to use the Gulf Stream to their advantage. Others, especially English ship owners, were slower to learn. Folger, himself a sea captain, wrote that Nantucket whalers ". . . in crossing it have sometimes met and spoke with those packets who were in the middle of and stemming it. We have informed them that they were stemming a current that was against them to the value of three miles an hour and advised them to cross it, but they were too wise to be counseled by simple American fishermen."

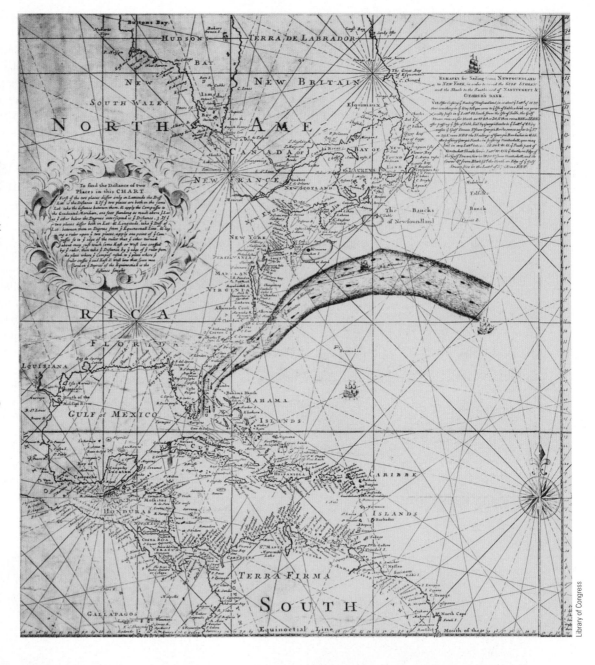

merchant named Tim Folger, noted Franklin's puzzlement and provided him with a rough chart of the "Gulph Stream" that he (Folger) had worked out. By staying within the stream on the outbound leg and adding its speed to their own, and by avoiding it on their return, captains could traverse the Atlantic much more quickly. It was Franklin who published, in 1769, the first chart of any current **(Figure 2.20)**.

But Maury was the first person to sense the worldwide pattern of surface winds and currents. Based on his analysis, he produced a set of directions for sailing great distances more efficiently. Maury's sailing directions quickly attracted worldwide notice: He had shortened the passage for vessels traveling

> **M**atthew Maury was perhaps the first person to undertake the systematic study of the ocean as a full-time occupation.

from the American East Coast to Rio de Janeiro by 10 days, and to Australia by 20. His work became famous in 1849 during the California gold rush—his directions made it possible to save 30 days on the voyage around Cape Horn to California. Applicable U.S. charts still carry the inscription "Founded on the researches of M. F. M. while serving as a lieutenant in the U.S. Navy." His crowning achievement, *The Physical Geography of the Seas*, a book explaining his discoveries, was published in 1855.

Maury, considered by many to be the father of physical oceanography, was perhaps the first person to undertake the systematic study of the ocean as a full-time occupation.

Lt. Pelham Aldrich, first lieutenant of HMS *Challenger*, kept a detailed journal of the *Challenger* Expedition. With accuracy and humor he kept this record in good weather and bad and had the patience and skill to include watercolors of the most exciting events. This is part of the first page of his journal.

The *Challenger* Expedition Was Organized from the First as a Scientific Expedition

The first sailing expedition devoted completely to marine science was conceived by Charles Wyville Thomson, a professor of natural history at Scotland's University of Edinburgh, and his Canadian-born student, John Murray. Stimulated by their own curiosity and by the inspiration of **Charles Darwin's** voyage in HMS *Beagle*, they convinced the Royal Society and British government to provide a Royal Navy ship and trained crew for a prolonged and arduous voyage of exploration across the oceans of the world. Thomson and Murray even coined a word for their enterprise: **oceanography**. Though the term literally implies only marking or charting, it has come to mean the science of the ocean. Prime Minister Gladstone's administration and the Royal Society agreed to the endeavor provided that a proportion of any financial gain from discoveries was handed over to the Crown. This arranged, the scientists made their plans.

HMS *Challenger*, a 2,306-ton steam corvette **(Figure 2.21a)**, set sail on 21 December 1872 on

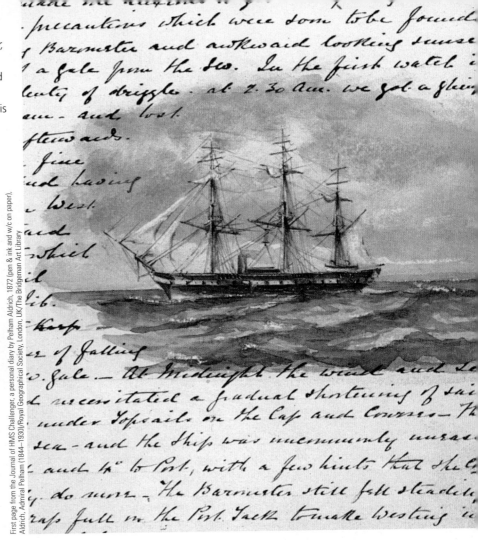

First page from the Journal of HMS Challenger, a personal diary by Pelham Aldrich, 1872 (pen & ink and w/c on paper). Aldrich, Admiral Pelham (1844–1930)/Royal Geographical Society, London, UK/The Bridgeman Art Library

Figure 2.21 The voyage of HMS *Challenger*.

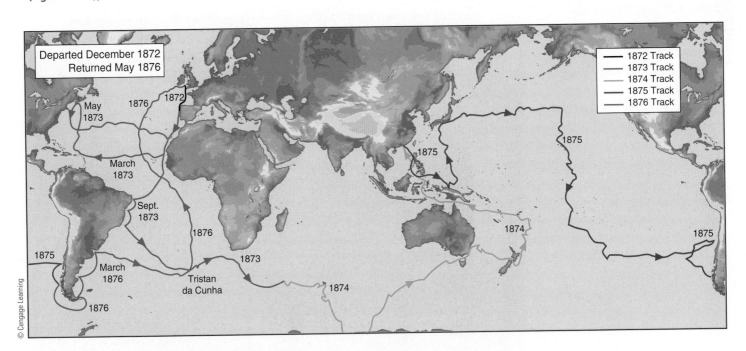

Departed December 1872
Returned May 1876

	1872 Track
	1873 Track
	1874 Track
	1875 Track
	1876 Track

May 1873 · 1876 · 1872 · 1875 · 1875

March 1873 · 1875

Sept. 1873 · 1875 · 1874

1876 · 1873 · 1875

1875 · 1873 · 1874

March 1876 · Tristan da Cunha · 1874

1876

© Cengage Learning

b HMS *Challenger's* track from December 1872 to May 1876. The Challenger Expedition remains the longest continuous oceanographic survey on record.

a 4-year voyage around the world, covering 127,600 kilometers (79,300 miles). Although the captain was a Royal Navy officer, the six-man scientific staff directed the course of the voyage. *Challenger's* track is shown in **Figure 2.21b.**

One important mission of the ***Challenger* expedition** was to investigate Edinburgh professor Edward Forbes's contention that life below 549 meters (1,800 feet) was impossible because of high pressure and lack of light. The steam winch on board made deep sampling practical, and samples from depths as great as 8,185 meters (26,850 feet) were collected off the Philippines. Through the course of 492 deep **soundings** with mechanical grabs and nets at 362 stations (including 133 dredgings), Forbes was proved resoundingly wrong. With each hoist, animals new to science were strewn on the deck; in all, staff biologists discovered 4,717 new species! **Figure 2.22** depicts researchers making some of these discoveries.

The scientists also took salinity, temperature, and water-density measurements during these soundings. Each reading contributed to a growing picture of the physical structure of the deep ocean. They completed at least 151 open-water trawls and stored 77 samples of seawater for detailed analysis ashore. The expedition collected new information on ocean currents, meteorology, and the distribution of sediments; the locations and profiles of coral reefs were charted. Thousands of pounds of specimens were brought to British museums for study. Manganese nodules, brown lumps of mineral-rich sediments, were discovered on the seabed, sparking interest in deep-sea mining. The work was agonizing and repetitive—a quarter of the 269 crew members eventually deserted!

In spite of the drudgery, this first pure oceanographic investigation was an unqualified success. The discovery of life in the depths of the oceans stimulated the new science of marine biology. The scope, accuracy, thoroughness, and attractive presentation of the researchers' written reports made this expedition a high point in scientific publication. The *Challenger Report*, the record of the expedition, was published between 1880 and 1895 by Sir John Murray in a well-written and magnificently illustrated 50-volume set. It is still used today. Indeed, it was the 50-volume *Report*, rather than the cruise, that provided the foundation for the new science of oceanography. The expedition's many financial spin-offs indicated that pure research was a good investment, and the British government realized quick profits from the exploitation of newly discovered mineral deposits on islands. The *Challenger* expedition remains history's longest continuous scientific oceanographic expedition.

With successes like these, the pace of exploration accelerated. American naturalist Alexander Agassiz, sailing in 1877 in the U.S. Coast and Geodetic Survey ship *Blake*, collected data corroborating the *Challenger* material at 355 deep-sea stations. The distribution of manganese nodules was found to be widespread. Further work by Agassiz and his students around the turn of the 20th century in the survey ship *Albatross* helped train a generation of influential American marine biologists. In 1886,

> The *Challenger* expedition was the first large research project devoted solely to marine science. It remains history's longest continuous scientific oceanographic expedition.

Figure 2.22 Scientists investigate specimens in the zoology laboratory aboard HMS *Challenger.*

THINKING BEYOND THE FIGURE

How do you suppose the scientists and the deck hands got along on this expedition?

Mary Evans/Science Source

Figure 2.23 Alfred Thayer Mahan, naval historian and strategist. Mahan served in the Union Navy in the American Civil War and was later appointed commander of the new U.S. Naval War College in 1886. He organized his lectures into his most influential book, *The Influence of Sea Power upon History, 1660–1783*. Received with great acclaim, his work was closely studied in Britain and Germany, influencing their buildup of forces in the years prior to World War I.

💬 **THINKING BEYOND THE FIGURE**

Consider the title of Mahan's work. Now think of the great sea battles of the First and Second World Wars. What point do you think Mahan was making between resource availability and distribution, and world power?

---

the Russians entered the field of marine exploration with the 3-year cruise of *Vitiaz* under the leadership of S. O. Makarov; their main contribution was a careful analysis of the salinity and temperature of North Pacific waters.

Ocean Studies Have Military Applications

Marine science is also applied to military interests. **Sea power** is the means by which a nation extends its military capacity onto the ocean. History has been greatly influenced by sea power—for example, the defeat of the Persian fleet by the Greeks at Salamis in 480 B.C.E. and the triumph of British Admiral Horatio Nelson over French forces at Trafalgar in 1805 led to eras of cultural and economic supremacy by both nations.

In 1892, **Alfred Thayer Mahan** (Figure 2.23), an American naval officer and historian, published *The Influence of Sea Power upon History, 1660–1783*. Based on his studies of the rise and fall of nation-states, this book had profound consequences for the development of the modern world. Mahan stressed the interdependence of military and commercial control of seaborne commerce and the ability of safe lines of transportation and communication to influence the outcomes of conflicts. Coming at a time of unprecedented technological improvements in shipbuilding, Mahan's work was read avidly in Great Britain, Germany, and the United States. For better or worse, the naval hardware, strategy, and tactics of the last century's greatest wars—along with their outcomes—was influenced by his clear analysis.

2.5 Contemporary Oceanography Makes Use of Modern Technology

In the 20th century, oceanographic voyages became more technically ambitious and expensive. Scientist-explorers sought out and investigated places that once had been too difficult to attain. Though the deep ocean floor was coming into reach, it was the forbidding polar ocean that attracted their first attentions.

Polar Exploration Advanced Ocean Studies

Polar oceanography began with the pioneering efforts of Fridtjof Nansen (Figure 2.24a). Nansen courageously allowed his specially designed ship *Fram* to be trapped in the Arctic ice, where he and his crew of 13 drifted with the pack for nearly 4 years (1893–1896), exploring to 85°57'N, a record for the time. The 1,650-kilometer (1,025-mile) drift of *Fram* proved that no Arctic continent existed. Nansen's studies of the drift, of meteorological and oceanographic conditions, of life at high latitudes, and of deep sounding and sampling techniques form the underpinnings of modern polar science.

Living up to its name—*Fram* means "forward" in Norwegian—Nansen's ship continued to play a pivotal role in exploration (Figure 2.24b). In 1910, Roald Amundsen, a student of Nansen's, set out in the sturdy little vessel for the coast of

Topical Press Agency/Hulton Archive/Getty Images

a Fridtjof Nansen, pioneering Norwegian ocean-ographer and polar explorer, looking every inch the Viking. In 1908, Nansen became the first professor of oceanography, a post created for him at Christiania University.

Figure 2.24 The rigors of polar exploration.

Alda Amundsen/Scott Polar Research Institute

b Nansen's 123-foot schooner *Fram* ("forward"). With 13 men, *Fram* sailed on 22 June 1893 to the high Arctic with the specific purpose of being frozen into the ice. *Fram* was designed to slip up and out of the frozen ocean and drifted with the pack ice to within about 4° of the North Pole. The whole harrowing adventure took nearly 4 years. The ship's 1,650-kilometer (1,025-mile) drift proved no Arctic continent existed beneath the ice. Living conditions aboard can be sensed from this recently rediscovered photograph.

Antarctica, the first leg of a journey to the South Pole. Nansen himself settled down to a long and distinguished career as an oceanographer, inventor, zoologist, artist, statesman, and professor. He was awarded the Nobel Peace Prize in 1922 for his unstinting work in worldwide humanitarian causes.

Modern technology has eased the burden of high-latitude travel. In 1958, under the command of Captain William Anderson, the U.S. nuclear submarine *Nautilus* sailed beneath the North Pole during a submerged transit beneath the Arctic pack from Point Barrow, Alaska, to the Norwegian Sea.

New Ships for New Tasks

In 1925, the German *Meteor* **expedition**, which crisscrossed the South Atlantic for 2 years, introduced modern optical and electronic equipment to oceanographic investigation. Its most

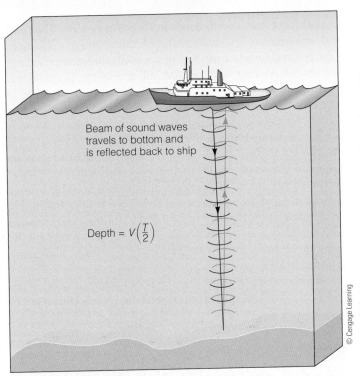

Figure 2.26 A model of R/V *Chikyu* ("Earth"), lead vessel in the 18-nation Integrated Ocean Drilling Program. *Chikyu* is 45% longer and 2.4 times the mass of *JOIDES Resolution,* the ship it replaces. Damaged in the 2011 tsunami, the ship has been repaired and is once again in operation.

Figure 2.25 Echo sounders sense the contour of the seafloor by beaming sound waves to the bottom and measuring the time required for the sound waves to bounce back to the ship. If the round-trip travel time and wave velocity are known, distance to the bottom can be calculated. This technique was first used on a large scale by the German research vessel *Meteor* in the 1920s.

💬 **THINKING BEYOND THE FIGURE**

Can you think of some factors that might affect the accuracy of an echo sounding?

important innovation was use of an **echo sounder**, a device that bounces sound waves off the ocean bottom, to study the depth and contour of the seafloor **(Figure 2.25)**. The echo sounder revealed to *Meteor* scientists a varied and often extremely rugged bottom profile rather than the flat floor they had anticipated.

In October 1951, a new HMS *Challenger* began a 2-year voyage that would make precise depth measurements in the Atlantic, Pacific, and Indian oceans and in the Mediterranean Sea. With echo sounders, measurements that would have taken the crew of the first *Challenger* nearly 4 hours to complete could be made in seconds. *Challenger II's* scientists discovered the deepest part of the ocean's deepest trench, naming it Challenger Deep in honor of their famous predecessor. In 1960, U.S. Navy lieutenant Don Walsh and Jacques Piccard descended into the Challenger Deep in *Trieste*, a Swiss-designed, blimplike bathyscaphe.[3]

In 1968, the drilling ship *Glomar Challenger* set out to test a controversial hypothesis about the history of the ocean floor. It was capable of drilling into the ocean bottom beneath more than 6,000 meters (20,000 feet) of water and recovering samples of seafloor sediments. These long and revealing plugs of seabed provided confirming evidence for seafloor spreading and plate

[3]More about this harrowing adventure will be found in Chapter 4.

tectonics. (The wonderful details will be found in Chapter 3.) In 1985, deep-sea drilling duties were taken over by the much larger and more technologically advanced ship *JOIDES Resolution*. Beginning in October 2003, deep-drilling responsibilities were passed to the Integrated Ocean Drilling Program (IODP), an international research consortium that operated a successor to *JOIDES Resolution* and an even larger drillship, R/V *Chikyu* ("Earth") **(Figure 2.26)**. The new Japanese ship, fully operational in 2007, contains equipment capable of drilling cores as much as 11 kilometers (7 miles) long! The vessel has equipment to control any flows of oil or gas, so it can safely drill deep into sedimentary basins on continental margins considered unsafe for *JOIDES Resolution*. This ship cost US$500 million and houses one of the most completely equipped geological laboratories ever put to sea.

The U.S. Navy operates five oceanographic survey ships, the newest of which is the USNS *Bruce C. Heezen*. Ships of this class crisscross the world ocean to map the seabed, take water samples, conduct acoustic tests, and launch and recover instrument packages. Joining these ships in 2010 was NOAA's new *Okeanos Explorer* **(Figure 2.27)**, the only research ship to have a dedicated ROV (remotely operated vehicle).

Oceanographic Institutions Arose to Oversee Complex Research Projects

The demands of scientific oceanography have become greater than the capability of any single voyage. Oceanographic institutions, agencies, and consortia evolved in part to ensure continuity of effort. The first of these coordinating bodies was founded by Prince Albert I of Monaco, who endowed his country's oceano-

Figure 2.27 NOAA's *Okeanos Explorer*, latest in a series of world-ranging oceanographic research vessels, and the only such ship to contain a dedicated ROV (remotely operated vehicle).

graphic laboratory and museum in 1906 **(Figure 2.28)**. The most famous alumnus of Albert's Institut Océanographique is Jacques Cousteau, co-inventor in 1943 of the scuba underwater breathing system. Monaco also became the site of the International Hydrographic Bureau, founded in 1921 as an association of maritime nations. This bureau published one of the first general charts of the ocean showing bottom contours.

In the United States, the three preeminent oceanographic institutions are the Woods Hole Oceanographic Institution on Cape Cod, founded in 1930 (and associated with the Massachusetts Institute of Technology and the neighboring Marine Biological Laboratory, founded in 1888); the Scripps Institution of Oceanography, founded in La Jolla, California, and affiliated with the University of California in 1912 **(Figure 2.29)**; and the Lamont–Doherty Earth Observatory of Columbia University, founded in 1949.

China is rapidly expanding its ocean research capabilities. Much of this effort is concentrated at the Ocean University of

a Founded by Prince Albert I of Monaco in 1908, the *Institut Océanographique* continues research and public-education programs in Monaco (seen here) and Paris. Jacques-Yves Cousteau, co-inventor of the SCUBA system, was director from 1957 to 1988.

b Scientists and technicians at work in the first autonomous oceanographic institution.

Figure 2.28 *Musée Océanographique,* Monaco.

a The Woods Hole Oceanographic Institution, Woods Hole, Massachusetts. Marine science has been an important part of this small Cape Cod fishing community since Spencer Fullerton Baird, then assistant secretary of the Smithsonian Institution, established the U.S. Commission of Fish and Fisheries there in 1871. The Marine Biological Laboratory was founded in 1888, the Oceanographic Institution in 1930. The institution buildings seen here surround calm Eel Pond.

Figure 2.29 Oceanographic Institutions.

b The Scripps Institution of Oceanography, La Jolla, California. Begun in 1892 as a portable laboratory-in-a-tent, Scripps was founded by William Ritter, a biologist at the University of California. Its first permanent buildings were erected in 1905 on a site purchased with funds donated by philanthropic newspaper owner E. W. Scripps and his sister, Ellen.

China in Qingado, an institution of some 19,000 undergraduate and graduate students dedicated almost exclusively to marine science.

The U.S. government has been active in oceanographic research. Within the Department of the Navy are the Office of Naval Research, the Office of the Oceanographer of the Navy, the Naval Oceanic and Atmospheric Research Laboratory, and the Naval Ocean Systems Command. These agencies are responsible for oceanographic research related to national defense. The **National Oceanic and Atmospheric Administration (NOAA)**, founded within the Department of Commerce in 1970, seeks to facilitate commercial uses of the ocean. NOAA includes the National Ocean Service, the National Weather Service, the National Marine Fisheries Service, and the Office of Sea Grant.

International consortia are playing a growing role in marine research. The IODP was noted earlier, and there are others. For example, the United States and Canada cooperate in the National Science Foundation's Ocean Observatories Initiative (OOI). In Project *Neptune*, the observatory's network of 900 kilometers (560 miles) of undersea cable now links 130 instruments and 400 sensors to the Internet, providing continuous monitoring of part of the northwestern Pacific Ocean system. The Census of Marine Life, a vast decade-long undertaking of 2,700 researchers in more than 80 nations, completed its scientific report in October 2010. Results of these endeavors will be discussed in later chapters.

Robot Devices Are Becoming More Capable

ROVs (remotely operated vehicles) are revolutionizing marine science. Telepresence systems now being perfected allow operators aboard ship or ashore to "feel" surface textures and forces

c The Ocean University in Qingdao, China. This is the world's largest institution dedicated exclusively to marine science education and research.

being applied at great depths and distances by a robot hand. Some robots are autonomous—they are not tethered to the ship and operate on their own, carrying out instructions programmed into them before they are released. Among the most capable of these is HROV *Nereus*, the deepest-diving vehicle now in operation (**Figure 2.30**). In May 2009, *Nereus* reached the bottom of the world's deepest ocean trench at a depth of 10,902 meters (35,768 feet).

The ability of remotely controlled robots to take samples and manipulate equipment in conditions far beyond the ability of humans to withstand was widely observed in the summer of 2010 during the *Horizon* Platform oil spill in the Gulf of

Woods Hole Oceanographic Institution

Figure 2.30 HROV *Nereus* in launch position. As a hybrid ROV, *Nereus* can act autonomously or under guidance from human operators aboard ship or ashore. The deepest-diving vehicle now in use, *Nereus* can reach depths of 11,000 meters (36,000 feet).

Mexico. Operated by technicians sometimes hundreds of miles from the site of the spill, robots brought eyes and energy to bear to turn valves, lift pipes, force plugs into place, and otherwise mitigate what would otherwise have been an even greater tragedy. A photo of one of these robots in action is included as **Figure 2.31**.

Satellites Have Become Important Tools in Ocean Exploration

The National Aeronautics and Space Administration (NASA), organized in 1958, has become an important institutional contributor to marine science. For 4 months in 1978, NASA's *Seasat*, the first oceanographic satellite, beamed oceanographic data to Earth. More recent contributions have been made by satellites beaming radar signals off the sea surface to determine wave height, variations in sea-surface contour and temperature, and other information of interest to marine scientists.

The first of a new generation of oceanographic satellites was launched in 1992 as a joint effort of NASA and the Centre National d'Études Spatiales (the French space agency). The centerpiece of **TOPEX/Poseidon**, as the project is known, is a satellite orbiting 1,336 kilometers (835 miles) above Earth in an orbit that allows coverage of 95% of the ice-free ocean every 10 days. The satellite's *TOPography EXperiment* uses a positioning device that allows researchers to determine its position to within 1 centimeter (½ inch) of Earth's center! The radars aboard can then determine the height of the sea surface with unprecedented accuracy. Other experiments in this 5-year program include sensing water vapor over the ocean, determining the precise location of ocean currents, and determining wind speed and direction.

Jason-1, NASA's ambitious follow-on to *TOPEX/Poseidon*, was launched in December 2001. Now flying 1 minute and 370 kilometers (230 miles) ahead of *TOPEX/Poseidon* on an identical ground track, its primary task is to monitor global climate interactions between the sea and the atmosphere. Its 5-year mission has been extended.

AQUA, one of three of NASA's next generation of Earth-observing satellites, was launched into polar orbit on 4 May 2002. It is the centerpiece of a project named for the large amount of information that will be collected about Earth's water cycle, including evaporation from the oceans; water vapor in the atmosphere; phytoplankton and dissolved organic matter in the oceans; and air, land, and water temperatures. *AQUA* flies in for-

HO/Reuters/Corbis

Figure 2.31 *Skandi Neptune,* a remotely operated vehicle (ROV), works near the *Horizon* platform's failed wellhead in the Gulf of Mexico on 21 July 2010.

NASA

Figure 2.32 NASA's "A-Train," a string of satellites that orbit Earth one behind the other on the same track. They are spaced a few minutes apart so their collective observations may be used to build three-dimensional images of Earth's atmosphere, ocean surface, and land topography.

mation with sisters *TERRA, AURA, PARASOL,* and *CloudSat* to monitor Earth and air **(Figure 2.32).**

A satellite system you can use every day? The U.S. Department of Defense has built the **Global Positioning System (GPS)**, a constellation of 24 satellites (21 active and 3 spare) in orbit 17,000 kilometers (10,600 miles) above Earth. The satellites are spaced so that at least four of them are above the horizon from any point on Earth. Each satellite contains a computer, an atomic clock, and a radio transmitter. On the ground, every GPS receiver contains a computer that calculates its own geographical position using information from at least three of the satellites. The longitude and latitude it reports are accurate to less than 1 meter (39.37 inches), depending on the type of equipment used. Handheld GPS receivers can be purchased for less than US$70—most "smartphones" contain this technology. The use of the GPS in marine navigation and positioning has revolutionized data collection at sea.

Satellite oceanography is an important frontier, and discoveries made by satellites are discussed in later chapters.

The remarkable progress of marine studies is shown in **Table 2.1.** Our predecessors would be proud, we think, of the progress we have made and excited about the discoveries that lie in the future.

CONCEPT CHECK

Before going on to the next section, check your understanding of some of the important ideas presented so far:

Why were oceanographic conditions at Earth's poles of interest to scientists?

How is the echo sounder an improvement over a weighted line in taking soundings? Which expedition first employed an echo sounder? Can you think of a few things that might cause echo sounding to give false information?

What stimulated the rise of oceanographic institutions?

Satellites orbit in space. How can a satellite conduct oceanography research?

What role does field research play in modern oceanography?

Table 2.1 Time Line for the History of Marine Science

Date	Event	Date	Event
4000 B.C.E.	Egyptian trade on Nile.	1742	Anders Celsius invents the centigrade temperature scale (see **Chapter 6**).
3800 B.C.E.	First maps showing water (river charts).	1758	Carolus Linnaeus publishes tenth edition of *Systema Naturae,* in which biological nomenclature is formalized (see **Chapter 13**).
1200 B.C.E.	Phoenicians trade from Mediterranean to Britain and West Africa.		
1000 B.C.E.	Polynesians first inhabit Tonga, Samoa.	1760	John Harrison's Number Four chronometer (see **Chapter 2**).
900 B.C.E.	Greeks first use the term *okeanos,* root of our word *ocean.*	1768	James Cook's first voyage of discovery (see **Chapter 2**).
800 B.C.E.	First graphic aids to marine navigation.	1769	Benjamin Franklin publishes first chart showing an ocean current (see **Chapter 2**).
600 B.C.E.	Greek Pythagoreans assume a spherical Earth.	1779	James Cook dies in Hawai'i.
325 B.C.E.	Pytheas voyages to Britain, links tides to movement of the moon (see **Chapter 11**); Chinese invent the compass.	1818	John Ross takes first deep-water and sediment samples.
300 B.C.E.	Library founded at Alexandria.	1831	Charles Darwin departs on 5-year voyage aboard HMS *Beagle* (see **Chapter 13**).
230 B.C.E.	Eratosthenes calculates circumference of Earth, invents latitude and longitude (see **Chapter 2**).		
127 B.C.E.	Hipparchus arranges latitude and longitude in regular grid by degrees.	1835	Gaspard Coriolis publishes first papers on an object's horizontal motion across Earth's surface (see **Chapter 8**).
A.D. 150	Claudius Ptolemy errs in estimating Earth's circumference.	1836	William Harvey devises a taxonomy of seaweeds (see **Chapter 14**).
A.D. 415	Library of Alexandria destroyed.	1838	Departure of the United States Exploring Expedition.
A.D. 500	Hawai'i colonized by Polynesians.	1847	Hans Christian Oersted observes plankton (see **Chapter 14**).
A.D. 780	Viking raids begin.		
1000	Norwegian colonies in North America.	1855	Matthew Maury publishes *Physical Geography of the Seas* (see **Chapter 2**).
1403	Chinese explorer Zheng He launches first global exploratory expedition.	1859	Darwin's *Origin of Species* published (see **Chapter 13**).
1460	Prince Henry the Navigator dies.	1872	Departure of *Challenger* expedition.
1492	Columbus's first voyage.	1877	Alexander Agassiz begins research in *Blake.*
1522	Magellan's crew completes first circumnavigation.	1880	William Dittmar determines major salts in seawater (see **Chapter 7**).
1609	Hugo Grotius publishes *Mare Liberum,* the foundation for all modern law of the sea (see **Chapter 18**).	1888	Marine Biological Laboratory founded at Woods Hole, Massachusetts.
1687	Isaac Newton's publication of *Principia Mathematica,* which includes an explanation of the operation of gravity (see **Chapter 11**).	1890	Alfred Thayer Mahan completes *The Influence of Sea Power upon History.*

(continued)

Table 2.1 Time Line for the History of Marine Science (*continued*)

Date	Event	Date	Event
1891	Sir John Murray and Alphonse Renard classify marine sediments (see **Chapter 5**).	1985	*JOIDES Resolution* replaces *Glomar Challenger* in Deep Sea Drilling Project (see **Chapters 2** and **3**).
1893	Fridtjof Nansen in Arctic in *Fram* (see **Chapter 2**).	1985	R. D. Ballard locates wreck of *Titanic*.
1900	Richard D. Oldham identifies P and S waves on seismograph (see **Chapter 3**).	1987	Observations of supernova 1987A confirm theories of the origin of elements (see **Chapter 1**).
1906	Prince Albert I of Monaco establishes the Institut Océanographique.	1991	*JOIDES Resolution* researchers bore to a depth of 2 kilometers (1.24 miles) beneath the seafloor near the Galápagos Islands (see **Chapter 3**).
1907	Bertram Boltwood calculates age of Earth by radioactive decay (see **Chapter 3**).	1992	U.S.–French *TOPEX/Poseidon* satellite launched.
1911	Roald Amundsen first at South Pole.	1995	*Kaiko*, a small remotely controlled Japanese submersible, sets a new depth record: 10,978 meters (36,008 feet) in the Challenger Deep.
1912	Alfred Wegener's Frankfurt lectures on continental drift (see **Chapter 3**).		
1912	Scripps Institution allied with the University of California.	1998	*Galileo* spacecraft finds possible evidence of an ocean on Jupiter's moon Europa (see **Chapter 1**).
1918	Vilhelm Bjerknes formulates theory of atmospheric fronts (see **Chapter 8**).	2000	*Mars Global Surveyor* photographs channels perhaps carved by flowing water (see **Chapter 1**).
1921	International Hydrographic Bureau founded.	2000	First comprehensive census of marine life undertaken.
1925	Departure of *Meteor* expedition; first echo sounder in operation (see **Chapters 2** and **3**).	2002	R/V *Chikyu*, lead ship of the Integrated Ocean Drilling Program, is launched (see **Chapters 2** and **3**).
1930	Woods Hole Oceanographic Institution founded.	2003	Inauguration of the Integrated Ocean Drilling Program (IODP) (see **Chapters 3** and **4**).
1931	*Atlantis* launched.		
1937	*E. W. Scripps* launched.	2004	Mars *Rover* explores Gusev crater to seek evidence of water. Lethal tsunami strikes the Indian Ocean. NOAA establishes GOESS (Global Earth Observation System of Systems).
1942	*The Oceans,* first modern reference text, published.		
1943	Jacques Cousteau and Emile Gagnan invent the scuba regulator and tank combination, the "aqualung."	2005	Researchers aboard *JOIDES Resolution* recover rocks more than 1,416 meters (4,644 feet) below the sea floor. Most active Atlantic hurricane season on record.
1949	Maurice Ewing forms the Lamont–Doherty Earth Observatory (see **Chapters 2** and **3**).	2006	President George W. Bush establishes the largest marine sanctuary in the tropical Pacific. Lakes of liquid methane found on Saturn's moon Titan.
1958	U.S. nuclear submarine *Nautilus* makes first submerged transit of the Arctic ice pack, passes through North Pole (see **Chapter 2**).	2007	R/V *Chikyu* drills first deep ocean cores (see **Chapter 3**).
1960	Bathyscaphe *Trieste* carrying Jacques Piccard and Don Walsh reaches bottom of deepest trench at 10,915 meters (35,801 feet).	2008	*Jason-2,* an ocean-sensing satellite, is launched.
		2008	Google Earth makes first three-dimensional seafloor images widely available.
1962	Rachel Carson's book *Silent Spring* initiates the U.S. environmental movement (see **Chapter 18**).	2009	European Space Agency launches the Gravity Field and Ocean Circulation Explorer satellite (GOCE).
1968	*Glomar Challenger* returns first cores, indicating the age of Earth's crust. The cores support the theory of plate tectonics (see **Chapters 3** and **5**).	2009	NOAA dispatches *Okeanos Explorer,* the first vessel with a permanent mission to explore the deepest reaches of the ocean.
1969	Santa Barbara, California, oil well blowout captures national attention (see **Chapter 18**).	August 2010	The Chinese announce *Jiaolong,* a submersible capable of diving to 7,000 meters—deeper than any human-carrying device currently in operation (see **Chapter 4**).
1970	National Oceanic and Atmospheric Administration (NOAA) established.		
1970	John Tuzo Wilson writes brief history of the tectonic revolution in geology in *Scientific American* (see **Chapter 3**).	October 2010	Culminating a 10-year exploration, 2,700 scientists from 80 nations report first Census of Marine Life.
1974	Project FAMOUS (French-American Mid-Ocean Undersea Study) maps and samples the Mid-Atlantic Ridge, a zone of seafloor spreading (see **Chapter 3**).	March 2011	9.0 earthquake and tsunami in central Japan (see **Chapters 3** and **10**).
		2011	*Aquarius* satellite launched.
1977	*Alvin* finds hydrothermal vents in Galápagos rift (see **Chapters 4** and **16**).	2012	James Cameron becomes the third person to reach the ocean's deepest spot (see **Chapter 4**).
1978	*Seasat,* the first satellite dedicated to ocean studies, is launched.	2014	DSRV *Alvin* returns to service after 3-year retrofit.

In this chapter you learned that science and exploration have gone hand-in-hand. Voyaging for necessity evolved into voyaging for scientific and geographical discovery. The transition to scientific oceanography was complete when the *Challenger Report* was published in 1895. The rise of the great oceanographic institutions quickly followed, and those institutions and their funding agencies today mark our path into the future.

In the next chapter you will learn how scientific exploration has discovered Earth's inner layers—layers that are density stratified. You'll find these layers to be heavier and hotter as depth increases, and you'll learn how we know what's inside our planet even though we've never been beneath the outermost layer. As you'll see, today's earthquakes and volcanoes, and the slow movement of continents, are all remnants of our distant cosmological past.

QUESTIONS FROM STUDENTS

1. **If the Alexandrian Library was so powerful and of such great value to learning and intelligent discourse, why was it so easy to turn local sentiment against its mission? Didn't the citizens of Alexandria appreciate the institution in their midst?**

Perhaps the Library was an easy target for destruction because there is no record of any of the researchers explaining or popularizing the monumental discoveries being made there. Scientific inquiry was the province of a privileged few, and—except for economic information available to traders—the librarians' intellectual achievements had little practical value. As Carl Sagan wrote, "Science never captured the imagination of the multitude. There was no counterbalance to stagnation, to pessimism, to the most abject surrenders to mysticism. When, at long last, the mob came to burn the Library down, there was nobody to stop them."[4]

2. **Did Columbus discover North America?**

No. He *never saw* North America.

3. **If Columbus was unsuccessful in his attempt to sail around the world and Magellan died without completing his circumnavigation, who was the first captain to complete the trip?**

Sir Francis Drake, of England, was the first captain to sail his own ship around the world. His expedition, begun in 1577, lasted 3 years. In the eastern Pacific, he raided Spanish merchants and captured a fortune in gold, silver, coins, and precious stones. He was the first European to sight the west coast of what is now Canada, and he claimed California for Queen Elizabeth I. His transit of the Pacific lasted 68 days, and on his return to England he concluded spice-trade negotiations with various heads of state; some of the agreements remain in effect to this day! On 26 September 1580 he returned to England a wealthy man, was knighted by the Queen, and went on in 1588 to help defeat the Spanish Armada.

Little was known of his explorations until comparatively recently. The trade and geographical information he brought home to England was considered so valuable that it was given the highest security classification—thus few people saw it or benefited from it!

4. **What was James Cook's motivation for those extraordinary voyages?**

One could say simply that he was a serving Royal Naval Officer and was ordered to go. But Beaglehole, Hough, and other biographers suggest the story is much more complex. How did a relatively unschooled man become leader of one of the first scientific oceanographic expeditions? Cook had the usual attributes of a successful person—intelligence, strength of character, meeting the right people at the right time, health, focus, luck—but he also had a driving intellectual curiosity and a rare (for that era) tolerance and respect for alien cultures. As Hough (1994) writes, "Cook stood out like a diamond amidst junk jewelery. . . ." It was no surprise that the Lords of the Admiralty settled upon this unique man to lead the adventure.

5. **What would you say is the most important "unknown" early marine oceanographic expedition?**

A French expedition into the Pacific occurred after Cook's last voyage but before Wilkes's. Backed by Napoléon and led by Nicolas Baudin, two lavishly equipped ships conveyed 23 scientists to study and map Australia from 1800 to 1804. They discovered 2,542 new species and returned the largest and most valuable natural history collection of its time—some 200,000 specimens and 30 live animals! Baudin's exploits are little known outside Australia because of British primacy in the area, but the French are rediscovering this part of their maritime history and voyage documents are being made available. Will we need to replace HMS *Challenger* as the first purely scientific oceanography expedition? Watch this space!

---

[4]Sagan, Carl. 1980. *Cosmos.* Random House. See especially pages 331–345.

6. How do modern navigators find their position at sea?

Very dull story. They push a few buttons on a small box and read their latitude and longitude directly on a screen. This is accomplished by analysis of radio transmissions from satellites. For about US$70, you can now buy a small, hand-held portable receiver capable of receiving GPS satellite signals. Most smart cell telephones also contain a GPS chip. The GPS system is accurate to about 1 meter (3 feet) and can even tell you which direction to go to get home (or anywhere else you want to go)! None of these methods is nearly as much fun as the old-fashioned sextant-and-chronometer method, but we suspect that any of the explorers mentioned in this chapter would be very impressed by our new tools.

7. What's this about a chronometer not having to keep perfect time? I thought you had to know exactly what time it is to be able to calculate your longitude.

Yes, you need accurate time. But a chronometer is valuable not because it necessarily keeps perfect time but because it loses or gains time at a constant, known rate. Each day, the navigator multiplies the number of seconds the clock is known to gain (or lose) by the number of days since the clock was last set—and then adds the total to the time shown on the chronometer's face to obtain the real time. The value of a chronometer lies entirely in its *consistency*.

8. You wrote that the future of oceanography lies in the big institutions. Is there a place for individual initiative in marine science?

Always. Every adventure begins with a person sitting quietly nurturing an idea. The notion may seem crazy at first, or it may seem impossible to prove or disprove, but the idea won't go away. He or she shares the idea with colleagues. If a research consensus is reached, plans are made, grants are proposed and funded, data flow. But the trail always begins with one person and his or her idea.

TERMS AND CONCEPTS TO REMEMBER

AQUA	echo sounder	longitude	*oceanus*
cartographer	Eratosthenes of Cyrene	Magellan, Ferdinand	Polynesians
Challenger expedition	Franklin, Benjamin	Mahan, Alfred Thayer	Prince Henry the Navigator
chart	government	marine science	sea power
Chinese navigators	GPS (Global Positioning System)	Maury, Matthew	sounding
chronometer		*Meteor* expedition	*TOPEX/Poseidon*
Columbus, Christopher	Harrison, John	National Oceanic and Atmospheric Administration (NOAA)	United States Exploring Expedition
compass	*Jason-1*		Vikings
Cook, James	latitude	oceanography	voyaging
Darwin, Charles Robert	Library of Alexandria		

STUDY QUESTIONS

Thinking Critically

1. How could you convince a 10-year-old child that Earth is round? What evidence would the child offer that it's flat? How can you counter those objections?
2. How did the Library of Alexandria contribute to the development of marine science? What happened to most of the information accumulated there? Why do you suppose the residents of Alexandria became hostile to the librarians and the many achievements of the library?
3. How did Eratosthenes calculate the approximate size of Earth? Which of his assumptions was the "shakiest"?
4. If Columbus didn't discover North America, then who did?
5. What were the contributions of Captain James Cook? Does he deserve to be remembered more as an explorer or as a marine scientist?
6. What was the first purely scientific oceanographic expedition, and what were some of its accomplishments?
7. Who was probably the first person to undertake the systematic study of the ocean as a full-time occupation? Are his contributions considered important today?
8. What famous American is also famous for publishing the first image of an ocean current? What was his motivation for studying currents?
9. What is an echo sounder? Can you think of some ways error could be introduced in an echo sounder's readings?
10. Sketch briefly the major developments in marine science since 1900. Do individuals, separate voyages, or institutions figure most prominently in this history?

Thinking Analytically

1. Imagine that you set your watch at local noon in Kansas City on Monday and then fly to the coast on Tuesday. You stick a pole into the ground on a sunny day at the beach, wait until its shadow is shortest, and look at your watch. The watch says 10:00 a.m. Are you on the East Coast or the West Coast? What is the difference in longitude? (Hint: 360° divided by 24 hours is 15°. The sun moves through the sky at a rate of 15° per hour.)
2. Look at "How Do We Know? 2.1: Where We Are: Latitude and Longitude." Provide a rough estimate of the latitude and longitude of your present position.
3. Magellan's crew kept very careful records of their circumnavigation, yet when they returned home, they were 1 day off. Why? Had they gained a day, or lost a day?
4. Replicate Eratosthenes's measurement of the diameter of Earth. Try this technique: Contact a friend who lives about 800 kilometers (500 miles) north or south of you (a distance comparable to the distance between Alexandria and Syene). Drive a tall pole into the ground at each location. Make sure the poles are vertical (using a weight on a string). Watch around noon, and when the pole casts the shortest shadow, measure the sun's angle of inclination from the shadow cast. Can you take it from there?
5. How do you think a satellite in space can study the ocean? It's up there, and the ocean is down here. . . .

GLOBAL GEOSCIENCE WATCH

Visit www.cengagebrain.com to access MindTap, a complete digital course which includes access to Global Geoscience Watch and more. Choose a historical figure, an expedition, or important research that has helped expand our understanding of marine science and research this topic within the GREENR database. Create an outline of the achievements, challenges, and consequences associated with your topic. Be prepared to discuss the achievements and importance of your topic's contributions to the field of marine science, and also if your topic has led to additional important discoveries.

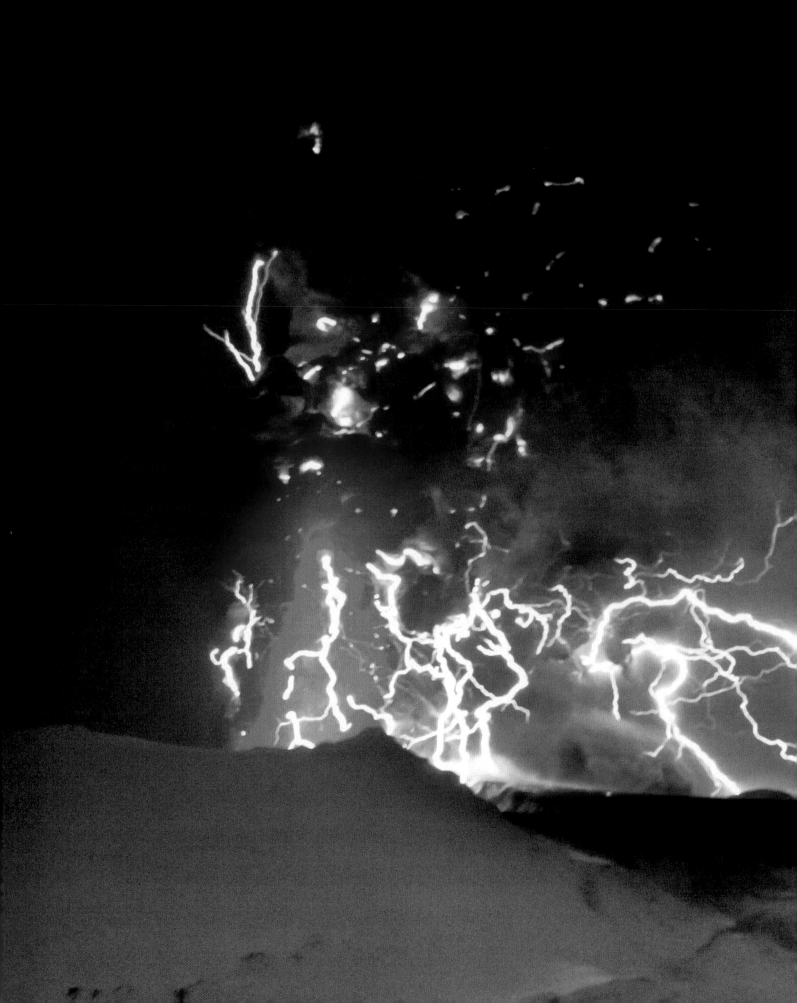

3 Earth Structure and Plate Tectonics

> ## KEY CONCEPTS

Earth's interior is layered, and the layers are arranged by density. Each deeper layer is denser than the layer above. © Cengage Learning

Continents rise above sea level because they float on a dense, deformable layer beneath them. © Cengage Learning

The brittle surface of Earth is fractured into about a dozen tilelike "plates." © Cengage Learning

Movement of the subterranean material on which these plates float moves them relative to one another. © Cengage Learning

Continents and oceans are formed and destroyed where the plates collide, flex, and sink. Patrick Taschler

Compelling evidence for plate movement is recorded in magnetic fields within the ocean floor. Cordelia Molloy/Science Source

Lightning bolts mix with lava and ash erupting from Eyjafjallajokull volcano in southern Iceland early on the morning of 18 April 2010. **Ash from the eruption closed Europe's airspace for days. Iceland lies on the Mid-Atlantic Ridge and is one of Earth's most geologically active places.**
David Jon/NordicPhotos/Getty Images

Media Connection

Start off this chapter by watching a video that models the plate movement that caused the tsunami of March 2011 in Japan. Visit www.cengagebrain.com to access MindTap, a complete digital course that includes this video and other resources.

©BBC

3.1 Pieces of Earth's Surface Look Like They Once Fit Together

In some places, the continents look as if they would fit together like jigsaw-puzzle pieces if the intervening ocean were removed. In 1620, Francis Bacon also wrote of a "certain correspondence" between shorelines on either side of the South Atlantic. **Figure 3.1a** shows this remarkable appearance. Could the continents have somehow been together in the distant past?

> **"**Nothing lasts long under the same form. I have seen what once was solid earth changed into sea, and lands created out of what once was ocean. Seashells lie far away from ocean waves. . .**"**
> —Ovid, *Metamorphoses*, Book XV

As they probed the submerged edges of the continents, marine scientists found that the ocean bottom nearly always sloped gradually out to sea for some distance and then dropped steeply to the deep-ocean floor. They realized that these shelflike continental edges were extensions of the continents themselves. Where they had measurements, researchers found that the fit between South America and Africa, impressive at the shoreline, was even better along the submerged edges of the continents. In an early use of computer graphics, researchers provided a best-fit view along these submerged edges **(Figure 3.1b)**.

Such an accurate fit almost certainly could *not* have occurred by chance.

A key to this curious puzzle had been provided 50 years earlier by **Alfred Wegener**, a busy German meteorologist and polar explorer **(Figure 3.2)**. In a lecture in 1912, he proposed a startling and original theory called **continental drift**. Wegener suggested that all Earth's land had once been joined into a single supercontinent surrounded by an ocean. He called the landmass **Pangaea** (*pan*, "all"; *gaea*, "Earth, land") and the

surrounding ocean **Panthalassa** (*pan*, "all"; *thalassa*, "ocean"). Wegener thought Pangaea had broken into pieces about 200 million years ago. Since then, he said, the pieces had moved to their present positions and were still moving.

Of course, Wegener's evidence included the apparent shoreline fit of continents across the North and South Atlantic, but he also commented on the alignment of mountain ranges of similar age, composition, and structure on both sides of the Atlantic. Sir Ernest Shackleton's 1908 discovery of coal, the fossilized remains of tropical plants, in frigid Antarctica did not escape Wegener's attention. Wegener was also aware of Edward Suess's discovery, in 1885, of the similarities of fossils found across these separated continents, especially fossils of the 1-meter (3.3-foot)-long reptile *Mesosaurus* and the seed fern *Glossopteris* **(Figure 3.3)**.

Wegener was not the first scientist to reassemble the continents according to geology and geometry (that honor probably goes to French geographer Antonio Snider-Pellegrini in 1858). But Wegener *was* first to propose a mechanism to account for the hypothetical drift. He believed that the heavy continents were slung toward the equator on the spinning

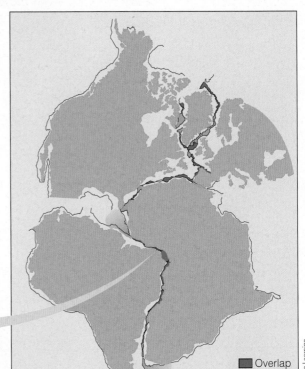

a From the time accurate charts became available in the late 1700s, observers noticed the remarkable coincidence of shape of the Atlantic coasts of Africa and South America.

b The fit of all the continents around the Atlantic at a water depth of about 137 meters (450 feet), as calculated by Sir Edward Bullard at the University of Cambridge in the 1960s. Note especially the relationship between Africa and South America. This early computer graphic was an effective stimulus to the tectonic revolution.

Overlap
Gap

© Cengage Learning

Figure 3.1 Corresponding coastlines around the South Atlantic.

Figure 3.2 Alfred Wegener studies his journal at the beginning of what would be his last expedition to Greenland, in 1930. His remarkable book *The Origin of Continents and Oceans* was published in 1915. In it, he outlined interdisciplinary evidence for his theory of continental drift. One year before this photo was taken, Wegener wrote: "It is as if we were to refit the torn pieces of a newspaper by matching their edges and then checking whether the lines of printing run smoothly across. If they do, there is nothing left but to conclude that the pieces were in fact joined in this way."

Earth by a centrifugal effect. This inertia, coupled with the tidal drag on the continents from the combined effects of sun and moon, would account for the phenomenon of drifting continents, he thought.

Wegener was dismissed as a crank. His detractors claimed, with some justification, that he had carefully selected only those data supporting his hypothesis, ignoring contrary evidence. Where, for instance, were the wakes or tracks through old seabed that the migrating continents would leave? But a few geologists sided with Wegener. These "drifters" were hesitant

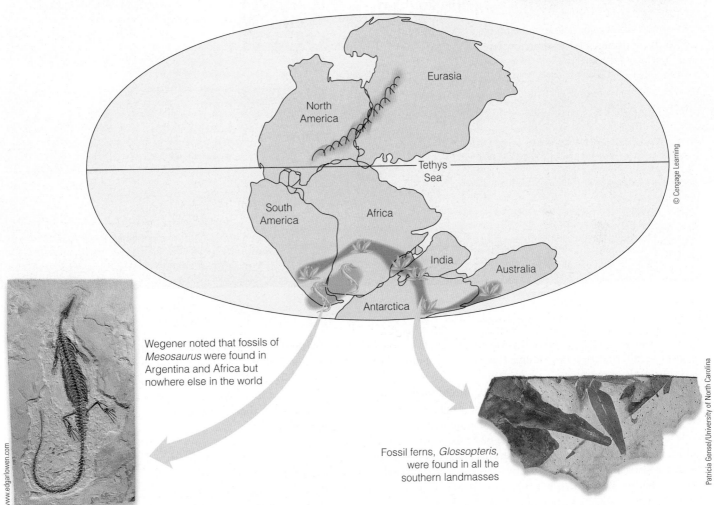

Wegener noted that fossils of *Mesosaurus* were found in Argentina and Africa but nowhere else in the world

Fossil ferns, *Glossopteris*, were found in all the southern landmasses

© Cengage Learning

Figure 3.3 Mountain ranges in Scandinavia, Scotland, and North America are now separated by the Atlantic but are remarkably similar in age and composition. Fossils of the reptile *Mesosaurus* were found in Argentina and Africa, but nowhere else. The seed fern *Glossopteris* was found in all the southern landmasses. *If the* continents were once joined, as shown here, these mountain ranges and fossil bands would have formed continuous chains.

Figure 3.4: Laborers in the TauTona Mine in South Africa hold the record for the deepest human penetration of Earth. In 2008 miners reached 3.9 kilometers (2.43 miles, 12,870 feet) beneath the surface, where the rock temperature is 55°C (131°F). They work in pairs, one chipping the rock and the other aiming cold air at his partner. After a time they trade places. Workers earn between 250 and 350 Ran (US$35–50) per 8-hour shift. Because of the extraordinarily high price of gold, the mine can make money by recovering just 10 grams (0.35 ounce) of gold from each metric ton of rock! © Tom Fox/Dallas Morning News/Corbis

to embrace the centrifugal force theory, yet they were unable to propose an alternative power source that could have moved the massive granitic continents.

The greatest block to the acceptance of continental drift was geology's view of Earth's mantle. The available evidence seemed to suggest that a deep, solid mantle supported the crust and mountains *mechanically*—like giant scaffolding—from below. Drift would be impossible with this kind of rigid subterranean construction. A few perceptive seismic researchers then noticed that the upper mantle reacted to earthquake waves as if it were a deformable mass, not a firm solid. Perhaps such a layer would resemble a slug of iron heated in a blacksmith's forge. If so, the upper mantle would deform with pressure and even flow slowly. Established geologists dismissed this interpretation, however, saying that the mountains would simply fall over or

sink without rigid underpinnings. By 1926 the "drifters" were in full retreat. When Wegener died on an expedition across Greenland in 1930, his theory was already in eclipse.

Does that mean continents don't move?

To answer that question, we need to investigate Earth's inner structure.

CONCEPT CHECK

Before going on to the next section, check your understanding of some of the important ideas presented so far:

What force did Wegener believe was responsible for the movement of continents?

What were the greatest objections to Wegener's hypothesis?

3.2 Earth's Interior Is Layered

As we saw in Chapter 1, Earth was formed by accretion from a cloud of dust, gas, ices, and stellar debris. Gravity later sorted the components by density, separating Earth into layers (see Figure 1.10). Because each deeper layer is denser than the layer above, we say Earth is **density stratified**. Remember, **density** is an expression of the relative heaviness of a substance; it is defined as the mass per unit volume, usually expressed in grams per cubic centimeter (g/cm^3). The density of pure water is $1\ g/cm^3$. Granite rock is about 2.7 times as dense, at $2.7\ g/cm^3$.

It might seem easy to satisfy our curiosity about the nature of inner Earth by digging or drilling for samples. People have descended to a depth of 3,900 meters (2.4 miles) in search of gold (**Figure 3.4**). The deepest hole drilled so far was bored by researchers on the Kola Peninsula in Russia. In 1994, they halted drilling at a depth of 12.262 kilometers (7.5 miles), short of their goal of 15 kilometers. High temperature (245°C, or 473°F) and pressure squeezed the hole closed. Drilling has also been conducted at sea. The oceanic drilling record is held by *JOIDES Resolution*. In 1991, after 12 years of intermittent effort, a drill aboard the ship penetrated 2 kilometers (1.2 miles) of seafloor beneath 2.5 kilometers (1.6 miles) of seawater. The R/V *Chikyu* (see Figure 2.26) can drill to a depth of 7 kilometers (4.4 miles) beneath the seafloor.

No matter where investigators drill, the samples of rock they recover are not exceptionally dense. The deepest probes have penetrated less than 1/500 of the radius of Earth, however. Should we assume that Earth consists of lightweight (less dense) rock all the way through?

Thanks to studies of Earth's orbit begun in the late 1700s, we know Earth's total mass. This mass is much greater than would have been predicted from even the deepest rocks ever collected. Earth's interior must therefore contain heavier (denser) substances than the rocks from the deep drill holes. We might expect materials from the interior to get denser gradually with increasing depth, but geologists have shown that the density of the materials increases abruptly at specific depths. Thus, geologists are convinced that Earth has *distinct interior layers*, somewhat resembling the inside of an onion (**Figure 3.5**).

CONCEPT CHECK

Before going on to the next section, check your understanding of some of the important ideas presented so far:

What do we mean when we say something is dense?

How is density expressed (units)?

Has anybody drilled into Earth's mantle or core to return samples of the densest interior layers?

3.3 The Study of Earthquakes Provides Evidence for Layering

Scientists have known since the mid-1800s that low-frequency waves can travel through the interior of Earth. The forces that cause **earthquakes** generate low-frequency waves called seis-

mic waves (*seismos*, "earthquake"). Some of these waves radiate through Earth, reflecting or bending as they travel, and eventually reappear at the surface. Careful study of the time and location of their arrival at the surface, along with changes in the frequency and strength of the waves themselves, has revealed information about the nature of Earth's interior. (We use the same kind of analysis to select a ripe watermelon. If we tap the outside and hear a *tick*, we suspect the melon isn't ripe. A *thunk* indicates a winner.)

Seismic Waves Travel through Earth and along Its Surface

Seismic waves form in two types: surface waves and body waves (**Figure 3.6**).

Surface waves move along Earth's surface. Like ocean waves, they ripple the free surface and can sometimes be seen as an undulating wavelike motion in the ground. Surface waves cause most of the property damage suffered in an earthquake.

Body waves are less dramatic, but they are very useful for analyzing Earth's interior structure. One kind of wave, the **P wave** (or primary wave), is a compressional wave similar in behavior to a sound wave. Rapidly pushing and pulling a very flexible spring (like a Slinky) generates P waves. The **S wave** (or secondary wave) is a shear wave like that seen in a rope shaken side to side. Both kinds of body waves are shown in Figure 3.6.

P waves and S waves are generated simultaneously at the source of an earthquake. P waves travel through Earth nearly twice as fast as S waves, so P waves arrive first at a distant **seismograph** (*seismos*, "earthquake"; *graphein*, "to write"), an instrument that senses and records earthquakes. Liquids are unable to transmit the side-to-side S waves but do propagate compressional P waves. Solid rock transmits both kinds of waves. Analysis of the characteristics of seismic waves returning to Earth's surface after passage through the interior suggests which parts of the interior are solid, liquid, or partially melted.

Earthquake Wave Shadow Zones Confirmed the Presence of Earth's Core

In 1900, the English geologist Richard Oldham first identified P and S waves on a seismograph. If Earth were perfectly homogeneous, body waves would travel at constant speeds from an earthquake, and their paths through the interior would be straight lines (**Figure 3.7a**, see p. 66). Oldham's investigations, however, showed that body waves were arriving *earlier* than expected at seismographs far from the quake. This meant that the waves must have traveled *faster* as they went down into Earth. They must also have been refracted—bent back toward the surface (**Figure 3.7b**, see p. 66).[1] Oldham reasoned that the waves were being influenced by passage through areas of Earth with different density and elastic properties from those seen at the surface. This reasoning showed that Earth is not homogeneous and that its properties vary with depth.

---

[1]The principle of refraction is explained and illustrated in Figure 6.21.

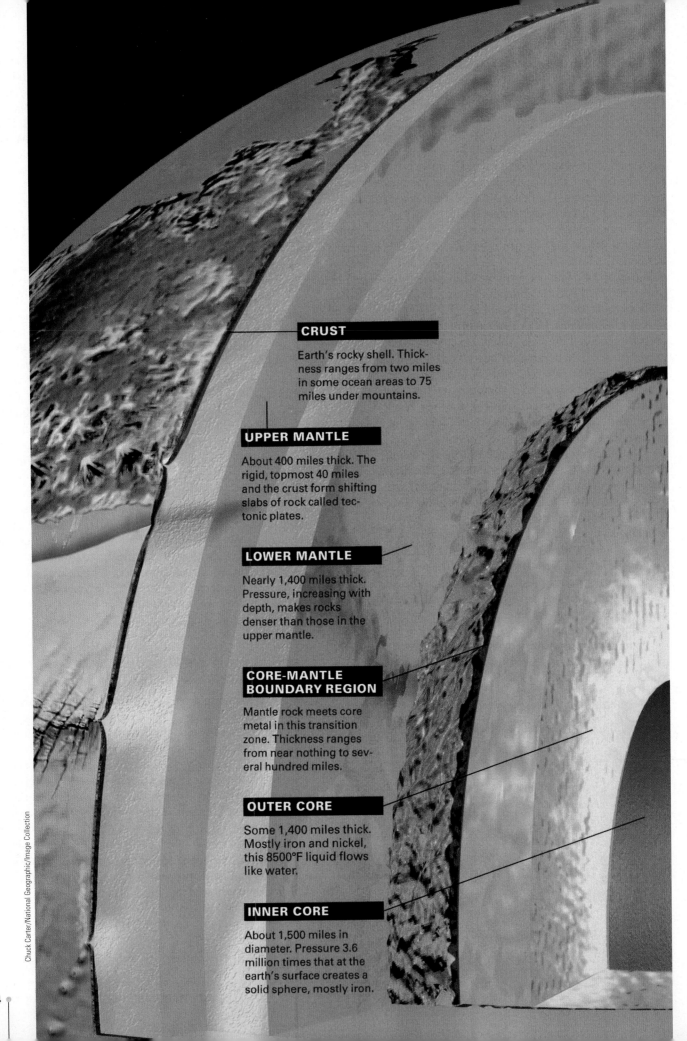

CRUST

Earth's rocky shell. Thickness ranges from two miles in some ocean areas to 75 miles under mountains.

UPPER MANTLE

About 400 miles thick. The rigid, topmost 40 miles and the crust form shifting slabs of rock called tectonic plates.

LOWER MANTLE

Nearly 1,400 miles thick. Pressure, increasing with depth, makes rocks denser than those in the upper mantle.

CORE-MANTLE BOUNDARY REGION

Mantle rock meets core metal in this transition zone. Thickness ranges from near nothing to several hundred miles.

OUTER CORE

Some 1,400 miles thick. Mostly iron and nickel, this 8500°F liquid flows like water.

INNER CORE

About 1,500 miles in diameter. Pressure 3.6 million times that at the earth's surface creates a solid sphere, mostly iron.

Figure 3.5 A cross section through Earth showing the internal layers.

💬 **THINKING BEYOND THE FIGURE**

Without looking ahead, can you think of a way scientists could have arrived at the conclusion that the inner Earth is layered?

In 1906, Oldham made a critical discovery: No S waves survived deep passage through Earth **(Figure 3.7d)**. Oldham deduced that a dense fluid structure, or core, must exist within Earth to absorb the S waves. He further predicted that a **shadow zone**, a wide band from which S waves were absent, would be found on the side of Earth opposite the location of an earthquake. The existence of the liquid core and shadow zone were verified by seismographic analysis in 1914.

What about the P waves—the compressional waves that can pass through liquid? Oldham found that P waves arrived at a seismograph farthest away from an earthquake (that is, on the opposite side of the globe) much more slowly than expected. They had been deflected but not stopped by Earth's core **(Figure 3.7c)**. Working from this information, later researchers were able to calculate that the mantle–core boundary is about 2,900 kilometers (1,800 miles) below the surface.

More sensitive seismographs were developed in the 1930s. In 1935, the Danish seismologist Inge Lehmann suggested that the very faint, very low frequency P waves discovered opposite earthquake sites had sped up as they passed through an inner core, indicating that it was a solid. Measurements of subtle differences in the pull of gravity, plus a more accurate estimate of Earth's mass (derived from precise timings of the orbits of artificial satellites), gave further clues to the layered structure of Earth.

Figure 3.6 Seismic waves travel through Earth.
a–c: © Cengage Learning

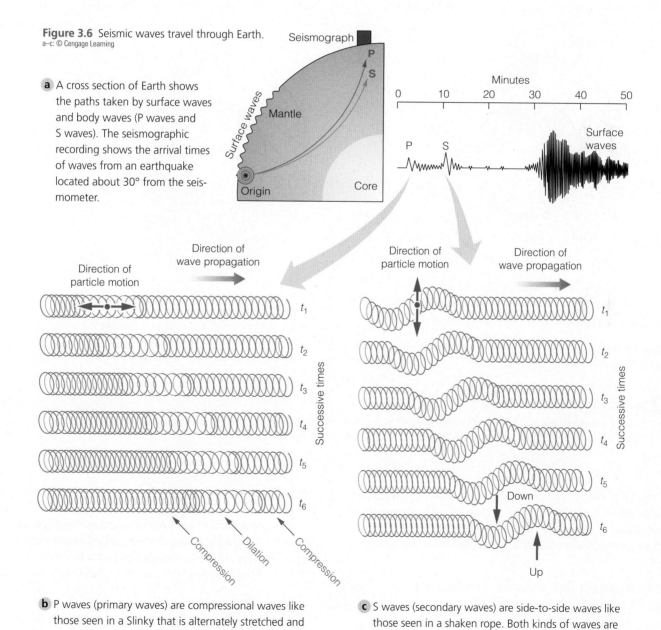

a A cross section of Earth shows the paths taken by surface waves and body waves (P waves and S waves). The seismographic recording shows the arrival times of waves from an earthquake located about 30° from the seismometer.

b P waves (primary waves) are compressional waves like those seen in a Slinky that is alternately stretched and compressed.

c S waves (secondary waves) are side-to-side waves like those seen in a shaken rope. Both kinds of waves are associated with earthquakes.

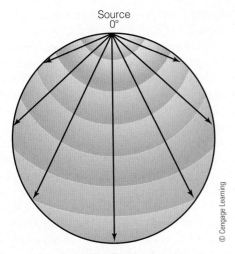

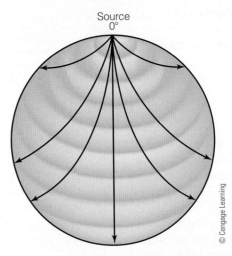

a Earthquake waves passing through a homogeneous planet would not be reflected or refracted (bent). The waves would follow linear paths (arrows).

b In a planet that becomes gradually denser and more rigid with depth, the waves would bend along evenly curved paths.

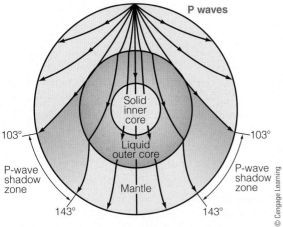

P waves

Solid inner core

Liquid outer core

Mantle

103° 103°

143° 143°

P-wave shadow zone

P-wave shadow zone

S waves

Solid inner core

Liquid outer core

Mantle

103° 103°

S-wave shadow zone

c P waves (compressional waves) can penetrate the liquid outer core but are bent in transit. A P-wave shadow zone forms between 103° and 143° from an earthquake's source.

d Our Earth has a liquid outer core through which the side-to-side S waves cannot penetrate, creating a large "shadow zone" between 103° and 180° from an earthquake's source. Very sensitive seismographs can sometimes detect weak P-wave signals reflecting off the solid inner core.

Figure 3.7 How earthquakes contributed to our model of the layered Earth.

THINKING BEYOND THE FIGURE

Inge Lehmann (see text) lived and worked in Denmark. What is special about Denmark's location and shadow zones? (Hint: Relative to Denmark, where do most great earthquakes occur?)

By the early 1960s another new generation of sensitive seismographs stood ready to provide geologists with an even better understanding of Earth's inner configuration. To confirm their theories, scientists needed data from a very large earthquake.

Data from an Earthquake Confirmed the Model of Earth Layering

They did not have to wait long.

On the last Friday of March 1964, at 5:36 p.m., one of the largest earthquakes ever recorded struck 144 kilometers (90 miles) east of Anchorage, Alaska **(Figure 3.8).**[2] The release of energy tore the surface of Earth for 800 kilometers (500 miles) between the small port of Cordova in the east and Kodiak Island in the west. In some places the vertical movement of the crust was 3.7 meters (12 feet); one small island was lifted 11.6 meters (38 feet). Horizontal movement caused the greatest damage:

[2]The magnitude of this earthquake—9.2—was the second greatest ever measured. (The greatest, at magnitude 9.5, struck Chile in 1960.) In comparison, the 2011 Tohoku earthquake centered west of Japan registered 9.0, tying three other quakes as the third largest in modern times.

Figure 3.8 Central Anchorage lies in ruins after the huge 1964 earthquake, one of the most powerful ever recorded.

WINFIELD PARKS/National Geographic Creative

65,000 square kilometers (25,000 square miles) of land abruptly moved west. In 4½ minutes of violent shaking, Anchorage had moved sideways 2 meters (6.6 feet), and the town of Seward had moved 14 meters (46 feet)! A seismic (earthquake-generated) sea wave destroyed two harbors. More than 75% of the state's commerce was disrupted, and thousands of people were made homeless. Damage exceeded US$750 million, and considering the violence of the earthquake, it is a wonder that only 115 lives were lost.

Seismic stations over much of the world saw extraordinarily large P waves arrive from Alaska. Many of the 800 seismographs online worldwide were physically damaged. The arrival times of the P and S waves at each station were carefully noted. When correlated with the frequency, intensity, and phase characteristics of the waves, this information helped confirm the models of Earth layering. The badly shaken citizens of Anchorage probably didn't derive much comfort from the knowledge gained from "their" earthquake, but this "natural experiment" confirmed theories of Earth's layering.

CONCEPT CHECK

Before going on to the next section, check your understanding of some of the important ideas presented so far:

What are the two kinds of seismic waves? Which causes most of the damage in an earthquake? Which are the more useful in determining the nature of Earth's interior?

Differentiate between P waves and S waves. Which can go through fluids? Through solids? How are "shadow zones" related?

How did the 1964 Alaska earthquake enhance our understanding of Earth's interior?

3.4 Earth's Inner Structure Was Gradually Revealed

Although researchers have never directly collected samples from below the outermost layer of Earth, they have indirect evidence about the chemical composition, density, temperature, and thickness of each layer of Earth's interior. This evidence was pieced together from measurements of earthquake shocks, volcanic gases, and variations in the pull of gravity.

Each of Earth's Inner Layers Has Unique Characteristics

Each layer inside Earth has different chemical and physical characteristics. One classification of Earth's interior emphasizes chemical composition. The uppermost layer is the lightweight, brittle, aptly named **crust**. The crust beneath the ocean differs in thickness, composition, and age from the crust of the continents. The thin **oceanic crust** is primarily **basalt**, a heavy dark colored rock composed mostly of oxygen, silicon, magnesium, and iron. By contrast, the most common material in the thicker **continental crust** is **granite**, a familiar speckled rock composed mainly of oxygen, silicon, and aluminum. The **mantle**, the layer beneath the crust, is thought to consist mainly of oxygen, iron, magnesium, and silicon. Most of Earth is mantle—it accounts for 68% of Earth's mass and 83% of its volume. The outer and inner **cores**, which consist mainly of iron and nickel, lie beneath the mantle at Earth's center.

Chemical makeup is not the only important distinction between layers. Different conditions of temperature and pressure occur at different depths, and these conditions influence the physical properties of the materials. The behavior of a rock is determined by three factors: temperature, pressure, and the rate at which a deforming force (stress) is applied. Geologists have therefore devised another classification of Earth's interior based on *physical* rather than *chemical* properties. These are shown in **Figure 3.9**.

- The **lithosphere** (*lithos* = rock)—Earth's cool, rigid outer layer—is 70–200 kilometers (44–125 miles) in thickness. It is composed of the continental and oceanic crusts *and* the uppermost cool and rigid portion of the mantle.
- The **asthenosphere** (*asthenes* = weak) is the hot, partially melted, slowly flowing layer of upper mantle below the lithosphere extending to a depth of about 350–650 kilometers (220–400 miles).
- The **lower mantle** extends to the core. The asthenosphere and the mantle below the asthenosphere (the lower mantle) have a

similar chemical composition. Although it is hotter, mantle below the asthenosphere does not melt because of rapidly increasing pressure. As a result, it is more dense and flows much more slowly.

- The **core** has two parts. The outer core is a dense, viscous liquid. The inner core is a solid with a maximum density about 16 g/cm^3, nearly six times the density of granite rock. Both parts are extremely hot, with an average temperature of about 5,500°C (9,900°F). Recent evidence indicates that the inner core may be as hot as 6,600°C (12,000°F) at its center, hotter than the surface of the sun! Curiously, the solid inner core also rotates eastward at a slightly faster rate than the mantle. The core accounts for only one-sixth of Earth's volume, but a third of its mass.

Figure 3.9 expands to show the lithosphere and asthenosphere in detail. *Note that the rigid sandwich of crust and upper mantle— the lithosphere—floats on (and is supported by) the denser deformable asthenosphere.* Note also that the structure of oceanic lithosphere differs from that of continental lithosphere. Because the thick granitic continental crust is not exceptionally dense, it can project above sea level. In contrast, the thin dense basaltic oceanic crust is almost always submerged.

Earth's Interior Is Heated by the Decay of Radioactive Elements

The temperature of Earth's core is about the same as the temperature of the surface of the sun! When research showed that Earth is about 4.6 billion years old, calculations of heat flow indicated that our planet should have cooled almost completely by now. Something must be heating the interior.

> **T**he temperature of Earth's core is about the same as the temperature of the surface of the sun.

In the late 1800s the British mathematician and physicist William Thomson (Lord Kelvin) calculated that Earth was about 80 million years old, an estimate he based on the rate at which the planet would have cooled from an original molten mass. Geologists tried to explain mountain building based on Lord Kelvin's assumption of progressive cooling. In their "drying-fruit" model, Earth was considered to have shrunk as it cooled. Mountains were thought to be shrinkage wrinkles—like those seen as a grape transforms into a raisin—and earthquakes were thought to be caused by jerks during this wrinkling. Earth's true age was not then known, and the "wrinkle" theory depended on the rapid cooling of a relatively young Earth. Earthquakes, active volcanoes, and hot springs clearly indicated Earth was still hot inside. There must be other sources of heat energy besides the trapped ancient heat of formation. Better explanations for mountain building and earthquakes were needed.

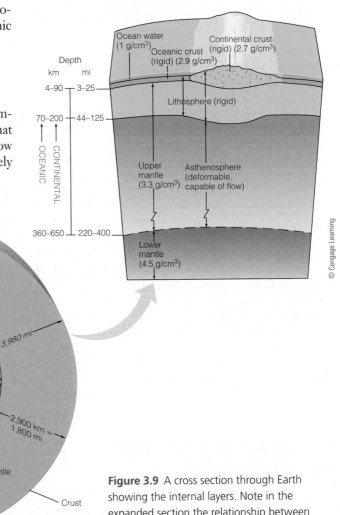

Figure 3.9 A cross section through Earth showing the internal layers. Note in the expanded section the relationship between lithosphere and asthenosphere, and between crust and mantle. This representation is not to scale.

The Age of Rocks

We can use the natural radioactivity of rocks.

Radioactive decay is the process by which unstable atomic nuclei break apart. As you have seen, radioactive decay is accompanied by the release of heat, some of which warms Earth's interior and helps drive the processes of plate tectonics.

Although it is impossible to predict exactly when *any one* unstable nucleus in a sample will decay, it is possible to discover the time required for one-half of *all* the unstable nuclei in a sample to decay. This time is called the half-life. Every radioactive element has its own unique half-life. For example, one of the radioactive forms of uranium has a half-life of 4.5 billion years, and a radioactive form of potassium has a half-life of 8.4 billion years. During each half-life, one-half of the remaining amount of the radioactive element decays to become a different element.

Radiometric dating is the process of determining the age of rocks by observing the ratio of unstable radioactive elements to stable decay products. Geologists consider radiometric dating a form of **absolute dating** because the age of a rock that contains a radioactive element may be determined with an accuracy of 1% to 2% of its actual age.

Figure A shows how samples may be dated by this means.

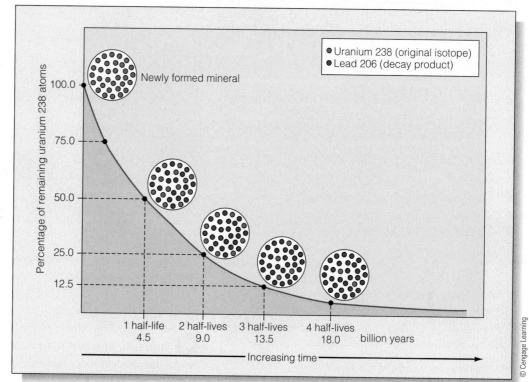

Figure A The rate of radioactive decay of a radioactive form of uranium (^{238}U) into lead. The half-life of ^{238}U is 4.5 billion years. During each half-life, one-half of the remaining amount of the radioactive element decays to become a different element. The assumption is that the system has remained closed—no radioactive atoms or decay products have been added or removed from the sample. Ages obtained from absolute and relative dating of continental rocks and rocks and sediments from the seabeds coincide with ages predicted by plate tectonic theory.

Relative dating is a method of dating a sample by comparing its position to the positions of other samples. Younger sediments are typically laid down over older deposits—events are placed in their proper sequence. If a group of rocks or fossilized remains contains no radioactive elements, researchers can determine whether the sample is older or younger than a different sample close by, but not the actual age of the assemblages. These two methods of dating can work together to determine the age of materials.

An important source of heat that was not recognized in Lord Kelvin's time is **radioactive decay**. Though most atoms are stable and do not change, some forms of elements are unstable and give off heat when their nuclei break apart (decay). Radioactive particles are ejected in the process. As we also saw in Chapter 1, radioactive decay within the newly formed Earth released heat that contributed to the melting of the original mass. Most of the melted iron sank toward the core, releasing huge amounts of energy. This residual heat combines with the much greater heat given off by the continuous decay of radioactive elements within the crust and upper mantle (primarily potassium, uranium, and thorium).

Some of Earth's internal heat journeys toward the surface by **conduction**, a process analogous to the slow migration of heat along a skillet's handle. Some heat also rises by **convection** in the asthenosphere and mantle. Convection occurs when a fluid or semisolid is heated, expands and becomes less dense, and rises. (Convection causes air to rise over a warm radiator—see, for example, Figure 8.9.)

So, even after 4.6 billion years, heat continues to flow from within Earth. As we shall see, this heat, not raisin-like global shrinkage, builds mountains and volcanoes, causes earthquakes, moves continents, and shapes ocean basins.

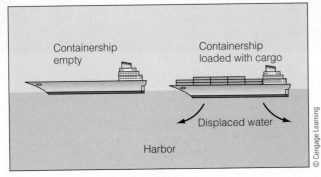

a A ship sinks until it displaces a volume of water equal in weight to the weight of the ship and its cargo.

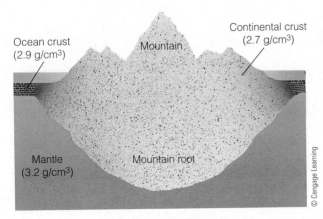

c Continents are supported in a similar way.

b Icebergs sink into water so that the same proportion of their volume (about 90%) is submerged. The more massive the iceberg, the greater this volume is. The large iceberg rides higher but also extends to a greater depth than the small one.

Figure 3.10 The principle of buoyancy.

 THINKING BEYOND THE FIGURE
Steel is heavier than water. How can a steel ship float?

Isostatic Equilibrium Supports Continents above Sea Level

Why do large regions of continental crust stand high above sea level? If the asthenosphere is partially melted and deformable, why don't mountains sink because of their mass and disappear? Another look at the expanded part of Figure 3.9 will help explain the situation. The mountainous parts of continents have "roots" extending into the asthenosphere. The continental crust and the rest of the lithosphere "float" on the denser asthenosphere. The situation involves buoyancy, the principle that explains why ships and icebergs float.

Buoyancy is the ability of an object to float in a fluid by displacing a volume of that fluid equal in weight to the floating object's own weight. *A steel ship floats because it displaces a volume of water equal in weight to its own weight plus the weight of its cargo.* An empty containership displaces a smaller volume of water than the same ship when fully loaded **(Figure 3.10a)**. The water that supports the ship is not *strong* in the mechanical sense; water does not support a ship the same way a steel bridge supports the weight of a car. Buoyancy, rather than mechanical strength, supports the ship and its cargo.

The buoyancy of an object depends on its density *and* its mass. Think of floating objects of different *sizes* but the same *density*—icebergs, for example. All icebergs float with about 10% of the volume exposed above sea level. Because the sub-

merged portion (the "root") of an iceberg is nine times as large as its exposed portion, larger icebergs ride higher and extend to greater depths than smaller ones **(Figure 3.10b)**. An iceberg 10 meters (33 feet) high will stick out of the water about 1 meter (3 feet), but an iceberg 100 meters (330 feet) high will stand 10 meters (33 feet) above sea level. *For an iceberg to stand high in the water, it must have a deep submerged portion to support it.*

Now think of continents. Any part of a continent that projects above sea level must be supported in the same way **(Figure 3.10c)**. Consider the continent that contains Mount Everest, the highest of Earth's mountains at 8.84 kilometers (29,007 feet) above sea level. Mount Everest and its neighboring peaks are not supported by the *mechanical* strength of the materials within Earth; nothing in our world is that strong. Over a long period, and under the tremendous weight of the overlying crust, the asthenosphere behaves like a dense, viscous, slowly moving fluid. *The continent's mountains float high above sea level because the lithosphere gradually sinks into the deformable asthenosphere until it has displaced a volume of asthenosphere equal in mass to the mountains' mass.* The mountains stand at great height, nearly in balance with their subterranean underpinnings but susceptible to rising or falling as erosion or crustal stresses dictate. Lower regions are supported by shallower "roots." In a slow-motion version of a ship floating in water, the entire continent stands in **isostatic equilibrium** (*isos*, "equal"; *stasis*, "standing").

Continental
crust

Mountains

Mantle

Low-density
mountain root

a

Transport   Erosion

Deposition

Subsidence

Uplift

b

Subsidence

c

© Cengage Learning

d

©Jarno Gonzalez Zarraonandia/Shutterstock.com

Figure 3.11 Erosion and isostatic readjustment can cause continental crust to become thinner in mountainous regions. As mountains are eroded over time (a–c), isostatic uplift causes their roots to rise. (The same thing happens when a ship is unloaded or an iceberg melts.) Further erosion exposes rocks that were once embedded deep within the peaks, sometimes exposing once-buried structures like Half-Dome in Yosemite Valley (d).

What happens when a mountain erodes? In much the same way as a ship rises when cargo is removed, Earth's crust will rise in response to the reduced load. Ancient mountains that have undergone millions of years of erosion often expose rocks that were once embedded deep within their roots. This kind of isostatic readjustment results in the thinning of the continental crust beneath the mountains and subsidence beneath areas of deposited sediments. This process is shown in **Figure 3.11**.

Unlike the asthenosphere on which the lithosphere floats, crustal rock does not slowly flow at normal surface temperatures. A ship or iceberg reacts to any small change in weight with a correspondingly small change in vertical position in the water, but an area of continent or ocean floor cannot react to every small weight change because the underlying rock is *not* liquid, the deformation does not occur rapidly, and the edges of the continent or seabed are mechanically bound to adjacent crustal masses. When the force of uplift or down-bending exceeds the mechanical strength of the adjacent rock, the rock will fracture along a plane of weakness—a **fault**. The adjacent crustal fragments will move vertically in relation to each other. This sudden adjustment of the crust to isostatic forces by fracturing, or faulting, is one cause of earthquakes. The lithosphere does not always behave as a rigid, brittle solid, however. Where forces

are applied slowly enough, some of this material may deform without breaking.

CONCEPT CHECK

Before going on to the next section, check your understanding of some of the important ideas presented so far:

How are Earth's inner layers classified?

What's the relationship between crust and lithosphere? Between lithosphere and asthenosphere?

Which part of Earth's interior is thought to be a liquid?

Why is the inside of Earth so hot?

How does heat move from the inner Earth to the surface?

How can something as heavy as a continent be so high? The Himalayas are more than 8,800 kilometers (29,000 feet) high—what holds them up there?

3.5 The New Understanding of Earth Evolved Slowly

Earth's layered internal structure—its brittle lithosphere floating on the hot and viscous asthenosphere—ensures that its surface will be geologically active. Earthquakes and volcanoes attest

Figure 3.12 Rocks at Siccar Point, Scotland, that helped convince James Hutton of Earth's great age. Both layers of rock, now at right angles, were laid down horizontally on the sea floor at different times. After the first layer was deposited, Earth's movements tilted it vertically and uplifted it to form mountains. Erosion wore down those mountains. When the lower layer was once again submerged, the upper layers were deposited horizontally on top. Now both sets of rock layers have been uplifted and eroded. Clearly this process required time—and lots of it.

to the commotion within. But how do internal layering and heat contribute to mountain building, the arrangement of continents, the nature of the seafloors, and the wealth of seemingly randomly distributed geological features found everywhere? Are there patterns and order in this apparent chaos?

Let's see how we achieved our current understanding of the answers to these questions.

The Age of Earth Was Controversial and Not Easily Determined

Geologists read Earth's history in its surface rocks and features. In the past the effort to understand geological processes was especially hindered by an incomplete understanding of Earth's age. Advances in geology had to wait—quite literally—for the time to be right.[3]

Interlinking lines of evidence now tell us the 4.6-billion-year age of Earth can be accepted with confidence. But before appropriate tools were developed, researchers were frustrated by tradition and seemingly contradictory data. At the end of the 18th century most European natural scientists believed in a young Earth, one that had formed only about 6,000 years ago. This age had been determined not by an analysis of rocks but through the genealogy of the Bible's Old Testament. Imaginative reading of the book of Genesis in 1654 had convinced Irish bishop James Ussher that the Creation had taken place on 26 October 4004 B.C.E.

[3]For information on geologic time and its divisions, please see Appendix 2.

James Hutton—a Scottish physician with an interest in geology—decided that the biblical account of the Creation was incorrect because it implied that the landscape was mostly stable and unchanging. Hutton, however, had measured geological changes: the rate at which streambeds eroded, the distribution of sediments by rivers, and the patterns of rocks in the Scottish countryside. From his observations, Hutton concluded that the rate of geological change today is not greatly different from the rate of change in the past. His principle of **uniformitarianism**, formalized in 1788, suggested that all of Earth's geological features and history could be explained by processes identical to ones acting today and that these processes must have been at work for a very long time **(Figure 3.12)**.

A few scientists agreed with Hutton. His detractors, however, asked some troubling questions: If Earth is very old, and if erosive forces have continued uniformly through time, why isn't Earth's surface eroded flat? Why isn't the ocean brimming with sediment? What post-Creation forces could build mountains?

Believers in another school of thought, **catastrophism**, were able to answer these objections by interpreting the biblical account of the Creation literally. The catastrophists were convinced that Earth was very young and that the biblical flood was responsible for the misleading appearance of Earth's great age. The flood, they maintained, had folded and exposed strata, toppled mountains, filled shallow ocean basins with sediments, and caused many plants and animals to become extinct. This hypothesis had the additional benefit of explaining how seashell fossils could be present on mountaintops.

A further complication was introduced in the late 1850s when Charles Darwin and Alfred Russell Wallace proposed a rational mechanism—natural selection—by which new kinds of living things might come about. Since natural selection required long periods of time to generate the overwhelming variety of life-forms on Earth, biological evidence also suggested an ancient Earth. The arguments intensified.

Scientists attempting to prove or disprove uniformitarianism, catastrophism, and biological evolution made a wealth of new discoveries during the last half of the 19th century. As we've seen, they improved the seismograph and discovered long-distance earthquake (seismic) waves. Others probed the ocean floors, drew more accurate charts, collected mineral samples from great heights and depths, measured the flow of heat from within Earth, and identified patterns in the worldwide distribution of fossils. All the theories were reassessed. *The evidence from these explorations convinced most researchers that Earth was truly of great age.* The stage was now set for a revolution in geology, the development of what we know today as the theory of plate tectonics. The first steps toward the theory were tentative, though, and as we have seen, some of its proponents were dismissed as lunatics.

CONCEPT CHECK

Before going on to the next section, check your understanding of some of the important ideas presented so far:

Why were traditional views of the age of Earth an impediment to early understanding of Earth's inner structure?

3.6 Wegener's Idea Is Transformed

Wegener's concept of continental drift had refused to die—those neatly fitted continents provided a haunting reminder of Wegener to anyone looking at an Atlantic chart.

In 1935, a Japanese scientist, Kiyoo Wadati, speculated that earthquakes and volcanoes near Japan might be associated with continental drift. In 1940, seismologist Hugo Benioff plotted the locations of deep earthquakes at the edges of the Pacific. His charts revealed the true extent of the **Pacific Ring of Fire**, a circle of violent geological activity surrounding much of the Pacific Ocean. Seismographs were now beginning to reveal a worldwide pattern of earthquakes and volcanoes. Deep earthquakes did not occur randomly over Earth's surface but rather were concentrated in zones that extended in lines along Earth's surface.

Benioff, Wadati, and others wondered what could cause such an orderly pattern of deep earthquakes. Many of the lines corresponded with a worldwide system of oceanic ridges, the first of which was plotted in 1925 by oceanographers aboard the research ship *Meteor* working in the middle of the North Atlantic. Benioff's sensitive seismographs also began to gather strong evidence for a deformable, nonrigid layer in the upper mantle. Could the continents somehow be sliding on that layer?

Other seemingly unrelated bits of information were accumulating. Radiometric dating of rocks, discussed earlier, was perfected after World War II. To the surprise of many geologists, the maximum age of the ocean floor and its overlying sediments was radiometrically dated to less than 200 million years—only about 4% of the age of Earth! The centers of the continents are *much* older. Some parts of the continental crust are more than 4 billion years old, about 90% of the age of Earth. *Why was oceanic crust so young?*

Attention had turned to the deep-ocean floors, the complex profiles of which were now being revealed by **echo sounders**—devices that measure depth by bouncing high-frequency sound waves off the bottom (see again Figure 2.25). In particular, scientists aboard the Lamont–Doherty Geological Observatory deep-sea research vessel *Vema* (a converted three-masted schooner) invented deep survey techniques as they went. After World War II, they probed the bottom with powerful echo sounders and looked beneath sediments with reflected pressure waves generated by surplus Navy depth charges dropped gingerly overboard. The overall shape of the Mid-Atlantic Ridge was slowly revealed. The ridge's conformance to shorelines on either side of the Atlantic raised many eyebrows. Ocean-floor sediments were thickest at the edge of the Atlantic and thinnest near this mid-ocean ridge.

Mantle studies were keeping pace. The first links in the Worldwide Standardized Seismograph Network, begun during the International Geophysical Year in 1957, were beginning to report data from seismic waves reflected and refracted through the planet's inner layers. This information verified the existence of a layer in the upper mantle that caused a decrease in the velocity of seismic waves. This finding strongly suggested that the layer was deformable. Perhaps the lithosphere was isostatically balanced in this partially melted layer, and perhaps continents could move around in it *if* a suitable power source existed.

CONCEPT CHECK

Before going on to the next section, check your understanding of some of the important ideas presented so far:

How did a careful plot of earthquake locations affect the discussion of the Theory of Continental Drift (as it was first called)? What *about* the jigsaw-puzzle-like fit of continents around the Atlantic?

How did echo sounding and an understanding of radiometric dating influence the debate?

3.7 The Breakthrough: From Seafloor Spreading to Plate Tectonics

In 1960, Professor Harry Hess of Princeton University and Robert Dietz of Scripps Institution of Oceanography proposed a radical idea to explain the features of the ocean floor and the "fit" of the continents. They suggested that new seafloor develops at the Mid-Atlantic Ridge (and the other newly discovered ocean ridges) and then spreads outward from this line of origin. Continents would be carried along by the same forces that cause the ocean to grow. This motion could be powered by **convection currents**, slow-flowing circuits of material within the mantle **(Figure 3.13)**.[4]

Seafloor spreading, as the new hypothesis was called, pulled many loose ends together. If the mid-ocean ridges were **spreading centers** and sources of new ocean floor rising from the asthenosphere, then they should be hot. They were—indeed, they were found to be lines of volcanoes! If the new oceanic crust cooled as it moved from the spreading center, then it should shrink in volume and become denser, and the ocean should be deeper farther from the spreading center. It was. Sediments at the edges of the ocean basin should be thicker than those near the spreading centers. They were, and they were also older.

Did this mean that Earth was continuously expanding? Since there was no evidence for a growing Earth, the creation of new crust at spreading centers would have to be balanced by the destruction of crust somewhere else. Then researchers discovered that the crust plunges down into the mantle along the periphery of the Pacific. The process is known as **subduction**, and these areas are called **subduction zones** (or Wadati–Benioff zones in honor of their discoverers).

In 1965, the ideas of continental drift and seafloor spreading were integrated into the overriding concept of **plate tec-**

> In 1965, the ideas of continental drift and seafloor spreading were integrated into the overriding concept of plate tectonics.

---

[4]Curiously, the English geologist Arthur Holmes first proposed the idea of mantle convection as Wegener's ideas were being ridiculed. This idea was swept away with little attention at the time.

tonics (*tekton*, "builder"; the English word *architect* has the same root), primarily by the work of **John Tuzo Wilson**, a geophysicist at the University of Toronto. In this theory Earth's outer layer consists of about a dozen separate major litho-

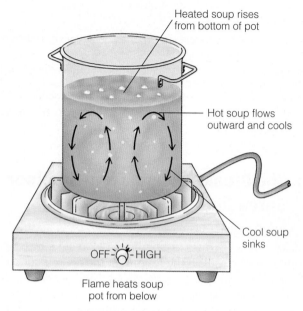

Heated soup rises
from bottom of pot

Hot soup flows
outward and cools

Cool soup
sinks

OFF — HIGH

Flame heats soup
pot from below

a A convection current forms when soup is heated from the bottom of the pot.

spheric **plates** floating on the asthenosphere. When heated from below, the deformable asthenosphere expands, becomes less dense, and rises **(Figure 3.14)**. It turns aside when it reaches the lithosphere, lifting and cracking the crust to form the plate edges. The newly forming pair of plates (one on each side of the spreading center) slide down the swelling ridges—they diverge from the spreading center. New seabed forms in the area of divergence. The large plates include both continental and oceanic crust. The major plates jostle about like huge slabs of ice on a warming lake. Plate movement is slow in human terms, averaging about 5 centimeters (2 inches) a year. The plates interact at converging, diverging, or sideways-moving boundaries, sometimes forcing one another below the surface or wrinkling into mountains.

Plate movement appears to be caused by two forces:

- Plates form and slide off the raised ridges of the spreading centers.
- Plates are pulled downward into the mantle by their cool, dense leading edges.

We now know that *through the great expanse of geologic time, this slow movement remakes the surface of Earth, expands and splits continents, and forms and destroys ocean basins.* The less dense, ancient granitic continents ride high in the lithospheric plates, rafting on the slowly moving asthenosphere below. This process has

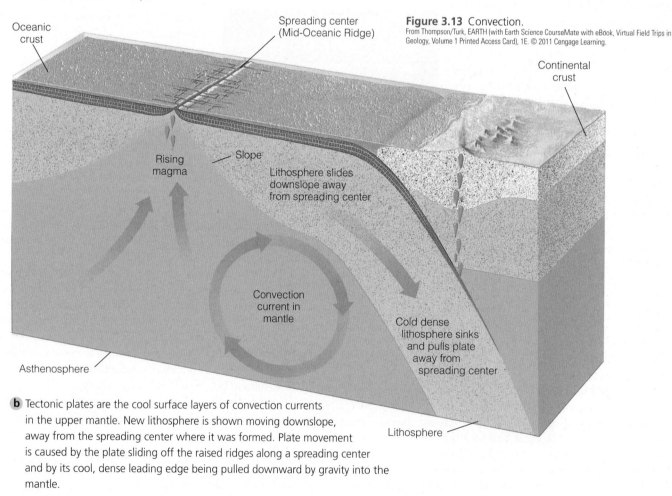

Oceanic
crust

Spreading center
(Mid-Oceanic Ridge)

Figure 3.13 Convection.
From Thompson/Turk, EARTH (with Earth Science CourseMate with eBook, Virtual Field Trips in Geology, Volume 1 Printed Access Card), 1E. © 2011 Cengage Learning.

Continental
crust

Rising
magma

Slope

Lithosphere slides
downslope away
from spreading center

Convection
current in
mantle

Cold dense
lithosphere sinks
and pulls plate
away from
spreading center

Asthenosphere

Lithosphere

b Tectonic plates are the cool surface layers of convection currents in the upper mantle. New lithosphere is shown moving downslope, away from the spreading center where it was formed. Plate movement is caused by the plate sliding off the raised ridges along a spreading center and by its cool, dense leading edge being pulled downward by gravity into the mantle.

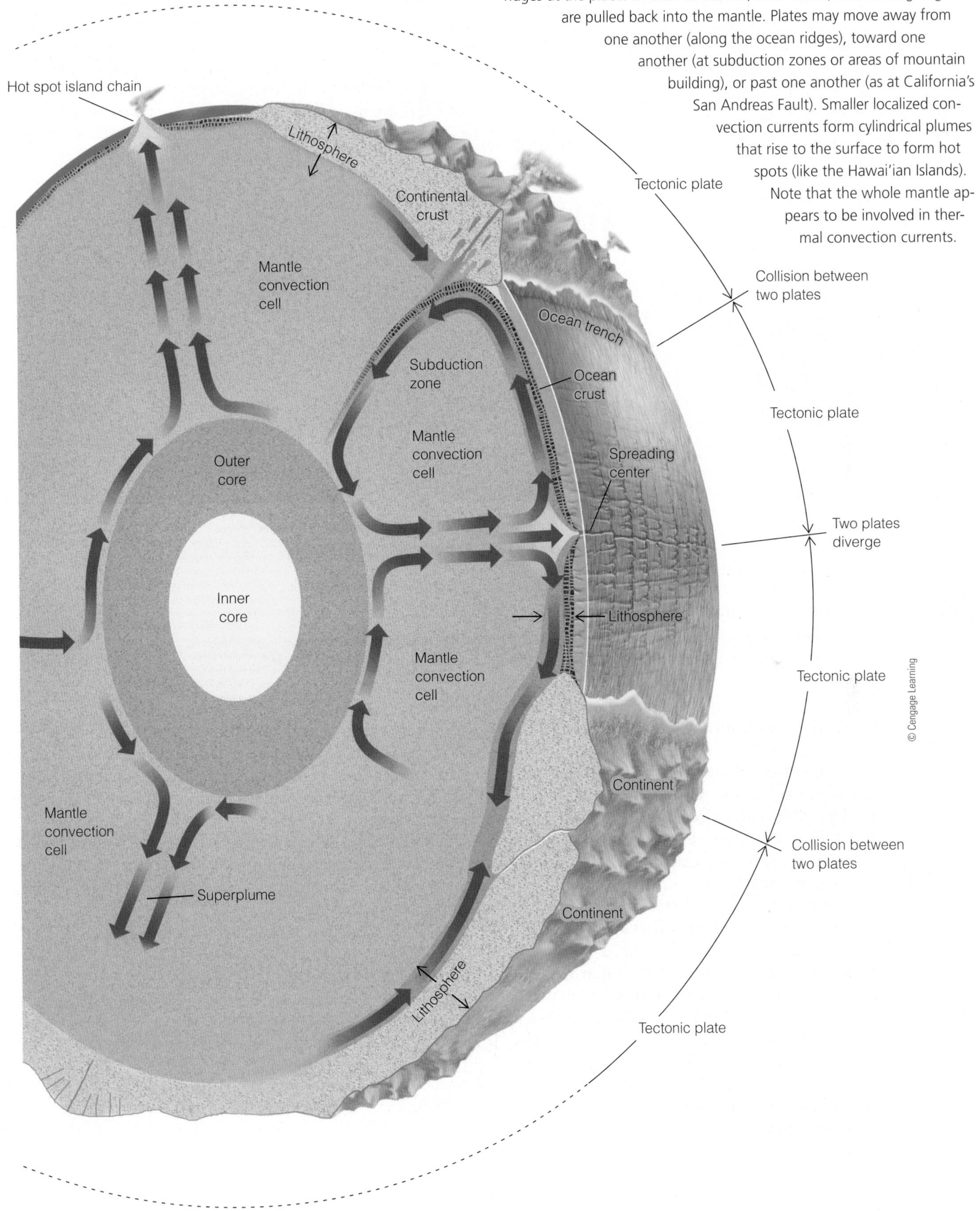

Figure 3.14 The tectonic system is powered by heat. Some parts of the mantle are warmer than others, and convection currents form when warm mantle material rises and cool material falls. Above the mantle floats the cool, rigid lithosphere, which is fragmented into plates. Plate movement is powered by gravity: The plates slide down the ridges at the places of their formation; their dense, cool leading edges are pulled back into the mantle. Plates may move away from one another (along the ocean ridges), toward one another (at subduction zones or areas of mountain building), or past one another (as at California's San Andreas Fault). Smaller localized convection currents form cylindrical plumes that rise to the surface to form hot spots (like the Hawai'ian Islands). Note that the whole mantle appears to be involved in thermal convection currents.

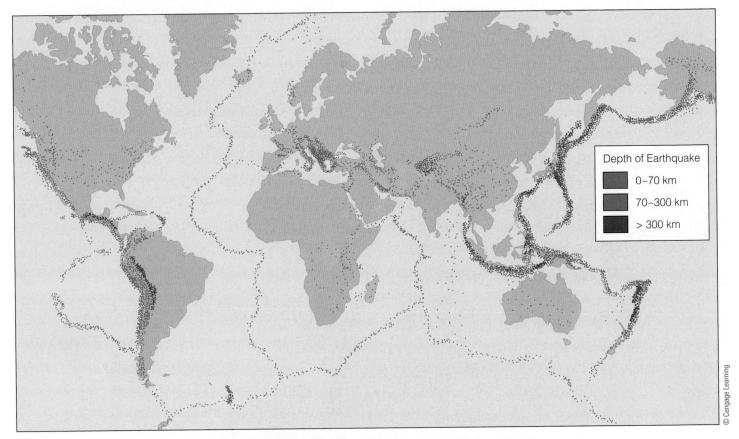

Figure 3.15 Seismic events worldwide, January 1977 through December 1986. The locations of about 10,000 earthquakes are colored red, green, and blue to represent event depths of 0–70 kilometers, 70–300 kilometers, and below 300 kilometers, respectively.

THINKING BEYOND THE FIGURE

Look at the pattern of earthquake depths along the western coast of South America. Without looking ahead, can you think why earthquakes occur at greater depths at greater distances from the Pacific Ocean?

progressed since Earth's crust first cooled and solidified. Figure 3.14 presents an overview of the whole tectonic system.

Literally and figuratively, it all fits; a cooling, shrinking, raisin-like wrinkling is no longer needed to explain Earth's surface features. This 20th-century understanding of the ever-changing nature of Earth has given fresh meaning to historian Will Durant's warning: "Civilization exists by geological consent, subject to change without notice."

After a series of raucous scientific meetings in 1966 and 1967, the revolution in geology entered a period of rapid consolidation. In 1968, *Glomar Challenger* drilled its first deep-ocean crustal cores and provided the confirmation of plate tectonics. Researchers found supporting data from many sources that tended to confirm Wilson's surprising synthesis. Every scientist had to re-examine his or her specialty in light of this new information. Zoologists found new explanations for the unusual animals of Australia. Biologists discovered a new cause of the isolation required for the formation of new species by natural selection. Paleontologists found an expla-

nation for similar fossils on different continents. Resource specialists could at last explain why coal deposits were buried in Antarctica. Some geologists were pleased; some were skeptical. All were eager to explore further to prove or disprove this new theory.

This historical overview aims to convey the sense of discovery and excitement surrounding the 20th-century revolution in geology. Now we can investigate the workings of plate tectonics in more detail.

CONCEPT CHECK

Before going on to the next section, check your understanding of some of the important ideas presented so far:

What was the key insight that Hess and Wilson brought to the discussion?

Can you outline—in very simple terms—the action of Earth's crust described by the theory of plate tectonics?

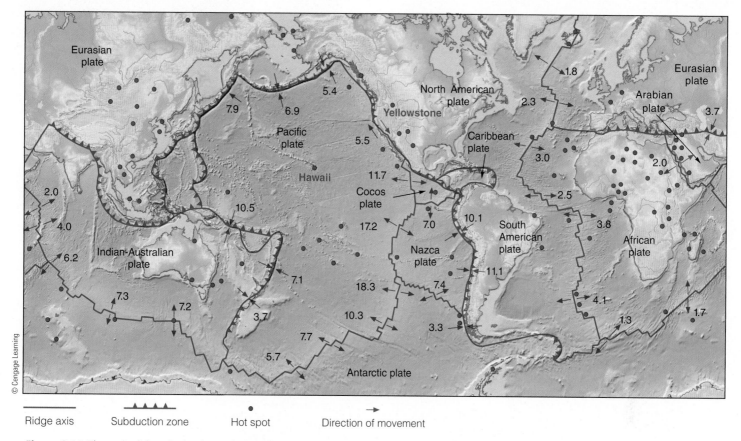

| Ridge axis | Subduction zone | Hot spot | Direction of movement |

Figure 3.16 The major lithospheric plates, showing their directions of relative movement and the location of the principal hot spots. Note the correspondence of plate boundaries and earthquake locations—compare this figure to Figure 3.15. Most of the million or so earthquakes and volcanic events each year occur along plate boundaries.

3.8 Plates Interact at Plate Boundaries

Figure 3.15 is a plot of about 10,000 earthquakes. Notice the odd pattern they form—almost as if Earth's lithosphere is divided into sections!

The lithospheric plates and their margins are shown in **Figure 3.16**. The plates float on a dense, deformable asthenosphere and are free to move relative to one another. Plates interact with neighboring plates along their mutual boundaries. In **Figure 3.17**, movement of Plate A to the left (west) requires it to slide along its north and south margins. An overlap is produced in front (to the west), and a gap is created behind (to the east). Different places on the margins of Plate A experience separation and extension, convergence and compression, and transverse movement (shear).

The three types of plate boundaries that result from these interactions are:

- Divergent plate boundaries (two plates move apart from each other).
- Convergent plate boundaries (two plates move toward each other and interact).

- Transverse (or transform) plate boundaries (two plates slide laterally past each other.

Look for these boundaries in Figure 3.17d.

Ocean Basins Form at Divergent Plate Boundaries

Imagine the effect a rising plume of heated mantle might have on overlying continental crust. Pushed from below, the relatively brittle continental crust would arch and fracture. The broken pieces would be pulled apart by the diverging asthenosphere, and spaces between the blocks of continental crust would be filled with newly formed (and relatively dense) oceanic crust. As the broken plate separated at this new spreading center, molten rock called **magma** would rise into the crustal fractures. (Magma is called *lava* when found aboveground. Magma becomes more fluid as it rises because the pressure near the surface is less than the pressure deeper within the crust.) Some of the magma would then cool and solidify in the fractures; some would erupt from volcanoes. A **rift valley** would form.

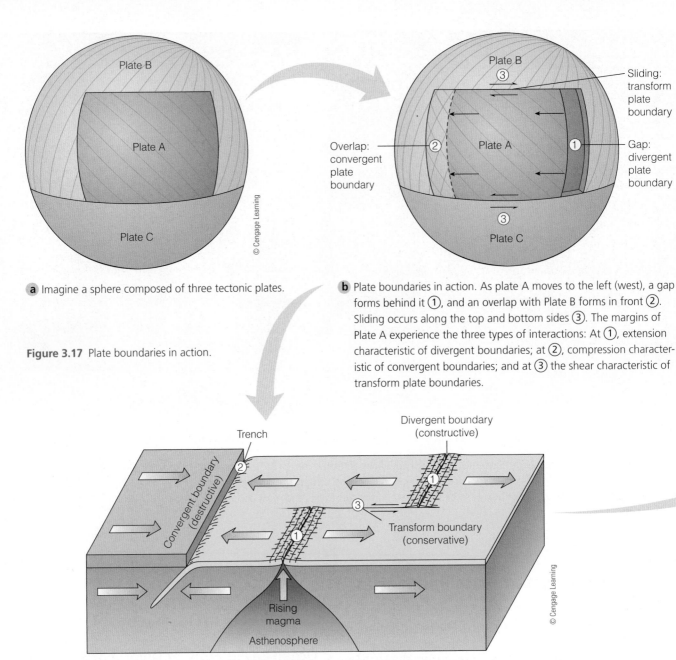

a Imagine a sphere composed of three tectonic plates.

Figure 3.17 Plate boundaries in action.

b Plate boundaries in action. As plate A moves to the left (west), a gap forms behind it ①, and an overlap with Plate B forms in front ②. Sliding occurs along the top and bottom sides ③. The margins of Plate A experience the three types of interactions: At ①, extension characteristic of divergent boundaries; at ②, compression characteristic of convergent boundaries; and at ③ the shear characteristic of transform plate boundaries.

c Plate boundaries on the surface of a sphere. The divergent ①, convergent ②, and transform ③ margins match those of the margins of Plate A in (b).

The East African Rift, one of the newest and largest of Earth's rift valley systems, was formed in this way **(Figure 3.18).** It extends from Ethiopia to Mozambique, a distance of nearly 3,000 kilometers (1,666 miles) **(Figure 3.19,** see p. 82). Long, linear depressions have partially filled with water to form large freshwater lakes. To the north, the rift widens to form the Red Sea and the Gulf of Aden. Between the freshwater lakes to the south and the gulfs to the north, seawater is leaking through the fractured crust to fill small depressions—the first evidence of an ocean-basin-to-be in central eastern Africa.

The Atlantic Ocean experienced a similar youth. Like the East African Rift, the spreading center of the Mid-Atlantic Ridge is a **divergent plate boundary,** a line along which two plates are moving apart and at which oceanic crust forms. The growth of the Atlantic began about 210 million years ago when heat caused the asthenosphere to expand and rise, lifting and fracturing the lighter, solid lithosphere above. **Figure 3.20** (see p. 83) shows the Atlantic, a large new ocean basin that formed between the diverging plates when the rift became deep enough for water to collect. A long mid-ocean ridge divided by a central rift valley traverses the ocean floor roughly equidistant from the shorelines in both the North and South Atlantic, terminating north of Iceland.

Plate divergence is not confined to East Africa or the Atlantic, nor has it been limited to the last 200 million years. As may also be

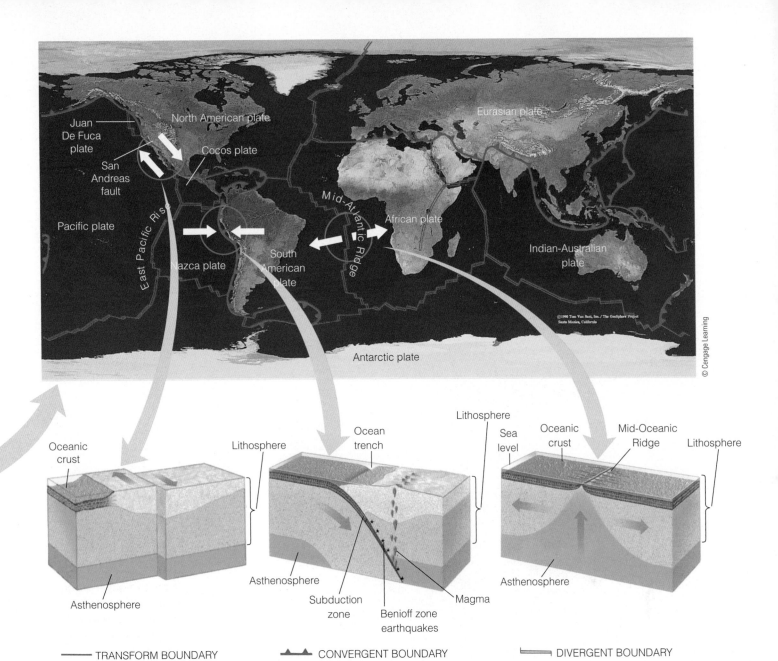

TRANSFORM BOUNDARY   CONVERGENT BOUNDARY   DIVERGENT BOUNDARY

d Lateral movement at *transform* boundaries causes shear, compression at *convergent* boundaries produces buckling and shortening, and extension of *divergent* boundaries causes splitting and rifting.

seen in Figure 3.17d, the Mid-Atlantic Ridge has counterparts in the Pacific and Indian oceans. The Pacific floor, for example, diverges along the East Pacific Rise and the Pacific Antarctic Ridge, spreading centers that form the eastern and southern boundaries of the great Pacific Plate. In East Africa, rift valleys have formed relatively recently as plate divergence begins to separate another continent. As happened in the Red Sea, the ocean will invade when the rift becomes deep enough.

Figure 3.21 (see p. 84) shows how divergence has formed other ocean basins (and uses Wegener's Pangaea as a model). About 20 cubic kilometers (4.8 cubic miles) of new ocean crust forms each year.

Island Arcs Form, Continents Collide, and Crust Recycles at Convergent Plate Boundaries

Since Earth is not getting larger, divergence in one place must be offset by convergence in another. Oceanic crust is destroyed at **convergent plate boundaries**, regions of violent geological activity where plates push together.

Ocean–Continent Convergence South America, embedded in the westward-moving South American Plate, encounters the Pacific's Nazca Plate as it moves eastward. The relatively thick and light continental lithosphere of South America rides up and over the heavier oceanic lithosphere of the Nazca Plate, which is

Pangaea warps and stretches
Volcanic activity
© Cengage Learning
Lithosphere
Rising magma
Continental crust

Figure 3.18 A model for the formation of a new plate boundary: the breakup of Pangaea and the formation of the Atlantic.

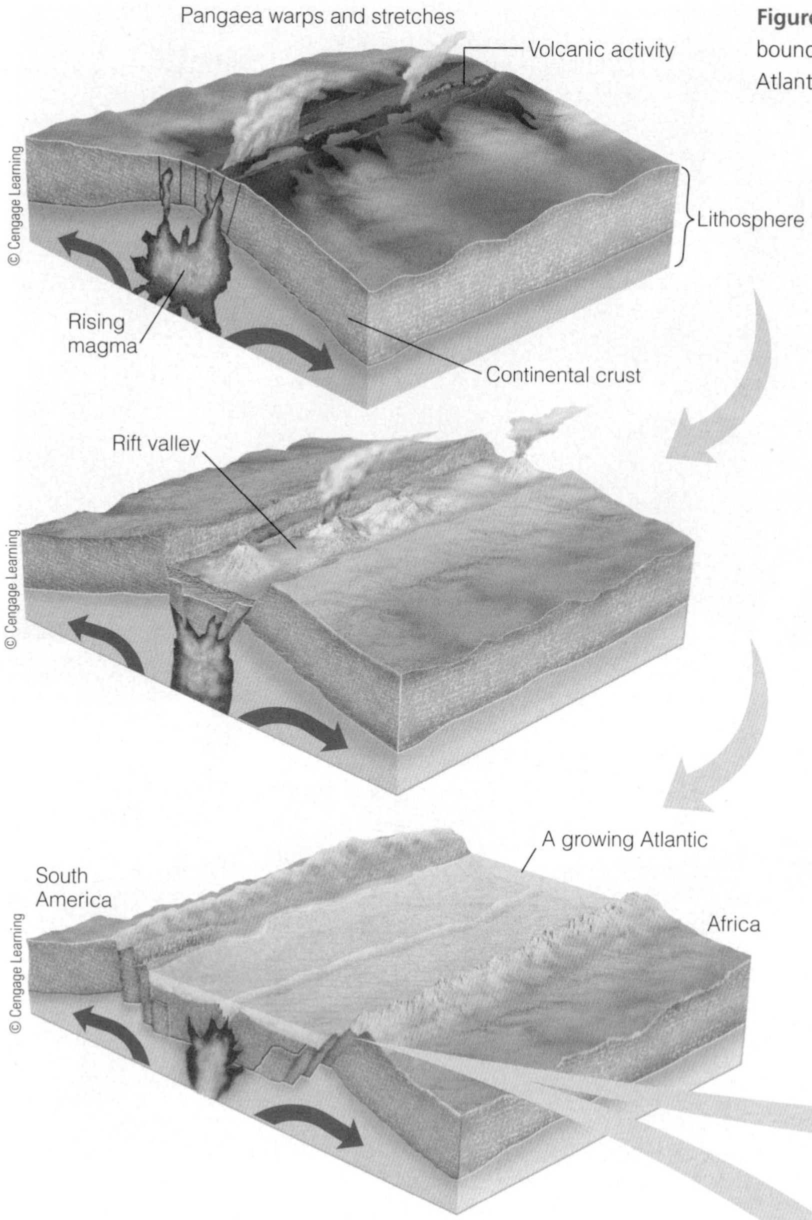

a As the lithosphere began to crack, a rift formed beneath the continent, and molten magma began to rise to form a new basaltic ocean floor.

Rift valley
© Cengage Learning

b As the rift continued to open, the two new continents were separated by a growing ocean basin. Volcanoes and earthquakes occur along the active rift area, which is the mid-ocean (mid-Atlantic) ridge. The East African Rift Valley, though not yet submerged, currently resembles this stage (see Figure 3.19).

A growing Atlantic
South America
Africa
© Cengage Learning

c A new ocean basin forms beneath a new ocean.

Jump to Figure 4.27 on pages 126–127 to see another example of the saw-tooth configuration of a mid-ocean ridge.

d The Red Sea currently resembles this stage. Note the remarkable saw-tooth configuration of the peaks on the horizon and their similarity to the diagram.

V. Courtillot

DR. ROBERT BALLARD recounts an expedition to a mid-ocean ridge.

At 12:13 we reach the deepest point in our dive: 9,100 feet. Pressure on the outside of our sphere is now more than two tons to the square inch.

No strain. *Alvin* is built to work safely as deep as 12,000 feet; an identical sphere has been tested to 22,500 feet without failing.

Our first encounter with a tectonic feature comes near the end of this first dive. As we cross the steep ridge that divides the central depression, we see a fault, or fracture, a few inches wide running southward down the axis of the rift valley.

The crack cuts across a lava flow, splitting individual rock feature. We follow it for about a hundred yards before losing it in a pile of lava fragments. Fresh, unbroken lava outpourings are seen only in a narrow strip along the central axis. Elsewhere, it is as if a massive wrenching, cracking, and grinding has worked and reshaped the rock floor.

Fissures like the one we have found at the very axis of the plate boundary are little more than hairline cracks. As we proceed, either east or west of the central region, we cross others increasingly larger in size. They range up to large fissures, that have dilated or opened as much as twenty to thirty feet, with vertical fault scarps sometimes towering more than 1,000 feet.

Many of us believed that the fissures might be a good place to look for hot-springs activity and possible deposits of minerals. Some geologists theorize that seawater may travel or percolate down these deep cracks, be heated by the hot magma somewhere deep below, then rise back to the surface carrying dissolved minerals. As the water cools, the minerals may be deposited in layers around open vents in the floor.

Source: National Geographic Magazine, May 1975, pp. 604–615.
The versatile deep submersible Alvin descends on another voyage of exploration.

Background photo: Emory Kristof/National Geographic Creative

O. LOUIS MAZZATENTA/National Geographic Creative

---

subducted along the deep trench that parallels the west coast of South America. **Figure 3.22** (see p. 85) is a cross section through these plates.

Some of the oceanic crust and its sediments will melt as the plate plunges downward and its temperature rises. Volatile components—mostly water and carbon dioxide—are driven off and rise toward the overriding plate. This in turn lowers the melting temperature of the surrounding mantle, forming a magma rich in dissolved gases. In places this magma then rises through overlying layers to the surface and causes volcanic eruptions. The high volumes of gas contained in these melts can be explosively released as the magma nears the surface. The violent volcanoes of Central America and South America's Andes Mountains are a product of this activity, as are the area's numerous earthquakes. The North American Cascade volcanoes, including Mount St. Helens, result from similar processes.

> **"Civilization exists by geological consent, subject to change without notice."** —WILL DURANT

Most of the subducted crust mixes with the mantle. As shown in Figures 3.22c and d, some of it continues downward through the mantle, eventually reaching the mantle–core boundary 2,800 kilometers (1,700 miles) beneath the surface! Subduction at converging oceanic plates was responsible for the great Alaska earthquake of 1964 and the devastating tsunami-generating Indian Ocean earthquake of 2004. Plate convergence (and divergence) is faster in the Pacific than in the Atlantic, in a few places reaching a rate of 18 centimeters (7 inches) a year. You can now clearly see the source of the Pacific Ring of Fire.

Ocean–Ocean Convergence In the previous example, continental crust met oceanic crust. What happens when two *oceanic* plates converge? One of the colliding plates will usually be older, and therefore cooler and denser, than the other. Pulled by gravity, this heavier plate will slip steeply below the lighter one into the

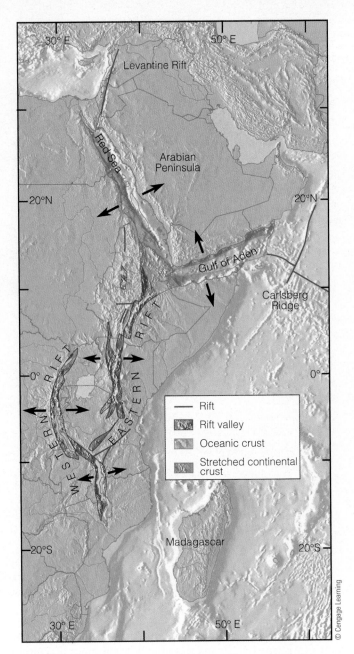

Figure 3.19 The East African rift system, a divergent boundary. East Africa is being pulled apart by tectonic forces thought to be driven by a superplume originating at the core–mantle boundary (see again Figure 3.14). The lithosphere in this region is relatively thin, and as the upward-arching lithosphere cracks and splits, long linear blocks have fallen along faults. Some of these blocks are overlain by freshwater lakes; some are dry and occasionally below sea level. Oceanic crust has been generated in the area to the north (the Red Sea, the Gulf of Aden) for about 5 million years—these are the freshest bits of a new ocean-to-be.

asthenosphere. The ocean bottom distorts in these areas to form deep trenches, the ocean's greatest depths. Again, the temperature of the descending plate rises, and water and carbon dioxide trapped with the melting rock of the subducting plate rise into the overlying mantle, lowering its melting point. As before, this fluid melt of magma and subducted material forms a relatively

light magma that powers vigorous volcanoes, but the volcanoes emerge from the seafloor rather than from a continent. These volcanoes appear in patterns of curves on the overriding oceanic crust; when they emerge above sea level, they form curving arcs of islands (**Figure 3.23**, see p. 86).

Convergent margins are vast "continent factories," where materials from the surface descend and are heated, compressed, partially liquefied, separated, mixed with surrounding materials, and recycled to the surface. Relatively light continental crust is the main product, and it is produced at a rate of about 1 cubic kilometer (0.24 cubic mile) per year. Some geophysicists believe all of Earth's continental crust may have originated from granitic rock produced in this way. The island arcs may have coalesced to form larger and larger continental masses.

Continent–Continent Convergence Two plates bearing continental crust can also converge. Since both plates are of approximately equal density, neither plate edge is being subducted; instead, both are compressed, folded, and uplifted, to form mountains, as **Figure 3.24** (see p. 86) shows. These mountains—Earth's largest land features—are composed of the remains of sedimentary rocks originally formed from seabed sediments. The most spectacular example of such a collision, between the India-Australian and Eurasian Plates some 45 million years ago, formed the Himalayas. The lofty top of Mt. Everest is made of rock formed from sediments deposited long ago in a shallow sea!

Crust Fractures and Slides at Transform Plate Boundaries

Remember, movement of lithospheric plates over the mantle is occurring on the surface of a sphere, not on a plane. The axis of spreading is not a smoothly curving line but a jagged trace abruptly offset by numerous faults. These features are called **transform faults** (see **Figure 3.25**, p. 87, and look for these features on the mid-ocean ridges of Figure 3.15). Transform faults are named from the fact that the relative plate motion is changed, or transformed, along them. We will discuss transform faults in more detail in Chapter 4 in our discussion of the mid-ocean ridge system (see, for example, Figure 4.24) but the concept is important in our discussion of plate boundaries because lithospheric plates shear laterally past one another at **transform plate boundaries**. Crust is neither produced nor destroyed at this type of junction.

The potential for earthquakes at transform plate boundaries can be great as the plate edges slip past each other. The eastern boundary of the Pacific Plate is a long transform fault system. As you can see in Figures 3.16 and 3.25, California's San Andreas Fault is merely the most famous of the many faults marking the junction between the Pacific and North American plates. The Pacific Plate moves steadily, but its movement is stored elastically at the North American Plate boundary until friction is overcome. Then the Pacific Plate lurches in abrupt jerks to the northwest along much of its shared border with the North American Plate, an area that includes the major population centers of California. These jerks cause California's famous earthquakes. Because of this movement, coastal southwestern California is gradually sliding north along the rest of North America; some 50 million years from now, it will encounter the Aleutian Trench.

CONCEPT CHECK

Before going on to the next section, check your understanding of some of the important ideas presented so far:

What kinds of plate boundaries exist?

What happens at each?

About how fast do plates move?

Which kind of plate movement is related to earthquakes and tsunami?

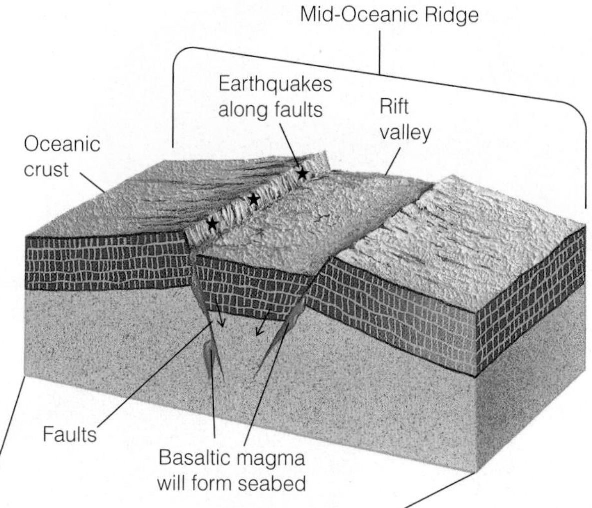

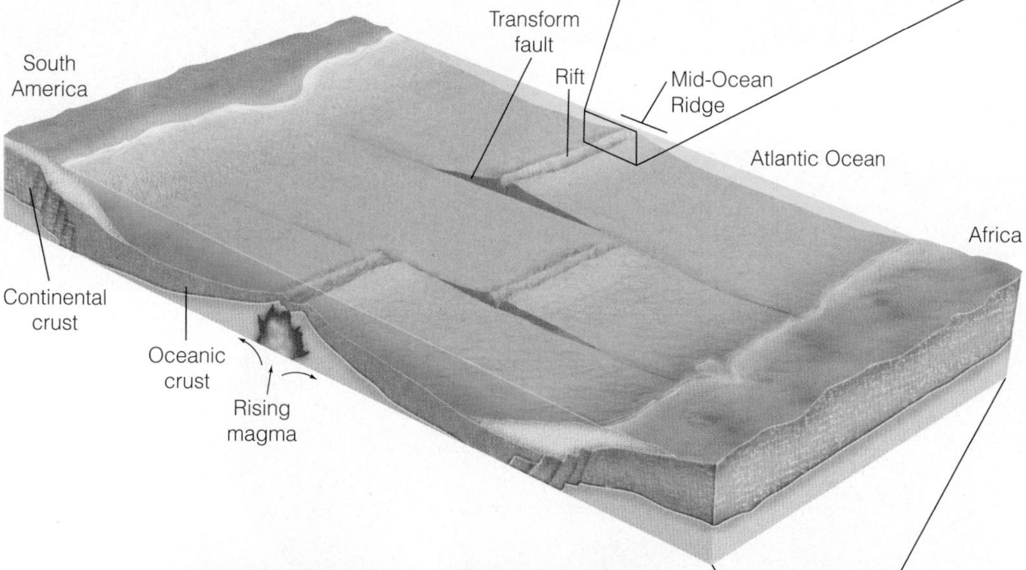

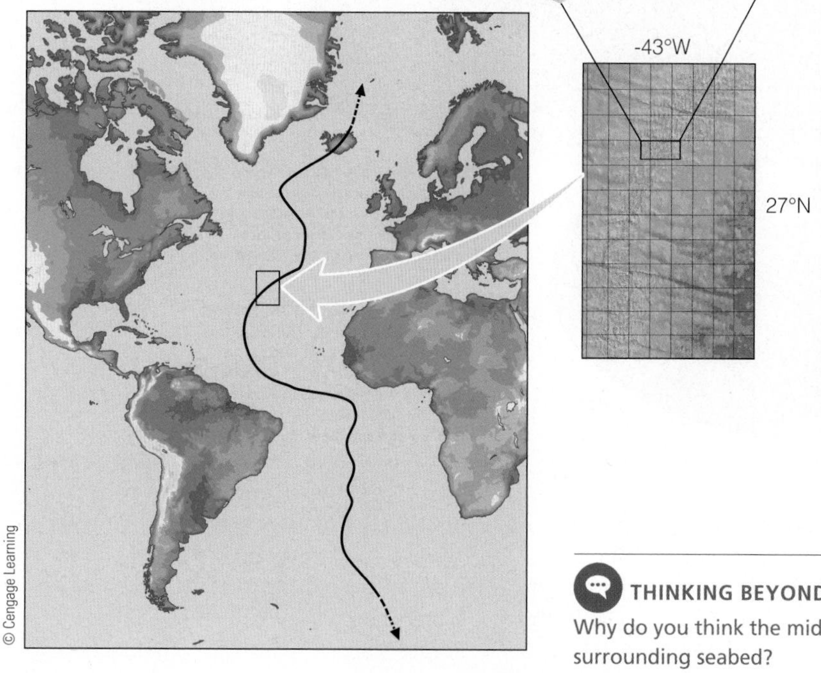

Figure 3.20 The Mid-Atlantic Ridge, showing its conformance to the coastlines of the adjacent continents. The first inset shows a detail of the ridge generated from side-scan sonar data—the central rift is clearly visible. Red and orange colors indicate the crest of the ridge; dark blue, the deeper seabed on either side. The second insert shows the location of the ridge and associated valley in the Atlantic. (Transform faults are explained in Figure 3.25)

THINKING BEYOND THE FIGURE

Why do you think the mid-ocean ridges are higher than the surrounding seabed?

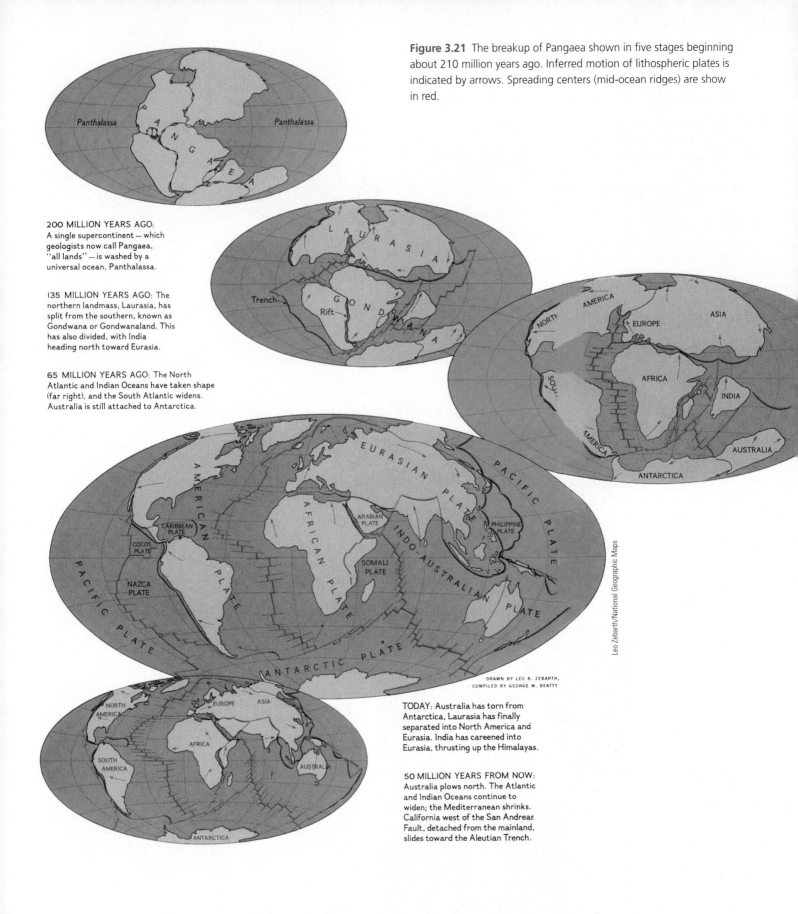

Figure 3.21 The breakup of Pangaea shown in five stages beginning about 210 million years ago. Inferred motion of lithospheric plates is indicated by arrows. Spreading centers (mid-ocean ridges) are show in red.

200 MILLION YEARS AGO: A single supercontinent — which geologists now call Pangaea, "all lands" — is washed by a universal ocean, Panthalassa.

135 MILLION YEARS AGO: The northern landmass, Laurasia, has split from the southern, known as Gondwana or Gondwanaland. This has also divided, with India heading north toward Eurasia.

65 MILLION YEARS AGO: The North Atlantic and Indian Oceans have taken shape (far right), and the South Atlantic widens. Australia is still attached to Antarctica.

TODAY: Australia has torn from Antarctica, Laurasia has finally separated into North America and Eurasia. India has careened into Eurasia, thrusting up the Himalayas.

50 MILLION YEARS FROM NOW: Australia plows north. The Atlantic and Indian Oceans continue to widen; the Mediterranean shrinks. California west of the San Andreas Fault, detached from the mainland, slides toward the Aleutian Trench.

Leo Zebarth/National Geographic Maps

DRAWN BY LEO B. ZEBARTH, COMPILED BY GEORGE W. BEATTY

Figure 3.22 Subduction.

b An Andean volcano in full blast in 2006. The 5,000-meter (16,400-foot) high volcano Tungurahua, located in Ecuador, becomes active roughly every 90 years.

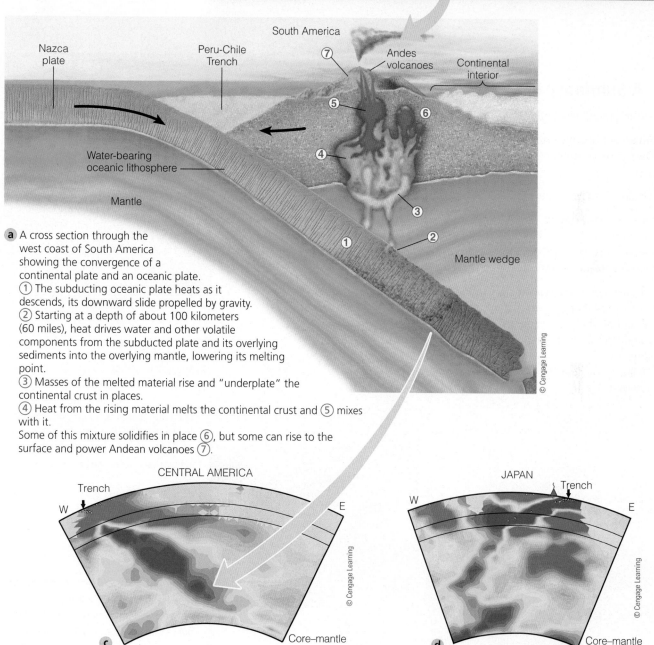

a A cross section through the west coast of South America showing the convergence of a continental plate and an oceanic plate.
① The subducting oceanic plate heats as it descends, its downward slide propelled by gravity.
② Starting at a depth of about 100 kilometers (60 miles), heat drives water and other volatile components from the subducted plate and its overlying sediments into the overlying mantle, lowering its melting point.
③ Masses of the melted material rise and "underplate" the continental crust in places.
④ Heat from the rising material melts the continental crust and ⑤ mixes with it.
Some of this mixture solidifies in place ⑥, but some can rise to the surface and power Andean volcanoes ⑦.

c and **d** Vertical slices through Earth's mantle beneath Central America and Japan showing the distribution of warmer (red) and colder (blue) material. The configuration of colder material suggests that the subducting slabs beneath both areas have penetrated to the core-mantle boundary, which is at a depth of about 2,900 kilometers (1,800 miles).

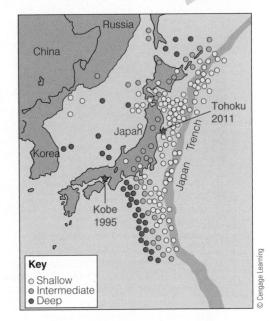

a The formation of an island arc along a trench as two oceanic plates converge. The volcanic islands form as masses of magma reach the seafloor. The Japanese islands were formed in this way.

Figure 3.23 The formation of island arcs.

3.9 A Summary of Plate Interactions

There are, then, two kinds of plate divergences:

- Divergent oceanic crust (such as in the Mid-Atlantic)
- Divergent continental crust (as in the Rift Valley of East Africa)

And there are three kinds of plate convergences:

- Oceanic crust toward continental crust (west coast of South America)
- Oceanic crust toward oceanic crust (northern Pacific)
- Continental crust toward continental crust (Himalayas)

Transform boundaries mark the locations at which crustal plates move past one another (San Andreas Fault).

Each of these movements produces a distinct topography, and each zone contains potential dangers for its human inhabitants.

Table 3.1 on page 88 summarizes the characteristics of plate boundaries.

b The distribution of shallow, intermediate, and deep earthquakes for part of the "Pacific Ring of Fire" in the vicinity of the Japan trench. Note that earthquakes occur only on one side of the trench, the side on which the plate subducts. The sites of the catastrophic 1995 Kobe and 2011 Tohoku subduction earthquakes are marked.

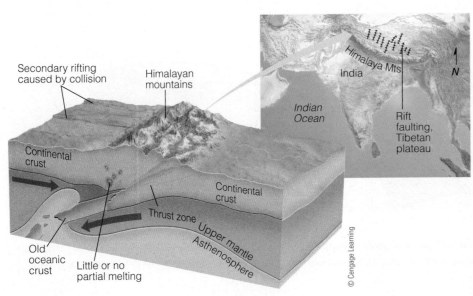

Figure 3.24 A cross section through the Himalayan plateau, showing the convergence of two continental plates. Neither plate is dense enough to subduct; instead, their compression and folding uplift the plate edges to form the Himalayan Mountains. Notice the massive supporting "root" beneath the emergent mountain needed for isostatic equilibrium.

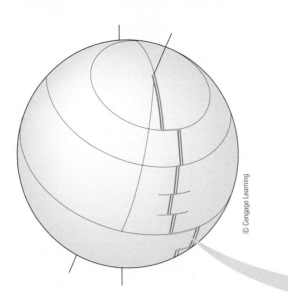

a Transform faults (orange) form because the axis of seafloor spreading on the surface of a sphere cannot follow a smoothly curving line. The motion of two diverging lithospheric plates (arrows) rotates about an imaginary axis extending through Earth.

© Cengage Learning

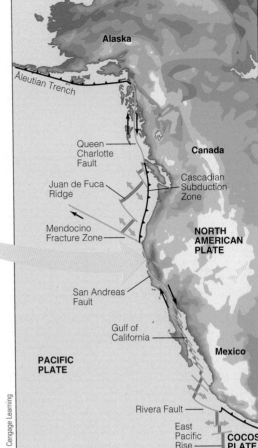

© Cengage Learning

b A long transform plate boundary, which includes California's San Andreas Fault. Note the offset plate boundaries caused by divergence on a sphere.

Tom Garrison

c California's San Andreas Fault, a transform fault. The fault trace is clearly visible between the arrows in this photograph taken south of San Francisco. The fault takes its name from San Andreas Lake, visible near the center of the photograph. The large landfill is the site of San Francisco International Airport.

Figure 3.25 Transform faults.

💬 **THINKING BEYOND THE FIGURE**

Figure (c) is a bit small, but perhaps you can see the marked difference between the terrain and vegetation on the west (top) side of the San Andreas Fault and the east (bottom) side. Why do you think that difference exists?

3.10 The Confirmation of Plate Tectonics

The theory of plate tectonics has had the same effect on geology that the theory of evolution has had on biology. In each case a catalog of seemingly unrelated facts was unified by a powerful central idea. As we will see in this section, many discoveries contributed to our present understanding of plate tectonics, but the most compelling evidence is locked within the floors of the young ocean basins themselves.

> The theory of plate tectonics has had the same unifying effect on geology that the theory of evolution has had on biology.

A History of Plate Movement Has Been Captured in Residual Magnetic Fields

Earth's persistent magnetic field is caused by the movement of molten metal in the outer core. A compass needle points toward the magnetic north pole because the needle aligns with Earth's magnetic field (**Figure 3.26**). Tiny particles of an iron-bearing magnetic mineral called magnetite occur naturally in basaltic magma. When this magma erupts at mid-ocean ridges, it cools to form solid rock. The magnetic minerals act

Table 3.1 Characteristics of Plate Boundaries

Plate Boundary		Plate Movement	Seafloor	Events Observed	Example locations
Divergent plate boundaries	Ocean–ocean	Apart	Forms by seafloor spreading.	Ridge forms at spreading center. Ocean basin expands; plate area increases. Many small volcanoes and/or shallow earthquakes.	Mid-Atlantic Ridge, East Pacific Rise
	Continent–continent		New ocean basin may form as continent splits.	Continent spreads; central rift collapses; ocean fills basin.	East African Rift Valley, Red Sea
Convergent plate boundaries	Continent–ocean	Together	Destroyed at subduction zones.	Dense oceanic lithosphere plunges beneath less dense continental. Earthquakes trace path of downmoving plate as it descends into asthenosphere. A trench forms. Subducted plate partially melts. Magma rises to form continental volcanoes.	Western South America, Cascade Mountains in western United States
	Ocean–ocean			Older, cooler, denser crust slips beneath less dense crust. Strong quakes. Deep trench forms in arc shape. Subducted plate heats in upper mantle; magma rises to form curving chains of volcanic islands.	Aleutians, Marianas
	Continent–continent		Closure of ocean basins.	Collision between masses of granitic continental lithosphere. Neither mass is subducted. Plate edges are compressed, folded, uplifted; one may move beneath the other.	Himalayas, Alps
Transform plate boundaries		Past each other	Neither created nor destroyed.	A line (fault) along which lithospheric plates move past each other. Strong earthquakes along fault.	San Andreas Fault; South Island, New Zealand
				Transform faults across spreading center.	Mid-ocean ridges

like miniature compass needles. As they cool below their *Curie point* (or Curie temperature)—about 580°C (1,080°F)—to form new seafloor, the magnetic minerals' magnetic fields align with Earth's magnetic field. Thus, the orientation of Earth's magnetic field at that particular time becomes frozen in the rock as it solidifies. Any later change in the strength or direction of Earth's magnetic field will not significantly change the characteristics of the field trapped within the solid rocks. **Figure 3.27** shows the process. The "fossil," or remanent, magnetic field of a rock is known as **paleomagnetism** (*palaios*, "ancient").

A **magnetometer** measures the amount and direction of residual magnetism in a rock sample. In the late 1950s, geophysicists towed sensitive magnetometers just above the ocean floor to detect the weak magnetism frozen in the rocks. When plotted on charts, the data revealed a pattern of symmetrical magnetic stripes or bands on both sides of a spreading center (**Figure 3.28a**). The magnetized minerals contained in the rocks in some bands add to Earth's present magnetic orientation to enhance the strength of the local magnetic field, but the magnetism in rocks in adjacent bands weakens it. What could cause such a pattern?

In 1963, geologists Drummond Matthews, Frederick Vine, and Lawrence Morley proposed a clever interpretation. They knew similar magnetic patterns had been found in layered lava flows on land that had been independently dated by other means. They also knew that Earth's magnetic field reverses at irregular intervals of a few hundred thousand years. In a time of reversal a compass needle would point south instead of north, and any particles of magnetic material falling below their Curie points in fresh seafloor basalt at a spreading center would be imprinted with the reversed field. The alternating magnetic bands represent rocks with alternating magnetic polarity—one band having normal polarity (magnetized in the same direction as today's magnetic field direction), and the next band having reversed polarity (opposite from today's direction). These researchers realized that the pattern of alternating weak and strong magnetic fields was symmetrical because freshly magnetized rocks born at the ridge are spread apart and carried away from the ridge by plate movement (**Figure 3.28b**).

By 1974, scientists had compiled charts showing the paleomagnetic orientation—and the age—of the seafloors of the eastern Pacific and the Atlantic for about the last 200 million years (**Figure 3.29**, see p. 92). Plate tectonics beautifully explains these patterns, and the patterns themselves are among the most compelling of all arguments for the theory.

Paleomagnetic data have recently been used to measure spreading rates, to calibrate the geologic time scale, and to reconstruct continents. Paleomagnetism has been among the

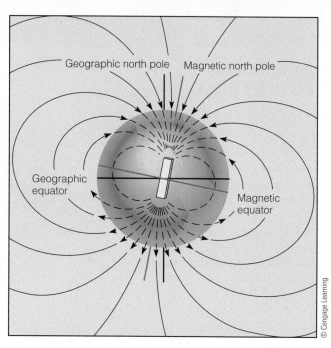

Figure 3.26 Earth's magnetism

Geographic north pole   Magnetic north pole

Geographic equator

Magnetic equator

a The magnetic field pattern of a simple bar magnet is revealed by iron filings that align themselves along the lines of magnetic force.

b Earth's magnetic field is thought to arise from fluid motions and electric currents in the outer core. The likely energy source is heat from the solid inner core causing convection currents in the outer core. These flowing currents of liquid iron, coupled with the rotation of Earth, form the magnetic field in a process analogous to the way a power station generator produces electricity. The axis of Earth's magnetic field is tilted about 11° from the axis of the geographical North and South poles. Compass needles point to magnetic north, not to geographical (true) north.

most productive specialties in geology for the past three decades, and other lines of paleomagnetic investigation have also shed light on the process of plate tectonics.

Geologists already knew that periodic magnetic field reversals were not the only unusual feature of Earth's magnetic field. A plot of the apparent position of the north magnetic pole, as measured by the magnetic orientation of rocks in North Amer-

ica, South America, Europe, and Africa, showed that the magnetic pole seemed to have moved—it appeared to have migrated to its present position from a point much farther south, in the Pacific Ocean.

Figure 3.27 Particles of iron-bearing magnetite occur naturally in basaltic magma. As this rock forms new seabed, the magnetic particles cool, "locking" their magnetic orientation to that of Earth's prevailing magnetic field. If Earth's magnetic field changes direction later, the "locked" particles will not respond, but magnetic particles in any new (hot) magma above will orient to the new field direction.

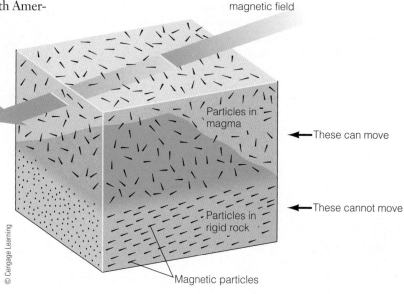

Direction of magnetic field

Particles in magma — These can move

Particles in rigid rock — These cannot move

Magnetic particles

💬 **THINKING BEYOND THE FIGURE**

If a continent twists as it moves over time (see Figure 3.21 for example), would magnetic particles locked in rocks still point north (or south)?

Patterns of paleomagnetism and their explanation by plate tectonic theory.

a When scientists conducted a magnetic survey of a spreading center, the Mid-Atlantic Ridge, they found bands of weaker and stronger magnetic fields locked in the rocks.

b – d The molten rocks forming at the spreading center take on the polarity of the planet when they are cooling and then move slowly in both directions from the center. When Earth's magnetic field reverses, the polarity of newly formed rocks changes, creating symmetrical bands of opposite polarity.

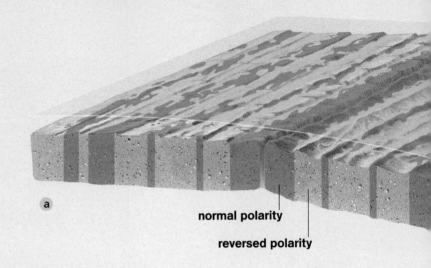

a

normal polarity

reversed polarity

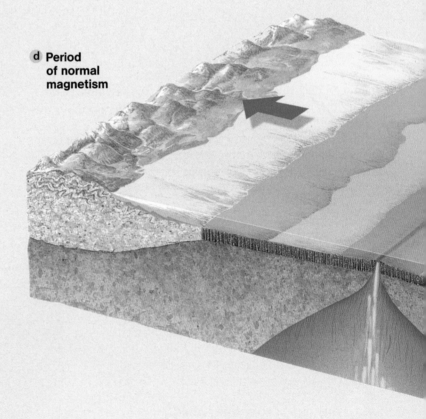

d Period of normal magnetism

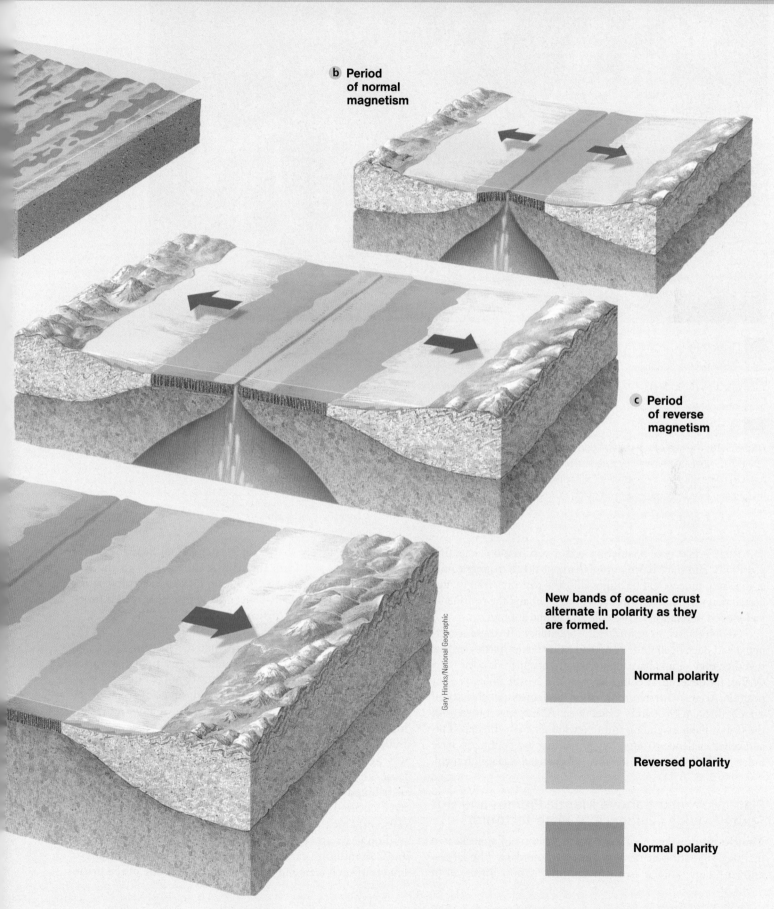

b Period
of normal
magnetism

c Period
of reverse
magnetism

New bands of oceanic crust
alternate in polarity as they
are formed.

Normal polarity

Reversed polarity

Normal polarity

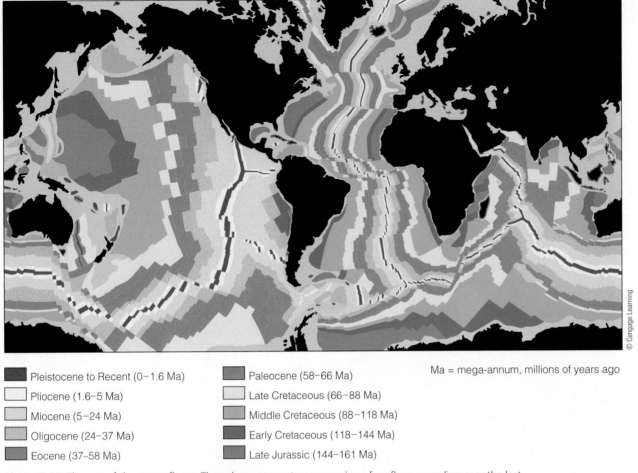

■ Pleistocene to Recent (0–1.6 Ma)	■ Paleocene (58–66 Ma)	Ma = mega-annum, millions of years ago
□ Pliocene (1.6–5 Ma)	■ Late Cretaceous (66–88 Ma)	
■ Miocene (5–24 Ma)	■ Middle Cretaceous (88–118 Ma)	
■ Oligocene (24–37 Ma)	■ Early Cretaceous (118–144 Ma)	
■ Eocene (37–58 Ma)	■ Late Jurassic (144–161 Ma)	

Figure 3.29 The age of the ocean floors. The colors represent an expression of seafloor spreading over the last 200 million years as revealed by paleomagnetic patterns. Note especially the relative symmetry of the Atlantic basin in contrast with the asymmetrical Pacific, where the spreading center is located close to the eastern margin and intersects the coast of California.

Since *actual* pole wandering was a very remote scientific possibility, and since having more than one north magnetic pole at the same time would be impossible, the pattern strongly suggested that the poles had probably stayed put and the continents had moved in relation to the pole and to one another.

Geophysicists examined this possibility. If these continents had once been united and had drifted over Earth's surface together, they should have identical paths of apparent polar wandering, terminating in coincidence at the North Pole's present position. Armed with this information, geologists quickly noticed a fault through Scotland's Caledonian Mountains joined with the Cabot Fault extending from Newfoundland to Boston. This and other evidence strongly suggested that the continents have indeed drifted apart, carrying their magnetized rocks with them.

Plate Movement above Mantle Plumes and Hot Spots Provides Evidence of Plate Tectonics

Mantle plumes are continent-sized columns of superheated mantle originating at the core–mantle boundary. The largest known plume, known as a **superplume**, is now lifting all of Africa. Its relatively sharp edges extend from Scotland to the Indian Ocean, and from the Mid-Atlantic Ridge to the Red Sea. As we saw in Figure 3.19, the center of Africa is fracturing and spreading, and what will eventually be new seabed is forming rapidly in the East African rift valleys.

Plumes and superplumes are conduits for heat from the core. Current research suggests that the heat in the asthenosphere that powers plate tectonics is resupplied from the core by superplumes. Indeed, in the not-so-distant past, superplume heat may have been responsible for some of the most dramatic events on Earth's surface. A huge outpouring of Earth's interior occurred over much of present-day India about 65 million years ago. The Indian subcontinent was deluged with more than 1 million cubic kilometers of lava! The stacks of lava are known as the Deccan Traps. If distributed evenly, this cataclysmic series of eruptions would have covered Earth's surface with a layer of lava 3 meters (10 feet) thick! Similar megaeruptions happened about 17 million years ago in what is now the U.S. Pacific Northwest, and 248 million years ago in Siberia. The toxic atmospheric effects of such tremendous upheavals led to one of the great-

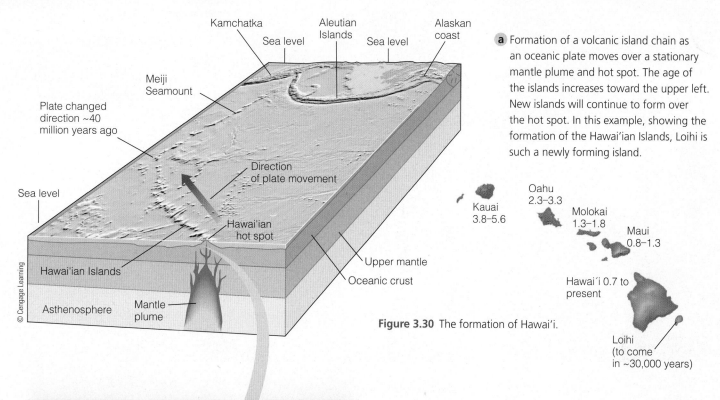

Kamchatka

Aleutian Islands

Sea level

Alaskan coast

Sea level

Sea level

Meiji Seamount

Plate changed direction ~40 million years ago

Direction of plate movement

Sea level

Hawai'ian hot spot

© Cengage Learning

Hawai'ian Islands

Upper mantle

Oceanic crust

Asthenosphere

Mantle plume

a Formation of a volcanic island chain as an oceanic plate moves over a stationary mantle plume and hot spot. The age of the islands increases toward the upper left. New islands will continue to form over the hot spot. In this example, showing the formation of the Hawai'ian Islands, Loihi is such a newly forming island.

Kauai 3.8–5.6

Oahu 2.3–3.3

Molokai 1.3–1.8

Maui 0.8–1.3

Hawai'i 0.7 to present

Loihi (to come in ~30,000 years)

Figure 3.30 The formation of Hawai'i.

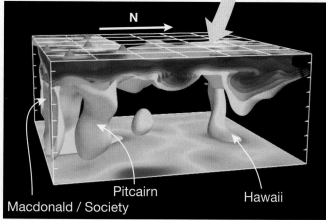

From Science Magazine Vol. 342 no. 6155 pp. 227–230. Reprinted with permission from AAAS.

N

Pitcairn

Hawaii

Macdonald / Society

b Tomographic images beneath the Pacific seabed down to the core–mantle boundary. Maps of earthquake wave velocity perturbations at different depths show a vast area of unusually high temperature (red and yellow areas) that extends up from the core–mantle boundary (about 2,800 kilometers [1,700 miles] from the surface). These areas of high temperatures are mantle plume. Activity atop one plume powers the Hawai'ian volcanoes. Plumes like these are thought to bring to the asthenosphere much of the heat needed to power plate tectonics.

est mass extinctions in Earth's history. You'll learn more about mass extinctions in Chapter 13.

Hot spots are one of the surface expressions of plumes of magma rising from relatively stationary sources of heat in the mantle. Hot spots are not always located at plate boundaries, and no one knows why their source of heat is localized or what anchors them in place. As lithospheric plates slide over these fixed locations, the plates are weakened from below by rising heat and magma. A volcano can form over the hot spot, but because the plate is moving, the volcano is carried away from its source of magma after a few million years and becomes inactive. It is replaced at the hot spot by a new volcano a short distance away. A chain of volcanoes and volcanic islands results.

Figure 3.30 shows the most famous of these "assembly-line" chains, which extends from the old eroded volcanoes of the Emperor Seamounts to the still-growing island of Hawai'i. In fact, the abrupt bend in the chain was caused by a change in the direction of movement of the Pacific Plate, from largely northward to more westward, about 40 million years ago. The next Hawai'ian island that will come into being—already named Loihi—is building on the ocean floor at the southeastern end of the chain. Now about 1,000 meters (3,300 feet) beneath the surface, Loihi will break the surface about 30,000 years from now.

There are other hot spots in the Pacific. The island chains formed by their activity also jog in the Hawai'ian pattern, indicating that they are positioned on the same lithospheric plate. Chains of undersea volcanoes in the Atlantic, centered on the Mid-Atlantic Ridge, suggest a similar process is at work there. Hot spots can exist beneath continental crust as well; Yellowstone National Park is believed to be over a huge hot spot beneath the westward-moving North American Plate **(Figure 3.31)**. Look once again at Figure 3.16 to see the locations of hot spots around the world. The configuration and length of all these chains of volcanoes and geothermal sites are consistent with the theory of plate tectonics.

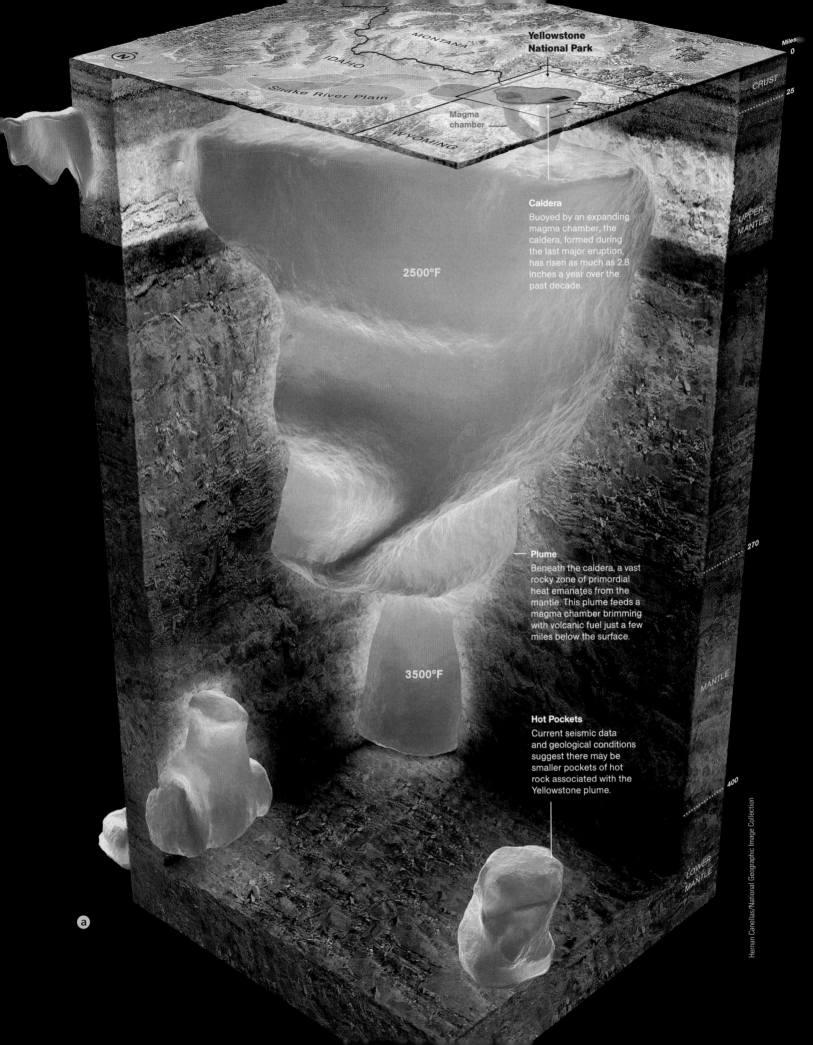

Yellowstone National Park

Snake River Plain

MONTANA

IDAHO

WYOMING

Magma chamber

CRUST

UPPER MANTLE

MANTLE

LOWER MANTLE

Miles
0

25

270

400

2500°F

3500°F

Caldera
Buoyed by an expanding magma chamber, the caldera, formed during the last major eruption, has risen as much as 2.8 inches a year over the past decade.

Plume
Beneath the caldera, a vast rocky zone of primordial heat emanates from the mantle. This plume feeds a magma chamber brimming with volcanic fuel just a few miles below the surface.

Hot Pockets
Current seismic data and geological conditions suggest there may be smaller pockets of hot rock associated with the Yellowstone plume.

a

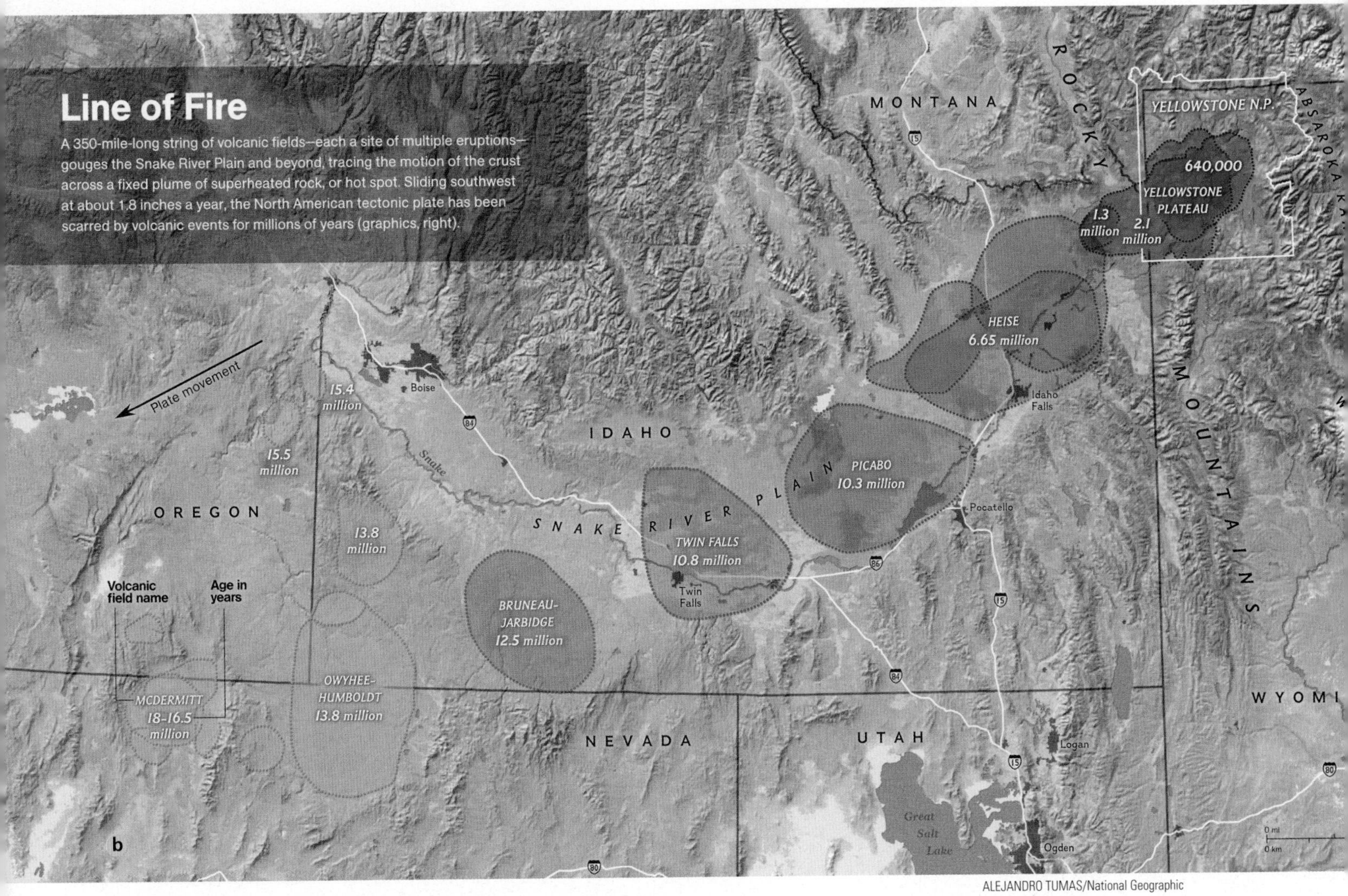
Line of Fire
A 350-mile-long string of volcanic fields—each a site of multiple eruptions—gouges the Snake River Plain and beyond, tracing the motion of the crust across a fixed plume of superheated rock, or hot spot. Sliding southwest at about 1.8 inches a year, the North American tectonic plate has been scarred by volcanic events for millions of years (graphics, right).
MONTANA
ROCKY MOUNTAINS
YELLOWSTONE N.P.
ABSAROKA
640,000
YELLOWSTONE PLATEAU
1.3 million
2.1 million
HEISE 6.65 million
Idaho Falls
Plate movement
15.4 million
Boise
IDAHO
15.5 million
Snake
PICABO 10.3 million
SNAKE RIVER PLAIN
Pocatello
OREGON
13.8 million
TWIN FALLS 10.8 million
Twin Falls
Volcanic field name
Age in years
BRUNEAU-JARBIDGE 12.5 million
OWYHEE-HUMBOLDT 13.8 million
MCDERMITT 18–16.5 million
NEVADA
UTAH
WYOMI
Logan
Great Salt Lake
Ogden
b
ALEJANDRO TUMAS/National Geographic

Figure 3.31 The Yellowstone caldera, perhaps Earth's largest volcano.

a The Yellowstone system in cross section. The vast reservoir of magma is heated by a hot spot on the mantle. The magma's expansion has lifted the floor of the caldera about 8 centimeters (3 inches) in the last decade.

b Just as the geological activity forming of the Hawai'ian Islands moved as the Pacific Plate drove northwest (Figure 3.29), the position of Yellowstone eruptions has moved as the American plate has moved westward across a mantle plume over the last few million years. The last major eruption was 640,000 years ago.

Sediment Age and Distribution, Oceanic Ridges, and Terranes Are Explained by Plate Tectonics

If the ocean basins are genuinely ancient, and if the processes that produce sediments have been operating for most or all of that time, both the thickness and age of sediments on the ocean floor should be great. They are not. The young spreading ridges are almost free of sediment, and the oldest edges of the basins support layers of sediment 15 to 20 times as thin as the age of the ocean itself would suggest. The oldest sediments of the ocean basins are rarely more than 180 million years old (see Figure 5.24). The reason is that sediments are subducted at a plate's leading edge.

The location and configuration of the oceanic ridges are clear evidence of past events. The volcanic nature of ridge islands like Iceland, the shape of the longitudinal rifts splitting the ridgetops, and the sinking of the seabed as new oceanic crust cools and travels outward are all consistent with the theory of plate tectonics. The distribution of transform faults and fracture zones along the oceanic ridges (features you'll learn about in Chapter 4) also supports the theory of plate tectonics, as do on-the-spot geological observations made by researchers in deep submersibles.

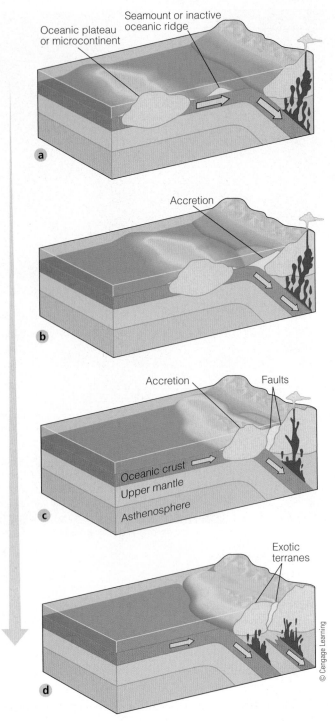

Figure 3.32 Terrane formation. Oceanic plateaus usually composed of relatively low-density rock are not subducted into the trench with the oceanic plate. Instead, they are "scraped off," causing uplifting and mountain building as they strike a continent (a–d). Though rare, assemblages of subducting oceanic lithosphere can also be scraped off (obducted) onto the edges of continents. Rich ore deposits are sometimes found in them.

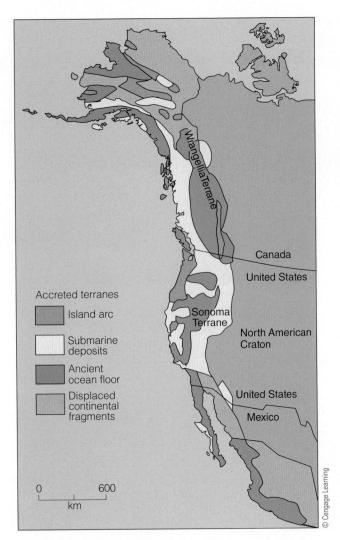

Figure 3.33 North American terranes. These fragments have differing histories and origins. Some have moved thousands of kilometers to be scraped off onto the North American core as their transporting plate subducted.

💬 **THINKING BEYOND THE FIGURE**
Do you live on a terrane?

Buoyant continental and oceanic plateaus (submerged small fragments of continents), island arcs, and fragments of granitic rock and sediments can be rafted along with a plate and scraped off onto a continent when the plate is subducted. This process is similar to what happens when a sharp knife is scraped across a tabletop to remove pieces of cool candle wax. The wax accumulates and wrinkles on the knife blade in the same way landmasses and ocean sediments accumulate against the face of a continent as the lithosphere in which they are embedded reaches a plate boundary. Plateaus, isolated segments of seafloor, ocean ridges, ancient island arcs, and parts of continental crust that are squeezed and sheared onto the face of a continent are called **terranes**. The thickness and low density of terranes prevent their subduction. A simplified account of terrane accumulation is diagrammed in **Figure 3.32**.

Terranes are surprisingly common. New England, much of the Pacific Northwest of North America, and most of Alaska appear to be composed of this sort of crazy-quilt assemblage of material, some of which has evidently arrived from thousands of kilometers away. For example, western Canada's Vancouver Island may have moved north some 3,500 kilometers (2,200 miles) in the last 75 million years **(Figure 3.33)**.

Curiously, terranes can also contain fragments of dense oceanic crust. Roughly 0.001% of oceanic lithosphere is not subducted, but rather *obducted*—scraped off—onto the edges of continents. The heavy wrinkled rocks contain pillow basalts and material derived from the upper mantle. Called **ophiolites** (*ophion*, "snake") because of their sinuous shape, these assemblages can contain metallic ores similar to those known to exist at the mid-ocean ridge spreading centers. Ophiolites are found on all continents, and as might be expected, those near the edges of the present continents are generally younger than those embedded deep within continental interiors **(Figure 3.34)**. Some of these ancient reminders of past episodes of plate tectonics are more than 1.2 billion years old.

Figure 3.34 Exposed ophiolites in Gros Morne National Park, Newfoundland, Canada.

CONCEPT CHECK

Before going on to the next section, check your understanding of some of the important ideas presented so far:

Is Earth's magnetic field a constant? That is, would a compass needle always point north?

How can Earth's magnetic field be "frozen" into rocks as they form?

Can you explain the matching magnetic alignments seen south of Iceland (see Figure 3.27)?

What is a hot spot? Where can you go to see hot-spot activity?

How does the long chain of Hawai'ian volcanoes seem to confirm the theory of plate tectonics?

Earth is 4,600 million years old, and the ocean is nearly as old. Why is the oldest ocean floor so young—rarely more than 200 million years old?

3.11 Scientists Still Have Much to Learn about the Tectonic Process

The theory of plate tectonics reveals much about the nature of Earth's surface. **Figure 3.35** summarizes surface tectonic activity.

In case you think that geophysicists have all the answers, however, consider just a few of the theory's unsolved problems:

- Why should long *lines* of asthenosphere be any warmer than adjacent areas?
- Is the increasing density of the leading edge of a subducting plate more important than the plate's sliding off the swollen mid-ocean ridge in making the plate move?
- Why do mantle plumes form? What causes a superplume? How long do they last?

- How far do most plates descend? Recent evidence suggests that much of the material spans the entire mantle, reaching the edge of the outer core.
- Has seafloor spreading always been a feature of Earth's surface? Has a previously thin crust become thicker with time, permitting plates to function in the ways described here?
- There is evidence of tectonic movement prior to the breakup of Pangaea. Will the process continue indefinitely, or are there cycles within cycles?
- When did plate tectonics begin **(Figure 3.36)**?

Though there is clearly much to learn, plate tectonics is already an especially powerful predictive theory. Discoveries and insights made by the researchers mentioned in this chapter, and hundreds of others, have borne out the intuition of Alfred Wegener. Our understanding of the process will evolve as more data become available, but there seems very little chance that geologists will ever return to the dominant pre-1960 view of a stable and motionless crust.

The theory of plate tectonics shows us the picture of an actively cycling Earth and an ever-changing surface, with *a single world ocean* changing shape and shifting position as the plates slowly move.

The configuration of the ocean basins—discussed in the next chapter—is the result of plate tectonic activity. The variety of these features will make more sense now that you are armed with an understanding of the theory.

CONCEPT CHECK

Before going on to the next section, check your understanding of some of the important ideas presented so far:

Has plate tectonics been a feature of Earth since its formation?

Can you suggest areas for future research in plate tectonics?

In your opinion, how has an understanding of plate processes revolutionized geology?

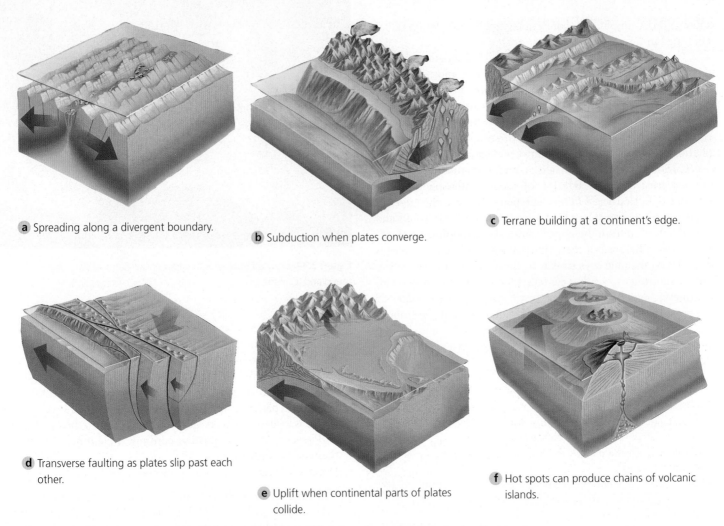

a Spreading along a divergent boundary.

b Subduction when plates converge.

c Terrane building at a continent's edge.

d Transverse faulting as plates slip past each other.

e Uplift when continental parts of plates collide.

f Hot spots can produce chains of volcanic islands.

Figure 3.35 The Wilson Cycle, named in honor of John Tuzo Wilson's synthesis of plate tectonics. Over great spans of time, ocean floors form and are destroyed. Mountains erode, sediments subduct, and continents rebuild. The world ocean moves from basin to basin.

a: Source: National Geographic Visual Atlas p. 32b; b: Source: National Geographic Visual Atlas p. 33a; c: Source: National Geographic Visual Atlas p. 32a; d: Source: National Geographic Visual Atlas p. 33a; e: Source: National Geographic Visual Atlas p. 33b; f: Source: National Geographic Visual Atlas p. 32a

Steve Richardson, Department of Geological Sciences, University of Cape Town, South Africa

Figure 3.36 The first evidence of crustal rocks mixing with the mantle at subduction zones—a signature of plate tectonics—has been found in tiny mineral inclusions in diamonds like this one. The oldest of these dates to between 3.0 and 3.2 billion years ago, marking the origin of the Wilson Cycle.

In this chapter you learned that Earth is composed of concentric spherical layers, with the least dense layer on the outside and the densest as the core. The layers may be classified by chemical composition into crust, mantle, and core or by physical properties into lithosphere, asthenosphere, mantle, and core. Geologists have confirmed the existence and basic properties of the layers by analysis of seismic waves, which are generated by the forces that cause large earthquakes.

The theory of plate tectonics explains the nonrandom distribution of earthquake locations, the curious jigsaw-puzzle fit of the continents, and the patterns of magnetism in surface rocks. The theory of plate tectonics suggests that Earth's surface is not a static arrangement of continents and ocean but a dynamic mosaic of jostling lithospheric plates. The plates have converged, diverged, and slipped past one another since Earth's crust first solidified and cooled, driven by slow, heat-generated currents rising and falling in the asthenosphere. Continental and oceanic crusts are generated by tectonic forces, and most major continental and seafloor features are shaped by plate movement. Plate tectonics explains why our ancient planet has surprisingly young seafloors; the oldest is only as old as the oldest dinosaurs, that is, about 1/23 the age of Earth.

In the next chapter you will learn about seabed features, which the slow process of plate movement has made and re-made on our planet. The continents are old; the ocean floors, young. The seabed bears the marks of travels and forces only now being understood. Hidden from our view until recent times, the contours and materials of the seabed have their own stories to tell.

QUESTIONS FROM STUDENTS

1. What is the difference between crust and lithosphere? Between lithosphere and asthenosphere?

Lithosphere includes crust (oceanic and continental) and rigid upper mantle down to the asthenosphere. The velocity of seismic waves in the crust is much different from that in the mantle. This suggests differences in chemical composition or crystal structure or both. The lithosphere and asthenosphere have different physical characteristics: The lithosphere is generally rigid, but the asthenosphere is capable of slow movement. The asthenosphere and lithosphere also transmit seismic waves at different speeds.

2. What are the most abundant elements inside Earth?

You may be surprised to learn that oxygen accounts for about 46% of the mass of Earth's crust. On an atom-for-atom basis the proportion is even more impressive: Of every 100 atoms of Earth's crust, 62 are oxygen. Most of this oxygen is not present as the gaseous element but is combined with other atoms into oxides and other compounds. Most of the familiar crustal rocks and minerals are oxides of aluminum, silicon, and iron (rust, for example, is iron oxide).

In Earth as a whole, iron is the most abundant element, making up 35% of the mass of the planet, and oxygen accounts for 30% overall. Remember that most of the mass of the universe is hydrogen gas. Oxygen and iron are abundant on Earth only because fusion reactions in stars can transform light elements like hydrogen into heavy ones.

3. How do geologists determine the location and magnitude of an earthquake?

Geologists use the time difference in the arrival of the seismic waves at their instruments to determine the distance to an earthquake. At least three seismographs in widely separated locations are needed to get a fix on the location.

The strength of the waves, adjusted for the distance, is used to calculate an earthquake's magnitude. Earthquake magnitude is often expressed on the **Richter scale**. Each full step on the Richter scale represents a 10-fold change in surface wave amplitude and a 32-fold change in energy release. Thus, an earthquake with a Richter magnitude of 6.5 releases about 32 times as much energy as an earthquake with a magnitude of 5.5, and about 1,000 times as much as a 4.5-magnitude quake. People rarely notice an earthquake unless the Richter magnitude is 3.2 or higher, but the energy associated with a magnitude-6 quake may cause significant destruction.

The energy released by the 1964 Alaska earthquake was more than a billion times as great as the energy released by the smallest earthquakes felt by humans. The energy release was equal to about twice the energy content of world coal and oil production for an entire year. Very low or very high Richter magnitudes are not easy to measure accurately. The Alaska earthquake's magnitude was initially calculated as between 8.3 and 8.6 on the Richter scale, but recent reassessment has yielded an extraordinary magnitude of 9.2. Earth rang like a great silent bell for 10 days after that earthquake.

4. What's the potential for serious loss of life and property damage due to tectonic plate movement?

Relatively great. About 40% of the world's largest cities lie within 160 kilometers (100 miles) of a plate boundary. By the year 2015, about 350 million people will be living in high-risk areas. About 80% of those at risk live in developing nations, where seismic safety is not a high priority in building design. Another 210 million people are also threatened by active volcanoes; most live along the subduction zones of the Pacific Ring of Fire, where 75% of Earth's 850 active volcanoes are located. Since 1600, there have been approximately 262,000

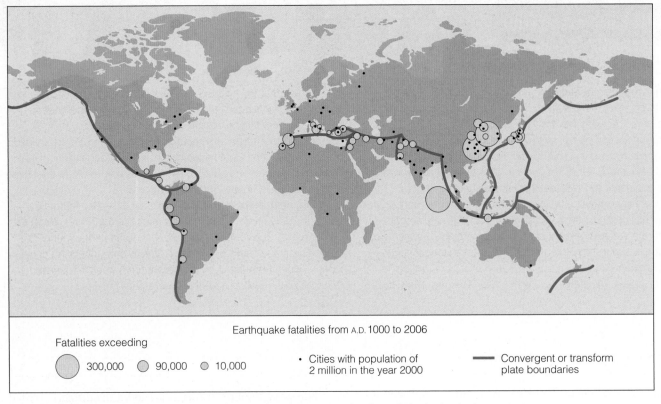

Earthquake fatalities from A.D. 1000 to 2006

Fatalities exceeding

◯ 300,000 ◯ 90,000 ◦ 10,000

• Cities with population of 2 million in the year 2000

━━━ Convergent or transform plate boundaries

Figure 3.37 Population centers and earthquake areas. Large cities near plate boundaries in developing countries are likely to have the highest number of fatalities in massive earthquakes.

deaths from volcanic eruptions, about 76,000 in the last century. See **Figure 3.37** for a graphic assessment of the risk.

5. How common are large earthquakes?

About every 2 days, somewhere in the world, there's an earthquake of 6.0–6.9 on the Richter scale—roughly equivalent to the quake that shook Northridge and the rest of southern California in January 1994, or Kobe, Japan, in January 1995. Once or twice a month, on average, there's a 7.0–7.9 quake somewhere. There is about one 8–9 earthquake—similar in magnitude to the 1964 earthquake in Alaska, the 1812 New Madrid quakes, and the 2011 Tohoku disaster in Japan—each year.

Northridge- and Kobe-sized quakes are moderate in size. Large losses of life and property can occur when these earthquakes occur in populated areas, however. Damage estimates from the Northridge earthquake exceeded US$40 billion. In Kobe, more than 5,000 people died, and more than 26,000 were injured. Some 56,000 buildings were destroyed; estimates of the cost of reconstruction exceeded US$400 billion.

Japan's great Tohoku magnitude 9.0 earthquake and associated tsunami of March, 2011, was among the greatest natural disasters of modern times. More than 20,000 people died, and 144,000 structures were destroyed (along with an estimated 230,000 cars and trucks). Property damage estimates currently exceed a quarter of a trillion U.S. dollars.

6. Is plate movement a new feature of Earth?

No. Multiple lines of evidence suggest we may be in the middle of the sixth or seventh major tectonic cycle since tectonic movement began about 3 billion years ago. Megacontinents like Pangaea appear to have formed, split, moved, and rejoined many times since Earth's crust solidified.

Have you ever wondered why the Mississippi River is where it is? It flows along a seam produced when Pangaea was assembled. Stress in that seam generated one of the largest earthquakes ever felt in North America: The great New Madrid (Missouri) earthquake of 1812 had a magnitude of about 8.0 on the not-yet-invented Richter scale. The devastating quake (and two that followed) could be felt over the entire eastern United States.

7. What about plate tectonic processes on other planets? Is there any evidence?

Earth remains tectonically "alive," but while there is evidence of past tectonic activity on Mars and the moon (and probably Mercury), it does not continue today. Volcanic activity and tectonic movement require heat, and the relatively small sizes of these bodies resulted in their relatively rapid loss of internal heat.

Venus, however, may tell a different story. High volumes of sulfur were noticed in Venus's atmosphere in 1979 and then decreased over the next few years. Perhaps the sulfur resulted from a large series of volcanic eruptions? Beginning in 1990, the Magellan spacecraft's radar altimeter revealed dramatic volcanic features and long deep valleys similar in size and shape to Earth's subduction zones **(Figure 3.38)**.

a The northward-moving Pacific Plate subducts beneath Alaska's Aleutian Islands.

NOAA

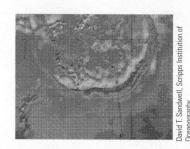

b Artemis Corona on Venus at the same scale.

David T. Sandwell, Scripps Institution of Oceanography

Figure 3.38 Plate tectonics on Earth and Venus.

TERMS AND CONCEPTS TO REMEMBER

absolute dating	density stratification	ophiolite	seismic waves
asthenosphere	divergent plate boundary	P wave (primary wave)	seismograph
basalt	earthquake	Pacific Ring of Fire	shadow zone
body wave	echo sounder	paleomagnetism	spreading center
buoyancy	fault	Pangaea	subduction
catastrophism	granite	Panthalassa	subduction zone
conduction	hot spot	plate	superplume
continental crust	isostatic equilibrium	plate tectonics	surface wave
continental drift	lithosphere	radioactive decay	terrane
convection	lower mantle	radiometric dating	transform fault
convection current	magma	relative dating	transform plate boundary
convergent plate boundary	magnetometer	Richter scale	uniformitarianism
core	mantle	rift valley	Wegener, Alfred
crust	mantle plume	S wave (secondary wave)	Wilson, John Tuzo
density	oceanic crust	seafloor spreading	

STUDY QUESTIONS

Thinking Critically

1. What difficulties complicate scientists' attempts to predict earthquakes?
2. Some earthquakes are linked to adjustments of isostatic equilibrium. How can this occur? Where would you be likely to experience such an earthquake?
3. Would the most violent earthquakes be associated with spreading centers or with subduction zones? Why?
4. Describe the mechanism that powers the movement of the lithospheric plates.
5. Where are the youngest rocks in the ocean crust? The oldest? Why?
6. Why did geologists have such strong objections to Wegener's ideas when he proposed them in 1912?
7. What biological evidence supports plate tectonics theory?
8. Imagine a tectonic plate moving westward. What geological effects would you expect to see on its northern edge? Western edge? Eastern edge?
9. What evidence can you cite to support the theory of plate tectonics? What questions remain unanswered? Which side would you take in a debate?

Thinking Analytically

1. How much farther would Columbus have to sail if he crossed the Atlantic today?
2. Look at Figure 3.28. Knowing the Atlantic's spreading rate, how far apart would the blue normal-polarity blocks be 3 million years in the figure?
3. Look again at Figures 3.30 and 3.31. What properties of Earth's interior make it possible to "see" this level of detail beneath the surface?

GLOBAL GEOSCIENCE WATCH

Visit www.cengagebrain.com to access MindTap, a complete digital course which includes access to Global Geoscience Watch and more. Research volcanic hot spots, mantle plumes, and superplumes in the GREENR database and learn about how they form and where they are thought to be located. Identify at least five different locations that are thought to originate from hot spots/mantle plumes and then summarize some of the current ideas and research addressing how they may form. Use this information to create a report on how you think hot spots have influenced marine geology and biology.

CENGAGEbrain.com   Visit www.cengagebrain.com to access course materials for this text, including interactive learning tools, videos, and more.

4 Ocean Basins

KEY CONCEPTS

Tectonic forces shape the seabed. © Cengage Learning

The ocean floor is divided into continental margins and deep-ocean basins. © Cengage Learning

The continental margins are seaward extensions of the adjacent continents and are usually underlain by granite; the deep seabeds have different features and are usually underlain by basalt. © Cengage Learning

The mid-ocean ridge system is perhaps Earth's most prominent feature. Gary Hincks/National Geographic

Most of the water of world ocean circulates through the hot oceanic crust of the ridges about every 10 million years. © Cengage Learning

---

Explorers investigate an ocean basin. The seabed comprises 70.8% of Earth's solid surface and reaches a depth of 10,920 meters (35,818 feet), much deeper than Mount Everest is high.
Brian J. Skerry/National Geographic Creative

Media Connection
Start off this chapter by listening to a podcast with Don Walsh, as he talks about being the first to reach the Mariana Trench. Visit www.cengagebrain.com to access MindTap, a complete digital course that includes this podcast and other resources.

4.1 The Ocean Floor Is Mapped by Bathymetry

Mars Reconnaissance Orbiter, an orbiting robot spacecraft, has mapped most of the surface of our planetary neighbor at a resolution that would reveal a dinner table resting on the sand. There are no oceans and few storms to spoil the view.

Mapping Earth is much more difficult because water and clouds hide more than three quarters of the surface. Until surprisingly recently we have known more about the global contours of the moon and the inner planets than we have known about our own home. Thanks to modern bathymetry, our view is clearing.

The discovery and study of ocean floor topography is called **bathymetry** (*bathy*, "deep"; *meter*, "measure"). The earliest known bathymetric studies were carried out in the Mediterranean by a Greek named Posidonius in 85 B.C.E. He and his crew let out nearly 2 kilometers (1.25 miles) of rope until a stone tied to the end of the line touched bottom. Bathymetric technology had not improved by the time Sir James Clark Ross obtained soundings of 4,893 meters (16,054 feet) in the South Atlantic in 1818. In the 1870s, the researchers aboard HMS *Challenger* added the innovation of a steam-powered winch to raise the line and weight, but the method was the same (Figure 4.1). The *Challenger* crew made 492 bottom soundings and confirmed Matthew Maury's earlier discovery of the Mid-Atlantic Ridge.

> **U**sing about a thousand radar pulses each second, the U.S. Navy's *Geosat* satellite measured its distance from the ocean surface to within 0.03 meter (1 inch)!

Echo Sounders Bounce Sound off the Seabed

The sinking of the RMS *Titanic* in 1912 stimulated research that finally ended slow, laborious weight-on-a-line efforts. By April 1914, one of Thomas Edison's former employees had developed the "Iceberg Detector and Echo Depth Sounder." The detector directed a powerful underwater sound pulse ahead of a ship and then listened for an echo from the submerged portion of an iceberg. It was easy to direct the beam downward to sense the distance to the bottom. It might take most of a day to lower and raise a weighted line, but echo sounders could take many bottom recordings in a minute.

In June 1922, an echo sounder based on his designs made the first continuous profile across an ocean basin aboard the USS *Stewart*, a U.S. Navy vessel. Using an improved echo sounder, the German research vessel *Meteor* made 14 profiles across the Atlantic from 1925 to 1927. The wandering path of the Mid-Atlantic Ridge was revealed, and its obvious coincidence with coastlines on both sides of the Atlantic stimulated the discussions that culminated in our present understanding of plate tectonics.

Echo sounding wasn't perfect. The ship's exact position was sometimes uncertain. The speed of sound through seawater varies with temperature, pressure, and salinity, and those variations

Figure 4.1 Seamen handling the steam winch aboard HMS *Challenger*. The winch was used to lower a weight on the end of a line to the seabed to find the ocean depth. The work was difficult and repetitive—a quarter of the 269 crew members eventually deserted during the 4½-year journey! This illustration is from the Challenger Report (1880).

 THINKING BEYOND THE FIGURE

Only 144 of the original crew of 216 returned with the ship. What do you suppose the crew found so objectionable about the voyage?

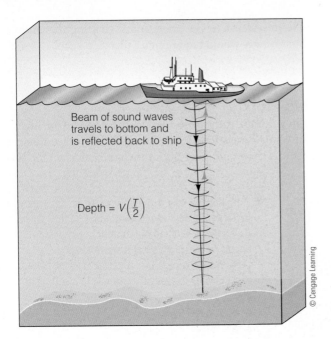

a The accuracy of an echo sounder can be affected by water conditions and bottom contours. The pulses of sound energy, or "pings," from the sounder spread out in a narrow cone as they travel from the ship. When depth is great, the sounds reflect from a large area of seabed, thus hiding details. As you'll see in Figure 4.4, there is a clever solution to this problem.

The equation shown in the figure is:

$$\text{Depth} = V\left(\frac{T}{2}\right)$$

Figure 4.3 Marie Tharp and geologist Bruce Heezen study the final version of Heinrich Berann's World Ocean Floor Panorama in Berann's studio, Austria, 1977. Berann was engaged by the U.S. Navy and the National Geographic Society to incorporate Tharp's work and data from other sources in a unified form. One of Berann's beautiful maps is included as Figure 4.22.

💬 **THINKING BEYOND THE FIGURE**

Why would the U.S. Navy have been interested in seeing such deep-water charts produced?

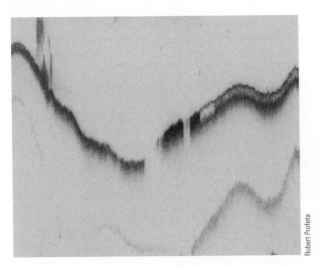

b An echo sounder trace. A sound pulse from a ship is reflected off the seabed and returns to the ship. Transit time provides a measure of depth. For example, it takes about 2 seconds for a sound pulse to strike the bottom and return to the ship when water depth is 1,500 meters (4,900 feet). Bottom topography is revealed as the ship sails a steady course. In this trace, the horizontal axis represents the course of the ship, and the vertical axis represents water depth. The ship has sailed over a small submarine canyon.

Figure 4.2 Echo sounding

made depth readings slightly inaccurate. Simple depth sounder images (such as that shown in **Figure 4.2**) were also unable to resolve the fine detail that oceanographers needed to explore seabed features. Even so, researchers using depth sounder tracks

had painstakingly compiled the first comprehensive charts of the ocean floor by 1959 (**Figure 4.3**).[1]

Since then, two new techniques—made possible by improved sensors and fast computers—have been perfected to minimize inaccuracies and speed the process of bathymetry. Multibeam echo sounder systems and satellite altimetry (as well as other systems) have been used to study the features discussed in this chapter. Any of them is surely an improvement over lowering rocks into the ocean on ropes!

Multibeam Systems Combine Many Echo Sounders

Like other echo sounders, a multibeam system bounces sound off the seafloor to measure ocean depth. Unlike a simple echo sounder, a multibeam system may have as many as 121 beams radiating from a ship's hull. Fanning out at right angles to the

[1]A portion of one of those beautiful hand-drawn charts is shown in Figure 4.24.

direction of travel, these beams can cover a 120° arc **(Figure 4.4a)**. Typically, a pulse of sound energy is sent toward the seabed every 10 seconds. Listening devices record sounds reflected from the bottom, but only from the narrow corridors corresponding to the outgoing pulse. Successive observations build a continuous swath of coverage beneath the ship. By "mowing the lawn"—moving the ship in a coverage pattern similar to one you would follow in cutting grass—researchers can build a complete map of an area **(Figure 4.4c)**. Fewer than 200 research vessels are equipped with

multibeam systems. At the present rate, charting the entire sea-floor in this way would require more than 125 years.

Satellites Can Be Used to Map Seabed Topography

Satellites cannot measure ocean depths directly, but they can measure small variations in the elevation of surface water. Using about a thousand radar pulses each second, the U.S. Navy's

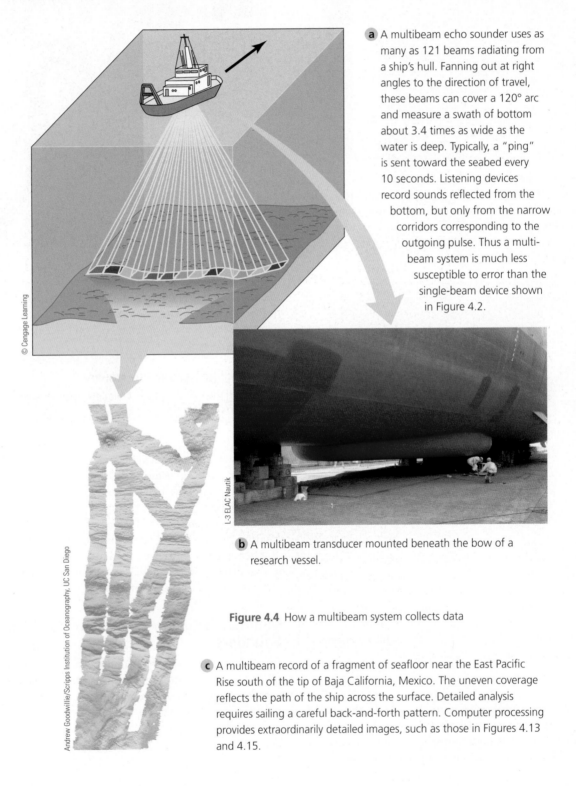

a A multibeam echo sounder uses as many as 121 beams radiating from a ship's hull. Fanning out at right angles to the direction of travel, these beams can cover a 120° arc and measure a swath of bottom about 3.4 times as wide as the water is deep. Typically, a "ping" is sent toward the seabed every 10 seconds. Listening devices record sounds reflected from the bottom, but only from the narrow corridors corresponding to the outgoing pulse. Thus a multi-beam system is much less susceptible to error than the single-beam device shown in Figure 4.2.

b A multibeam transducer mounted beneath the bow of a research vessel.

Figure 4.4 How a multibeam system collects data

c A multibeam record of a fragment of seafloor near the East Pacific Rise south of the tip of Baja California, Mexico. The uneven coverage reflects the path of the ship across the surface. Detailed analysis requires sailing a careful back-and-forth pattern. Computer processing provides extraordinarily detailed images, such as those in Figures 4.13 and 4.15.

Mark Thiessen/National Geographic Creative

DR. SYLVIA EARLE discusses Sir John Murray and modern developments in basin exploration.

I wonder what John Murray, who sailed aboard HMS *Challenger* between 1872 and 1876, edited the more than 50 volumes that resulted from that great voyage of discovery, and received the Royal Geographic Society's Founders medal in 1895, might think if he could go aboard a 21st century ocean research vessel.

What would impress him most? The ability to pinpoint exactly where in the ocean he was using satellites beaming data from high in the sky? The ability to hear the voice and see the image of a scientist half a world away and compare notes in real time? Or to call his family from thousands of miles away, see their faces, and get an instant update about what was happening on the homefront?

Surely he would marvel at maps, not on paper, but on large, illuminated screens portraying great mountain chains that curve like giant backbones down the major ocean basins; of sea mounts, valleys, and broad plains that hold most of Earth's water—and most of Earth's life.

Murray was the first to note the existence of the Mid-Atlantic Ridge and of oceanic trenches, but he did not know, could not know, about plate tectonics, the underlying processes that cause continents to move and oceans to shrink and expand.

He did not and could not know about hydrothermal vents and chemosynthesis as a mechanism for powering deep sea communities of life. The technology did not yet exist to take observers miles under the sea to record what was there—and safely return them to the surface.

He might be astonished that humans have walked on the moon and now inhabit a station in space, live underwater, and journey across continents and oceans in hours, not months or years.

Source: **Remarks for Royal Geographic Society, June 6, 2011, upon Dr. Earle's acceptance of the Patron's Medal of the Royal Geographical Society of the United Kingdom.**

A section of the outer barrier reef, Australia.

Background photo: Image Source/Corbis

Geosat satellite **(Figure 4.5a)** measured its distance from the ocean surface to within 0.03 meter (1 inch)! Because the precise position of the satellite can be calculated, the average height of the ocean surface can be known with great accuracy.

Disregarding waves or tides or currents, researchers have found the ocean surface can vary from the ideal smooth (ellipsoid) shape by as much as 200 meters (660 feet). The reason is that the pull of gravity varies across Earth's surface depending on the nearness (or distance away) of massive parts of Earth. The gravity from an undersea mountain or ridge "pulls" water toward it from the sides, forming a mound of water over itself **(Figure 4.5b)**. For example, a typical undersea volcano,

> **T**he great density of the seabed partly explains why more than half of Earth's solid surface is at least 3,000 meters (10,000 feet) below sea level.

with a height of 2,000 meters (6,600 feet) above the seabed and a radius of 20 kilometers (32 miles), would produce a 2-meter (6.6-foot) rise in the ocean surface. (This mound cannot be seen with the unaided eye because the slope of the surface is very gradual.) The large features of the seabed are amazingly and accurately reproduced in the subtle standing irregularities of the sea surface **(Figure 4.5d)**!

Geosat and its successors, *TOPEX/ Poseidon* and *Jason-1* and *Jason-2*, have allowed the rapid mapping of the world ocean floor from space. Hundreds of previously unknown features have been discovered through the data they have provided.

Figure 4.5 Measuring the
seabed from space

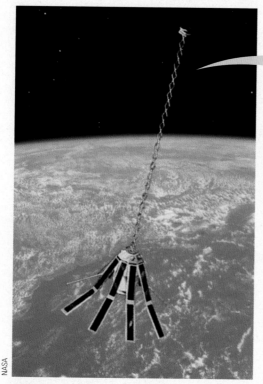

NASA

a *Geosat,* a U.S. Navy satellite that operated from
1985 through 1990, provided measurements of
sea surface height from orbit. Moving above the
ocean surface at 7 kilometers (4 miles) a second,
Geosat bounced 1,000 pulses of radar energy
off the ocean every second. Height accuracy was
within 0.03 meter (1 inch)! Other satellites have
taken its place.

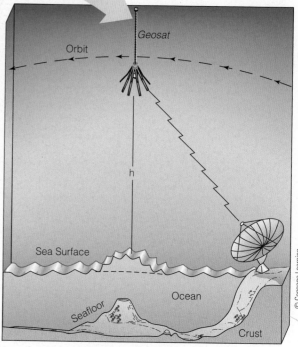

© Cengage Learning

b Distortion of the sea surface above a seabed feature
occurs when the extra gravitational attraction of
the feature "pulls" water toward it from the sides,
forming a mound of water over itself.

Robots Descend to Observe the Details

Small robot submersibles were much in the news during the
Gulf oil spill in the summer of 2010. These nimble, capable devices can manipulate valves, lift and reposition equipment, and
act as remote sets of eyes for decision makers.[2] Scientists use
these devices to probe submerged geological features, examine
shipwrecks, and measure water characteristics **(Figure 4.6)**.

Perhaps the most imaginative new technology being incorporated into these devices is *telepresence*, the extension of a person's
senses by remote manipulators. A scientist might wear a helmet
containing small stereo television screens and earphones, and
place his hands in special gloves equipped with tactile feedback
units. Movements of his head and hands would be duplicated by
a robot on the seafloor, and sensations "felt" by the robot would
be relayed back to the scientist through the TVs, earphones, and
gloves. He would thus have the sensation of being on the seafloor
and could take samples, manipulate tools, or just look around.

---

[2]What's the difference between a submarine and a submersible? A *submarine* is
self-contained and can operate independently; a *submersible* relies on another
vessel or facility for support. Submersibles can carry humans or be robotic.

Other researchers could watch or participate at distant locations
via a high-speed data link.

CONCEPT CHECK

Before going on to the next section, check your understanding of
some of the important ideas presented so far:

How was bathymetry accomplished in years past? How do scientists
do it now?

Echo sounders bounce sound off the seabed to measure depth. How
does that work?

Satellites orbit in space. How can a satellite conduct oceanographic
research? Why does the surface of the ocean "bunch up" over
submerged mountains and ridges?

4.2 Ocean-Floor Topography Varies with Location

Most people think an ocean basin is shaped like a giant bathtub.
They imagine that the continents drop off steeply just beyond
the surf zone and that the ocean is deepest somewhere out in the

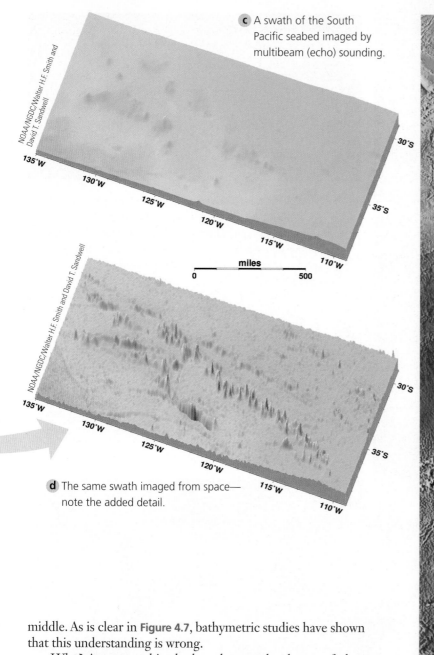

c A swath of the South Pacific seabed imaged by multibeam (echo) sounding.

NOAA/NGDC/Walter H.F. Smith and David T. Sandwell

NOAA/NGDC/Walter H.F. Smith and David T. Sandwell

miles
0   500

d The same swath imaged from space—note the added detail.

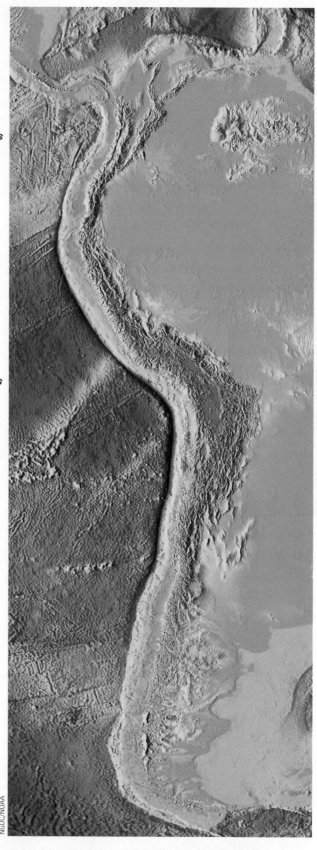

NGDC/NOAA

middle. As is clear in **Figure 4.7**, bathymetric studies have shown that this understanding is wrong.

Why? As you read in the last chapter, the theory of plate tectonics suggests that Earth's surface is not a static arrangement of continents and ocean but a dynamic mosaic of jostling lithospheric plates. The lighter continental lithosphere floats in isostatic equilibrium above the level of the heavier lithosphere of the ocean basins. The great density of the seabed partly explains why more than half of Earth's solid surface is at least 3,000 meters (10,000 feet) below sea level **(Figure 4.8)**. Continental crust is thicker than oceanic crust, and continental lithosphere is less dense than oceanic lithosphere. Therefore, the less dense lithosphere containing the continents floats in isostatic equilibrium above the level of the denser lithosphere containing ocean basins. The seabed topography we observe is the result of this dynamic balance *and* the jostling of tectonic plates.

Notice in **Figure 4.9** (see p. 112) the transition between the thick (and less dense) granitic rock of the continents and the

e South America viewed by satellite altimetry. Note the high Andes mountains, the Peru–Chile trench running the length of the continent's active west coast, the transform faults and fracture zones of the Chile Rise (at lower left), and the very large continental shelf on the passive (trailing) edge of the southern part of the continent.

b Operated by the Mystic Aquarium's Institute for Exploration, in conjunction with the University of Rhode Island and NOAA, ROV *Hercules* is photographed from the ROV *Argo*, a sled slowly towed by a mothership. *Hercules* is tethered to *Argo* by a 30-meter (100-foot) cable. *Argo* was used to discover the remains of the RMS *Titanic* in 1985.

a The HROV *Nereus,* named after a Greek god who could change himself into any shape, is capable of operating in two modes: free-swimming (autonomous) and tethered. (HROV stands for hybrid remotely operated vehicle.) In either mode, *Nereus* can spend up to 36 hours working in the ocean's deepest recesses. On 31 May 2009, *Nereus* successfully reached the deepest part of the ocean—a depth of 10,920 meters (6.8 miles) in the western Pacific's Mariana Trench.

Figure 4.6 Remote sensing moves to the future.

💬 **THINKING BEYOND THE FIGURE**

What (if any) advantages would a human in a submarine bring to a research situation that a robot equipped with telepresence could not provide?

Figure 4.7 Cross sections of the Atlantic Ocean basin and the continental United States, showing the range of elevations. The vertical exaggeration is 100:1. Although ocean depth is clearly greater than the average elevation of the continent, the general range of contours is similar.

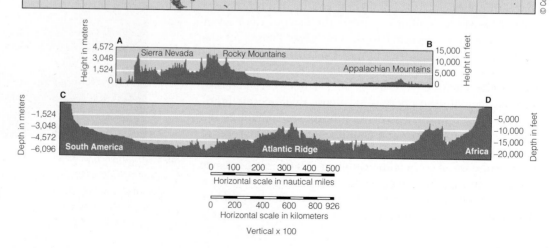

Figure 4.8 A graph showing the distribution of elevations and depths on Earth. This curve is not a land-to-sea profile of Earth but rather a plot of the area of Earth's surface above any given elevation or depth below sea level. Note that more than half of Earth's solid surface is at least 3,000 meters (10,000 feet) below sea level. The average depth of the ocean is much greater than the average elevation of the continents: The average depth of the world ocean (3,796 meters or 12,451 feet) is much greater than the average height of the continents (840 meters or 2,760 feet).

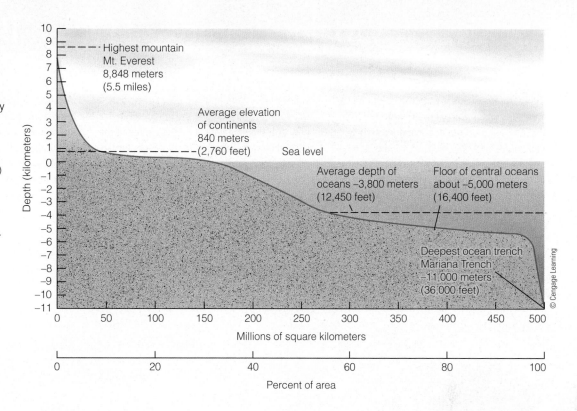

relatively thin (and denser) basalt of the deep-sea floor. Near shore the features of the ocean floor are similar to those of the adjacent continents because they share the same granitic basement. The transition to basalt marks the *true* edge of the continent and divides ocean floors into two major provinces. The submerged outer edge of a continent is called the **continental margin**. The deep-sea floor beyond the continental margin is properly called the **ocean basin**. **Figure 4.10** (see p. 114) shows the relative amounts of Earth's surface composed of continental and oceanic structures and the important subdivisions of each.

4.3 Continental Margins May Be Active or Passive

You learned in Chapter 3 that lithospheric plates converge, diverge, or slip past one another. As you might expect, the submerged edges of continents—continental margins—are greatly influenced by this tectonic activity. Continental margins facing the edges of *diverging* plates are called **passive margins**

because relatively little earthquake or volcanic activity is now associated with them. Because they surround the Atlantic, passive margins are sometimes referred to as *Atlantic-type* margins. Continental margins near the edges of *converging* plates (or near places where plates are slipping past one another) are called **active margins** because of their earthquake and volcanic activity. Because of their prevalence in the Pacific, active margins are sometimes referred to as *Pacific-type* margins.

Figure 4.11 (see p. 114) shows active and passive margins west and east of South America. Note that active margins coincide with plate boundaries but passive margins do not. Passive margins are also found outside the Atlantic, but active margins are confined mostly to the Pacific.

Continental margins have three main divisions: a shallow, nearly flat continental *shelf* close to shore; a more steeply sloped continental *slope* seaward; and an apron of sediment—the continental *rise*—that blends the continental margins into the deep-ocean basins.

Continental Shelves Are Seaward Extensions of the Continents

The shallow submerged extension of a continent is called the **continental shelf**. Continental shelves, an extension of the adjacent continents, are underlain by granitic continental crust. They are much more like the continent than like the deep-ocean floor, and they may have hills, depressions, sedimentary rocks, and mineral and oil deposits similar to those on the dry land nearby. Taken together, the area of the continental shelves is 7.4% of Earth's ocean area.

Submarine canyons, cut by rivers when sea level was lower, carry material eroded from the land out onto the deep seafloor. These sediments, plus the shells of trillions of microscopic creatures from the surface waters, cover the rough seabed terrain with vast flat abyssal plains. Seafloor spreading ridges create new oceanic crust from molten mantle material that emerges through the central rift valley of the ridge and cools into new rock. Convection cells deep in the mantle then force new crust to move away on both sides of the spreading ridge. Stationary hot spots within the mantle create straight lines of volcanoes and island chains as the crust moves over them. Oceanic crust is destroyed when it is subducted—pushed and pulled down under another plate and melted back into the mantle. As the oceanic crust is forced down, it creates deep-sea trenches, the deepest places in the ocean. Some of the melted material escapes back to the surface in volcanic island arcs formed on the other side of the trenches. (The vertical scale has been greatly exaggerated to emphasize basin topography.)

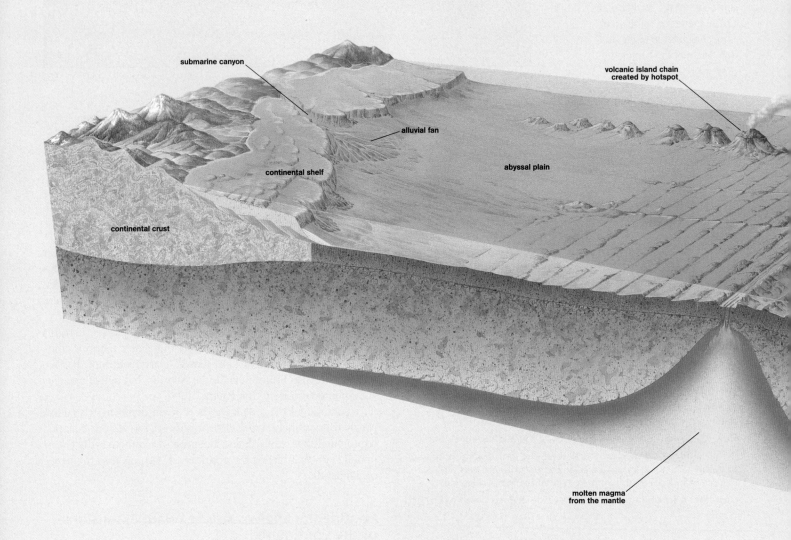

submarine canyon

volcanic island chain created by hotspot

alluvial fan

continental shelf

abyssal plain

continental crust

molten magma from the mantle

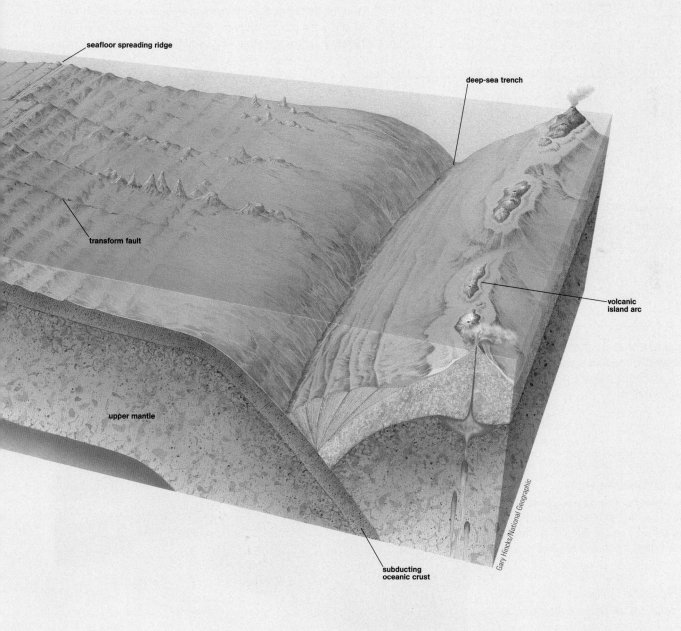

seafloor spreading ridge

deep-sea trench

transform fault

volcanic
island arc

upper mantle

Gary Hincks/National Geographic

subducting
oceanic crust

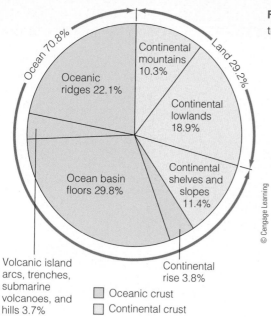

Figure 4.10 Features of Earth's solid surface shown as percentages of the planet's total surface.

Ocean 70.8%

Oceanic ridges 22.1%

Continental mountains 10.3%

Land 29.2%

Continental lowlands 18.9%

Ocean basin floors 29.8%

Continental shelves and slopes 11.4%

Volcanic island arcs, trenches, submarine volcanoes, and hills 3.7%

Continental rise 3.8%

☐ Oceanic crust
☐ Continental crust

Figure 4.12 shows a passive-margin continental shelf characteristic of Atlantic Ocean edges. The broad shelf extends far from shore in a gentle incline, typically 1.7 meters per kilometer (0.1°, or about 9 feet per mile), much more gradual than the slope of a well-drained parking lot. Shelves along the margin of the Atlantic Ocean often reach 350 kilometers (220 miles) in width and end at a depth of about 140 meters (460 feet), where a steeper drop-off begins.

The passive-margin shelves of the Atlantic Ocean formed as the fragments of Pangaea were carried away from each other by seafloor spreading. The oceanic lithosphere, thinned during initial rifting, cooled and contracted as it moved away from the spreading center, submerging the trailing edges of the continents and forming the shelves.

Most of the material composing a shelf comes from erosion of the adjacent continental mass. Rivers assist in passive shelf building by transporting huge amounts of sediments to the

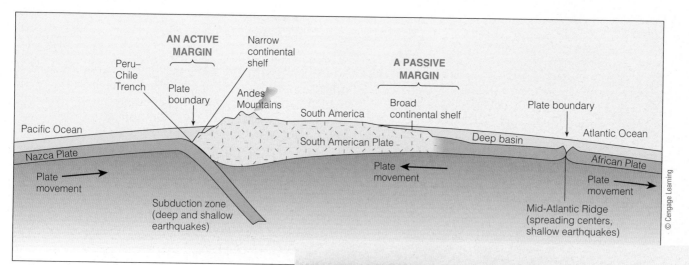

a Typical continental margins bordering the tectonically *active* (Pacific-type) and tectonically *passive* (Atlantic-type) edges of a moving continent. (The vertical scale has been exaggerated.)

Figure 4.11 Active and passive margins.

💬 **THINKING BEYOND THE FIGURE**
One of the most violent earthquakes recorded in North America occurred near New Madrid, Missouri, in 1811. Is Missouri at an active or passive margin? No margin? Can you propose a cause for this earthquake?

b A more detailed view of subduction and earthquakes at an active margin; in this case, the American Pacific Northwest.

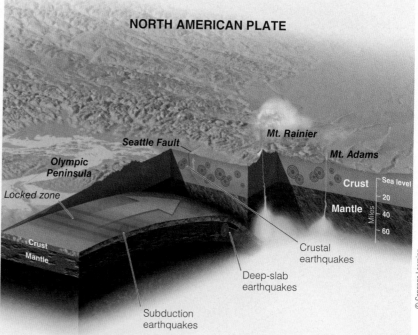

Figure 4.12 The features of a passive continental margin.

a Vertical exaggeration 50:1

b With no vertical exaggeration

© Cengage Learning

shore from far inland. In some places the sediments accumulate behind natural dams formed by ancient reefs or ridges of granitic crust. The weight of the sediment isostatically depresses the continental edges and allows the sediment load to grow even thicker. Sediment at the outer edge of a shelf can be up to 15 kilometers (9 miles) thick and 150 million years old at the bottom.

The width of a shelf is usually determined by its proximity to a plate boundary. You can see in Figure 4.12 that the shelf at the *passive* margin (east of South America) is broad, but the shelf at the *active* margin (west of South America) is very narrow. The widest shelf, 1,280 kilometers (800 miles) across, lies north of Siberia in the tectonically quiet Arctic Sea. Shelf width depends not only on tectonics but also on marine processes: fast-moving ocean currents can sometimes prevent sediments from accumulating. For example, the east coast of Florida has a very narrow shelf because there is no natural offshore dam formed by ridges of granitic crust and because the swift current of the nearby Gulf Stream scours surface sediment away. Florida's west coast, however, has a broad shelf with a relatively steep terminating slope **(Figure 4.13)**.

Figure 4.13 A relatively steep slope more than 1.6 kilometers (1 mile) high marks the edge of the continental shelf west of central Florida. The steep continental slope seen here is unusual. Perhaps fresh water seeping from the adjacent land has undermined the slope and caused it to collapse. Currents have removed much of the material that would otherwise be found at the base of the slope.

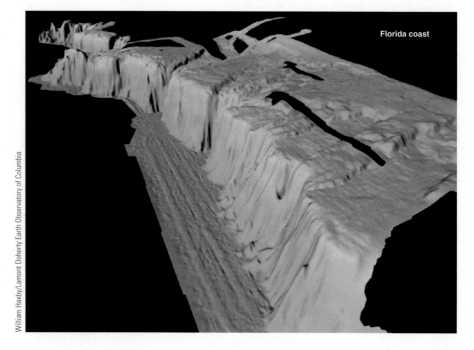

William Haxby/Lamont Doherty Earth Observatory of Columbia

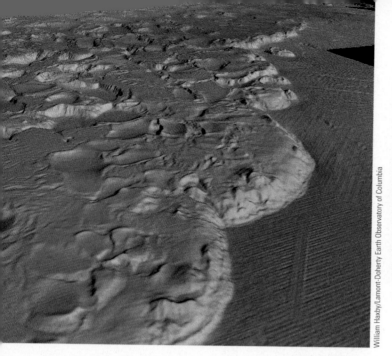

William Haxby/Lamont-Doherty Earth Observatory of Columbia

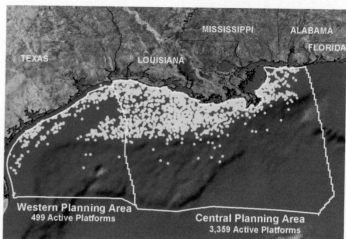

Expedition to Deep Slope 2006, NOAA-OER

a The broad continental shelf south of Texas and Louisiana (looking east). Sediments carried by the Mississippi have overlain ancient salt deposits. The weight of the sediments causes salt domes to form and dissolve, leaving pockmarks in the continental shelf.

b The demand for petroleum products has greatly affected continental shelves. Nearly 4,000 active oil and gas drilling platforms were active off the Texas, Louisiana, Mississippi, and Alabama coasts in 2009. The dangers inherent in the exploitation of these resources were made clearer by the destruction of an offshore oil-drilling platform (and subsequent release of oil into the Gulf of Mexico) in the summer of 2010. More on this subject will be found in Chapters 17 and 18.

Figure 4.14 The continental shelf along the northern edge of the Gulf of Mexico.

A broad shelf has resulted from sediment accumulation along the sheltered edges of the Gulf of Mexico **(Figure 4.14)**. Sediment eroded from the Rocky Mountains has been carried by the Mississippi River and deposited south of Texas and Louisiana. This sediment overlies salt deposited about 180 million years ago when much of the water in the Gulf evaporated. The weight of the overlying sediment causes domes of salt to rise, spread out, and dissolve; collapsed salt domes create the pockmarked bottom characteristic of the area.

The shelves of the active Pacific margins are generally not as broad and flat as the Atlantic shelves. An example is the abbreviated shelf off the west coast of South America, where the steep western slope of the Andes Mountains continues nearly uninterrupted beneath the sea into the depths of the Peru–Chile Trench (see again Figure 4.5e). Active-margin shelves have more varied topography than passive-margin shelves; the character of continental shelves at an active margin may be determined more by faulting, volcanism, and tectonic deformation than by sedimentation **(Figure 4.15)**.

A few Pacific shelves are broad, however. As in the Atlantic, natural offshore dams trap sediments and form shelves; but in the Pacific the sediment-trapping dams are more commonly offshore chains of volcanoes or lines of coral reefs. Volcanic activity east of China and Southeast Asia has formed a broad basin that is now filling with sediments and is one of the largest shelves in the Pacific.

Because of their gentle slope, continental shelves are greatly influenced by changes in sea level. Around 18,000 years ago—at the height of the last **ice age** (period of wide-spread glaciation)—massive ice caps covered huge regions of the world's continents. The water that formed these thick ice sheets came from the ocean, and sea level fell about 125 meters (410 feet) below its present position **(Figure 4.16)**.[3] The continental shelves were almost completely exposed, and the surface area of the continents was about 18% greater than it is today. Rivers and waves cut into the sediments that had accumulated during periods of higher sea level, and they transported some coarse sediments to their present locations at the shelves' outer edges. Sea level began to rise again when the ice caps melted, and sediments again began to accumulate on the shelves. More on the history and effects of sea-level change will be found in the discussion of coasts in Chapter 12 and of environmental issues in Chapter 18.

The continental shelves have been the focus of intense exploration for natural resources. Because shelves are the submerged margins of continents, any deposits of oil or minerals along a coast are likely to continue offshore. Water depth over shelves averages only about 75 meters (250 feet), so large areas of the shelves are accessible to mining and drilling activities. Many of the techniques used to find and exploit natural resources on land can also be used on the continental shelves.

---

[3]Note that sea level has been considerably below its present position for nearly all of the past quarter-million years. Consider the implications for coastal civilizations. During the time between 18,000 years ago and 8,000 years ago, sea level rose over 100 meters (330 feet) at a rate of about 1 centimeter (½ inch) a year; over 0.5 meters in a human lifetime. Could this have given rise to the flood legends common to many religions? And what will be the implications of a future rise?

Figure 4.15
The complex continental shelf off southern California, typical of an active margin. Compare to Figure 4.14a. USGS Pacific Sea-Floor Mapping Project

💬 **THINKING BEYOND THE FIGURE**

What connection might there be between this complex seabed and the formation of terranes (see Figures 3.32 and 3.33)?

Figure 4.16 Changes in sea level over the past 250,000 years, as traced by data taken from ocean floor cores. The rise and fall of sea level is due largely to the coming and going of ice ages— periods of increased and decreased glaciation, respectively. Water that formed the ice-age glaciers came from the ocean, and this caused sea level to drop. Point ⓐ indicates a low stand of −125 meters (−410 feet) at the climax of the last ice age some 18,000 years ago. Point ⓑ indicates a high stand of +6 meters (+19.7 feet) during the last interglacial period about 120,000 years ago. Point ⓒ shows the present sea level. Sea level continues to rise as we emerge from the last ice age and enter an accelerating period of global warming. For more detail of the past 25,000 years, please look ahead to Figure 12.2 and Chapter 18's discussion of Doggerland.

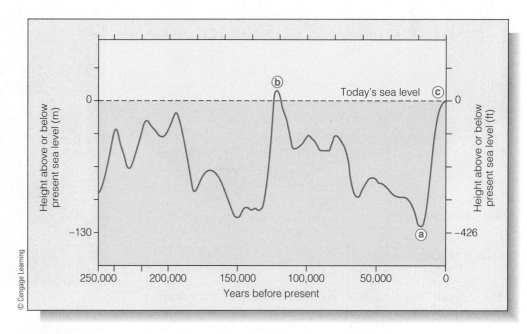

Figure 4.17 Platform *Brent Bravo* rides out a North Atlantic storm. The Brent oil field lies on the continental shelf beneath 140 meters (460 feet) of seawater between Scotland and Norway.

Resource development requires intensive scientific investigations, and our understanding of the geology of the shelves has benefited greatly from the search for offshore oil and natural gas **(Figure 4.17).**[4]

---

[4]The economic, environmental, and legal implications of the exploitation of these riches are touched on in Chapter 17's discussion of the Law of the Sea and the establishment of exclusive economic zones.

Continental Slopes Connect Continental Shelves to the Deep-Ocean Floor

The **continental slope** is the transition between the gently descending continental shelf and the deep-ocean floor. Continental slopes are formed of sediments that reach the built-out edge of the shelf and are transported over the side. At active margins a slope may also include marine sediments scraped off a descending plate during subduction. The inclination of a typical continental slope is about 4° (70 meters per kilometer, or 370 feet per mile), slightly steeper than the steepest road slope allowed on the interstate highway system. As Figure 4.12b implied, even the steepest of these slopes is not precipitous: a 25° slope is the greatest incline yet discovered. In general, continental slopes at active margins are steeper than those at passive margins. Continental slopes average about 20 kilometers (12 miles) wide and end at the continental rise, usually at a depth of about 3,700 meters (12,000 feet). The bottom of the continental slope is the true edge of a continent.

The **shelf break** marks the abrupt transition from continental shelf to continental slope (see once again Figure 4.12). The depth of water at the shelf break is surprisingly constant—about 140 meters (460 feet) worldwide—but there are exceptions.

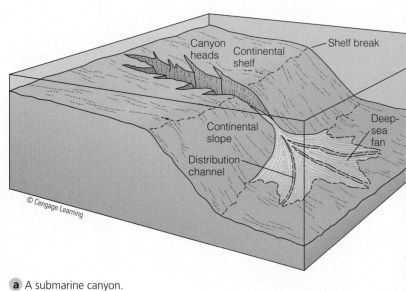

© Cengage Learning

a A submarine canyon.

Figure 4.18 Submarine canyons

b A multibeam image of Hudson canyon east of New Jersey. The shelf in this area is broad. The canyon can be seen nicking the shelf-slope junction and then continuing toward the abyssal plain to the southeast. The underwater topography has been exaggerated by a factor of 5 relative to the land topography. The fine black line marks sea level.

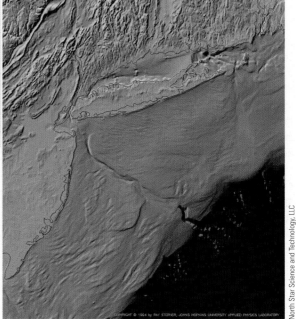

North Star Science and Technology, LLC

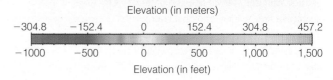

Elevation (in meters)					
−304.8	−152.4	0	152.4	304.8	457.2
−1000	−500	0	500	1,000	1,500

Elevation (in feet)

118 • CHAPTER 4

Figure 4.19 A turbidity current flowing down a submerged slope off the island of Jamaica. A turbidity current is not propelled by the water within it, but by gravity. The propeller of a submarine caused the turbidity current by disturbing sediment along the slope.

The great weight of ice on Antarctica, for example, has isostatically depressed that continent, and the depth of water at the shelf break is 300–400 meters (1,000–1,300 feet). The shelf break in Greenland is similarly depressed.

Submarine Canyons Form at the Junction between Continental Shelf and Continental Slope

Submarine canyons cut into the continental shelf and slope, often terminating on the deep-sea floor in a fan-shaped wedge of sediment (**Figure 4.18a**). Hundreds of submarine canyons nick the edge of nearly all of Earth's continental shelves. The canyons generally trend at right angles to the shoreline (and shelf edge), sometimes beginning very close to shore. Congo Canyon actually extends into the African continent as a deep estuary at the mouth of the Congo River. These enigmatic features can be quite large. In fact, submarine canyons similar in size and profile to Arizona's Grand Canyon have been discovered!

Hudson Canyon, a typical large canyon on a passive margin, is shown in **Figure 4.18b**. Like many submarine canyons, Hudson Canyon is located just offshore of the mouth of a river or stream—in this case, New York's Hudson River. Because of their similarity to canyons on land, submarine canyons appear to have been created by erosion, so marine geologists initially thought the canyons were carved into the shelves by stream erosion at times of lower sea level. But most researchers agree that sea level has never fallen more than 200 meters (660 feet) below its present level in the past 600 million years. Stream erosion could account for the shape of the uppermost parts of the canyons. However, since submarine canyons can be traced to depths in excess of 3,000 meters (10,000 feet) below sea level, stream erosion could not have played a direct role in cutting their lower depths.

What, then, caused the submarine canyons to form? Local landslides or sediment liquefaction triggered by earthquakes sometimes causes an abrasive underwater "avalanche" of sediments. These mass movements of sediment, called **turbidity currents**, occur when turbulence mixes sediments into water above a sloping bottom. The sediment-filled water is denser than the surrounding water, so the thick muddy fluid runs down the slope at speeds up to 27 kilometers (17 miles) per hour. **Figure 4.19** is a rare photo of a turbidity current.

What is the connection between turbidity currents and submarine canyons? Sediments may cascade continuously down the canyons (**Figure 4.20**), but earthquakes can shake loose huge masses of boulders and sand that rush down the edge of the shelf, scouring the canyon deeper as they go. Most geologists believe that the canyons have been formed by abrasive turbidity currents plunging down the canyons. In this way the canyons can be cut to depths far below the reach of streams even during the low sea levels of the ice ages.

Figure 4.20 A continuous cascade of sediment at the head of San Lucas submarine canyon (off the coast of Baja California, Mexico), which may be eroding the narrow gorge in conjunction with occasional turbidity currents. About 100,000 cubic meters of sand slip down this canyon every year. U.S. Department of the Navy

Continental Rises Form as Sediments Accumulate at the Base of the Continental Slope

Along passive margins, the oceanic crust at the base of the continental slope is covered by an apron of accumulated sediment called the **continental rise** (see again Figure 4.12). Sediments from the shelf slowly descend to the ocean floor along the whole continental slope, but most of the sediments that form the continental rise are transported to the area by turbidity currents. The width of the rise varies from 100 to 1,000 kilometers (63 to 630 miles), and its slope is gradual—about one-eighth that of the continental slope. One of the widest and thickest continental rises has formed in the Bay of Bengal at the mouths of the Ganges–Brahmaputra River, the most sediment-laden of the world's great rivers.

Deep-ocean currents are an important factor in shaping continental rises, especially along the western boundaries of most ocean basins. Deep boundary currents held against the continental slopes by Coriolis effect (news of which awaits you in Chapter 8) pick up volcanic debris and sediments transported by turbidity currents and sweep it along the ocean floor. As the currents are forced around bends and across depressions, they slow and deposit the suspended material on the rises in the form of ridges and mud.

> " The wonders of the sea are as marvelous as the glories of the heavens. . . Could the waters of the Atlantic be drawn off so as to expose to view this great sea gash. . . [it] would present a scene most rugged, grand and imposing. The very ribs of the solid Earth, with the foundations of the sea, would be brought to light. "
> —MATTHEW MAURY, *THE PHYSICAL GEOGRAPHY OF THE SEA*

The deep-ocean floor consists mainly of oceanic ridge systems and the adjacent sediment-covered plains. Deep basins may be rimmed by trenches or by masses of sediment. Flat expanses are interrupted by islands, hills, active and extinct volcanoes, and active zones of seafloor spreading. The sediments on the deep-ocean floor reflect the history of the surrounding continents, the biological productivity of the overlying water, and the ages of the basins themselves.

Oceanic Ridges Circle the World

If the ocean evaporated, the oceanic ridges would be Earth's most remarkable and obvious feature. An **oceanic ridge** is a mountainous chain of young basaltic rock at the active spreading center of an ocean. Stretching 65,000 kilometers (40,000 miles), more than 1½ times Earth's circumference, oceanic ridges girdle the globe like seams surrounding a softball **(Figure 4.21a)**. The rugged ridges, which often are devoid of sediment, rise about 2 kilometers (1.25 miles) above the seafloor. In places they project above the surface to form islands such as Iceland, the Azores, and Easter Island. Oceanic ridges and their associated structures account for 22% of the world's solid surface area (all the land above sea level accounts for 29%). Although these features are often called mid-ocean ridges, less than 60% of their length actually exists along the centers of ocean basins.

As we saw in our discussion of plate tectonics, the rift zones associated with oceanic ridges are sources of new ocean floor where lithospheric plates diverge. The oceanic ridges are widest where they are most active. The youngest rock is located at the active ridge center, and rock becomes older with distance from the center. As the lithosphere cools, it shrinks and subsides. Slowly spreading ridges have a steeper profile than rapidly spreading ones because slowly diverging seafloor cools and shrinks closer to the spreading center. **Figure 4.21b** shows these distinctly different ridge profiles.

Figure 4.22, a beautiful hand-drawn bathymetric map of the North Atlantic, clearly shows the great extent of the Mid-Atlantic Ridge, a typical oceanic ridge. **Figure 4.23a**, a multibeam image, provides a detailed look at the young central rift. **Figure 4.23b** shows the mountainous rift away from the spreading center being buried by falling sediments.

As could be seen in Figure 4.22, the Mid-Atlantic Ridge does not run in a straight line. It is offset at more or less regular intervals by transform faults. A *fault*, recall, is a fracture in the lithosphere along which movement has occurred, and **transform faults** are fractures along which lithospheric plates slide horizontally **(Figure 4.24**, see p. 124). As you saw in Figure 3.17c and 3.25, when segments of a ridge system are offset, the fault connecting the axis of the ridge is a transform fault. Shallow

CONCEPT CHECK

Before going on to the next section, check your understanding of some of the important ideas presented so far:

What are the features of the continental margins?

How is an active tectonic margin different from a passive tectonic margin?

How do the widths of continental shelves differ between active margins and passive margins?

How has sea level varied with time? Is sea level unusually high or low at present?

What are submarine canyons? Where are they found, and how are they thought to have been formed?

Where would you look for a continental rise? What forms continental rises?

4.4 The Topography of Deep-Ocean Basins Differs from That of the Continental Margin

Away from the margins of continents the structure of the ocean floor is quite different. Here the seafloor is a blanket of sediment up to 5 kilometers (3 miles) thick overlying basaltic rocks. Deep-ocean basins constitute more than half of Earth's surface.

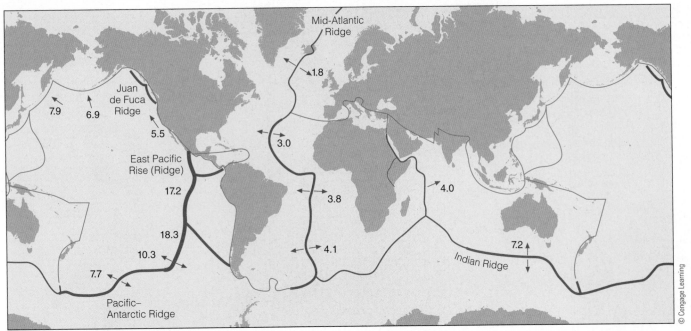

a The oceanic ridge system (in red) stretches some 65,000 kilometers (40,000 miles) around Earth. The thickness of the red lines indicates the rate of spreading for some of the most rapidly spreading sections, and the numbers give spreading rates in centimeters per year. The East Pacific Rise typically spreads about six times faster than the Mid-Atlantic Ridge. The thin green lines represent subduction zones.

b Rapid spreading at the East Pacific Rise spreads ridge features over a greater area. The slower spreading along the Mid-Atlantic Ridge (upper right) concentrates the features in a smaller area with a more vertical topography and a pronounced central rift. The relatively slow-spreading ridge is shown in more detail in Figure 4.23a.

Figure 4.21 The oceanic ridge system.

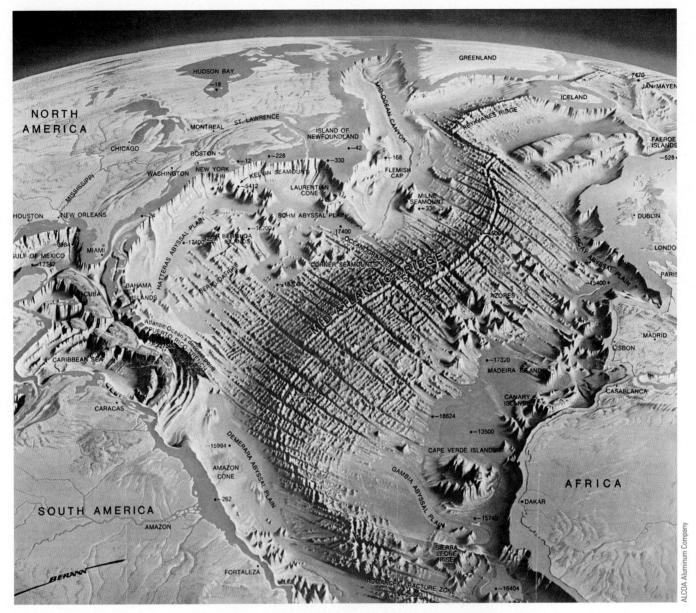

Figure 4.22 Heinrich Berann's hand-drawn map of a portion of the Atlantic Ocean floor showing some major oceanic features: mid-ocean ridge, transform faults, fracture zones, submarine canyons, seamounts, continental rises, trenches, and abyssal plains. If the ocean evaporated, the ridge system would be Earth's most remarkable and obvious feature. Depths are in feet. The map is vertically exaggerated.

earthquakes are common along transform faults. Since the ocean floor cannot expand evenly on the surface of a sphere, plate divergence on the spherical Earth can only be irregular and asymmetrical, and transform faults and fracture zones result.

Transform faults are the active part of **fracture zones.** Extending outward from the ridge axis, fracture zones are seismically inactive areas that show evidence of past transform fault activity. While segments of a lithospheric plate on either side of a transform fault move in *opposite* directions from each other, the plate segments adjacent to the outward segments of a fracture zone move in the *same* direction, as Figure 4.24 shows.

Hydrothermal Vents Are Hot Springs on Active Oceanic Ridges

Some of the most exciting features of the ocean basins are the **hydrothermal vents.** In 1977 Robert Ballard and J. F. Grassle of the Woods Hole Oceanographic Institution discovered hot springs on oceanic ridges. Diving in *Alvin* at 3 kilometers (1.9 miles) near the Galápagos Islands along the East Pacific Rise (an oceanic ridge), they came across rocky chimneys up to 20 meters (66 feet) high, from which dark, mineral-laden water was blasting at 350°C (660°F) **(Figure 4.25).** Only the great pressure at this depth prevented the escaping water from flashing to steam. These *black smokers*, as they were quickly nicknamed, fasci-

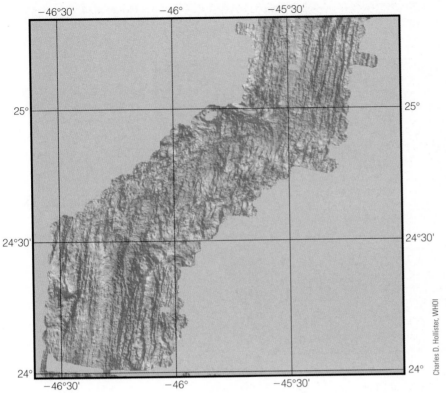

Figure 4.23 A mid-ocean ridge in detail.

a The fine structure of the central portion of the mid-Atlantic ridge between Florida and western Africa. The depressed central valley (the spreading center, shown in blue) is clearly visible in this computer-generated multi-beam image.

Charles D. Hollister, WHOI

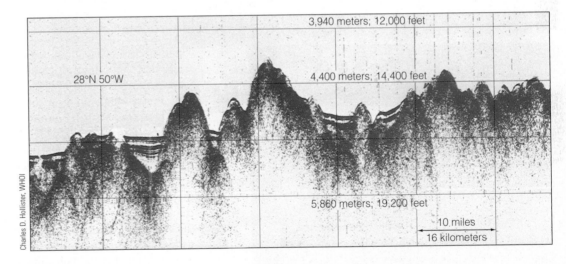

b As this seismic profile shows, the rugged relief of the North Atlantic's oceanic ridge is gradually being buried by slowly accumulating sediments.

Charles D. Hollister, WHOI

nate marine geologists. It is believed that water descends through fissures and cracks in the ridge floor until it comes into contact with very hot rocks associated with active seafloor spreading. There the superheated, chemically active water dissolves minerals and gases and escapes upward through the vents by convection **(Figure 4.26)**. As we will see in Chapter 7, this contact between water and rocks has great implications for ocean chemistry.

Since that first discovery, vents have been found on the Mid-Atlantic Ridge east of Florida, in the Sea of Cortez east and south of Baja California, and on the Juan de Fuca Ridge off the coast of Washington and Oregon. Scientists now believe that hydrothermal vents may be very common on oceanic ridges, especially in zones of rapid seafloor spreading. In July 1990 vents were discovered in fresh water, at the bottom of

Lake Baikal in southern Siberia. This discovery suggests that the world's oldest and deepest lake may someday become part of the ocean as Asia slowly breaks apart.

In Iceland these vents may be seen on dry land! As you may recall, the country of Iceland rests uneasily on a mid-ocean ridge lifted above sea level (see again Chapter 3's opener). **Figure 4.27** (see p. 126). is an extraordinary view of the central rift of that ridge (similar to that seen as the blue-colored contour in Figure 4.23a. The rift rises from an Icelandic lake at the left, traverses to the right, and supports many thermal vents visible as jets of steam. Notice the hills paralleling the rift on the far side. Crustal spreading in this area averages about 10 centimeters (4 inches) a year.

Not all vents form chimneys of mineral deposits—some are simply cracks in the seabed, or porous mounds, or broad seg-

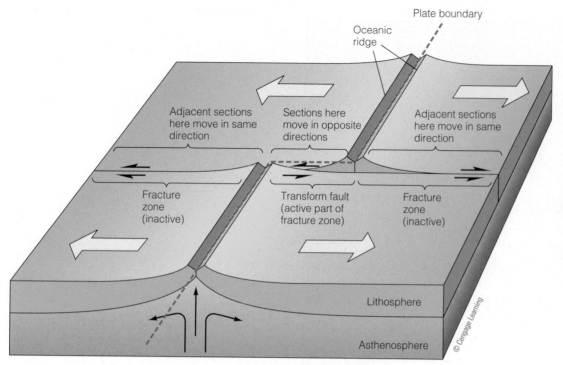

© Cengage Learning

Figure 4.24 Transform faults and fracture zones along an oceanic ridge. Transform faults are fractures along which lithospheric plates slide horizontally past one another. Transform faults are the active part of fracture zones. For a review of this process in larger scale, please see Figure 3.25.

💬 **THINKING BEYOND THE FIGURE**

How is California's famous San Andreas Fault related to transform faults and fracture zones?

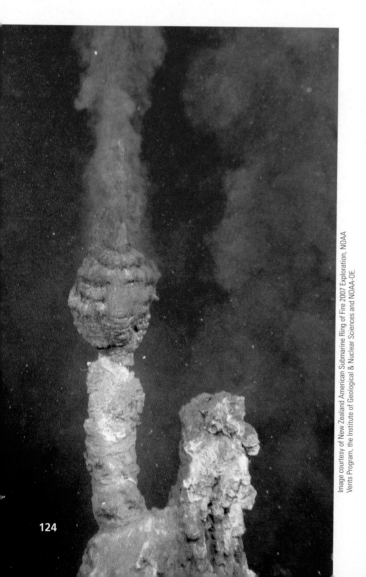

Image courtesy of New Zealand American Submarine Ring of Fire 2007 Exploration, NOAA Vents Program, the Institute of Geological & Nuclear Sciences and NOAA-OE.

ments of ocean ridge floor through which warm, mineral-laden water percolates upward. Cooler vents result when hot, rising water mixes with cold bottom water before reaching the surface. Water temperature in the vicinity of most hydrothermal vents averages 8°–16°C (46°–61°F), much warmer than usual for ocean-bottom water, which has an average temperature of 3°–4°C (37°–39°F). We will study the unusual communities of marine animals that populate these vents in Chapter 16.

A volume of water equal to the volume of the world ocean is thought to circulate through the hot oceanic crust at spreading centers about every 10 million years! The water coming from the vents and seeping from the floor is more acidic, is enriched with metals, and has higher concentrations of dissolved gas than seawater. The heat and chemicals issuing from these structures may play important roles in the chemical composition of seawater and the atmosphere, and in the formation of mineral deposits.

Abyssal Plains and Abyssal Hills Cover Most of Earth's Surface

A quarter of Earth's surface consists of abyssal plains and abyssal hills. *Abyssal* is an adjective derived from a Greek word meaning "without bottom." Although this is obviously not literally true, you can appreciate how the term came into use following the

Figure 4.25 A black smoker discovered at a depth of about 2,800 meters (9,200 feet) along the East Pacific Rise.

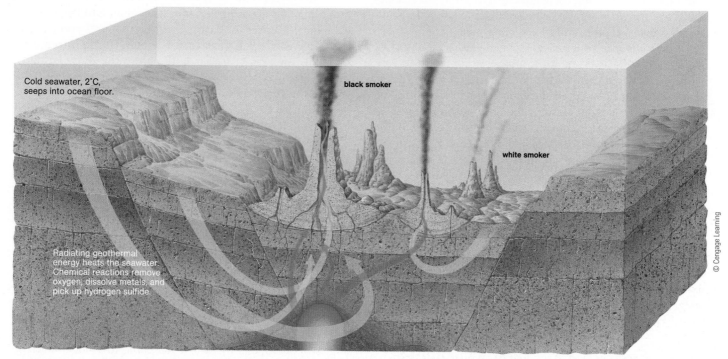

Cold seawater, 2°C, seeps into ocean floor.

black smoker

white smoker

Radiating geothermal energy heats the seawater. Chemical reactions remove oxygen, dissolve metals, and pick up hydrogen sulfide.

© Cengage Learning

Figure 4.26 A cross-section of the central part of a mid-ocean ridge—similar to that shown in Figures 4.23a and 4.27—showing the origin of hydrothermal vents. Cool water (blue arrows) is heated as it descends toward the hot magma chamber, leaching sulfur, iron, copper, zinc, and other materials from the surrounding rocks. The heated water (red arrows) returning to the surface carries these elements upwards, discharging them at the hydrothermal springs on the seafloor. The areas around the vents support unique communities of organisms (see Chapter 16).

Challenger expedition's laborious soundings of these extremely deep areas!

Abyssal plains are flat, featureless expanses of sediment-covered ocean floor found on the periphery of all oceans. They are most common in the Atlantic, less so in the Indian Ocean, and relatively rare in the active Pacific, where peripheral trenches trap most of the sediments flowing from the continents. They lie between the continental margins and the oceanic ridges about 3,700 to 5,500 meters (12,000 to 18,000 feet) below the surface (see again Figure 4.24). The Canary Abyssal Plain, a huge plain west of the Canary Islands in the North Atlantic, has an area of about 900,000 square kilometers (350,000 square miles).

Abyssal plains are extraordinarily flat. A 1947 survey by the Woods Hole Oceanographic Institution ship *Atlantis* found that a large Atlantic abyssal plain varies no more than a few meters in depth over its entire area. Such flatness is caused by the smoothing effect of the layers of sediment, which often exceed 1,000 meters (3,300 feet) in thickness. Most of the sediment that forms the abyssal plains appears to be of terrestrial or shallow-water origin, not derived from biological activity in the ocean above. Some of it may have been transported to the plains by winds or turbidity currents. These deep sediment layers mask irregularities in the underlying ocean crust, but a powerful type of echo sounder can "see" through this sediment to reveal the complex topography of the basaltic basin floor below **(Figure 4.28)**. The broad basaltic shoulders of the Mid-Atlantic Ridge extend beneath this cloak of sediment almost as far as the bordering continental slopes.

Abyssal plain sediments may not be thick enough to cover the underlying basaltic floor near the edges bordering the oceanic ridges. Here the plains are punctuated by **abyssal hills**—small, sediment-covered extinct volcanoes or intrusions of once-molten rock, usually less than 200 meters (650 feet) high (one is seen extending above the flat surface in Figure 4.30). These abundant features are associated with seafloor spreading; they form when newly formed crust moves away from the center of a ridge, stretches, and cracks. Some blocks of the crust drop to form valleys, and others remain higher as hills. Lava erupting from the ridge flows along the fractures, coating the hills. This helps explain why abyssal hills occur in lines parallel to the flanks of the nearby oceanic ridge and why they occur most abundantly in places where the rate of seafloor spreading is fastest. Abyssal plains and hills account for nearly all the area of deep-ocean floor that is not part of the oceanic ridge system. They are Earth's most common "landform."

Volcanic Seamounts and Guyots Project above the Seabed

The ocean floor is dotted with thousands of volcanic projections that do not rise above the surface of the sea. These projections are called **seamounts**. Seamounts are circular or elliptical, more than 1 kilometer (0.6 mile) in height, with relatively steep slopes of 20° to 25°. (Abyssal hills, in contrast, are much more abundant, less than a kilometer high, and not as steep.)

Seamounts may be found alone or in groups of 10 to 100. Though many form at hot spots, most are thought to be submerged inactive volcanoes that formed at spreading centers (**Figure 4.29**). Movement of the lithosphere away from spreading centers has carried them outward and downward to their present positions and will eventually cause them to bond with the edge of a continent or disappear into a trench (**Figure 4.30**). As many as 10,000 seamounts are thought to exist in the Pacific, about half the world total.

Guyots are flat-topped seamounts that once were tall enough to approach or penetrate the sea surface. Generally they are confined to the west-central Pacific. The flat top suggests that they were eroded by wave action when they were near sea level. Their plateau-like tops eventually sank too deep for wave erosion to continue wearing them down. Like the more abundant seamounts, most guyots were formed near spreading centers and transported outward and downward as the seafloor moved away from a spreading center and cooled.

Trenches and Island Arcs Form in Subduction Zones

A **trench** is an arc-shaped depression in the deep-ocean floor. These creases in the seafloor occur where a converging oceanic plate is subducted. The water temperature at the bottom of a trench is slightly cooler than the near-freezing temperatures of the adjacent flat ocean floor, reflecting the fact that trenches are underlain by old, relatively cold ocean crust sinking into the upper mantle and that cold bottom wa-

ter is flowing into them. Trenches (and their associated island arcs topped by erupting volcanoes) are among the most active geological features on Earth. Great earthquakes and tsunami (huge waves we will discuss in Chapter 10) often originate in them. **Figure 4.31** (see p. 130) shows the distribution of the ocean's major trenches. It is not surprising that most are around the edges of the active Pacific.

Trenches are the deepest places in Earth's crust, 3 to 6 kilometers (1.9 to 3.7 miles) deeper than the adjacent basin floor. The ocean's greatest depth is the Mariana Trench of the western Pacific, where the ocean bottom is 11,022 meters (36,163 feet) below sea level, 20% deeper than Mount Everest is high (**Figure 4.32**, see p. 131). The Mariana Trench is about 70 kilometers (44 miles) wide and 2,550 kilometers (1,600 miles) long, typical dimensions for these structures.

Trenches are curving chains of V-shaped indentations. The trenches are curved because of the geometry of plate interactions on a sphere. The convex sides of these curves generally face the open ocean (see again Figure 4.31). The trench walls on the island side of the depressions are steeper than those on the seaward side, indicating the direction of plate subduction. The sides of trenches become steeper with depth, normally reaching angles of about 10°–16° before flattening to a floor underlain by thick sediment. (Parts of the concave wall of the Kermadec–Tonga Trench are the world's steepest at 45°.) No continental rise occurs along coasts with trenches because the sediment that would form the rise ends up at the bottom of the trench.

Island arcs, curving chains of volcanic islands and seamounts, are almost always found parallel to the concave edges of trenches. As you may remember from Chapter 3, trenches and island arcs are formed by tectonic and volcanic activity associated with subduction. The descending lithospheric plate contains some mate-

(Continues from Figure 3.18)

Tom Garrison

Figure 4.28 The deep, smooth sediments of the Atlantic's Northern Madeira Abyssal Plain bury 100-million-year-old mountains. Note the one lonesome seamount emerging from the muck. This image was generated by a powerful echo sounder.

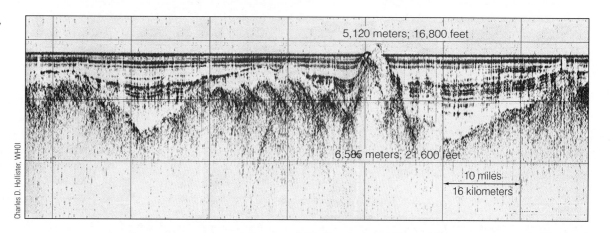

5,120 meters; 16,800 feet

6,585 meters; 21,600 feet

10 miles
16 kilometers

Charles D. Hollister, WHOI

rials that melt as the plate sinks into the mantle. These materials rise to the surface as magmas and lavas that form the chain of islands behind the trench. The Aleutian Islands, most Caribbean islands, and the Mariana Islands are island arcs. (See Figure 3.23 for a review of the processes involved in their construction.)

CONCEPT CHECK

Before going on to the next section, check your understanding of some of the important ideas presented so far:

What are typical features of deep-ocean basins?

What is the extent of the mid-ocean ridge system? Are mid-ocean ridges always literally in mid-ocean?

Draw a cross section through an active mid-ocean ridge. Where are the hydrothermal vents located? Where is new seabed being formed?

What are fracture zones? What causes these lateral breaks?

What are abyssal plains? What is unique about them?

Why are abyssal plains relatively rare in the Pacific?

How do guyots form? How were lines of guyots and seamounts important in deciphering plate tectonics?

How are the ocean's trenches formed? How are earthquakes related to their formation?

Figure 4.27 The Mid-Atlantic ridge comes ashore in southwestern Iceland. The central rift (similar to that seen as the blue-colored contour in Figure 4.23a) is the valley in the middle distance. Notice the linear hills paralleling the rift and the steam issuing from thermal vents. If this part of the rift were submerged, these fumaroles would be hydrothermal vents similar to the one seen in Figure 4.26. Reykjavik, Iceland's largest city, is supplied with domestic hot water, hot water for space heating, and geothermally generated electricity from this valley.

 THINKING BEYOND THE FIGURE

Why can't geothermal steam be produced on, say, the east coast of the United States?

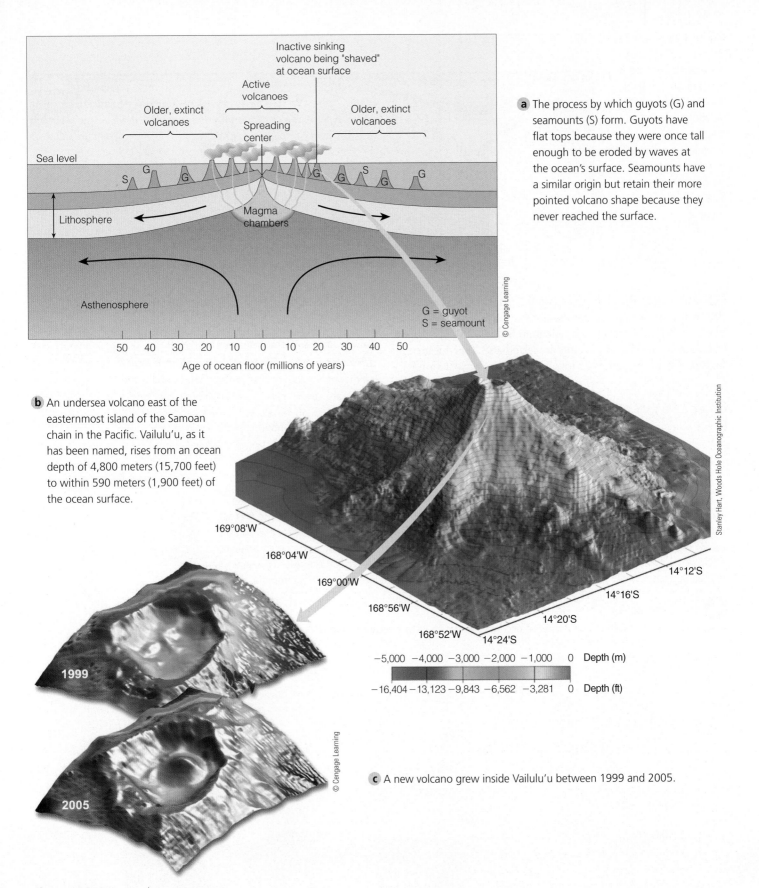

a The process by which guyots (G) and seamounts (S) form. Guyots have flat tops because they were once tall enough to be eroded by waves at the ocean's surface. Seamounts have a similar origin but retain their more pointed volcano shape because they never reached the surface.

b An undersea volcano east of the easternmost island of the Samoan chain in the Pacific. Vailulu'u, as it has been named, rises from an ocean depth of 4,800 meters (15,700 feet) to within 590 meters (1,900 feet) of the ocean surface.

c A new volcano grew inside Vailulu'u between 1999 and 2005.

Figure 4.29 Guyots and seamounts.

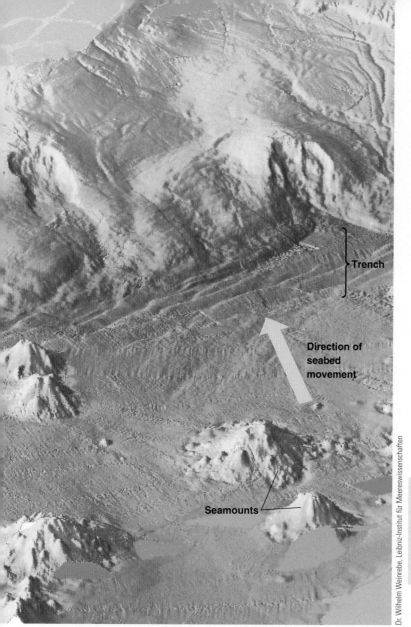

Trench

Direction of
seabed
movement

Seamounts

Dr. Wilhelm Weinrebe, Leibniz-Institut für Meereswissenschaften

Figure 4.30 Seamounts slowly approach a trench in the Pacific Basin west of Coast Rica. They will collapse into the trench and may trigger earthquakes in the process.

4.5 The Marine Environment Is Classified in Distinct Zones

Scientists have found it useful to divide the marine environment into **zones**—areas with homogeneous physical features. We can make divisions based on light, temperature, salinity, depth, latitude, water density, or almost any of the other physical dimensions we have discussed. Some classifications, such as the classifications by light and location, are particularly useful, however. These classifications are shown in **Figure 4.33**.

We've already described classification by light level, in which the primary division is between the aphotic and photic

zones. It's particularly valuable in studying marine life because light powers photosynthesis and thus primary productivity.

If we classify by location, the primary division is between water and ocean bottom. Open water is called the **pelagic zone**, and we divide it into two subsections: the **neritic zone**, near shore over the continental shelf, and the deep-water **oceanic zone**, beyond the continental shelf.

We divide the oceanic zone further by depth into zones. The *epipelagic zone* corresponds to the lighted photic zone. In the aphotic depths are layered the *mesopelagic, bathypelagic,* and *abyssopelagic* zones. Abyssopelagic water is water in the deep trenches.

Bottom divisions are labeled **benthic** and begin with the intertidal **littoral zone**, the band of coast alternately covered and uncovered by tidal action. (The *supralittoral zone,* the splash zone *above* the high intertidal, is not technically part of the ocean bottom.) Past the littoral is the **sublittoral zone**, which is further divided into inner and outer segments: The *inner sublittoral* is ocean bottom near shore, and the *outer sublittoral* is ocean floor out to the edge of the continental shelf.[5] The **bathyal zone** covers seabed on the slopes and down to great depths, where the **abyssal zone** begins. The **hadal zone** (*Hades* = underworld) is the deepest seabed of all, the trench walls and floors.

Oceanographers most often use classification by location as the basic, standard system to describe everything from the position of their physical and chemical measurements to the realms where specific organisms are found.

4.6 The Grand Tour

Researchers at the National Oceanic and Atmospheric Administration have generated a map of the world ocean floor based on satellite observations of the shape of the sea surface (**Figure 4.34**, p. 134). The graphic shows all the features discussed in this chapter. These features—and a basic understanding of the geological reasons for their existence—will help you recall the dramatic nature and history of the seafloor that we have discussed in the past two chapters.

---

[5]Many workers use the words *supertidal, intertidal,* and *subtidal* instead of the *littoral* terms.

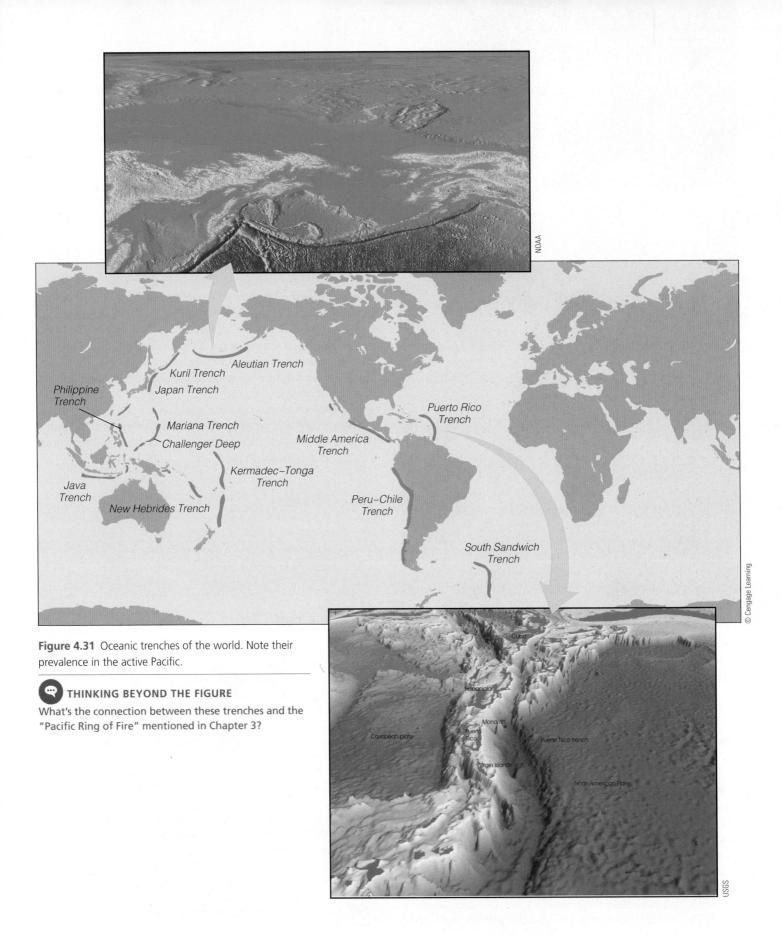

Figure 4.31 Oceanic trenches of the world. Note their prevalence in the active Pacific.

THINKING BEYOND THE FIGURE

What's the connection between these trenches and the "Pacific Ring of Fire" mentioned in Chapter 3?

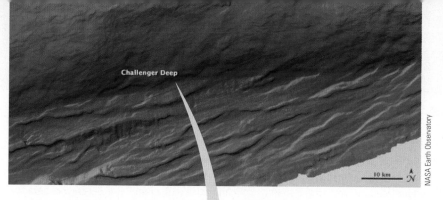

a A multibeam scan of the area of the ocean's deepest trench. Deeper areas are shown in darker shades of blue.

NASA Earth Observatory

Figure 4.32 The Mariana Trench.

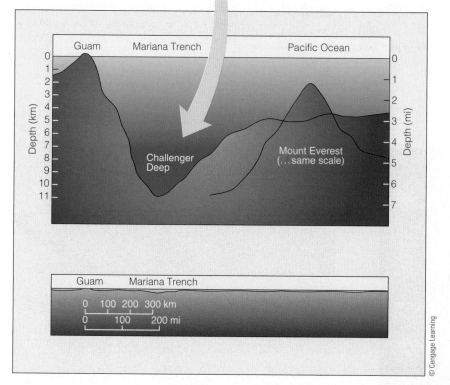

b Comparing the Challenger Deep and Mount Everest at the same scale shows that the deepest part of the Mariana Trench is about 20% deeper than the mountain is high.

c The Mariana Trench shown without vertical exaggeration.

© Cengage Learning

Figure 4.33 Classification of marine environments. This diagram is designed to show divisions; its proportions are exaggerated.

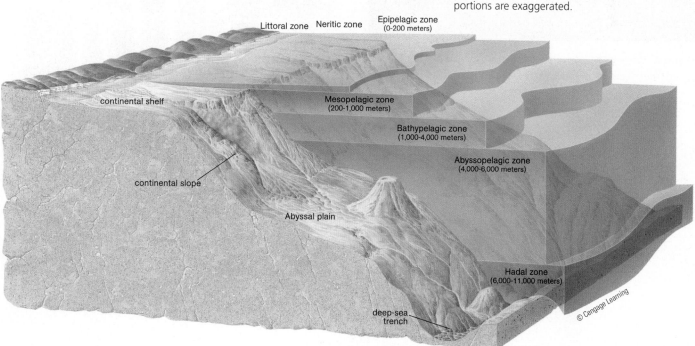

© Cengage Learning

What It's Like at the Ocean's Deepest Spot

Studying deep-ocean basins has always presented daunting obstacles. Amazing as robots and satellites and multibeam systems are, though, sometimes there is no substitute for actually *seeing*—focusing a well-trained set of eyes on the ocean floor.

The most difficult problem is to reach extreme depths, but, amazingly, scientists have visited the bottom of the deepest ocean basin. On 23 January 1960, U.S. Navy Lieutenant Don Walsh and Dr. Jacques Piccard descended to a depth of 11,022 meters (6.85 miles) into the Challenger Deep, an area of the Mariana Trench discovered in 1951 by the British oceanographic research vessel *Challenger II*.[6]

---

[6]The position of the Challenger Deep is marked in Figure 4.31.

The vehicle used in the descent was *Trieste* **(Figures A and B)**, a deep-diving submersible designed like a blimp with a very strong and thick (and cramped) steel crew sphere suspended below. A blimp uses helium gas for buoyancy, but a gas would be compressed by water pressure, so gasoline, which is relatively incompressible, was used to provide lift. With the possible exception of the moon, no place

Dr. Don Walsh, International Maritime, Inc.

Figure A *Trieste* being loaded aboard ship for the journey to the south Pacific. The personnel sphere is visible beneath the flotation envelope.

Naval Historical Center

Figure B A close-up view of the front of *Trieste's* personnel sphere, which varies from 10 to 18 centimeters (5–7 inches) in thickness. The Plexiglas window and instrument leads are visible in this 1959 photo along with part of the striped blimplike flotation structure. The sphere had to withstand the pressure of the deepest spot in the world.

visited by humans has been more hazardous to explore.

We have come a long way since the 1960s. *Alvin,* the best-known and oldest of the deep-diving manned research submarines now in operation, has made more than 4,500 dives since its commissioning in 1964 **(Figure C)**. The 6.7-meter (22-foot) truck-sized submersible carries three people in a sealed titanium sphere and is capable of diving to 4,000 meters (13,120 feet). The most famous of the research submersibles, *Alvin* has explored the Mid-Atlantic Ridge near the Azores at 2,700 meters (9,000 feet), measuring rock temperatures, collecting water samples for chemical analysis, and taking photographs. Many of the features discussed in this chapter were first observed by researchers in this trusty little submersible.

Alvin's abilities have been surpassed by a new class of manned submersibles, the most capable of which is Japan's *Shinkai 6500* and China's *Jiaolong* **(Figure D)**. *Jiaolong* is 8.2 meters (27 feet) long, weighs nearly 22 metric tons (24 tons), and has a depth capability of 7,000 meters (22,960 feet)—enough to reach 99.8% of the seabed. The crew of three can conduct research for up to 9 hours at a time.

Image courtesy of John Paul Tolson, NOAA Ocean Service

Figure C *Alvin* on the deck of the mothership *Lulu.*

Imaginechina/Corbis

Figure D China's manned submersible *Jiaolong* is hoisted aboard its mothership after a successful dive to 5,039 meters (16,530 feet) on 26 July 2011.

Figure E The tracked robot *Nereus IV* is lowered from the cable ship *Teneo* off Santiago de Compostela, Galicia, at the start of exploration works following the oil spill caused by sunken oil tanker *Prestige.*

X. REY/EPA/Newscom

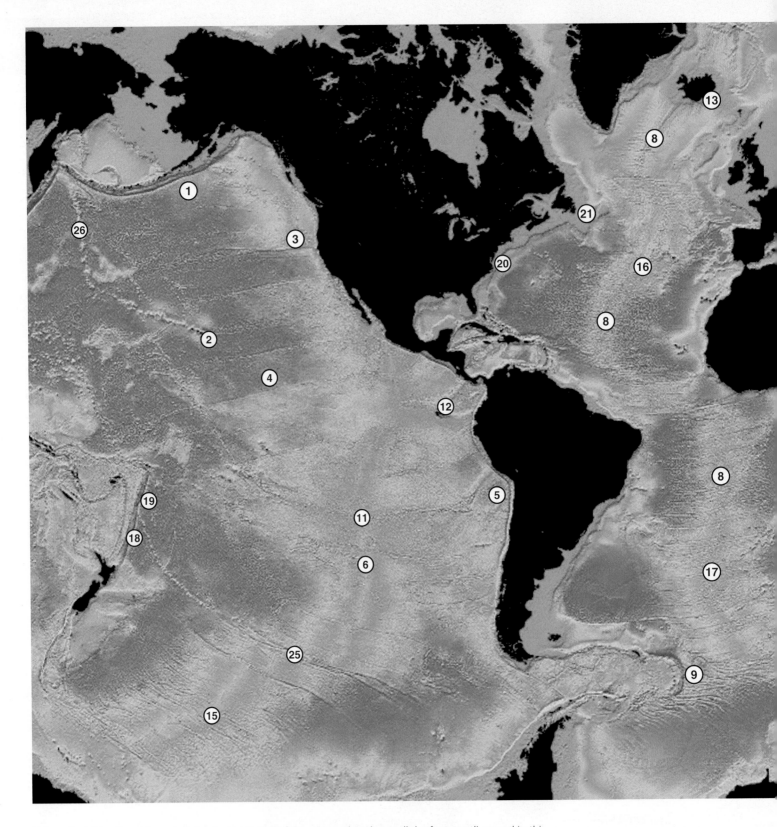

Figure 4.34 The Grand Tour. A technological *tour de force:* a map that shows all the features discussed in this chapter, derived from data provided to the National Geophysical Data Center from satellites and shipborne sensors. These features—and a basic understanding of the geological reasons for their existence—will help you recall the dramatic nature and history of the seafloor that we have discussed in the past two chapters.

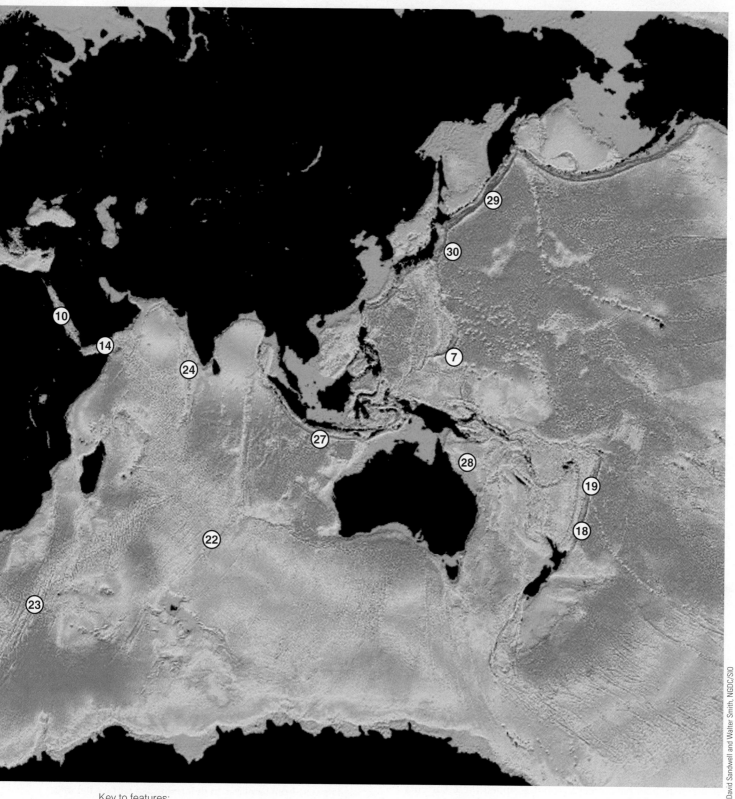

Key to features:
(1) Aleutian Trench
(2) Hawai'ian islands
(3) Juan de Fuca Ridge
(4) Clipperton Fracture Zone
(5) Peru–Chile Trench
(6) East Pacific Rise
(7) Mariana Trench, Challenger Deep
(8) Mid–Atlantic Ridge

(9) South Sandwich Trench
(10) Red Sea
(11) Easter Island
(12) Galápagos Rift
(13) Iceland
(14) Gulf of Aden
(15) Pacific–Antarctic Ridge
(16) Azores

(17) Tristan de Cunha
(18) Kermadec Trench
(19) Tonga Trench
(20) Hatteras Abyssal Plain
(21) Grand Bank
(22) Mid–Indian Ridge
(23) Atlantic–Indian Ridge
(24) Maldive Islands

(25) Eltanin Fracture Zone
(26) Emperor Seamounts
(27) Java Trench
(28) Great Barrier Reef
(29) Kuril Trench
(30) Japan Trench

David Sandwell and Walter Smith, NGDC/SIO

In this chapter you learned how difficult it has been to discover the shape of the seabed. Even today, the surface topography of Mars is better known than that of our ocean floor.

We now know that seafloor features result from a combination of tectonic activity and the processes of erosion and deposition. The ocean floor can be divided into two regions: continental margins and deep-ocean basins. The continental margin, the relatively shallow ocean floor nearest the shore, consists of the continental shelf and the continental slope. The continental margin shares the structure of the adjacent continents, but the deep-ocean floor away from land has a much different origin and history. Prominent features of the deep-ocean basins include rugged oceanic ridges, flat abyssal plains, occasional deep trenches, and curving chains of volcanic islands. The processes of plate tectonics, erosion, and sediment deposition have shaped the continental margins and ocean basins.

In the next chapter you will learn that nearly all the ocean floor is blanketed with sediment. Except for the spreading centers themselves, the broad shoulders of the oceanic ridge systems are buried according to their age—the older the seabed, the greater the sediment burden. Some oceanic crust near the trailing edges of plates may be overlain by sediments more than 1,500 meters (5,000 feet) thick. Sediments have been called the "memory of the ocean." The memory, however, is not a long one. Before continuing, can you imagine why that is so?

QUESTIONS FROM STUDENTS

1. Shouldn't the ocean be deeper in the middle? And why is Iceland one of only a few places in the world where oceanic crust is found above sea level?

Understanding *why* there are ocean basins explains their shapes. A basin usually contains an expanding ridge that is higher than the surrounding bed. Oceanic crust is thin and dense. Because of isostasy, the ocean floor lies at a lower elevation than the thicker, less dense, higher continents. Water filled the lower elevations first, submerging nearly all the basaltic basement we now call ocean floor. In areas of rapid seafloor spreading, or at hot spots, peaks are occasionally pushed toward the ocean surface by the large volume of mantle material rising in plumes from below. Large quantities of erupted magma (lava) then build the crests above sea level to form islands, as in Iceland. The Azores is another place on the Mid-Atlantic Ridge where this is happening.

2. Turbidity currents seem important in forming canyons and distributing deep sediments over abyssal plains. Has anybody ever seen a turbidity current in action?

Yes, surprisingly. In the late 1940s, the Dutch geologist Philip Kuenen produced turbidity currents in his laboratory by pouring muddy water into a trough with a sloping bottom. His observations confirmed 19th-century reports that the muddy Rhône River continued to flow in a dense stream along the bottom of Lake Geneva. In the 1960s, Robert Dill and Francis Shepard viewed sandfalls in Scripps Canyon from the diving saucer, and French researchers have recently photographed these currents in the Mediterranean.

3. How deep can a person dive (without a submarine, that is)?

The record free dive (no scuba gear) was set at 214 meters (702 feet) by Austrian diver Herbert Nitsch in 2007. At this depth, his lungs were collapsed by the pressure. Using scuba, South African diver Nuno Gomes set the record at 318 meters (1,044 feet) and nearly perished in the attempt. His descent took 14 minutes.

Given the choice, we'll take the sub!

4. The pressure at the bottom of a trench must be crushing. How do animals withstand this force?

At the bottom of the deepest trench, the water column above exerts a pressure of 1,086 bars (15,750 pounds per square inch—nearly 8 tons). This is more than a thousand times the average atmospheric pressure at sea level. At this pressure, the density of water is increased by 4.96%—95 liters of water from the Challenger Deep contains the same mass as 100 liters at the surface.

Now think of our situation as residents beneath an ocean of air. From sea level to the top of the atmosphere, a 1-centimeter square column of air weighs 1.03 kilograms (a 1-inch square weighs about 15 pounds). If the palm of your hand has an area of 12 square inches, the expected force bearing down on it would equal about 180 pounds. Does it *feel* like 180 pounds? No.

Now imagine placing your hand over an empty jar, making a good seal with your palm against the rim. If the air were removed from the jar, your hand would be "sucked" into the jar.[7] This imbalance is not present in animals in nature—there is as much air pressure *inside* your lungs pushing *out* as there is air pressure *outside* your lungs pushing *in*. If you were instantly transported into the vacuum of space, you'd blow up. If you were instantly transported to great depth, you'd be mashed.

So, marine animals living at great depth are in balance with ambient pressure, just as we are here at sea level.

5. Wouldn't wave action and tides hopelessly clutter the radar signals sent from satellites to determine sea surface height?

Satellite altimetry is one of the most sophisticated oceanographic uses of high-speed computer processing. Imagine the processing power needed to reduce the data generated by more than 1,000 radar pulses from orbit each second! Programmers

---

[7] Actually, of course, your hand would be "pushed." There is no such thing as "suck."

subtract predicted tidal height from the measurements and then use algorithms to average and cancel wave crests and troughs. The remaining sea surface height—determined to perhaps 2 centimeters (slightly less than 1 inch)—is due to gravitational variations caused by submerged features. Still more processing is needed to generate graphic images like that of Figure 4.5e.

The procedures were pioneered by the U. S. Navy and the Office of Naval Research. The initial goal was to provide detailed seafloor maps for use in anti-submarine defense.

6. **How far away is the horizon when I stand on the beach and look out to sea?**

That depends on how tall you are. If you're about six feet tall, the distance to the horizon is about 3 miles (4.8 kilometers). For a more precise estimate, subtract 4 inches (a third of a foot) to find the height of your eyes (in feet) above the ground, then divide that number by 0.5736. Take the square root of the result, and there's the number of miles to the horizon. Lifeguards standing on a 10-foot tower see a horizon roughly 8.5 kilometers (5.3 miles) distant.

TERMS AND CONCEPTS TO REMEMBER

abyssal hill	continental margin	hydrothermal vent	oceanic ridge	sublittoral zone
abyssal plain	continental rise	ice age	oceanic zone	submarine canyon
abyssal zone	continental shelf	island arc	passive margin	transform fault
active margin	continental slope	littoral zone	pelagic zone	trench
bathyal zone	fracture zone	neritic zone	seamount	turbidity current
bathymetry	guyot	ocean basin	shelf break	zone
benthic zone	hadal zone			

STUDY QUESTIONS

Thinking Critically

1. Why did people think an ocean was deepest at its center? What changed their minds?
2. What do the facts that (a) granite underlies the edges of continents and (b) basalt underlies deep-ocean basins, suggest? (Hint: Consider thicknesses and densities.)
3. The terms *leading* and *trailing* are also used to describe continental margins. How do you suppose these words relate to *active* and *passive*, or *Atlantic-type* and *Pacific-type* used in the text?
4. What forces control the shape of a continental shelf? A continental slope? A continental rise?
5. Answer this question if you have already read Chapter 3: Your time machine has been programmed to deliver you to Frankfurt, Germany, on a chilly evening in January 1912, to hear Wegener's lectures on continental drift. What two illustrations from this chapter would you take with you to cheer him up after the lecture? Why did you select those particular illustrations? Did you enjoy the beer and wurst?

Thinking Analytically

1. The speed of sound through seawater is about 1,500 meters per second. If a ship equipped with a multibeam mapping system is surveying a feature 3,500 meters below the surface, and if the researchers wish to obtain an image of the feature at a resolution of 10 meters, what is the maximum speed the ship can steam?
2. Review the speed of a turbidity current. In the unlikely event that a fast-running current formed near the shoreline of a trailing edge coast, how long would it take for the current to traverse a typical continental shelf and arrive at the shelf break? Would you expect the current to move at a constant speed during this traverse?
3. How much wider has Iceland become because of seafloor spreading since the last sequence of major eruptions in the 1500s? [Hint: Use the data in Figure 4.21a.]
4. What would be the *approximate* age of a seamount produced at the East-Pacific Rise at the position marked ⑥ in Figure 4.34 when it collapses into the Peru–Chile Trench (marked ⑤)?

GLOBAL GEOSCIENCE WATCH

Visit www.cengagebrain.com to access MindTap, a complete digital course which includes access to Global Geoscience Watch and more. Within the GREENR database, search for "hydrothermal vents" and focus on recent articles under the "News" heading. Select at least three news articles that describe new findings and make sure to record your references. Be prepared to report back to the class regarding what you have learned and participate in a class current events discussion focused on hydrothermal vents.

CENGAGE brain.com   Visit www.cengagebrain.com to access course materials for this text, including interactive learning tools, videos, and more.

5 Sediments

KEY CONCEPTS

Sediments are loose accumulations of particulate material. Charles D. Hollister/Woods Hole Oceanographic Institution

The depth and composition of marine sediments tell us of relatively recent events in the ocean basin above. NASA/BSFC, ORBIMAGE, SeaWiFS

The most abundant sediments are terrigenous (from land) and biogenous (from once-living things). Susumu Nishinaga/Science Source

Marine sediments have been uplifted and exposed on land. Arizona's Grand Canyon is made of marine sediment, as is the top of the world's highest mountain. Craig Kassover/National Geographic

Because marine sediments are usually subducted along with the seabed on which they lie, the oldest sediments are relatively young—rarely older than 180 million years. © Cengage Learning

---

Arizona's Grand Canyon is a slice through an ancient ocean floor. The rocks exposed at the canyon rim formed from sediments deposited in a shallow ocean basin about 270 million years ago. Younger layers originally stacked above the present rim have already been removed and deposited by the Colorado River into the Gulf of California.
© Scott Stulberg/Corbis

Media Connection

Start off this chapter by listening to a podcast with author McKenzie Funk, as he discusses the war over the energy resources beneath the Arctic seabed. Visit www.cengagebrain.com to access MindTap, a complete digital course that includes this podcast and other resources.

5.1 Ocean Sediments Vary Greatly in Appearance

Sediment is particles of organic or inorganic matter that accumulate in a loose, unconsolidated form. The particles originate from the weathering and erosion of rocks, from the activity of living organisms, from volcanic eruptions, from chemical processes within the water itself, and even from space. Most of the ocean floor is being slowly dusted by a continuing rain of sediments. Accumulation rates on the deep seafloor vary from a few centimeters per year to the thickness of a dime every thousand years.

Marine sediments occur in a broad range of sizes and types. Beach sand is

> **B**ecause they can provide a record of conditions in the past, marine sediments have been called the "memory of the ocean."

sediment; so are the muds of a quiet bay and the mix of silt and tiny shells found on the continental margins. Less familiar sediments are the fine clays of the deep-ocean floor, the biologically derived oozes of abyssal plains, and the nodules and coatings that form around hard objects on the seafloor. The origin of these materials—and the distribution and sizes of the particles—depends on a combination of physical and biological processes.

What do sediments look like? That depends on where you look. **Figure 5.1** shows a sea anemone on the Mid-Atlantic Ridge. The young rocky outcrop on which it rests is only lightly powdered with sediment. Contrast that rough ridge with the smooth seafloor shown in **Figure 5.2**. The sediment there is about 35 meters (116 feet) thick and marked by the tracks of brittle stars. These widely distributed organisms feed on surface bacteria and fallen particles of organic sediment. Note that the surface of the sediment is not always smooth. Where bottom currents are swift and persistent, they can cause ripples like those on a streambed **(Figure 5.3)**.

The extraordinary thickness of some layers of marine sediment can be seen in **Figure 5.4**, a seismic profile of the eastern edge of a seamount in the North Atlantic's Sohm Abyssal Plain south of Nova Scotia. The sediment at the eastern boundary of this profile covers the oceanic crust to a depth of more than 1.8 kilometers (1.1 miles).

The colors of marine sediments are often striking. Sediments of biological origin are white or cream colored, with deposits high in silica tending toward gray. Some deep-sea clays—though tradition-

Figure 5.2 Brittle stars and their tracks on the continental slope off the coast of New England. The depth here is 1,476 meters (4,842 feet).

 THINKING BEYOND THE FIGURE

Why is the sediment in Figure 5.1 very thin and the sediment layer in Figure 5.2 very thick?

Figure 5.1 Sediment near the crest of the Mid-Atlantic Ridge. A sea anemone clings to newly formed rock outcrops only lightly dusted with sediment.

Figure 5.3 Ripples on the sediment beneath the swift Antarctic Circumpolar Current in the northern Drake Passage. The depth here is 4,010 meters (13,153 feet).

Charles D. Hollister/Woods Hole Oceanographic Institution

ally termed "red clays" from the rusting (oxidation) of iron within the sediments to form iron oxide—can range from tan to chocolate brown. Other clays are shades of green or tan. Nodular sediments are a dark sooty brown or black. Some nearshore sediments contain decomposing organic material and smell of hydrogen sulfide, but most are odorless.

Very few areas of the seabed are altogether free of overlying sediments. The water over these areas is not completely sediment free, but for some reason sediment does not collect on the bottom. Strong currents may scour the sediments away, or the seafloor may be too young in these areas for sediments to have had time to accumulate, or hot water percolating upward through a porous seafloor may dissolve the material as fast as it settles.

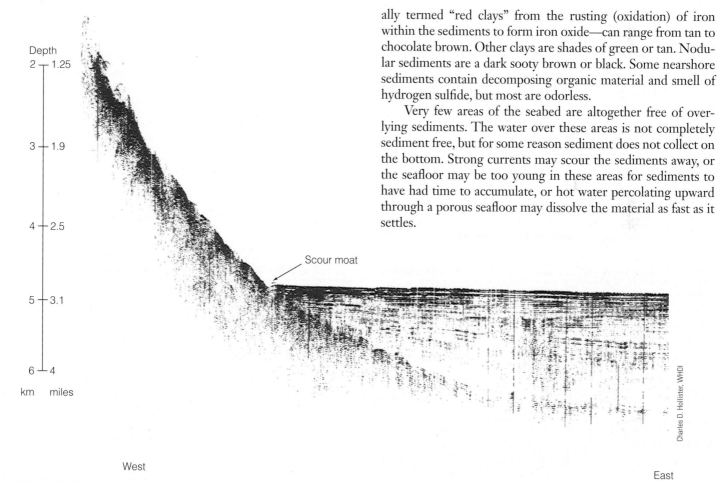

West

East

Charles D. Hollister, WHOI

Figure 5.4 The deep sediments of the Sohm Abyssal Plain in the North Atlantic south of Nova Scotia have buried the base of this seamount. This seismic profile shows the depth of the sediments above the geological base of the seamount to be more than 1.8 kilometers (1.1 miles). Note the scour moat—a depression along the boundary of seamount and sediment—caused by a persistent deep boundary current in the area. The vertical exaggeration in this figure is about 12:1.

CONCEPT CHECK

Before going on to the next section, check your understanding of some of the important ideas presented so far:

What is sediment?

Why are very few areas of the seabed completely free of sediments?

The ocean is more than 4 billion years old, yet marine sediments are rarely older than about 180 million years. How can that be?

5.2 Sediments Are Classified by Particle Size

Particle size is frequently used to classify sediments. The scheme shown in **Table 5.1** was formalized in 1922 and has been used by geologists, soil scientists, and oceanographers ever since. In this classification the coarsest particles are boulders, which are more than 256 millimeters (about 10 inches) in diameter. Although boulders, cobbles, and pebbles occur in the ocean, most marine sediments are made of finer particles: **sand**, **silt**, and **clay**. The particles are defined by their size.

Generally, the smaller the particle, the more easily it can be transported by streams, waves, and currents. As sediment is transported it tends to be sorted by size; coarser grains, which are moved only by turbulent flow, tend not to travel as far as finer grains, which are more readily moved. The clays, particles less than 0.004 millimeter in diameter, can remain suspended for very long periods and may be transported great distances by ocean currents before they are deposited. As indicated in **Figure 5.5**, cohesiveness of smaller particles can be as important as grain size in determining whether they will be eroded and transported. Once in suspension, the finest clays may circulate in the ocean for decades.

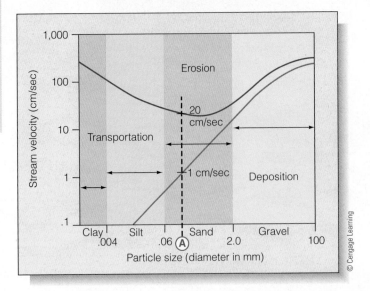

Figure 5.5 The velocities of currents required for erosion, transportation, and deposition (sedimentation) of sediment particles of different sizes. To dislodge and carry a particle of size Ⓐ, the speed of a current must exceed 20 centimeters per second (8 inches per second). When the current falls below 1 centimeter per second (½ inch per second), the particle will be deposited.

A layer of sediment can contain particles of similar size, or it can be a mixture of different-sized particles. Sediments composed of particles of mostly one size are said to be **well-sorted sediments**. Sediments with a mixture of sizes are **poorly sorted sediments**. Sorting is a function of the energy of the environment—the exposure of that area to the action of waves, tides, and currents. Well-sorted sediments occur in an environment where energy fluctuates within narrow limits. Sediments of the

Table 5.1 Particle Sizes and Settling Rate in Sediment

	Size Class	Size	Size Comparison	Settling Velocity in Still Water	Time to Settle 4 Kilometers (2.5 Miles)
Gravel	Boulder	>256 mm	Basketball		
	Cobble	64–256 mm	Potato to grapefruit		
	Pebble	4–64 mm	Throwing and skipping size		
	Granule	2–4 mm	Pea		
Sand	Very coarse sand	1–2 mm	Peppercorns		
	Coarse sand	0.5–1 mm	Coarse sugar	2.5 cm/sec (1 inch/sec)	1.8 days
	Medium sand	0.25–0.5 mm	Granulated sugar		
	Fine sand	0.125–0.25 mm	Confectioners' sugar		
	Very fine sand	0.0625–0.125 mm	Visible to the eye		
Mud	Coarse silt	0.0310–0.0625 mm	Barely visible to the eye	0.025 cm/sec (1/100 inch/sec)	6 months
	Medium to very fine silt	0.0039–0.0310 mm	Microscopic		
	Clay	<0.0039 mm	Microscopic	0.00025 cm/sec	50 years

Based on the Udden–Wentworth Sediment Grain Size Scale.

calm deep-ocean floor are typically well sorted (see again Figure 5.2). Poorly sorted sediments form in environments where energy fluctuates over a wide spectrum. The mix of rubble at the base of a rapidly eroding shore cliff is a good example of poorly sorted sediment.

CONCEPT CHECK

Before going on to the next section, check your understanding of some of the important ideas presented so far:

What types of particles compose most marine sediments?

Which particles are most easily transported by water?

How do well-sorted sediments differ from poorly sorted sediments?

5.3 Sediments May Be Classified by Source

Another way to classify marine sediments is by their origin. Such a scheme was first proposed in 1891 by Sir John Murray and A. F. Renard after a thorough study of sediments collected during the *Challenger* expedition. A modern modification of their organization is shown in **Table 5.2**. This scheme separates sediments into four categories by source: terrigenous, biogenous, hydrogenous (also called authigenic), and cosmogenous.

Terrigenous Sediments Come from Land

Terrigenous sediments (*terra*, "Earth"; *generare*, "to produce") are the most abundant. As the name implies, they originate on the continents or islands from erosion, volcanic eruptions, and blown dust.

The rocks of Earth's crust are made up of **minerals**, inorganic crystalline materials with specific chemical compositions. The texture of igneous rocks—rocks that crystallize from molten material—is determined by how rapidly they cool. Igneous rocks that cool rapidly, such as the basalt that forms the ocean floor at spreading centers or pours from volcanic vents on land, solidify so quickly that obvious crystals do not have a chance to form. Slower cooling produces the most commonly encountered crystals, those about the size of a grain of rice or the head of a pin. Nearly all terrigenous sediments are derived directly

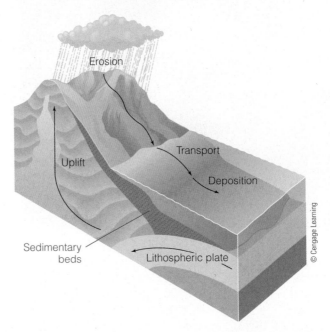

Figure 5.6 A simplified sediment cycle. Over geologic time, mountains rise as lithospheric (crustal) plates collide, fuse, and subduct. Water and wind erode the mountains and transport resulting sediment to the sea. The sediments are deposited on the seafloor, where they travel with the plate and are either uplifted or subducted. The material is made into mountains again.

💬 THINKING BEYOND THE FIGURE

Think again about the Grand Canyon in Arizona, an old seabed being eroded by the Colorado River. Old mountains eroded, and their sedimentary remains formed that seabed. Do you see why the term "deep time" is so apt?

or indirectly from these crystals. You have probably seen the crystals in granite, the most familiar continental igneous rock. Granite is the source of quartz and clay, the two most common components of terrigenous marine sediments.

Terrigenous sediments are part of a slow and massive cycle **(Figure 5.6)**. Over the great span of geologic time, mountains rise as plates collide, fuse, and subduct. The mountains erode. The resulting sediments are transported to the sea by wind and water,

Table 5.2 Classification of Marine Sediments by Source of Particles

Sediment Type	Source	Examples	Distribution	Percent of All Ocean Floor Area Covered
Terrigenous	Erosion of land, volcanic eruptions, blown dust	Quartz sand, clays, estuarine mud	Dominant on continental margins abyssal plains, polar ocean floors	~45%
Biogenous	Organic; accumulation of hard parts of some marine organisms	Calcareous and siliceous oozes	Dominant on deep-ocean floor (siliceous ooze below about 5 km)	~55%
Hydrogenous (authigenic)	Precipitation of dissolved minerals from water, often by bacteria	Manganese nodules, phosphorite deposits	Present with other, more dominant sediments	1%
Cosmogenous	Dust from space, meteorite debris	Tektite spheres, glassy nodules	Mixed in very small proportion with more dominant sediments	1%

Sources: Kennett, *Marine Geology*, 1982; Weihaupt, *Exploration of the Oceans*, 1979; Sverdrup, Johnson, and Fleming, *The Oceans: Their Physics, Chemistry, and General Biology*, 1942.

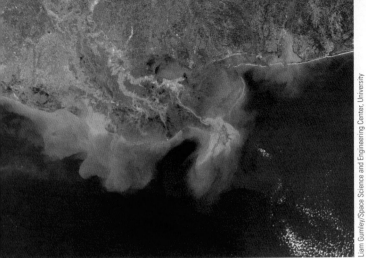

Liam Gumley/Space Science and Engineering Center, University of Wisconsin-Madison/MODIS Science Team/NASA

a Rivers are the main source of terrigenous sediments. This photo, taken from space, shows sediment entering the Gulf of Mexico from the Mississippi River.

Figure 5.7 Sources of terrigenous sediments.

💬 **THINKING BEYOND THE FIGURE**

Where do you think most terrigenous sediment ends up—close to the shore or far away from the shore?

where they collect on the seafloor. The sediments travel with the plate and are either uplifted or subducted. The material is made into mountains. The cycle begins anew.

Although estimates vary, it appears that about 15 billion metric tons (16.5 billion tons) of terrigenous sediments are transported in rivers to the sea each year, with an additional 100 million metric tons transported annually from land to ocean as fine airborne dust and volcanic ash **(Figure 5.7)**.[1]

Biogenous Sediments Form from the Remains of Marine Organisms

Biogenous sediments (*bio*, "life"; *generare*, "to produce") are the next most abundant marine sediment. The siliceous (silicon-containing) and calcareous (calcium-carbonate–containing) compounds that make up these sediments of biological origin were originally brought to the ocean in solution by rivers or dissolved in the ocean at oceanic ridges. The siliceous and calcareous materials were then extracted from the seawater by the normal activity of tiny plants and animals to build protective shells and skeletons. Some of this sediment derives from larger mollusk shells or from stationary colonial animals such as corals, but most of the organisms that produce biogenous sediments drift free in the water as plankton (about which you'll learn in Chapters 14 and 15). After the death of their owners, the hard structures fall to the bottom and accumulate in layers. Biogenous sediments are most abundant where ample nutrients encourage high biological productivity, usually near continental margins and areas of upwelling. Over millions of years, organic molecules within these sediments can form oil and natural gas (see Chapter 17 for details).

Note in Table 5.2 that biogenous sediments cover a larger percentage of the *area* of the ocean floor than terrigenous sediments do, but the terrigenous sediments dominate in total *volume*.

[1]An important effect of this injection of ash into the atmosphere is discussed in Questions from Students #4 on page 161.

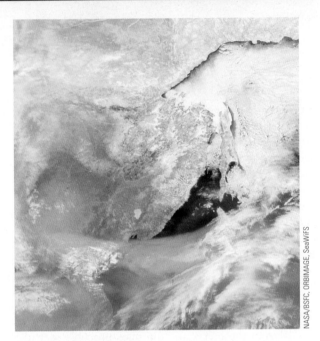

NASA/BSFC, ORBIMAGE, SeaWiFS

b Dust from the Gobi Desert blows eastward across the Pacific on 18 March 2002. The particles will fall to the ocean surface and descend slowly to the bottom to end up as terrigenous sediments.

NASA/MODIS RapidResponse Team

c The ash cloud caused by the April 2010 eruption of Eyjafjallajökull, a volcano in Iceland. The wind transported ash from the eruption for hundreds of kilometers and greatly disrupted aviation in Europe. Much of the ash ended up in the North Atlantic.

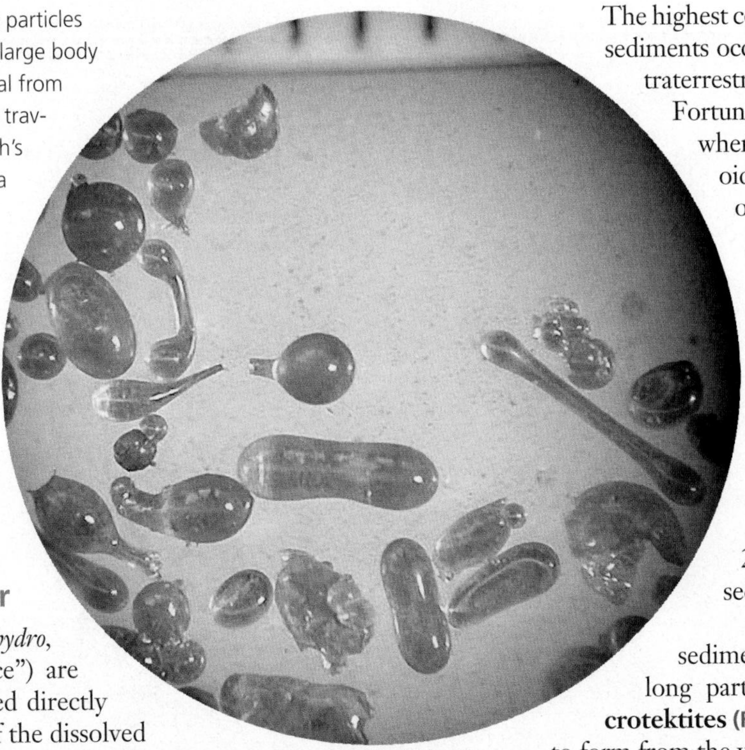

Figure 5.8 Microtektites, very rare particles that began a long journey when a large body impacted Earth and ejected material from Earth's crust. Some of this material traveled through space, reentered Earth's atmosphere, melted, and took on a rounded or teardrop shape. These specimens of sculptured glass range from 0.2 to 0.8 millimeter in length. Glassy dust much finer in size, as well as nut-sized chunks, have also fallen on Earth.

Image courtesy of Michael Daniels

Hydrogenous Sediments Form Directly from Seawater

Hydrogenous sediments (*hydro*, "water"; *generare*, "to produce") are minerals that have precipitated directly from seawater. The sources of the dissolved minerals include submerged rock and sediment, fresh crust leaching at oceanic ridges, material issuing from hydrothermal vents, and substances flowing to the ocean in river runoff. As we shall see, the most obvious hydrogenous sediments are manganese nodules, which litter some deep seabeds, and phosphorite nodules, seen along some continental margins. Hydrogenous sediments are also called **authigenic sediments** (*authis*, "in place, on the spot") because they were formed in the place they now occupy.

Though they usually accumulate very slowly, rapid deposition of hydrogenous sediments is possible—in a rapidly drying lake, for example.

Cosmogenous Sediments Come from Space

Cosmogenous sediments (*cosmos*, "universe"; *generare*, "to produce"), which are of extraterrestrial origin, are the least abundant. These sediments are typically greatly diluted by other sediment components and rarely constitute more than a few parts per million of the total sediment in any layer. Scientists believe that cosmogenous sediments come from two major sources: interplanetary dust that falls constantly into the top of the atmosphere and rare impacts by large asteroids and comets.

Interplanetary dust consists of silt- and sand-sized micrometeoroids that come from asteroids and comets or from collisions between asteroids. The silt-sized particles settle gently to Earth's surface, but larger, faster-moving dust is heated by friction with the atmosphere and melts, sometimes glowing as the meteors we see in a dark night sky. Though much of this material is vaporized, some may persist in the form of iron-rich cosmic spherules. Most of these dissolve in seawater before reaching the ocean floor. About 15,000 to 30,000 metric tons (16,500–33,000 tons) of interplanetary dust enters Earth's atmosphere every year.

The highest concentrations of cosmogenous sediments occur when large volumes of extraterrestrial matter arrive all at once. Fortunately, this happens only rarely, when Earth is hit by a large asteroid or comet. Very few examples of this are known, but most geologists believe that an impact like the one described in "How Do We Know? 5.1" on page 158 would have blown vast amounts of debris into space around Earth. Much of it would fall back and be deposited in layers. Cosmogenous components may make up between 10% and 20% of these extraordinary sediments!

Occasionally cosmogenous sediment includes translucent oblong particles of glass known as **microtektites (Figure 5.8)**. Tektites are thought to form from the violent impact of large meteors or small asteroids on the crust of Earth. The impact melts some of the crustal material and splashes it into space; the material melts again as it rushes through the atmosphere, producing the various shapes shown in the photo. Tektites do not dissolve easily and usually reach the ocean floor. Most are smaller than 1.5 millimeters ($\frac{1}{16}$ inch) long.

Marine Sediments Are Usually Combinations of Terrigenous and Biogenous Deposits

Sediments on the ocean floor only rarely come from a single source; most sediment deposits are a mixture of biogenous and terrigenous particles, with an occasional hydrogenous or cosmogenous supplement. The patterns and composition of sediment layers on the seabed are of great interest to researchers studying conditions in the overlying ocean. Different marine environments have characteristic sediments, and these sediments preserve a record of past and present conditions within those environments.

The sediments on the continental margins are generally different in quantity, character, and composition from those on the deeper basin floors. Continental shelf sediments—called **neritic sediments** (*neritos*; of the coast)—consist primarily of terrigenous material. Deep-ocean floors are covered by finer sediments than those of the continental margins, and a greater proportion of deep-sea sediment is of biogenous origin. Sediments of the slope, rise, and deep-ocean floor that originate in the ocean are called **pelagic sediments** (*pelagios*; of the sea).

The average thickness of the marine sediments in each oceanic region is shown in **Figure 5.9** and **Table 5.3**. Note that 72% of the total volume of all marine sediment is associated with continental slopes and rises, which constitute only about 12%

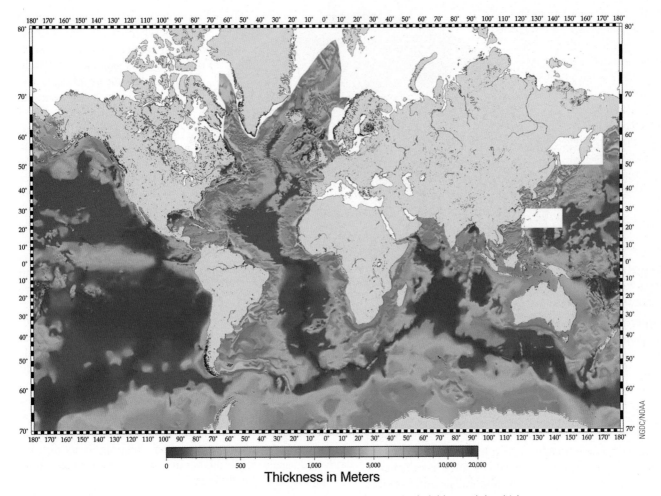

Thickness in Meters

0   500   1,000   5,000   10,000   20,000

NGDC/NOAA

Figure 5.9 Total sediment thickness of the ocean floor, with the thinnest deposits in dark blue and the thickest in red. Note the abundant deposits along the east and Gulf coast of North America, in the South China Sea, and in the Bay of Bengal east of India.

of the ocean's area. **Figure 5.10** shows the worldwide distribution of marine sediment types. Put a bookmark in this page—you'll want to refer to these images as our discussion continues.

CONCEPT CHECK

Before going on to the next section, check your understanding of some of the important ideas presented so far:

What are the four main types of marine sediments?

Which type of sediment is most abundant?

Which type of sediment covers the greatest seabed area?

Which type of sediment is rarest? Where does this sediment originate?

Do most sediments consist of a single type? (That is, are terrigenous deposits made exclusively of terrigenous sediments?)

How do neritic sediments differ from pelagic ones?

Table 5.3 The Distribution and Average Thickness of Marine Sediments

Region	Percent of Ocean Area	Percent of Total Volume of Marine Sediments	Average Thickness
Continental shelves	9%	15%	2.5 km (1.6 mi)
Continental slopes	6%	41%	9 km (5.6 mi)
Continental rises	6%	31%	8 km (5 mi)
Deep-ocean floor	78%	13%	0.6 km (0.4 mi)

Sources: Emery in Kennett, *Marine Geology*, 1982 (Table 11-1); Weihaupt, *Exploration of the Oceans*, 1979; Sverdrup, Johnson, and Fleming, *The Oceans: Their Physics, Chemistry, and General Biology*, 1942.

5.4 Neritic Sediments Overlie Continental Margins

Most neritic sediments are *terrigenous;* they are eroded from the land and carried to streams, where they are transported to the ocean. Currents distribute sand and larger particles along the coast; wave action carries the silts and clays to deeper water. When the water is too deep to be disturbed by wave action, the finest sediment may come to rest or continue to be transported by the turbulence of deep currents toward the deeper ocean floor. Ideally, these processes produce an orderly sorting of particles by size from relatively large grains near the coast to relatively small grains near the shelf break.

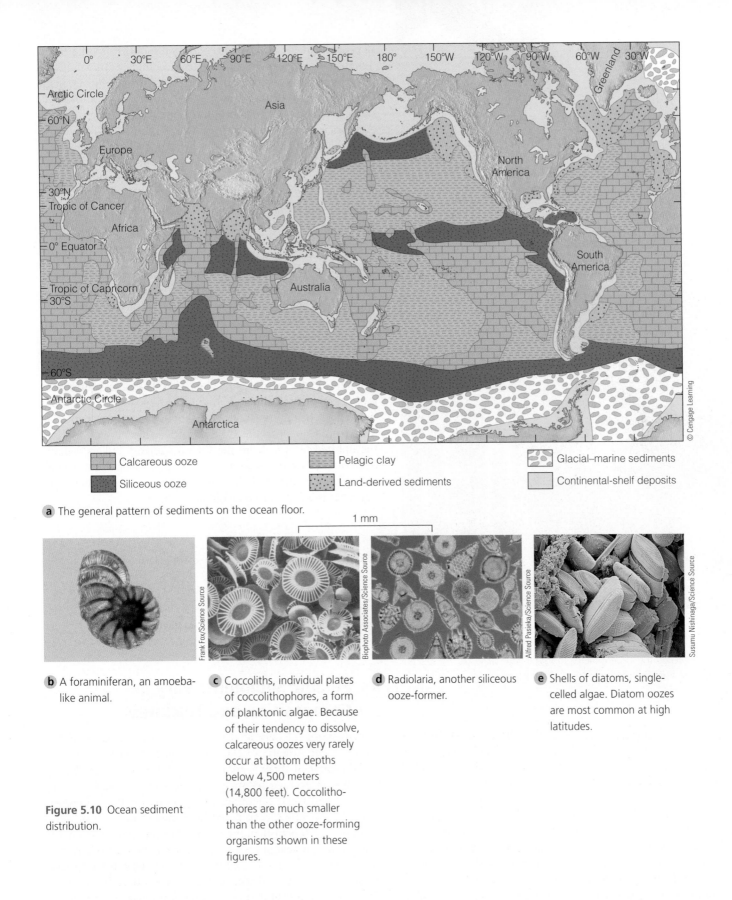

Calcareous ooze	Pelagic clay	Glacial–marine sediments
Siliceous ooze	Land-derived sediments	Continental-shelf deposits

a The general pattern of sediments on the ocean floor.

1 mm

b A foraminiferan, an amoeba-like animal.

c Coccoliths, individual plates of coccolithophores, a form of planktonic algae. Because of their tendency to dissolve, calcareous oozes very rarely occur at bottom depths below 4,500 meters (14,800 feet). Coccolithophores are much smaller than the other ooze-forming organisms shown in these figures.

d Radiolaria, another siliceous ooze-former.

e Shells of diatoms, single-celled algae. Diatom oozes are most common at high latitudes.

Figure 5.10 Ocean sediment distribution.

There are exceptions, however. Shelf deposits are subject to further modification and erosion as sea level fluctuates: Larger particles may be moved toward the shelf edge when sea level is low, as it was during periods of widespread glaciation—ice ages.

Poorly sorted sediments are also found as glacial deposits. In polar regions, glaciers and ice shelves give rise to icebergs. These carry particles of all sizes, and when they melt, they distribute their mixtures of rocks, gravel, sand, and silt onto high-latitude deposition in the deep ocean. Near the mouths of large rivers, 1 meter (about 3 feet) of sediment may accumulate every thousand years. Along the east coast of the United States, however, many large rivers terminate in estuaries, which trap most of the sediment brought to them. The continental shelf of eastern North America is therefore covered mainly by sediments laid down during the last period of glaciation, when sea level was lower.

In addition to terrigenous material, neritic sediments almost always contain biological material. Biological productivity in coastal waters is often quite high, and biogenous sediments—the skeletal remains of creatures living on the bottom or in the water above—mix with the terrigenous sediments and dilute them.

Sediments can build to impressive thickness on continental shelves. In some cases, neritic sediments undergo **lithification**: they are converted into sedimentary rock by pressure-induced compaction or by cementation. If these lithified sediments are thrust above sea level by tectonic forces, they can form mountains or plateaus as noted in this chapter's opener. The top of Mount Everest, the world's highest peak, is a shallow-water biogenic marine limestone (a calcareous rock). Much of the Colorado Plateau, with its many stacked layers, was formed by sedimentary deposition and lithification beneath a shallow continental sea beginning about 570 million years ago. As noted in this chapter's opener, the Colorado River has cut and exposed the uplifted beds to form the Grand Canyon. Hikers walking from the canyon rim down to the river pass through spectacular examples of continental-shelf sedimentary deposits. Their journey takes them deep into an old ocean floor!

Figure 5.11 Researchers suspended over an iceberg take samples of the dust and gravel scraped off a nearby continent by the iceberg's parent glacier. When the iceberg melts, this sediment will fall to the seabed at a distance from the continent from which it came.

Mark Drinkwater/European Space Agency, ESTEC

CONCEPT CHECK

Before going on to the next section, check your understanding of some of the important ideas presented so far:

Are neritic sediments generally terrigenous or biogenous?

What is lithification? How is sedimentary rock formed?

Can you think of an example of lithified sediment on land?

continental margins and deep-ocean floor **(Figure 5.11)**. Turbidity currents also disrupt the orderly sorting of sediments on the continental margin by transporting coarse-grained particles away from coastal areas and onto the deep-ocean floor.

Ice ages have other effects on sediment deposition. Note in Figures 4.12 and 4.16 that continental shelves were almost completely exposed by the lowered sea level during times of widespread glaciation. Rivers carry their sediment right to the shelf edge, and it goes straight to the continental slope and deep seabed, mostly in turbidity currents.

Between ice ages, when the shelves are covered with water, the rate of sediment deposition on continental shelves is variable, but it is almost always greater than the rate of sediment

5.5 Pelagic Sediments Vary in Composition and Thickness

The thickness of pelagic sediments is highly variable. When averaged, the Atlantic Ocean bottom is covered by sediments to a thickness of about 1 kilometer (3,300 feet), and the Pacific floor has an average sediment thickness of less than 0.5 kilometer (1,650 feet). There are two reasons for this difference. First, the Atlantic Ocean is fed by a greater number of rivers laden with sediment than the Pacific, but the Atlantic is smaller in area; thus, it gets more sediment for its size than the Pacific. Second, in the Pacific Ocean many oceanic trenches trap sediments moving toward basin centers. Beyond this, the composition and thickness of pelagic sediments also vary with location, being thickest on the abyssal plains and thinnest (or absent) on the oceanic ridges.

Turbidites Are Deposited on the Seabed by Turbidity Currents

Dilute mixtures of sediment and water periodically rush down the continental slope in turbidity currents (**Figure 5.12**). A turbidity current is not propelled by the water within it but by gravity (the water suspends the particles, and the mixture is denser than the surrounding seawater). As we have seen, the erosive force of turbidity currents is thought to help cut submarine canyons (see again Figure 4.20). These underwater avalanches of thick, muddy fluid can reach the continental rise and often continue moving onto an adjacent abyssal plain before eventually coming to rest. The resulting deposits are called **turbidites**, graded layers of terrigenous sand interbedded with the finer pelagic sediments typical of the deep-sea floor. Each distinct layer consists of coarse sediment at the bottom with finer sediment above, and each graded layer is the result of sediment deposited by one turbidity-current event.

Clays Are the Finest and Most Easily Transported Terrigenous Sediments

About 38% of the deep seabed is covered by clays and other fine terrigenous particles. As we have seen, the finest terrigenous sediments are easily transported by wind and water currents. Microscopic waterborne particles and tiny bits of windborne dust and volcanic ash settle slowly to the deep-ocean floor, forming fine brown, olive-colored, or reddish clays. As Table 5.1 shows, the velocity of particle settling is related to particle size, and clay particles usually fall very slowly indeed. Terrigenous sediment accumulation on the deep-ocean floor is typically about 2 millimeters (⅛ inch) every thousand years.

Oozes Form from the Rigid Remains of Living Creatures

Seafloor samples taken farther from land usually contain a greater proportion of biogenous sediments than those obtained near the continental margins. The reason is not that biological productivity is higher farther from land (the opposite is usually true) but that there is less terrigenous material far from shore, and thus pelagic deposits contain a greater proportion of biogenous material.

Deep-ocean sediment containing at least 30% biogenous material is called an **ooze** (surely one of the most descriptive terms in the marine sciences). Oozes are named after the dominant remnant organism constituting them. The organisms that contribute their remains to deep-sea oozes are small, single-celled, drifting, plantlike organisms and the single-celled animals that feed on them. The hard shells and skeletal remains of these creatures are composed of relatively dense glasslike silica or calcium carbonate. When these organisms die, their shells settle slowly toward the bottom, mingle with fine-grained ter-

rigenous silts and clays, and accumulate as ooze. The silica-rich residues give rise to **siliceous ooze**; the calcium-containing material to **calcareous ooze**.

Oozes accumulate slowly, at a rate of about 1 to 6 centimeters (½–2½ inches) per thousand years. But they collect more than 10 times as quickly as deep-ocean terrigenous clays. The accumulation of any ooze therefore depends on a delicate balance between the abundance of organisms at the surface, the rate at which they dissolve once they reach the bottom, and the rate of accumulation of terrigenous sediment.

Calcareous ooze forms mainly from shells of the amoeba-like **foraminifera** (Figure 5.10b), small drifting mollusks called pteropods, and tiny algae known as **coccolithophores** (Figure 5.10c). When conditions are ideal, these organisms generate prodigious volumes of sediment. The remains of countless coccolithophores have been compressed and lithified to form the impressive White Cliffs of Dover in southeastern England (**Figure 5.13**). Though formed at moderate ocean depth about 100 million years ago, tectonic forces have uplifted Dover's chalk cliffs to their present prominent position.

Although foraminifera and coccolithophores live in nearly all surface ocean water, calcareous ooze does not accumulate everywhere on the ocean floor. Shells are dissolved by seawater at great depths because it contains more carbon dioxide than seawater near the surface and thus becomes slightly acid. This acidity, combined with the increased solubility of calcium carbonate in cold water under pressure, dissolves the shells more rapidly, as you will see in Figure 7.13. At a certain depth, called the **calcium carbonate compensation depth (CCD)**, the rate at which calcareous sediments are supplied to the seabed equals the rate at which those sediments dissolve. Below this depth, the tiny skeletons of calcium carbonate dissolve on the seafloor, so no calcareous oozes accumulate. Calcareous sediment dominates the deep-sea floor at depths of less than about 4,500 meters (14,800 feet), the usual calcium carbonate compensation depth. Sometimes a line analogous to a snow line on a terrestrial mountain can be seen on undersea peaks: above the line, the white sprinkling of calcareous ooze is visible; below it, the "snow" is absent (**Figure 5.14**). About 48% of the surface of deep-ocean basins is covered by calcareous oozes.

Siliceous (silicon-containing) ooze predominates at greater depths and in colder polar regions. Siliceous ooze is formed from the hard parts of another amoeba-like animal, the beautiful glassy **radiolarian** (Figure 5.10d), and from single-celled algae called **diatoms** (Figure 5.10e). After a radiolarian or diatom dies, its shell will also dissolve back into the seawater, but this dissolution occurs much more slowly than the dissolution of calcium carbonate. Slow dissolution at all depths, combined with very high diatom productivity in some surface waters, leads to the buildup of siliceous ooze. Diatom ooze is most common in the deep-ocean basins surrounding Antarctica because strong ocean currents and seasonal upwelling in this area support large

Hikers descending to the Colorado River in Arizona's Grand Canyon walk through deeper and deeper sedimentary layers of an ancient ocean floor.

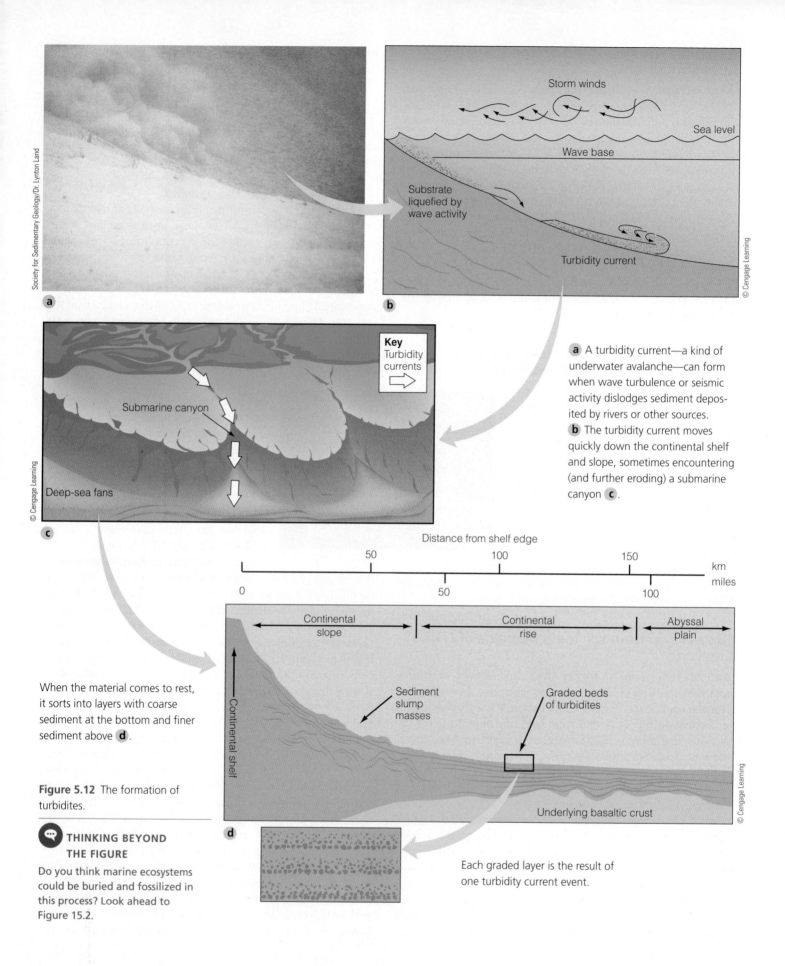

Society for Sedimentary Geology/Dr. Lynton Land

a

Storm winds

Sea level

Wave base

Substrate liquefied by wave activity

Turbidity current

© Cengage Learning

b

Key
Turbidity currents

Submarine canyon

Deep-sea fans

© Cengage Learning

c

a A turbidity current—a kind of underwater avalanche—can form when wave turbulence or seismic activity dislodges sediment deposited by rivers or other sources.
b The turbidity current moves quickly down the continental shelf and slope, sometimes encountering (and further eroding) a submarine canyon **c**.

Distance from shelf edge

50   100   150

km
miles

0   50   100

Continental slope | Continental rise | Abyssal plain

Continental shelf

Sediment slump masses

Graded beds of turbidites

Underlying basaltic crust

© Cengage Learning

d

When the material comes to rest, it sorts into layers with coarse sediment at the bottom and finer sediment above **d**.

Figure 5.12 The formation of turbidites.

💬 **THINKING BEYOND THE FIGURE**

Do you think marine ecosystems could be buried and fossilized in this process? Look ahead to Figure 15.2.

Each graded layer is the result of one turbidity current event.

Figure 5.13 Dover's famous white cliffs are uplifted masses of lithified coccolithophores. This chalklike material was deposited on the seabed around 100 million years ago, overlain by other sediments, and transformed into soft limestone by heat and pressure.

Aflo Co. Ltd./Alamy

populations of diatoms. Radiolarian oozes occur in equatorial regions, most notably in the zone of equatorial upwelling west of South America (as was seen in Figure 5.10a). About 14% of the surface of the deep-ocean floor is covered by siliceous oozes.

The very small particles that make up most of these pelagic sediments would need between 20 and 50 years to sink to the bottom. By that time they would have drifted a great lateral distance from their original surface position. But researchers have noted that the composition of pelagic sediments is usually similar to the particle composition in the water directly above. How could such tiny particles fall quickly enough to avoid great horizontal displacement? The answer appears to involve their compression into fecal pellets (Figure 5.15). While still quite small, the fecal pellets of small animals are much larger than

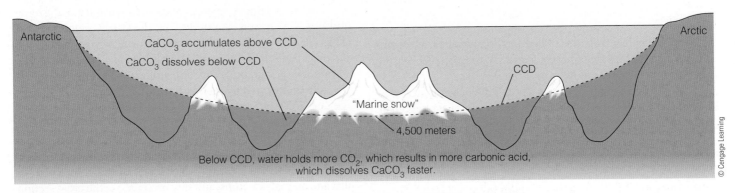

© Cengage Learning

Figure 5.14 The dashed line shows the calcium carbonate ($CaCO_3$) compensation depth (CCD). At this depth, usually about 4,500 meters (14,800 feet), the rate at which calcareous sediments accumulate equals the rate at which those sediments dissolve.

 THINKING BEYOND THE FIGURE

Is there a silicon dioxide compensation depth?

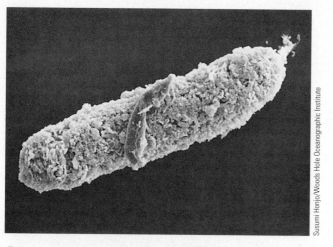

Susumi Honjo/Woods Hole Oceanographic Institute

a The compressed pellet is about 80 micrometers long.

Susumi Honjo/Woods Hole Oceanographic Institute

b Enlargement of the pellet's surface, magnified about 2,000 times. The pellet consists of the indigestible remains of small microscopic plantlike organisms, mostly coccolithophores. Unaided by pellet packing, these remains might take months to reach the seabed, but compressed in this way, they fall much faster and can be added to the ooze in perhaps 2 weeks.

Figure 5.15 A fecal pellet of a small planktonic animal.

the tiny individual skeletons of diatoms, foraminifera, and other plantlike organisms that they consumed, so they fall much faster, reaching the deep-ocean floor in about 2 weeks.

Some deep-sea oozes have been uplifted by geological processes and are now visible on land. (The white calcareous chalk cliffs of Dover are partially lithified deposits composed largely of foraminifera and coccolithophores.) Fine-grained siliceous deposits called *diatomaceous earth* are mined from other deposits. This fossil material is a valued component in flat paints, pool and spa filters, and mildly abrasive car and tooth polishes.

Hydrogenous Materials Precipitate out of Seawater Itself

Hydrogenous sediments also accumulate on deep-sea floors. They are associated with terrigenous or biogenous sediments and rarely form sediments by themselves. Most hydrogenous

a A broken manganese nodule shows concentric layers of manganese and iron oxides. This nodule is about 7 centimeters (3 inches) long, a typical size.

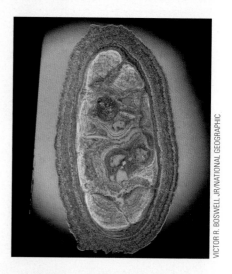

VICTOR R. BOSWELL JR/NATIONAL GEOGRAPHIC

Institute of Oceanographic Sciences/NERC/Science Source

b Lemon-sized manganese nodules littering the abyssal Pacific.

Figure 5.16 Manganese nodules.

sediments originate from chemical reactions that occur on particles of the dominant sediment.

The most famous hydrogenous sediments are manganese **nodules**, which were discovered by the hardworking crew of HMS *Challenger*. The nodules consist primarily of manganese and iron oxides but also contain small amounts of cobalt, nickel, chromium, copper, molybdenum, and zinc. They form in ways not fully understood by marine chemists, "growing" at an average rate of 1 to 10 millimeters (0.04–0.4 inch) per *million* years, one of the slowest chemical reactions in nature. Though most are irregular lumps the size of a potato, some nodules exceed 1 meter (3.3 feet) in diameter. Manganese nodules often form around nuclei such as sharks' teeth, bits of bone, microscopic alga and animal skeletons, and tiny crystals—as the cross section of a manganese nodule in **Figure 5.16a** shows. Bacterial activity may play a role in the development of a nodule. Between 20% and 50% of the Pacific Ocean floor may be strewn with nodules **(Figure 5.16b)**.

Why don't these heavy lumps disappear beneath the constant rain of accumulating sediment? Possibly the continuous churning of the underlying sediment by creatures living there

Figure 5.17 Geology students inspect the base of a thick layer of rock gypsum in Colorado. This rock probably formed by the lithification of evaporites left behind as a shallow inland sea dried up.

Stan Finney

keeps the dense lumps on the surface, or perhaps slow currents in areas of nodule accumulation waft particulate sediments away.

Evaporites Precipitate as Seawater Evaporates

Evaporites are an important group of hydrogenous deposits that include many salts important to humanity. These salts precipitate as water evaporates from isolated arms of the ocean or from landlocked seas or lakes. For thousands of years people have collected sea salts from evaporating pools or deposited beds. Evaporites are forming today in the Gulf of California, the Red Sea, and the Persian Gulf. The first evaporites to precipitate as water's salinity increases are the carbonates, such as calcium carbonate (from which limestone is formed). Calcium sulfate, which gives rise to gypsum, is next. Crystals of sodium chloride (table salt) will form if evaporation continues.

Figure 5.17 shows a thick deposit of rock gypsum within sedimentary rocks in the Rocky Mountains. Deposition of such a thick evaporite layer would have required evaporation of an arm of the ocean over a long period of time.

Oolite Sands Form When Calcium Carbonate Precipitates from Seawater

Not all hydrogenous calcium carbonate deposits are caused by evaporation, however. A small decrease in the acidity of seawater, or an increase in its temperature, can cause calcium carbonate to precipitate from water of normal salinity. In shallow areas of high biological productivity where sunlight heats the water, microscopic plants use up dissolved carbon dioxide, making seawater slightly less acidic (see Figure 7.13). Molecules of calcium carbonate then may precipitate around shell fragments or other particles. These white, rounded grains are called ooliths (*oon*; egg) because they resemble fish eggs **(Figure 5.18)**. **Oolite sands**—sands composed of ooliths—are abundant in many warm, shallow waters such as those of the Bahama Banks.

5.6 Researchers Have Mapped the Distribution of Deep-Ocean Sediments

Look again at the types and distribution of marine sediments in Figure 5.10a. Notice especially the lack of radiolarian deposits in much of the deep North Pacific; the strand of siliceous oozes extending west from equatorial South America; and the broad expanses of the Atlantic, South Pacific, and Indian ocean floors covered by calcareous oozes. The broad, deep, relatively old Pacific contains extensive clay deposits, most delivered in the form

Tom Garrison

Figure 5.18 Oolite sand. Note the uniform rounded shape reminiscent of fish eggs.

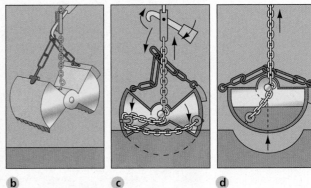

Figure 5.19 Sampling sediments aboard the research vessel *Robert Gordon Sproul*.

a A scoop of muddy ocean-bottom sediments collected with a clamshell sampler is dumped onto the deck for study.

b Before sampling, **c** during sampling, and **d** after the sample has been taken. Note that the sample is relatively undisturbed.

of airborne dust. Why? Though some of the world's largest and muddiest rivers empty into the Pacific, most of their sediments are trapped in the peripheral trenches and cannot reach the mid-basins. And as you might expect, the poorly sorted glacial deposits are found only at high latitudes.

Figures 5.9 and 5.10 summarize more than a century of effort by marine scientists. Studies of sediments will continue because of their importance to natural resource development and because of the details of Earth's history that remain locked beneath their muddy surfaces.

CONCEPT CHECK

Before going on to the next section, check your understanding of some of the important ideas presented so far:

Why are Atlantic sediments generally thicker than Pacific sediments?

How do turbidity currents distribute sediments? What do these sediments (turbidites) look like?

What is the origin of oozes? What are the two types of oozes?

What is the CCD? How does it affect ooze deposition at great depths?

How do hydrogenous materials form? Give an example of hydrogenous sediment.

How do evaporites form?

5.7 Geologists Use Specialized Tools to Study Ocean Sediments

Deep-water cameras have enabled researchers to photograph bottom sediments. The first of these cameras was simply lowered on a cable and triggered by a trip wire. Other more elaborate cameras have been taken to the seafloor on towed sleds or deep submersibles.

Actual samples usually provide more information than photographs do. HMS *Challenger* scientists used weighted, wax-tipped poles and other tools attached to long lines to obtain samples, but today's oceanographers have more sophisticated equipment. Shallow samples may be taken using a **clamshell**

sampler (named because of its method of operation, not its target; **Figure 5.19**).

These sediments can be separated by a series of ever-finer sieve screens (**Figure 5.20**). The relative amount of material trapped by each screen can provide information on the composition and recent history of the sample.

Deeper samples are taken by a **piston corer** (**Figure 5.21**), a device capable of punching through as much as 25 meters

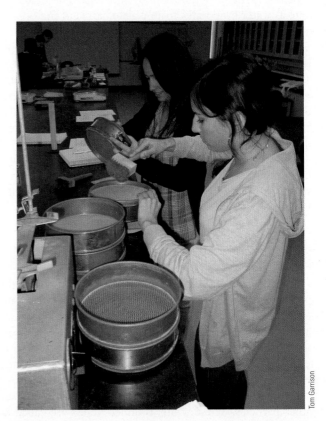

Figure 5.20 Students clean sieve screens before assembling them into a coarse-to-fine sequence that will separate sediment particles of different sizes. A coarse screen is seen in the foreground.

(a)

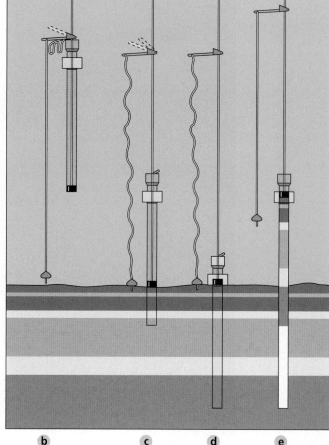

(b)   **(c)**   **(d)**   **(e)**

Figure 5.21 Sampling using a piston corer.
(a) The piston corer. **(b)** The corer is allowed to fall toward the bottom. **(c)** The corer reaches the bottom and continues, forcing a sample partway into the cylinder. **(d)** Tension on the cable draws a small piston within the corer toward the top of the cylinder, and the pressure of the surrounding water forces the corer deeper into the sediment. **(e)** The corer and sample being hauled in.

(82 feet) of sediment and returning an intact plug of material. Using a rotary drilling technique similar to that used to drill for oil, the drilling ship *JOIDES Resolution* **(Figure 5.22)** returned much longer core segments, some more than 1,100 meters (3,600 feet) long![2] These cores are stored in core libraries, a valuable scientific resource **(Figure 5.23)**. Analysis of sediments and fossils from the Deep Sea Drilling Project cores helped verify the theory of plate tectonics. It has also shed light on the evolution of lifeforms and helped researchers decipher the history of changes in Earth's climate over the past 100,000 years.

CONCEPT CHECK

Before going on to the next section, check your understanding of some of the important ideas presented so far:

How are sediments studied?

How have studies of marine sediments advanced our understanding of plate tectonics?

---

[2]R/V *Chikyu* (Figure 2.26) assumed these duties in 2005.

5.8 Sediments Are Historical Records of Ocean Processes

In 1899, the British geologist W. J. Sollas theorized that deep-sea deposits could reveal much of the planet's history. In the era before plate tectonics theory, this certainly seemed reasonable—the deep-ocean bottom was thought to be a calm, changeless place, where an unbroken accumulation of sediment could be probed to discover the entire history of the ocean. Unfortunately for this promising idea, difficulties began to crop up almost immediately. For one thing, the sediments should have been much *thicker* than early probes indicated. If Earth's ocean is truly older than a few hundred thousand years, and if life has existed within it for most of that time, the sediment layer should be thicker than had been observed. Another difficulty lay in the uneven distribution of sediments. Sollas thought that the center of an ocean basin should contain the thickest layers of sediment, yet the ridged mid-Atlantic bottom was nearly naked. There didn't seem to be any difference in the nature of the overlying seawater that could account for the variations in thickness and composition of the sediments across the bottom of the Atlantic. Oozes were especially puzzling: The organisms that form oozes grow well at the surface of the middle Atlantic, yet the mid-Atlantic floor seemed to bear little ooze.

Figure 5.22

a *JOIDES Resolution,* the deep-sea drilling ship operated by the Joint Oceanographic Institutions for Deep Earth Sampling. The vessel is 124 meters (407 feet) long, with a displacement of more than 16,000 tons. The rig can drill to a depth of 9,150 meters (30,000 feet) below sea level.

Figure 5.23 Sediment cores in storage. Cores are sectioned longitudinally, placed in trays, and stored in hermetically sealed cold rooms. The Gulf Coast Repository of the Ocean Drilling Program, located at Texas A&M University (pictured here), stores about 75,000 sections taken from more than 80 kilometers (50 miles) of cores recovered from the Pacific and Indian Oceans. Smaller core libraries are maintained at the Scripps Institution in California (Pacific and Indian Oceans) and at the Lamont–Doherty Earth Observatory in New York state (Atlantic Ocean).

b The difficulty of deep-sea drilling can be sensed from this scale drawing: The length of the drill ship is 120 meters (394 feet); the depth of water through which the drill string must pass to reach the bottom is 5,500 meters (18,000 feet)!

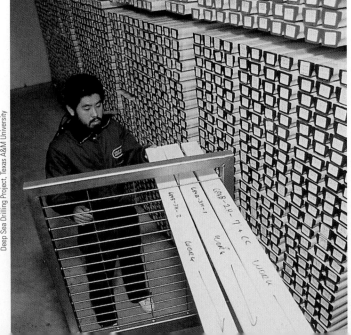

Turn-of-the-century geologists were understandably confused, but today we know the tectonic reasons for these discrepancies. Because the deep-sea sediment record is ultimately destroyed in the subduction process, the ocean's sedimentary "memory" does not start with the ocean's formation as originally reasoned by early marine scientists. But modern studies of deep-sea sediments using seafloor samples, cores obtained by deep drilling, and continuous seismic profiling have demonstrated that these deposits contain a remarkable record of relatively recent (that is, about the past 180 million years) ocean history. However, these same data have also shown that the record is not uninterrupted, as early workers had originally assumed. In fact, some of the gaps in the deep-sea deposits represent erosional events and constitute evidence of changes in deep-sea circulation, and hence are valuable in their own right. The analysis of layered sedimentary deposits, whether in the ocean or on land, represents the discipline of **stratigraphy** (*stratum;* layer, *graph;* a drawing). Deep-sea stratigraphy utilizes variations in the composition of rocks, microfossils, depositional patterns, geochemical character, and physical character (density and such) to trace or correlate distinctive sedimentary layers from place to place, establish the age of the deposits, and interpret changes in ocean and atmospheric circulation, productivity, and other aspects of past ocean behavior. In turn, these sorts of studies and the advent of deep-sea drilling have given rise to the emerging science of **paleoceanography** (*palaios;* ancient), the study of the ocean's past.

Early attempts to interpret ocean and climate history from evidence in deep-sea sediments occurred in the 1930s through 1950s as cores became available. These initial studies relied primarily on identifying variations in the abundance and distribution of glacial marine sediments, carbonate and siliceous oozes, and temperature-sensitive microfossils in the cores. Modern paleoceanographic studies continue to utilize these same features, but researchers have much greater understanding of their significance and are aided by seismic imaging of the deposits over large areas. In addition, newer and more precise methods of dating deep-sea sediments have enabled them to place events in a proper time context. Finally, scientists now have instruments capable of analyzing very small variations in the relative abundances of the stable isotopes of oxygen preserved within the carbonate shells of microfossils found in deep-sea sediments; these instruments allow them to interpret changes in the temperature of surface and deep water over time. These same data are also used to estimate variations in the volume of ice stored in continental ice sheets, and thus to track the ice ages. Other geochemical evidence contained in the shells of marine microfossils, including variations in carbon isotopes and trace metals such as cadmium, provide insights into ancient patterns of ocean circulation, productivity of the marine biosphere, and upwelling. These sorts of data have already provided quantitative records of the glacial-interglacial climatic cycles of the past 2 million years. Future drilling and analysis of deep-sea sediments are poised to extend our paleoceanographic perspective much further back in time.

> **I**n 2010, an estimated 39% of the world's crude oil and 35% of its natural gas was extracted from the sedimentary deposits of continental shelves and continental rises.

Figure 5.24 shows the age of regions of the Pacific Ocean floor using data obtained largely from analyses of the overlying sediment. Note that sediments get older with increasing distance from the East Pacific Rise spreading center.

Earth might not be the only planet where marine sediments have left historical records. As you read in Chapter 1, Mars probably had an ocean between 3.2 and 1.2 billion years ago. In November of 2004, NASA's Mars Exploration Rover *Opportunity* photographed lithified sediments that look suspiciously marine in origin (**Figure 5.25**, see p. 160). One can only wonder what stories they will tell.

CONCEPT CHECK

Before going on to the next section, check your understanding of some of the important ideas presented so far:

Would you say the "memory" of the sediments is long or short (in geologic time)?

How might past climate be inferred from studies of marine sediments?

5.9 Marine Sediments Are Economically Important

Study of sediments has brought practical benefits. You probably have more daily contact with marine sediment than you think. Components of the building materials for roads and structures, toothpaste, paint, and swimming pool filters come directly from sediments. In 2010, an estimated 39% of the world's crude oil and 35% of its natural gas were extracted from the sedimentary deposits of continental shelves and continental rises (**Figure 5.26**, see p. 160). Offshore hydrocarbons currently generate annual revenues in excess of $200 billion. Deposits within the sediments of continental margins account for about one third of the world's estimated oil and gas reserves.

In addition to oil and gas, in 2005, sand and gravel valued at more than $550 million were taken from the ocean. This is about 1% of world needs. Commercial mining of manganese nodules has also been considered. In addition to manganese, these nodules contain substantial amounts of iron and other industrially important chemical elements. The high iron content of these nodules has prompted a proposal to rename them *ferromanganese* nodules. We will investigate these resources in more detail in Chapter 17.

CONCEPT CHECK

Before going on to the next section, check your understanding of some of the important ideas presented so far:

What percentages of the total production of petroleum and natural gas are extracted from the seabed?

What products containing marine sediments have you used today?

Other than petroleum and natural gas, what is the most valuable material taken from marine sediments?

About Past Climates and Catastrophes

Figure A An artist's conception of a catastrophic asteroid strike. The 10-kilometer object would have vaporized above Earth's surface and struck with catastrophic force. The energy of the collision (imagined here about 45 seconds after impact) would have sent shock waves and debris around Earth.

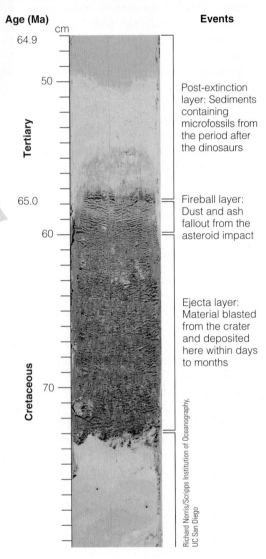

Age (Ma)   **Events**

Post-extinction layer: Sediments containing microfossils from the period after the dinosaurs

Fireball layer: Dust and ash fallout from the asteroid impact

Ejecta layer: Material blasted from the crater and deposited here within days to months

Richard Norris/Scripps Institution of Oceanography, UC San Diego

Figure B This cross section of a seabed core shows clear evidence of the impact and its aftermath.

Could sediment cores tell us something about Earth's history and thus offer insight into future change?

Researchers employing a variety of techniques have long known about two relatively recent and profound changes to our planet's ocean and atmosphere. One of these occurred about 65 million years ago and resulted in (among other things) the extinction of nearly all of the dinosaurs. The other happened 56 million years ago when Earth's climate warmed suddenly and powerfully. Analysis of seabed sediment cores was the key to understanding these two dramatic events.

Our relatively short life spans don't allow much of a cosmological perspective; most of us believe Earth to be a safe and relatively benign place for living things. Not so. Since 1992, about 800 objects—some larger than Pluto and most much smaller—have been found orbiting the sun in the dim reaches past the planet Nep-

tune. Once in a while one of these bodies sets off on a voyage to the inner solar system. The great majority of them return to the outer darkness without incident.

But even comparatively small visitors are still capable of considerable mayhem if they strike a planet. Earth was apparently struck 65 million years ago by an asteroid about 10 kilometers (6 miles) across **(Figure A)**. The cataclysmic collision is thought to have propelled shock waves and huge clouds of seabed and crust all over Earth, producing a time of cold and dark that contributed to the extinction of vast numbers of marine and terrestrial species. The accompanying photograph of a deep-sea core **(Figure B)** shows evidence of this disaster, about which you'll learn more in Chapter 13.

We can't do much about asteroids, but the global warming event that began 56 million years ago might offer some insight and perhaps

a valuable lesson. During the Paleocene-Eocene Thermal Maximum (or PETM), atmospheric temperature skyrocketed by more than 5°C (9°F), and polar ice disappeared. The cause is thought to be a sudden and massive release of carbon (mostly in the form of carbon dioxide and methane) into the atmosphere. The source of the carbon is uncertain, but Earth was already warming gradually at the beginning of the PETM, so methane hydrate (an ice-like compound discussed in Chapters 17 and 18) may have escaped from submarine deposits and melting permafrost to accumulate in the atmosphere and ocean. More warming released more carbon, and temperature spiraled out of control. Droughts, floods, plagues of insects, and mass extinctions followed. Ocean life forms were dramatically altered **(Figure C)**.

The PETM lasted more than 150,000 years—until the excess carbon was absorbed by organisms and geological processes. The total amount of carbon injected into the atmosphere during this period was roughly the amount that would be released if all Earth's reserves of coal, natural gas, and oil were burned. We're working hard on doing exactly that. Reminds us of an old Chinese curse: "May you live in interesting times."

Figure C This sediment core shows an abrupt change in the Atlantic Ocean 56 million years ago at the onset of the Paleocene-Eocene thermal maximum. The shells of small marine organisms (white) vanished from the seafloor sediment, shifting its color from white to red. As planet-warming CO_2 surged into the atmosphere, it seeped into the ocean, acidifying the water and dissolving the shells.

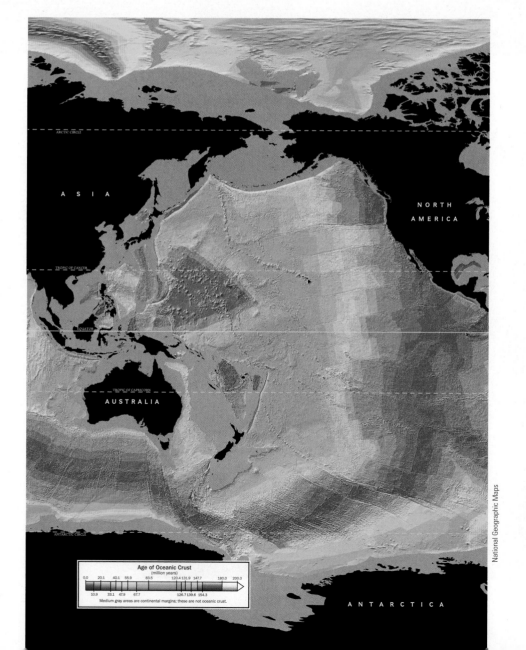

Age of Oceanic Crust
(million years)
0.0 20.1 40.1 55.9 83.5 120.4 131.9 147.7 180.0 200.0
   10.9   33.1 47.9 67.7   126.7 139.6 154.3
Medium gray areas are continental margins; these are not oceanic crust.

Figure 5.24 The ages of portions of the Pacific Ocean floor, based on core samples of sediments just above the basalt seabed, in millions of years ago (Ma, mega-annum). The youngest sediments are found near the East Pacific Rise and the oldest close to the eastern side of the trenches.

💬 **THINKING BEYOND THE FIGURE**
How far back in time can the "memory of the ocean" (sediments) provide information?

Figure 5.25
Ancient marine sediments on Mars? This photo taken in August 2012 by NASA's Mars Exploration Rover *Curiosity* shows an eroded area on the slope of nearby Mt. Sharp. Their stepped configuration suggests the layers' deposition by water with subsequent erosion by wind. NASA/JPL-Caltech/MSSS

Figure 5.26 Natural gas is burned off during exploration in the North Sea. Marine oozes are often rich in trapped hydrocarbons.

Philip Stephen/Bluegreen Pictures/Alamy

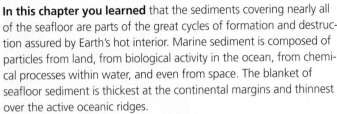

Chapter in Perspective

In this chapter you learned that the sediments covering nearly all of the seafloor are parts of the great cycles of formation and destruction assured by Earth's hot interior. Marine sediment is composed of particles from land, from biological activity in the ocean, from chemical processes within water, and even from space. The blanket of seafloor sediment is thickest at the continental margins and thinnest over the active oceanic ridges.

Sediments may be classified by particle size, source, location, or color. Terrigenous sediments, the most abundant, originate on continents or islands. Biogenous sediments are composed of the remains of once-living organisms. Hydrogenous sediments are precipitated directly from seawater. Cosmogenous sediments, the ocean's rarest, come to the seabed from space.

The position and nature of sediments provide important clues to Earth's recent history, and valuable resources can sometimes be recovered from them.

In the next chapter you'll learn about our story's main character—water itself. You know something about how water molecules were formed early in the history of the universe, something about the inner workings of our planet, and something about the nature of the ocean's "container." Now let's fill that container with water and see what happens.

QUESTIONS FROM STUDENTS

1. The question of sediment age seems to occupy much of sedimentologists' time. Why?

The dating of sediments has been a central problem in marine science for many years. In 1957, during the International Geophysical Year, sedimentologists designed a coordinated effort to determine sediment age, which included plans for the *Glomar Explorer* and *Glomar Challenger* drilling surveys. Their primary interest was to seek evidence for the hypothesis of the then-new idea of seafloor spreading. Cores returned by the Deep Sea Drilling Project in 1968 enabled researchers, including J. Tuzo Wilson, Harry Hess, and Maurice Ewing, to put the evidence together. Much of the proof for plate tectonics rests on the interpretation of sediment cores.

2. Where are sediments thickest?

Sediments are thickest close to eroding land and beneath biologically productive neritic waters and thinnest over the fast-spreading oceanic ridges of the eastern South Pacific. The thickest accumulations of sediment may be found along and beneath the continental margins (especially on continental rises). Some are typically more than 1,500 meters (5,000 feet) thick. Remember that much of the rocky material of the Grand Canyon was once marine sediment atop an isostatically depressed ancient seabed. The Grand Canyon is nearly 2 kilometers (1¼ miles) deep, and the uppermost layer of sedimentary rock has already been eroded completely away!

3. What's the relationship between deep-sea animals and the sediments on which they live?

Though microscopic bacteria and benthic foraminifera may be very abundant on the seabed, visible life is not abundant on the bottom of the deep ocean. There are no plants at great depths because there is no light, but animals do live there. Some, like brittle stars, move slowly along the surface searching for bits of organic matter to eat. Others burrow through the muck in search of food particles. Worms eat quantities of sediment to extract any nutrients that may be present and then deposit strings of fecal material as they move forward. The deeps are uninviting places, but life is tenacious and survives even in this hostile environment.

4. What's the deal with airliners and volcanic eruptions? Why are flights cancelled when there's just a hint of volcanic smoke in the air?

The combination of finely divided silicate rock and finely engineered turbofan engines is not a happy one. Airline routes from the United States to Asia often pass over the Aleutian Islands west of Alaska where active subduction powers near-continuous volcanism. Iceland's brace of active volcanoes is perched inconveniently on the North America–Europe track. Thousands of passengers and tens of millions of dollars of cargo pass over these volcanoes every day. Airborne ash can erode cockpit windows, damage flight control systems, clog cabin air filters, and cause jet engines to fail. In the past 30 years, more than 100 commercial jet airliners have encountered clouds of volcanic ash and suffered damage (see **Figure**)

The eruption in 2010 of Iceland's Eyjafjallajökull volcano repeatedly injected glass-rich ash into the eastward-moving jet stream (see Figure 5.7c). European airspace closed on 14 April 2010 and was disrupted for weeks afterward. This was the largest disruption to air travel since World War II. An estimated US$1.7 billion of airline revenue was lost, along with the incalculable value of delays and disruptions to personal plans, goods, and services.

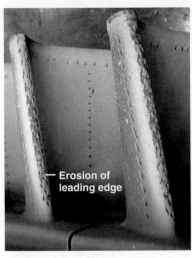

The aerodynamic shape of jet turbine blades can be eroded by volcanic ash.

— Erosion of leading edge

NASA

TERMS AND CONCEPTS TO REMEMBER

authigenic sediment	diatom	nodule	sand
biogenous sediment	evaporite	oolite sand	sediment
calcareous ooze	foraminiferan	ooze	siliceous ooze
calcium carbonate compensation depth (CCD)	hydrogenous sediment	paleoceanography	silt
clamshell sampler	lithification	pelagic sediment	stratigraphy
clay	microtektite	piston corer	terrigenous sediment
coccolithophore	mineral	poorly sorted sediment	turbidite
cosmogenous sediment	neritic sediment	radiolarian	well-sorted sediment

STUDY QUESTIONS

Thinking Critically

1. Is the thickness of ooze always an accurate indication of the biological productivity of surface water in a given area? (Hint: See next question.)
2. What is the calcium carbonate compensation depth? Is there a compensation depth for the siliceous components of once-living things?
3. What sediments accumulate most rapidly? Least rapidly?
4. Can marine sediments tell us about the history of the ocean from the time of its origin?
5. What problems might arise when working with deep-ocean cores? (Imagine the process of taking a core sample and think of what can go wrong!)

Thinking Analytically

1. Given an average rate of accumulation, how much time would it take to build an average-size manganese nodule?
2. Assuming an average rate of sediment accumulation and seafloor spreading, how far from the Mid-Atlantic Ridge would one need to travel before encountering a layer of sediment 1,000 meters thick?
3. Microtektites are often found in "fields"—elongated zones of relative concentration a few hundred kilometers long. Why do you suppose that is?
4. How much faster do fragments of diatoms fall when they are compacted into fecal pellets than when they are not? (Hint: See Table 5.1)

GLOBAL GEOSCIENCE WATCH

Visit www.cengagebrain.com to access MindTap, a complete digital course which includes access to Global Geoscience Watch and more. Go to the "Gulf of Mexico Oil Spill 2010" topic portal in the GREENR database to learn more about the relationship among marine sediments, energy, and pollution from oil spills. Write a 2–3 page report that describes how oil was formed in the Gulf of Mexico, with emphasis on its relationship to biogenous marine sediments. Explain what impacts the oil spill had on the seafloor and local marine life.

CENGAGE brain.com   Visit www.cengagebrain.com to access course materials for this text, including interactive learning tools, videos, and more.

SEDIMENTS   163

6 Water and Ocean Structure

KEY CONCEPTS

Heat is not the same as temperature. Temperature is an object's response to an input (or removal) of heat. Not all substances respond in the same way. MARK THEISSAN/NATIONAL GEOGRAPHIC

Water resists rising in temperature as heat is added. Water gives off heat when it freezes and absorbs heat as it thaws. These properties of liquid water moderate Earth's surface temperatures. © Cengage Learning

The ocean is density stratified. Dense cold and salty water underlies less dense warm and fresher water. © Cengage Learning

Light is quickly extinguished by passage through water. Sound is not. © Cengage Learning

Light and sound can be refracted (bent) by passage between water masses whose physical characteristics differ. © Cengage Learning

"There is nothing in the world more soft and weak than water, yet for attacking things that are hard and strong there is nothing that surpasses it. Nothing can take its place."—Lao Tzu, sixth century B.C.E.
© Ocean/Corbis

Media Connection

Start off this chapter by listening to a podcast with David Braun, as he reports on the effects sounds have on sea creatures and what's being done to help them. Visit www.cengagebrain.com to access MindTap, a complete digital course that includes this podcast and other resources.

6.1 A Note to the Reader

This chapter and the next work together as a pair. In this chapter we'll first consider the physics of pure water—mainly its response to the input and removal of heat energy. Then we'll add some dissolved solids (salts) and see how properties change, and then discuss density stratification—an old friend from previous chapters. Last, we'll look at light and sound in the ocean.

In the **next** chapter, we'll investigate salinity in more detail and add concepts involving dissolved gases and the ocean's acid-base balance.

Together, these two chapters set the stage for the air–ocean interactions in the two chapters that follow.

6.2 Familiar, Abundant, and Odd

Water is so familiar and abundant that we don't always appreciate its unusual characteristics **(Figure 6.1)**. Here you'll meet the water you never knew. This chapter introduces the characteristics that make water unusual—the molecule's polarity and the bonds that hold it together, the large amount of heat needed to

change its temperature, and the heat needed to change its physical state. And, no, heat and temperature are not the same thing.

Two big lessons follow.

One lesson is the influence of water on global temperatures. Liquid water's thermal characteristics prevent broad swings of temperature during day and night, and, through a longer span, during winter and summer. Heat is stored in the ocean during the day and released at night. A much greater amount of heat is stored through the summer and given off during the winter. Liquid water has an important thermostatic balancing effect—an oceanless Earth would be much colder in winter and much hotter in summer than the moderate temperatures we experience.

The other lesson is the influence of density on ocean structure. You'll see that the ocean's structure and large-scale movement depend on changes in the density of seawater, with density dependent on temperature and salt content (salinity).

6.3 The Water Molecule Is Held Together by Chemical Bonds

Pure water is a **compound**—that is, a substance that contains two or more different elements in a fixed proportion. The familiar chemical formula for water, H_2O, shows that it is composed of the elements hydrogen (H) and oxygen (O) in a fixed proportion of two to one. An **element** is a substance composed of identical particles called **atoms** that cannot be broken into simpler substances by chemical means.

Water is a **molecule**, a group of atoms held together by chemical bonds. **Chemical bonds**, the energy relationships between atoms that hold them together, are formed when **electrons**—tiny negatively charged particles found toward the outside of an atom—are shared between atoms or moved from one atom to another. A water molecule forms when electrons are

Figure 6.1 Liquid water covers most of Earth's surface. Water's properties are familiar to us, but it is, in fact, an unusual compound. Its odd thermal properties are responsible for Earth's moderate climate, and its ability to dissolve nearly any substance was critical to the appearance and maintenance of life on Earth.

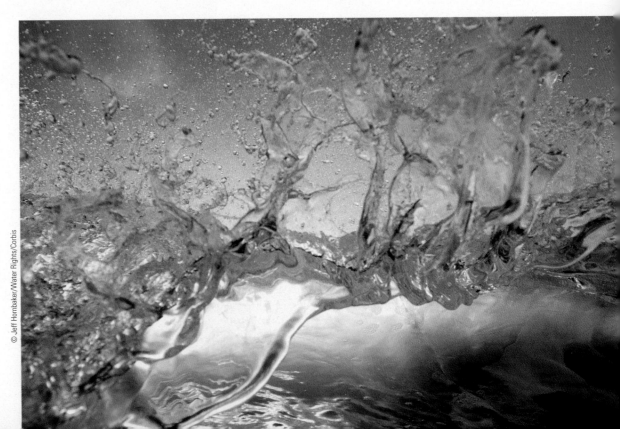

© Jeff Hornbaker/Water Rights/Corbis

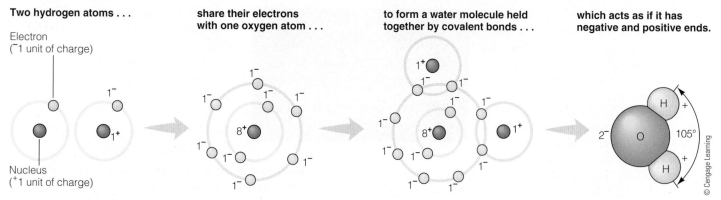

Electron
(⁻1 unit of charge)

Nucleus
(⁺1 unit of charge)

share their electrons
with one oxygen atom . . .

to form a water molecule held
together by covalent bonds . . .

which acts as if it has
negative and positive ends.

Figure 6.2 The formation of a water molecule.

shared between two hydrogen atoms and one oxygen atom (**Figure 6.2**). The bonds formed by shared pairs of electrons are known as **covalent bonds**. Covalent bonds hold together many familiar molecules besides water, including CO_2 (carbon dioxide), CH_4 (methane gas), and O_2 (atmospheric oxygen).

Because of the way a water molecule's oxygen electrons are distributed, the overall geometry of the molecule is a bent or angular shape. The angle formed by the two hydrogen atoms and the central oxygen atom is about 105°. The angular shape of the water molecule makes it electrically asymmetrical. Each water molecule can be thought of as having a positive (+) end and a negative (−) end. This is because **protons** of the hydrogen atoms—the positively charged particles in the **nucleus** (center)—are left partially exposed when the negatively charged electrons bond more closely to oxygen. The water molecule behaves something like a magnet: its positive end attracts particles that have a negative charge, and its negative end (or pole) attracts particles that have a positive charge. For this reason, water is called a **polar molecule**. When water comes into contact with compounds whose elements are held together by the attraction of opposite electrical charges (most salts, for example), the polar water molecule will separate that compound's component elements from each other. This explains why water can easily dissolve so many other compounds.

The polar nature of water also permits it to attract other water molecules. When a hydrogen atom (the positive end) in one water molecule is attracted to the oxygen atom (the negative end) of an adjacent water molecule, a **hydrogen bond** forms. A hydrogen bond *between* molecules is about 5% to 10% as strong as a covalent bond *within* a molecule. Hydrogen bonds link water molecules together by electrostatic forces. The resulting loosely held webwork of water molecules is shown in **Figure 6.3**. Hydrogen bonds greatly influence the properties of water by allowing individual water molecules to stick to each other, a property called **cohesion**. Cohesion gives water an unusually high **surface tension**, which results in a surface "skin" capable

> "Standing on the shore and looking out to sea, the boy said, 'There's a lot of water out there.' And the wise old oceanographer responded, 'And that's only the top of it.'" —RICHARD ELLIS
> *SINGING WHALES AND FLYING SQUID*

of supporting needles, razor blades, and even walking insects. Surface tension allows a clean water glass to be filled slightly above its brim. The formation of hydrogen bonds between water molecules allows the water to bulge above the container. (Add too much water, though, and gravity wins.)

Adhesion, the tendency of water to stick to other materials, enables water to adhere to solids—that is, to make them wet. Cohesion and adhesion are the causes of capillary action, the tendency of water to spread through a towel when one corner is dipped in water.

Hydrogen bonds are also what give pure water its pale blue hue. When water molecules vibrate, adjacent molecules tug

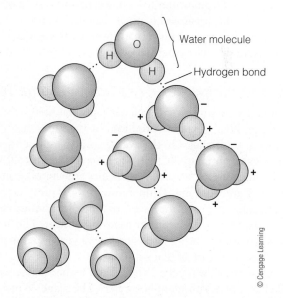

Water molecule

Hydrogen bond

Figure 6.3 Hydrogen bonds in liquid water. The attractions between adjacent polar water molecules form a webwork of hydrogen bonds. These bonds are responsible for surface tension and cohesion, the properties of water that cause surface tension and wetting. Hydrogen bonds among water molecules also make it difficult for individual molecules to escape from the surface.

and push against their hydrogen-bonded neighbors. This action absorbs a small amount of red light, leaving proportionally more blue light to scatter back to our eyes. The same blue color is seen in ice formations.

If hydrogen bonds did not hold water molecules together, they would form a gas rather than a liquid. The compound hydrogen sulfide (H_2S) is chemically similar to water but lacks water's ability to form networks of hydrogen bonds. H_2S is therefore a gas (with individual molecules separated from each other) at normal temperatures and pressures.

> **W**ater has an extraordinarily high heat capacity—it resists changing *temperature* when *heat* is added or removed.

CONCEPT CHECK

Before going on to the next section, check your understanding of some of the important ideas presented so far:

How are atoms different from molecules?

What holds molecules together?

Why is water a polar molecule? What properties of water derive from its polar nature?

What familiar property of water is due to its cohesion? What kinds of bonds are involved?

6.4 Water Has Unusual Thermal Characteristics

Perhaps the most important physical properties of water are related to its behavior as it absorbs or loses heat. Water's unusual thermal characteristics prevent wide temperature variation from day to night and from winter to summer; permit vast amounts of heat to flow from equatorial to polar regions; and power Earth's great storms, wind waves, and ocean currents. To see why, we first need to examine water's thermal characteristics in some detail.

Heat and Temperature Are Not the Same Thing

Heat and temperature are related concepts, but they are not the same thing. **Heat** is energy produced by the random vibration of atoms or molecules. On the average, water molecules in hot water vibrate more rapidly than water molecules in cold water. Heat is a measure of *how many* molecules are vibrating and *how rapidly* they are vibrating. Temperature records only *how rapidly* the molecules of a substance are vibrating. **Temperature** is an object's response to an input (or removal) of heat. The amount of heat required to bring a substance to a certain temperature varies with the nature of that substance.

This example will help: Which has a higher temperature: a candle flame or a bathtub of hot water? The flame. Which contains more heat? The tub. The molecules in the flame vibrate very rapidly, but there are relatively few of them. The molecules of water in the tub vibrate more slowly, but there are a great many of them, so the total amount of heat energy in the tub is greater.

Temperature is measured in **degrees**. One degree Celsius (°C) = 1.8 degrees Fahrenheit (°F). Though Americans are more familiar with the older Fahrenheit scale, Celsius degrees are more useful in science because they are based on two of pure water's most significant properties: its freezing point (0°C) and its boiling point (100°C).

Not All Substances Have the Same Heat Capacity

Heat capacity is a measure of the heat required to raise the temperature of 1 gram (0.035 ounce) of a substance by 1°C (1.8°F). Different substances have different heat capacities. *Not all substances respond to identical inputs of heat by rising in temperature the same number of degrees* **(Table 6.1)**. Heat capacity is measured in calories per gram. A **calorie** is the amount of heat required to raise the temperature of 1 gram of pure water by 1°C.[1]

Because of the great strength and large number of hydrogen bonds between water molecules, more heat energy must be added to speed up molecular movement and raise water's temperature than would be necessary in a substance held together by weaker bonds. Liquid water's heat capacity is therefore among the highest of all known substances. This means that *water can absorb (or release) large amounts of heat while changing relatively little in temperature.*

Anyone who waits by a stove for water to boil knows a lot about water's heat capacity—it seems to take a very long time to warm water for soup or coffee. Compared with water, ethyl alcohol has a much lower heat capacity. If both liquids absorb heat from identical stove burners at the same rate, pure ethyl alcohol (the active ingredient in alcoholic beverages) will rise in temperature about three times as fast as an equal mass of water.

---

[1]A nutritional Calorie, the unit we see on cereal boxes, also known as a kilocalorie, equals 1,000 of these calories. A gram is about 10 drops of seawater. One calorie = 4.184 kilojoules.

Table 6.1 Heat Capacity of Common Substances

Substance	Heat Capacity[a] in calories/gram/°C
Silver	0.06
Granite	0.20
Aluminum	0.22
Alcohol (ethyl)	0.30
Gasoline	0.50
Acetone	0.51
Pure water	**1.00**
Ammonia (liquid)	1.13

[a] Heat capacity is a measure of the heat required to raise the temperature of 1 gram (0.035 ounce) of a substance by 1°C (1.8°F). Different substances have different heat capacities. *Not all substances respond to identical inputs of heat by rising in temperature the same number of degrees.* Notice how little heat is required to raise the temperature of 1 gram of silver 1°.

Because of the great strength and large number of the hydrogen bonds between water molecules, water can gain or lose large amounts of *heat* with very little change in *temperature*. This *thermal inertia* moderates temperatures worldwide. Of all common substances, only liquid ammonia has a higher heat capacity than liquid water.

Beach sand has an even lower heat capacity. A gram of sand requires as little as 0.2 calories to rise 1°C (1.8°F). So, on sunny days beaches can get too hot to stand on with bare feet, while the water remains pleasantly cool.

As we will soon see, the concept of heat capacity is very important in oceanography. But for now, remember this: Water has an extraordinarily high heat capacity—it resists changing *temperature* when *heat* is added or removed.

Water's Temperature Affects Its Density

The uniqueness of water becomes even more apparent when we consider the effect of a temperature change on water's **density** (its mass per unit of volume). You may recall from Chapter 3 that the density of pure water is 1 gram per cubic centimeter (1 g/cm³). Granite rock is heavier, with a density of about 2.7 g/cm³, and air is lighter, with a density of about 0.0012 g/cm³. Most substances become denser (weigh more per unit of volume) as they get colder. Pure water generally becomes denser as heat is removed and its temperature falls, but water's density behaves in an unexpected way as the temperature approaches the freezing point.

A **density curve** shows the relationship between the temperature (or salinity) of a substance and its density. Most substances become progressively denser as they cool; their temperature–density relationships are linear (that is, appear as a straight line on graphs). But **Figure 6.4** shows the unusual temperature–density relationship of pure water. Imagine heat being removed from some water placed in a freezer. Initially,

the water is at room temperature (20°C, or 68°F), point Ⓐ on the graph. As expected, the density of water increases as its temperature drops along the line from point Ⓐ toward point Ⓑ. As the temperature approaches point Ⓑ, the density increase slows, reaching a maximum at point Ⓑ of 1 g/cm³ at 3.98°C (39.16°F). As the water continues to cool, its framework of hydrogen bonds becomes more rigid, which causes the liquid to expand slightly because the molecules are held slightly farther apart. So water becomes slightly less dense as cooling continues, until point Ⓒ (0°C, or 32°F) is reached. At point Ⓒ the water begins to freeze—to change state by crystallizing into ice.

State is an expression of the internal form of a substance (**Figure 6.5**). Changes in state are accompanied by either an input or an output of energy. Water exists on Earth in three physical states: liquid, gas (water vapor), and solid (ice). If the freezer continues to remove heat from the water at point Ⓒ in Figure 6.4, the water will change from liquid to solid state. Through this transition from water to ice—from point Ⓒ to point Ⓓ—the density of the water *decreases* abruptly. Ice is therefore lighter than an equal volume of water. Ice increases in density as it gets colder than 0°C. No matter how cold it gets, however, ice never reaches the density of liquid water. Being less dense than water, ice "freezes over" as a floating layer instead of "freezing under" like the solid forms of virtually all other liquids.

As we'll see in a moment, the implications of water's high heat capacity and the ability of ice to float are vital in maintaining Earth's moderate surface temperature. First, however, we look at the transition from point Ⓒ to point Ⓓ in Figure 6.4.

Figure 6.4 The relationship of density and temperature for pure water. Note that points Ⓒ and Ⓓ both represent 0°C (32°F) but show different densities, and thus different states of water. Ice floats because the density of ice is lower than the density of liquid water.

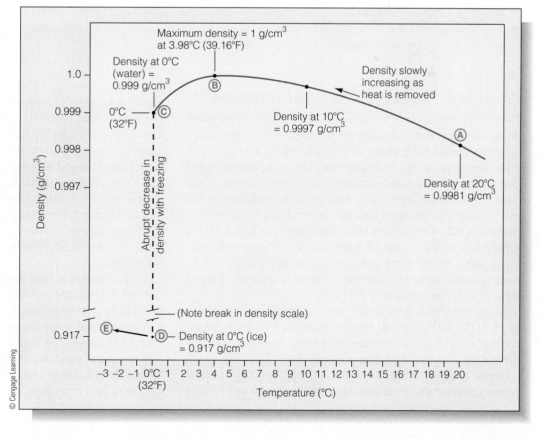

WATER AND OCEAN STRUCTURE • **169**

GAS

Fills closed container uniformly

Molecules in high-speed motion

Collisions and rebounds occur

Density very low

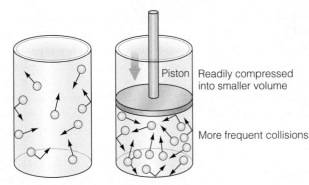

Piston | Readily compressed into smaller volume

More frequent collisions

© Cengage Learning

Figure 6.5 The three common states of matter—solid, liquid, and gas. On Earth, water can occur in all three states.

a A *gas* is a substance that can expand to fill any empty container. Atoms or molecules of gas are in high-speed motion and move in random directions.

LIQUID

Free upper surface

Flows freely to lower level

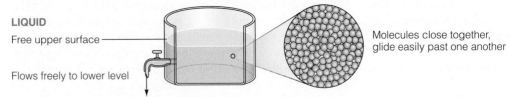

Molecules close together, glide easily past one another

b A *liquid* is a substance that flows freely in response to unbalanced forces but has a free upper surface in a container it does not fill. Atoms or molecules of a liquid move freely past one another as individuals or small groups. Liquids compress only slightly under pressure. Gases and liquids are classed as *fluids* because both substances flow easily.

SOLID
(crystalline)

Strong, rigid

Fractures when sudden, strong stress is applied

Density high

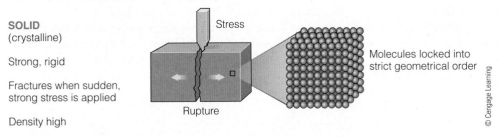

Stress

Rupture

Molecules locked into strict geometrical order

c A *solid* is a substance that resists changes of shape or volume. A solid can typically withstand stresses without yielding permanently. A solid usually breaks suddenly.

Water Becomes Less Dense When It Freezes

During the transition from liquid to solid state at the **freezing point**, the bond angle between the oxygen and hydrogen atoms in water widens from about 105° to slightly more than 109°. This change allows the hydrogen bonds in ice to form a crystal lattice (**Figure 6.6**). The space taken by 27 water molecules in the liquid state will be occupied by only 24 water molecules in the solid lattice. The resulting structure leaves a small gap between the atoms that wasn't there when the water was liquid. Water expands about 9% as the crystal forms. Ice is less dense than liquid water, so it floats. A cubic centimeter of ice at 0°C (32°F) has a mass of only 0.917 gram, but a cubic centimeter of liquid water at 0°C has a mass of 0.999 gram.

The transition from liquid water to ice crystal (point Ⓒ to point Ⓓ in Figure 6.4) requires continued removal of heat energy; the change in state does not occur instantly throughout the mass when the cooling water reaches 0°C (32°F). Again, consider water in a freezer. **Figure 6.7**, a plot of heat removal versus temperature, illustrates the water's progress to ice. As in

Figure 6.4, point Ⓐ represents 20°C (68°F) water just placed in the freezer. The removal of heat does not stop when the water reaches point Ⓒ, *but the decline in temperature stops*. Even though heat continues to be removed, the water will not get colder until all of it has changed state from liquid (water) to solid (ice). Heat may therefore be removed from water when it is changing state (that is, when it is freezing) without the water dropping in temperature. Indeed, the continued removal of heat is what makes the change in state possible. Heat is released as hydrogen bonds form to make ice, and that heat must be removed to allow more ice to form.

The removal of heat from point Ⓐ to point Ⓒ in Figures 6.4 and 6.7 produces a *measurable* lowering of temperature detectable by a thermometer. Removing just 1 calorie of heat from a gram of liquid water causes its temperature to drop 1°C. This detectable decrease in heat is called **sensible heat** loss. But the loss of heat as water freezes between points Ⓒ and Ⓓ is not measurable (that is, not sensible) by a thermometer. Removing a calorie of heat from freezing water at 0°C (32°F) won't change

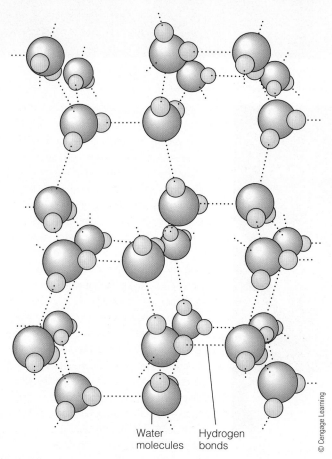

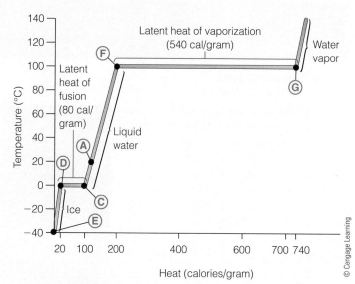

© Cengage Learning

Figure 6.6 The lattice structure of an ice crystal, showing its hexagonal arrangement at the molecular level. The space taken by 24 water molecules in the solid lattice could be occupied by 27 water molecules in liquid state, so water expands about 9% as the crystal forms. Because of the way molecules are arranged during freezing, ice is less dense than liquid water and floats.

💬 **THINKING BEYOND THE FIGURE**

Would ice still float if water molecules in ice occupied the same space as water molecules in liquid water?

its temperature at all; 80 calories of heat energy must be removed per gram of pure water at 0°C (32°F) to form ice. This heat is called the **latent heat of fusion** (*latere*, "to be hidden"). The straight line between points Ⓒ and Ⓓ in Figure 6.7 represents water's latent heat of fusion.

No more ice crystals can form when all the water in the freezer has turned to ice. If the removal of heat continues, the ice will get colder and will soon reach the temperature inside the freezer, point Ⓔ in Figures 6.4 and 6.7.

Latent heat of fusion is also a factor during thawing. When ice melts, it *absorbs* large quantities of heat (the same 80 calories per gram), but it does not change in temperature until all the ice has turned to liquid. This explains why ice is so effective in cooling drinks.

Figure 6.7 A graph of temperature versus heat as water freezes, melts, and vaporizes. The horizontal line between points Ⓒ and Ⓓ represents the latent heat of fusion, when heat is being added or removed but the temperature is not changing. The horizontal line between points Ⓖ and Ⓕ represents the latent heat of vaporization, when heat is being added or removed but the temperature is not changing. (Note that points Ⓐ–Ⓔ on this graph are the same as those in Figure 6.4.)

💬 **THINKING BEYOND THE FIGURE**

"Latent" means hidden. Do you see why the horizontal lines between points Ⓒ and Ⓓ and points Ⓖ and Ⓕ are called "latent heats"?

Water Removes Heat from Surfaces As It Evaporates

Let's reverse the process now and warm the ice. Imagine the water resting at −40°C, point Ⓔ at the lower left of Figure 6.7. Add heat, and the ice warms toward point Ⓓ. It begins to melt. The horizontal line between point Ⓓ and point Ⓒ represents the latent heat of fusion: Heat is absorbed but temperature does not change as the ice melts. All liquid now at point Ⓒ, the water warms past our original point Ⓐ and arrives at point Ⓕ. It begins to boil—it *vaporizes*.

When water vaporizes (or evaporates), individual water molecules diffuse into the air (**Figure 6.8**). Since each water molecule is hydrogen-bonded to adjacent molecules, heat energy is required to break those bonds and allow the molecule to fly away from the surface. Evaporation cools a moist surface because departing molecules of water vapor carry this energy away with them. (This is how perspiring cools us when we're hot. The heat energy required to evaporate water from our skin is taken away from our bodies, cooling us.)

Hydrogen bonds are quite strong, and the amount of energy required to break them—known as the **latent heat of vaporization**—is very high. The long horizontal line between points Ⓕ and Ⓖ in Figure 6.7 represents the latent heat of vaporization. As before, the term *latent* applies to heat input that

How Water Freezes

Every semester our students ask the same question: "Does hot water freeze faster than cold?" Every semester we patiently answer "No." Why should it? If two equal amounts of water are placed in a freezer and one is warmer than the other, the cooler one will always freeze first. After all, the hot water must lose heat to come down to the starting temperature of the cool water, and by that time the water that started out cool might be near the freezing point.

Right?

Well, yes and no.

Yes, because the logic above makes complete thermodynamic sense!

No, because other factors can intervene!

Curious? So was G. S. Kell, who wrote a paper on the subject for the *American Journal of Physics* (May 1969). He lived in Canada, where many people apparently believe that, if left outside on a cold night, a bucket of hot water will freeze more quickly than a bucket of cold water. Kell tested this belief using covered buckets. The freezing went exactly as predicted above: The warmer water must first fall to the starting temperature of the cold water, and then its cooling curve (a graph of the fall in temperature with time) will follow the cooling curve already taken by the cooler bucket. The bucket containing cooler water always froze first. *Good!*

So why do many Canadians (not to mention American students) persist in their belief? Might

Figure A Lab students compare the freezing points of fresh water and seawater.

there be a grain of truth here? Kell tried the experiment using buckets without lids. Now the water can evaporate, and hot water evaporates much more rapidly than cold water, especially in cold, dry winter air. He demonstrated that a bucket of water cooling from the boiling point

to the freezing point would lose about 16% of its mass. Now there's less water to freeze. A smaller volume of water initially at 0°C (32°F) will freeze faster than a larger volume *if* the same surface area is exposed to the cold.

And there's more. As you know, the latent heat of vaporization for water is very high (540 calories per gram). Water molecules escaping the surface of the hot water take away a very large amount of heat. The hot water's cooling curve plummets. All of these conditions together could make the hot water seem to freeze faster than the cold.

Even so, if the buckets are made of metal, much heat would be lost through the container walls to the cold air. It is unlikely that hot water in a metal bucket would freeze first. But what if the buckets are made of wood, an excellent insulator? Then most energy loss would occur through evaporation, and the initially hot water might actually freeze faster. (But this is not a variable that Kell tested for his paper.)

So should you make ice cubes with warm water? No, never. First you pay the electric (or gas) company to warm the water. Then you pay the electric company to run the refrigerator to bring the water down to tap temperature. Now you freeze the water. If you started with warm water, you'd have less ice. Would it really freeze faster? Re-read the discussion of the scientific method in Chapter 1, and then check it out for yourself!

does not cause a temperature change but does produce a change of state—in this case from liquid to gas. Even though more heat is applied, the water cannot get warmer until all of it has vaporized. At 540 calories per gram, water has the highest latent heat of vaporization of any known substance.

About 1 meter (3.3 feet) of water evaporates each year from the surface of the ocean, a volume of water equivalent to 334,000 cubic kilometers (80,000 cubic miles). The great quantities of solar energy that cause this evaporation are carried from the ocean by the escaping water vapor. When a gram of water vapor condenses back into liquid water, the same 540 calories is again available to do work. As we shall see, winds, storms, ocean currents, and wind waves are all powered by that heat.

Why the big difference between water's latent heat of *fusion* (80 calories per gram) and its latent heat of *vaporization* (540 cal-

ories per gram)? Only a small percentage of hydrogen bonds are broken when ice melts, but *all* of them must be broken during evaporation. Breaking these bonds requires additional energy in proportion to their number.

A summary of water's unusual properties is provided in **Figure 6.9** and **Table 6.2**.

Seawater and Pure Water Have Slightly Different Thermal Properties

Seawater is about 96.5% pure water and 3.5% dissolved solids and gases. The solids dissolved in seawater change its thermal characteristics, lowering its latent heat by about 4%. Only 0.96 calorie of heat energy is needed to raise the temperature of 1 gram of seawater by 1°C.

Figure 6.8 Water vapor is invisible, but as it evaporates from the sea surface and rises into cool air, it can condense into tiny droplets that form clouds and fog. The low-lying fog bank often seen obscuring the Golden Gate Bridge at the entrance to San Francisco Bay forms when warm, moist air passes over the cold waters of the California Current. The air is cooled to the saturation point, and some of its water vapor condenses into fog droplets.

© Photri Inc./age fotostock Spain S.L./Corbis

The dissolved solids also interfere with the formation of the ice lattice, acting as "antifreeze" to lower the freezing point. The saltier the water, the lower the freezing point. The temperature of maximum density moves toward the freezing point as salinity (salt content) increases (see **Figure 6.10**), finally coinciding with the freezing point at a salinity of 24.7‰ ($-1.33°C$, 29.61°F).[2] Seawater at 35‰, typical ocean salinity, freezes at $-1.91°C$ (28.6°F). Seawater's density simply increases smoothly with decreasing temperature until it freezes. The crystals that form are pure water ice, with the seawater salts excluded. The leftover cold, salty water is very dense. Some of this water may be trapped among the ice crystals, but most is free to fall toward the seabed, pulled rapidly downward by its great density. As we'll see later, this is an important cause of deep ocean currents.

Seawater evaporates more slowly than freshwater under identical circumstances because the dissolved salts tend to attract and hold water molecules. The **latent heat of evaporation**, however, is essentially the same for both freshwater and seawater. Salts are left behind as seawater evaporates. The remaining cool, salty water is also very dense, and it, too, sinks toward the ocean floor.

[2]The symbol % (percent) represents parts-per-*hundred*. Oceanographers use the symbol ‰ to represent parts-per-*thousand*. So 2.47%; 24.7‰, and saying that water has a salinity of 24.7‰, for example, means that there are 24.7 grams of salts dissolved in 1,000 grams of water.

Figure 6.9 We must add 80 calories of heat energy to change a gram of ice to liquid water. After the ice is melted, about 1 calorie of heat is needed to raise each gram of water by 1°C. But 540 calories must be added to each gram of water to vaporize it—to boil it away. The process is reversed for condensation and freezing.

 **THINKING BEYOND THE FIGURE**

How does the condensation of water vapor on the outside of a soda can warm the contents inside?

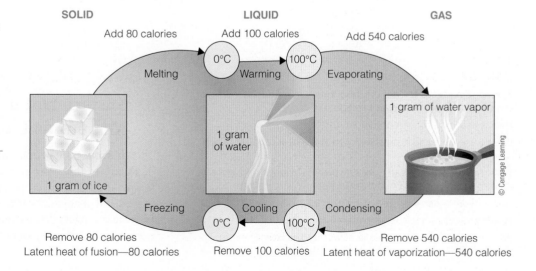

© Cengage Learning

Table 6.2 Properties of Water

Property	Remarks	Importance to the Ocean Environment
Physical state	Only substance occurring naturally in all three phases as solid, liquid, and gas (vapor) on Earth's surface	Transfer of heat between ocean and atmosphere by phase change
Dissolving ability	Dissolves more substances in greater quantities than any other common liquid	Important in chemical, physical, and biological processes
Density: mass per unit volume	Density determined by (1) temperature, (2) salinity, and (3) pressure, in that order of importance. The temperature of maximum density for pure water is 4°C. For seawater, the freezing point decreases with increasing salinity.	Controls oceanic vertical circulation, aids in heat distribution, and allows seasonal stratification
Surface tension	Highest of all common liquids	Controls drop formation in rain and clouds; important in cell physiology
Conduction of heat	Highest of all common liquids	Important on the small scale, especially on cellular level
Heat capacity: quantity of heat required to raise the temperature of 1 g of a substance 1°C	Highest of all common solids and liquids	Prevents extreme range in Earth's temperatures; great heat moderator
Latent heat of fusion: quantity of heat gained or lost per unit mass by a substance changing from a solid to a liquid or a liquid to a solid phase without an accompanying rise in temperature	Highest of all common liquids and most solids (80 cal/g)	Thermostatic heat-regulating effect due to the release of heat on freezing and absorption on melting
Latent heat of vaporization: quantity of heat gained or lost per unit mass by a substance changing from a liquid to a gas or a gas to a liquid phase without an increase in temperature	Highest of all common substances (540 cal/g)	Immense importance: a major factor in the transfer of heat in and between ocean and atmosphere, driving weather and climate
Refractive index	Increases with increasing salinity and decreases with increasing temperature	Objects appear closer than in air
Transparency	Relatively great for visible light; absorption high for infrared and ultraviolet	Important in photosynthesis
Sound transmission	Good compared with other fluids	Allows sonar and precision depth recorders to rapidly determine water depth, detect subsurface features and animals; sounds can be heard great distances underwater
Compressibility	Only slight	Density changes only slightly with pressure/depth
Boiling and melting points	Unusually high	Allows water to exist as a liquid on most of Earth

Sources: Sverdrup et al., 1942; Ingmanson and Wallace, 1995.

CONCEPT CHECK

Before going on to the next section, check your understanding of some of the important ideas presented so far:

How is heat different from temperature?

What is meant by heat capacity? Why is the heat capacity of water unique?

What factors affect the density of water? Why does cold air or water tend to sink? How does salinity affect density?

How is water's density affected by freezing? Why does ice float?

What is the difference between sensible and nonsensible heat?

What's the latent heat of fusion of water? The latent heat of vaporization? Why do we use the term "latent"?

How do the properties of seawater differ from those of fresh water?

6.5 Surface Water Moderates Global Temperature

The **thermostatic properties** (*therme*; heat, *stasis*; standing still) of water are those properties that act to moderate changes in temperature. Water temperature rises as the sun's energy is absorbed and changed to heat, but as we've seen, water has a very high heat capacity, so its temperature will not rise very much even if a large quantity of heat is added. This tendency of a substance to resist a change in temperature with the gain or loss of heat energy is called **thermal inertia**.

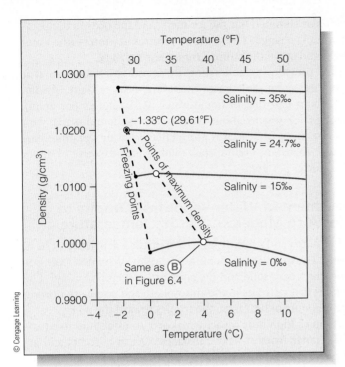

Figure 6.10 The dependence of freezing temperature and temperature of maximum density upon salinity (salt content). As we saw in Figure 6.4, pure water is densest at 3.98°C (39.2°F) (point B), and its freezing point is 0°C (32°F). Seawater with 15‰ salinity is densest at 0.73°C (33.31°F), and its freezing point is −0.80°C (30.56°F). The temperature of maximum density and the freezing point coincide at −1.33°C (29.61°F) in seawater with a salinity of 24.7‰. At salinities greater than 24.7‰ the density of water always decreases as temperature increases. Note that the symbol ‰ represents parts-per-thousand, so 2.47%; 24.7‰.

Remember the hot sand and cool water on a hot summer afternoon? Think about Earth as a whole. The highest temperatures on land, in the north African desert, exceed 50°C (122°F); the lowest, on the Antarctic continent, drop below −90°C (−129°F). That's a difference of 140°C (or 250°F)! On the ocean surface, however, the range is from −2°C (29°F) where sea ice is forming to about 32°C (90°F) in the tropics—a difference of only 34°C (61°F). Consisting of water, the ocean rises very little in temperature as it absorbs heat. The ocean's thermal inertia is much greater than the land's.

A practical example of thermal inertia can be seen in **Figure 6.11**. San Francisco, California, and Norfolk, Virginia, are

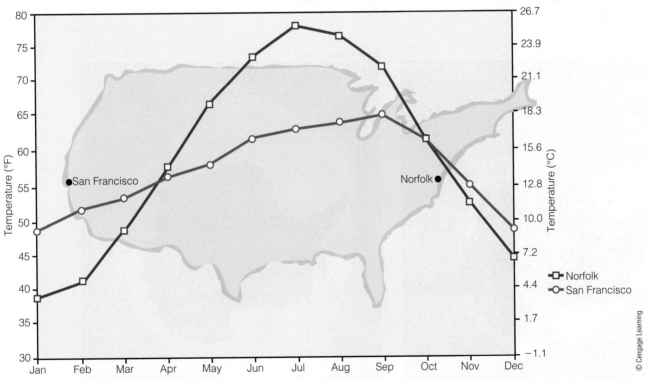

Figure 6.11 San Francisco, California, and Norfolk, Virginia, are on the same line of latitude, yet San Francisco is warmer in the winter and cooler in the summer than Norfolk. Part of the reason is that wind tends to flow from west to east at this latitude. Thus, air in San Francisco has moved over the ocean while air in Norfolk has approached over land. Water doesn't warm as much as land in the summer, nor cool as much in winter—a demonstration of thermal inertia.

💬 **THINKING BEYOND THE FIGURE**

Look up the places in the continental United States where the hottest and coldest temperature records have been set. Do you see any relationship with thermal inertia?

on the same line of latitude—each is the same distance from the equator. Wind tends to flow from west to east at this latitude (for reasons discussed in Chapter 8). Compared to Norfolk, San Francisco is warmer in the winter and cooler in the summer, in part because air in San Francisco has moved over the ocean, while air in Norfolk has approached over land.[3]

Annual Freezing and Thawing of Ice Moderates Earth's Temperature

As you know, removing a calorie of heat from freezing pure water at 0°C (32°F) won't change its temperature at all—80 calories of heat energy must be removed per gram of liquid water to form ice. More than 18,000 cubic kilometers (4,300 cubic miles) of polar ice, covering as many as 20 million square kilometers (7.7 million square miles) of surface, thaws and refreezes in the Southern Hemisphere each year—an area of ocean larger than South America (**Figure 6.12**)! The annual change in sea-ice cover is less in the Arctic, averaging about 5 million square kilometers (2 million square miles). Incoming solar heat melts ice in the

[3]Mark Twain was said to have remarked that the coldest winter he spent in his life was a summer in San Francisco!

local polar summer, but the ice melts and the ocean's temperature doesn't change. The situation reverses in winter—the water freezes and again the temperature doesn't change.

Ice provides a moderating thermostatic effect even if it doesn't get warm enough to melt. The heat capacity of solid ice is about half that of liquid water (0.51 calories per gram). Although ice warms about twice as fast as liquid water with the same input of heat, it is more effective at moderating temperature than, say, granite rock (with a heat capacity of about 0.20 calories per gram).

Movement of Water Vapor from Tropics to Poles Also Moderates Earth's Temperature

Earth's north and south poles have a marked deficiency of heat, and the equator has a pronounced surplus. Why don't the polar oceans freeze solid and the equatorial ocean boil away? The reason is that currents in the atmosphere and ocean are moving huge amounts of heat from the tropics toward the poles.

Water's high heat capacity makes it an ideal fluid to equalize the polar-tropical heat imbalance. Ocean currents and atmospheric weather result from the response of water and air to unequal solar heating. Although weather and currents are discussed in more detail in the next two chapters, here's a brief preview.

March 2007

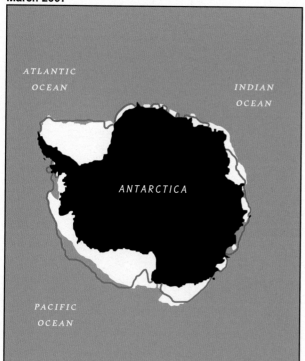

September 2007

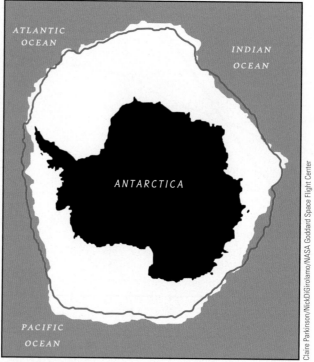

Claire Parkinson/NickDiGirolamo/NASA Goddard Space Flight Center

Figure 6.12 About 20 million square kilometers (7.7 million square miles) of ocean surface thaws and refreezes in the Southern Hemisphere each year—an area of ocean larger than South America! The autumn cooling of the atmosphere is delayed because heat energy is released as masses of water turn to ice. Heat is absorbed during ice melt in the spring. Seasonal extremes are moderated by the absorption and release of heat energy as ice thaws and refreezes. (Remember, the seasons are reversed in the Southern Hemisphere.) The scale shows the percent of ocean surface completely covered by ice.

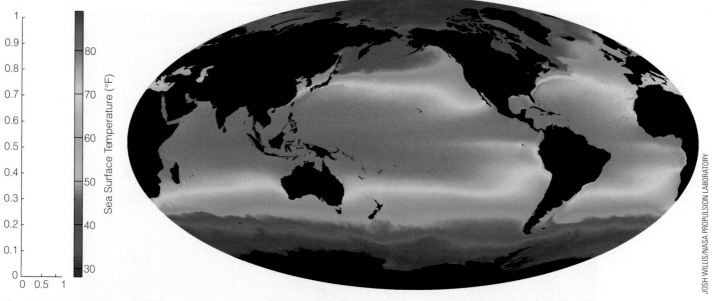

Figure 6.13 Typical ocean surface temperature. This image was derived from satellite data and is updated daily (see http://ourocean.jpl.nasa.gov). Blue, purple, red, yellow, and white represent progressively warmer water. Warm water is shown streaming northward in the Gulf Stream along the east coast of the United States. In a week's time this water will have a moderating effect on the European winter climate.

Ocean currents carry heat from the tropics (where incoming energy exceeds outgoing) to the polar regions (where outgoing energy exceeds incoming). The amount of heat transferred in this way is astonishing **(Figure 6.13)**. For example, "outbound" water in the warm Gulf Stream (a large northward-flowing ocean current just offshore of the eastern United States) is about 10°C (18°F) warmer than "inbound" water coming from the central eastern Atlantic to replace it, meaning that about 10 million calories are transported per cubic meter. Since the flow rate of the Gulf Stream is about 55 million cubic meters per second, some 550 trillion calories are being transported northward in the western North Atlantic each *second*! Nearly half of these calories reach the high latitudes above 40°N. This warmth has a dramatic moderating influence on the winter climate of northwestern Europe.

As impressive as the figures are for ocean currents, the amount of heat transported by water vapor in the atmosphere is even greater. About half of the solar energy entering water results in evaporation. The solar energy required for this evaporation is later surrendered during condensation and cloud formation (and rain), but usually at a distance from where the initial evaporation occurred. So, the ocean surface near Cuba may be cooled by evaporation today, the water vapor may then be moved north by winds, and eastern Canada may be warmed by condensation of the same water in a rainstorm later in the week.

Both atmosphere and ocean transfer heat by movement, but water's exceptionally high latent heat of vaporization means that water vapor transfers much more heat (per unit of mass) than liquid water. Masses of moving air account for about two thirds of the poleward transfer of heat; ocean currents move the other third.

Global Warming May Be Influencing Oceanic Surface Temperature and Salinity

Ocean surface temperature and salinity are changing relatively rapidly, probably in response to accelerating greenhouse warming. The year 2010 was the warmest year yet measured, the latest in a series of record-setting years extending back to the early 1980s. The ocean warms at its surface **(Figure 6.14a)**. Researchers at the Bedford Institute of Oceanography in Nova Scotia, Canada, compared conditions in the Atlantic for two 14-year periods centered on 1962 and 1992 **(Figure 6.14b)**. Over that 30-year interval, the tropical ocean shallower than 1,000 meters (3,300 feet) had become warmer and saltier, while water in the far north and south had become fresher. The world's heat-driven cycle of evaporation and precipitation seems to have become between 5% and 10% faster during that time, increasing both the rate of water evaporation in the tropics and the amount of precipitation in the polar regions. The implications of these changes will be explored in Chapter 18.

Ocean Surface Conditions Depend on Latitude, Temperature, and Salinity

Table 6.3 summarizes a few important properties of seawater at polar, temperate, and tropical latitudes. The temperature data show the effects of solar radiation at various latitudes, along with the ratio of evaporation to precipitation. Notice that the temperature of ocean surface water is more variable through the year in the temperate zone than in either the polar or tropical areas, but that temperate zone salinity stays relatively constant. Notice also that evaporation generally exceeds precipitation in the tropics, but that precipitation dominates in temperate and

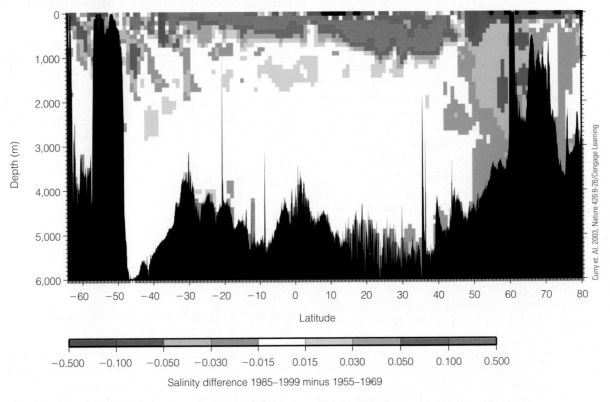

Atlantic Ocean Temperature Transect

Temperature readings in °C

National Geographic Maps

a Most seawater lies in cold darkness. In this image, a small and thin warm layer, concentrated near the equator, brightens an otherwise bleak view. As Earth warms, this layer will extend poleward. The irregular black shapes at the bottom of the graphs represent the ocean floor at the latitude shown.

Curry et. Al. 2003, Nature 426:8-26/Cengage Learning

−0.500 −0.100 −0.050 −0.030 −0.015 0.015 0.030 0.050 0.100 0.500

Salinity difference 1985–1999 minus 1955–1969

b The difference in average ocean salinity measured from 1985 to 1999 *minus* the years 1955 to 1969. Over the past 40 years the tropical ocean shallower than 1,000 meters (3,300 feet) has become warmer and saltier, while water in the far north and south has become fresher. The world's heat-driven cycle of evaporation and precipitation seems to have become between 5% and 10% faster during that time, increasing both the rate of water evaporation in the tropics and the amount of precipitation in the polar regions.

Figure 6.14 Temperature and salinity across the Atlantic ocean basin at about 30° west longitude.

Table 6.3 Some Characteristics of the World Ocean Surface, by Latitude

Characteristic	Tropical Oceanic Waters	Temperate Oceanic Waters	Polar Oceanic Waters
Winter temperature	20–25°C (68–77°F)	5–20°C (41–68°F)	About –2°C (28°F)
Annual variation of temperature	Less than 5°C (9°F)	About 10°C (18°F)	Less than 5°C (9°F)
Average salinity	35‰–37‰	About 35‰	28‰–32‰
Annual variation of air temperature	Less than 5°C (9°F)	About 10°C (18°F)	Up to 40°C (72°F)
Precipitation–evaporation balance	P exceeds E	E exceeds P	P exceeds E

Source: H. Charnock, 1971. Table created by Tom Garrison.

polar zones. The relationships between latitude, temperature, and salinity are shown graphically in **Figure 6.15**.

Ocean surface temperature is highest in the Pacific, north and east of Borneo, where westward-flowing currents funnel seawater warmed by long exposure to tropical sunlight. The highest open ocean surface temperatures range to 32°C (90°F). Surface salinity is especially high in the warm central North and South Atlantic, where evaporation rates are high and surface water is isolated by the currents flowing around the ocean's pe-

riphery. Some of this information is summarized graphically in Figure 6.15 and **Figure 6.16** showing typical world ocean surface temperatures and salinities.

Remember, though, that variations in temperature, salinity, and density between high and low latitudes are usually confined to the uppermost 2,000 meters (6,500 feet) of water. Below this depth conditions are similar at all latitudes. The deep ocean is nearly as cold, salty, and dense in the equatorial Pacific as it is off the Siberian coast.

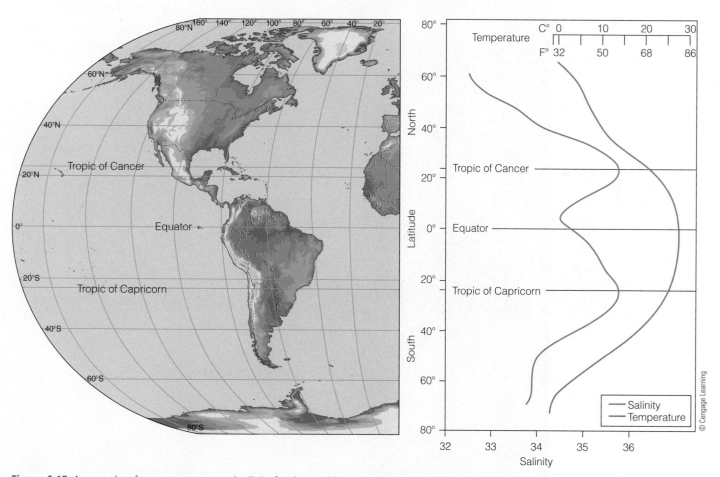

Figure 6.15 Averaged surface temperature and salinity for the world ocean. As you would expect, temperatures are lowest in the polar regions and highest near the equator. Heavy rainfall in the equatorial regions "freshens" the ocean near the equator, while hot and dry conditions near the tropic lines (Tropic of Capricorn and Tropic of Cancer) result in higher surface salinity in those areas.

Sea Surface Average Salinities

This image was assembled from data collected by NASA's *Aquarius* satellite from December 2011 through December 2012. Red colors represent the areas of high salinity, while blue shades represent areas of low salinity.

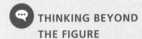 **THINKING BEYOND THE FIGURE**

Why do you think salinity is so high in the northeastern Atlantic?

NASA/GSFC/JPL–Caltech

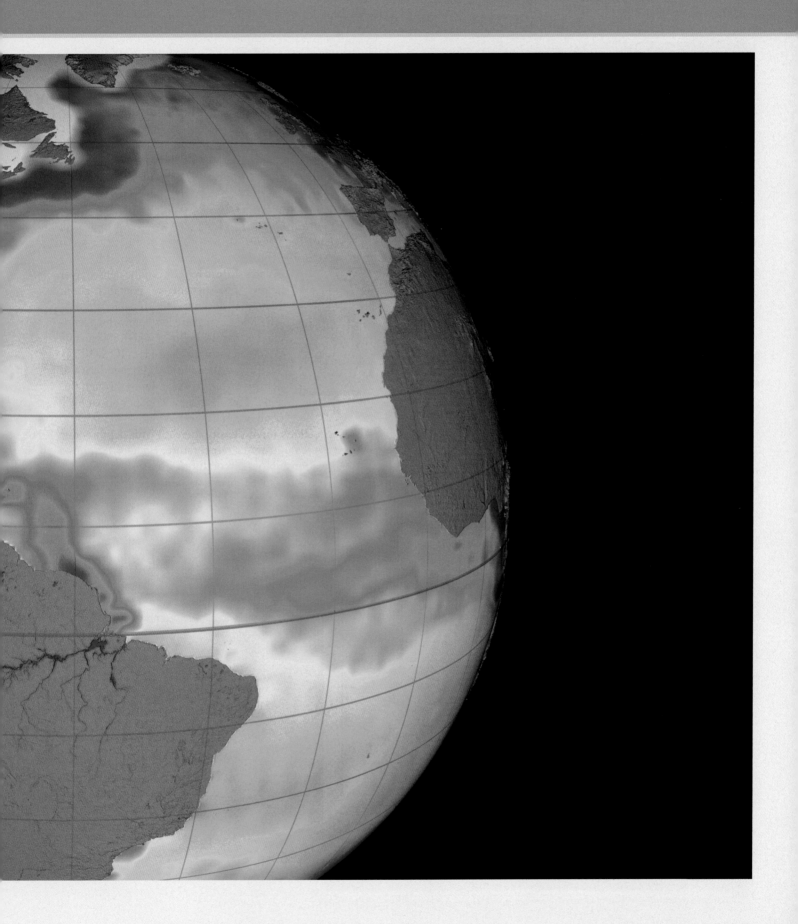

6.6 The Ocean Is Stratified by Density

As we've said, the density of water is mainly a function of its temperature and salinity. We will learn more about the specifics of salinity in Chapter 7, but for now you need to know that a liter of seawater weighs between 2% and 3% more than a liter of pure water because of the solids (often called salts) dissolved in seawater. The density of seawater is thus between 1.020 and 1.030 g/cm^3 compared with 1.000 g/cm^3 for pure water at the same temperature. *Cold, salty water is more dense than warm, less salty water.* Seawater's density increases with increasing salinity, increasing pressure, and decreasing temperature. **Figure 6.17** shows the relationship among temperature, salinity, and density. Notice that two samples of water can have the *same* density at *different combinations* of temperature and salinity.[4]

> **S**eawater's density increases with increasing salinity, increasing pressure, and decreasing temperature.

---

[4]This fact has fascinating implications for deep-ocean circulation, as Figure 9.30 will show.

The Ocean Is Stratified into Three Density Zones by Temperature and Salinity

Much of the ocean is divided into three density zones: the surface zone, the pycnocline, and the deep zone. The **surface zone**, or **mixed layer**, is the upper layer of ocean **(Figure 6.18a)**. Temperature and salinity are relatively constant with depth in the surface zone because of the action of waves and currents. The surface zone consists of water in contact with the atmosphere and exposed to sunlight; it contains the ocean's least dense water and accounts for only about 2% of total ocean volume. The surface zone (or mixed layer) typically extends to a depth of about 150 meters (500 feet), but depending on local conditions, it may reach a depth of 1,000 meters (3,300 feet) or be absent entirely.

The **pycnocline** (*pyknos;* strong, *clinare;* slope, to lean), is a zone in which density increases with increasing depth. This zone isolates surface water from the denser layer below. The pycnocline contains about 18% of all ocean water.

The **deep zone** lies below the pycnocline at depths below about 1,000 meters (3,300 feet) in mid-latitudes (40°S–40°N). There is little additional change in water density with increasing depth through this zone. This deep zone contains about 80% of all ocean water.

The pycnocline's rapid density increase with depth is due mainly to a decrease in water temperature. **Figure 6.18b** shows the general relationship of temperature with depth in the open sea. The surface zone is well mixed, with little decrease in temperature with depth. In the next layer, temperature drops rapidly

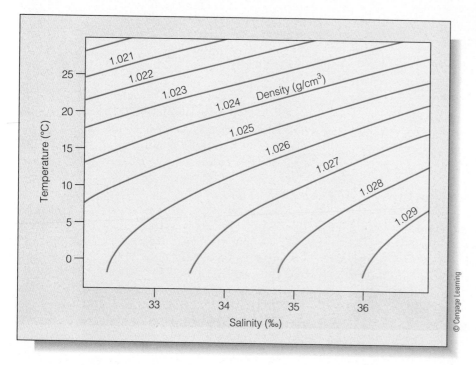

Figure 6.17 The complex relationship among the temperature, salinity, and density of seawater. Note that two samples of water can have the *same* density at *different* combinations of temperature and salinity.

© Cengage Learning

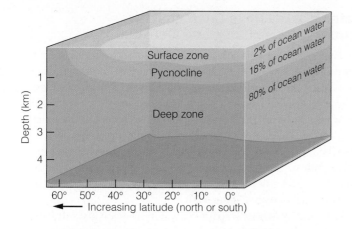

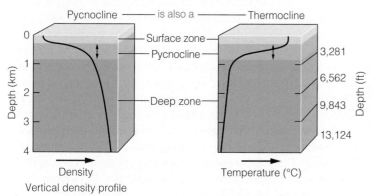

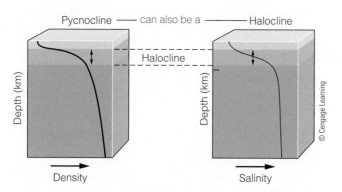

a In most of the ocean, a surface zone (or mixed layer) of relatively warm, low-density water overlies a layer called the *pycnocline*. Density increases rapidly with depth in the pycnocline. Below the pycnocline lies the deep zone of cold, dense water—about 80% of total ocean volume.

b The rapid density increase in the pycnocline is mainly due to a decrease in temperature with depth in this area—the *thermocline*.

c In some regions, especially in shallow water near rivers, a pycnocline may develop in which the density increase with depth is due to vertical variations in salinity. In this case, the pycnocline is a *halocline*.

Figure 6.18 Density stratification in the ocean.

with depth. Beneath it lies the deep zone of cold, stable water. The middle layer, the zone in which temperature changes rapidly with depth, is called the **thermocline** (*therm;* heat), and falling temperature is the major contributor to the formation of the pycnocline.

Thermoclines are not identical in form in all areas or latitudes. Surface temperature is proportional to available sunlight. More solar energy is available in the tropics than in the polar regions, so the water there is warmer. The ocean's sunlit upper layer is thicker in the tropics, both because the solar angle there is more nearly vertical and because water in the open tropical ocean contains fewer suspended particles (and is therefore clearer than water in open temperate or polar regions). Because the ocean is heated to a greater depth, the tropical thermocline is deeper than thermoclines at higher latitudes. It is also much

more pronounced: The transition to the colder, denser water below is more abrupt in the tropics than at high latitudes.

Polar waters, which receive relatively little solar warmth, are not stratified by temperature and generally lack a thermocline because surface water in the polar regions is nearly as cold as water at great depths.

Figure 6.19 contrasts polar, tropical, and temperate thermal profiles, showing that the thermocline is primarily a mid- and low-latitude phenomenon. Thermocline depth and intensity also vary with season, local conditions (storms, for example), currents, and many other factors.

Below the thermocline, water is very cold, ranging from −1°C to 3°C (30.5°F–37.5°F). Because this deep and cold layer contains the bulk of ocean water, the average temperature of the world ocean is a chilly 3.9°C (39°F). Low salinity can also con-

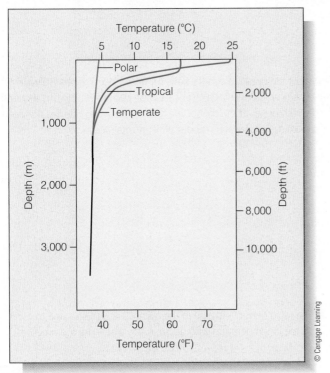

Figure 6.19 Typical temperature profiles at polar, tropical, and middle (temperate) latitudes. Note that polar waters lack a strong thermocline.

💬 **THINKING BEYOND THE FIGURE**

Knowing what you do about the factors controlling the density of ocean water, can you suggest in which zone it would be easiest to lift nutrient-rich bottom water toward the ocean's sunlit surface? Why would this be important?

tribute to the pycnocline, especially in cool regions where precipitation is high or along coasts where freshwater runoff mixes with surface water. Wherever precipitation exceeds evaporation, salinity will be low. These differences produce the **halocline** (*halos*; salt), a zone of rapid salinity increase with depth **(Figure 6.18c)**. The halocline often coincides with the thermocline, and the combination produces a pronounced pycnocline.

Water Masses Have Characteristic Temperature, Salinity, and Density

The layers we have been discussing are distinct water masses. A **water mass** is a body of water with characteristic temperature and salinity and therefore density. Curiously, even the deepest of these layers originates at the ocean's surface. Very cold and salty water produced during the formation of sea ice at the polar ocean surface is denser than the surrounding water and sinks to the seabed. In a few marginal basins, the most notable being the Mediterranean Sea, evaporation can also produce salty, dense water that sinks toward the bottom until it reaches a layer of water of equal density **(Figure 6.20)**.

Layering by density traps dense water masses at great depths, where they are not exposed to daily heating and cooling, to surface circulation driven by winds and storms, or to light. The pycnocline effectively isolates 80% of the world ocean's water from the 20% involved in surface circulation. Dense water masses form near polar continental shelves (as cold water freezes and excludes salt) or in enclosed areas such as the Mediterranean Sea (where evaporation exceeds precipitation and river input, raising salinity). These heavy water masses sink, sometimes overlapping one another and often retaining their identity for long periods. Separate water masses below the pycnocline tend not to merge because little energy is available for mixing in these quiet depths. (Density differences in water masses power the deep ocean currents, as we will see in Chapter 9.)

Density Stratification Usually Prevents Vertical Water Movement

The vertical movement of large volumes of water from the surface to great depths (and vice versa) is possible only where surface water density is similar to deep-water density. The great difference in temperature, and therefore density, between surface water and deep water in the tropics makes the water column very stable and prevents an exchange of surface and deep water. This stability is maintained despite the fact that the surface of the tropical ocean is in constant horizontal motion, churned by tropical cyclones and stirred by currents.

Figure 6.20 Sound waves change in subtle ways as they pass through ocean layers of different temperatures and salinities. Using sensitive new acoustical detectors, researchers have recently been able to visualize the swirls and strata of water in fine detail.

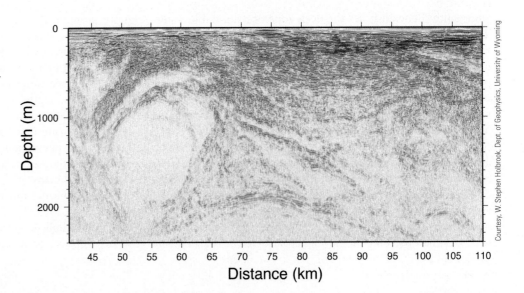

Vertical movement of water in the northern polar ocean is also limited. There, however, the stratification is caused largely by a salinity difference between surface water and water at great depths. The surface of the Arctic Ocean receives a large volume of freshwater runoff from Siberian and Canadian rivers. Continental masses block the formation of large currents, and the landlocked northern ocean communicates sluggishly with other ocean areas; so the surface water tends not to mix with deeper water or to flow to lower latitudes.

In contrast, the southern polar ocean is only weakly stratified. The cold temperature of southern ocean surface water closely matches that of deep water, so no thermocline divides surface water from deep water (see again Figure 6.19). The absence of confining continental margins and mixing at the boundaries of the Antarctic Circumpolar Current minimize salinity differences. Turbulence and weak stratification encourage a huge volume of deep-water upwelling, which contributes to high surface nutrient levels and high biological productivity.

CONCEPT CHECK

Before going on to the next section, check your understanding of some of the important ideas presented so far:

How is the ocean stratified by density? What names are given to the ocean's density zones?

What, generally, are the water characteristics of the *surface* zone? Do these conditions differ significantly between the polar regions and the tropics?

What, generally, are the water characteristics of the *deep* zone? Do these conditions differ significantly between the polar regions and the tropics?

How is the pycnocline related to the thermocline and halocline?

How is a water mass defined?

How does the ocean's density stratification limit the vertical movement of seawater?

6.7 Refraction Can Bend the Paths of Light and Sound through Water

The ring of light sometimes seen around the moon and the safe concealment of a submarine may not seem related, but both events depend on **refraction**, the bending of waves. Light and sound are both wave phenomena. When a light wave or a sound wave leaves a medium of one density—such as air—and enters a medium of a different density—such as water—at an angle other than 90°, it is bent from its original path. The reason for this bending is that light or sound waves travel at different speeds in the different media.

The situation is analogous to a line of people marching along a desert highway with their arms linked over each other's shoulders. The marchers can walk faster if they stay on the pavement than if they walk in the sand next to the highway. Their speed on the pavement, then, is greater than their speed in the sand. As long as they stay on the pavement, they won't change direction. But if their marching angle gradually takes them off the edge (into a medium in which their speed is *lower*), the people who reach the sand first will suddenly slow

down, and the line will pivot quickly off the highway. They have been *refracted*. Their progress is depicted in **Figure 6.21a**. Note that the transition from one medium to another must occur at an angle other than 90° for refraction to occur; our marchers will not change direction if they march straight off the asphalt into the sand. They will still slow down, however (see **Figure 6.21b**).

The refraction of light waves by water happens in much the same way. The speed of light in water is only about three-quarters its speed in air, so water effectively refracts light. (Glass bends light even more.) The degree to which light is refracted from one medium to another is expressed as a ratio called the **refractive index**. The higher the refractive index, the greater the bending of waves between media. The refractive index of water increases with increasing salinity. A beautiful example of the difference between the refractive index of freshwater and seawater is seen in **Figure 6.22**.

Examples of the refraction of light by water are all around you. A pencil sticking out of a glass of water looks bent because of refraction; the submerged steps of a swimming pool ladder look closer than they are because of refraction; and refraction magnifies objects and causes fishermen to exaggerate the size of the fish that got away.

CONCEPT CHECK

Before going on to the next section, check your understanding of some of the important ideas presented so far:

What is refraction?

What is refractive index?

6.8 Light Does Not Travel Far through the Ocean

Light is a form of electromagnetic radiation, or radiant energy, that travels as waves through space, air, and water. The visible spectrum—the wavelengths of light that human eyes can detect—is only a small part of the electromagnetic spectrum, which also includes radio waves, infrared, ultraviolet, and X-rays, for example. The wavelength of light determines its color: Shorter wavelengths are bluer, longer wavelengths are redder. Except for very long radio waves, water rapidly absorbs nearly all electromagnetic radiation. Only blue and green wavelengths pass through water in any appreciable quantity or distance.

The Photic Zone Is the Sunlit Surface of the Ocean

Sunlight has a difficult time reaching and penetrating the ocean; clouds and the sea surface reflect light while atmospheric gases and particles scatter and absorb it. Once past the sea surface, light is rapidly attenuated by scattering and absorption. **Scattering** occurs as light is bounced between air or water molecules, dust particles, water droplets, or other objects before being absorbed. The greater density of water (along with the greater number of suspended and dissolved

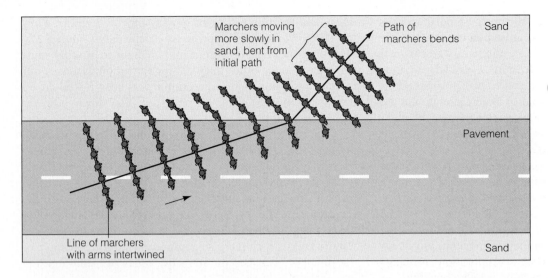

a If the marchers head off the pavement at an angle other than 90°, their path will bend (refract) as they hit the sand because some will be walking more slowly than others.

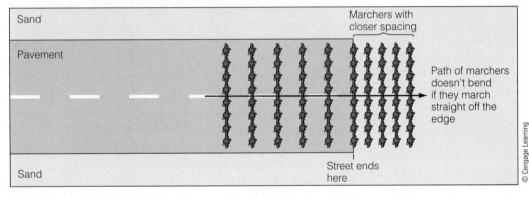

b If they march straight off the pavement, the ranks will slow down but not bend as they hit the sand.

Figure 6.21 An analogy for refraction. The ranks of marchers represent light or sound waves; the pavement and sand represent different media. The marchers can walk faster if they stay on the pavement than if they walk in the sand next to the highway.

💬 **THINKING BEYOND THE FIGURE**

Imagine light striking the edge of a prism. Draw the path of the light through the prism using the model of the marchers.

Figure 6.22 A broken rainbow, a beautiful demonstration of the difference between the refractive index of fresh water and that of seawater. Broken rainbows occur when an atmospheric rainbow (caused by raindrops of pure water) meets a surf-droplet rainbow. The rainbow angle for rainwater is 0.8° greater than for seawater.

particles) makes scattering more prevalent in water than in air. The **absorption** of light is governed by the structure of the water molecules it happens to strike. When light is absorbed, molecules vibrate and the light's electromagnetic energy is converted to heat.

Even the clearest seawater is not perfectly transparent. If it were, the sun's rays would illuminate the greatest depths of the ocean, and seaweed forests would fill its warmed basins. The thin film of lighted water at the top of the surface zone is called the **photic zone** (*photo*; light). In the very clearest tropical waters, the photic zone may extend to a depth of 600 meters (2,000 feet), but a more typical value for the open ocean is 100 meters (330 feet). In contrast, light typically penetrates the coastal waters in which we swim only to about 40 meters (130 feet). All the production of food by photosynthetic marine organisms takes place in this thin, warm surface layer. Here, water is heated by the sun, heat is transferred from the ocean into the atmosphere and space, and gases are exchanged with the atmosphere. The thermostatic effects we've discussed function largely within this zone. Most of the ocean's life is found here. The photic zone may be extraordinarily thin, but it is also extraordinarily important.

The ocean below the photic zone lies in blackness. Except for light generated by certain living organisms, the region is perpetually dark. This dark water beneath the photic zone is called the **aphotic zone** (*a*; without, *photo*; light).

> **O**nly blue and green wavelengths pass through water in any appreciable quantity or distance.

Water Transmits Blue Light More Efficiently Than Red

The energy of some colors of light is converted into heat—that is, its wavelengths are absorbed—nearer to the surface than the energy of other colors. **Figure 6.23** shows this differential absorption by color. The top meter (3.3 feet) of the ocean absorbs nearly all the infrared radiation that reaches the ocean surface, significantly contributing to surface warming. The top meter also absorbs 71% of red light. The dimming light becomes bluer with depth because the red, yellow, and orange wavelengths are being absorbed. By 300 meters (1,000 feet), even the blue light has been converted into heat.

From above, clear ocean water looks blue because blue light can travel through water far enough to be scattered back through the surface to our eyes. Divers near the surface see an even brighter blue color. Because nearly all red light is converted to heat in the first few meters of ocean water, red objects a short distance beneath the surface look gray. If you were a diver working at a depth of 10 meters (33 feet) and you cut your hand, you would see gray blood rather than red, because there is not enough red light at that depth to reflect from blood's red pigment and stimulate your eye. The underwater pictures of red organisms you've seen are possible only because the photographer has brought along a source of white light (which contains all colors). **Figure 6.24** suggests colors at depth and shows some organisms characteristic of those depths.

Absorption of Light in Different Wavelengths (Colors) by Seawater

Color	Wavelength (nm)	% Absorbed in 1 m of Water	Depth in Which 99% Is Absorbed (m)
Ultraviolet (UV)	310	14.0	31
Violet (V)	400	4.2	107
Blue (B)	475	1.8	254
Green (G)	525	4.0	113
Yellow (Y)	575	8.7	51
Orange (O)	600	16.7	25
Red (R)	725	71.0	4
Infrared (IR)	800	82.0	3

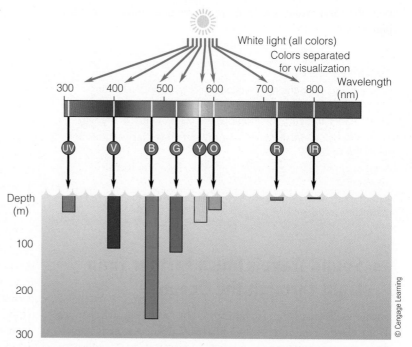

a The table shows the percentage of light absorbed in the uppermost meter of the ocean, and the depths at which only 1% of the light of each wavelength remains.

b The bars show the depths of penetration of 1% of the light of each wavelength (as in the last column of the table).

© Cengage Learning

Figure 6.23 Only a thin film of seawater is illuminated by the sun. Except for light generated by living organisms, the vast majority of the ocean lies in complete blackness.

Figure 6.24 Light and color fade rapidly as depth increases. This illustration suggests this and shows some organisms characteristic of various depths. You'll meet these communities in Chapter 16.

© Cengage Learning

Suspended particles scatter some colors of light and absorb others. Some sediments reflect yellow light, giving the ocean a yellow cast. The Red Sea gets its name from its abundance of cyanobacteria, small plantlike organisms that contain a reddish pigment.

CONCEPT CHECK

Before going on to the next section, check your understanding of some of the important ideas presented so far:

What factors influence the intensity and color of light in the sea?

Intensities being equal, which color of light moves farthest through seawater. Least far?

What happens to the energy of light when light is absorbed in seawater?

What factors affect the depth of the photic zone? Could there be a photocline in the ocean?

6.9 Sound Travels Much Farther Than Light through the Ocean

Sound is a form of energy transmitted by rapid pressure changes in an elastic medium. Sound intensity decreases as it travels through seawater because of spreading, scattering, and absorption. Intensity loss due to spreading is proportional to the square of the distance from the source. Scattering occurs as sound bounces off bubbles, suspended particles, organisms, the surface, the bottom, or other objects. Eventually sound is absorbed and converted by molecules into a very small amount of heat. The absorption of sound is proportional to the square of the frequency of the sound: Higher frequencies are absorbed sooner. Sound waves can travel for much greater distances through water than light waves can before being absorbed. Because sound travels through water so efficiently, many marine animals use sound rather than light to "see" in the ocean.

The speed of sound in seawater of average salinity is about 1,500 meters per second (3,345 miles per hour) at the surface, almost five times the speed of sound in air. The speed of sound in seawater increases as temperature and pressure increase. Sound travels faster at the warm ocean surface than it does in deeper, cooler water. Its speed decreases with depth, eventually reaching a minimum at about 1,000 meters (3,300 feet). Below that depth, however, the effect of increasing pressure offsets the effect of decreasing temperature, so speed increases again. Near the bottom of an ocean basin, the speed of sound may actually be higher than at the surface. Though these variations are important in the behavior of oceanic sound, they amount to only 2% or 3% of the average speed of sound in seawater. The relationship between depth and sound speed is shown in **Figure 6.25**.

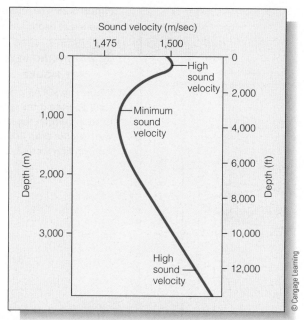

Figure 6.25 The relationship between water depth and the speed of sound.

Refraction Causes SOFAR Layers and Shadow Zones

The depth at which the speed of sound reaches its minimum varies with conditions, but it is usually near 1,200 meters (3,900 feet) in the North Atlantic or about 600 meters (2,000 feet) in the North Pacific. Although its speed is relatively slow, the transmission of sound in this minimum-velocity layer is very efficient because refraction tends to cause sound energy to remain within the layer. The outer edges of sound waves escaping from this layer will enter water in which the speed of sound is higher. This will cause the wave to speed up but then pivot back into the minimum-velocity layer, as shown in **Figure 6.26**. Upward-traveling sound waves that are generated within the minimum-velocity layer will tend to be refracted downward, and downward-traveling sound waves will tend to be refracted upward. In short, sound waves bend *toward* layers of lower sound velocity and so tend to stay within the zone. Therefore, loud noises made at this depth can be heard for thousands of kilometers.

Navy depth charges detonated in the minimum-velocity layer in the Pacific have been heard 3,680 kilometers (2,290 miles) away from the explosion. In the early 1960s, the U.S. Navy experimented with the use of sound transmission in the minimum-velocity layer as a lifesaving tool. Survivors in a life raft would drop a small charge into the water that was set to explode at the proper depth. A number of widely spaced listening stations ashore would compare the differences in the arrival times of the signal and then compute the position of the raft. The project—which has since been abandoned in favor of radio beacons—was called **SOFAR** (for *s*ound *f*ixing *a*nd *r*anging). The minimum-velocity layer has come to be known as the **SOFAR layer**.

Sound travels slowly in the SOFAR layer, but it moves rapidly near the bottom of the well-mixed surface layer. Temperature and salinity conditions are homogeneous there, so they do not produce any refraction. Pressure still increases with depth, however, causing a thin high-velocity layer at around 80 meters (260 feet), just above the pycnocline.

Some ships project pulses of sound into the water to search for animals, submarines, or hazards to navigation. Depending on the angle at which the sound waves arrive at the high-velocity layer, they will sometimes split and refract to the surface or bend into the depths. An object beyond the area of divergence may be undetectable; it would be within a **shadow zone**, a region into which very little sound energy penetrates. Shadow zones are of particular interest to submariners and to the ship captains who use sound to hunt them. As depicted in **Figure 6.27**, a smart submarine captain can use the shadow zone to hide from a pursuer.

Sonar Systems Use Sound to Detect Underwater Objects

Crews aboard surface ships and submarines employ **active sonar** (*s*ound *n*avigation *a*nd *r*anging), the projection and return through water of short pulses (*pings*) of high-frequency sound to search for objects in the ocean **(Figure 6.28)**. In a modern system, electrical current is passed through crystals to produce powerful sound pulses pitched above the limit of human hearing. (Though this high-frequency sound is absorbed rapidly in the ocean, its use greatly improves the resolution of images.) Some of the sound from the transmitter bounces off any object larger than the wavelength of sound employed and returns to a microphone-like sensor. Signal processors then amplify the echo and reduce the frequency of the sound to within the range of human hearing. An experienced sonar operator can tell the direction of the contact, its size and heading, and even something about its composition (whale or submarine or school of fish) by analyzing the characteristics of the returned ping.

Side-scan sonar is a type of active sonar. Operating with as many as 121 transmitter/receivers tuned to high sound frequencies, side-scan systems towed in the quiet water beneath a ship are sometimes capable of near-photographic resolution (see **Figure 6.29**, p. 192). Side-scan systems are used for geological investigations, archaeological studies, and locating downed ships and airplanes. The multibeam system you read about in Chapter 4 is a form of side-scan sonar (see again Figure 4.4).

For deeper soundings, or to "see" into sediment layers below the surface, geologists use *seismic reflection profilers* employing powerful electrical sparks, explosives, or compressed air to generate a very energetic low-frequency sound pulse. Again, the round-trip travel time of the sound waves is crucial. The low-frequency sound cannot resolve great detail, but the echo can usually provide an image of the sedimentary layers beneath the surface (see Figure 4.28). Low-frequency sound also has the advantage of efficient travel with less absorption.

The first human use of sea sound was passive: Mariners listened through their hulls to the whistles and clicks of whales and other animals. When submarine warfare emerged during World War I, the British invented a simple underwater listening device to detect noises made by enemy submarines. Operators would listen through a sensitive directional microphone for the telltale sounds of a propeller, a torpedo, or even a dropped wrench or slammed hatch cover. The first systems were primitive, but

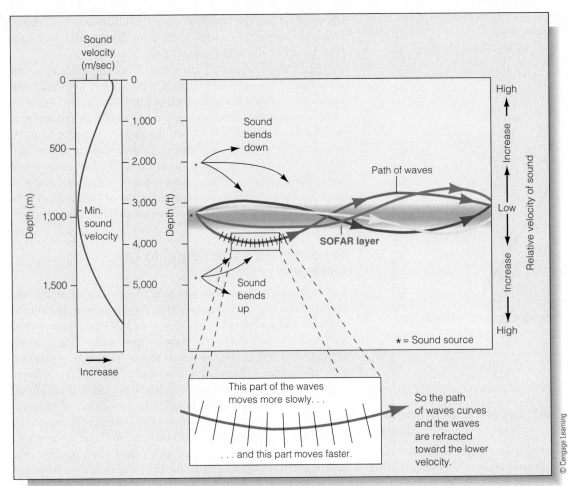

© Cengage Learning

Figure 6.26 The refraction of sound at depth.

💬 **THINKING BEYOND THE FIGURE**

Would the SOFAR principle operate if the speed of sound within the layer were *higher* than in the surrounding water?

a The depth of minimum sound velocity (see again Figure 6.25).

b The SOFAR layer, in which sound waves travel at minimum speed. Sound transmission is particularly efficient—that is, sounds can be heard for great distances—because refraction tends to keep sound waves within the layer.

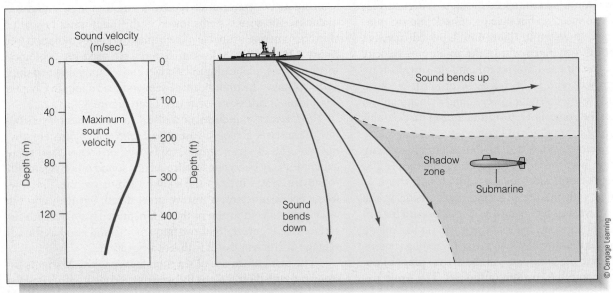

© Cengage Learning

a The depth of maximum sound velocity (see again Figure 6.25).

b The shadow zone. The thin, high-sound-velocity layer, which forms at a depth of about 80 meters (260 feet), deflects sound. The shadow zone creates a good place for submarines to hide from sonar. (Note the difference in depth scales from Figures 6.25 and 6.26.)

Figure 6.27 Another form of sound refraction.

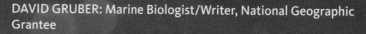

DAVID GRUBER: Marine Biologist/Writer, National Geographic Grantee

While exploring the Bloody Bay Wall off Little Cayman Island in 2012 to chronicle its biofluorescent creatures, marine biologist Gruber felt the strange sensation of stepping into his backyard and beholding an unknown and surprising universe.

Just a few hours after this photo was taken, darkness fell and I descended down the sheer face of the coral wall, along with teammates Jim Hellemn and John Sparks. The wall drops 1,000 feet, but we hovered at just 90, capturing the large biofluorescent mural that now appears in the traveling exhibit "Creatures of Light." On that evening, we unwittingly photographed many cryptic biofluorescent animals whose secret light had never before been glimpsed by human eyes.

Fluorescent proteins from marine species have already provided science with one of its most valuable tools for illuminating processes in living cells and neurons. So when it was announced that President Obama's administration is planning to launch a decade-long initiative to map the human brain, I was inspired by new possibilities. Just as biofluorescence plays many roles for corals and fishes in their dim, blue-lit world, it may also come to play a greater role in understanding how our own brain cells communicate.

For example, as a thought pops into my head, my neurons fire (a process illuminated to science by biofluorescent tags decoded from the DNA of marine organisms). I imagine a role reversal: a pair of Warteye stargazers (*Gillellus uranidea*) exploring Greenwich Village with special lights and cameras, trying to discover how we communicate so as to better understand themselves. This planet is full of unsuspected biological connections and we are just beginning to explore the bioluminescent and biofluorescent universe under the sea.

Natural fluorescence brightens this coral colony in a reef off Little Cayman Island.

Background photo: ©D. Gruber/V. Pieribone

passive sonar, as these listening-only devices are now called, is currently undergoing a renaissance. Modern passive devices are much more sophisticated and sensitive than their World War I predecessors. Passive sonar confers the benefit of surprise—unlike active sonar, in which a listener can hear the loud *ping* long before an operator aboard the sending vessel can hear his faint echo. Usually, it's safer just to listen. The combination of computerized signal processing, microphones towed at a distance from the listener's noisy ship, and better knowledge of ocean physics will improve the usefulness of both kinds of sonar.

Humans are not the only organisms to use sonar. Whales and other marine mammals use clicks and whistles to find food and avoid obstacles. We'll discuss biological active sonar in Chapter 15.

Ocean Sound Is Used to Monitor Climate Change

In 1993, nearly 100 researchers at 13 institutions began a 5-year US$40 million experiment to measure ocean temperature within a few thousandths of a degree using sound.

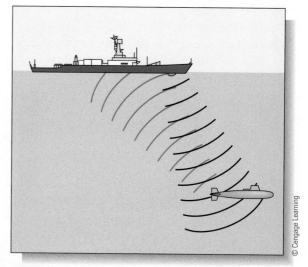

Figure 6.28 The principle of active sonar. Pulses of high-frequency sound are radiated from the sonar array of the sending vessel. Some of the energy of this ping reflects from the submerged submarine and returns to the sending vessel. The echo is analyzed to plot the position of the submarine.

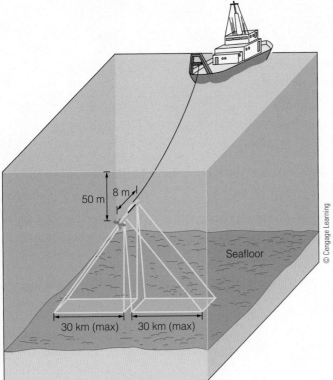

a One form of side-scan sonar in action.

50 m   8 m

30 km (max)   30 km (max)

Seafloor

© Cengage Learning

Figure 6.29 Side-scan sonar.

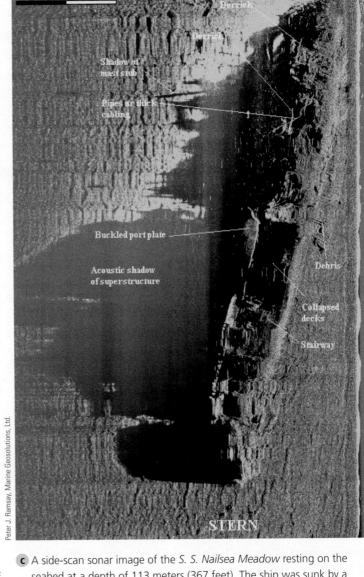

500 kHz side-scan sonar image

± 40 m

BOW

Derrick

Derrick

Shadow of mast stub

Pipes or thick cabling

Buckled port plate

Acoustic shadow of superstructure

Debris

Collapsed decks

Stairway

STERN

Peter J. Ramsay, Marine Geosolutions, Ltd.

b Sound pulses leave the submerged towed array, bounce off the bottom, and return to the device. Computers process the impulses into images.

c A side-scan sonar image of the *S. S. Nailsea Meadow* resting on the seabed at a depth of 113 meters (367 feet). The ship was sunk by a German U-boat in 1942 off the coast of South Africa. Many fine details of the wreck are clearly visible.

Peter J. Ramsay, Marine Geosolutions, Ltd.

Since sound travels more rapidly through warmer water, water temperature differences can be calculated by recording the time taken for sound to reach receivers around the world. The temperature of the ocean is an important measure of potential global warming. Computer models of Earth's climate suggest that greenhouse gases could cause the ocean to warm by about 0.005°C (0.009°F) per year at a depth of 1,000 meters (3,300 feet). Such a subtle temperature change would be very difficult to detect by conventional monitoring, but over a decade this warming would cut a few seconds from the travel time of a sound signal.[5]

Investigators constructed and tested sound sources capable of projecting a 75-cycle-per-second signal with 260 watts of acoustic power. Sound was generated and trapped in the Indian Ocean's minimum-velocity (SOFAR) layer, in which it traveled both into the Atlantic and around Australia into the Pacific. The sound was detected by a ship equipped with sensitive hydrophones lowered to the SOFAR depth off the California coast, near Point Conception. It took about 3½ hours for the sound to travel that far, nearly halfway around the world. Similarly equipped ships and offshore stations detected sound from other transmitters at other locations (**Figure 6.30**).

By January 1998, 15 months of data had been collected by some of these receivers. Key results include confidence that travel time is clearly related to known ocean processes and that the data are consistent with and complementary to related estimates derived from measurements of sea surface height made by the *TOPEX/Poseidon* satellite altimeter. (Sea surface height is related to ocean temperature because of thermal expansion.)

As experiments proceed, the extent of ocean warming—and thus global warming—may become clear.

---

[5]More information on the implications and consequences of global warming may be found in Chapter 18.

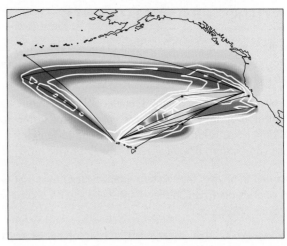

Figure 6.30 A map of changes in the long-range speed of sounds in the North Pacific Ocean. This map is based on data measuring the speed of sound through water to a depth of 1,000 meters (3,300 feet). Variations from normal speed range from −0.5 meter per second (dark blue) to +0.5 meter per second (dark red). Such variations suggest the temperature of the seawater through which the sound is passing and suggest the extent of global warming. More on this topic will be found in Chapter 18.

CONCEPT CHECK

Before going on to the next section, check your understanding of some of the important ideas presented so far:

How much faster is the speed of sound in water than in air?

Is the speed of sound the same at all ocean depths?

What's a SOFAR layer? A shadow zone?

How does sonar work? What kinds of sonar systems are used in oceanographic research?

How can sound be used to measure the temperature of an ocean?

Chapter in Perspective

In this chapter you learned of the polar nature of the water molecule. This polarity and the hydrogen bonds that form between water molecules result in water's unexpected thermal properties. You found that liquid water is remarkably resistant to temperature change with the addition or removal of heat, and ice, with its large latent heat of fusion and low density, melts and re-freezes over large areas of the ocean to absorb or release heat with no change in temperature. These thermostatic effects, combined with the mass movement of water and water vapor, prevent large swings in Earth's surface temperature. And, of course, you also learned that the words *heat* and *temperature* are not interchangeable.

Changes in temperature and salinity greatly influence water density. Ocean water is usually layered by density, with the densest water on or near the bottom. The physical characteristics of the world ocean are largely determined by the physical properties of seawater. These properties include water's heat capacity, density, salinity, and its ability to transmit light and sound.

In the next chapter you'll learn what happens when solids and gases dissolve in seawater. Most of the properties of seawater are different from those of pure water because of the substances dissolved in the seawater.

QUESTIONS FROM STUDENTS

1. If water has the highest latent heat of vaporization, why does a drop of alcohol make my hand feel colder when it evaporates than a drop of water does?

Heat versus temperature again. The alcohol makes your hand *colder*, but it evaporates much *more quickly*. The water stays around longer (takes longer to evaporate), and although it may not cause your skin to become as *cold*, it will remove more *heat*.

2. I learned in physics that water's latent heat of evaporation is 585 calories per gram. In this chapter you say it's 540 calories per gram. What's up?

Actually, we wrote that water's latent heat of *vaporization* is 540 calories per gram. The latent heat of vaporization is the amount of heat that must be added to 1 gram of a substance, at its boiling point, to change it completely from a liquid to a gas. But water doesn't have to be at its boiling point to vaporize. When water molecules evaporate at temperatures below the boiling point (as from your skin during a workout), more heat must be supplied to evaporate each molecule than is needed to vaporize it at the boiling point. The latent heat of vaporization increases with decreasing temperature—at 20°C (68°F), the latent heat of vaporization is 585 calories per gram. This is sometimes called the latent heat of *evaporation*.

3. Liquid methane appears to be the only other substance found in quantity in a liquid state in the solar system. If our ocean were of liquid methane rather than liquid water, how would conditions here be different?

Liquid methane freezes at −183°C (−272°F) and boils at −162°C (−234°F). Methane has a much lower heat capacity than liquid water because it does not form a lattice of hydrogen bonds as water does and consequently does not require a large energy input to release molecules. Because of methane's low heat capacity, the difference between Earth's polar and tropical daytime temperatures would be drastically greater unless the circulation rate of liquid and vapor currents accelerated to keep pace. Computer modeling yields a very unattractive vision of a planet with a boiling equatorial ocean, crushing atmospheric pressures from the greater amount of vapor in the air, torrential polar methane rainfall, cataclysmic cyclonic storms, and average wind speeds at mid-latitudes of hundreds of kilometers per hour. We miss all this excitement because our ocean is made of water.

4. Input of solar radiation in the Northern Hemisphere peaks on 22 June and reaches a minimum on 22 December, because of Earth's orbital tilt (more about that in Chapter 8). Why, then, do our warmest days occur in August or September and our coldest days in January or February?

Because of thermal inertia. There is a lag between maximum sunlight and maximum warmth because of water's great heat ca-pacity. The sun must shine on this watery planet for many weeks to raise the summer hemisphere's temperature. Of course, water also retains heat well, so the coldest days in the winter hemisphere come well after the darkest ones.

5. What about pressure? Is seawater compressible? Does compression have anything to do with water density?

Seawater is nearly incompressible—it doesn't squeeze very much as pressure builds. Still, there is some effect. Pressure in the deepest ocean trench is equal to the weight of 1,086 atmospheres (about 8½ tons per square inch), and a cubic centimeter of water lowered to that level will lose about 2% of its volume. The average pressure acting on the world ocean's volume compresses water enough to lower sea level by about 37 meters (121 feet). If gravity (and therefore pressure) were released, sea level would rise that much. Still, the effect of pressure on density is very small in comparison to the effects of temperature and salinity.

6. Why don't sounds seem to travel easily from air to ocean, or vice versa?

Sound waves can make the transition from one medium to another with little energy loss only when the speeds of sound waves in the two different media are similar. The speeds of sound in water and air, however, are too different for an efficient transition to be made. Too great a contrast in speed produces reflection (not refraction) of the sound waves at the junction. This is why you can't hear people shouting from the edge of the pool while you're underwater, even though the weak sound of a submerged pebble clicking against the side is very clear and sharp.

If you place a solid medium in which the speed of sound is intermediate between air and water, the sound can move across one junction and then the other, for a more efficient total transition. Wood works well for this, which explains why ocean noises are easy to hear in wooden boats. Even if the speed of sound in the intermediate medium is higher than in water (as in steel, for instance), some sound will be audible simply because the hard surface provides a good radiating surface for noises coming from the water.

7. I can't tell where sound is coming from when I'm underwater. It seems to be coming from inside my head. Why is that?

Our normal sensation of stereo hearing depends in part on the difference in the arrival times of sound from one ear to the other. The speed of sound in water, however, is more than four times greater than its speed in air; so our brains are unable to sense arrival-time differences from sounds originating nearby. You'd need to be more than four times as sensitive—or have a head more than four times as large—to hear stereophonically underwater.

TERMS AND CONCEPTS TO REMEMBER

absorption	density curve	molecule	side-scan sonar
active sonar	electron	nucleus	**SOFAR**
adhesion	element	passive sonar	**SOFAR layer**
aphotic zone	freezing point	photic zone	sound
atom	halocline	polar molecule	state
calorie	heat	proton	surface tension
chemical bond	heat capacity	pycnocline	surface zone
cohesion	hydrogen bond	refraction	temperature
compound	latent heat of evaporation	refractive index	thermal inertia
covalent bond	latent heat of fusion	scattering	thermocline
deep zone	latent heat of vaporization	sensible heat	thermostatic property
degree	light	shadow zone	water mass
density	mixed layer		

STUDY QUESTIONS

Thinking Critically

1. What is the latent heat of vaporization? Of fusion? Which requires more heat to transform water's physical state? Why?
2. How does water's high latent heat influence the ocean? Leaving aside its effect on beach parties, how do you think conditions on Earth would differ if our ocean consisted of ethyl alcohol?
3. How does the seasonal freezing and thawing of the polar ocean areas influence global temperatures?
4. How does refraction permit sound to be transmitted in the ocean for thousands of miles?
5. What factors influence the density of seawater?

Thinking Analytically

1. How many Calories (note capital "C") are required to raise the temperature of a can of diet cola (12 fluid ounces) from refrigerator temperature to body temperature? What percentage of the energy (Calories) in the can of diet cola is required to do this? Assume, for the moment, that the diet cola has the same thermal properties as pure water. Hint: See footnote 1 on page 168.
2. How much heat is required to melt a quart of ice resting at 0°C?
3. How much heat is required to vaporize a quart of pure water that has just come to a boil?

GLOBAL GEOSCIENCE WATCH

Visit www.cengagebrain.com to access MindTap, a complete digital course which includes access to Global Geoscience Watch and more. Within the GREENR database, search for "sonar." You may wish to further narrow your results by searching for "side scan" or "multibeam" sonar within your results. Utilize as many different types of sources as possible. Research the different types of sonar within the database. Use this information to write a short report which addresses the following: (1) What are the major types of sonar used today? (2) How do they each work? (3) Describe at least three (3) different applications that modern sonar is used for.

7 Ocean Chemistry

KEY CONCEPTS

Water is a powerful solvent. The concentration of dissolved inorganic solids in water is its salinity. © Cengage Learning

Though salinity may vary with location, the ratio of dissolved solids in seawater is constant. NASA/GSFC/JPL-Caltech

Gases dissolve in seawater. Cold water can hold more gas in solution than warm water. © Micha Pawlitzki/Terra/Corbis

The ocean is a vast reservoir of carbon. The dynamics of carbon exchange between ocean and atmosphere affect Earth's climate. U.S. Coast Guard photo

The ocean's acid–base (pH) balance varies with depth and dissolved components. Carbonate chemistry serves to moderate (buffer) wide swings in oceanic pH. David Littschwager/National Geographic Creative

"The finest workers in stone are not copper or steel tools, but the gentle touches of air and water working at their leisure with a liberal allowance of time."—Henry David Thoreau
Radius Images/Corbis

Media Connection
Start off this chapter by watching a video on how an increase in carbon dioxide can impact our oceans and lead to disruption of life systems. Visit www.cengagebrain.com to access MindTap, a complete digital course that includes this video and other resources.

7.1 Water Is a Powerful Solvent

In Chapter 1 you saw that most of Earth's surface waters are thought to have been delivered by comets or escaped from the crust and mantle through the process of outgassing. Outgassing of other substances, and water's ability to dissolve crustal material as it cycles from the ocean to the atmosphere and the land and back again, have added solids and gases to the ocean (**Figure 7.1**).

All this water is in constant motion, circulating between the ocean, the atmosphere, and the land. About 1.37 billion cubic kilometers (329 million cubic miles) of water exist at Earth's surface, the vast majority (97.5%) in the ocean. About 1.8% is bound in glaciers and the great ice caps of Greenland and Antarctica. Only about 0.64% is fresh, temporarily residing in rivers, lakes, wetlands, and below the surface as ground water.

Water cycles continuously. The **hydrologic cycle**, as the process is called, is powered by solar radiation (with a small assist from geothermal energy) (**Figure 7.2**). About 85% of all water entering the atmosphere evaporates from the ocean (the remaining 15% comes from water on land). Water vapor rising from the surface condenses into clouds and forms rain and snow. About 80% of this precipitation falls back into the ocean, and the 20% that falls on land eventually finds its way there after spending varying amounts of time moving through plants, masses of ice, sediments, and streams. As we'll see in a moment, the average

> **"The fall of dropping water wears away the stone."** —Lucretius

time water stays in the ocean (before being evaporated) is about 4,100 years. Once in the air, its residence time is only 9 days!

Water is remarkably effective in dissolving crustal material as it cycles—indeed, water can dissolve more substances than almost any other liquid.

Why?

As we have learned, water is a polar molecule. It can be thought of as having a positive (+) end and a negative (−) end (look again at Figures 6.2 and 6.3). In the polar water molecule, opposites attract: Its positive end attracts particles having a negative charge, and its negative end attracts particles having a positive charge. When water comes into contact with compounds whose elements are held together by the attraction of opposite electrical charges (most salts, for example), the polar water molecule will separate that compound's component elements from each other.

No wonder, then, that seawater and most other liquids in nature are water solutions. A **solution** is made of two components: The **solvent**, usually a liquid, is always the more abundant constituent; the **solute**, often a dissolved solid or gas, is the less abundant. In a true solution (sugar in well-stirred coffee, for example), the molecules of the solute are homogeneously dispersed among the molecules of solvent; that is, the solution has uniform properties throughout. In a **mixture**, different substances are closely intermingled but retain separate identities. The properties of a mixture are heterogeneous; they may vary from place to place within the mixture. Think of noodle soup as a mixture of noodles and liquid.

Water's dissolving power results from the polar nature of the water molecule. Consider how water dissolves sodium chloride (or NaCl), the most common salt, into its constituents, sodium ions (Na^+) and chloride ions (Cl^-).[1] An **ion** is an atom (or small group of atoms) that has an unbalanced electrical charge

---

[1]Na, the chemical symbol for sodium, is derived from *natrium*, the Latin name for sodium.

Figure 7.1 Water has been called the "universal solvent" because, given time, it will dissolve nearly any substance. Even the hardest rock is susceptible to its slow, relentless attack.

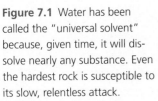

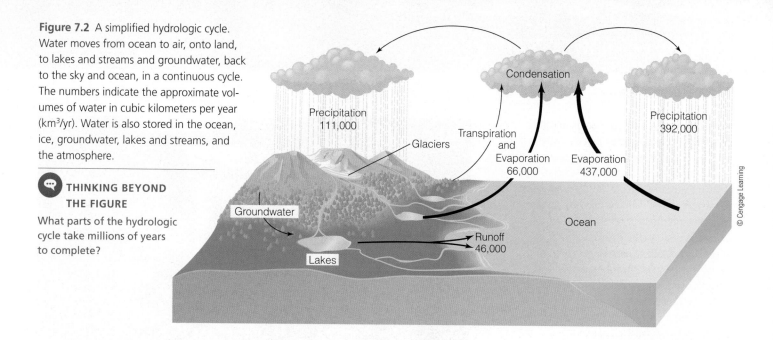

Figure 7.2 A simplified hydrologic cycle. Water moves from ocean to air, onto land, to lakes and streams and groundwater, back to the sky and ocean, in a continuous cycle. The numbers indicate the approximate volumes of water in cubic kilometers per year (km³/yr). Water is also stored in the ocean, ice, groundwater, lakes and streams, and the atmosphere.

💬 **THINKING BEYOND THE FIGURE**

What parts of the hydrologic cycle take millions of years to complete?

because it has gained or lost one or more electrons. Unlike the electron *sharing* found in covalently bonded molecules such as water, the sodium atoms in NaCl have *lost* electrons, and chlorine atoms have *gained* them. The resulting ions are linked in a salt by the mutual attraction of their opposite electrical charges. The ions of sodium and chloride in NaCl are said to be held together by **ionic bonds**, electrostatic attraction that exists between ions that have opposite charge. When NaCl dissolves in water (**Figure 7.3**), the polarity of water reduces the electrostatic attraction (ionic bonding) between the Na⁺ and Cl⁻ ions. This causes the two ions to separate. They move away from the salt crystal, permitting water to attack the next layer of NaCl.

Note that NaCl does not exist as "salt" in seawater; its components are separated when salt crystals dissolve in water, but they are joined when crystals re-form as water evaporates.

Some other salts form ions in which groups of atoms bond together and carry an electrical charge. The sulfate ion (SO_4^{2-}), for example, carries two extra electrons. The charge on the overall ion is therefore -2.

By contrast, oil doesn't dissolve in water even if the two are thoroughly shaken together. When oil is dispersed in water, it forms a mixture because molecules of oil are nonpolar in character. This means that oil has no positive or negative charges to attract the polar water molecule. In a way this is fortunate: Liv-

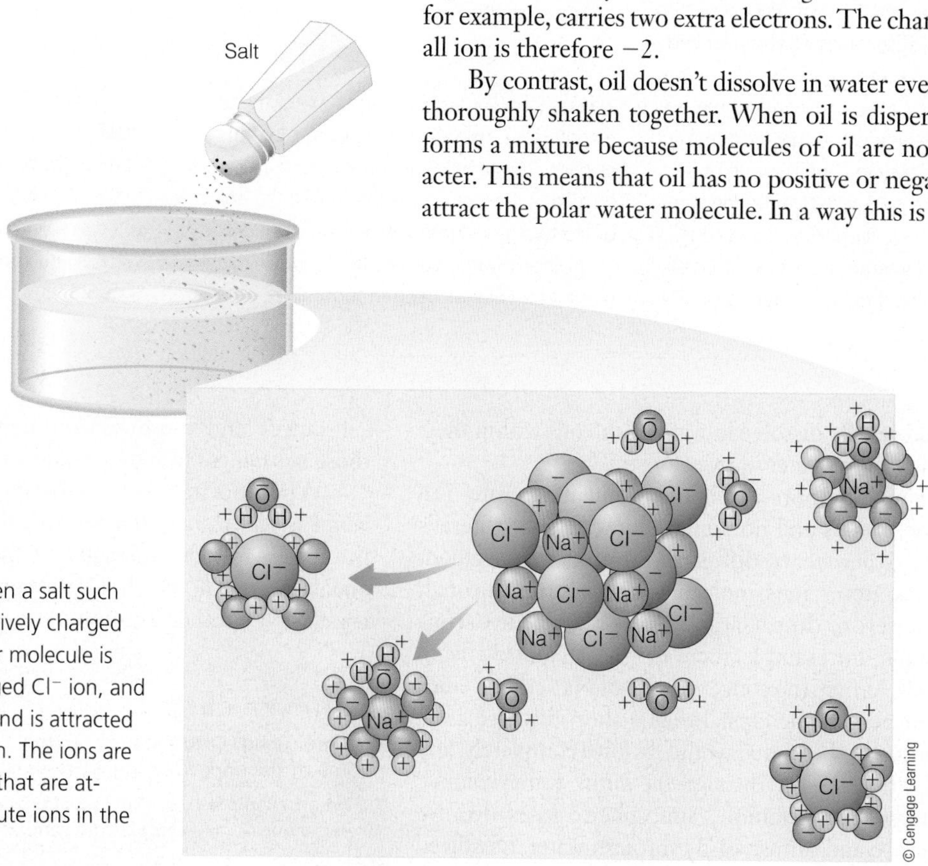

Figure 7.3 Salt in solution. When a salt such as NaCl is put in water, the positively charged hydrogen end of the polar water molecule is attracted to the negatively charged Cl⁻ ion, and the negatively charged oxygen end is attracted to the positively charged Na⁺ ion. The ions are surrounded by water molecules that are attracted to them and become solute ions in the solvent.

The Nature of Water

In the fifth century B.C.E., the Greek philosopher Empedocles suggested that all matter was composed of four primary substances in various combinations: earth, air, fire, and water. Divide anything into small enough bits, he wrote, and you would end up with one or more of these four elements, none of which could be further divided. A century later this concept was endorsed by Aristotle, a philosopher considered by many to be the father of modern science. Aristotle's stature lent credence to the idea.

By the 1700s, the four-element theory was being seriously challenged. Chemists were making many discoveries that could not adequately be explained using Empedocles's theory. A test that would unequivocally disprove the four-element theory would be to separate one of the four elements into smaller components, and water became the target. On 24 June 1783, the French chemist Antoine-Laurent Lavoisier discovered that water could be subdivided, a landmark event in the history of chemistry.

Lavoisier burned two gases in a special vessel. One gas was produced by dripping acid onto iron or copper. This clear, odorless gas was called "inflammable air" because it burned when ignited. The other gas—also colorless and odorless—was produced by heating a red precipitate of mercury. This second gas was known as the "acid maker" because of its tendency to combine with other substances to make mild acids. During the combustion of these gases, droplets of liquid condensed on the cool interior of an attached flask. Tests showed this liquid to be "as pure as distilled

Figure A Antoine-Laurent Lavoisier is credited with the "discovery" of water in 1783. Though a pioneering chemist and one of the founders of the metric system of measurements, Lavoisier could not escape the terrors of the French Revolution. He was sentenced to death by a revolutionary tribunal and guillotined on 8 May 1794.

The Art Archive/Palais de Compiegne/CCI/Picture Desk

water." Lavoisier argued that in chemistry, as in mathematics, the whole is equal to the sum of its parts. "Since only water and nothing else was formed. . . the water was the sum of the [masses] of the two gases from which it was produced."

In short, water was not the irreducible element it had always been thought to be. It was a combination—a compound—of the acid maker (*oxys;* acid, *gen;* making) and the newly named "begetter of water" (*hydro;* water, *gen;* making). Oxygen and hydrogen were found to be simpler substances than water. Elements were clearly more numerous than Aristotle had thought, and water more complex.

---

ing tissues would readily dissolve in water if the oils within their membranes didn't blunt water's powerful attack.

As we'll examine in more detail in a moment, seawater is a complex solution of ions and nonionic solutes. These materials can move through still water by **diffusion**, the random movement of materials (gases, atoms, ions, molecules, and so forth) through a solution. If you were to drop a large crystal of NaCl into a container of pure water, for example, the water would quickly begin to dissolve the salt. At first, the concentration of Na^+ and Cl^- ions would be greater next to the crystal than a short distance away from it, but eventually, the crystal would dissolve completely, and the ions would diffuse evenly through the entire water volume, forming a homogeneous solution. Atmospheric gases dissolve in the ocean surface and diffuse easily through water. Dissolved substances tend to diffuse from regions of high concentration of those substances to regions of low concentration.

When no more of a substance will dissolve in water, the water is said to be *saturated* with that substance. At **saturation**, the rate at which molecules of the solute are being dissolved equals the rate at which they are **precipitating** (re-forming into crystals) at another location in the solution.

CONCEPT CHECK

Before going on to the next section, check your understanding of some of the important ideas presented so far:

Give an example of water in a very short, rapid part of a hydrologic cycle. What is an example of a long and slow hydrologic cycle?

7.2 Seawater Consists of Water and Dissolved Solids

About 97.2% of the 1,370 million cubic kilometers (329 million cubic miles) of Earth's surface water is marine. By weight, seawater is about 96.5% water and 3.5% dissolved substances, most of which are salts of various kinds. The world ocean contains some 5,000 trillion kilograms (5.5 trillion tons) of salts. If the ocean's water evaporated completely, leaving its salts behind, the dried residue could cover the entire planet with an even layer 45 meters (150 feet) thick.

Salinity Is a Measure of Seawater's Total Dissolved Inorganic Solids

The total quantity (or concentration) of dissolved inorganic solids in water is its **salinity**. The ocean's salinity varies from about 3.3% to 3.7% by mass, depending on such factors as evaporation, precipitation, and freshwater runoff from the continents, but the average salinity is usually given as 3.5%. Most of the dissolved solids in seawater are salts that have been separated into ions. Sodium and chloride are the most abundant of these.

The many ions present in seawater react with each other (and with water molecules) in complex ways to modify the physical properties of pure water. Consider the following:

- The heat capacity of water decreases with increasing salinity; that is, less heat is necessary to raise the temperature of seawater by 1° than is required to raise the temperature of freshwater by the same amount.
- Dissolved salts disrupt the webwork of hydrogen bonding in water. As salinity increases, the freezing point of water becomes lower; the salts act as a sort of antifreeze. Sea ice therefore forms at a lower temperature than ice in freshwater lakes.
- Because dissolved salts tend to attract water molecules, seawater evaporates more slowly than freshwater. Swimmers usually notice that freshwater evaporates quickly and completely from their skin, but seawater lingers.
- Osmotic pressure, the pressure exerted on a biological membrane when the salinity of the environment is different from that within the cells, rises with increasing salinity. This is a key factor in transmitting water into and out of cells.

These four properties, which vary with the quantity of solutes dissolved in the water, are called water's **colligative properties** (*colligatus;* to bind together). Because colligative properties are the properties of *solutions*, the more concentrated (saline) water is, the more important these properties become. (Because it is not a solution, pure water has no colligative properties.)

A Few Ions Account for Most of the Ocean's Salinity

Because about 3.5% of seawater consists of dissolved substances, boiling away 100 kilograms of seawater theoretically produces a residue with a mass of 3.5 kilograms. Because variations of 0.1% are significant, however, oceanographers prefer to use parts-per-thousand notation (‰) rather than percent (%, parts per hundred) in discussing these materials.[2] The seven ions listed below oxygen and hydrogen in **Table 7.1** make up more than 99% of this residual material—sodium and chloride make up 85% of the total. When seawater evaporates, its ionic components combine in many different ways to form table salt ($NaCl$), epsom salts ($MgSO_4$), and other mineral salts. **Figure 7.4** shows the proportions of ions in seawater.

Seawater also contains minor constituents. The ocean is sort of an "Earth tea"—Nearly every element present in the crust and atmosphere is also present in the oceans, though sometimes in extremely small amounts. Only 14 elements have concentrations in seawater greater than one part per million (ppm). Elements present in amounts less than 0.001‰ (1 ppm) are known as **trace elements**. These are more easily given in parts per billion (ppb). **Table 7.2** lists some of the minor and trace elements, many of which are crucial to life processes.[3]

The Components of Ocean Salinity Came from, and Have Been Modified by, Earth's Crust

Remembering the effectiveness of water as a solvent, you might think that the ocean's saltiness has resulted from the ability of rain, groundwater, or crashing surf to dissolve crustal rock. Much of the sea's dissolved material originated in that way, but is crustal rock the source of all the ocean's solutes? An easy way to find out would be to investigate the composition of salts in river water and compare these figures to those of the ocean as a whole. If crustal rock is the only source, then the salts in the ocean should be like those of concentrated river water. But they are not. River water is usually a dilute solution of bicarbonate and calcium ions, while the principal ions in seawater are chloride and sodium (**Figure 7.5a**, p. 204) The magnesium content of seawater would also be higher if seawater were simply concentrated river water. The proportions of salts in isolated salty inland lakes, such as Utah's Great Salt Lake or the Dead Sea, are much different from the proportions of salts in the ocean. So weathering and erosion of crustal rocks cannot be the only source of sea salts.

The components of ocean water whose proportions are *not* accounted for by the weathering of surface rocks are called **excess volatiles**. To find the source of these excess volatiles, we must look to Earth's deeper layers. The upper mantle appears to contain more of the substances found in seawater (including the water itself) than are found in surface rocks, and their proportions are about the same as found in the ocean. As you read

---

[2]Note that 3.5% = 35‰. If you begin with 1,000 kilograms of seawater, you would expect 35 kilograms of residue.

[3]A periodic table of the elements, Appendix 7, provides more information on individual elements.

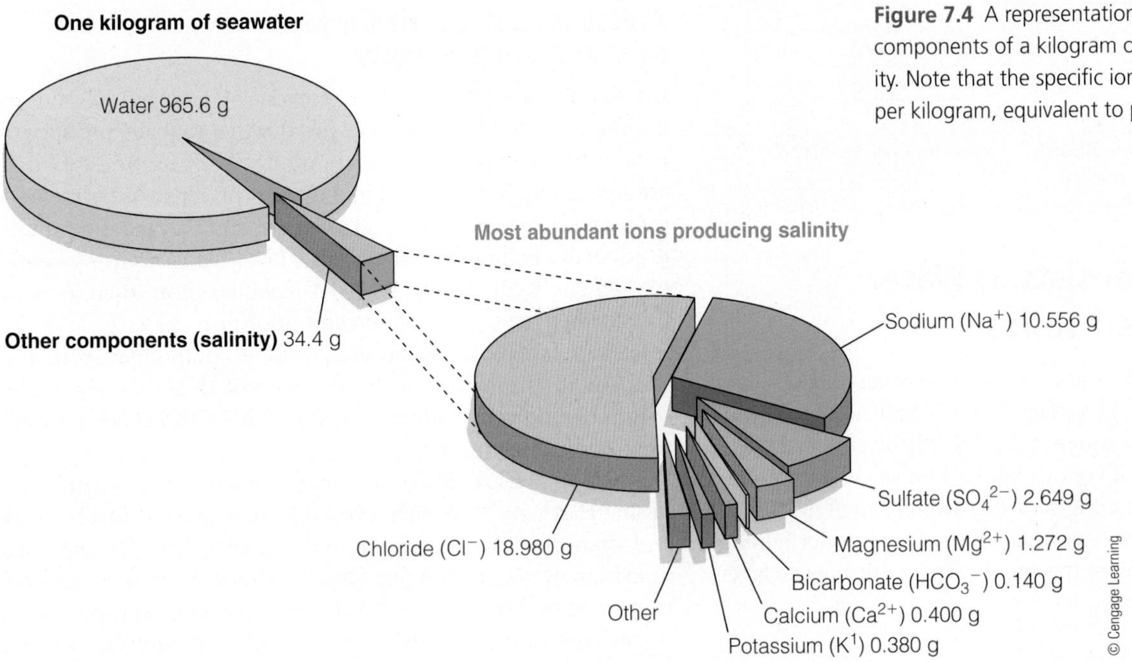

One kilogram of seawater

Figure 7.4 A representation of the most abundant components of a kilogram of seawater at 34.4‰ salinity. Note that the specific ions are represented in grams per kilogram, equivalent to parts per thousand (‰).

Water 965.6 g

Most abundant ions producing salinity

Other components (salinity) 34.4 g

Sodium (Na$^+$) 10.556 g

Sulfate (SO$_4^{2-}$) 2.649 g

Chloride (Cl$^-$) 18.980 g

Magnesium (Mg^{2+}) 1.272 g

Bicarbonate (HCO$_3^-$) 0.140 g

Other

Calcium (Ca^{2+}) 0.400 g

Potassium (K^1) 0.380 g

© Cengage Learning

in Chapter 3, convection currents slowly churn Earth's mantle, causing the movement of tectonic plates. Because of this activity, some deeply trapped volatile substances escape to the exterior, outgassing through volcanoes and rift vents. These excess volatiles include carbon dioxide, chlorine, sulfur, hydrogen, fluorine, nitrogen, and, of course, water vapor. This material, along with residue from surface weathering, accounts for the chemical constituents of today's ocean.

Some of the ocean's solutes are hybrids of the two processes of weathering and outgassing (see **Figure 7.5b**). Table salt, or so-

dium chloride, is an example. The sodium ions come from the weathering of crustal rocks, while the chloride ions come from the mantle by way of volcanic vents and outgassing from mid-ocean rifts. As for the lower-than-expected quantity of magnesium and sulfate ions in the ocean, research at a spreading center east of the Galápagos Islands suggests that the chemical composition of seawater percolating through mid-ocean rifts is altered by contact with fresh crust. The water that circulates through new ocean floor at these sites is apparently stripped of magnesium and a few other elements. The magnesium seems to be

Table 7.1 Major Constituents of Seawater at 34.4‰ Salinity

Constituent	Concentration in Parts per Thousand (‰), or Grams per Kilogram (g/kg)	Percent by Mass	Percent of Total
Water Itself			
Oxygen	857.8	85.8	96.5
Hydrogen	107.2	10.7	
Most Abundant Ions			
Chloride (Cl$^-$)	18.980	1.9	
Sodium (Na$^+$)	10.556	1.1	
Sulfate (SO$_4^{2-}$)	2.649	0.3	
Magnesium (Mg^{2+})	1.272	0.1	3.4
Calcium (Ca^{2+})	0.400	0.04	
Potassium (K$^+$)	0.380	0.04	
Bicarbonate (HCO$_3^-$)	0.140	0.01	
Total	999.377 g/kg	99.99%	

Source: Data from *CRC Handbook of Marine Science*, CRC Press, Edited by Walton-Smith, 1994. Table created by Tom Garrison.

Table 7.2 Minor and Trace Elements in Seawater at 35‰ Salinity

Constituent	Concentration in Parts per Thousand (‰), or g/kg	Concentration in Parts per Million (ppm), or mg/kg (mg/L)	Concentration in Parts per Billion (ppb), or mg/1,000 kg
Minor Elements			
Bromine (Br)	0.065	65	
Strontium (Sr)		8	
Boron (B)		4	
Silicon (Si)		3	
Fluorine (F)		1	
Important Trace Elements			
Nitrogen (N)[a]		0.28	280
Lithium (Li)			125
Iodine (I)			60
Phosphorus (P)			30
Zinc (Zn)			110
Iron (Fe)			6
Aluminum (Al)			2
Manganese (Mn)			2
Lead (Pb)			0.04
Mercury (Hg)			0.03
Gold (Au)			0.0013

Source: Data from M.J. Kennish, 1994, cited in *CRC Handbook of Marine Science,* CRC Press, Edited by Walton-Smith, 1994. Table created by Tom Garrison.
[a]Refers to nitrogen available as a nutrient, not as dissolved gas.

incorporated into mineral deposits, but calcium is added as hot water dissolves adjacent rocks **(Figure 7.5c)**.

Recent research has shown that temperature and density gradients inside seamounts also drive great quantities of water into close association with hot geological bits. The ocean contains about 15,000 seamounts, and the volume of seawater circulated through them may exceed the amounts associated with ridges.

Astonishingly, all the water in the ocean is thought to cycle through the seabed at rift zones every 1 to 2 million years (see "How Do We Know? 7.2")!

The Ratio of Dissolved Solids in the Ocean Is Constant

In 1865, the chemist Georg Forchhammer noted that although the total *amount* of dissolved solids (salinity) might vary among samples, the *ratio* of major salts was constant in samples of seawater from many locations. In other words, the percentage of various salts in seawater is the same in samples from many places, regardless of how salty the water is. For example, when the solids are isolated from any seawater sample, whether from the high-salinity North Atlantic or low-salinity Arctic oceans, 55.04% of those sol-

ids will be chloride ions. This constant ratio is known as **Forchhammer's principle**, or the **principle of constant proportions**. Forchhammer was also the first to observe that seawater contains fewer silica and calcium ions than concentrated river water—and the first to realize that removal of these compounds by marine animals and plants to form shells and other hard parts might account for part of the difference. The English chemist William Dittmar, working with HMS *Challenger* samples 10 years after they had been collected, confirmed Forchhammer's principle of constant proportions. Building on Forchhammer's and Dittmar's work, and taking advantage of improved analytical techniques, researchers have established a reliable way to determine salinity.

Salinity Is Calculated by Seawater's Conductivity

Water's salinity by weight would seem an easy property to measure. Why not simply evaporate a known weight of seawater and weigh the residue? This simple method yields imprecise results because some salts will not release all the molecules of water associated with them. If these salts are heated to drive off the water, other salts (carbonates, for example) will decompose to

a Seawater is not concentrated river water. The chemical constituents in landlocked lakes with no outlets (Utah's Great Salt Lake; Mono Lake in California) are much different than those of the ocean.

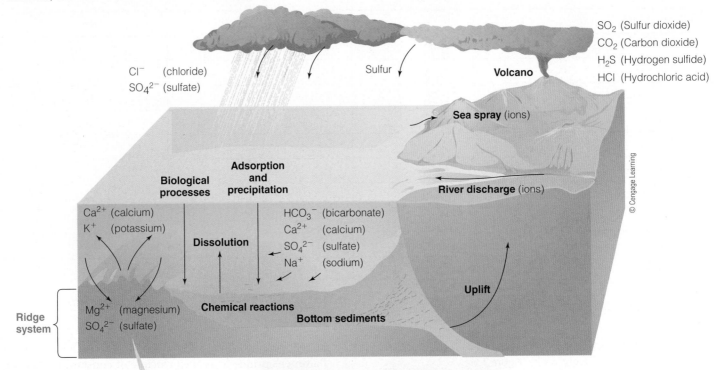

SO_2 (Sulfur dioxide)
CO_2 (Carbon dioxide)
H_2S (Hydrogen sulfide)
HCl (Hydrochloric acid)

Cl^- (chloride)
SO_4^{2-} (sulfate)

Sulfur

Volcano

Sea spray (ions)

River discharge (ions)

Biological processes

Adsorption and precipitation

Ca^{2+} (calcium)
K^+ (potassium)

Dissolution

HCO_3^- (bicarbonate)
Ca^{2+} (calcium)
SO_4^{2-} (sulfate)
Na^+ (sodium)

Uplift

Ridge system

Mg^{2+} (magnesium)
SO_4^{2-} (sulfate)

Chemical reactions
Bottom sediments

b Ions are *added* to seawater by rivers running off crustal rocks, volcanic activity, groundwater, hydrothermal vents and cold springs, and the decay of once-living organisms. Ions are *removed* from the ocean by chemical entrapment as water percolates through the mid-ocean ridge systems and seamounts, sea spray, uptake by living organisms, incorporation into sediments, and ultimately by subduction.

Figure 7.5 Processes that regulate the major constituents in seawater.

c An active hydrothermal vent spouts superheated seawater. Recent research at a spreading center east of the Galápagos Islands suggests that mid-ocean rifts strip the water that circulates through new ocean floor of magnesium and a few other elements, while calcium is added as hot water dissolves adjacent rocks.

Recycling on a Grand Scale

Ocean water has a residence time. Scientists estimate that about 334,000 cubic kilometers (80,000 cubic miles) of pure water evaporates from the ocean each year. Assuming a total ocean volume of 1,370 million cubic kilometers (329 million cubic miles), we can calculate that about 0.024% of the water in the ocean vaporizes each year. Thus, if it were not replenished by precipitation and runoff from land, the ocean would evaporate completely in about 4,100 years.

For a complete picture, more than oceanic water should be considered. Earth's total surface water volume—including the ocean, lakes, rivers, groundwater, ice caps, and so on—is about 1,399 million cubic kilometers (336 million cubic miles). The total yearly evaporation/precipitation estimate is 396,000 cubic kilometers (95,000 cubic miles) of pure water. From these figures we again calculate a recycling time of about 4,100 years. With the exception of water subducted into the mantle along with sediments at subduction zones, then, water molecules have a residence time in the ocean of about 4,100 years.

Earth's ocean is more than 4 billion years old. Thus, on average, individual water molecules have evaporated from and returned to the ocean *more than a million times* since the world ocean formed in its basins!

Zuckerstaetter/Stock4B/Alloy/Corbis

Figure A Water molecules cycle between sea and atmosphere. The average time a water molecule spends in the ocean before being evaporated and precipitated is about 4,100 years.

form gases and solid compounds not originally present in the water sample.

Until recently, oceanographers preferred to use parts-per-thousand notation (‰) rather than percent (%, parts-per-hundred) in discussing these materials. In 1978, however, oceanographers redefined salinity in the *Practical Salinity Scale* (PSS), a ratio of the conductivity of a seawater sample to a standard solution of potassium chloride. This is measured by an electronic device called a **salinometer**. Conductivity varies with the concentration and mobility of ions present and with water temperature. Circuits in the salinometer adjust for water temperature, convert conductivity to salinity, and then display salinity (or the PSS ratio itself, expressed as the dimensionless unit "**S**").

Salinometers are calibrated against a sample of known conductivity and salinity. The best salinometers can determine salinity to an accuracy of 0.001%. Some salinometers are designed for remote sensing—the electronics stay aboard ship while the sensor coil is lowered over the side.

Seawater samples can be obtained by methods ranging from tossing a clean bucket over the side of the ship to elaborate tube-and-pump systems. Typically water samples are collected using a group of sampling bottles (as in **Figure 7.6**). The bottles are lowered from a ship and triggered to close at specific depths by an electronic signal. The bottles are hauled to the surface and their contents analyzed. **Figure 7.7** shows the results of these measurements. Note the areas of high salinity in the central Atlantic and Mediterranean Sea, and areas of low salinity in the polar regions and near the mouths of large rivers.

The Ocean Is in Chemical Equilibrium

If outgassing and the chemical weathering of rock are continuing processes, shouldn't the ocean become progressively saltier with age? Landlocked seas and some lakes usually become saltier as they grow older, but the ocean does not. The ocean appears to be in **chemical equilibrium**; that is, the proportion *and amounts*

Figure 7.6 A "rosette" of sampling bottles. Each of these 10-liter bottles may be separately closed by a signal from the research ship when the array reaches predetermined depths. The bottles are hauled to the surface and their contents analyzed by a salinometer.

Photo courtesy of William Cochlan, Ph.D., Hancock Institute for Marine Science, University of Southern California

of dissolved salts per unit volume of ocean are nearly constant and have been so for millions of years. Evidently, whatever goes in must come out somewhere else.

Geologists in the 1950s developed the concept of a *steady state ocean*. The idea suggests that ions are added to the ocean at the same rate as they are being removed. This theory helps explain why the ocean is not growing saltier. The idea was quantified by T. F. W. Barth, who in 1952 devised the concept of **residence time**, the average length of time an atom of an element spends in the ocean. Residence time for a particular element may be calculated by this equation:

$$\text{Residence Time} = \frac{\text{Amount of element in the ocean}}{\begin{array}{c}\text{Rate at which the element is added}\\ \text{to (or removed from) the ocean}\end{array}}$$

Additions of salts from the mantle or from the weathering of rock are balanced by subtractions of minerals being bound into sediments. Dissolved salts precipitate out of the water, and the hard parts of living organisms containing silicon and calcium carbonate drift slowly down to the seabed. Some of these sediments are removed from the ocean and drawn into the mantle at subduction zones by the cycling of crustal plates. Input (from runoff and outgassing) equals outfall (binding into sediments) for each dissolved component.

The residence time of an element depends on its chemical activity. Atoms (or ions) of some elements, such as aluminum and iron, remain in seawater for a relatively short time before becoming incorporated in sediments; others, such as chloride, sodium, and magnesium, remain in water for millions of years. The approximate residence times for the major constituents of

seawater are shown in **Table 7.3**. Even seawater itself has a residence time (see Box 7.2).

If constituent minerals remain in ocean water longer than the ocean's mixing time, they will become evenly distributed throughout the ocean. Because of the vigorous activity of currents, the **mixing time** of the ocean is thought to be on the order of 1,600 years, so the ocean has been mixed hundreds of thousands of times during its long history. The relatively long residence times of seawater's major constituents assure thorough mixing, the foundation of Forchhammer's principle of constant proportions.

Seawater's Constituents May Be Conservative or Nonconservative

As we have seen, the major constituent salts maintain constant ratios in seawater, but the quantity and proportions of minor and trace elements can change because living things take them up or excrete them, or because of local geological events.

Those constituents of seawater that occur in constant proportion or change very slowly through time are **conservative constituents**. Conservative elements have long residence times. Not surprisingly, these are the most abundant dissolved constituents of water—the ones that constitute the bulk of the ocean's dissolved material, seen at the top of Table 7.3. The inert gases dissolved in the ocean (and the water of the ocean itself) are also conservative constituents.

Nonconservative constituents are those substances dissolved in seawater that are tied to biological or seasonal cycles or to very short geological cycles. They have short residence times.

Biologically important nonconservative constituents include dissolved oxygen produced by plants, carbon dioxide produced by animals, silica and calcium compounds needed for plant and animal shells, or the nitrates and phosphates needed for production of protein and other biochemicals. Aluminum, with a residence time of only 600 years, is rapidly removed by adsorption onto clay sediment particles and other objects, so it is a nonconservative element. Aluminum is very rare in seawater (10 parts per billion). Many trace elements are in great biological demand or tend to form insoluble compounds in water.

Oceanic residence times calculated to be less than 1,600 years—the ocean's mixing time—are mainly a function of strong local demand variations. The definition of residence time assumes the existence of steady-state conditions and a well-mixed ocean, so a residence time less than the ocean's mixing time is mainly an indication of an element's reactivity.

> **U**nlike solids, gases dissolve most readily in *cold* water.

CONCEPT CHECK

Before going on to the next section, check your understanding of some of the important ideas presented so far:

How is seawater's salinity expressed?

What are some of seawater's colligative properties? Does pure water have colligative properties?

Figure 7.7 An average of worldwide ocean surface salinity. Note the high-salinity areas corresponding to high rates of evaporation in the central Atlantic Ocean, the Mediterranean Sea, and Arabian Sea west of India. The polar regions and ocean areas adjacent to Earth's great rivers (the Ganges-Brahmaputra system, the Amazon, and the Mississippi) show correspondingly lower salinities.

Sea Surface Salinity

5   14   25   40
practical salinity units

Other than hydrogen and oxygen, what are the most abundant ions in seawater?

What are the sources of the ocean's dissolved solids?

What is the Principle of Constant Proportions?

How is salinity determined?

What is meant by "residence time"? Does seawater itself have a residence time?

7.3 Gases Dissolve in Seawater

Most gases in the air dissolve readily in seawater at the ocean's surface. Plants and animals living in the ocean require these dissolved gases to survive. No marine animal has the ability to break down water molecules to obtain oxygen directly, and no marine plant can manufacture enough carbon dioxide to support its own metabolism. In order of their relative abundance, the major gases found in seawater are nitrogen, oxygen, and carbon dioxide (see **Table 7.4**). The proportions of dissolved gases in the ocean are very different from the proportions of the same gases in the atmosphere because of differences in their solubility in water and air.

Unlike solids, gases dissolve most readily in *cold* water. A cubic meter of chilly polar water usually contains a greater volume of dissolved gases than a cubic meter of warm tropical water.

Table 7.3 Approximate Residence Times for Constituents of Seawater

Constituent	Residence Time (years)
Chloride (Cl^-)	100,000,000
Sodium (Na^+)	68,000,000
Magnesium (Mg^{2+})	13,000,000
Potassium (K^+)	12,000,000
Sulfate (SO_4^{2-})	11,000,000
Calcium (Ca^{2+})	1,000,000
Carbonate (CO_3^{2-})	110,000
Silicon (Si)	20,000
Water (H_2O)	4,100
Manganese (Mn)	1,300
Aluminum (Al)	600
Iron (Fe)	200

Sources: Data from Broecker and Peng, 1982; Bruland, 1983; Riley and Skirrow, 1975.

Table 7.5 shows the relation between temperature and the saturation solubility of oxygen and nitrogen in seawater. Note that the dissolved oxygen concentrations in parts of the tropical ocean may be so low that animals may be unable to survive. Tropical waters can be further stressed by pollutants that consume oxygen, such as sewage or agricultural runoff.

Dissolved nitrogen gas is a conservative component of seawater, while biologically active oxygen and carbon dioxide are nonconservative components.

Nitrogen Is the Most Abundant Gas Dissolved in Seawater

About 48% of the dissolved gas in seawater is nitrogen. (In contrast, the atmosphere is slightly more than 78% nitrogen by volume.) The upper layers of ocean water are usually saturated with nitrogen gas; that is, additional nitrogen gas will not dissolve. Living organisms require nitrogen to build proteins and other important biochemicals, but the vast majority of them cannot use the nitrogen gas in the atmosphere and ocean directly. It must first be bound, or *fixed*, into usable chemical forms by specialized organisms. Though some species of bottom-dwelling bacteria can manufacture usable nitrates from the nitrogen gas dissolved in seawater, most of the nitrogen compounds needed by living organisms must be recycled among the organisms themselves.

Dissolved Oxygen Is Critical to Marine Life

About 36% of the gas dissolved in the ocean is oxygen, but there is about a hundred times more gaseous oxygen in Earth's atmosphere than is dissolved in the whole ocean. An average of 6 milligrams of oxygen is dissolved in each liter of seawater (that is, 6 parts per million parts of gaseous oxygen per liter of seawater, by weight). Yet this small amount of oxygen is a vital resource for animals that extract oxygen with gills. The sources of the ocean's dissolved oxygen are the photosynthetic activity of plants and plant-like organisms and the diffusion of oxygen from the atmosphere.

The Ocean Is a Vast Carbon Reservoir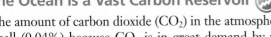

The amount of carbon dioxide (CO_2) in the atmosphere is very small (0.04%) because CO_2 is in great demand by photosynthetic plants as a source of carbon for growth. Carbon dioxide is very soluble in water, though; the proportion of dissolved CO_2 in water is about 15% of all dissolved gases. Because CO_2 combines chemically with water to form a weak acid (carbonic acid, H_2CO_3), water can hold perhaps a thousand times more carbon dioxide than either nitrogen or oxygen at saturation. Carbon dioxide is quickly used for marine photosynthesis, however; dissolved quantities of CO_2 are almost always much less than this theoretical maximum. Even so, there is about 60 times as much CO_2 dissolved in the ocean as in the atmosphere.

CO_2 moves quickly from atmosphere to ocean, but more slowly from ocean to atmosphere. This is because some dissolved CO_2 forms carbonate ions, which combine with calcium ions in seawater to form the calcium carbonate ($CaCO_3$) used by many marine organisms to build shells and skeletons. When these organisms die, their coverings and bones sink to form sediments that may, in time, become limestone rock. Most of Earth's surface carbon—40,000 times that in the total mass of all life on Earth—is stored in sediments. This carbon slowly re-enters the cycle as sediments dissolve and re-form CO_2 that can enter the atmosphere. Geological uplift can also bring sediments to the surface, exposing the carbonate rock to chemical attack by oxygen and conversion to CO_2. Acidic rain falling on exposed limestone rock also releases some CO_2 back into the atmosphere. (This cycle will be seen in Figure 7.10.)

TABLE 7.5 The Solubility of Gases Decreases as Temperature Increases

Temperature	Solubility (mL/L at atmosphere pressure and salinity of 33‰)[a]		
	N_2	O_2	CO_2
0°C (32°F)	14.47	8.14	8,700.0
10°C (50°F)	11.59	6.42	8,030.0
20°C (68°F)	9.65	5.26	7,350.0
30°C (86°F)	8.26	4.41	6,600.0

[a]Figures are given at *saturation*, the maximum amount of gas held in solution before bubbling begins.

Source: F.G. Walton-Smith, *CRC Handbook of Marine Science* (Cleveland, OH: CRC Press, 1974). Table created by Tom Garrison.

Table 7.4 **Major Gases in the Atmosphere and Ocean**

Gas	Percent of Gas in Atmosphere, by Volume	Percent of Dissolved Gas in Seawater at Ocean Surface, by Volume	Concentration in Seawater in Parts per Million, by Mass for Ocean as a Whole
Nitrogen (N_2)	78.08	48	10–18
Oxygen (O_2)	20.95	36	0–13
Carbon dioxide (CO_2)	0.038 in 2008	15[a]	64–107

Sources: Data from Weihaupt, 1979; Hill, 1963.

[a]Also present in seawater as carbonic acid, carbonate ions, and bicarbonate ions.

Gas Concentrations Vary with Depth

Figure 7.8 illustrates how CO_2 and oxygen concentrations vary with depth. Oxygen is abundant near the surface because of the activity of marine photosynthesizers (plants and plantlike organisms). The oxygen concentration decreases below the sunlit layer because of the respiration of marine animals and bacteria, and because of the oxygen consumed by the decay of tiny dead organisms slowly sinking through the area. In contrast, plants and plantlike organisms use carbon dioxide during photosynthesis, so surface levels of CO_2 are low. Because photosynthesis cannot take place in the dark, CO_2 given off by animals and bacteria tends to build up at depths below the sunlit layer. Levels of CO_2 also increase with depth because its solubility increases as pressure increases and temperature decreases.

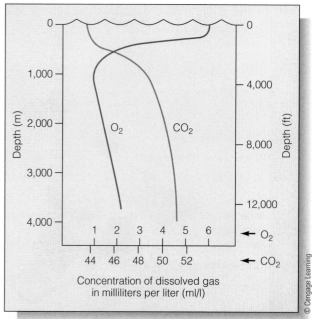

Figure 7.8 How concentrations of oxygen and carbon dioxide vary with depth. High concentrations of oxygen at the surface are usually by-products of photosynthesis in the ocean's brightly lit upper layer. Since plants and plantlike organisms require carbon dioxide for photosynthesis, surface CO_2 concentrations tend to be low. A decrease in oxygen below the sunlit upper layer usually results from the respiration of bacteria and marine animals, activity that tips the balance in favor of carbon dioxide. Oxygen levels are slightly higher in deeper water because fewer animals are present to take up the oxygen that reaches these depths.

CONCEPT CHECK

Before going on to the next section, check your understanding of some of the important ideas presented so far:

Solids dissolve more readily in warm water than in cool. Does this also hold true for gases?

What dissolved gases are most abundant in the ocean?

Why is oxygen sometimes in short supply in the tropical ocean? Might this shortage affect marine life?

Can carbon dioxide be stored in the ocean? How does the amount of carbon dioxide in the ocean compare to the amount in the atmosphere?

7.4 The Ocean's Acid–Base Balance Varies with Dissolved Components and Depth

As you saw in Chapter 6, water can separate to form hydrogen ions (H^+) and hydroxide ions (OH^-). In pure water, these two ions are present in equal concentrations. An imbalance in the ions produces either an acidic or a basic solution. An **acid** is a substance that *releases* a hydrogen ion in solution. A **base** is a substance that *combines with* a hydrogen ion in solution. A solution containing a base is also called an **alkaline** solution.

The acidity or alkalinity of a solution is measured in terms of the **pH scale**, which measures the concentration of hydrogen ions in a solution. An excess of hydrogen ions (H^+) in a solution makes that solution acidic. An excess of hydroxide ions (OH^-) makes a solution alkaline. **Figure 7.9** shows a pH scale and the pH values of a few familiar solutions. The scale is logarithmic, which means that a change of one pH unit represents a 10-fold change in the hydrogen ion concentration. A modern nonphosphate detergent is a thousand times more alkaline than seawater, and black coffee is a hundred times more acidic than pure water. Pure water, which is neutral (neither acidic nor basic), has a pH of 7; lower numbers indicate greater acidity (more H^+), and higher numbers indicate greater alkalinity (fewer H^+).

Seawater is slightly alkaline; its average pH is about 8.0. This seems odd because of the large amount of CO_2 dissolved in the ocean. If dissolved CO_2 combines with water to form carbonic

THINKING BEYOND THE FIGURE

How do you think oxygen reaches the deep ocean for use by the animals living there?

acid, why is the ocean mildly alkaline and not slightly acidic? When dissolved in water, however, CO_2 is actually present in several different forms. Carbonic acid (H_2CO_3) is only one of these. In water solutions, some carbonic acid breaks down to produce the hydrogen ion (H^+), the bicarbonate ion (HCO_3^-), and the carbonate ion (CO_3^{2-}).

It works like this:

Step One:	$CO_2 + H_2O \Leftrightarrow H_2CO_3$
Step Two:	$H_2CO_3 \Leftrightarrow HCO_3^- + H^+$
Step Three:	$HCO_3^- + H^+ \Leftrightarrow CO_3^{2-} + 2H^+$

The double-headed arrows indicate the reactions can proceed in either direction.

In Step One, carbon dioxide and water combine to form carbonic acid. Carbonic acid rapidly dissociates into a bicarbonate ion (HCO_3^-) and a hydrogen ion (H^+) in Step Two. Most of the CO_2 dissolved in seawater ends up as HCO_3^-. Some of the bicarbonate and hydrogen ions will then combine to form carbonate (CO_3^{2-}) ions and two hydrogen ions ($2H^+$) in Step

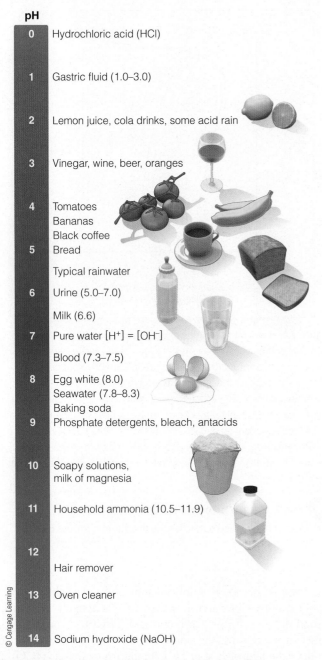

pH

0 — Hydrochloric acid (HCl)

1 — Gastric fluid (1.0–3.0)

2 — Lemon juice, cola drinks, some acid rain

3 — Vinegar, wine, beer, oranges

4 — Tomatoes
Bananas
Black coffee

5 — Bread

Typical rainwater

6 — Urine (5.0–7.0)

Milk (6.6)

7 — Pure water [H$^+$] = [OH$^-$]

Blood (7.3–7.5)

8 — Egg white (8.0)
Seawater (7.8–8.3)
Baking soda

9 — Phosphate detergents, bleach, antacids

10 — Soapy solutions,
milk of magnesia

11 — Household ammonia (10.5–11.9)

12 —

Hair remover

13 — Oven cleaner

14 — Sodium hydroxide (NaOH)

© Cengage Learning

Figure 7.9 The pH scale. pH 7 is neutral; higher numbers represent bases and lower numbers represent acids. The ocean's pH is slightly alkaline at ~8.

Three. At any given pH CO$_2$, H$_2$CO$_3$, HCO$_3$$^-$, CO$_3$$^{2-}$, and H$^+$ exist in equilibrium with each other.

But what if the pH changes?

If acid is added to seawater (perhaps at a sewage outfall or volcanic eruption), and the pH of the ocean drops (becomes more acidic), the reaction proceeds toward Step One (to the left), raising the pH by removing the excess H$^+$ ions. If an alkaline solution is added, and the pH of the ocean rises (becomes more alkaline), the reaction proceeds toward Step Three (to the right), lowering the pH by releasing H$^+$ ions. This behavior acts

to **buffer** the water, preventing broad swings of pH when acids or bases are introduced. Rivers and lakes usually have a much smaller buffering capacity since they have lower amounts of total dissolved inorganic carbon. This makes them more susceptible to pH swings when acids or bases are added at waste outfalls or form acid rain.

Figure 7.10 shows the progress of these reactions in the ocean. At the normal pH of seawater (slightly above 8.0), about 80% of the carbon compounds are in the form of HCO$_3$$^-$. Carbon dioxide is used in photosynthesis by plantlike organisms for the production of sugars. A decrease in dissolved CO$_2$ will cause bicarbonate to dissociate and form more CO$_2$. Carbonate ions combine with hydrogen ions to form more bicarbonate. Calcium carbonate (CaCO$_3$) can then dissolve from the carbonate shells of marine organisms or from seabed sediments.

Though seawater remains slightly alkaline, it is subject to some variation. In areas of rapid photosynthesis, for example, pH will rise because CO$_2$ is used by plants and plantlike organisms. The reactions move toward Step One, removing free H$^+$ ions. Because temperatures are generally warmer at the surface, less CO$_2$ can dissolve in the first place. So surface pH in warm productive water is usually around 8.5. **Figure 7.11** shows the relative abundance of carbonic acid, bicarbonate, and carbonate in seawater as a function of pH. Seawater has an average pH of about 8, and the bicarbonate ion is most prevalent. As the graph shows, carbonic acid dominates at low pH and the carbonate ion at higher pH.

At middle depths and in deep water, more CO$_2$ may be present than at the surface. Its source is the respiration of animals and the decay of the remains of organisms falling from the sunlit layer above (bacterial respiration). With cold temperatures, high pressure, and no photosynthetic plants to remove it, this CO$_2$ will lower the pH of water, making it less alkaline with depth (see **Figure 7.12**). Deep, cold seawater below 4,500 meters (15,000 feet) has a pH of around 7.5. This lower pH can dissolve calcium-containing marine sediments. A drop to pH 7 can occur at the deep ocean floor when bottom bacteria consume oxygen and produce hydrogen sulfide. You may recall from Chapter 5 that sediments containing calcium carbonate are rarely found in deep water. Now does Figure 5.14, the calcium carbonate compensation depth—CCD—make more sense?

As you will see in Chapter 18, burning fossil fuels to provide energy for transportation and industry produces a large amount of carbon dioxide. The ocean serves as a natural sink for this excess carbon dioxide—about 25% of the CO$_2$ emitted by human activity in the period 2000 to 2006 was taken up by the ocean. The CO$_2$ combines with water to become carbonic acid. The additional carbonic acid produced has made the ocean less alkaline—indeed, since the industrial revolution began, it is estimated that surface ocean pH has dropped by about 0.1 units (on the logarithmic pH scale), and it is estimated that it will drop by an additional 0.3 to 0.5 units by 2100 as the ocean absorbs more anthropogenic (human-generated) carbon dioxide (**Figure 7.13**). Increasing acidity decreases the concentration of calcium carbonate in the water, making it unavailable to organisms for

Figure 7.10 Carbon dioxide (CO_2) combines readily with seawater to form carbonic acid (H_2CO_3). Carbonic acid can then lose a H^+ ion to become a bicarbonate ion (HCO_3^-), or two H^+ ions to become a carbonate ion (CO_3^{2-}). Some bicarbonate ions dissociate to form carbonate ions, which combine with calcium ions in seawater to form calcium carbonate ($CaCO_3$), used by some organisms to form hard shells and skeletons. When their builders die, these structures may fall to the seabed as carbonate sediments, eventually to be redissolved. As the double arrows indicate, all these reactions may move in either direction.

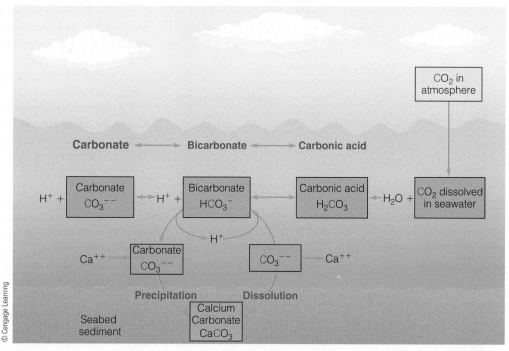

the construction of some hard parts (shells, skeletons, rigid coverings, and so forth).

The ocean's extensive and productive coral reefs are in particular danger from accelerating ocean acidification. In 2014, the atmospheric concentration of CO_2 was about 398 parts per million by volume—at no time in the past 10 million years has the concentration been so high. Since 1990, the growth of coral in the Australian Barrier Reef has slowed to its lowest rate in at least 400 years, possibly as a result of the warmer and more acidic ocean. Additionally, coral skeletons appear to have lost some of their density, and thus are thinner and more brittle. The ability of coral colonies to withstand wave stress may be compromised, and their resistance to predation by grazing animals (parrotfish, urchins, and other gnawing species) lowered. Also, as coral species exert more energy to maintaining skeletal integrity, cellular resources are diverted from other essential processes such as reproduction and fighting disease. Coral reefs are endangered worldwide.

Figure 7.11 The relative abundance of carbonic acid, bicarbonate, and carbonate in seawater is a function of pH. Seawater has an average pH of about 8, at which the bicarbonate ion is most prevalent. As the graph shows, carbonic acid dominates at low pH and the carbonate ion at higher pH.

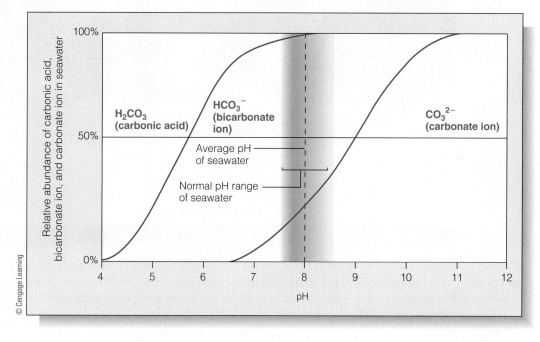

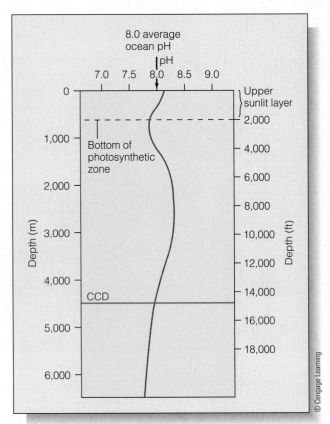

Figure 7.12 The variation in pH with depth. The average calcium carbonate compensation depth (CCD) is represented by a red line. For a brief review of the CCD, please see Figure 5.14.

Apart from the obvious loss of species diversity and extinctions that a rapid decline in coral reefs would involve, there are other considerations. People, infrastructure, lagoon and estuarine ecosystems, mangrove forests, fisheries, aquaculture operations, tourism, sand generation, bird populations, and a host of other interconnected systems would be negatively affected.

Indeed, the whole ocean could undergo rapid change: Could the CCD rise toward the surface, changing the ecological balance of organisms in the upper productive layer of the ocean? Our inadvertent global experiment continues.[4]

[4]A glance ahead at Figures 13.20c and 18.23 might be of interest.

CONCEPT CHECK

Before going on to the next section, check your understanding of some of the important ideas presented so far:

How is pH expressed?

What happens when carbon dioxide dissolves in seawater?

What factors affect seawater's pH? How does the pH of seawater change with depth? Why?

What's a buffer? Why is seawater's buffering capacity important to marine life?

What might be some of the consequences of ocean acidification?

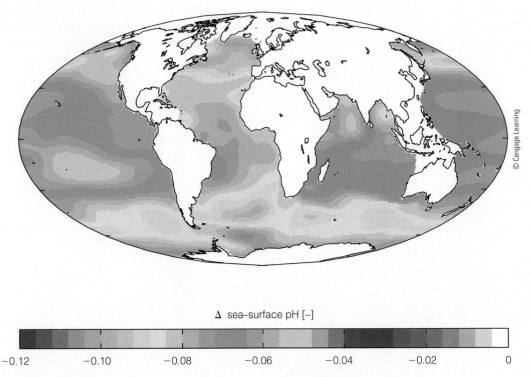

Δ sea–surface pH [–]

−0.12   −0.10   −0.08   −0.06   −0.04   −0.02   0

a The ocean is becoming more acidic as it absorbs additional carbon dioxide from the atmosphere. A less alkaline environment will make it more difficult for organisms to build hard structures containing calcium (shells, coral, and so forth). The chart shows changes in sea surface pH between the 1700s and the late 1990s.

Figure 7.13 The ocean's pH is changing.

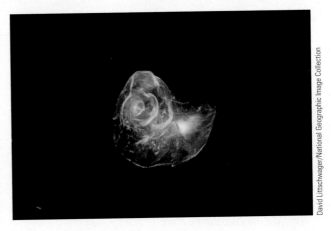

b, **c** Progressive dissolution of a pteropod (a small animal with a calcareous shell) in acidified seawater.

Figure 7.13 The ocean's pH is changing.

💬 **THINKING BEYOND THE FIGURE**

Pteropods are fairly delicate. Do you think ocean acidification could affect more robust shells—clams, for example?

Chapter in Perspective

In this chapter you learned that water has the remarkable ability to dissolve more substances than any other natural solvent. Though most solids and gases are soluble in water, the ocean is in chemical equilibrium, and neither the proportion nor amount of most dissolved substances changes significantly through time. Most of the properties of seawater are different from those of pure water because of the substances dissolved in the seawater.

About 3.5% (35‰) of seawater consists of dissolved substances. These almost always exist as ions—"salts" don't exist in the ocean. The most abundant ions dissolved in seawater are chloride, sodium, and sulfate. Seawater is not concentrated river water or rain water—its chemical composition has been altered by circulation through the crust at oceanic spreading centers and by other chemical and biological processes.

Most gases in the air dissolve readily in seawater at the ocean's surface. Plants and animals living in the ocean require these dissolved gases to survive. In order of their relative abundance, the major gases found in seawater are nitrogen, oxygen, and carbon dioxide. The proportions of dissolved gases in the ocean are very different from the proportions of the same gases in the atmosphere because of differences in their solubility in water and air.

When you've finished reading, consider Chapters 6 and 7 together. Your appreciation of the uniqueness of water will grow as your oceanographic perspective expands.

In the next chapter you'll learn how atmosphere and ocean work together to shape Earth's weather, climate, and natural history. Our past discussions of heat, temperature, convection, dissolved gases, and geological cycles will be put to use. You'll find surprising oceanic connections among storms, tropical organisms, deserts, and balloon races!

QUESTIONS FROM STUDENTS

1. If outgassing continues today, why aren't the oceans getting bigger? By now shouldn't they have covered the continents?

Well, yes, outgassing does continue, but the water is being drawn from the mantle. Some marine geologists believe that the mantle shrinks a little, making the ocean basins a bit deeper, which accommodates the ocean's greater volume. Also, geological processes have increased the amount of continental crust, and water is drawn back into the mantle at subduction zones. Though some of this water eventually reappears at the surface in volcanic vapor, some is incorporated back into mantle material. It probably just about balances out.

2. I see that there is 100 times more oxygen in the atmosphere than in the ocean. How can that be? Isn't water 86% oxygen by weight?

Yes, but the oxygen of water (H_2O) is bonded tightly to hydrogen atoms. Unlike atmospheric oxygen, the oxygen in water molecules cannot be used by organisms for respiration. Fish can't extract this oxygen with their gills. Marine animals must depend on dissolved oxygen, paired molecules of oxygen (O_2) present in the water and free to move through the gill membranes. Compared to the atmosphere, very little of this free oxygen is present in the ocean.

3. Why does the ocean smell like the ocean? I can sometimes tell I'm near the coast before I see it.

Salt doesn't have a smell, so microscopic crystals of salt are not responsible. A gaseous compound normally emitted by an abundant drifting photosynthetic organism (one of the phytoplankters you'll meet in Chapter 14) may be responsible for the characteristic scent. Dimethyl sulfide given off by *E. huxleyi* combines with algae pheromones to make the sea smell like the sea.

4. There is more than $28 million worth of gold (at 2014 prices) in each cubic kilometer of seawater. Why don't we remove some of it and pay off the national debt?

Easy to say, difficult to do. For one thing, extracting the gold will require sophisticated chemical treatment. But for the sake of argument, let's assume you perfect a magic method of precipitating gold from seawater with a simple wave of your hand. All you have to do is pump water out of the ocean into a large holding tank to do your trick. Unfortunately the energy cost of lifting the water just a few millimeters into the tank would eat up all the profits, and then some. German economists made a study after the First World War to see if this method could be used to retire their war debt. It was shown to be fruitless. So much for get-rich-quick schemes then and now.

5. Some essential resources are relatively abundant in ocean water. Are there any prospects for chemical "mining" of seawater?

Yes. Commercial extraction of magnesium from seawater began in 1940. Seawater is treated with calcium hydroxide and hydrochloric acid to form magnesium chloride. This substance is dried and electrolytically separated into chlorine gas (which can be used to make more hydrochloric acid) and magnesium metal. Magnesium is an essential component of the aluminum alloys from which aircraft, beverage cans, and automobile parts are manufactured.

Many nonmetal resources are also obtained from seawater. For example, bromine—an element important in the production of motor fuels, photographic film, dyes, and insecticides—is extracted from concentrated brines or from crystallized sea salts. And, as we'll see in Chapter 17, the salts themselves are among the most important of the ocean's physical resources.

TERMS AND CONCEPTS TO REMEMBER

acid	diffusion	nonconservative constituent	salinity
alkaline	excess volatiles	pH scale	salinometer
base	Forchhammer's principle	precipitate	saturation
buffer	hydrologic cycle	principle of constant	solute
chemical equilibrium	ion	proportions	solution
chlorinity	ionic bond	residence time	solvent
colligative properties	mixing time	S	trace element
conservative constituent	mixture		

STUDY QUESTIONS

Thinking Critically

1. Where did the water of the ocean come from? Go ahead—try the "long version," starting with stars.
2. How are chemical methods of determining salinity dependent on the principle of constant proportions?
3. What was the earthly origin of the sodium and chloride ions in common table salt?
4. Technically, there are no "salts" in seawater. How can that be?
5. How are seawater's conservative constituents different from its nonconservative constituents? Give an example of each.
6. Which dissolved gas is present in the ocean in much greater proportion than in the atmosphere? Why the disparity?
7. There's lots of oxygen in water (H_2O). Why can't fish breathe that? Why do they have to breathe oxygen dissolved in the water?

8. Seawater is denser than freshwater. A ship moving from the Atlantic Ocean into the Great Lakes goes from seawater to freshwater. Will the ship sink farther into the water during the passage, stay at the same level, or rise slightly?

Thinking Analytically

1. Imagine you have two small containers (say 10 milliliters each). Fill one with seawater and the other with pure water. Now, drop by drop, add dilute hydrochloric acid to each container and swirl the solution. After each drop, check the pH. What do you suppose a graph of pH versus the number of drops would look like for pure water? For seawater? Explain the difference.
2. Why is the amount of dissolved chloride ion (Cl^-) a constant in the ocean, but the amount of dissolved carbon dioxide (CO_2) greatly variable?
3. What is the salinity of a sample in parts-per-thousand if its **chlorinity** measures 13.4‰?

GLOBAL GEOSCIENCE WATCH

Visit www.cengagebrain.com to access MindTap, a complete digital course which includes access to Global Geoscience Watch and more. Within the GREENR database, search "ocean acidification" and focus on recent articles on this topic. Select at least three news, magazine, or journal articles that detail specific consequences of ocean acidification. Be prepared to report back to the class what you have learned and to participate in a class current events discussion focused on ocean acidification.

8 Circulation of the Atmosphere

KEY CONCEPTS

Earth's ocean and atmosphere are unevenly heated by the sun—more solar energy is absorbed near the equator than near the poles. The atmosphere circulates in response to this difference in heating. NASA

Moving objects tend to move to the right of their initial course in the northern hemisphere (and to the left in the southern hemisphere). This tendency is called the *Coriolis effect.* © Cengage Learning

The atmosphere circulates in six large circuits (three in each hemisphere). NASA

Storms can form between two air masses (frontal storms) or within one air mass (tropical cyclones). NASA

Earth Observatory image by Robert Simmon with data courtesy of the NASA/NOAA GOES Project Science team

A storm broods over Marco Island's Tigertail beach, Florida. Violent thunderstorms are a daily summer occurrence in this part of the world.

RAUL TOUZON/National Geographic Creative

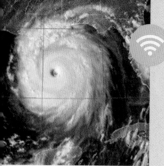
Naval Research Laboratory-Monterey

8.1 The Atmosphere and Ocean Interact with Each Other

The **atmosphere** is the volume of gases, water vapor, and airborne particles enveloping Earth. Earth's atmosphere and ocean are intertwined, their gases and waters freely exchanged. Gases entering the atmosphere from the ocean have important effects on climate, and gases entering the ocean from the atmosphere can influence sediment deposition, the distribution of life, and some of the physical characteristics of the seawater itself.

The atmosphere is surprisingly thin—about 80% of its mass lies within 15 kilometers (9 miles) of Earth's surface. While the ocean is heated from above by the sun (and is thus relatively stable), the atmosphere is heated from below as sunlight is absorbed into land and ocean. Because warm air rises, the lower atmosphere is consequently *un*stable, its conditions changing rapidly from day to day and place to place (**Figure 8.1**).

Figure 8.2 illustrates our atmosphere's layers. The **troposphere** (*trope;* overturn), the turbulent layer in which nearly all weather occurs, begins at Earth's surface and extends to 9 kilometers (30,000 feet) at the poles and 17 kilometers (56,000 feet) at the equator. The calmer *stratosphere* extends to about 51 kilometers (32 miles or 170,000 feet). Atmospheric pressure at the top of the stratosphere is about 1/1000th that of

Figure 8.1 A waterspout touches down near Tampa Bay, Florida. Heated from below, the lower atmosphere is inherently unstable. Storms and winds result when heat is redistributed.

Joey Mole

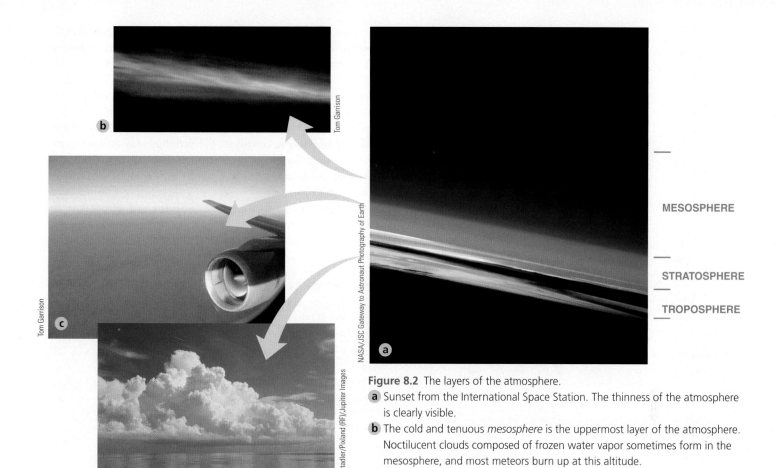

Figure 8.2 The layers of the atmosphere.

a Sunset from the International Space Station. The thinness of the atmosphere is clearly visible.

b The cold and tenuous *mesosphere* is the uppermost layer of the atmosphere. Noctilucent clouds composed of frozen water vapor sometimes form in the mesosphere, and most meteors burn up at this altitude.

c Pilots and passengers appreciate the calm *stratosphere* where airflow is much less turbulent than in the troposphere.

d Most weather occurs in the unsettled *troposphere,* the layer nearest Earth's surface that contains most of the atmosphere's mass.

sea level. The cold and tenuous *mesosphere* lies above the stratosphere. Its upper reaches gradually give way to space.

Water evaporated from the ocean surface and moved by **wind**, the mass movement of air, helps minimize worldwide extremes of surface temperature; through rain, it provides moisture for agriculture. The weather that so profoundly affects our daily lives is shaped at the junction of wind and water, and the flow of air greatly influences the movement of water in the ocean. **Weather** is the state of the atmosphere at a specific time and place; **climate** is the long-term statistical sum of weather in an area. To understand the interactions of atmosphere and ocean that produce weather and climate, we begin by examining the composition and properties of the atmosphere.

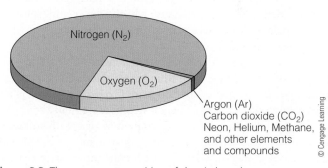

Figure 8.3 The average composition of dry air, by volume.

CONCEPT CHECK

Before going on to the next section, check your understanding of some of the important ideas presented so far:

What atmospheric layer comprises most of the atmosphere's mass and generates the weather?

How is weather different from climate?

8.2 The Atmosphere Is Composed Mainly of Nitrogen, Oxygen, and Water Vapor

The lower atmosphere is a nearly homogeneous mixture of gases, most plentifully nitrogen (78.1%) and oxygen (20.9%). Other elements and compounds, as shown in **Figure 8.3**, make up less than 1% of the composition of the lower atmosphere.

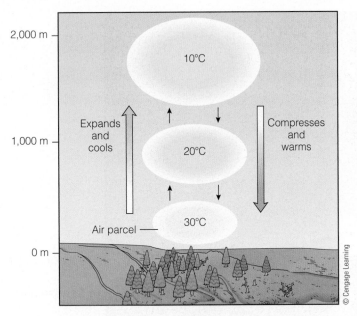

2,000 m

10°C

1,000 m

Expands
and
cools

20°C

Compresses
and
warms

Air parcel

30°C

0 m

© Cengage Learning

Figure 8.4 Expansion and cooling (or compression and warming) occur as an air parcel moves up (or down). Descending air warms as it compresses—the droplets (clouds) may evaporate.

 THINKING BEYOND THE FIGURE

If you exhale slowly on a cold day, your breath becomes visible. Why? It's not rising or falling a great distance.

Air is never completely dry; **water vapor**, the gaseous form of water, can occupy as much as 4% of its volume. Sometimes liquid droplets of water are visible as clouds or fog, but more often the water is simply there—invisible in vapor form, having entered the atmosphere from the ground, plants, and sea surface. The residence time of water vapor in the lower atmosphere is about 10 days. Water leaves the atmosphere by condensing into dew, rain, or snow.

Air has mass. A 1-square-centimeter (0.16-square-inch) column of dry air, extending from sea level to the top of the atmosphere, weighs about 1.04 kilograms (2.3 pounds). A 1-square-foot column of air the same height weighs more than a ton.

The temperature and water content of air greatly influence its density. Because the molecular movement associated with heat causes a mass of warm air to occupy more space than an equal mass of cold air, warm air is less dense than cold air. But contrary to what we might guess, humid air is *less* dense than dry air at the same temperature—because molecules of water vapor have less mass than the nitrogen and oxygen molecules that the water vapor displaces.

Near Earth's surface, air is packed densely by its own weight. Air lifted from near sea level to a higher altitude is subjected to less pressure and will expand. As anyone knows who has felt the cool air rushing from an open tire valve, air becomes *cooler* when it *expands*. The opposite effect is also familiar: Air *compressed* in

a tire pump becomes *warmer*. Air descending from high altitude warms as it is compressed by the higher atmospheric pressure near Earth's surface.

Warm air can hold more water vapor than cold air can. Water vapor in rising, expanding, cooling air will often condense into clouds (aggregates of tiny droplets) because the cooler air can no longer hold as much water vapor (**Figure 8.2d**). If rising and cooling continue, the droplets may coalesce into raindrops or snowflakes. The atmosphere will then lose water as **precipitation**, liquid or solid water that falls from the air to Earth's surface. These rising–expanding–cooling and falling–compressing–heating relationships are important in understanding atmospheric circulation, weather, and climate (**Figure 8.4**).

CONCEPT CHECK

Before going on to the next section, check your understanding of some of the important ideas presented so far:

What is the composition of air?

Can more water vapor be held in warm air or cool air?

Which is denser at the same temperature and pressure: humid air or dry air?

How does air's temperature change as it expands? As it is compressed?

What happens when air containing water vapor rises?

8.3 The Atmosphere Moves in Response to Uneven Solar Heating and Earth's Rotation

Atmospheric circulation is powered by sunlight. Only about 1 part in 2.2 billion of the sun's radiant energy is intercepted by Earth, but that amount averages 7 million calories per square meter per day at the top of the atmosphere—or, for Earth as a whole, an impressive 17 trillion kilowatts (23 trillion horsepower)! As may be seen in **Figure 8.5**, on a global basis about 51% of the incoming energy is absorbed by Earth's land and water surface. How much light penetrates the ocean depends greatly on several factors: the angle at which it approaches, the sea state (surface turbulence), the presence of an ice covering or light-colored foam, and others.

The 51% of short-wave solar energy (light) striking land and sea is converted to longer-wave infrared radiation (heat) and then transferred into the atmosphere by conduction, radiation, and evaporation. The atmosphere, like the land and ocean, eventually radiates this heat back into space in the form of long-wave infrared radiation. The heat-input and heat-outflow "account" for Earth is its **heat budget**. As in your personal financial budget, income must eventually equal outgo. Over long periods of time, the total incoming heat (plus that from earthly sources) equals the total heat radiating into the cold of space. So, Earth

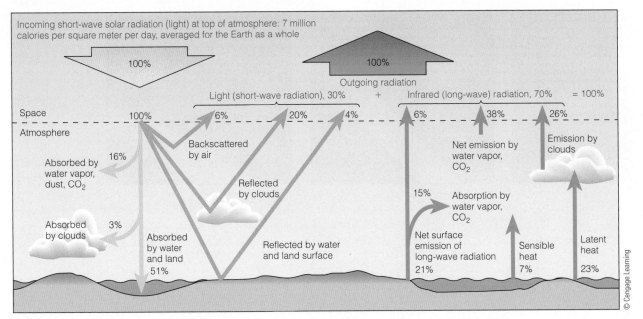

Figure 8.5 An estimate of the heat budget for Earth. On an average day, about half of the solar energy arriving at the upper atmosphere is absorbed at Earth's surface. Light (short-wave) energy absorbed at the surface is converted into heat. Heat leaves Earth as infrared (long-wave) radiation. Since input equals output over long periods of time, the heat budget is balanced.

 THINKING BEYOND THE FIGURE
How do you think our present period of global warming relates to Earth's overall heat budget?

is in **thermal equilibrium**: it is growing neither significantly warmer nor colder.[1]

The Solar Heating of Earth Varies with Latitude

Although the heat budget for Earth *as a whole* is in balance, the heat budget for its *different latitudes* is not. As can be seen in **Figure 8.6**, sunlight striking polar latitudes spreads over a greater area (that is, polar areas receive less radiation per unit area) than sunlight at tropical latitudes. Near the poles, light also filters through more atmosphere and approaches the surface at a low angle, favoring reflection. Polar regions receive no sunlight at all during the depths of local winter. By contrast, at tropical latitudes, sunlight strikes at a more nearly vertical angle, which distributes the same amount of sunlight over a much smaller area. The light passes through less atmosphere and minimizes reflection. Tropical latitudes thus receive significantly more solar energy than the polar regions, and mid-latitude areas receive more heat in summer than in winter.

Figure 8.7 shows heat received versus heat re-radiated at different latitudes. Near the equator, the amount of solar energy received by Earth greatly exceeds the amount of heat radiated into space. In the polar regions the opposite is true.

So why doesn't the polar ocean freeze solid and the equatorial ocean boil away? The reason is that water itself is moving huge amounts of heat between tropics and poles. Water's thermal properties make it an ideal fluid to equalize the polar-tropical heat imbalance. Water's heat capacity moves heat poleward in ocean currents, but water's exceptionally high latent heat of vaporization (540 calories per gram) means that water vapor transfers much more heat (per unit of mass) than liquid water. Masses of moving air account for about two-thirds of the poleward transfer of heat; ocean currents move the other third.

The Solar Heating of Earth Also Varies with the Seasons

At mid-latitudes the Northern Hemisphere receives about three times as much solar energy per day in June as it does in December. This difference is due to the 23½ tilt of Earth's rotational axis relative to the plane of its orbit around the sun (**Figure 8.8**, see p. 224) As Earth revolves around the sun, the constant tilt of its rotational axis causes the Northern Hemisphere to lean *toward* the sun in June but *away* from it in December. The sun

---

[1]Changes in heat balance do occur over short periods of time. Increasing amounts of carbon dioxide and methane in Earth's atmosphere may contribute to an increase in surface temperature called *global warming*. More on this subject may be found in Chapter 18.

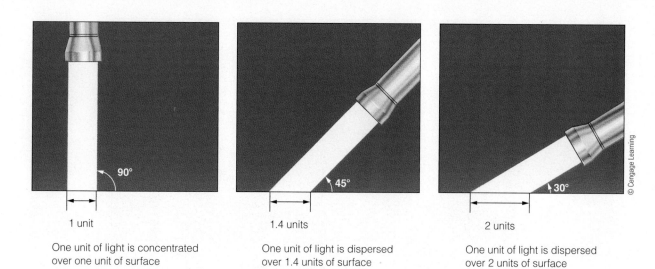

One unit of light is concentrated over one unit of surface

One unit of light is dispersed over 1.4 units of surface

One unit of light is dispersed over 2 units of surface

This ray strikes parallel to surface.

Rays strike surface at angle.

This ray strikes perpendicular to surface.

Ray strikes surface at angle.

This ray strikes parallel to surface.

Figure 8.6 How solar energy input varies with latitude. Equal amounts of sunlight are spread over a greater surface area near the poles than in the tropics. Ice near the poles reflects much of the energy that reaches the surface there.

therefore appears higher in the sky in the summer but lower in winter. The inclination of Earth's axis also causes days to become longer as summer approaches but shorter with the coming of winter. Longer days mean more time for the sun to warm Earth's surface. The tilt causes the seasons.

Mid-latitude heating is strongly affected by season: The mid-latitude regions of the Northern Hemisphere receive about three times as much light energy in June as in December.

Earth's Uneven Solar Heating Results in Large-Scale Atmospheric Circulation

The concentration of solar energy near the equator affects the atmosphere. We know that warm air rises and that cool air sinks. Think of air circulation in a room with a hot radiator opposite a cold closed window **(Figure 8.9)**. Air warms, expands, becomes less dense, and *rises* over the radiator. Air cools, contracts, becomes more dense, and *falls* near the cold glass window. The circular current of air in the room, a **convection current**, is caused by the difference in air density resulting from the temperature difference between the ends of the room.

A similar process occurs over the surface of Earth. As we have seen, surface temperatures are higher at the equator than at the poles, and air can gain heat from warm surroundings. Since air is free to move over Earth's surface, it would be reasonable to assume that an air circulation pattern like the one shown in **Figure 8.10** would develop over Earth. In this ideal model, air heated in the tropics would expand and become less dense, rise to high altitude, turn poleward, and "pile up" as it converged near the poles. The air would then cool by radiating heat into

a *Terra* satellite data for the month of July 2000 shows areas on Earth where solar heat input exceeds the radiation of heat back into space (red, orange) and areas where the radiation of heat into space exceeds heat input from the sun (blue). The ocean does not boil away near the equator or freeze solid near the poles because heat is transferred by winds and ocean currents from equatorial areas to the polar regions.

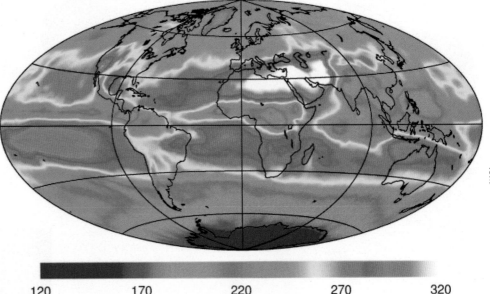

All-sky Longwave Flux from CERES/Terra
July 2000

NASA

Figure 8.7 Areas of heat gain and loss on Earth's surface.

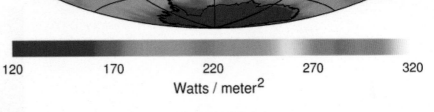

120   170   220   270   320

Watts / meter^2

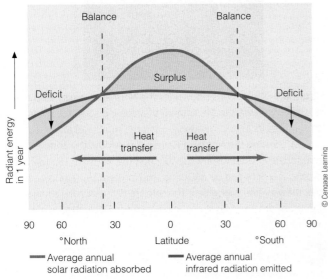

b The average annual incoming solar radiation (red line) *absorbed* by Earth is shown along with the average annual infrared radiation (blue line) *emitted* by Earth. As in **(a)**, note that polar latitudes lose more heat to space than they gain, and tropical latitudes gain more heat than they lose. Only at about 38°N and 38°S latitudes does the amount of radiation received equal the amount lost. Since the area of heat gained (orange area) equals the area of heat lost (blue areas), Earth's total heat budget is balanced over time.

space, become more dense, sink to the surface, and turn equatorward, flowing along the surface back to the tropics to complete the circuit.

But this is *not* what happens. Global circulation of air is governed by two factors: uneven solar heating *and* Earth's rotation. The eastward rotation of Earth on its axis deflects the moving air or water (or any moving object that has mass) away from its initial course. This deflection is called the **Coriolis effect** in honor of **Gaspard Gustave de Coriolis**, the French scientist who worked out its mathematics in 1835.

An understanding of the Coriolis effect is important to an understanding of atmospheric and oceanic circulation.

The Coriolis Effect Deflects the Path of Moving Objects

To an earthbound observer, any object moving freely across the globe appears to curve slightly from its initial path. In the Northern Hemisphere, this curve is to the right, or *clockwise*, from the expected path; in the Southern Hemisphere, it is to the left, or *counterclockwise*. To earthbound observers the deflection is very real; it isn't caused by some mysterious force, and it isn't an optical illusion or some other trick caused by the shape of the globe itself. *The observed deflection is caused by the observer's moving frame of reference on the spinning Earth.*

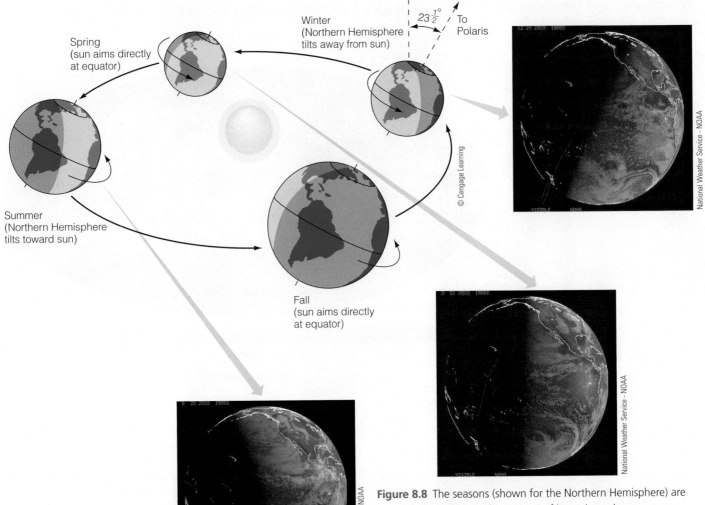

Spring
(sun aims directly
at equator)

Winter
(Northern Hemisphere
tilts away from sun)

$23\frac{1}{2}°$ To
Polaris

Summer
(Northern Hemisphere
tilts toward sun)

Fall
(sun aims directly
at equator)

© Cengage Learning

National Weather Service - NOAA

National Weather Service - NOAA

National Weather Service - NOAA

Figure 8.8 The seasons (shown for the Northern Hemisphere) are caused by variations in the amount of incoming solar energy as Earth makes its annual rotation around the sun on an axis tilted by 23½°. During the Northern Hemisphere winter, the Southern Hemisphere is tilted toward the sun, and the Northern Hemisphere receives less light and heat. During the Northern Hemisphere summer, the situation is reversed. The satellite images clearly show the significant difference in illumination angles in December, September, and June.

💬 **THINKING BEYOND THE FIGURE**
Why are the seasons reversed in the Southern Hemisphere?

The influence of this deflection can be illustrated by performing a mental experiment involving concrete objects—in this case, cities and cannonballs—and then applying the principle to atmospheric circulation.

Let's pick as examples for our experiment the equatorial city of Quito, the capital of Ecuador, and Buffalo, New York. Both cities are on almost the same line of longitude (79°W), so Buffalo is almost exactly north of Quito (as **Figure 8.11** shows). Like everything else attached to the rotating Earth, both cities make one trip around the world each 24 hours. Through one day, the north–south relationship of the two cities never changes: Quito is *always* due south of Buffalo.

A complete trip around the world is 360°, so each city moves eastward at an angular rate of 15° per hour (360°/24 hours = 15°/hour). Even though their angular rates are the same, the two cities move eastward at different speeds. Quito is on the

equator, the "fattest" part of Earth. Buffalo is farther north at a "skinnier" part. Imagine both cities isolated on flat disks, and imagine Earth's sphere being made of a great number of these disks strung together on a rod connecting the North Pole to the South Pole. From Figure 8.11, you can see that Buffalo's disk has a smaller circumference than Quito's. Buffalo doesn't have as far to go in one day as Quito because Buffalo's disk is not as large. So, Buffalo must move eastward *more slowly* than Quito to maintain its position due north of Quito.

Look at Earth from above the North Pole, as shown in **Figure 8.12.** The Quito disk and the Buffalo disk must turn through

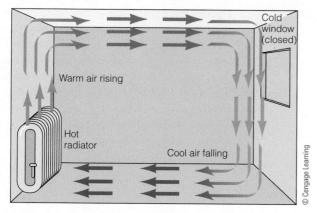

Figure 8.9 A convection current forms in a room when air flows from a hot radiator to a cold closed window and back. (For a practical oceanic application of this principle, look ahead to Figure 8.20.)

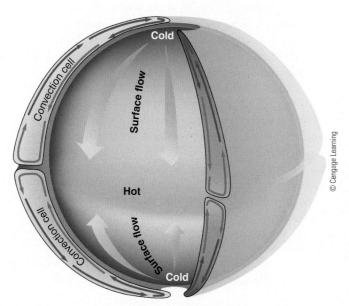

Figure 8.10 A hypothetical model of Earth's air circulation if uneven solar heating were the only factor to be considered. (The thickness of the atmosphere is greatly exaggerated.)

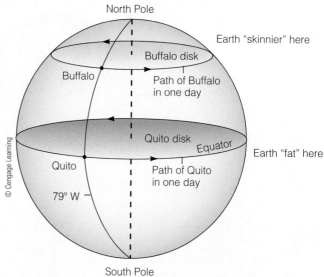

Figure 8.11 A sketch of the thought experiment in the text, showing that Buffalo travels a *shorter* path on the rotating Earth than Quito does.

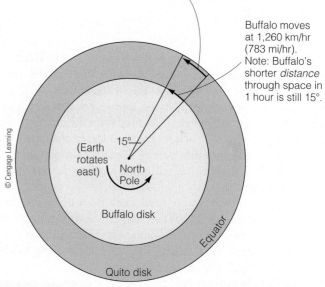

Figure 8.12 A continuation of the thought experiment. A look at Earth from above the North Pole shows that Buffalo and Quito move at different speeds.

15° of longitude each hour (or Earth would rip itself apart), but the city on the equator must move faster to the east to turn its 15° each hour because its "slice of the pie" is larger. Buffalo must move at 1,260 kilometers (783 miles) per hour to go around the world in one day, while Quito must move at 1,668 kilometers (1,036 miles) per hour to do the same.

Now imagine a massive object moving between the two cities. A cannonball shot north from Quito toward Buffalo would carry Quito's eastward speed as it goes; that is, regardless of its northward speed, the cannonball is also moving *east* at 1,668 kilometers (1,036 miles) per hour. The fact of being fired northward by the cannon does not change its eastward movement in the least. As the cannonball streaks north, an odd thing happens. The cannonball veers from its northward path, angling

slightly to the right (east) **(Figure 8.13)**. Actually, this first cannonball is moving just as an observer from space would expect it to, but to those of us on the ground, the cannonball "gets ahead of Earth." As cannonball 1 moves north, the ground beneath it is no longer moving eastward at 1,668 kilometers per hour. During the ball's time of flight, Buffalo (on its smaller disk) *has not moved eastward enough to be where the ball will hit*. If the time of flight for the cannonball is 1 hour, a city 408 kilometers east of Buffalo

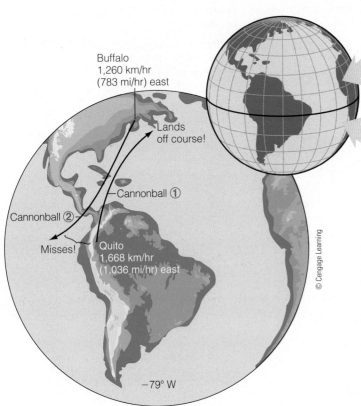

Buffalo
1,260 km/hr
(783 mi/hr) east

Lands off course!

—Cannonball ①

Cannonball ②

Misses!

Quito
1,668 km/hr
(1,036 mi/hr) east

−79° W

© Cengage Learning

Right
Clockwise

N

S

Left
Counter-
Clockwise

Figure 8.13 The final step in the thought experiment. As observed from space, cannonball ① (shot northward) and cannonball ② (shot southward) move as we might expect; that is, they travel straight away from the cannons and fall to Earth. Observed from the ground, however, cannonball ① veers slightly east and cannonball ② veers slightly west of their intended targets. The effect depends on the observer's frame of reference.

(1,668 [Quito's speed] − 1,260 [Buffalo's speed] = 408) will have an unexpected surprise. Albany may be in for some excitement!

Still having trouble? Try this: Remember that the *northward*-moving cannonball, in a sense, brought with it the *eastward* velocity it had before it was fired from the muzzle of the cannon back in Quito. When it gets to its target, the target has lagged behind (because that part of Earth is moving eastward more slowly). The cannonball will strike Earth to the right of its aiming position. Better?

Now if a cannonball were fired south from Buffalo toward Quito, the situation would be reversed. This second cannonball has an eastward speed of 1,260 kilometers (783 miles) per hour even while it sits in the muzzle. Once fired and moving southward, cannonball 2 travels over portions of Earth that are moving ever faster in an eastward direction. The ball again appears to veer off course to the right of its initial direction of travel (see Figure 8.13 again), falling into the Pacific to the west of Ecuador. Don't be deceived by the word *appears* in the last sentence. The cannonballs really do veer to the right, or *clockwise*. Only to an observer in space would they appear to go straight, and points on Earth would appear to move out from underneath them.

The Coriolis effect is a real effect dependent on our rotating frame of reference. Part of that frame of reference involves the direction from which you view the problem. Thus, in Figure 8.13, can-

To an earthbound observer, any object moving freely across the globe appears to curve slightly from its initial path. In the Northern Hemisphere, this curve is to the right, or *clockwise*, from the expected path; in the Southern Hemisphere, it is to the left, or *counterclockwise*.

nonball 2 looks to you as if it is veering left, but to the citizens of Buffalo facing south to watch the cannonball disappear, it moves to the right (west). Coriolis deflection works *counterclockwise* in the Southern Hemisphere because the frame of reference there is reversed. Also, except at the equator (where the Coriolis effect is nonexistent), the Coriolis effect influences the path of objects moving from east to west, or west to east.

An easy way to remember the Coriolis effect: In the Northern Hemisphere, moving objects veer off course clockwise, to the right; in the Southern Hemisphere, moving objects veer off course counterclockwise, to the left (see the inset in Figure 8.13).

Because the Coriolis effect influences any object with mass—*as long as that object is moving*—it plays a large role in the movements of air and water on Earth. The Coriolis effect is most apparent in mid-latitude situations involving the almost frictionless flow of fluids: between layers of water in the ocean and in the circuits of winds. Does the Coriolis effect influence the directions of cars and airplanes? Yes, but in these cases friction (of tires on pavement, of wings on air) is much greater than the influence of the Coriolis effect, so the deflection is not observable.

The Coriolis Effect Influences the Movement of Air in Atmospheric Circulation Cells

We can now modify our original model of atmospheric circulation (Figure 8.10) into the more correct representation. Yes, air does warm, expand, and rise at the equator; and air does cool, contract, and fall at the poles. But instead of continuing all the way from equator to pole in a continuous loop in each hemisphere, air rising from the equatorial region moves poleward and is gradually deflected eastward; that is, it turns to the *right* in the Northern Hemisphere and to the *left* in the Southern Hemisphere. This eastward deflection is caused by the Coriolis effect. (Note that the Coriolis effect does not *cause* the wind; it only *influences* the wind's direction.)

Remember, the atmosphere is very thin **(Figure 8.14)**—nothing like the exaggerated

view in Figure 8.10. There is not room for a single uninterrupted loop of wind from equator to pole and back again. As air rises at the equator, it loses moisture by precipitation (rainfall) caused by expansion and cooling. This drier air now grows denser in the upper atmosphere as it radiates heat to space and cools. When it has traveled *about a third of the way* from the equator to the pole—that is, to about 30°N and 30°S latitudes—the air becomes dense enough to fall back toward the surface. Most of the descending air turns back toward the equator when it reaches the surface. In the Northern Hemisphere the Coriolis effect again deflects this surface air to the right, and the air blows across the ocean or land from the northeast. Though it has been heated by compression during its descent, this air is generally still colder than the surface over which it flows. The air soon warms as it moves equatorward, however, evaporating surface water and becoming humid. The warm, moist, less dense air then begins to rise as it approaches the equator and completes the circuit.

What would that global pattern look like?

Three Atmospheric Circulation Cells Circulate in Each Hemisphere

A pair of cells like the ones just described exists in the tropics, one on each side of the equator. They are known as **Hadley cells** in honor of George Hadley, the London lawyer and philosopher who worked out an overall scheme of wind circulation in 1735. Look for them in **Figure 8.15**. This Figure shows the large circuits of air called **atmospheric circulation cells**, and there are six of these cells in our world model.

Figure 8.14 Earth's atmosphere viewed from space. Note how thin it is. If Earth were to shrink to the size of a large beach ball, its inhabitable atmosphere would be thinner than a piece of paper. Compare this image with Figures 8.10 and 8.15.

NASA

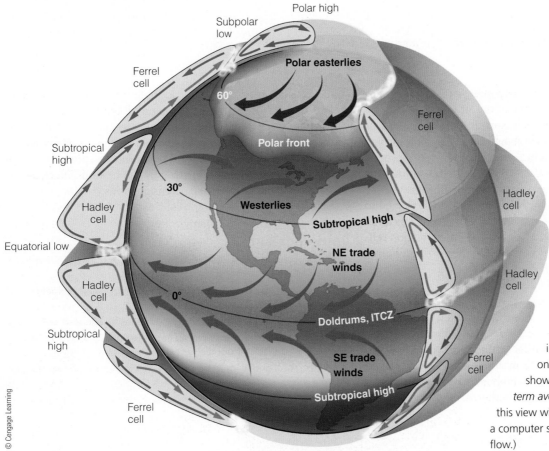

Figure 8.15 Global air circulation as described in the six-cell circulation model. As shown in Figure 8.10, air rises at the equator and falls at the poles. But instead of one great circuit in each hemisphere from equator to pole, there are *three in each hemisphere*. Note the influence of the Coriolis effect on wind direction. The circulation shown here is ideal—that is, a *long-term average* of wind flow. (Contrast this view with Figure 8.18, an excerpt from a computer simulation of true atmospheric flow.)

© Cengage Learning

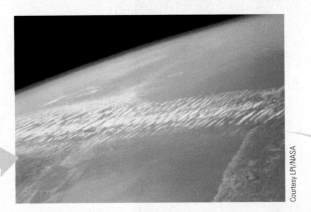

b A polar jet stream flowing across the north Pacific. Polar jets form at around 7 to 12 kilometers (23,000–52,000 feet) and are of great importance to commercial aviation.

Courtesy LPI/NASA

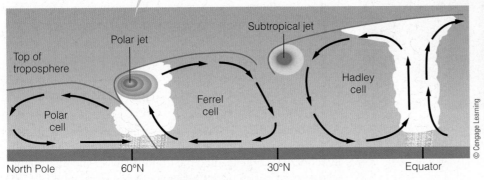

Top of troposphere

Polar jet

Subtropical jet

Polar cell

Ferrel cell

Hadley cell

North Pole   60°N   30°N   Equator

© Cengage Learning

Tail Wind	225 mph
Ground Speed	810 mph
Altitude	33000 feet
Outside Air Temperature	-40 °F

Tom Garrison

a The polar, Ferrel, and Hadley Cells shown in detail. Note that the troposphere is higher in the tropics than in the polar regions. Jet streams—rivers of fast-moving air—form at cell junctions.

c Airliners can save fuel by riding in (or avoiding) jet streams. This 747 has added the 225-mile-per-hour speed of a well-placed northern Pacific jet stream to its normal cruising speed and is traveling eastward at a ground speed of 810 miles per hour!

Figure 8.16 A cross-section through the lower atmosphere from the North Pole to the Equator.

A more complex pair of circulation cells operates at mid-latitudes in each hemisphere. Some of the air descending at 30° latitude turns poleward rather than equatorward. Before this air descends to the surface, it is joined at high altitude by air returning from the north. As can be seen in Figure 8.15, a loop of air forms a mid-latitude cell between 30° and about 50° to 60° latitude. As before, the air is driven by uneven heating and influenced by the Coriolis effect. Surface wind in this circuit is again deflected to the right, this time flowing from the west to complete the circuit (the westerlies in Figure 8.15). The mid-latitude circulation cells of each hemisphere are named **Ferrel cells** after William Ferrel, the American who discovered their inner workings in the mid-19th century. They, too, can be seen in Figure 8.15.

Meanwhile, air that has grown cold over the poles begins blowing toward the equator at the surface, turning to the west as it does so. Between 50° and 60° latitude in each hemisphere, this air has taken up enough heat and moisture to ascend. However, this polar air is denser than the air in the adjacent Ferrel cell and does not mix easily with it. The unstable zone between these two cells generates most mid-latitude weather. At high altitude, the ascending air from 50° to 60° latitude turns poleward to complete a third circuit. These are the **polar cells**.

Three large atmospheric circulation cells—a Hadley cell, a Ferrel cell, and a polar cell—exist in each hemisphere (**Figure 8.16**). Air circulation within and between cells, powered by uneven solar heating and influenced by the Coriolis effect, generates predictable long-term patterns of winds.

CONCEPT CHECK

Before going on to the next section, check your understanding of some of the important ideas presented so far:

What is meant by thermal equilibrium? Is Earth's heat budget in balance?

How does solar heating vary with latitude? With the seasons?

What is a convection current? Can you think of any examples of convection currents around your house?

Describe the Coriolis effect to the next person you meet. Go ahead—give it a try!

If all of Earth rotates eastward at 15° an hour, why does the eastward speed of locations on Earth vary with their latitude?

How many atmospheric circulation cells exist in each hemisphere?

How does the Coriolis effect influence atmospheric circulation?

8.4 Atmospheric Circulation Generates Large-Scale Surface Wind Patterns

The model of atmospheric circulation described earlier has many interesting features. Look once more at Figure 8.15. At the boundaries between circulation cells, the air is moving *vertically* and surface winds are weak and erratic. Such conditions exist at the equator (where air rises and atmospheric pressure is generally low) or at 30° latitude in each hemisphere (where air falls and atmospheric pressure is generally high). Places within these circulation cells where air moves rapidly *horizontally* across the surface from zones of high pressure to zones of low pressure are characterized by strong, dependable winds.

Sailors have a special term for the calm equatorial areas where the surface winds of the two Hadley cells converge: the equatorial low called the **doldrums**. The word has come to be associated with a gloomy, listless mood, perhaps reflecting the sultry air and variable breezes found there. Scientists who study the atmosphere call this area the **intertropical convergence zone (ITCZ)** to reflect the influence of wind convergence on conditions near the equator. Strong heating in the ITCZ causes surface air to expand and rise. The humid, rising, expanding air loses moisture as rain, some of which contributes to the success of tropical rain forests.

Sinking air, in contrast, is generally arid. The great deserts of both hemispheres, dry bands centered around 30° latitude, mark the intersection of the Hadley and Ferrel cells. Air falls toward Earth's surface in these areas, causing compressional heating. Because evaporation is higher than precipitation in these areas, ocean-surface salinity tends to be highest at these latitudes (as can be seen in Figure 6.16). At sea, these areas of high atmospheric pressure and little surface wind are called the subtropical high, or **horse latitudes**. Spanish ships laden with supplies for the New World were often becalmed there, sometimes for weeks on end. When the mariners ran out of water and feed for their livestock, they were forced to eat the animals or throw them over the side.

Of much more interest to sailing masters were the bands of dependable surface winds *between* the zones of ascending and descending air. Most constant of these are the persistent **trade winds**, or easterlies, centered at about 15°N and 15°S latitudes. The trade winds are the surface winds of the Hadley cells as they move from the horse latitudes to the doldrums. In the Northern Hemisphere, they are the northeasterly trade winds; the south-easterly trade winds are the Southern Hemisphere counterpart.[2] The **westerlies**, surface winds of the Ferrel cells centered at about 45°N and 45°S latitudes, flow between the horse latitudes and the boundaries of the polar cells in each hemisphere. The westerlies, then, approach from the southwest in the Northern Hemisphere and from the northwest in the Southern Hemisphere. Sailors outbound from Europe to the New World learned to drop south to catch the trade winds and to return home by a more northerly route to take advantage of the westerlies. Trade winds and westerlies are shown in Figure 8.15.

> The Coriolis effect is most pronounced in mid-latitude situations involving the almost frictionless flow of fluids: between layers of water in the ocean and in the circuits of winds.

Cell Circulation Centers on the Meteorological (Not Geographical) Equator

The six-cell model of atmospheric circulation discussed here represents an average of airflow through many years over the planet as a whole. Though the model is accurate in a general sense, local details of cell circulation vary because surface conditions are different at different longitudes. The ocean's thermostatic effect is the major factor reducing irregularities in cell circulation over water.

For example, the equator-to-pole patterns of airflow within circulation cells along the 20°E line of longitude, a meridian crossing Africa and Europe, are much more complex than the patterns of flow along 170°W longitude, which is almost exclusively over the ocean. The ITCZ is also much narrower and more consistent over the ocean than over land. Because the Northern Hemisphere contains much less ocean surface than the Southern Hemisphere and because landmasses have a lower specific heat than the ocean, seasonal differences in temperature and cell circulation are more extreme in the north. Cell circulation is also much more symmetrical in the Southern Hemisphere.

Another consequence of the markedly different proportions of land to ocean surface in the two hemispheres is the position of the ITCZ. The convergence zone does not coincide with the **geographical equator** (0° latitude). Instead, it lies at the **meteorological equator** (or *thermal equator*), an irregular imaginary line of thermal equilibrium between the hemispheres, situated about 5° north of the geographical equator. The positions of the meteorological equator and the ITCZ generally coincide. They change with the seasons, moving slightly farther north in the northern summer and returning toward the equator in the northern winter **(Figure 8.17)**. Atmospheric and oceanic circulation in the two hemispheres is approximately symmetrical about the meteorological equator, not the geographical equator. Thus,

---

[2]Winds are named by the direction *from* which they blow. A west wind blows *from* the west toward the east; a northeast wind blows *from* the northeast toward the southwest.

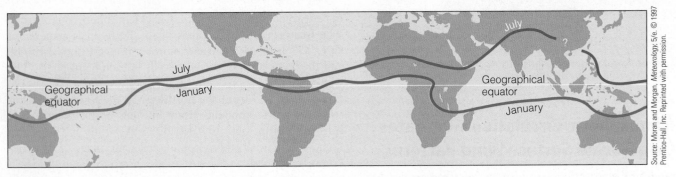

Figure 8.17 Seasonal changes in the position of the intertropical convergence zone (ITCZ). The zone reaches its most northerly location in July and its most southerly location in January. Because of the thermostatic effect of water, the seasonal north–south movement is generally less over the ocean than over land.

the doldrums, trade winds, horse latitudes, and westerlies shift north in the northern summer and south in the northern winter.

There are also east–west variations in the expected patterns of the circulation cells. In the northern winter, air above the chilled continental masses of North America and Siberia becomes very cold and dense. This air sinks and forms zones of high atmospheric pressure over the continents. Air over the relatively warmer waters near the Aleutians and Iceland rises and forms zones of low atmospheric pressure. Air flows from the high-pressure zones toward low-pressure zones, modifying the flow of air within the cells. In summer the situation is reversed: Low pressure forms over the heated landmasses, and higher pressure forms over the cooler ocean. These effects are most pronounced in the middle latitudes of the Northern Hemisphere, where land and water are present in near-equal amounts.

Figure 8.18 is an excerpt from a computer simulation of worldwide winds. As you can see, the patterns can depart from what we would expect based on the six-cell model in Figure 8.15. Most of the difference is caused by the geographical distribution of landmasses, the different responses of land and ocean to solar heating, and chaotic flow. But, as noted above, over long periods of time (many years), *average* flow looks remarkably like what we would expect from that model.

The major surface wind and pressure systems of the world, and their prevailing weather conditions, are summarized in **Table 8.1**. These great wind patterns are responsible for about two-thirds of the heat transfer from the tropics to the polar regions. (Ocean currents account for the other third.)

Monsoons Are Wind Patterns That Change with the Seasons

A **monsoon** is a pattern of wind circulation that changes with the season. (The word *monsoon* is derived from *mausim*, the Arabic word for "season.") Areas subject to monsoons generally have wet summers and dry winters.

Monsoons are linked to the different specific heats of land and water and to the annual north–south movement of the ITCZ. In the spring, land heats more rapidly than the adjacent ocean. Air above the land becomes warmer and so rises. Rela-

tively cool air flows from over the ocean to the land to take the warmer air's place. Continued heating causes this humid air to rise, condense, and form clouds and rain. In autumn, the land cools more rapidly than the adjacent ocean. Air cools and sinks over the land, and dry surface winds move seaward. The intensity and location of monsoon activity depend on the position of the ITCZ. Note that the monsoons follow the ITCZ south in the Northern Hemisphere's winter **(Figure 8.19a)** and north in its summer **(Figure 8.19b)**.

In Africa and Asia, more than 2 billion people depend on summer monsoon rains for drinking water and agriculture. The most intense summer monsoons occur in Asia. Heating of the great landmass of Asia draws vast quantities of warm, moist air from the Indian Ocean **(Figure 8.19c)**. Winds from the south drive this moisture toward Asia, where it rises and condenses to produce a months-long deluge. The volume of water vapor carried inland is astonishing—the town of Cherrapunji, on the slopes of the Khasi Hills in northeastern India, receives about 10 *meters* (425 inches) of rain each year, most of it between April and October! Much smaller monsoons occur in North America as warming and rising air over the South and West draws humid air and thunderstorms from the Gulf of Mexico.

Sea Breezes and Land Breezes Arise from Uneven Surface Heating

Land breezes and sea breezes are small, daily mini-monsoons. Morning sunlight falls on land and adjacent sea, warming both. The temperature of the water doesn't rise as much as the temperature of the land, however. The warmer inland rocks transfer heat to the air, which expands and rises, creating a zone of low atmospheric pressure over the land. Cooler air from over the sea then moves toward land; this is the **sea breeze (Figure 8.20a)** The situation reverses after sunset, with land losing heat to space and falling rapidly in temperature. After a while, the air over the still-warm ocean will be warmer than the air over the cooling land. This air will then rise, and the breeze direction will reverse, becoming a **land breeze (Figure 8.20b)**. Land breezes and sea breezes are common and welcome occurrences in coastal areas.

Source: Moran and Morgan, *Meteorology*, 5/e. © 1997 Prentice-Hall, Inc. Reprinted with permission.

Figure 8.18 An excerpt from a 2-year "nature run" using a NASA supercomputer to model atmospheric movement. Winds are colored by speed, with red indicating the fastest (at 175 meters per second or 390 miles per hour). Tropical cyclones are seen as white circles. Although short-term views such as this one depart substantially from wind flow predicted in the six-cell model developed in Figure 8.15, the *average* wind flow over many years looks remarkably like what we would expect from that model. William Putman/NASA Goddard Space Flight Center

Table 8.1 **Major Wind and Pressure Systems and Related Weather**

Region	Name	Pressure	Surface Winds	Weather
Equator (0°)	Doldrums (ITCZ) (equatorial low)	Low	Light, variable winds	Cloudiness, abundant precipitation in all seasons; breeding ground for hurricanes. Relatively low sea-surface salinity because of rainfall (see Figure 6.16)
0°–30°N and S	Trade winds (easterlies)	—	Northeast in Northern Hemisphere, southeast in Southern Hemisphere	Summer wet, winter dry; pathway for tropical disturbances
30°N and S	Horse latitudes (sub-tropical high)	High	Light, variable winds	Little cloudiness; dry in all seasons. Relatively high sea-surface salinity because of evaporation
30°–60°N and S	Prevailing westerlies	—	Southwest in Northern Hemisphere, northwest in Southern Hemisphere	Winter wet, summer dry; pathway for subtropical high and low pressure
60°N and S	Polar front	Low	Variable	Stormy, cloudy weather zone; ample precipitation in all seasons
60°–90°N and S	Polar easterlies	—	Northeast in Northern Hemisphere, southeast in Southern Hemisphere	Cold polar air with very low temperatures
90°N and S	Poles	High	Southerly in Northern Hemisphere, northerly in Southern Hemisphere	Cold, dry air; sparse precipitation in all seasons

Note: Compare to Figure 8.13. (From *Earth in Crisis: An Introduction to the Earth Sciences,* 2/e, Thomas L. Burrus, Herbert J. Spiegel, 1980. C. V. Mosby Co. Reprinted by permission of the authors.)

El Niño, La Niña

Sometimes cell circulation doesn't seem to play by the rules. In 3- to 8-year cycles, atmospheric circulation changes significantly from the patterns shown in Figure 8.15. A reversal in the distribution of atmospheric pressure between the eastern and western Pacific causes the trade winds to weaken or reverse. The trade winds normally drag huge quantities of water westward along the ocean's surface near the equator, but without the winds, these equatorial currents crawl to a stop. Warm water that has accumulated at the western side of the Pacific can build eastward along the equator toward the coast of Central and South America, greatly changing ocean conditions there.

El Niño and La Niña are primarily ocean current phenomena, so we will study them in Chapter 9's discussion of ocean currents.

CONCEPT CHECK

Before going on to the next section, check your understanding of some of the important ideas presented so far:

What happens to airflow *between* circulation cells? (Hint: What causes Earth's desert climates?)

Draw the general pattern for the atmospheric circulation of the Northern Hemisphere (without looking at Figure 8.15). Now locate these features: the doldrums (or ITCZ), the horse latitudes, the prevailing westerlies, and the trade winds.

Why is atmospheric circulation between the two hemispheres centered about the meteorological equator, not the geographical equator? Why is there a difference? (Hint: Think of the heat capacity of water and which hemisphere contains more surface water.)

What's a monsoon? Do we experience monsoons in the continental United States?

How do sea breezes and land breezes form?

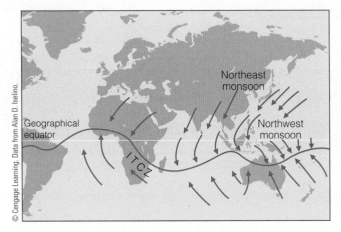

a January

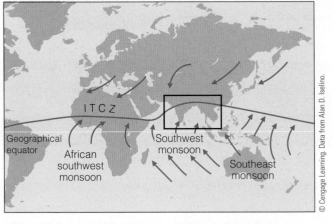

b July

a During the monsoon circulations of January **a** and July **b**, surface winds are deflected to the right in the Northern Hemisphere and to the left in the Southern Hemisphere.

c The summer monsoon in Kolkata (Calcutta), India, one of the world's wettest places. Rainfall in the region can exceed 10 meters (425 inches) per year!

Figure 8.19 Monsoon patterns.

THINKING BEYOND THE FIGURE

Do monsoons form outside Asia?

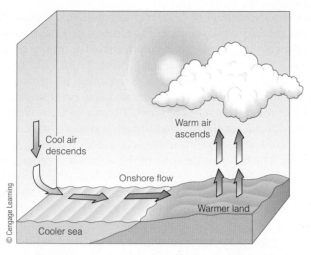

a In the afternoon, the land is warmer than the ocean surface, and warm air rising from the land is replaced by an onshore sea breeze.

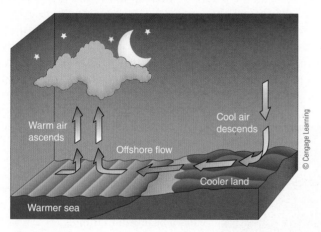

b At night, as the land cools, the air over the ocean is now warmer than the air over the land. The ocean air rises. Air flows offshore to replace it, generating an offshore flow (a land breeze).

Figure 8.20 The flow of air in coastal regions during stable weather conditions.

8.5 Storms Are Variations in Large-Scale Atmospheric Circulation

Storms are regional atmospheric disturbances characterized by strong winds, often accompanied by precipitation. Few natural events underscore human insignificance like a great storm. When powered by stored sunlight, the combination of atmosphere and ocean can do fearful damage.

In Bangladesh, on 13 November 1970, a tropical cyclone (a hurricane) with wind speeds of more than 200 kilometers (125 miles) per hour roared up the mouths of the Ganges–Brahmaputra River, carrying with it masses of seawater up to 12 meters (40 feet) high. Water and wind clawed at the aggrega-

tion of small islands, most just above sea level, that makes up this impoverished country. In only 20 minutes, at least 300,000 lives were lost, and estimates ranged up to a million dead! Property damage was essentially absolute. Photographs taken soon after the storm showed a horizon-to-horizon morass of flooded, deep-gashed ground tortured by furious winds. There was almost no trace of human inhabitants, farms, domestic animals, or villages. Another great storm that struck in May 1991 killed another 200,000 people. The economy of the shattered country may not recover for decades.

A much different type of storm hammered the U.S. East Coast in March 1993. Mountainous snows from New York to North Carolina (1.3 meters, or 50 inches, at Mount Mitchell), winds of 175 kilometers (109 miles) per hour in Florida, and record cold in Alabama ($-17°C$, or $2°F$, in Birmingham) were elements of a 4-day storm that spread chaos from Canada to Cuba. At least 238 people died on land; another 48 were lost at sea. At one point more than 100,000 people were trapped in offices, factories, vehicles, and homes; 1.5 million were without electricity. Damage exceeded US$1 billion.

These great storms are examples of *tropical cyclones* and *extratropical cyclones* at their worst. As the name implies, tropical cyclones like the hurricane that struck Bangladesh are primarily a tropical phenomenon. Extratropical cyclones—the winter weather disturbances with which residents of the U.S. eastern seaboard and other mid-latitude dwellers are most familiar—are found mainly in the Ferrel cells of each hemisphere. (Note that the prefix *extra-* means "outside" or "beyond," so *extratropical* refers to the location of the storm, not its intensity.)

Both kinds of storms are **cyclones**, huge rotating masses of low-pressure air in which winds converge and ascend. The word *cyclone*, derived from the Greek noun *kyklon* (which means "an object moving in a circle"), underscores the spinning nature of these disturbances. (Don't confuse a cyclone with a **tornado**, a much smaller funnel of fast-spinning wind associated with severe thunderstorms.)

Storms Form within or between Air Masses

Cyclonic storms form within or between air masses. An **air mass** is a large body of air with nearly uniform temperature, humidity, and therefore density throughout. Air pausing over water or land will tend to take on the characteristics of the surface below. Cold, dry land causes the mass of air above to become chilly and dry. Air above a warm ocean surface will become hot and humid. Cold, dry air masses are dense and form zones of high atmospheric pressure. Warm, humid air masses are less dense and form zones of lower atmospheric pressure.

Air masses can move within or between circulation cells. Density differences, however, will prevent the air masses from mixing when they approach one another. Energy is required to mix air masses. Since that energy is not always available, a dense air mass may slide beneath a lighter air mass, lifting the lighter one and causing its air to expand and cool. Water vapor in the rising air may condense. All of these effects contribute to turbulence at the boundaries of the air masses.

Figure 8.21 Vilhelm Bjerknes, the Norwegian meteorologist who, with his son Jacob, formulated the air-mass theories on which our understanding of weather is based.

The boundary between air masses of different density is called a **front**. The term was coined by **Vilhelm Bjerknes (Figure 8.21)**, a pioneering Norwegian meteorologist who saw a similarity between the zone where air masses meet and the violent battle fronts of World War I.

Extratropical cyclones and weather fronts form between *two* air masses. Tropical cyclones form from disturbances within *one* warm and humid air mass.

Extratropical Cyclones Form between Two Air Masses

Extratropical cyclones form at the boundary between each hemisphere's polar cell and its Ferrel cell—the **polar front**. These great storms occur mainly in the winter hemisphere when temperature and density

differences across the polar front are most pronounced. Remember that the cold wind poleward of the front is generally moving from the east; the warmer air equatorward of the front is generally moving from the west (see again Figure 8.15). The smooth flow of winds past each other at the front may be interrupted by zones of alternating high and low atmospheric pressure that bend the front into a series of waves. Because of the difference in wind direction in the air masses north and south of the polar front, the wave shape will enlarge, and a twist will form along the front. The different densities of the air masses prevent easy mixing, so the cold, dense air mass will slide beneath the warmer, lighter one. Formation of this twist in the Northern Hemisphere, as seen from above, is shown in **Figure 8.22**. The twisting mass of air becomes an extratropical cyclone.

The twist that generates an extratropical cyclone circulates counterclockwise in the Northern Hemisphere, seemingly in opposition to the Coriolis effect. The reasons for this paradox become clear, however, when we consider the wind directions and the nature of interruption of the airflow between the cells. (In fact, the counterclockwise motion of the cyclone *is* Coriolis-driven because the large-scale airflow pattern at the edges of the cells is generated in part by the Coriolis effect.) Wind speed increases as the storm "wraps up," in much the same way that a spinning skater increases rotation speed by pulling in his or her arms close to the body. Air rushing toward the center of the spinning storm rises to form a low-pressure zone at the center. Extratropical cyclones are embedded in the westerly winds and thus move eastward. They are typically 1,000 to 2,500 kilometers (620–1,600 miles) in diameter and last from 2 to 5 days. **Figure 8.23** provides a beautiful example.

Precipitation can begin as the circular flow develops. The sequence of events shown in Figure 8.22 shows why. Precipitation is caused by the lifting and consequent expansion and cooling of the mass of mid-latitude air involved in the twist.

> **F**ew natural events underscore human insignificance like a great storm. When powered by stored sunlight, the combination of atmosphere and ocean can do fearful damage.

Figure 8.22 Formation of a mid-latitude cyclone. These cyclonic (spinning) storms form between air masses and are responsible for most of the weather in the temperate zones.

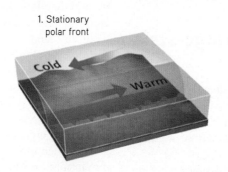

1. Stationary polar front

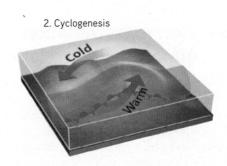

2. Cyclogenesis

Mid-latitude cyclones are found between 35° and 70° of latitude in the zone of the westerly winds. Most are occluded fronts. (1) Characterized by

intense, heavy precipitation, cold polar air—with a boundary known as a front—meets warm tropical air. (2) A wave develops along the frontal

boundary as the opposing air masses interact. Cyclogenesis (the birth of a cyclone) begins. (3) The faster-moving cold air forces the warm air

Figure 8.23 A well-developed extratropical cyclone swirls over the northeastern Pacific on 27 October 2000. Looking like a huge comma, the cloud-dense cold front extends southward and westward from the storm's center. Spotty cumulus clouds and thunderstorms have formed in the cold, unstable air behind the front. This picture was taken by the *GOES-10* satellite in visible light.

 **THINKING BEYOND THE FIGURE**

Would residents of California be expecting rain from this weather system?

As it rises and cools, this air can no longer hold all of its water vapor, so clouds and rain result. When *cold* air advances and does the lifting, a *cold front* occurs. A *warm front* happens when *warm* air is blown on top of the retreating edge of cold air. The wind and precipitation associated with these fronts are sometimes referred to as **frontal storms**. They are the principal cause of weather in the mid-latitude regions, where most of the world's people live.

North America's most violent extratropical cyclones are the **nor'easters**

> **"You could be a meteorologist all your life. . . and never see something like this. It would be a disaster of epic proportions. It would be. . . the perfect storm."**
> —METEOROLOGIST DESCRIBING A VAST EXTRATROPICAL CYCLONE IN THE 2000 FILM *THE PERFECT STORM*.

(northeasters) that sweep the eastern seaboard in winter. The name indicates the direction from which the storm's most powerful winds approach. About 30 times a year, nor'easters moving along the mid-Atlantic and New England coasts generate wind and waves with enough force to erode beaches and offshore barrier islands, disrupt communication and shipping schedules, damage shore and harbor installations, and break power lines. About every hundred years a nor'easter devastates coastal settlements. In spite of a long

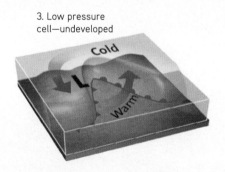

3. Low pressure cell—undeveloped

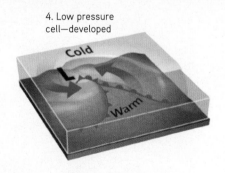

4. Low pressure cell—developed

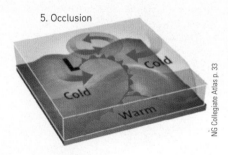

5. Occlusion

to lift above the cold. (4) Full rotation develops, counter-clockwise in the Northern Hemisphere and clockwise in the Southern Hemisphere. (5) Complete

occlusion occurs as the warm air, fully caught-up by the cold air, has been lifted away from the surface. Because the warm air is completely separated

from the surface, the characteristics of the cold air are felt on the ground in the form of unsteady, windy, and wet weather.

Figure 8.24 Waves crash over the seawall and threaten homes in Hull, Massachusetts, during a furious nor'easter on 6 March 2001. Nor'easters are North America's most violent extratropical cyclones.

history of destruction from nor'easters, people continue to build on unstable, exposed coasts **(Figure 8.24)**.

Tropical Cyclones Form in One Air Mass

Tropical cyclones are great masses of warm, humid, rotating air. They occur in all tropical oceans but are extremely rare in the equatorial South Atlantic. Large tropical cyclones are called **hurricanes** (*Huracan* is the god of the wind of the Caribbean Taino people) in the North Atlantic and eastern Pacific, *typhoons* (*Tai-fung* is Chinese for "great wind") in the western Pacific, *tropical cyclones* in the Indian Ocean, and *willi-willis* in

the waters near Australia. To qualify formally as a hurricane or typhoon, the tropical cyclone must have winds of at least 119 kilometers (74 miles) per hour. About 50 tropical cyclones grow to hurricane status each year. A very few of these develop into superstorms, with winds near the core that exceed 250 kilometers (155 miles) per hour![3] Tropical cyclones containing winds less than hurricane force are called *tropical storms* and *tropical depressions*.

---

[3] To imagine what winds in such a storm might feel like, picture yourself clinging to the wing of a twin-engine private airplane in flight!

From above, tropical cyclones appear as circular spirals (Figure 8.25). They may be 1,000 kilometers (620 miles) in diameter and 15 kilometers (9.3 miles, or 50,000 feet) high. The calm center, or *eye* of the storm—a zone some 13 to 16 kilometers (8–10 miles) in diameter—is sometimes surrounded by clouds so high and dense that the daytime sky above looks dark. Farther out, churned by furious winds, the rainband clouds condense huge amounts of water vapor into rain. A mature tropical cyclone is diagrammed in **Figure 8.26**.

Unlike extratropical cyclones, these greatest of storms are generated within *one* warm, humid air mass that forms between 10° and 25° latitude in either hemisphere **(Figure 8.27)**. (Though air conditions would be favorable, the Coriolis effect closer to the equator is too weak to initiate rotary motion.)

You may have noticed that tropical cyclones turn *counterclockwise* in the Northern Hemisphere and *clockwise* in the Southern Hemisphere. Does this mean that the Coriolis effect does not apply to tropical cyclones? No. This apparent anomaly is caused by the Coriolis deflection of winds approaching the center of a low-pressure area from great distances. In the Northern Hemisphere there is rightward deflection of the *approaching air*. The edge spin given by this approaching air causes the storm to spin counterclockwise in the Northern Hemisphere **(Figure 8.28)**.

The origins of tropical cyclones are not well understood. A tropical cyclone usually develops from a small tropical depression. Tropical depressions form in easterly waves, areas of lower pressure within the easterly trade winds that are thought to originate over a large, warm landmass. When air containing the disturbance is heated over tropical water with a temperature of about 26°C (79°F) or higher, circular winds begin to blow in the vicinity of the wave, and some of the warm, humid air is forced upward. Condensation begins, and the storm takes shape.

Although its birth process is somewhat mysterious, the source of the storm's power is well understood. Its strength comes from the same seemingly innocuous process that warms a chilled soft-drink can when water from the atmosphere condenses on its surface. As you may recall from Chapter 6, it takes quite a bit of energy to break the bonds that hold water molecules together and evaporate water into the atmosphere—water's latent heat of vaporization is very high. That heat energy is released when the water vapor recondenses as liquid. It tends to warm your drink very quickly, and the more humid the air, the faster the condensing and warming. Think of the situation in relation to a hair dryer. The heat energy generated by the dryer causes water to evaporate rapidly from your hair. When that water recondenses to liquid (on a nearby can of cold soda, for instance), the original heat used to evaporate the water is released. The cycle of evaporation and condensation has carried heat from the hair dryer to your soda. In tropical cyclones the condensation energy generates air movement (wind), not more heat. Fortunately, only 2% to 4% of this energy of condensation is converted into motional energy!

A tropical cyclone is an ideal machine for "cashing in" water vapor's latent heat of vaporization. Warm, humid air forms in great quantity only over a warm ocean. As already noted, tropical cyclones originate in ocean areas having surface temperatures in excess of 26°C (79°F) (see again Figure 8.24). When hot, humid tropical air rises and expands, it cools and is unable to contain the moisture it held when warm. Rainfall begins. The rainfall rate in some parts of the storm routinely exceeds 2.5 centimeters (1 inch) per hour, and 20 billion metric tons of water can fall from a large tropical cyclone in a day! Tremendous energy is released as this moisture changes from water vapor to liquid. In one day, a large tropical cyclone generates about 2.4 trillion kilowatt-hours of power, equivalent to the electrical energy needs of the entire United States for a year! So, solar energy ultimately powers the storm in a cycle of heat absorption, evaporation, condensation, and conversion of heat energy to kinetic energy. This energy is available as long as the storm stays over warm water and has a ready source of hot, humid air.

Under ideal conditions the embryo storm reaches hurricane status—that is, with wind speeds in excess of 119 kilometers (74 miles) per hour—in 2 to 3 days. The spray kicked up

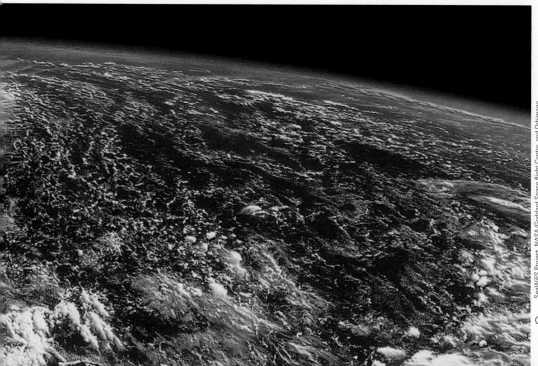

Figure 8.25 Hurricane Alberto spins in the North Atlantic east of Bermuda. Note the thinness of Earth's atmosphere in this oblique view.

SeaWiFS Project, NASA/Goddard Space flight Center, and Orbimage

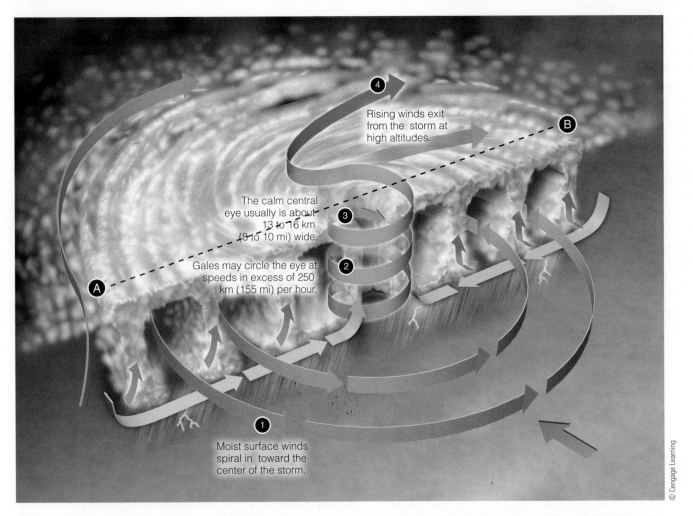

The calm central eye usually is about 13 to 16 km (8 to 10 mi) wide.

Gales may circle the eye at speeds in excess of 250 km (155 mi) per hour.

Rising winds exit from the storm at high altitudes.

Moist surface winds spiral in toward the center of the storm.

© Cengage Learning

a The internal structure of a mature northern hemisphere tropical cyclone, or hurricane. In this diagram the storm is moving to the northwest. (The vertical dimension is greatly exaggerated.)

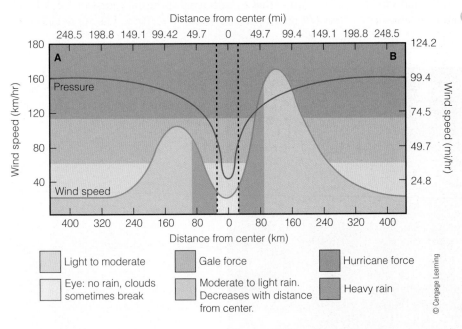

Light to moderate

Eye: no rain, clouds sometimes break

Gale force

Moderate to light rain. Decreases with distance from center.

Hurricane force

Heavy rain

© Cengage Learning

b A cross section of the storm showing wind speed and atmospheric pressure. Because the storm is moving northwestward *and* is rotating counterclockwise, the greatest destruction from wind and storm surge will occur to the east of its center.

Figure 8.26 Tropical cyclone structure.

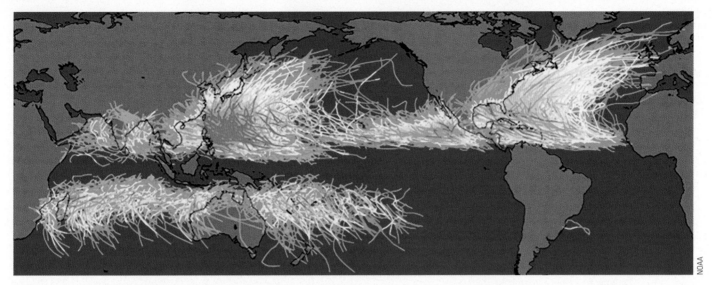

Figure 8.27 Colored lines indicate the paths of tropical storms (hurricanes, typhoons). Storm intensity is indicated by color: weaker storms in blue, category 5 storms in red. Cyclones are very rare over the South Atlantic or the southeast Pacific because their waters are too chilly. They are also rarely seen in the doldrums, the zone of still air within a few degrees of the equator.

by growing winds lubricates the junction between air and sea, effectively "decoupling" the wind from the friction of the rough ocean surface. Under ideal conditions, the storm is free to grow to enormous proportions (Table 8.2). The centers of most tropical cyclones move westward and poleward at 5 to 40 kilometers (3–25 miles) per hour.

Three aspects of a tropical cyclone can cause property damage and loss of life: wind, rain, and storm surge. The destructive force of winds of 250 kilometers (155 miles) per hour or more is self-evident. Rapid rainfall can cause severe flooding when the storm moves onto land. But the most danger lies in a **storm surge**, a mass of water driven by the storm. The low atmospheric pressure at the storm's center produces a dome of seawater that can reach a height of 1 meter (3.3 feet) in the open sea. The water height increases when waves and strong hurricane winds ramp the water mass ashore. If a high tide coincides with the arrival of all this water at a coast or if the coastline converges (as is the case at the mouths of the Ganges–Brahmaputra River in Bangladesh), rapid and catastrophic flooding will occur. Storm surges of up to 12 meters (40 feet) were reported at Bangladesh in 1970. Much of the catastrophic damage done by Hurricanes Katrina and Rita to the Mississippi Gulf Coast in 2005 was caused by an 8-meter (25-foot) storm surge arriving near high tide. You'll learn more about storm surges later in this chapter and in the discussion of large waves in Chapter 10.

Tropical cyclones last from 3 hours to 3 weeks; most have lives of 5 to 10 days. They eventually run down when they move over land or over water too cool to supply the humid air that sustains them. The friction of a land encounter rapidly drains a

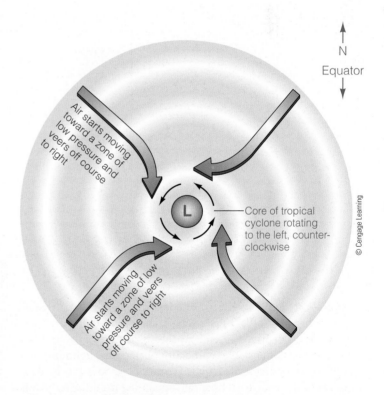

Figure 8.28 The dynamics of a tropical cyclone, showing the influence of the Coriolis effect. Note that the storm turns the "wrong" way (that is, counterclockwise) in the Northern Hemisphere, but for the "right" reasons.

 **THINKING BEYOND THE FIGURE**

Which way do tropical cyclones spin in the Southern Hemisphere?

Table 8.2 Classification of Tropical Cyclones

Storm	Description		Maximum Sustained Wind Speed	Storm Surge	Damage
Tropical Depression	An organized system of clouds and thunderstorms with defined surface circulation but no eye. Lacks the spiral shape of more powerful storms. It is already becoming a low-pressure system, a "depression."		Less than 62 km/h (38 mph)	None	None
Tropical Storm	An organized system of strong thunderstorms with a defined surface circulation and cyclonic shape, though usually lacking an eye. Government weather services assign first names to systems that reach this intensity (they become "named storms").		62–117 km/h (39–73 mph)	Little or none	Little or none
Tropical Cyclone	Intense cyclonic winds exceeding 119 km/h (74 mph). Tropical cyclones consist of a tightly organized band of thunderstorms surrounding a central eye.	Category 1	119–153 km/h (74–95 mph)	1.2–1.5 m (4–5 ft)	Damage to trees, shrubs, and unanchored mobile homes.
		Category 2	154–177 km/h (96–110 mph)	1.8–2.4 m (6–8 ft)	Some trees blown down; major damage to exposed mobile homes; roof damage to permanent structures.
		Category 3	178–209 km/h (111–130 mph)	2.7–3.7 m (9–12 ft)	Foliage removed from large trees; large trees blown down; mobile homes destroyed; some structural damage to permanent buildings.
		Category 4	210–249 km/h (131–155 mph)	4.0–5.5 m (13–18 ft)	All signs blown down; complete roof structure failure on small residences; extensive damage to windows and doors. Major erosion of beach areas. Terrain may be flooded well inland.
		Category 5	250 km/h (156 mph)	5.5 m (19 ft)	Complete roof failure on many residences and industrial buildings. Some complete building failures with small buildings blown over or away. Flooding causes major damage to lower floors of all structures near the shore. Evacuation of residential areas usually required.

© Cengage Learning

tropical cyclone of its energy, and a position above ocean water cooler than 24°C (75°F) is a sure harbinger of the storm's demise. When deprived of energy, the storm "unwinds" and becomes a mass of unstable humid air pouring rain, lightning, and even tornadoes from its clouds. Tropical cyclones can be dangerous to the end: Torrential rain streaming from the remnants of Hurricane Agnes in 1972 caused more than US$2 billion in damage, mostly to Pennsylvania. Chesapeake and Delaware bays were flooded with freshwater and sediments, destroying much of the shellfish industry there.

Tropical cyclones are nature's escape valves, flinging solar energy poleward from the tropics. They are beautiful, dangerous examples of the energy represented by water's latent heat of evaporation.

CONCEPT CHECK

Before going on to the next section, check your understanding of some of the important ideas presented so far:

What are the two kinds of large storms? How do they differ? How are they similar?

What is an air mass? How do air masses form?

Is a cyclone the same thing as a tornado?

What causes an extratropical cyclone? How are air masses involved?

What's a weather front? Is it typical of tropical or extratropical cyclones?

Why do extratropical cyclones rotate counterclockwise in the Northern Hemisphere?

What causes the greatest loss of life and property when a tropical cyclone reaches land?

8.6 Katrina and Sandy

Earth is warming. Worldwide, the year 2012 was one of the warmest on record.[4] In the continental United States, it was the hottest ever recorded—the previous record (set in 1998) was eclipsed by 0.6°C (1°F). The 2012 records were impressive: 34,008 daily highs were set at weather stations across the country, compared with only 6,664 lows. In Australia, The Bureau of Meteorology added a new color to their weather maps (purple) to indicate temperatures in excess of 50°C (122°F).

As the temperature of the atmosphere increases, the capacity to hold water increases, leading to stronger storms and higher rainfall amounts. Is there a connection between the warming ocean and the increasing intensity of tropical cyclones? Although the *number* of storms has remained relatively constant over the past 35 years, meteorologists have reported a striking 80% increase in the abundance of the *most powerful* (category 4 and 5) tropical cyclones in the past 35 years. Mathematical models suggest there is a relationship between the warming climate and intensification of tropical cyclones.

Hurricane Katrina, which made landfall in August 2005, was the cause of the most costly natural disaster to befall the United States.[5] Katrina's 200-kilometer-per-hour (125-mile-per-hour) winds and torrential rains pounded the Gulf Coast for most of a day. A storm surge, a huge dome of seawater driven by the storm, crested at a staggering 10.4 meters (34 feet) and essentially erased the Mississippi towns of Gulfport and Biloxi. (You'll learn more about storm surges in Chapter 10.)

The city of New Orleans, slightly west of the point of landfall, survived the initial blow. Though badly buffeted by winds and rain, the pumps that drain the city largely performed as designed. But several levees protecting New Orleans failed the next day, and the city, 80% of which lies below sea level, flooded rapidly **(Figure 8.29)**.

Some of Katrina's intensity was due to chance. The position and unusual warmth of the Gulf Stream loop contributed to the storm's rapid intensification. The lack of shearing winds aloft allowed the storm to form and grow without interference. The storm's energy rapidly increased to category 5 with maximum sustained winds of 280 kilometers (175 miles) per hour and then weakened as it passed over cooler water before again striking land along the central Gulf Coast near Buras-Triumph, Louisiana. With winds of 200 kilometers (125 miles) per hour, Katrina was one of the largest and strongest hurricanes ever recorded **(Figure 8.30a, b)**. Katrina's rain was intense (averaging 25 centimeters, or 10 inches, an hour at landfall), but it was the huge storm surge generated by wind and the

Figure 8.29 Hurricane Katrina was the most costly natural disaster in U.S. history. Houses were flooded when levees were overwhelmed and pumps failed. Downtown New Orleans, built on higher land, rises in the distance.

TYRONE TURNER/National Geographic Creative

storm's exceedingly low central pressure that caused the greatest immediate loss of life and property.

In contrast, Superstorm Sandy was the result of a rare and violent collision between a classic nor'easter (a frontal storm) and a tropical storm (Hurricane Sandy).

Sandy became a category 1 tropical storm on 22 October 2012 near Jamaica. As it moved north along the U.S. east coast, the hurricane intensified over the Gulf Stream, largely because of additional energy injected into the storm from its abnormally warm water. A blocking high to the northeast then caused the hurricane to veer westward and collide with a linear mass of cold air moving eastward. The comparative result can be seen from the wind charts in **Figures 8.30c** and **d**. For Katrina, winds over 65 kilometers per hour (40 miles per hour) stretched about 500 kilometers (300 miles) from edge to edge. But for Sandy, winds of that intensity stretched 1,500 kilometers (900 miles)! Because of its odd hybrid nature (circular tropical cyclone blending with a linear frontal storm), Sandy became the largest Atlantic hurricane on record (measured by diameter) with hurricane-force winds spanning 1,800 kilometers (1,100 miles). A buoy in New York harbor recorded a wave 10 meters (32.5 feet) high on 30 October! Sandy affected 24 states, including the whole east-

> **M**athematical models suggest a relationship between the warming climate and the intensification of tropical cyclones.

---

[4]Dependable temperature data have been available since 1880.

[5]At least 1,836 people died in Hurricane Katrina, and property damage has been estimated at US$81 billion. Estimates for superstorm Sandy place property damage at US$66 billion; 253 lives were lost.

Comparing Hurricane Katrina

a Katrina as a category 5 storm hours before landfall on 28 August 2005. The storm is relatively compact with strong winds confined to about 500 kilometers (300 miles) from edge to edge. Katrina's overall shape and well-formed eye is seen in **b**.

Sandy's wind field was much larger **c** and stretched across 1,500 kilometers (900 miles) on 30 October 2012. In **d**, the huge size and hybrid nature of the storm are clearly visible. The nor'easter is seen swept around and into the tropical cyclone from the west and south.

Hurricane Katrina August 28, 2005

a

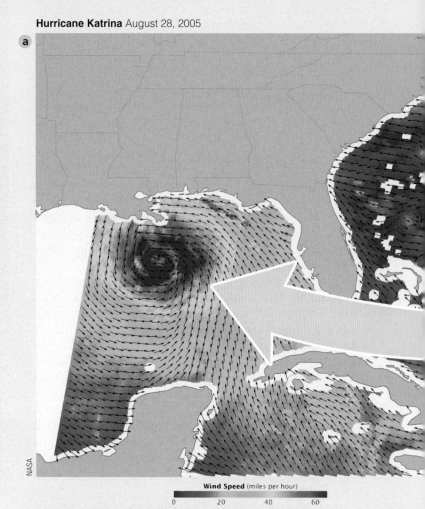

NASA

Wind Speed (miles per hour)

0   20   40   60

Hurricane Sandy October 30, 2012

c

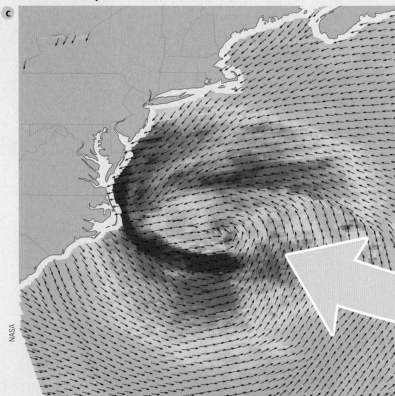

NASA

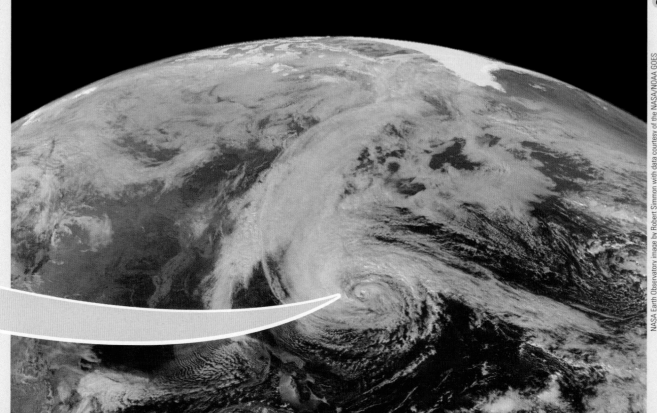

ern seaboard from Florida to Maine and into maritime Canada, and across the Appalachian Mountains as far as Michigan and Wisconsin.

Curiously, dust storms in the distant Sahara desert may have also played a role. Dust in air acts as condensation nuclei for raindrops, accelerating their formation and speeding the release of latent heat. Layers of dry, dusty air blowing westward into the tropical Atlantic appear to influence the intensity of tropical cyclones. As Earth warms and deserts spread, this influence could grow (Figure 8.31).

Figure 8.31 Dust blows off the African continent into the tropical Atlantic in October 2007. Dust in air acts as condensation nuclei for raindrops, accelerating their formation and speeding the release of latent heat. As Earth warms and deserts spread, hurricanes are expected to become stronger and more frequent.

Jeff Schmaltz, MODIS rapid response team—GSFC/NASA

CONCEPT CHECK

Before going on to the next section, check your understanding of some of the important ideas presented so far:

What things were unique about the 2005 Atlantic hurricane season?

Of a tropical cyclone's three most dangerous properties (wind, rain, storm surge), which of Katrina's characteristics caused the greatest loss of life? Of property?

How do large tropical cyclones affect the human-built coastal zone? The natural coastal zone?

Is there a proven link between global warming and the apparent growing intensity of Atlantic hurricanes?

Chapter in Perspective

In this chapter you learned that Earth and ocean are in continuous contact, and conditions in one are certain to influence conditions in the other. The interaction of ocean and atmosphere moderates surface temperatures, shapes Earth's weather and climate, and creates most of the sea's waves and currents.

The atmosphere responds to uneven solar heating by flowing in three great circulating cells over each hemisphere. This circulation of air is responsible for about two-thirds of the heat transfer from tropical to polar regions. The flow of air within these cells is influenced by Earth's rotation. To observers on the surface, Earth's rotation causes moving air (or any moving mass) in the Northern Hemisphere to curve to the right of its initial path, and in the Southern Hemisphere to the left. The apparent curvature of path is known as the Coriolis effect.

Uneven flow of air within cells is one cause of the atmospheric changes we call *weather*. Large storms are spinning areas of unstable air that occur between or within air masses. Extratropical cyclones originate at the boundary between air masses; tropical cyclones, the most powerful of Earth's atmospheric storms, occur within a single humid air mass. The immense energy of tropical cyclones is derived from water's latent heat of vaporization.

In the next chapter you will learn how movement of the atmosphere can cause movement of ocean water. Wind blowing over the ocean creates surface currents, and deep currents form when the ocean surface is warmed or cooled as the seasons change. Currents join with the atmosphere to form a giant heat engine that moves energy from regions of excess (tropics) to regions of scarcity (poles). This energy keeps the tropical seas from boiling away and the polar ocean from freezing solid in its basins.

The ocean's surface currents are governed by some of the principles you've learned here—the Coriolis effect and uneven solar heating continue to be important concepts in our discussion. Taken together, an understanding of air and water circulation is at the heart of physical oceanography.

QUESTIONS FROM STUDENTS

1. **Earth's orbit brings it closer to the sun in the Northern Hemisphere's winter than in its summer. Yet it's warmer in summer. Why?**

Earth's orbit around the sun is elliptical, not circular. The whole Earth receives about 7% more solar energy through the half of the orbit during which we are closer to the sun than through the other half. The time of greater energy input comes during our winter, but the entire Northern Hemisphere is tilted toward the sun during the summer, which results in much more light reaching it in the summer. Three times as much energy enters the Northern Hemisphere each day at midsummer as at midwinter.

2. **How did the trade winds get their name? Is their name a reminder of the assistance they provided to shipboard traders interested in selling their wares in distant corners of the world?**

The trade winds are not named after their contribution to commerce in the days of sail. This use of the word *trade* derives from an earlier English meaning equivalent to our adverbs *steadily* or *constantly*. These persistent winds were said to "blow trade."

3. **If the Coriolis effect acts on all moving objects, why doesn't it pull my car to the right? And does the Coriolis effect really make explorers wander to the left in the snows of Antarctica, or tree trunks grow in rightward spirals in Canadian forests, or water swirl clockwise down a toilet in Springfield?**

The Coriolis effect depends on the speed, mass, and latitude of the moving object. Your car's motion is affected. When you drive along the road at 70 miles per hour in an average-sized car, Coriolis "force" would pull your car to the right about 460 meters (1,500 feet) for every 160 kilometers (100 miles) you travel *if* it were not for the friction between your tires and the road surface.

Small, lightweight objects moving slowly are subject to many forces and conditions (such as wind currents, natural variations in basin shape, and friction) that overwhelm Coriolis acceleration. For example, think of how *very* small the difference in eastward speed of the northern edge of a toilet is in comparison to the southern edge. Any small irregularity in the toilet's shape will be hundreds of times as important as the Coriolis effect in determining whether water will exit in a rightward spin or a leftward spin! Explorers and trees aren't massive enough and don't move quickly enough to be affected by the Coriolis effect.

But if the moving object is at mid-latitude, is heavy, and is moving quickly, it *will* be deflected to the right of its intended path. The computers aboard jetliners subtly nudge the flight path in the appropriate direction, but unguided devices (artillery shells and so on) move noticeably.

4. **Does the ocean affect weather at the centers of continents?**

Absolutely. In a sense, *all* large-scale weather on Earth is oceanically controlled. The ocean acts as a solar collector and heat sink, storing and releasing heat. Most great storms (tropical and extratropical cyclones alike) form over the ocean and then sweep over land.

5. **On average, the meteorological equator lies about 5° north of the geographical equator. Why the displacement?**

The meteorological equator lies in the Northern Hemisphere because, overall, that hemisphere is slightly warmer than the Southern Hemisphere. At least three factors are responsible for this. First, the unbroken ice covering of Antarctica reflects more light back into space than the surface of the Arctic Ocean, which contains occasional light-absorbing patches of open water. Second, there is more land surface in the tropical latitudes of the Northern Hemisphere than in the Southern Hemisphere. Since land warms more than water with the same input of solar energy, tropical latitudes are warmer in the Northern Hemisphere. And third, ocean currents transport more warm water toward the north than toward the south.

6. **Are any of the results of large-scale atmospheric circulation apparent to the casual observer?**

Yes. The view of atmospheric circulation developed in this chapter explains some phenomena you may have experienced. For instance, flying from Los Angeles to New York takes about 40 minutes less than flying from New York to Los Angeles because of westerly headwinds (indicated in Figure 8.15).

Because of these same prevailing westerlies, most storms travel over the United States from west to east. Weather prediction is based on observations and samples taken from the air masses as they move. Forecasting is often easier in the East and Midwest than in the West because more data are available from an air mass when it's over land.

7. **Was Katrina the most powerful storm ever recorded?**

No. That honor belongs to Typhoon Haiyan. That storm tore across the Philippines in November 2013 with winds speeds of up to 314 kilometers (195 miles) per hour. A storm in 1979, Typhoon Tip, produced somewhat slower winds of 306 kilometers (190 miles) per hour but generated the lowest sea-level pressure ever recorded on Earth: 870 millibars, or 25.69 inHg!

8. **Has anything similar to Earth's weather patterns been seen on other planets?**

Yes, indeed. Saturn, Jupiter, and Neptune have tremendous cyclonic storms. A few on Jupiter are large enough to be seen from Earth through small telescopes. Tracks of tornadoes have recently been identified on Mars, and a huge cyclonic storm was photographed there in 1999. Venus has huge cloud banks suggestive of polar fronts and extratropical cyclones. The Coriolis effect and uneven solar heating are common to all planets with atmospheres in this solar system, so we shouldn't be surprised by similarities.

TERMS AND CONCEPTS TO REMEMBER

air mass
atmosphere
atmospheric circulation cell
Bjerknes, Vilhelm
climate
convection current
Coriolis, Gaspard Gustave de
Coriolis effect
cyclone
doldrums
extratropical cyclone

Ferrel cell
front
frontal storm
geographical equator
Hadley cell
heat budget
horse latitudes
hurricane
intertropical convergence
    zone (ITCZ)

land breeze
meteorological equator
monsoon
nor'easter (northeaster)
polar cell
polar front
precipitation
sea breeze
storm
storm surge

thermal equilibrium
tornado
trade winds
tropical cyclone
troposphere
water vapor
weather
westerlies
wind

STUDY QUESTIONS

Thinking Critically

1. What factors contribute to the uneven heating of Earth by the sun?

2. How does the atmosphere respond to uneven solar heating? How does the rotation of Earth affect the resultant circulation?

3. Why doesn't the ocean boil away at the equator and freeze solid near the poles?

4. Describe the atmospheric circulation cells in the Northern Hemisphere. At which latitudes does air move vertically? Horizontally? What are the trade winds? The westerlies? Where are deserts located? Why? What is ocean-surface salinity like in these desert bands?

5. How are the geographical equator, meteorological equator, and ITCZ related? What happens at the ITCZ?

6. What is a monsoon? How is monsoon circulation affected by the position of the ITCZ?

7. If the Coriolis effect causes the clockwise deflection of moving objects in the Northern Hemisphere, why does air rotate counterclockwise around zones of low pressure in that hemisphere?

Thinking Analytically

1. There is no such thing as "suck." Imagine the palm of your hand covering an open and empty peanut butter jar. Now imagine the air being pumped out of the jar. Your hand is not being "sucked" into the jar. Your hand is forced tightly onto the jar by differential pressure—the air pressure outside the jar is greater than the air pressure inside. Knowing the mass of a 1-square-centimeter column of air, calculate the total inward force exerted on the palm of your hand if you were at sea level and your hand were covering a jar with a diameter of 7.5 centimeters and all the air could be removed from inside the jar.

2. Why does water drip from the bottom of your car when the air conditioner is being used on an especially humid day?

3. Why is the longest day of the year almost never the hottest day of the year?

Visit www.cengagebrain.com to access MindTap, a complete digital course which includes access to Global Geoscience Watch and more. Within the GREENR database, search for "hurricane." On this page, look for "Special Libraries and Research Centers" in the right-hand column and click "VIEW ALL." Then click on "Florida International University - International Hurricane Research Center (IHRC)" and use this link to learn more information about the formation and impacts of tropical cyclones and hurricanes. Summarize some of the current ideas and research addressing how they form and what impact they have on coasts, marine life, and humans. Use this information to create a report on tropical cyclones and their effect on coastal and open ocean environments and research one particular hurricane to include in your report as a case study.

9 Circulation of the Ocean

KEY CONCEPTS

Ocean circulation is driven by winds and by differences in water density. Colin Monteath/Minden Pictures

Along with the winds, ocean currents distribute tropical heat worldwide. NASA/JPL

Surface currents are wind-driven movements of water at or near the ocean's surface. *Thermohaline* currents are the slow, deep currents that affect the vast bulk of seawater beneath the pycnocline. NASA/Goddard Space Flight Center Scientific Visualization Studio

Surface currents move in circular circuits—gyres—around the peripheries of major ocean basins. © Cengage Learning

El Niño and La Niña affect ocean and atmosphere. El Niño is an exception to normal wind and current flow. NASA

Storm winds blasting at a steady 80 knots (148 kilometers per hour, or 92 miles per hour) drive a surface current in the Drake Passage in the Southern Ocean beyond Cape Horn, Chile.
Colin Monteath/Minden Pictures

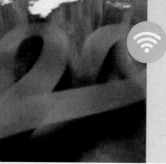

9.1 Mass Flow of Ocean Water Is Driven by Wind and Gravity

Sailors have long known that the ocean is on the move. The first traders to sail courageously out of the Mediterranean at Gibraltar noticed a persistent southerly set—they would often drift down the African coast despite the direction of the winds. Pytheas of Massalia, a Greek ship's captain who explored the northeastern Atlantic in the fourth century B.C.E., was the first observer to record this slow, continuous movement and to estimate its speed.

By the early 17th century, the massed fishing fleets of Japan were using a northward drift to their advantage to reach rich hauls off the Kamchatka Peninsula. (They returned home by sailing close to shore where the drift was weak.)

More recently, Sir John Murray noted in the *Challenger Report* that the temperature of the ocean's surface water was almost always higher than the temperature of deep water and that a zone of rapid temperature change (which you know as a thermocline) existed in most areas sampled.

Still later, in a pivotal 1961 paper, Klaus Wyrtki answered a persistent question: What keeps the thermocline up? Because of contact conduction of heat, shouldn't water temperature drop *gradually* and *continuously* as depth increases? Cold water is somehow rising from below to lift the warm water toward the ocean surface. Where does this cold water come from?

These ideas are related. The horizontal drift of ships and the vertical movement of cold water are caused by the mass flow of water—a phenomenon we know as ocean **currents**.

Surface currents are wind-driven movements of water at or near the ocean's surface, and *thermohaline currents* (so named because they depend on density differences caused by variations in water's temperature and salinity) are the slow, deep currents that affect the vast bulk of seawater beneath the pycnocline. Both have very important influences on Earth's temperature, climate, and biological productivity and will change as Earth's climate varies.

CONCEPT CHECK

Before going on to the next section, check your understanding of some of the important ideas presented so far:

What causes the two major types of ocean currents?

9.2 Surface Currents Are Driven by the Winds

About 10% of the water in the world ocean is involved in **surface currents**, water flowing horizontally in the uppermost 400 meters (1,300 feet) of the ocean's surface, driven mainly by wind friction. Most surface currents move water above the pyc-

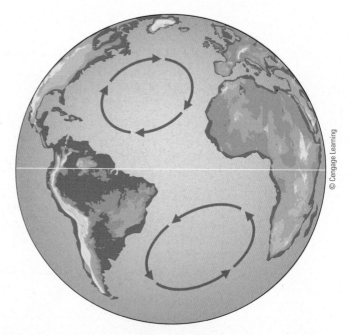

Figure 9.1 Winds, driven by uneven solar heating and Earth's spin, drive the movement of the ocean's surface currents. The prime movers are the powerful westerlies and the persistent trade winds (easterlies).

Figure 9.2 A combination of four forces—surface winds, the sun's heat, the Coriolis effect, and gravity—circulates the ocean surface clockwise in the Northern Hemisphere and counterclockwise in the Southern Hemisphere, forming gyres.

nocline, the zone of rapid density change with depth described in detail in Chapter 6.

The primary force responsible for surface currents is wind. As you read in Chapter 8, surface winds form global patterns within latitude bands (**Figure 9.1**; also see Figure 8.15. Most of Earth's surface wind energy is concentrated in each hemisphere's trade winds (easterlies) and westerlies. Waves on the sea surface transfer some of the energy from the moving air to the water by friction. This tug of wind on the ocean surface begins a mass flow of water. The water flowing beneath the wind forms a surface current.

The moving water "piles up" in the direction the wind is blowing. Water pressure is higher on the "piled up" side, and the force of gravity pulls the water down the slope—against the *pressure gradient*—in the direction from which it came. But the Coriolis effect intervenes. Because of the Coriolis effect (discussed in Chapter 8), Northern Hemisphere surface currents flow to the *right* of the wind direction. Southern Hemisphere currents flow to the *left*. Continents and basin topography often block continuous flow and help deflect the moving water into a circular pattern. This flow around the periphery of an ocean basin is called a **gyre** (*gyros*, "a circle"). Two gyres are shown in **Figure 9.2**.

Surface Currents Flow around the Periphery of Ocean Basins

Figure 9.3 shows the North Atlantic gyre in more detail. Though the gyre flows continuously without obvious places where one current ceases and another begins, oceanographers subdivide the North Atlantic gyre into four interconnected currents be-

cause each has distinct flow characteristics and temperatures. (Gyres in other ocean basins are similarly divided.) Notice that the east–west currents in the North Atlantic gyre flow to the right of the driving winds; once initiated, water flow in these currents continues in a roughly east–west direction. Where their flow is blocked by continents, the currents turn clockwise to complete the circuit.

Why does water flow around the *periphery* of the ocean basin instead of spiraling to the center? After all, the Coriolis effect influences any moving mass *as long as it moves*, so water in a gyre might be expected to curve to the center of the North Atlantic and stop. To understand this aspect of current movement, imagine the forces acting on the surface water at 45°N latitude (point ① in **Figure 9.4**). Here the westerlies blow from the southwest, so initially the water will move toward the northeast. The rightward Coriolis deflection then causes the water to flow almost due east. A particle at 15°N latitude (point ②) responds to the push of the trade winds from the northeast, however, and with Coriolis deflection it will flow almost due west. Water at the surface can flow at a velocity no greater than about 3% of the speed of the driving wind.

When driven by the wind, the topmost layer of ocean water in the Northern Hemisphere flows at about 45° to the right of the wind direction, a flow consistent with the arrows leading away from points ① and ② in Figure 9.4. But what about the water in the next layer down? It can't "feel" the wind at the surface; it "feels" only the movement of the water immediately above. This deeper layer of water moves *at an angle to the right* of the overlying water. The same thing happens in the layer below that, and the next layer, and so on, to a depth of about 100 meters (330 feet) at mid-latitudes. Each layer slides horizon-

Figure 9.3 The North Atlantic gyre, a series of four interconnecting currents with different flow characteristics and temperatures.

Figure 9.4 Surface water blown by the winds at point ① will veer to the right of its initial path and continue eastward. Water at point ② veers to the right and continues westward.

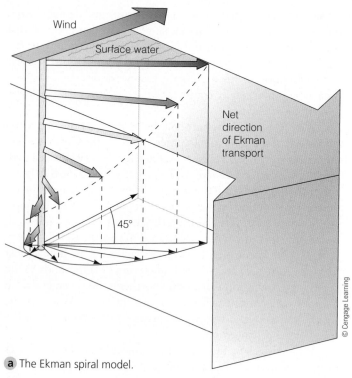

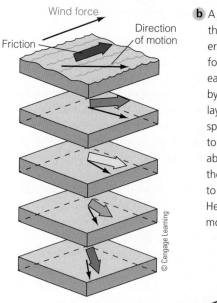

a The Ekman spiral model.

Figure 9.5 The Ekman spiral and the mechanism by which it operates. The length of the arrows in the diagrams is proportional to the speed of the current in each layer.

💬 **THINKING BEYOND THE FIGURE**

Is there any depth at which water flows in exactly the opposite direction from the direction of the surface wind?

b A body of water can be thought of as a set of layers. The top layer is driven forward by the wind, and each layer below is moved by friction. Each succeeding layer moves with a slower speed and at an angle to the layer immediately above it—to the right in the Northern Hemisphere, to the left in the Southern Hemisphere—until water motion becomes negligible.

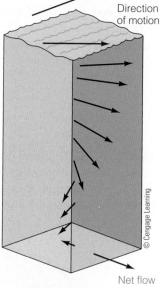

c Though the direction of movement varies for each layer in the stack, the theoretical net flow of water in the Northern Hemisphere is 90° to the right of the prevailing wind force.

tally over the one beneath it like cards in a deck, with each lower card moving at an angle slightly to the right of the one above. Because of frictional losses, each lower layer also moves more slowly than the layer above. The resulting situation, portrayed in **Figure 9.5**, is known as **Ekman spiral** after the Swedish oceanographer who worked out the mathematics involved.

The term *spiral* is somewhat misleading; the water itself does not spiral downward in a whirlpool-like motion like water going down a drain. Rather, the spiral is a way of conceptualizing the horizontal movements in a layered water column, each layer moving in a slightly different horizontal direction.

This may help: Imagine a massive deck of cards. A gust of wind moves the top card an inch to the north. In response to the Coriolis effect, this top card moves slightly to the right (to the northeast) of the wind direction. The card *beneath* it in the deck feels this card's movement and begins to move. The Coriolis effect shifts its direction a bit to the right of the direction of the topmost card (the second card is now moving to the east). Remember, the second card in the deck *cannot feel the wind directly*—it's only moving in response to the movement of the topmost card. The third card feels the second card, begins to move, and is further deflected to the right. And so on.

An unexpected result of the Ekman spiral is that at some depth (known as the friction depth), water will be flowing in the *opposite direction* from the surface current!

The *net* motion of the water down to about 100 meters, after allowance for the summed effects of Ekman spiral (the sum of all the arrows indicating water direction in the affected layers), is known as **Ekman transport** (or *Ekman flow*). In theory, the direction of Ekman transport is 90° to the *right* of the wind direction in the Northern Hemisphere and 90° to the *left* in the Southern Hemisphere.

Armed with this information, we can look in more detail at the area around point ② in Figure 9.4, which is enlarged in **Figure 9.6**. In nature, Ekman transport in gyres is less than 90°; in most cases the deflection barely reaches 45°. This deviation from theory occurs because of an interaction between the Coriolis effect and the pressure gradient. Some flowing Atlantic water has turned to the right and forms a hill of water; it follows the rightward dotted-line arrow in Figure 9.6.

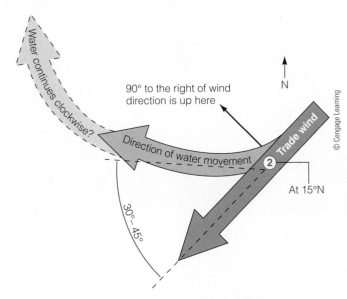

Figure 9.6 The movement of water away from point ② in Figure 9.4 is influenced by the rightward tendency of the Coriolis effect and the gravity-powered movement of water down the pressure gradient.

Why does the water now go straight west from point ② without turning? Because, as **Figure 9.7a** shows, to turn further *right*, the water would have to move uphill against the pressure gradient (and in defiance of gravity), but to turn *left* in response to the pressure gradient would defy the Coriolis effect. So the water continues westward and then clockwise around the whole North Atlantic gyre, dynamically balanced between the downhill urge of the pressure gradient and the uphill tendency of Coriolis deflection. The hill has consequences for deeper water as well. As **Figure 9.7b** shows, water flowing inward to form the hill sinks and depresses the thermocline.

Yes, *there really is* a hill near the middle of the North Atlantic, centered in the area of the Sargasso Sea; satellite images provide the evidence **(Figure 9.7c)**. This hill is formed of surface water gathered at the ocean's center of circulation. It is not a steep mountain of water—its maximum height is an unspectacular 2 meters (6.5 feet)—but rather a gradual rise and fall from coastline to open ocean and back to opposite coastline. Its slope is so gradual you wouldn't notice it on a transatlantic crossing.

The hill is maintained by wind energy. If the winds did not continuously inject new energy into currents, then friction within the fluid mass and with the surrounding ocean basins would slow the flowing water, gradually converting its motion into heat. The balance of wind energy and friction, and of the Coriolis effect and the pressure gradient (through the effect of gravity), propels the currents of the gyre and holds them along the outside edges of the ocean basin.

Currents are the very heart of physical oceanography. Their global effects, great masses of water, complex flow, and possible influence on human migrations make their study of particular importance.

Seawater Flows in Six Great Surface Circuits

Gyres in balance between the pressure gradient and the Coriolis effect are called **geostrophic gyres** (*Geos*, "Earth"; *strophe*, "turning"), and their currents are *geostrophic currents*. Because of the patterns of driving winds and the present positions of continents, the geostrophic gyres are largely independent of one another in each hemisphere.

There are six great current circuits in the world ocean, two in the Northern Hemisphere and four in the Southern Hemisphere **(Figure 9.8)**. Five are geostrophic gyres: the North Atlantic gyre, the South Atlantic gyre, the North Pacific gyre, the South Pacific gyre, and the Indian Ocean gyre. Though it is a closed circuit, the sixth and largest current is technically not a gyre because it does not flow around the periphery of an ocean basin. The **West Wind Drift**, or **Antarctic Circumpolar Current**, as this exception is called, flows endlessly eastward around Antarctica, driven by powerful, nearly ceaseless westerly winds. This greatest of all the surface ocean currents is never deflected by a continent.

We might expect the two gyres in the North and South Pacific (and the two gyres in the North and South Atlantic) to converge exactly at the geographical equator. However, as Figure 9.8b shows, the junction of equatorial currents lies a few degrees north of the geographical equator, at the meteorological equator. As noted in Chapter 8, the meteorological equator and the intertropical convergence zone (the band at which the trade winds converge) are displaced 5° to 8° northward primarily because of the heat accumulated in the Northern Hemisphere's greater tropical land-surface area. Ocean circulation, like atmospheric circulation, is balanced around the meteorological equator.

Boundary Currents Have Different Characteristics

Because of the different factors that drive and shape them, the currents that form geostrophic gyres have different characteristics. Geostrophic currents may be classified by their position within the gyre as western boundary currents, eastern boundary currents, or transverse currents.

Western Boundary Currents The fastest and deepest geostrophic currents are found at the western boundaries of ocean basins (that is, off the east coast of continents). These narrow, fast, deep currents move warm water poleward in each of the gyres. As you can see in Figure 9.8b, there are five large **western boundary currents**: the Gulf Stream (in the North Atlantic), the Japan or Kuroshio Current (in the North Pacific), the Brazil Current (in the South Atlantic), the Agulhas Current (in the Indian Ocean), and the East Australian Current (in the South Pacific).

The **Gulf Stream** is the largest of the western boundary currents. Studies of the Gulf Stream have revealed that off Mi-

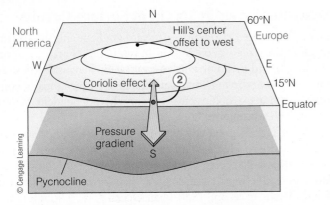

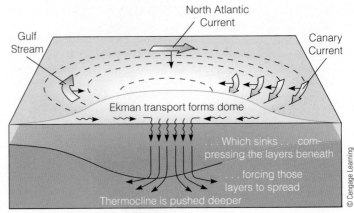

a The surface of the North Atlantic is raised through wind motion and Ekman transport to form a low hill. Water from point ② (see also Figures 9.4 and 9.6) turns westward and flows along the side of this hill. The westward-moving water is balanced between the Coriolis effect (which would turn the water to the right) and flows down the pressure gradient, driven by gravity (which would turn it to the left). Thus, water in a gyre moves along the outside edge of an ocean basin.

b The hill is formed by Ekman transport. Water turns clockwise (inward) to form the dome, then descends, depressing the thermocline.

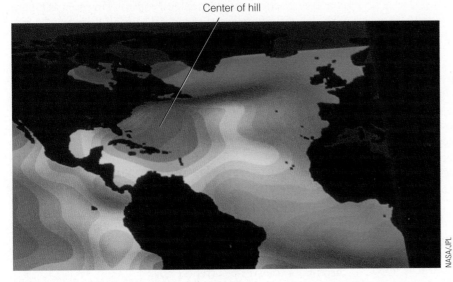

c The average height of the surface of the North Atlantic is shown in color in this image derived from data taken in 1992 by the *TOPEX/Poseidon* satellite. Red indicates the highest surface; green and blue, the lowest. Note that the measured position of the hill is offset to the west, as seen in **a**. (The westward offset is explained in Figure 9.13.) The gradually sloping hill is only 2 meters (6.5 feet) high and would not be apparent to anyone traveling across the ocean.

Figure 9.7 The hill of water in the North Atlantic.

 THINKING BEYOND THE FIGURE

Without looking ahead, can you think why the peak of the hill is offset to the west? Would it also be offset to the west in the south Atlantic?

ami, the Gulf Stream moves at an average speed of 2 meters per second (5 miles per hour) to a depth of more than 450 meters (1,500 feet). Water in the Gulf Stream can move more than 160 kilometers (100 miles) in a day. Its average width is about 70 kilometers (43 miles).

The volume of water transported in western boundary currents is extraordinary. The unit used to express volume transport in ocean currents is the **sverdrup (sv)**, named in honor of Harald Sverdrup, one of the last century's pioneering oceanographers.

A sverdrup equals 1 million cubic meters per second.[1] The Gulf Stream flow is at least 55 sv (55 million meters per second), about 300 times the usual flow of the Amazon, the greatest of rivers. In **Figure 9.9** the surface currents of the North Atlantic gyre are shown with their volume transport (in sverdrups) indicated.

---

[1]One million cubic meters is about one-half the volume of the Louisiana Superdome.

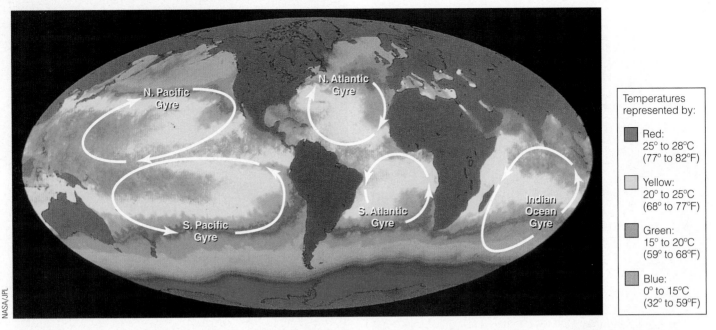

a An illustration of sea-surface temperature showing the general direction and pattern of surface current flow. Sea-surface temperatures were measured by a radiometer aboard *NOAA-7* in July 1984. The purple color around Antarctica and west of Greenland indicates water below 0°C, the freezing point of freshwater. Note the distortion of the temperature patterns we might expect from the effects of solar heating alone—the patterns twist clockwise in the Northern Hemisphere, counterclockwise in the Southern.

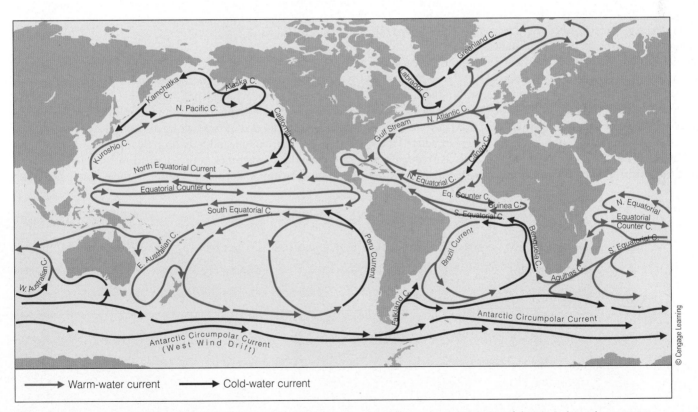

b A chart showing the names and usual direction of the world ocean's major surface currents. The powerful western boundary currents flow along the western boundaries of ocean basins in *both* hemispheres.

Figure 9.8 Two ways of viewing the major surface currents of the world ocean.

💬 **THINKING BEYOND THE FIGURE**

How may gyres exist in the world ocean?

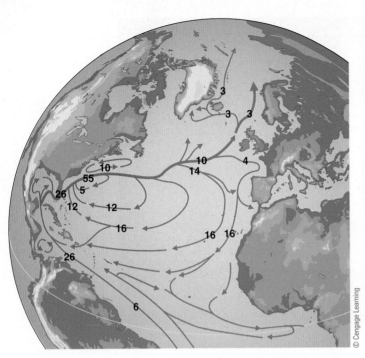

Figure 9.9 The general surface circulation of the North Atlantic. The numbers indicate flow rates in sverdrups (1 sv = 1 million cubic meters of water per second).

© Cengage Learning

NASA

Figure 9.10 Moving at a speed of about 10 kilometers (6 miles) per hour, the Gulf Stream departs the coast at Cape Hatteras, its warm, clear, blue water contrasting with the cooler, darker, more productive water to the north and west. Clouds form over the warm current as water vapor evaporates from the ocean surface. Look for this area in Figures 9.11 and 9.18.

Water in a current, especially a western boundary current, can move for surprisingly long distances within well-defined boundaries, almost as if it were a river. In the Gulf Stream, the current-as-river analogy can be startlingly apt: the western edge of the current is often clearly visible. Water within the current is usually warm, clear, and blue, often depleted of nutrients and incapable of supporting much life. By contrast, water over the continental slope west of the current is often cold, green, and teeming with life. **Figure 9.10** shows sun glinting off surface irregularities in the Gulf Stream as it passes the coast of Newfoundland.

Long, straight edges are the exception rather than the rule in western boundary currents, however. Unlike rivers, ocean currents lack well-defined banks, and friction with adjacent water can cause a current to form waves along its edges. Western boundary currents meander as they flow poleward. The looping meanders sometimes connect to form turbulent rings, or **eddies**, that trap cold or warm water in their centers and then

separate from the main flow. For example, *cold-core eddies* form in the Gulf Stream as it meanders eastward upon leaving the coast of North America off Cape Hatteras (**Figure 9.11**). *Warm-core eddies* can form north of the Gulf Stream when the warm current loops into the cold water lying to the north. When the loops are cut off, they become freestanding spinning masses of water. Warm-core eddies rotate clockwise, and cold-core eddies rotate counterclockwise.

The slowly rotating eddies move away from the current and are distributed across the North Atlantic. Some may be 1,000 kilometers (620 miles) in diameter and retain their identity for more than 3 years. In mid-latitudes, as much as one fourth of the surface of the North Atlantic may consist of old, slow-moving, cold-core eddy remnants! Eddies are visible in the satellite image shown in **Figure 9.12**. Recent research suggests that their influence reaches to the seafloor. Warm- and cold-core eddies may be responsible for slowly moving *abyssal storms*, which leave ripple marks that have been observed in deep sediments. Nutrients brought toward the surface by turbulence in eddies sometimes stimulate the growth of tiny marine plantlike organisms, seen here as green swirls.

Eastern Boundary Currents As shown in Figure 9.8b, there are five **eastern boundary currents** at the eastern edge of ocean basins (that is, off the west coast of continents): the Canary Current (in the North Atlantic), the Benguela Current (in the South Atlantic), the California Current (in the North Pacific), the West Australian Current (in the Indian Ocean), and the Peru or Humboldt Current (in the South Pacific).

Eastern boundary currents are the opposite of their western boundary counterparts in nearly every way: They carry cold water equatorward; they are shallow and broad, sometimes more than 1,000 kilometers (620 miles) across; their boundaries are not well defined; and eddies tend not to form. Their total flow is less than that of their western counterparts. The Canary Current in the North Atlantic carries only 16 sv of water at about

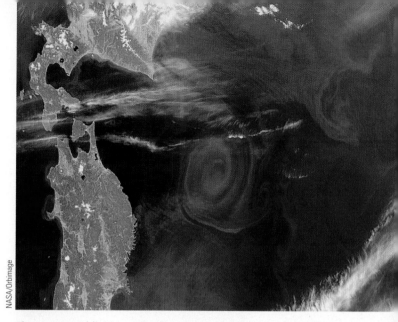

NASA/Orbimage

Figure 9.12 Eddies in another western boundary current, the Kuroshio, off Japan's east coast. The green color in this natural-color photograph indicates areas in which the growth of small plantlike organisms has been stimulated by nutrients brought to the surface by turbulence.

2 kilometers (1.2 miles) per hour. The current is so shallow and broad that sailors may not notice it. Contrast the flow rates of the North Atlantic's western and eastern boundary currents in Figure 9.9. **Table 9.1** summarizes the major differences between boundary currents in the Northern Hemisphere.

Transverse Currents As we have seen, most of the power for ocean currents is derived from the trade winds at the fringes of the tropics and from the mid-latitude westerlies. The stress of winds on the ocean in these bands gives rise to the **transverse currents**—currents that flow from east to west and west to east, linking the eastern and western boundary currents.

The trade-wind–driven North Equatorial Current and South Equatorial Current in the Atlantic and Pacific are moderately shallow and broad, but each transports about 30 sv westward. Because of the thrust of the trades, Atlantic water at Panama is usually 20 centimeters (8 inches) higher, on average, than water across the isthmus in the Pacific. The Pacific's greater expanse of water at the equator and stronger trade winds develop more powerful westward-flowing equatorial currents, and the height differential between the western and eastern Pacific is thought to approach 1 meter (3.3 feet)!

Westerly winds drive the eastward-flowing transverse currents of the mid-latitudes. Because they are not shepherded by the trade winds, eastward-flowing currents are wider and flow more slowly than their equatorial counterparts. The North Pacific and North Atlantic currents are Northern Hemisphere examples.

As can be seen in Figure 9.8b, the westward flow of the transverse currents near the equator proceeds unimpeded for great distances, but the eastward flow of transverse currents at middle and high latitudes in the northern ocean basins is interrupted by continents and island arcs. In the far south, however, eastward flow is almost completely free. Intense westerly winds

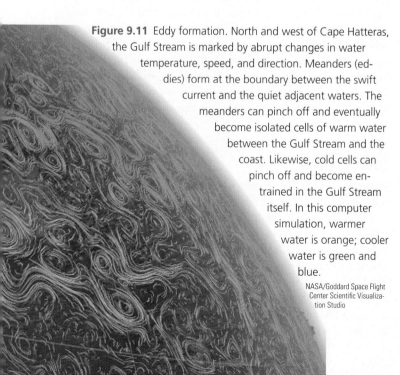

Figure 9.11 Eddy formation. North and west of Cape Hatteras, the Gulf Stream is marked by abrupt changes in water temperature, speed, and direction. Meanders (eddies) form at the boundary between the swift current and the quiet adjacent waters. The meanders can pinch off and eventually become isolated cells of warm water between the Gulf Stream and the coast. Likewise, cold cells can pinch off and become entrained in the Gulf Stream itself. In this computer simulation, warmer water is orange; cooler water is green and blue.

NASA/Goddard Space Flight Center Scientific Visualization Studio

Table 9.1 Boundary Currents in the Northern Hemisphere

Type of Current (example)	General Features	Speed	Transport (millions of cubic meters per second)	Special Features
Western Boundary Currents	**Warm**			
Gulf Stream, Kuroshio (Japan) Current	Narrow, <100 km; deep—substantial transport to depths of 2 km	Swift, hundreds of kilometers per day	Large, usually 50 sv or greater	Sharp boundary with coastal circulation system; little or no coastal upwelling; waters tend to be depleted in nutrients, unproductive; waters derived from trade-wind belts
Eastern Boundary Currents	**Cold**			
California Current, Canary Current	Broad, ~1,000 km; shallow, <500 m.	Slow, tens of kilometers per day	Small, typically 10–15 sv	Diffuse boundaries separating from coastal currents; coastal upwelling common; waters derived from mid-latitudes

Source: GROSS, M. GRANT, OCEANOGRAPHY: VIEW OF THE EARTH, 5th Edition, © 1990, p. 173. Reprinted by permission of Pearson Education, Inc., Upper Saddle River, NJ.

over the southern ocean drive the greatest of all ocean currents, the unobstructed West Wind Drift (or Antarctic Circumpolar Current). This current carries more water than any other—at least 100 sv west to east in the Drake Passage between the tip of South America and the adjacent Palmer Peninsula of Antarctica.

Westward Intensification

Why should western boundary currents be concentrated and eastern boundary currents be diffuse? The reasons are complex, but as you might expect, the Coriolis effect is involved. Due to the Coriolis effect—which increases as water moves farther from the equator—eastward-moving water on the north side of the North Atlantic Gyre is turned sooner and more strongly toward the equator than westward-flowing water at the equator is turned toward the pole. So, the peak of the hill described in Figure 9.7 is not in the center of the ocean basin but closer to its western edge. Its slope is steeper on the western side. If an equal volume of water flows around the gyre, the current on the eastern boundary (off the coast of Europe) is spread out and slow, and the current on the western boundary (off the U.S. east coast) is concentrated and rapid. Western boundary currents are faster (up to 10 times), deeper, and narrower (up to 20 times) than eastern boundary currents. The effect on current flow is known as **westward intensification (Figure 9.13)**, a phenomenon clearly visible in Figures 9.7c and 9.9.

Westward intensification doesn't happen just in the North Atlantic. The western boundary currents in the gyres of both hemispheres are more intense than their eastern counterparts.

Countercurrents and Undercurrents Are Submerged Exceptions to Peripheral Flow

Equatorial currents are typically accompanied by **countercurrents**, which flow on the surface in the opposite direction from the main current. As you may remember, at the meteorological equator, air is rising and the trade winds do not blow across the boundary. Without persistent winds to drive water to the west, some backward flow of water occurs at the meteorological equator. Some water flows away from the main equatorial currents (which flow westward a bit north and south of the meteorological equator) to return on the surface to the east. Look for this in the Pacific in Figure 9.8b.

Countercurrents, sometimes referred to as **undercurrents**, can also exist *beneath* surface currents. The first undercurrent was discovered in 1951 in the central Pacific by Townsend Cromwell, a researcher employed by the U.S. Fish and Wildlife Service to investigate deep, long-line fishing techniques pioneered by Japanese fishermen. The Pacific Equatorial Undercurrent, or Cromwell Current, flows eastward beneath the North Equatorial Current with an average velocity of 5 kilometers (3 miles) per hour at a depth of 100 to 200 meters (330–660 feet). It is about 300 kilometers (190 miles) wide and carries a volume equivalent to about half the Gulf Stream. It has been traced for more than 14,000 kilometers (8,700 miles), from New Guinea to Ecuador. Undercurrents have since been found under most major currents. They can be very large—their volumes sometimes approach the volume of the current above them.

Calm Centers

As we have seen, the fastest water movement in a gyre occurs at its periphery. Current flow at the centers of gyres is minimal. Water in these areas tends to be relatively high in salinity **(Figure 9.14)** and supports relatively little life. Winds are light, and floating material tends to collect in the low energy center of the gyre. There are usually few islands on which the floating material can beach, so it stays there in the gyre.

Much of this material is natural. The most famous example is *Sargassum*, a large marine alga (seaweed) that can grow unattached to a solid object **(Figure 9.15)**.

Without the Coriolis effect, ocean gyres would look like this:

With the Coriolis effect, they look like this:

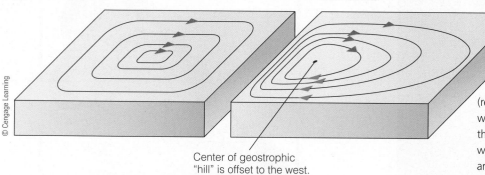

Center of geostrophic "hill" is offset to the west.

© Cengage Learning

a Without the Coriolis effect, water currents would form a regular and symmetrical gyre. However, because the Coriolis effect is strongest near the poles, water flowing eastward at high latitudes (red arrows) turns sooner to the right (clockwise), "short circuiting" the gyre. And because the Coriolis effect is nonexistent at the equator, water flowing westward near the equator (green arrows) tends not to turn clockwise until it encounters a blocking continent. Western boundary currents are therefore faster and deeper than eastern boundary currents, and the geostrophic hill is offset to the west.

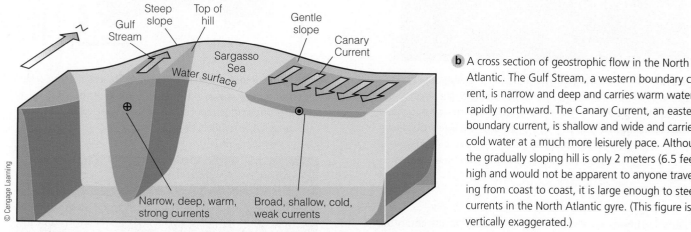

Gulf Stream • Steep slope • Top of hill • Sargasso Sea • Water surface • Gentle slope • Canary Current

Narrow, deep, warm, strong currents

Broad, shallow, cold, weak currents

© Cengage Learning

b A cross section of geostrophic flow in the North Atlantic. The Gulf Stream, a western boundary current, is narrow and deep and carries warm water rapidly northward. The Canary Current, an eastern boundary current, is shallow and wide and carries cold water at a much more leisurely pace. Although the gradually sloping hill is only 2 meters (6.5 feet) high and would not be apparent to anyone traveling from coast to coast, it is large enough to steer currents in the North Atlantic gyre. (This figure is vertically exaggerated.)

Figure 9.13 The influence of the Coriolis effect on westward intensification.

 THINKING BEYOND THE FIGURE

Would you expect westward intensification in the southern Hemisphere, or would the situation be reversed?

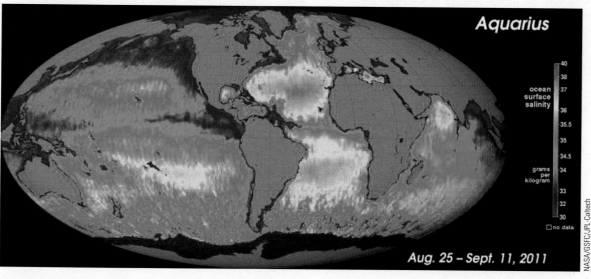

Figure 9.14 The *Aquarius* satellite, launched in 2011, visualizes gyres based on salinity.

ocean surface salinity

grams per kilogram

40
38
37
36
35.5
35
34.5
34
33
30

□ no data

Aquarius

Aug. 25 – Sept. 11, 2011

NASA/GSFC/JPL–Caltech

Figure 9.15 The calm centers of gyres tend to be relatively high in salinity and support relatively little life. Winds are light, and floating material tends to collect in the low energy centers of the gyres. Nearly all seaweeds are anchored to a hard, shallow seafloor. When Columbus encountered *Sargassum* (seen here) on his way across the Atlantic in 1492, he and his crew were convinced—wrongly—that the ocean was shallow and that land must lie close by.

More ominous are the growing vortices of plastic and other debris that circulate in the centers of gyres. As can be seen in Figure 9.14, the North Pacific subtropical gyre covers a large area of the Pacific. This area, about the size of Texas, has been dubbed "the Asian Trash Trail" the "Trash Vortex," or the "Eastern Garbage Patch." (A smaller western Pacific equivalent has formed midway between San Francisco and Hawaii; another lies off the east coast of the United States—**Figure 9.16**.) One researcher estimates the weight of the debris trapped in gyres to be about 3 million metric tons, comparable to a year's deposition at Los Angeles's largest landfill.[2] More on the topic of plastics in the ocean will be found in Chapter 18.

A Final Word on Gyres

Although we have stressed individual currents in our discussion, remember that gyres consist of currents that blend into one another. Flow is continuous without obvious places where one current ceases and another begins. The *balance* of wind energy, friction, the Coriolis effect, and the pressure gradient propels gyres and holds them along the outside of ocean basins.

---

[2]Contrary to popular belief, this plastic debris does not form vast mats that one could easily see or even walk across. The particles are small—sometimes too small to be seen from the deck of a ship—and widely spaced.

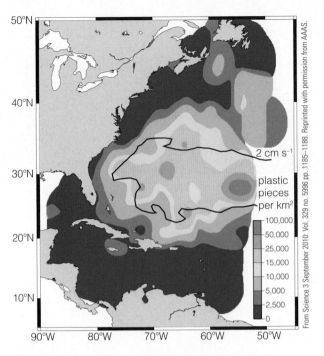

a Plastic pieces in the North Atlantic gyre. The estimated total amount of plastic in the area shown (most of which consists of particles about 1 millimeter—¹⁄₁₆ of an inch—in size) is 1,100 metric tons (1,210 tons).

b A clump of plastic debris near the center of the North Pacific gyre.

Figure 9.16 Waste plastic accumulates at the centers of gyres.

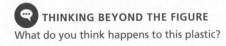

THINKING BEYOND THE FIGURE
What do you think happens to this plastic?

9.3 Surface Currents Affect Weather and Climate

Along with the winds, surface currents distribute tropical heat worldwide. Warm water flows to higher latitudes, transfers heat to the air and cools, moves back to low latitudes, and absorbs heat again; then the cycle repeats. The greatest amount of heat transfer occurs at mid-latitudes, where about 10 million billion calories of heat are transferred each second—a million times as much power as is consumed by all the world's human population in the same length of time! This combination of water flow and heat transfer from and to water influences climate and weather in several ways.

In winter, for example, Edinburgh, Dublin, and London are bathed in eastward-moving air only recently in contact with the relatively warm North Atlantic Current. Scotland, Ireland, and England have a maritime climate. They are warmed in part by the energy of tropical sunlight transported to high latitudes by the Gulf Stream. The power of the Gulf Stream to affect weather is clearly evident in **Figures 9.17** and **9.18**.

At lower latitudes on an ocean's eastern boundary the situation is often reversed. Mark Twain is supposed to have said that the coldest winter he ever spent was a summer in San Francisco. Summer months in that West Coast city are cool, foggy, and mild, while Washington, D.C., on nearly the same line of latitude (but on the *western* boundary of an ocean basin), is known for its all-but-intolerable August heat and humidity. Why the difference? Look at Figure 9.8b, and follow the currents responsible. The California Current, carrying cold water from the north, comes close to the coast at San Francisco. Wind approaching the California coast loses heat to the cold sea and comes ashore to chill San Francisco. On the other coast, summer air often flows around a similar high off the East Coast (the Bermuda High). Winds approaching Washington, D.C., therefore blow from the south and east. Heat and moisture from the Gulf Stream contribute to the capital's oppressive summers. (In winter, on the other hand, Washington, D.C., is colder than San Francisco because westerly winds approaching Washington are chilled by the cold continent they cross.)

9.4 Wind Can Cause Vertical Movement of Ocean Water

The wind-driven *horizontal* movement of water can sometimes induce *vertical* movement in the surface water. This movement is called **wind-induced vertical circulation**. Upward movement of water is known as **upwelling**; the process brings deep, cold, usually nutrient-laden water toward the surface. Downward movement is called **downwelling**.

Nutrient-Rich Water Rises Near the Equator

The South Equatorial Currents of the Atlantic and Pacific straddle the equator. Though the Coriolis effect is weak near the equator (and absent *at* the equator), water moving in the currents on either side of the equator is deflected slightly poleward and replaced by deeper water **(Figure 9.19)**. Thus, **equatorial upwelling** occurs in these westward-flowing equatorial surface currents. Upwelling is an important process because this water from within and below the pycnocline is often rich in the nutrients needed by marine organisms for growth. The long, thin band of upwelling and biological productivity extending along the equator westward from South America is clearly visible in Figures 9.23b. The layers of ooze on the equatorial Pacific seabed (Figure 5.10) are testimony to the biological productivity of surface water there. By contrast, generally poor conditions for growth prevail in most of the open tropical ocean, because strong layering isolates deep, nutrient-rich water from the sunlit ocean surface.

Wind Can Induce Upwelling Near Coasts

Wind blowing parallel to shore or offshore can cause **coastal upwelling**. The friction of wind blowing along the ocean surface causes the water to begin moving, the Coriolis effect deflects it to the right (in the Northern Hemisphere), and the resultant Ekman transport moves it offshore. As shown in **Figure 9.20a**, coastal upwelling occurs when this surface water is replaced by water rising along the shore. Again, because the new surface water is often rich in nutrients, prolonged wind can result in increased biological productivity. Coastal upwelling along the coast of California is visible in **Figure 9.20b**.

Figure 9.17 Only 48 kilometers (30 miles) off the coast of Cornwall at 50°N, these "tropical" gardens on Britain's Scilly Isles lie in the path of the warm waters of the Gulf Stream. The cities of western Europe and Scandinavia are warmed by the energy of tropical sunlight transported to their northern latitudes by winds and by moving masses of water called currents. Ireland and England therefore have a mild maritime climate.

Gary Blake/Alamy

NASA Earth Observatory

Figure 9.18 A cold, dry wind blowing to the southeast off the U.S. east coast on 7 January 2014 warms, picks up moisture, and forms clouds as it passes across the warm Gulf Stream.

a The South Equatorial Current, especially in the Pacific, straddles the geographical equator (see again Figure 9.8b). Water north of the equator veers to the right (northward), and water to the south veers to the left (southward). Surface water therefore diverges, causing upwelling. Most of the upwelled water comes from the area above the equatorial undercurrent, at depths of 100 meters (330 feet) or less.

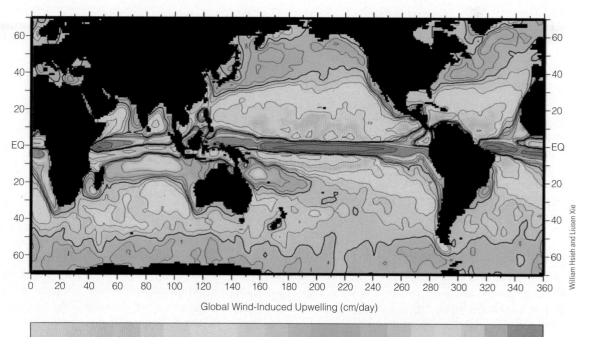

b The phenomenon of equatorial upwelling is worldwide but most pronounced in the Pacific. The red, orange, and yellow colors mark the areas of greatest upwelling.

Global Wind-Induced Upwelling (cm/day)

William Hsieh and Liusen Xie

Figure 9.19 Equatorial upwelling.

Figure 9.20 Coastal upwelling.

💬 **THINKING BEYOND THE FIGURE**
What season of the year results in the greatest upwelling—thus the greatest growth of small, plantlike organisms, off the coast of California?

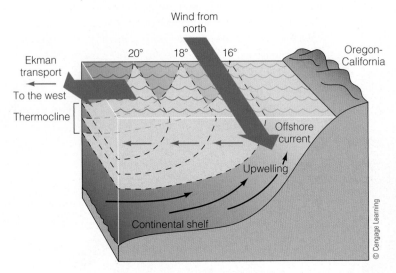

a In the Northern Hemisphere, coastal upwelling can be caused by winds from the north blowing along the west coast of a continent. Water moved offshore by Ekman transport is replaced by cold, deep, nutrient-laden water. In this diagram, temperature of the ocean surface is shown in degrees Celsius. (Vertical exaggeration ~100×.)

b A satellite view of the U.S. west coast shows (in artificial color) the growth of small plantlike organisms stimulated by upwelled nutrients. The color bar indicates the concentration of chlorophyll in milligrams per cubic meter of seawater. Notice that chlorophyll concentration—and biological productivity—is highest near the coast.

Upwelling can also influence weather. Wind blowing from the north along the California coast causes offshore movement of surface water and subsequent coastal upwelling. The overlying air becomes chilled, contributing to San Francisco's famous fog banks and cool summers. Wind-induced upwelling is also common in the Peru Current, along the west coast of Antarctica's Palmer Peninsula, in parts of the Mediterranean, and near some large Pacific islands.

Wind Can Also Induce Coastal Downwelling

Water driven toward a coastline will be forced downward, returning seaward along the continental shelf. This downwelling **(Figure 9.21)** helps supply the deeper ocean with dissolved gases and nutrients, and it assists in the distribution of living organisms. Unlike upwelling, downwelling has no direct effect on the climate or productivity of the adjacent coast.

Langmuir Circulation Affects the Ocean Surface

Winds that blow steadily across the ocean, and the small waves that such winds generate, can induce long sets of counter-rotating vortices (or cells) in the surface water. These slowly twisting vortices align in the direction of the wind **(Figure 9.22)**. It takes about an hour for a particle in a vortex to complete one revolution. Streaks of foam or seaweed or debris, known as *windrows*, collect in areas where adjacent vortices converge, while regions of divergence remain relatively clear. Observed by mariners for hundreds of years, these windrow lines (and the underlying vortices) were first explained in 1938 by Irving Langmuir, who observed them while crossing the calm Sargasso Sea in the center of the North Atlantic gyre. Named in his honor, **Langmuir circulation** rarely disturbs the ocean below a depth of about 20 meters (66 feet). Unlike the equatorial and coastal upwellings mentioned previously, Langmuir circulation

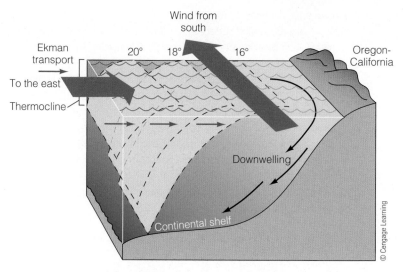

Figure 9.21 Wind blowing from the south along a Northern Hemisphere west coast for a prolonged period can result in downwelling. Areas of downwelling are often low in nutrients and therefore relatively low in biological productivity. (Vertical exaggeration ~100×.)

operates within the surface layer and thus does not lift nutrients trapped within or below the pycnocline.

CONCEPT CHECK

Before going on to the next section, check your understanding of some of the important ideas presented so far:

How can wind-driven horizontal movement of water induce vertical movement in surface water?

How is the Coriolis effect involved in equatorial upwelling?

How is upwelling linked to biological productivity?

a Winds that blow steadily across the ocean, and the small waves that such winds generate, can induce long sets of counter-rotating vortices in the surface water. If the sun angle is just right (as it was when this photo was taken while flying southwest across the Trade Winds over the South China Sea), the resulting small differences in ocean surface height can be seen.

Figure 9.22 Langmuir circulation.

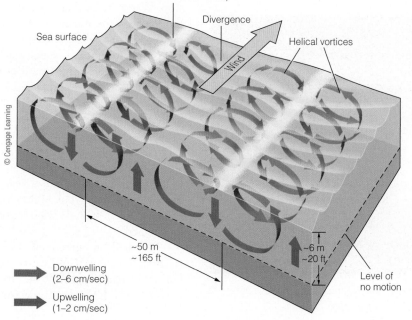

b These shallow, slowly twisting cells of surface water are known as Langmuir circulation in honor of the researcher who explained their motion.

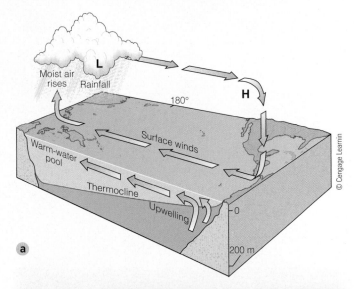

a

© Cengage Learnin

Figure 9.23 A non–El Niño year.

a Normally the air and surface water flow westward, the thermocline rises, and upwelling of cold water occurs along the west coast of Central and South America.

b This map from satellite data shows the temperature of the equatorial Pacific on 31 May 1988. The warmest water is indicated by dark red; and progressively cooler upwelled water, by yellow and green. Note the coastal upwelling along the coast at the lower right of the map and the tongue of recently upwelled water extending westward along the equator from the South American coast.

c A vertical section through the equatorial Pacific in a non–El Niño year (January 1997) shows warmer water to the west and cooler water to the east.

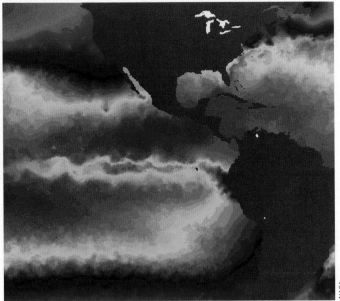

b

NASA

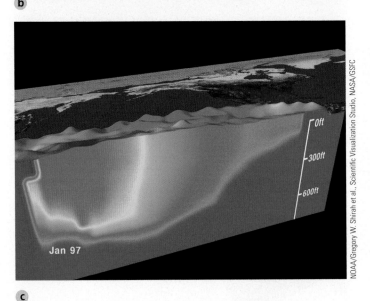

c

NOAA/Gregory W. Shirah et al., Scientific Visualization Studio, NASA/GSFC

9.5 El Niño and La Niña Are Exceptions to Normal Wind and Current Flow

Surface winds across most of the tropical Pacific normally move from east to west (review Figure 8.15). The trade winds blow from the normally high-pressure area over the eastern Pacific (near Central and South America) to the normally stable low-pressure area over the western Pacific (north of Australia). However, for reasons that are still unclear, these pressure areas change places at irregular intervals of roughly 3 to 8 years: high pressure builds in the western Pacific, and low pressure dominates the eastern Pacific. Winds across the tropical Pacific then reverse direction and blow from west to east—the trade winds weaken or reverse. This change in atmospheric pressure (and thus in wind direction) is called the **Southern Oscillation**. Fifteen of these attention-getting oscillations have occurred since 1950.

The trade winds normally drag huge quantities of water westward along the ocean's surface on each side of the equator, but as the winds weaken, these equatorial currents crawl to a stop. Warm water that has accumulated at the western side of the Pacific—the warmest water in the world ocean—can then build to the east along the equator toward the coasts of Central and South America. The eastward-moving warm water usually arrives near the South American coast around Christmastime. In the 1890s it was reported that Peruvian fishermen were using the expression *Corriente del Niño* ("current of the Christ Child") to describe the flow; that's where the current's name, **El Niño**, came from. The phenomena of the Southern Oscillation and El Niño are coupled, so the terms are often combined to form the acronym **ENSO**, for El Niño/Southern Oscillation. An ENSO event typically lasts about a year, but some have persisted for more than 3 years. The effects are felt not only in the Pacific; all ocean areas at trade wind latitudes in both hemispheres can be affected.

Normally, a current of cold water, rich in upwelled nutrients, flows north and west away from the South American continent **(Figure 9.23)**. When the propelling trade winds falter during an ENSO event, warm equatorial water that would

Figure 9.24 An El Niño year.

a When the Southern Oscillation develops, the trade winds diminish and then reverse, leading to an eastward movement of warm water along the equator. The surface waters of the central and eastern Pacific become warmer, and storms over land may increase.

b Sea-surface temperatures on 13 May 1992, a time of El Niño conditions. The thermocline was deeper than normal, and equatorial upwelling was suppressed. Note the absence of coastal upwelling along the coast and the lack of the tongue of recently upwelled water extending westward along the equator.

c A vertical section through the equatorial Pacific in an El Niño year (November 1997) shows warmer water spreading toward the east.

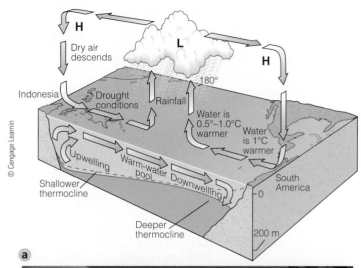

a

b

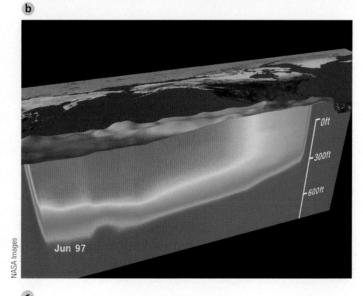

c

normally flow westward in the equatorial Pacific backs up to flow east **(Figure 9.24)**. The normal northward flow of the cold Peru Current is interrupted or overridden by the warm water. Upwelling within the nutrient-laden Peru Current is responsible for the great biological productivity of the ocean off the coasts of Peru and Chile. Although upwelling may continue during an ENSO event, the source of the upwelled water is nutrient-depleted water in the thickened surface layer approaching from the west **(Figure 9.25)**. When the Peru Current slows and its upwelled water lacks nutrients, fish and seabirds dependent on the abundant life it contains die or migrate elsewhere. Peruvian fishermen are never cheered by this Christmas gift!

During major ENSO events, sea level rises in the eastern Pacific, sometimes by as much as 20 centimeters (8 inches) in the Galápagos. Water temperature also increases by up to 7°C (13°F). The warmer water causes more evaporation, and the area of low atmospheric pressure over the eastern Pacific intensifies. Humid air rising in this zone, centered some 2,000 kilometers (1,200 miles) west of Peru, causes high precipitation in normally dry areas. The increased evaporation intensifies coastal storms, and rainfall inland may be much higher than normal. Marine and terrestrial habitats and organisms can be affected by these changes.

The two most severe ENSO events of this century occurred in 1982–1983 and 1997–1998 **(Figures 9.26** and **9.27,** see p. 270). In both cases, effects associated with El Niño were spectacular over much of the Pacific and some parts of the Atlantic and Indian oceans. In February 1998, 40 people were killed and 10,000 buildings damaged by a "wall" of tornadoes advancing over the southeastern United States. This record-breaking tornado event was spawned by the collision of warm, moist air that had lingered over the warm Pacific and a polar front that dropped from the north. In the eastern Pacific, heavy rains throughout the 1997–1998 winter in Peru left at least 250,000 people homeless, destroyed 16,000 dwellings, and closed every port in the country for at least 1 month. Hawai'i, however, experienced record drought, and some parts of southwestern Africa and Papua New Guinea received so

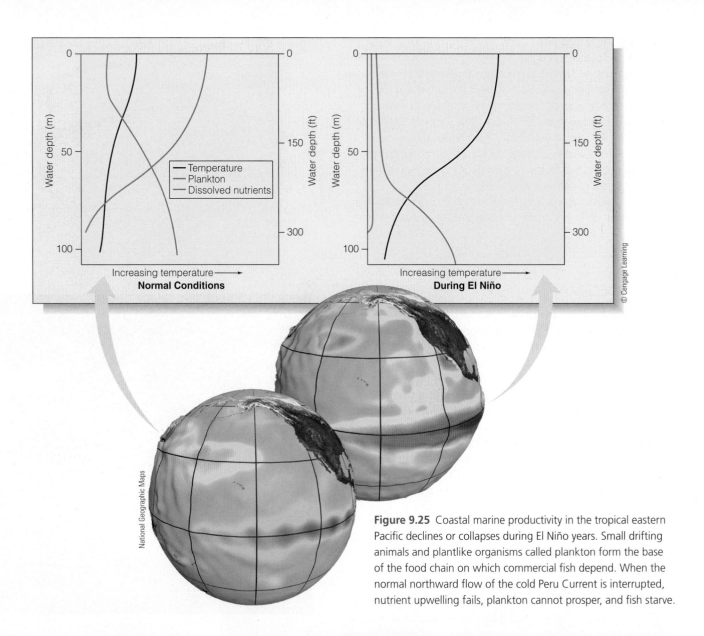

Normal Conditions

During El Niño

National Geographic Maps

Figure 9.25 Coastal marine productivity in the tropical eastern Pacific declines or collapses during El Niño years. Small drifting animals and plantlike organisms called plankton form the base of the food chain on which commercial fish depend. When the normal northward flow of the cold Peru Current is interrupted, nutrient upwelling fails, plankton cannot prosper, and fish starve.

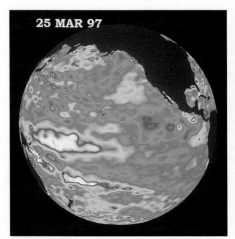

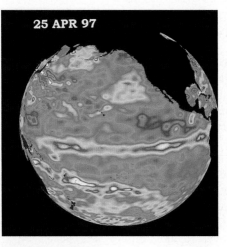

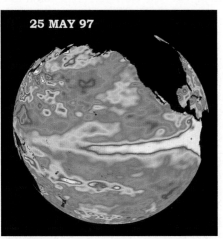

a March 1997. The slackening of the trade winds and westerly wind bursts allow warm water to move away from its usual location in the western Pacific Ocean. Red and white indicate sea level above average height.

b April 1997. About a month after it begins to move, the leading edge of the warm water reaches South America.

c May 1997. Warm water piles up against the South American continent. The white area of sea level is 13 to 30 centimeters (5–12 inches) above normal height, and 1.6° to 3°C (3°–5°F) warmer.

little rain that crops failed completely and whole villages were abandoned because of starvation.

Most of the United States escaped serious consequences—indeed, the midwestern states, Pacific Northwest, and eastern seaboard enjoyed a relatively mild fall, winter, and spring. But California's trials were widely reported. Greater evaporation of water from the warm ocean surface (**Figure 9.28**), combined with an increased number of winter storms steered into the area by the southward-trending jet stream, doubled rainfall amounts in most of the state. Landslides, avalanches, and other weather-related disasters crowded the evening news.

Conditions did not return to near normal until the late spring of 1998. Estimates of worldwide 1997–1998 ENSO-related damage exceed 23,000 deaths and US$33 billion.

Normal circulation sometimes returns with surprising vigor, producing strong currents, powerful upwelling, and chilly and dry conditions along the South American coast. These contrasting colder-than-normal events are given a contrasting name: **La Niña** ("the girl"). As conditions to the east cool off, the ocean to the west (north of Australia) warms rapidly. The renewed thrust of the trade winds piles this water upon itself, depressing the upper curve of the thermocline to more than 100 meters (328 feet). In contrast, the thermocline during a La Niña event in the eastern equatorial Pacific rests at about 25 meters (82 feet). A vigorous La Niña followed the 1997–1998 El Niño and persisted for nearly a year (Figure 9.26e). **Figure 9.29** contrasts how North American weather differs between El Niño and La Niña years.

Studies of the ocean and atmosphere in 1982–1983 and 1997–1998 have given researchers new insight into the behavior and effects of the Southern Oscillation. Some researchers believe that the 1982–1983 event was triggered by the violent 1982 eruption of El Chichón, a Mexican volcano, which injected huge quantities of sun-obscuring dust and sulfur-rich gases into the atmosphere. No similar trigger occurred before the 1997–1998 ENSO, however. Though the exact cause or causes of the Southern Oscillation are not yet understood, subtle changes in the atmosphere permit meteorologists to predict a severe El Niño nearly a year in advance of its most serious effects.

CONCEPT CHECK

Before going on to the next section, check your understanding of some of the important ideas presented so far:

Which way does wind typically blow over the tropical Pacific? How does this flow change during an El Niño event?

What is the Southern Oscillation? How is this related to El Niño?

Why do Peruvian fisheries decline—often dramatically—in El Niño years?

How might weather in the western United States be affected by El Niño?

How is La Niña different from El Niño?

d October 1997. By October, sea level is as much as 30 centimeters (12 inches) lower than normal near Australia. The bulge of warm water has spread northward along the coast of North America from the equator to Alaska. Fisheries in Peru are severely affected because the warm water prevents upwelling of cold, nutrient-rich water necessary for the support of large fish populations.

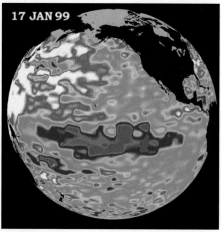

e Normal circulation sometimes returns with surprising vigor after an El Niño event, producing strong currents, powerful upwelling, and chilly and stormy conditions along the South American coast. This image was prepared from data for 15 February 1999. Note the mass of cold surface water and relatively low sea level (purple). Such cold water tends to deflect winds around it, changing the course of weather systems locally and the nature of weather patterns globally.

TOPEX/Poseidon Team, CNES, NASA

Figure 9.26 Development of the 1997–1998 El Niño, observed by the *TOPEX/Poseidon* satellite.

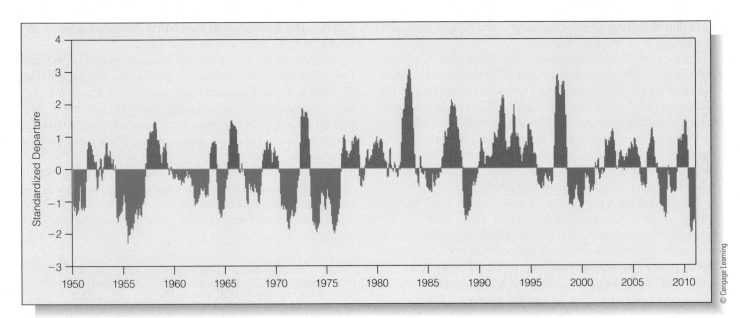

Figure 9.27 El Niño and La Niña events since 1950.

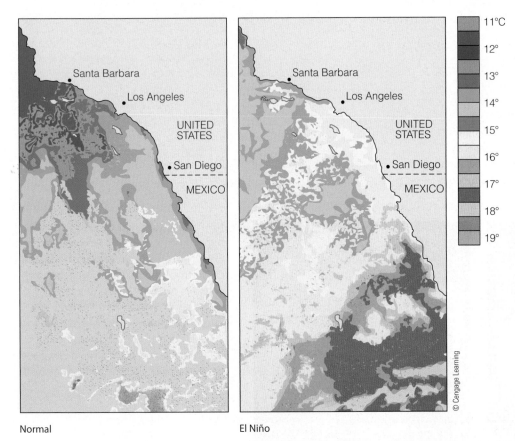

Normal

El Niño

Figure 9.28 Surface temperatures for southern California in January of the normal year of 1982 (left) and in the same month of an El Niño year (1983, right).

💬 **THINKING BEYOND THE FIGURE**

Look again at Figure 9.20. What do you think happens to biological productivity off the coast of southern California during an El Niño year?

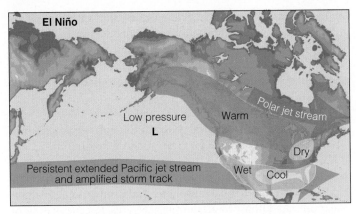

a During an El Niño, low atmospheric pressure south of Alaska allows storms to move unimpeded to the Pacific coast of North America. The resulting weather is wet and cool to the south and warm and dry in the north.

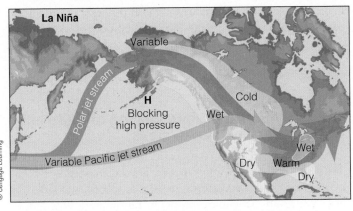

b In La Niña years high atmospheric pressure south of Alaska blocks the storm track. Winds veer north, lose their warmth over Canada, and sweep down as cold blasts. The Pacific Northwest gets its usual rain, but the southwest suffers drought.

Figure 9.29 El Niño changes atmospheric circulation and weather patterns.

9.6 Thermohaline Circulation Affects All the Ocean's Water

The surface currents we have discussed affect the uppermost layer of the world ocean (about 10% of its volume), but horizontal and vertical currents also exist below the pycnocline in the ocean's deeper waters. Because density is largely a function of water temperature and salinity, the movement of water due to differences in density is called **thermohaline circulation** (*therme*, "heat"; *halos*, "salt"). The whole ocean is involved in slow thermohaline circulation, a process responsible for the large-scale vertical movement of ocean water and the circulation of the global ocean as a whole.

Water Masses Have Distinct, Often Unique Characteristics

As you may recall from Chapter 6, the ocean is density stratified, with the densest water near the seafloor and the least dense near the surface. Each water mass has specific temperature and salinity characteristics. Density stratification is most pronounced at temperate and tropical latitudes because the temperature difference between surface water and deep water is greater there than near the poles.

The water masses possess distinct, identifiable properties. Like air masses, water masses don't often mix easily when they meet, because of their differing densities; instead, they usually flow above or beneath each other. Water masses can be remarkably persistent and will retain their identity for great distances and long periods of time. Oceanographers name water masses according to their relative position. In temperate and tropical latitudes, there are five common water masses:

- *Surface water*, to a depth of about 200 meters (660 feet)
- *Central water*, to the bottom of the main thermocline (which varies with latitude)
- *Intermediate water*, to about 1,500 meters (5,000 feet)
- *Deep water*, water below intermediate water but not in contact with the bottom, to a depth of about 4,000 meters (13,000 feet)
- *Bottom water*, water in contact with the seafloor

Surface currents move in the relatively warm upper environment of surface and central water. The boundary between central water and intermediate water is the most abrupt and pronounced.

The densest (and deepest) masses were formed by surface conditions that caused the water to become very cold and salty. Water masses near the surface can be warmer and less saline; they may have formed in warm areas where precipitation exceeded evaporation. Water masses at intermediate depths are intermediate in density.

In spite of this differentiation, the relatively cold water masses lying beneath the thermocline exhibit smaller variations in salinity and temperature than the water in the currents that move across the ocean's surface.

Different Combinations of Water Temperature and Salinity Can Yield the Same Density

Perhaps the best way to visualize ocean layering is with a **temperature–salinity (T–S) diagram** like the one in **Figure 9.30**. The S-shaped curve through the center of the figure shows the temperature and salinity of water at each depth indicated in this area of ocean. Note that many *combinations* of temperature and salinity can yield the *same* density and that the density of the water tends to increase with depth. The shape of the S curve is governed by the position and nature of water masses.

In some cases, two distinct water masses with the same *density* but with different *temperatures and salinities* will combine at a convergence to produce a new water mass of greater

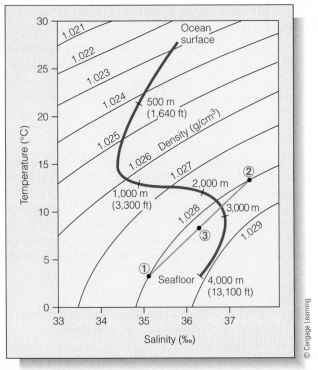

Figure 9.30 A general temperature–salinity (T–S) diagram, with density curves added. The S curve tracks the combination of temperature and salinity (and therefore density) with increasing depth. Note that the top of the S curve represents the temperature and salinity (and therefore density) of water at the ocean surface; the bottom of the curve represents the density of water at the ocean floor. Intermediate depths are shown on the curve.

Points ① and ② are on the same gently curving *isopycnal*—line of equal density. Note that seawater at points ① and ② has *different* temperatures and salinities but the *same* density. If these two masses of water were to merge, their combined density would be greater than either of their separate densities—a situation represented at point ③. This process is known as *caballing* and leads to the formation of deep water.

density. This mixing-and-sinking process is called **caballing**. To understand how caballing works, imagine the two water masses indicated by point ① and point ② in Figure 9.30. Though each has a different temperature and salinity, they lie on the same density line—they are equally dense. Now imagine the waters at these points mixing together. The density of the new water mass could be represented by a point halfway along a straight line connecting them—point ③. The combined water mass is denser than 1.0280—perhaps 1.0283—and will tend to sink. North Atlantic Intermediate Water, Antarctic Intermediate Water, and some Antarctic Bottom Water are produced by caballing.

Thermohaline Flow and Surface Flow: The Global Heat Connection

As we have seen, swift and narrow surface currents along the western margins of ocean basins carry warm, tropical surface waters toward the poles. In a few places, the water loses heat to the atmosphere and sinks to become deep water and bottom water. This sinking is most pronounced in the North Atlantic. The cold, dense water moves at great depths toward the Southern Hemisphere and eventually wells up into the surface layers of the Indian and Pacific oceans. Almost a thousand years are required for this water to make a complete circuit.

The transport of tropical water to the polar regions is part of a global conveyor belt for heat. A simplified outline of the global circuit, the result of three decades of concentrated effort to understand deep circulation, is shown in **Figure 9.31**. This slow circulation straddles the hemispheres and is superimposed upon the more rapid flow of water in surface gyres. Recent analysis of this global circuit suggests that some of the heat warming the coasts of Europe enters the ocean in the vicinity of Indonesia and Australia, travels to the Indian Ocean, and enters the Gulf Stream by way of the Agulhas Current rounding the southern tip of Africa. The surface water that leaves the Pacific is driven in part by excess rainfall and river runoff throughout the Pacific basin. The slow, steady, three-dimensional flow of water in the conveyor belt distributes dissolved gases and solids, mixes nutrients, and transports the juvenile stages of organisms among ocean basins.

The Formation and Downwelling of Deep Water Occurs in Polar Regions

Antarctic Bottom Water **Antarctic Bottom Water**, the most distinctive of the deep-water masses, is characterized by a salinity of 34.65‰, a temperature of −0.5°C (30°F), and a density of 1.0279 grams per cubic centimeter. This water is noted for its extreme density (the densest in the world ocean), for the great amount of it produced near Antarctic coasts, and for its ability to migrate north along the seafloor.

Most Antarctic Bottom Water forms near the Antarctic coast south of South America during winter (**Figure 9.32**, see p. 278). Salt is concentrated in pockets between crystals of pure water and then squeezed out of the freezing mass to form a frigid brine. Between 20 *million* and 50 *million* cubic meters of this brine form every second! The water's great density causes it to sink toward the continental shelf, where it mixes with nearly equal parts of water from the southern Antarctic Circumpolar Current.

The mixture settles along the edge of Antarctica's continental shelf, descends along the slope, and spreads along the deep seabed, creeping north in slow sheets. Antarctic Bottom Water flows many times as slowly as the water in surface currents: in the Pacific it may take a thousand years to reach the equator. Six hundred years later it may be as far away as the Aleutian Islands at 50°N! Antarctic Bottom Water also flows into the Atlantic Ocean basin, where it flows north at a faster rate than in the Pacific. Antarctic Bottom Water has been identified as high as 40° *north* latitude on the Atlantic floor, a journey that has taken some 750 years.

North Atlantic Deep Water Some dense bottom water also forms in the northern polar ocean, but the topography of the

Arctic Ocean basin prevents most of the bottom water from escaping, except in the deep channels formed in the submarine ridges separating Scotland, Iceland, and Greenland. These channels allow the cold, dense water formed in the Arctic to flow into the North Atlantic to form **North Atlantic Deep Water** (look ahead to Figure 9.34).

North Atlantic Deep Water forms when the relatively warm and salty North Atlantic Ocean cools as cold winds from northern Canada sweep over it. Exposed to the chilled air, water at the latitude of Iceland releases heat, cools from 10° to 2°C (50°–36°F), and sinks. (Transferred to the air, this bonus heat makes northern Europe far warmer than its high latitude suggests.) Gulf Stream water that sinks in the north is replaced by warm water flowing clockwise along the U.S. east coast in the North Atlantic gyre.

Other Deep-Water Masses Other distinct deep-water masses exist. Their positions in relation to each other are always determined by their relative densities. Consider the enclosed Mediterranean Sea, where surface water is made more saline by the excess of evaporation over freshwater input. About 3,000 cubic kilometers (720 cubic miles) more water evaporates annually from the Mediterranean than is replaced by river runoff or precipitation. In the cool winter months, Mediterranean water with a salinity of about 38‰ flows past the lip of Gibraltar and spreads into the Atlantic as Mediterranean Deep Water (**Figure 9.33**, see p. 279). Mediterranean Deep Water underlies much of the central water mass in the Atlantic, and some of this water can be traced as far south as the basins of the Antarctic. Though saltier than Antarctic Bottom Water or Atlantic Deep Water, Mediterranean Deep Water is considerably warmer and therefore not as dense. It will lie atop the layers of denser water at high southern latitudes.

Water Masses May Converge, Fall, Travel across the Seabed, and Slowly Rise

The great quantities of dense water sinking at polar ocean basin edges must be offset by equal quantities of water rising elsewhere. **Figure 9.34** (see p. 279) shows an idealized model of thermohaline flow. Note that water sinks relatively rapidly in a small area where the ocean is very cold, but it rises much more gradually across a very large area in the warmer temperate and tropical zones. It then slowly returns poleward near the surface to repeat the cycle. The continual diffuse upwelling of deep water maintains the existence of the permanent thermocline found everywhere at low- and midlatitudes. This slow upward movement is estimated to be about 1 centimeter (½ inch) per day over most of the ocean. If this rise were to stop, downward movement of heat would cause the thermocline to descend and would reduce its steepness. In a sense, the thermocline is "held up" by the continual slow upward movement of water.

Most features of this ideal circulation pattern exist in nature. The water masses, each of distinct density and sandwiched in layers, are slowly propelled by gravity. The water masses butt against one another in **convergence zones**, and the heavier water can slide beneath the lighter water (as Figure 9.32 shows).

Hundreds of years may pass before water masses complete a circuit or blend to lose their identities. Remember that Antarctic Bottom Water in the Pacific retains its character for up to 1,600 years! The residence time of most deep water is less, however; it takes about 200 to 300 years to rise to the surface. (By contrast, a bit of surface water in the North Atlantic gyre may take only a little more than a year to complete a circuit.)

Not all thermohaline circulation is so sedate. Ripple marks in sediments, scour lines, and the erosion of rocky outcrops on deep-ocean floors are evidence that relatively strong, localized bottom currents exist (see Figure 5.3). Some of these currents may move as rapidly as 60 centimeters (24 inches) per second. These relatively fast currents are strongly influenced by bottom topography, and they are sometimes called **contour currents** because their dense water flows around (rather than over) seafloor projections. Bottom currents generally move equatorward at or near the western boundaries of ocean basins (below the western boundary surface currents). The subtle density differences between deep-water masses are not capable of moving water at the speed of the wind-driven surface currents. Water in some of these currents may move only 1 to 2 meters (3–7 feet) per day. Even at that slow speed the Coriolis effect modifies their pattern of flow.

Currents are the very heart of physical oceanography. Their global effects, great masses of water, complex flow, and possible influence on human migrations make their study of particular importance.

> **B**etween 20 million and 50 million cubic meters of Antarctic Bottom Water forms every second.

CONCEPT CHECK

Before going on to the next section, check your understanding of some of the important ideas presented so far:

What drives the vertical movement of ocean water? What is the general pattern of thermohaline circulation?

What are water masses? What determines their relative position in the ocean?

Where are distinct water masses formed?

What happens in convergence zones? How is caballing associated with convergence zones?

How does thermohaline circulation force the thermocline toward the ocean's surface?

How does the length of time required for completion of a circuit of surface circulation compare with that needed for thermohaline circulation?

Traditional methods of studying currents are being replaced with high-tech devices. How do some of these work?

How can chlorofluorocarbons (CFCs) be used as such tracers? Would CFC-based methods be equally suitable for analysis of surface currents and thermohaline circulation?

The global pattern of deep circulation resembles a vast "conveyor belt" that carries surface water to the depths and back again. Begin with the formation of North Atlantic Deep Water north of Iceland. This water mass flows south through the Atlantic and then flows over (and mixes with) deep water formed near Antarctica. The combined mass circumnavigates Antarctica and then moves north into the Indian and Pacific ocean basins. Diffuse upwelling in all of the ocean returns some of this water to the surface. Water in the conveyor gradually warms and mixes upward to be returned to the North Atlantic by surface circulation. The whole slow-moving system is important in transporting water and heat.

 THINKING BEYOND THE FIGURE
How would western Europe's climate change if the Gulf Stream slowed?

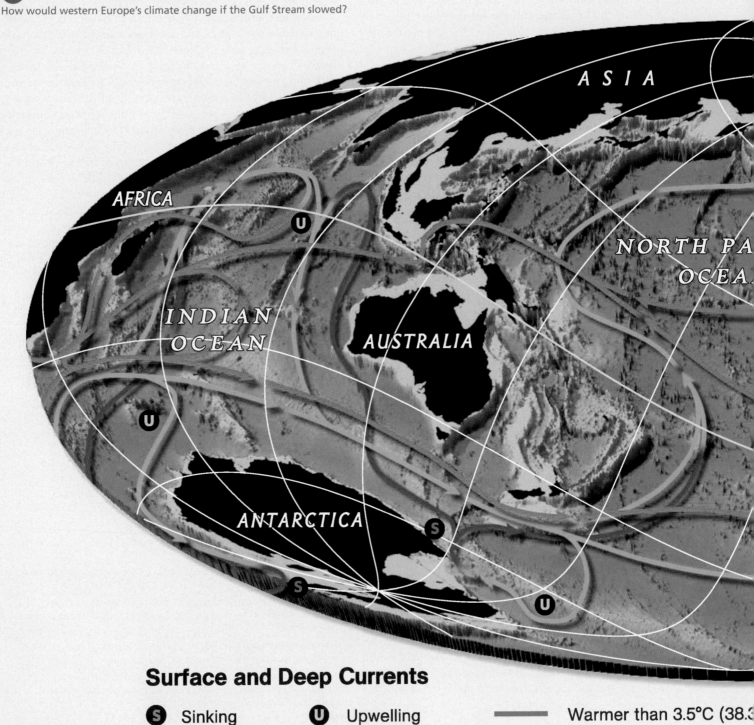

Surface and Deep Currents

S Sinking   **U** Upwelling   —— Warmer than 3.5°C (38.3

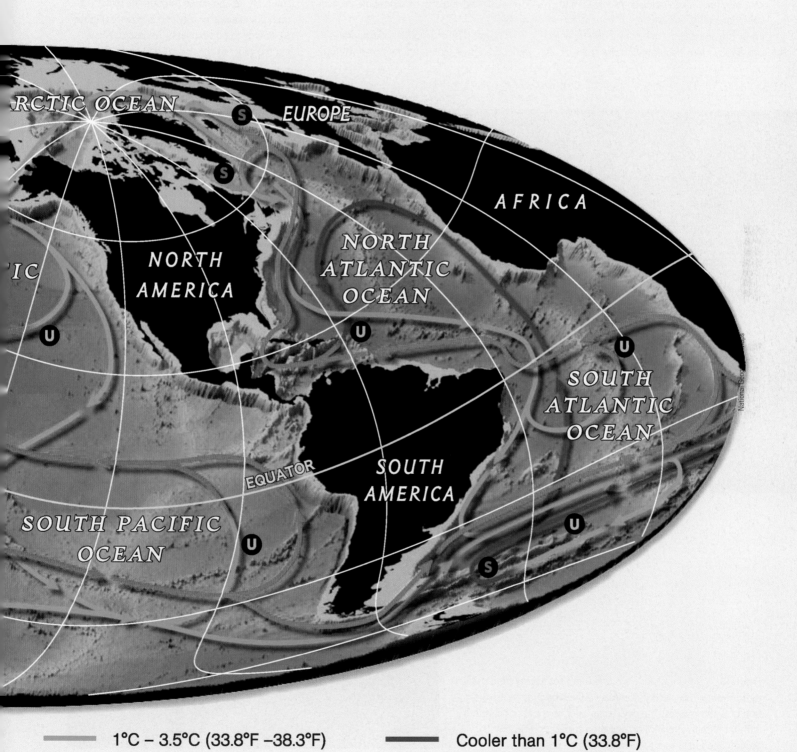

ARCTIC OCEAN ⓢ EUROPE

AFRICA

NORTH
ATLANTIC
OCEAN

NORTH
AMERICA

ⓤ

ⓤ

ⓤ

SOUTH
ATLANTIC
OCEAN

EQUATOR SOUTH
AMERICA

SOUTH PACIFIC
OCEAN

ⓤ

ⓤ

ⓢ

———— 1°C – 3.5°C (33.8°F –38.3°F)   ———— Cooler than 1°C (33.8°F)

How Ocean Currents Work

Surface currents can be traced with drift bottles or drift cards. These tools are especially useful in determining coastal circulation, but they provide no information on the path the drift bottle or card may have taken between its release and collection points. Researchers who want to know the precise track taken by a drifting object can deploy more elaborate drift devices, such as the buoy arrangement in **Figure A**. These buoys can be tracked continuously by radio direction finders or radar. Surface currents can also be tracked by noting the difference between the daily expected and observed positions of ships at sea.

A newer class of research tools operates autonomously—that is, on their own without human guidance. The first of these, initially deployed in 2003, is the Slocum, named after Joshua Slocum, first person to circumnavigate the globe alone. These little gliders "fly" smoothly up and down through the water column powered by gravity and buoyancy **(Figure B)**. Energy to pump ballast overboard, allowing the glider to rise, is provided by a simple

Chris Linder/Woods Hole Oceanographic Institution

Figure A A buoy carrying temperature probes is placed off Cape Hatteras to measure conditions where the Gulf Stream meets a cold, fresh, southward-moving coastal current.

Figure B A Slocum glider—a probe that uses energy from gravity, buoyancy, heat, and batteries to power long-range exploration of water masses. The device can operate for about a month without human intervention and then surface to transmit its accumulated data to a satellite for relay to scientists.

Webb Research

heat engine powered by the difference in temperature between ocean surface and great depths. The gliders can fall again as seawater is pumped aboard. Small fleets of Slocum gliders map the ocean's thermohaline depth profiles, chlorophyll content, and other parameters for years and, on their occasional visits to the surface, transmit data to satellites.

More numerous but less maneuverable than Slocum gliders are the 3,000+ floats of the **Argo system**. These smaller floats move vertically in the water column (to depths of 2 kilometers, about 7,000 feet). The floats return to the surface once every 10 days, measuring temperature and conductivity as they move. Data are uploaded to satellites and used to calculate salinity. Argo floats work with the satellite *Jason 1* as a part of the Integrated Ocean Observing System to measure ocean topography and worldwide climate.

The batteries that power each Argo float's vertical voyages last for around 5 years **(Figure C)**. About 750 new floats are deployed each year to replace floats that expire or are lost. The floats have an average spacing of 300 kilometers (190 miles), but the exact spacing depends on the random nature of float drift. **Figure D** shows the location of the system's 3,561 functioning floats in February of 2014.

Yet another method, developed for studying thermohaline circulation, senses the presence in seawater of chemical tracers—artificial substances with known histories of production or release. Because they dissolve easily in seawater, chlorofluorocarbons (CFCs) can be used as such tracers. A totally artificial chemical first produced in the 1930s for use as refrigerants, aerosol propellants, and blowing agents for foam, CFCs spread through the ocean like a dye, following oceanic circulation. The speed of deep currents has been measured by careful analysis of their CFC content.

Figure C A freshly deployed *Argo* float ready to carry out its mission.

Figure D The World Ocean Circulation Experiment, a consortium of research institutions formed in the late 1990s, planned a world-wide fleet of small drifting robot sensors. The target number of probes (about 3,000) was reached during 2007. About 750 floats are added each year to replace those that are lost or have expired. Each *Argo* float drifts at a depth of about 1,000 meters (3,300 feet) for 10 days. The float then descends to 2,000 meters (6,600 feet), then rises, collecting data of temperature, pressure, density, salinity and its position to transmit to satellites. The float then sinks to repeat the process. This figure shows the location of 3,561 *Argo* floats in February, 2014.

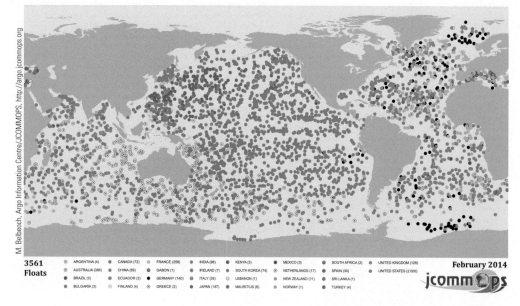

3561 Floats	ARGENTINA (4)	CANADA (72)	FRANCE (256)	INDIA (96)

February 2014

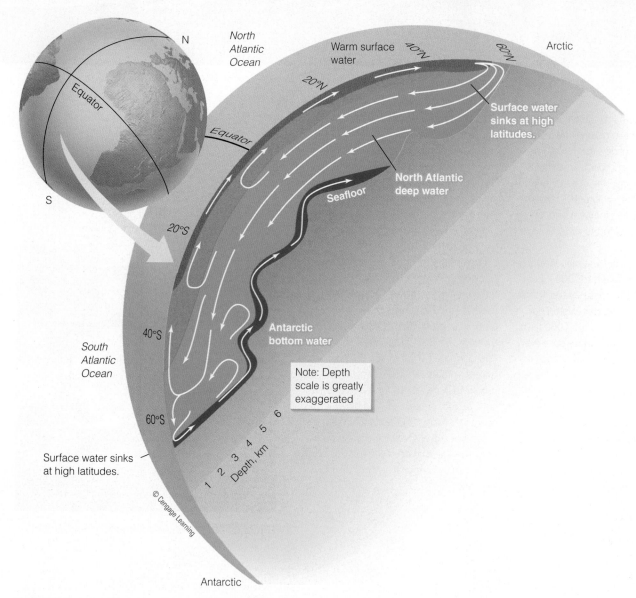

Figure 9.32 A simplified view of thermohaline circulation in the Atlantic. Surface water becomes dense and sinks in the north and south polar regions. Being denser, Antarctic Bottom Water slips beneath North Atlantic Deep Water. The water then gradually rises across a very large area in the tropical and temperate zones, then flows poleward to repeat the cycle.

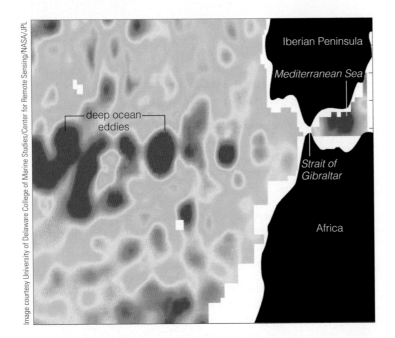

Figure 9.33 Warm water flowing out of the Mediterranean Sea past the Straits of Gibraltar flows down the southern margin of Portugal until it reaches its point of neutral buoyancy. It then breaks free of the slope and forms a layer between 800 and 1,500 meters (2,600 and 5,000 feet) deep. The subtle differences in the waters' temperature and salinity (therefore density) are imaged here by acoustical tomography.

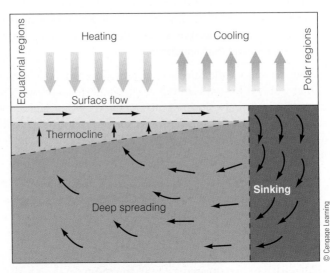

Figure 9.34 The classic model of a pure thermohaline circulation, caused by heating in lower latitudes and cooling in higher latitudes. Compare this figure to the convection cell shown in Figure 8.9.

Chapter in Perspective

In this chapter you learned that ocean water circulates in currents. Surface currents affect the uppermost 10% of the world ocean. The movement of surface currents is powered by the warmth of the sun and by winds. Water in surface currents tends to flow horizontally, but it can also flow vertically in response to wind blowing near coasts or along the equator. Surface currents transfer heat from tropical to polar regions, influence weather and climate, distribute nutrients, and scatter organisms. They have contributed to the spread of humanity to remote islands, and they are important factors in maritime commerce.

Circulation of the 90% of ocean water beneath the surface zone is driven by the force of gravity, as dense water sinks and less dense water rises. Since density is largely a function of temperature and salinity, the movement of deep water due to density differences is called *thermohaline circulation*. Currents near the seafloor flow as slow, river-like masses in a few places, but the greatest volumes of deep water creep through the ocean at an almost imperceptible pace. The Coriolis effect, gravity, and friction shape the direction and volume of surface currents and thermohaline circulation.

In the next chapter you will learn about ocean waves. The traveling crests produce the appearance of movement we see in a wave. In an ocean wave, a ribbon of *energy* is moving at the speed of the wave, but *water* is not. In a sense, an ocean wave is an illusion. How can you be knocked off your surfboard by an illusion? Well, there's much to learn!

1. **If the Gulf Stream warms Britain during the winter and keeps Baltic ports free of ice, why doesn't it moderate New England winters? After all, Boston is much closer to the warm core of the Gulf Stream than London is.**

Yes, but remember the direction of prevailing winds in winter. Winter winds at Boston's latitude are generally from the west, so any warmth is simply blown out to sea. On the other side of the Atlantic the same winds blow toward London. It does get cold in London, but generally winters in London are much milder than those in Boston.

2. **Could currents be used as a source of electrical power? With the Gulf Stream so close to Florida, it seems that some way could be devised to take advantage of all that water flow to turn a turbine.**

It's been considered. The total energy of the Gulf Stream flowing off Miami has been estimated at 25,000 megawatts! A Woods Hole Oceanographic Institution team has proposed a honeycomb-like array of turbines for the layer between 30 and 130 meters (100 and 430 feet) across 20 kilometers (13 miles) of the current. They estimate a power output of around 1,000 megawatts, equal to the generation potential of two large nuclear power plants. Engineering difficulties would be complex, however. To see the current (forgive the pun) state of the art, see **Figure 9.35**.

Photo by Jeff J Mitchell/Getty Images

Figure 9.35 A large turbine designed to generate electrical power from the flow of ocean currents is being prepared for placement among the Orkney Islands in northern Scotland. Figure 11.22 shows even larger turbines for use in generating electrical power from the rise and fall of the tides.

3. **Are there any *nongeostrophic* currents? Are any currents not noticeably influenced by gravity, the Coriolis effect, uneven solar heating, planetary winds, and so forth?**

Yes, there are small-scale currents that are not noticeably affected. Currents of freshwater from river mouths, rip currents in surf, and tidal currents in small harbors are much more affected by basin and bottom topography than by the Coriolis effect and gravity.

4. **Why are western boundary currents strong in *both* hemispheres? I thought things went the other way (counterclockwise) in the Southern Hemisphere. Shouldn't the *eastern* boundary currents be stronger down there?**

Western boundary currents are strong in part because the "Coriolis hill" is offset to the west, forcing water to move in a relatively narrow path along the ocean's western boundary. Truly, Coriolis effect works in a clockwise direction in the Northern Hemisphere and counterclockwise in the Southern Hemisphere, so the "hill" is offset to the west in both hemispheres. Thus, western boundary currents are strong in both hemispheres.

5. **Which takes the lead in producing the hill in the middle of a geostrophic gyre—the pressure gradient or the Coriolis effect?**

The question reminds us of the "chicken-and-egg" question—which came first? In ocean currents, both act together, in balance, to form both the hill and the circular flow around its crest. Imagine the situation some 150 million years ago when the Atlantic was first forming—Pangaea was splitting and the rift began to fill with water. Driven by winds, a small amount of water would have turned right to begin forming the hill. A pressure gradient was immediately formed, and water would have been forced back downhill by gravity. On its way back down, that water would achieve a balance with Coriolis effect and come to a "compromise" position of clockwise flow around the apex.

6. **A north wind comes from the north, but a north current is going north. Why the difference?**

Traditions die hard, it seems. For thousands of years winds have been named by where they come *from*. A north wind comes from the north, and a west wind comes from the west. Currents, though, are named by where they are *going*. A southern current is headed south; a western current is moving west. An exception is the Antarctic Circumpolar Current, or West Wind Drift, which moves eastward. This current, however, is named after the wind that drives it, the powerful polar westerlies.

It may also be a matter of perspective. Ancient peoples took shelter *from* winds (and referred to the wind's place of origin). But early oceanic travelers were aware of where the currents were carrying them *to*.

7. **Has anything been learned by sealing a message in a bottle and setting it adrift?**

Individual releases don't tell us much. Statistically, sample sizes need to be relatively large in order to provide meaningful data.

Bottle, card, or buoy studies are almost always carefully planned and executed, but not all surface drift releases have been intentional. In May 1990 a violent storm struck the containership *Hansa Carrier* en route between Korea and Seattle. Twenty-one boxcar-sized cargo containers were lost overboard, among them containers holding 30,910 pairs of Nike athletic shoes. About 6 months later, shoes from the broken containers began to wash up on beaches from British Columbia to Oregon. Because the shoes were not tied together in pairs, beachcombers placed advertise-ments in local newspapers and held swap meets to exchange the shoes (which were in surprisingly good condition despite having been exposed to the ocean). Oceanographers noticed the ads and asked the media to request that individuals let them know where and when they found shoes. By knowing the place where the shoes were lost and the places where they were found, researchers have been able to refine their computer models of the North Pacific gyre. Some of the shoes are completing their third full circuit of the North Pacific.

TERMS AND CONCEPTS TO REMEMBER

Antarctic Bottom Water	downwelling	gyre	transverse current
Antarctic Circumpolar Current (West Wind Drift)	eastern boundary current	La Niña	undercurrent
Argo system	eddy	Langmuir circulation	upwelling
caballing	Ekman spiral	North Atlantic Deep Water	West Wind Drift (Antarctic Circumpolar Current)
coastal upwelling	Ekman transport	Southern Oscillation	western boundary current
contour current	El Niño	surface current	westward intensification
convergence zone	ENSO	sverdrup (sv)	wind-induced vertical circulation
countercurrent	equatorial upwelling	temperature–salinity (T–S) diagram	
current	geostrophic gyre	thermohaline circulation	
	Gulf Stream		

STUDY QUESTIONS

Thinking Critically

1. Why does water tend to flow around the periphery of an ocean basin? Why are western boundary currents the fastest ocean currents? How do they differ from eastern boundary currents?

2. What causes El Niño? How does an El Niño situation differ from normal current flow? What are the usual consequences of El Niño?

3. What are countercurrents? Undercurrents? How might El Niño be related to these currents?

4. What is the role of ocean currents in the transport of heat? How can ocean currents affect climate? Contrast the climate of a mid-latitude coastal city at a western ocean boundary with a mid-latitude coastal city at an eastern ocean boundary.

5. Can you think of ways ocean currents have (or *might* have) influenced history?

6. What holds up the thermocline? Wouldn't water slowly warm in an even gradient all the way to the bottom? (Hint: See Figure 9.34.)

Thinking Analytically

1. Look again at Figure 9.7b and consider the descending water. First, why does it stop descending? Second, which way is it most likely to go when it stops descending? (Hint: The Coriolis effect is greatest near the poles and nonexistent at the equator.)

2. Calculate the time it would take for an ideally situated rubber running shoe to make a loop of the North Atlantic. Now calculate the time it would take for an ideally situated shoe to make a loop of the North Pacific. Why the difference?

GLOBAL GEOSCIENCE WATCH

Visit www.cengagebrain.com to access MindTap, a complete digital course which includes access to Global Geoscience Watch and more. El Niño can cause serious impacts on currents, weather, and marine life. Research El Niño in the GREENR database and use this information to learn about the causes, characteristics, and effects of El Niño. Write a two- to three-page report detailing your findings.

CENGAGEbrain.com   Visit www.cengagebrain.com to access course materials for this text, including interactive learning tools, videos, and more.

10 Waves

KEY CONCEPTS

Waves transmit energy, not water mass, across the ocean's surface. MANE/Barcroft Media/Landov

Ocean waves are classified by the disturbing force that creates them, the extent to which the disturbing force continues to influence them once they are formed, and by their wavelength. Tom Garrison

The speed (celerity) of an ocean wave is proportional to its wavelength. Tom Garrison

The orbits of water molecules in waves moving through water deeper than about half the wavelength are unaffected by the bottom. © Cengage Learning

The wavelength of tsunami and tides are so great that they are always in shallow water. National Oceanic and Atmospheric Administration (NOAA)

Garrett McNamara breaks his own world record for the largest wave ever surfed. Thought to be 30 meters (100 feet) high, this wave formed in November 2012 when swell from a storm west of Ireland encountered a submarine canyon off the coast of Nazaré, Portugal. MANE/Barcroft Media/Landov

10.1 Ocean Waves Move Energy across the Sea Surface

To most people an ocean wave in deep water appears to be a massive moving object—a ridge of water traveling across the sea surface (**Figure 10.1**). An ocean wave is one of several kinds of **waves**, all of which are disturbances caused by the movement of energy from a source through some medium (solid, liquid, or gas). As the energy of the disturbance travels, the medium through which it passes moves in specific ways. Sometimes this movement is visible to us as crests or ridges in the medium. The traveling crests produce the appearance of movement we see in a wave. In an ocean wave, a ribbon of *energy* is moving at the speed of the wave, but *water* is not. In a sense, the wave is an illusion.

Picture a resting seagull as it bobs on the wavy ocean surface far from shore. The gull moves in *circles*—up and forward as the tops of the waves move to its position, down and backward as the tops move past. Each circle is equal in diameter to the wave's height. As can be seen in **Figure 10.2**, energy in waves flows past the resting bird, but the gull and its patch of water move only a very short distance forward in each up-and-forward, down-and-back wave cycle. The water on which the bird rests does not move continuously across the sea surface as the wave illusion suggests.[1]

---

[1]To clarify the important idea of wave-as-illusion, imagine yourself at a sports stadium where spectators are doing "the wave." Your role in wave propagation is simple: You stand up and sit down in precise synchronization with your neighbors. Though you move only a few feet vertically, the wave of which you were a part circles the arena at high speed. You and all the other participants stay in place, but the wave moves faster than anyone can run.

The transfer of energy from water particle to water particle in these circular paths, or **orbits**, transmits wave energy across the ocean surface and causes the wave form to move. This kind of wave is known as an **orbital wave**—a wave in which particles of the medium (water) move in closed circles as the wave passes. Orbital ocean waves occur at the boundary between two fluid media (between air and water) and between layers of water of different densities. Because the wave form moves forward, these waves are a type of **progressive wave**.

The progressive wave that moved the gull was probably caused by wind. Other forces can generate much greater progressive waves in which water molecules move through much larger circular or elliptical orbits. Some of these waves are so large that they do not appear to us as waves at all but rather as the slow sloshing of water in a harbor or bay, as dangerous flooding surges of water, or as rhythmic and predictable ocean tides.

Ocean waves have distinct parts. The **wave crest** is the highest part of the wave above average water level; the **wave trough** is the valley between wave crests below average water level. **Wave height** is the vertical distance between a wave crest and the adjacent trough, and **wavelength** is the horizontal distance between two successive crests (or troughs). The relationship between these parts is shown in **Figure 10.3**. The time it takes for a wave to move a distance of one wavelength is known as the **wave period**. **Wave frequency** is the number of waves passing a fixed point per second.

The circular motion of water particles at the surface of a wave continues underwater. As **Figure 10.4** shows, the diameter of the orbits through which water particles move diminishes rapidly with depth. For all practical purposes, wave motion is negligible below a depth of one-half the wavelength where the circles are only 1/23 the diameter of those at the surface. So, divers in 20 meters of water would not notice the passage of a wind wave of 30-meter wavelength and might barely notice the wave if they were at a depth of 15 meters. Since most ocean waves have moderate wavelengths, the circular disturbance of the ocean that propagates these waves affects only the uppermost layer of water. *Note that the movement of water in circles doesn't resemble interlocking mechanical gears. Instead, there is a coordinated,*

Figure 10.1 Think of the ocean, and you think of waves. Waves move energy, not mass, across great distances.

uniform circular movement of water molecules in one direction as the waves pass.

Figure 10.4 indicates that water molecules in the crest of a passing wave move in the same direction as the wave, but molecules in the trough move in the opposite direction. If particle speed decreases with depth, molecules in the top half of the orbit will move farther forward in the direction the wave is moving than molecules in the bottom half of the orbit will move backward **(Figure 10.5)**. This flow, known as **Stokes drift**, is important in driving the ocean surface currents you studied in Chapter 9.

CONCEPT CHECK

Before going on to the next section, check your understanding of some of the important ideas presented so far:

We wrote that an ocean wave is, in a sense, an illusion. What's actually moving in an ocean wave?

Draw an ocean wave and label its parts. Include a definition of *wave period*.

10.2 Waves Are Classified by Their Physical Characteristics

Ocean waves are classified by the *disturbing force* that creates them, the extent to which the disturbing force *continues* to influence the waves once they are formed, the *restoring force* that tries to flatten them, and their *wavelength*. (Wave height is not often used for classification because it varies greatly depending on water depth, interference between waves, and other factors.)

Ocean Waves Are Formed by a Disturbing Force

Energy that causes ocean waves to form is called a **disturbing force**. Wind blowing across the ocean surface provides the disturbing force for *capillary waves* and *wind waves*. The arrival of a storm surge or seismic sea wave in an enclosed harbor or bay, or a sudden change in atmospheric pressure, is the disturbing force for the resonant rocking of water known as a *seiche*. Landslides, volcanic eruptions, and faulting of the seafloor associated with earthquakes are the disturbing forces for seismic sea waves (also known as *tsunami*). The disturbing forces for *tides* are changes in the magnitude and direction of gravitational forces among Earth, moon, and sun, combined with Earth's rotation. **Table 10.1** summarizes the characteristics of these waves.

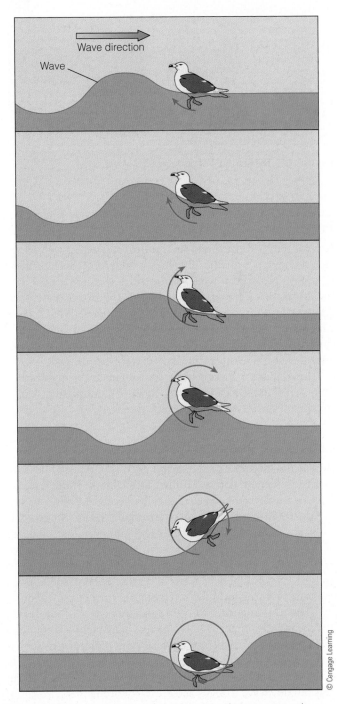

Figure 10.2 This floating seagull demonstrates that waves travel ahead, but the water itself does not. In this sequence, a wave moves from left to right as the gull (and the water in which it is resting) revolves in an imaginary circle, moving slightly to the left up the front of an approaching wave, then to the crest, and finally moving to the right down the back of the wave.

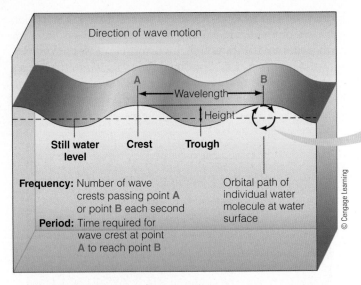

Figure 10.3 The anatomy of a progressive wave.

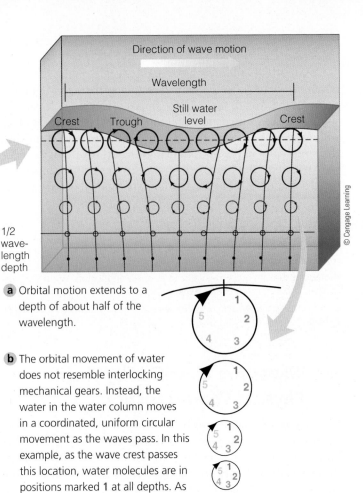

a Orbital motion extends to a depth of about half of the wavelength.

b The orbital movement of water does not resemble interlocking mechanical gears. Instead, the water in the water column moves in a coordinated, uniform circular movement as the waves pass. In this example, as the wave crest passes this location, water molecules are in positions marked **1** at all depths. As the crest passes, all molecules move to position **2**, then **3**, and so forth.

Figure 10.4 The orbital motion of water particles in a wave.

 THINKING BEYOND THE FIGURE

How deep would a submarine need to dive for its occupants to feel no motion from surface waves with a wavelength of 200 meters (660 feet)?

Free Waves and Forced Waves

A wave that is formed and then propagates across the sea surface without the further influence of the force that formed it is known as a **free wave**. When wind waves move away from the storm that created them, or when the storm ceases, they continue without the injection of additional wind energy. Likewise, tsunami—waves caused by submerged landslides or earthquakes—continue to move across the ocean surface long after the landslide or earthquake has stopped moving.

In contrast, a **forced wave** is maintained by its disturbing force. Tides are forced waves dependent on the gravitational attraction of the moon and sun.

Waves Are Reduced by a Restoring Force

Restoring force is the dominant force that returns the water surface to flatness after a wave has formed in it. If the restoring force of a wave were quickly and fully successful, a disturbed sea surface would immediately become smooth, and the energy of the embryo wave would be dissipated as heat. But that isn't what happens. Waves continue after they form because the restoring force overcompensates and causes oscillation. The situation is analogous to a weight bobbing at the bottom of a very flexible spring, constantly moving up and down past its normal resting point.

The restoring force for very small water waves—those with wavelengths less than 1.73 centimeters (0.68 inch)—is cohesion, the property that enables individual water molecules to stick to each other by means of hydrogen bonds (see Figure 6.3). The same force that makes the tea creep up on the sides of a teacup tugs the tiny wave troughs and crests toward flatness.

All waves with wavelengths greater than 1.73 centimeters depend mostly on gravity to provide the restoring force. Gravity pulls the crests downward, but the inertia of the water causes the crests to overshoot and become troughs. The repetitive nature of this movement, like the spring weight moving up and down, gives rise to the circular orbits of individual water molecules in an ocean wave. These larger waves are called **gravity waves**. Because the circular motion of water molecules in a wave is nearly friction-free, gravity waves can travel across thousands of miles of ocean surface without disappearing, eventually to break on a distant shore.

Wavelength Is the Most Useful Measure of Wave Size

Wavelength is an important measure of wave size. Table 10.1 lists the causes and typical wavelengths of capillary waves, wind waves, seiches, seismic sea waves, and tides. **Figure 10.6** shows the relationships between disturbing and restoring forces, period, and relative amount of energy present in the ocean's surface for each wave type. Note that more energy is stored in wind waves than in any of the other wave types.

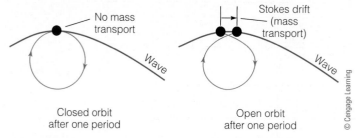

No mass transport

Closed orbit after one period

Stokes drift (mass transport)

Open orbit after one period

© Cengage Learning

Figure 10.5 Stokes drift, the small net amount of mass transport of water in the direction of a wave, contributes to the movement of the wind-driven surface currents discussed in Chapter 9. Although the orbits in Figures 10.1 and 10.3 are drawn as shown in the left figure, actually they are not completely closed. Instead, each particle moves forward a very small distance (exaggerated in the figure on the right) during each orbit.

Table 10.1 Disturbing Forces, Wavelength, and Restoring Forces for Ocean Waves

Wave Type	Disturbing Force	Restoring Force	Typical Wavelength
Capillary wave	Usually wind	Cohesion of water molecules	Up to 1.73 cm (0.68 in.)
Wind wave	Wind over ocean	Gravity	60 to 150 m (200–500 ft)
Seiche	Change in atmospheric pressure, storm surge, tsunami	Gravity	Large, variable; a function of ocean basin size
Seismic sea wave (tsunami)	Faulting of seafloor, volcanic eruption, landslide	Gravity	200 km (125 mi)
Tide	Gravitational attraction, rotation of Earth	Gravity	Half Earth's circumference

© Cengage Learning

CONCEPT CHECK

Before going on to the next section, check your understanding of some of the important ideas presented so far:

What are the types of ocean waves, arranged by disturbing force and wavelength?

How is a free wave different from a forced wave?

What is restoring force?

10.3 The Behavior of Waves Is Influenced by the Depth of Water through Which They Are Moving

Most characteristics of ocean waves depend on the relationship between their wavelength and water depth. Wavelength determines the *size* of the orbits of water molecules within a wave, but water depth determines the *shape* of the orbits. The paths of water molecules in a wind wave are circular only when the wave is traveling in deep water. A wave cannot "feel" the bottom when it moves through water deeper than half its wavelength because too little wave energy is contained in the small circles below that depth. Waves moving through water deeper than half their wavelength are known as **deep-water waves**. A wave has no way of "knowing" how deep the water is, only that it is in water deeper than about half its wavelength. For example, a wind wave of 20-meter wavelength will act as a deepwater wave if it is passing through water more than 10 meters deep **(Figure 10.7)**.

The situation is different for wind-generated waves close to shore. The orbits of water molecules in waves moving through shallow water are flattened by the proximity of the bottom. Water just above the seafloor cannot move in a circular path,

I stand on my dune top watching a great wave coursing in from sea, and know that I am watching an illusion, that the distant water has not left its place in the ocean to advance upon me, but only a force shaped in water, a bodiless pulse beat, a vibration.
—**HENRY BESTON**

only forward and backward. Waves in water shallower than $\frac{1}{20}$ their original wavelength are known as **shallow-water waves**. A wave with a 20-meter wavelength will act as a shallow-water wave if the water is less than 1 meter deep.

Of the five wave types listed in Table 10.1, only capillary waves and wind waves can be deep-water waves. To understand why, remember that most of the ocean floor is deeper than 125 meters (400 feet), half the wavelength of very large wind waves. The wavelengths of the larger waves are *much* longer: the wavelength of seismic sea waves usually exceeds 100 kilometers (62 miles). No ocean is 50 kilometers (31 miles) deep, so seiches, seismic sea waves, and tides are forever in water that to them is shallow or transitional in depth. Their huge orbit circles flatten against a distant bottom that is always less than half a wavelength away.

In general, the longer the wavelength, the faster the wave energy will move through the water. For *deep-water* waves, this relationship is shown in the formula

$$C = \frac{L}{T}$$

in which C represents speed (celerity), L is wavelength, and T is time, or period (in seconds).

The speed of all ocean waves is controlled by gravity, wavelength, and water depth. The speed of a deep-water wave may also be approximated by

$$C = \sqrt{\frac{gL}{2\pi}}$$

where g is the acceleration due to gravity, 9.8 meters per second per second. Because g and π (3.14) are constants,

$$C = 1.249\sqrt{L}$$

when C is measured in meters per second and L in meters. Note in both instances that wave speed is proportional to wavelength.

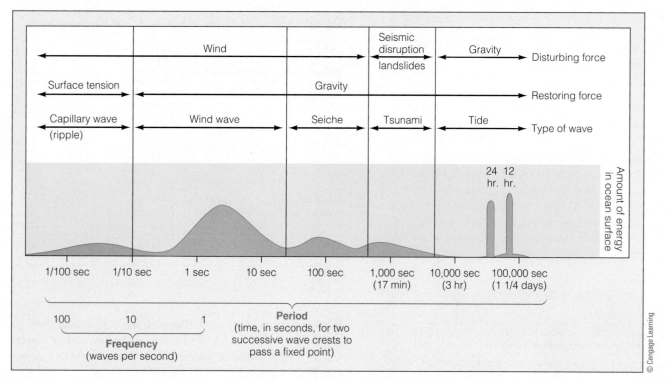

Figure 10.6 Wave energy in the ocean as a function of the wave period. As the graph shows, most wave energy is typically concentrated in wind waves. However, large tsunami, rare events in the ocean, can transmit more energy than all wind waves for a brief time. Notice that tides are waves—their energy is concentrated at periods of 12 and 24 hours.

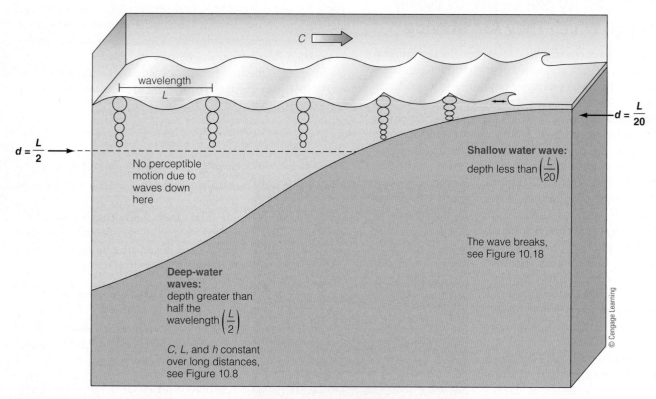

Figure 10.7 Progressive waves. Classification depends on their wavelength relative to the depth of water through which they are passing. The diagram is not to scale.

Now look at **Figure 10.8**, a diagram showing the relationship between wavelengths of deep-water waves and their speed and period. In the example shown by the red lines, the speed of a wave with a wavelength of 233 meters and a period of 12 seconds is

$$C = \frac{L}{T} = \frac{233 \text{ meters}}{12 \text{ seconds}} = \frac{19.4 \text{ meters}}{\text{seconds}}$$

Wavelength is difficult to determine at sea, but period is comparatively easy to find—for example, an observer simply times the movement of waves past the bow of a stopped ship. If period *(T)* is known, speed *(S)* can be calculated from the relationship

$$C(\text{in meters per second}) = \frac{gT}{2\pi}$$

$$= \frac{9.8 \text{ meters/sec}^2 \times T \text{ (in seconds)}}{2 \times (3.14)}$$

$$= 1.56T$$

where g is acceleration due to gravity.

The speed of *shallow-water* waves is described by a different equation that may be written as

$$\boldsymbol{C = \sqrt{gd} \text{ or } C = 3.1\sqrt{d}}$$

where *C* is speed (in meters per second), g is the acceleration due to gravity (9.8 meters per second per second), and *d* is the depth of the water (in meters). The *period* of a wave remains unchanged

> **O**nly capillary waves and wind waves can be deep-water waves.

regardless of the depth of water through which it is moving. As deep-water waves enter the shallows and feel bottom, however, their speed is reduced and their crests "bunch up," so their wavelength shortens.

Comparing deep-water wind waves and shallow-water seismic sea waves is like comparing apples and oranges, but the list below demonstrates the *general relationship* between wavelength and wave speed: the longer the wavelength, the greater the speed.

Wind Waves (Deep-Water Waves)

- Period to about 20 *seconds*
- Wavelength to perhaps 600 meters (2,000 feet) in extreme cases
- Speed to perhaps 112 kilometers (70 miles) per hour in extreme cases

Seismic Sea Waves (Shallow-Water Waves)

- Period to perhaps 20 *minutes*
- Wavelength typically 200 kilometers (125 miles)
- Speeds of 760 kilometers per hour (470 miles per hour)

Remember that *energy*—not the water *mass* itself—is moving through the water at the astonishing speed of 760 kilometers (470 miles) per hour (the speed of a jet airliner!) in seismic sea waves.

CONCEPT CHECK

Before going on to the next section, check your understanding of some of the important ideas presented so far:

What defines a deep-water wave? Are there any waves that can never be in deep water?

What is the mathematical relationship between celerity (speed), wavelength, and wave period for deep-water waves? For shallow-water waves?

10.4 Wind Blowing over the Ocean Generates Waves

Wind waves are gravity waves formed by the transfer of wind energy into water. Most wind waves are less than 3 meters (10 feet) high. Wavelengths from 60 to 150 meters (200–500 feet) are most common in the open ocean.

Wind waves grow from **capillary waves** (**Figure 10.9**). Capillary waves form as wind friction stretches the water surface and as surface tension tries to restore it to smoothness. These small ripples are important in transferring energy from air to water to drive ocean currents, but they are of little consequence in the overall picture of ocean waves because they are tiny and carry very little energy.

Capillary waves are nearly always present on the ocean. A capillary wave interrupts the smooth sea surface, deflecting the surface wind upward, slowing it, and causing some

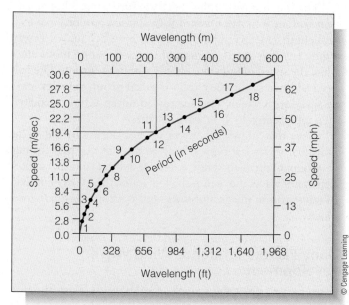

© Cengage Learning

Figure 10.8 The approximate relationship among speed, wavelength, and period in deep-water waves. Speed is equal to wavelength divided by period. If one characteristic of a wave can be measured, the other two can be calculated. The easiest characteristic to measure exactly is period. In the example shown in red, the speed of a wave with a wavelength of 233 meters and a period of 12 seconds is 19.4 meters per second.

Figure 10.9 Capillary waves form ahead of a slow moving hand in a swimming pool. Each has a wavelength less than 2 centimeters (about ⅔ of an inch).

of the wind's energy to be transferred into the water to drive the capillary wave crest forward **(Figure 10.10a)**. The wind may eddy briefly downwind of the tiny crest, creating a slight partial vacuum there. Atmospheric pressure pushes the trailing crest forward (downwind) toward the trough, adding still more energy to the water surface. The increasing energy in the water surface expands the circular orbits of water particles in the direction of the wind, enlarging the small wave's size. The capillary wave becomes a wind wave when its wavelength exceeds 1.73 centimeters (0.68 inch), the wavelength at which gravity supersedes capillary action as the dominant restoring force.

If the wind wave remains in water deeper than half its wavelength and the wind continues to blow, the wave becomes larger. Its crest is thrust higher into faster wind, extracting even more energy from the moving air. The circular orbits of water particles within the wave grow larger with more energy input; height, wavelength, and period increase proportionally. The irregular peaked waves in the area of wind wave formation are called **sea**; the chaotic surface is formed by simultaneous wind waves of many wavelengths, periods, and heights. When the wind slows or ceases, as it does away from a storm, the wave crests become rounded and regular. This process is shown in **Figure 10.10b**.

During their formation, moderate-sized wind waves in the open ocean exhibit a maximum 1:7 ratio of wave height to wavelength **(Figure 10.10c)**; this ratio is the **wave steepness**. Waves 7 meters long will not be more than 1 meter high, and waves with a 70-meter wavelength will not exceed 10 meters of height. The angle at their crest will not exceed 120°. A peaked appearance usually indicates the continuing injection of wind energy. If a wave gets any higher than the 1:7 ratio for its wave-

length, it will break, and excess energy from the wind will be dissipated as turbulence—hence the *whitecaps* or *combers* associated with a fully developed sea.

Larger Swell Move Faster Than Small Swell

Because they move faster, waves with longer wavelengths leave the area of wave formation sooner.[2] They outrun their smaller relatives. Mature waves from a storm sort themselves into groups with similar wavelengths and speeds. The process of wave separation, or **dispersion**, produces the familiar smooth undulation of the ocean surface called **swell** (see **Figure 10.11**). Swell often move thousands of kilometers from a storm to a shore, announcing the storm's impending arrival.

Contrary to what you might expect, observers far from the storm would first encounter large, quick-moving waves of long wavelength, then middle-sized waves, and then slow, small ones. Because the circular movement of water particles in deep-water waves is virtually friction free, the waves will continue until they break upon a shore. There they release their absorbed wind energy as random movement, heat, and sound.

Progressing groups of swell with the same origin and wavelength are called **wave trains**. The leading waves in the wave train are drained of energy because they must begin the circular movement of the undisturbed water into which they are intruding. These leading waves gradually disappear, but after the wave train has passed, some energy remains behind in the circles to form new waves. New waves thus form behind as the leading waves disappear at the front of the wave train. This process is shown in **Figure 10.10d**.

The implications of this detail are surprising. Though each *individual* wave moves forward with a speed proportional to its wavelength in deep water (**C**), the *wave train itself* moves forward at only *half* that speed. Groups of waves therefore move ahead at half the speed of individual waves within the group. The half-speed advance of the wave train is called **group velocity** and is the speed with which wave energy advances. Group velocity is often represented by **V** in wave equations.

Note that individual waves do not persist in the ocean. Individual waves last only as long as they take to pass through the group. Only deep-water waves are subject to dispersion. As deep-water waves move into shallow water, the speed of individual waves within the group slows until wave speed equals group velocity.

Many Factors Influence Wind Wave Development

Three factors affect the growth of wind waves. First, the wind must be moving faster than the wave crests for energy transfer from air to sea to continue, so the mean speed, or **wind strength**, of the wind is clearly important to wind wave development. A second factor is the length of time the wind blows, or **wind duration**; high winds that blow only a short time will not generate

---

[2] Remember that speed is directly proportional to wavelength in deep-water waves: $C = \dfrac{L}{T}$.

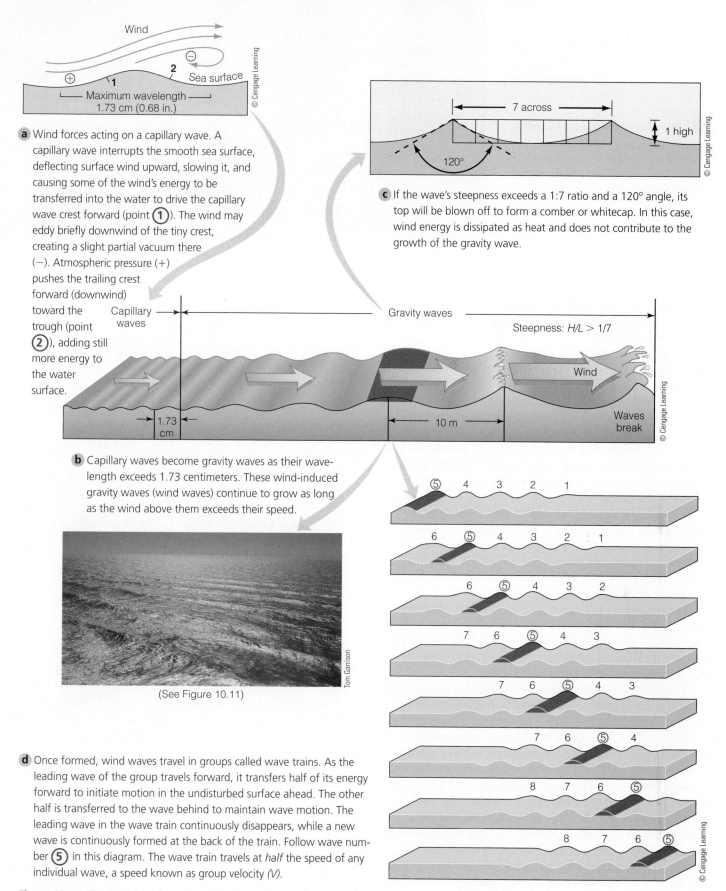

a Wind forces acting on a capillary wave. A capillary wave interrupts the smooth sea surface, deflecting surface wind upward, slowing it, and causing some of the wind's energy to be transferred into the water to drive the capillary wave crest forward (point ①). The wind may eddy briefly downwind of the tiny crest, creating a slight partial vacuum there (−). Atmospheric pressure (+) pushes the trailing crest forward (downwind) toward the trough (point ②), adding still more energy to the water surface.

c If the wave's steepness exceeds a 1:7 ratio and a 120° angle, its top will be blown off to form a comber or whitecap. In this case, wind energy is dissipated as heat and does not contribute to the growth of the gravity wave.

b Capillary waves become gravity waves as their wavelength exceeds 1.73 centimeters. These wind-induced gravity waves (wind waves) continue to grow as long as the wind above them exceeds their speed.

(See Figure 10.11)

d Once formed, wind waves travel in groups called wave trains. As the leading wave of the group travels forward, it transfers half of its energy forward to initiate motion in the undisturbed surface ahead. The other half is transferred to the wave behind to maintain wave motion. The leading wave in the wave train continuously disappears, while a new wave is continuously formed at the back of the train. Follow wave number ⑤ in this diagram. The wave train travels at *half* the speed of any individual wave, a speed known as group velocity *(V)*.

Figure 10.10 The growth and progress of wind waves.

Figure 10.11 Swell—mature, regular wind waves sorted by dispersion—off the Oregon coast. Small waves superimposed on the large swell are the result of local wind conditions.

Tom Garrison

💬 **THINKING BEYOND THE FIGURE**

Are swell like these influenced by the Coriolis effect? Have they come directly to this coast from their origin without changing direction?

large waves. The third factor is the uninterrupted distance over which the wind blows without significant change in direction, the **fetch** (Figure 10.12).

A strong wind must blow continuously in one direction for nearly 3 days for the largest waves to develop fully. A **fully developed sea** is the maximum wave size theoretically possible for a wind of a specific strength, duration, and fetch. Longer exposure to wind at that speed will not increase the size of the waves because energy is lost due to the breaking of wave tops and the formation of whitecaps. The first three columns of **Table 10.2** give examples of the combinations of wind speed, fetch, and duration required to form waves of the fully developed sea described in the remaining columns. Note that waves in a fully developed sea are small if wind speed is low. However, if the wind speed is 74 kilometers (46 miles) per hour through 1,313 kilometers (816 miles) for 42 hours—conditions not wholly unrealistic in the Pacific—waves that average 8.5 meters (28 feet) high can result. The highest 10% of the waves in this sea will exceed 17.2 meters (56.6 feet) in height!

The combination of factors required to produce truly great seas doesn't occur very often. In the example given at the bottom of Table 10.2, it seems unlikely that a wind would blow steadily at 92 kilometers (58 miles) per hour in one direction for 69 hours over 2,627 kilometers (1,633 miles).

The greatest potential for large waves occurs beneath the strong and nearly continuous winds of the West Wind Drift surrounding Antarctica. The early 19th-century French explorer of the South Seas, Jules Dumont d'Urville, encountered a wave train with heights estimated "in excess" of 30 meters (100 feet) in Antarctic waters. In 1916 Ernest Shackleton contended with occasional waves of similar size in the West Wind Drift during a heroic voyage to remote South Georgia Island in an open boat.[3] Satellite observations (like those of **Figure 10.13**) have shown that wave heights to 11 meters (36 feet) are fairly common in the West Wind Drift.

In zones of high winds a less than fully developed sea can also demand attention. Though wind speed within cyclonic

---

[3]This is one of the great adventure stories of all time. For a gripping read, find a copy of F. A. Worsley's 1933 book *Shackleton's Boat Journey*.

storm

wind

Figure 10.12 The fetch, the uninterrupted distance over which the wind blows without significant change in direction. Wave size increases with increased wind speed, duration, and fetch. A strong wind must blow continuously in one direction for nearly 3 days for the largest waves to develop fully.

 THINKING BEYOND THE FIGURE

Which ocean basin has the greatest potential fetch?

Table 10.2 Conditions Necessary for a Fully Developed Sea at Given Wind Speeds and the Parameters of the Resulting Waves

Wind Conditions			Wave Size		
Wind Speed in One Direction	Fetch	Wind Duration	Average Height	Average Wavelength	Average Period
19 km/hr (12 mi/hr)	19 km (12 mi)	2 hr	0.27 m (0.9 ft)	8.5 m (28 ft)	3.0 sec
37 km/hr (23 mi/hr)	139 km (86 mi)	10 hr	1.5 mi (4.9 ft)	33.8 m (111 ft)	5.7 sec
56 km/hr (35 mi/hr)	518 km (322 mi)	23 hr	4.1 m (13.6 ft)	76.5 m (251 ft)	8.6 sec
74 km/hr (46 mi/hr)	1,313 km (816 mi)	42 hr	8.5 m (27.9 ft)	136 m (446 ft)	11.4 sec
92 km/hr (58 mi/hr)	2,627 km (1,633 mi)	69 hr	14.8 m (48.7 ft)	212.2 m (696 ft)	14.3 sec

Source: Data from U.S. Army Corps of Engineers Coastal Research Center, Richmond, Virginia. Table created by Tom Garrison.

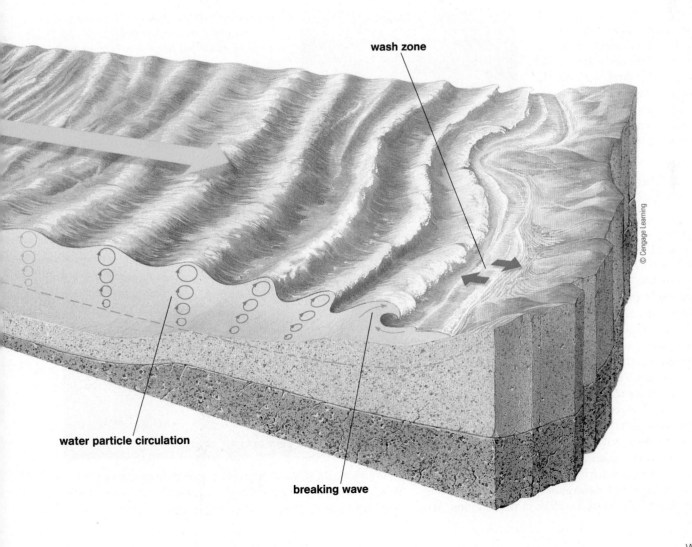

© Cengage Learning

Figure 10.13 Global wave height acquired by a radar altimeter aboard the *TOPEX/Poseidon* satellite in October 1992. In this image, the highest waves occur in the southern ocean, where waves more than 6 meters (19.8 feet) high (represented in white) were recorded. The lowest waves (indicated by dark blue) are found in the tropical and subtropical ocean, where wind speed is lowest.

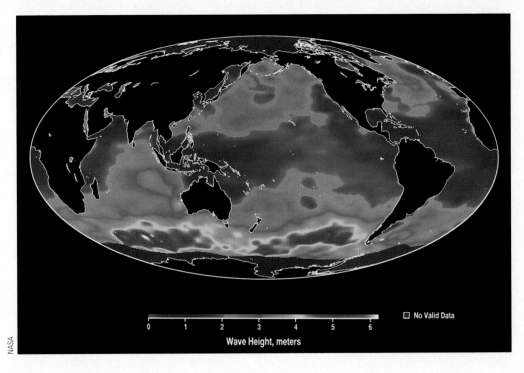

storms is often very strong, the circular motion of air doesn't allow long fetches, and fully developed seas rarely occur beneath such storms. Officers standing deck watches during storms rarely quibble with theoretical maximum height versus observed height, however. Wind waves can be overwhelming even if they are not fully developed **(Figure 10.14)**.

Wind Waves Can Grow to Enormous Size

How big can wind waves get?

The highest wave ever directly measured was sighted on the night of 7 February 1933 by Lt. Frederick Marggraff, a watch officer aboard the U.S. Navy tanker *Ramapo*. USS *Ramapo* was steaming from Manila to San Diego through a furious storm—a storm made more intense by the coalescence of three low-pressure centers. For days a steady wind had blown at 107 kilometers per hour (67 miles per hour), and gusts to 126 kilometers per hour (78 miles per hour) often lashed the decks. But the wind blew persistently from one direction, and though the monstrous waves it generated dwarfed the tanker, they were surprisingly orderly in form.

"The conditions for observing the seas from the ship were ideal," wrote *Ramapo's* executive officer. "We were running directly down the wind with the sea. There were no cross seas and therefore no peaks along wave crests. There was practically no rolling, and the pitching motion was easy because of the fact that the sides of the waves were much longer than the ship. The moon was out astern and facilitated observations during the night. The sky was partly cloudy." At about 3 o'clock in the morning, Mr.

The greatest potential for large waves occurs beneath the strong and nearly continuous winds of the West Wind Drift surrounding Antarctica.

Marggraff observed a train of tremendous waves looming in the moonlight. As the trough of the first wave approached the ship, he noted that its distant crest was on a level with the crow's nest on the mainmast. At that instant the ship's stern sank into the bottom of the onrushing trough. The next two waves were about the same size. Not surprisingly, such immense waves made an indelible impression on all who witnessed them.

Figure 10.14 The NOAA research ship *Discoverer* encounters a huge wave in the Bering Sea. (*Note:* When the NOAA photo library is asked to find an image for "seasick," this image is selected. Indeed!)

How big were these waves? When the executive officer had some time to spare, he did some calculations. **Figure 10.15** illustrates the position of the ship when the largest waves were measured. The height of the waves was determined by using a set of the ship's plans, a calculation of the height of the observer above the sea surface, the draft of the ship, and a sight to the horizon. The largest wave for which a dependable on-site observation had been made was 34 meters (112 feet) high, still a record!

We can personally attest to the fact that smaller waves can also add *much* excitement to a deck officer's watch!

CONCEPT CHECK

Before going on to the next section, check your understanding of some of the important ideas presented so far:

How are wind waves formed? What's a fetch?

How does the wavelength of a wind wave affect its speed?

What is a fully developed sea?

How is group velocity different from the velocity of an individual wave within the group?

10.5 Interference Produces Irregular Wave Motions

The real situation of wind waves at sea is not as simple as has been suggested earlier. The ideal vision of one set of waves moving in one direction at one speed across an otherwise smooth surface is almost never observed in the ocean.

Independent wave trains exist simultaneously in the ocean most of the time. Since long waves outrun shorter ones, wind waves from different storm systems can overtake and interfere with one another. One wave doesn't crawl over the others when they meet; instead, they add to or subtract from one another. Such interaction is known as **interference**. In **Figure 10.16a**, one wave is represented as a red line and a second wave with a slightly longer wavelength as a blue line. In the sea surface where these waves coincide (shown by the green line in **Figure 10.16b**), you can see the alternation between addition (large crests and troughs) and subtraction (almost no waves at all). The cancellation effect of subtraction is termed **destructive interference**— not because of harm to lives or property but because wave

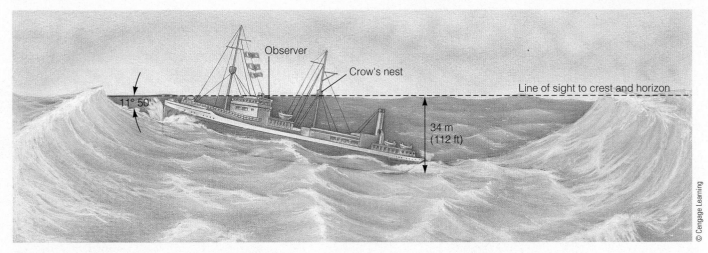

Figure 10.15 How the great wave observed from the USS *Ramapo* was measured. An officer on the bridge was looking toward the stern and saw the crow's nest in his line of sight to the crest of the wave, which had just come in line with the horizon. Wave height was later calculated based on the ship's design plans and the geometry of the situation.

interference destroys or cancels waves. **Constructive interference** is the additive formation of large crests or deep troughs, the size of which exceeds the size of each participating wave (as can be seen in the line in **Figure 10.16c**).

You have probably noticed that the surf along a coast seems to rise to a few big waves, diminish, and then build again. Surfers wait for the big "sets" to arrive, ride toward the shore, and then use the relatively calm interval to swim out into position for the next big set of waves. Constructive and destructive interferences explain this behavior, called **surf beat**. Constructive interference between waves of different wavelengths creates the sought-after big waves; destructive interference diminishes the waves and makes it easier to swim back out. The characteristics of surf beat explain why in some instances every ninth wave might be quite large. As wavelengths and interference change, though, every seventh wave might be large, or every fifth, or twelfth. Contrary to folklore, there is no set ratio.

Interference can have sudden unpleasant consequences on the open sea. In or near a large storm, wind waves at many wavelengths and heights may approach a single spot from different directions. If such a rare confluence of crests occurred at your position, a huge wave crest would suddenly erupt from a moderate sea to threaten your ship. The freak wave—called a **rogue wave**—would be much larger than any noticed before or after, and it would be higher than the theoretical maximum wave capable of being sustained in a fully developed sea. In such conditions, one wave in about 1,175 is more than three times average height, and one in every 300,000 is more than four times average height! This chapter's opener is an artist's conception of a catastrophic rogue wave that sank a large ship in 1978. **Figure 10.17** shows another harrowing example.

Currents can contribute to the formation of a kind of rogue wave that does not depend on interference. These waves are formed by the interaction of a wind wave and a swift surface current. The southeastern tip of Africa seems to breed a disproportionately large number of these giant waves. The Agulhas Current, a strong western boundary current, flows close to shore there, moving in the opposite direction to wind waves generated by high-latitude southwesterly gales. When large waves suddenly hit the 5-kilometer-per-hour (3-mile-per-hour) current, they can "trip" on it, rise suddenly to double their original height, and break precipitously. Mathematical modeling suggests that a swift current or large fields of random eddies can act like a lens to refract and focus wave trains at a point. Waves at that location would have a steep forward face preceded by a deep trough. Mariners who have experienced these kinds of rogue waves describe them as "holes in the sea." Such waves have broken tankers in half; many lives have been lost.

CONCEPT CHECK

Before going on to the next section, check your understanding of some of the important ideas presented so far:

Can constructive or destructive interference ever be seen on a casual visit to the beach?

What's a rogue wave? Are rogue waves potentially dangerous?

10.6 Deep-Water Waves Change to Shallow-Water Waves as They Approach Shore

Most wind waves eventually find their way to a shore and break, dissipating all their order and energy. The process begins with the transition of our now familiar deep-water wave to a transitional wave in water less than half a wavelength deep. **Figure 10.18** (see page 298) outlines the events that lead to the break.

① The wave train moves toward shore. When the depth of the water is less than half the wavelength, the wave "feels" bottom.

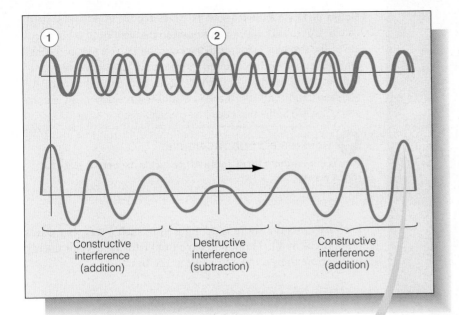

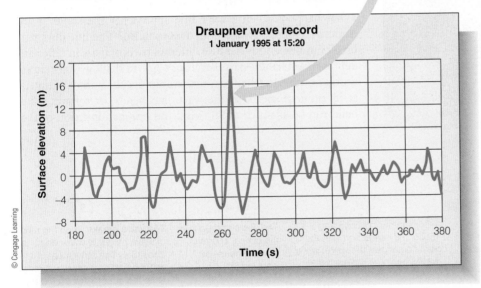

Draupner wave record
1 January 1995 at 15:20

a Two overlapping waves of different wavelength are shown, one in blue and one in red. Note that the wave shown in blue has a slightly longer wavelength.

b If both are present in the ocean at the same time, they will interfere with each other to form a composite wave. At the position of line ①, the two waves in **a** will constructively interfere to form very large crests and troughs, as shown in **b**. At the position of line ②, the two waves will destructively interfere, and the crests and troughs will be very small (again shown in **b**).

c A sea-level trace from the Draupner oil platform in the North Sea on 1 January 1995. Constructive interference was responsible for a maximum wave height of 18.5 meters (60 feet). Waves significantly larger than those seen before or after are sometimes called *rogue waves*.

Figure 10.16 Constructive and destructive interference.

② The circular motion of water molecules in the wave is interrupted. Circles near the bottom flatten to ellipses. The wave's energy must now be packed into less water depth, so the wave crests become peaked rather than rounded.

③ Interaction with the bottom slows the wave. Waves behind it continue toward shore at the original rate. Wavelength therefore decreases, but period remains unchanged.

④ The wave becomes too high for its wavelength, approaching the critical 1:7 ratio.

⑤ As the water becomes even shallower, the part of the wave below average sea level slows because of the restricting effect of the ocean floor on wave motion. When the wave was in deep water, molecules at the top of the crest were supported by the molecules ahead (thus transferring energy forward). This is now impossible because the *water* is moving faster than the *wave*. As the crest moves ahead of its supporting base, the wave breaks. The break occurs at about a 3:4 ratio of wave height to water depth (that is, a 3-meter wave will break in 4 meters of water). The turbulent mass of agitated water rushing shoreward during and after the break is known as **surf**. The **surf zone** is the region between the breaking waves and the shore.

Waves break against the shore in different ways, depending in part on the slope of the bottom. The break can be violent and toppling, leaving an air-filled channel (or *tube*) between the falling crest and the foot of the wave. These **plunging waves (Figure 10.19)** form when waves approach a steeply sloping bottom.

Slope alone does not determine the position and nature of the breaking wave. The contour and composition of the bottom can also be important. Gradually shoaling bottoms can sap waves of their strength because of prolonged interaction against the bottom of the lowest elliptical water orbits. Energy may be lost even more rapidly if the bottom is covered with loose gravel or irregular growths of coral. Masses of moving seaweed or jostling chunks of sea ice can also extract energy from a wave. In a

Figure 10.17 An artist imagines the cargo ship *München* immediately before its encounter with a rogue wave on the night of 12 December 1978. The ship sank quickly with the loss of all 27 crew members. One unused lifeboat had been stowed 20 meters (66 feet) above the waterline, yet one of its attachment pins had been bent by extreme force. A Maritime Court concluded that bad weather had caused "...an unusual event" leading to the ship's demise—probably a rogue wave.

💬 **THINKING BEYOND THE FIGURE**

Why is the southern tip of South Africa such a dangerous place for rogue waves?

few rare cases the shore is configured in such a way that waves don't break at all: The waves have lost virtually all their energy by the time their remnants arrive at the beach.

Waves Refract When They Approach a Shore at an Angle

What happens when a wave line approaches the shore at an angle, as it almost always does (**Figure 10.20**)? The line does not break simultaneously because different parts of it are in different depths of water. The part of the wave line in shallow water slows down, but the attached segment still in deeper water continues at its original speed, so the wave line bends, or refracts. The bend can be as much as 90° from the original direction of the

① The swell "feels" bottom when the water is shallower than half the wavelength.

② The wave crests become peaked because the wave's energy is packed into less water depth.

③ Constraint of circular wave motion by interaction with the ocean floor slows the wave, while waves behind it maintain their original speed. Therefore, wavelength shortens, but period remains unchanged.

④ The wave approaches the critical 1:7 ratio of wave height to wavelength.

⑤ The wave breaks when the ratio of wave height to water depth is about 3:4. The movement of water particles is shown in red. Note the change from a deep-water wave—through the transitional wave stage—to a shallow-water wave.

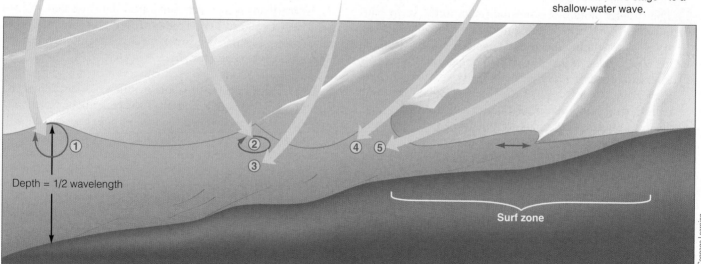

Figure 10.18 How a wave train breaks against the shore.

Figure 10.19
Garrett McNamara rides a big wave at Avalanche, off the O'ahu coast, Hawai'i. His record-breaking ride in Portugal is shown in this chapter's opener. Sean Davey/Corbis

💬 **THINKING BEYOND THE FIGURE**
What technical innovation made surfing waves of this magnitude possible?

wave train. This slowing and bending of waves in shallow water is called **wave refraction**.[4] The refracted waves break in a line almost parallel to the shore.

Wave refraction can produce some odd surf patterns. A swell with a wavelength of 600 meters "feels" bottom at 300 meters (1,000 feet), whereas a swell with a wavelength of 200 meters "feels" bottom at 100 meters (330 feet). You may remember that the average depth at the outer continental shelves is only about 150 meters (500 feet). As these two wave trains approach shore, the longer swell will be slowed by the shelf for perhaps 60 kilometers (37 miles) before the shorter swell is slowed by the bottom. With one wave bending and the other heading in its original direction, the potential for complex interference is great. Waves may break for a time at one spot, cease, and then begin breaking at another spot a few hundred meters down the beach. The ocean surface might have a checkerboard appearance from the constructive and destructive interference of these crests (**Figure 10.21**).

Waves Can Diffract When Wave Trains Are Interrupted

Wave diffraction is the propagation of a wave around an obstacle. Unlike wave refraction (which depends on a wave's response to a change in speed), wave diffraction depends on the

---

[4]To review the concept of refraction, please see Figure 6.21 on page 186.

interruption of the wave trains by an obstacle to provide a new point of departure for the waves. This is illustrated by the gap in the breakwater in **Figure 10.22**. Wave crests excite the water in the gap. Water moving in the gap generates smaller waves in the harbor, which radiate from the gap and disturb the quiet water. The diffracted waves are much smaller than those in the open ocean, but boats in the harbor might still be jostled.

Figure 10.23 shows a more complex case of diffraction in which waves are interrupted by a chain of islands. Some of the wave energy "squeezes" through the spaces between the islands. These spaces act as if they were new sources of waves. The waves spread into the space past the island, creating areas of reinforcement and cancellation. Polynesian navigators learned early in their training to sense the disturbances in wave patterns caused by diffraction. The best among them could feel the presence of islands beyond the horizon by subtle irregularities in the rocking of their canoe.

Waves Can Reflect from Large Vertical Surfaces

All of the waves discussed so far in this chapter—the familiar ocean waves in which the disturbance travels in one direction along the surface of the transmission medium—are progressive waves. A vertical barrier such as a seawall, large ship hull, or smooth jetty will reflect progressive waves with little loss of energy. If the waves approach the obstruction straight on, the reflected waves will move away from the obstruction in the direction from which they came. This **wave reflection** will cause

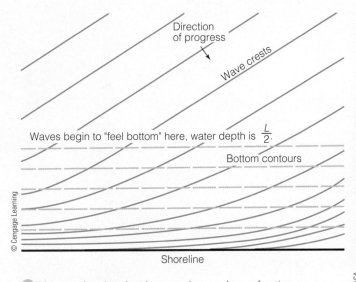

a Diagram showing the elements that produce refraction.

Figure 10.20 Wave refraction.

b Wave refraction around Maili Point, O'ahu, Hawai'i. Note how the wave crests bend almost 90° as they move around the point.

interference in the form of vertical oscillations called **standing waves**. As their name suggests, standing waves do not progress but appear as alternating crests and troughs at a fixed position. Figure 10.24 shows how a standing wave oscillates in a motion that resembles water sloshing back and forth in a half-filled bathtub. Because of constructive interference between crests (and troughs), these waves can be dangerous to boats or swimmers near the obstruction. As you will soon see, standing waves are important in the physics of tsunami and tides.

Waves approaching the obstruction at an angle can also reflect; the sea surface near the reflector will not form standing waves, but complicated sea-surface motions develop. Watch for these effects the next time you visit a solid breakwater or a steep beach.

CONCEPT CHECK

Before going on to the next section, check your understanding of some of the important ideas presented so far:

When does a wind wave become a shallow-water wave as it approaches shore?

What factors influence the breaking of a wind wave?

What might cause waves approaching a shore at an angle to bend to break nearly parallel with the shore?

a A "checkerboard" sea surface beyond the surf zone off La Jolla, California, the result of wave interference.

Figure 10.21 Wave interference.

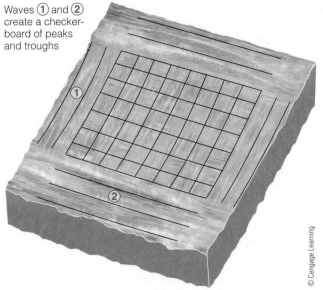

Waves ① and ② create a checker-board of peaks and troughs

b Intersecting wave trains generate the patterns of peaks and troughs seen in **a**.

Figure 10.22 Diffraction of waves at a breakwater gap at Morro Bay, California.

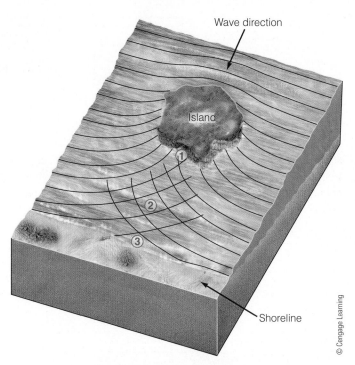

Figure 10.23 Wave diffraction past an island. At ①, waves are diminished in the lee of the island. At ②, the bending waves intersect, sometimes cancelling crests with troughs. At ③, crests and troughs reinforce each other to pound a section of coast. Polynesian navigators used diffraction patterns to sense the presence of islands out of sight over the horizon.

10.7 Internal Waves Can Form between Ocean Layers of Differing Densities

Progressive waves can occur at the junction between air and water, as we have seen, or they can form at the boundary between water layers of different densities. These sub-surface waves are called **internal waves**. As is the case with ocean waves at the air–ocean interface, internal waves possess troughs, crests, wavelength, and period. "Desktop-ocean" devices, which are sold at some gift shops, demonstrate internal waves. These sealed bottles contain nonmixing liquids of contrasting colors and slightly different densities. When you tilt the bottle, a very slow internal wave forms at the junction of the liquids, moves slowly to the low end, and breaks.

Normal ocean waves move rapidly because the difference in density between air and ocean is relatively great. Internal waves usually move very slowly because the density difference between the joined media is very small. Internal waves occur in the ocean at the base of the pycnocline, especially at the bottom edge of a steep thermocline **(Figure 10.25a)**. The wave height of internal waves may be greater than 30 meters (100 feet), causing the pycnocline to undulate slowly through a considerable depth **(Figure 10.25b)**. Their wavelength often exceeds 0.8 kilometer (0.5 mile); periods are typically 5 to 8 minutes. Internal waves are generated by wind energy, tidal energy, and ocean currents. Surface manifestations of internal waves have been photographed from space **(Figure 10.25c)**.

Are internal waves important? They may mix nutrients into surface water and trigger plankton blooms. They can also affect submarines and oil platforms. In 1963 the nuclear-powered USS *Thresher*, a fast attack submarine, was lost off the coast of Massachusetts with all hands. Running at high speed, *Thresher* may have encountered an internal wave and have been forced beyond its test depth. In 1980 a production oil platform was rotated nearly 90° from its original orientation by a series of internal waves. Also, the slow-motion breaking of internal waves against a shore may occasionally exaggerate tidal height.

We now turn our attention to the longer waves generated by the low atmospheric pressure of large storms, by the sloshing of water in enclosed spaces, and by the sudden displacement of ocean water.

CONCEPT CHECK

Before going on to the next section, check your understanding of some of the important ideas presented so far:

Are the wave speed and period of an internal wave comparable to those of a wind wave? A tsunami?

Are internal waves dangerous?

10.8 "Tidal Waves" Are Probably Not What You Think

The popular media and general public tend to label *any* unusually large wave a "tidal wave" regardless of its origin, and the waves described below are prime candidates for this error. Press accounts of storms at sea usually list a rogue wave as a tidal wave, and very large sets of wind waves are called tidal waves by some yachtsmen or surfers. The sea waves associated with earthquakes are almost always called tidal waves in media damage reports.

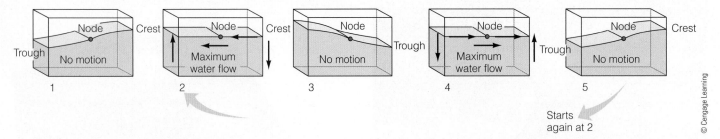

Figure 10.24 A standing wave in a basin. The wave oscillates about a node, a location at which there is no vertical movement. The rocking movement of water generates alternating crests and troughs at fixed distances from the node. As their name suggests, standing waves do not progress, and there is no net movement of water in them. (See also Figure 10.28.)

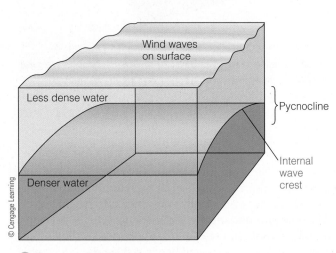

a The crest of an internal wave.

Figure 10.25 Internal waves can form between masses of water with different densities, especially at the base of the pycnocline.

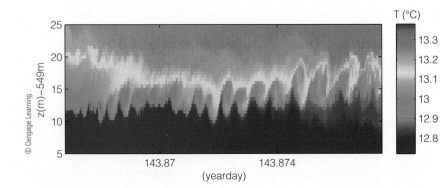

c The sun glints off surface manifestations of internal waves approaching the Baja California peninsula near Point Prieta. Patterns of constructive and destructive interference can be seen.

b Giant internal waves move down the side of a submerged mountain on the mid-Atlantic Ridge. Detected by temperature differences between water layers, these huge waves have a wavelength of 75 meters (250 feet) and a period of about 50 seconds.

What Waves Do in the Open Water

The complexity and expense of research ship operations makes it impossible for most of the ocean to be under direct observation at any given time. Fortunately, as you've seen in previous chapters, remote sensing—usually from satellites—is easing the burden of human observers needing to be present to take measurements. Wave height is routinely measured from space.

Now an ingenious new way of *directly* measuring wave activity (and other oceanic data) is coming online. Robots propelled by wave motion are being deployed to navigate huge ocean distances and relay information to scientists. The first four of this new generation of autonomous sensors, known as wave gliders (see **Figure A**), were launched on 17 November 2011. The gliders traveled together from San Francisco, California, to Hawai'i, and then took separate routes across the Pacific to Japan and Australia. The journey took about 300 days, traversed 40,000 kilometers (25,000 miles), and returned more than 2 million discrete data points. A second generation of gliders is now in production.

Liquid Robotics

Figure A Wave gliders move at about 3 kilometers (2 miles) per hour, powered by a set of underwater blades connected to the surface vessel by a long strap. The blades resemble window blinds. When a wave crest lifts the glider, the wings tilt up, rise through the water, and pull the craft forward. When the craft encounters a wave trough, the wings sink and tilt downward, pulling it forward. If the waves are very large, the submerged blade array pulls the whole craft underwater, mimicking a surfer's duck-dive. The full force of a large breaking wave is thus avoided. The solar panels power transmitters and on-board instrumentation.

The waves caused by the approach of a tropical cyclone to land may also incorrectly be termed tidal waves. The term *tidal wave* is *not* synonymous with *large wave*, however. As we will see in the next chapter, the only true tidal waves are relatively harmless waves associated with the tides themselves.

CONCEPT CHECK

Before going on to the next section, check your understanding of some of the important ideas presented so far:

Is there really any such thing as a true "tidal wave"?

10.9 Storm Surges Form beneath Strong Cyclonic Storms

The abrupt bulge of water driven ashore by a tropical cyclone (hurricane) or frontal storm is called a **storm surge (Figure 10.26)**. Its crest can temporarily add up to 7.5 meters (25 feet) to coastal sea level. Water can reach even greater heights when the surge is funneled into a confined bay or estuary.

Many factors contribute to the severity of a storm surge. The most important factor is the strength of the storm generating the surge. The low atmospheric pressure associated with a great storm will draw the ocean surface into a broad dome as much as 1 meter (about 3 feet) higher than average sea level. This dome of water accompanies the storm to shore, becoming much higher as the water gets shallower at the coast. There the water ramps ashore, driven forward by large storm-generated wind waves. A storm surge is a short-lived phenomenon. Technically it is not a progressive wave because it is only a crest; wavelength and period cannot be assigned to it.

Water in a storm surge does not come ashore as a single breaking wave but rushes inland in what looks like a sudden, very high, wind-blown tide. Indeed, storm surges are sometimes called *storm tides* because the volume of water they force onshore is greatly increased if the surge arrives at the same time as a high tide. The wrong combination of low atmospheric pressure, strong onshore winds, high tide, and bottom contour can be especially dangerous if estuaries in the area have been swollen by heavy rainfall preceding the storm.

Storm surges have had catastrophic consequences. The frightful tropical storm of November 1970 in Bangladesh (described in Chapter 8) generated a storm surge up to 9 meters (30 feet) high, which caused the death of more than 300,000 people. Storm surges associated with extratropical cyclones (frontal storms) can also do tremendous damage. On 1 February 1953, a storm surge and high tide arrived simultaneously against the

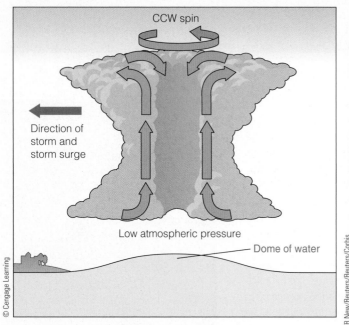

a The low pressure and high winds generated within a hurricane can produce a storm surge up to 9 meters (30 feet) high.

Figure 10.26 A storm surge.

b Low-lying southern Burma (Myanmar) was overwhelmed by tropical cyclone Nargis on 2 May 2008. Winds exceeding 165 kilometers (105 miles) per hour propelled a 5-meter (16-foot) storm surge across rice-growing areas west of the capital of Rangoon. The overwashed areas will take years to recover. More than 50,000 people perished immediately—many more later died of starvation and disease.

Dutch coast. Wind waves breached the dikes and flooded the low country, covering more than 3,200 square kilometers (800,000 acres) and drowning 1,783 people. The Dutch anticipate this coincidence of events will occur only about once in 400 years. The dikes have been rebuilt and the land reclaimed from the North Sea. On the opposite side of the North Sea, Londoners have spent US$1.3 billion on a flood defense system at the mouth of the River Thames. The centerpiece of the project is an immense barrier against storm surges **(Figure 10.27)**. Experts expect the barrier to prevent a devastating flood on an average of once every 50 years.

The U.S. coast is also at risk. In 1900 a storm surge topped the Galveston seawall, swept into the city, and killed more than 6,000 area residents. (In contrast, no lives were lost during a similar Texas storm in 1961 because coastal barriers had been constructed and advance warning allowed preparation and evacuation.) As you read in Chapter 8, Hurricane Katrina's 2005 assault on the coasts of Louisiana and Mississippi and Superstorm Sandy's devastation along the U.S. northeast coast were made even more lethal by their immense storm surges. Anyone living in a low-lying coastal area frequented by violent storms should be aware of the potential danger of storm surge.

CONCEPT CHECK

Before going on to the next section, check your understanding of some of the important ideas presented so far:

What causes a storm surge? Why is a storm surge so dangerous?

Can a storm surge be predicted?

10.10 Water Can Rock in a Confined Basin

When disturbed, water confined to a small space (such as a bucket, a bathtub, or a bay) will slosh back and forth at a specific resonant frequency. The frequency changes with different amounts of water or with different sizes or shapes of containers. If you carry a shallow container of water (like an ice cube tray) from one place to another, you're careful not to move the tray at its resonant frequency to avoid a spill. Most of the water's sloshing motion quickly settles down after you place the tray on a tabletop, but the water in the tray may rock gently for some seconds at this one resonant frequency. That rocking is a **seiche** (pronounced "saysh").

The seiche phenomenon was first studied in Switzerland's Lake Geneva by 19th-century researchers curious about why the water level at the ends of the long, narrow lake rises and falls at regular intervals after windstorms. They found that constant breezes tend to push water into the downwind end of the lake. When the wind stops, the water is released to rock slowly back and forth at the lake's resonant frequency, completing a crest–trough–crest cycle in a little more than an hour. At the ends of the lake the water rises and falls a foot or two; at the center it moves back and forth without changing height **(Figure 10.28)**. This kind of wave is called a

Figure 10.27 The Thames tidal barrier in London. The natural funnel shape of the Thames estuary concentrates storm surges as they near London. The barrier sections can be raised to prevent flooding upstream. Each barrier section is controlled by hydraulic arms within towers. The project cost US$1.3 billion and was completed in 1982. From its completion to February, 2014, the barrier has been raised 167 times to prevent flooding.

Tom Garrison

standing wave because it oscillates vertically with no forward movement. The point (or line) of no vertical wave action in a standing wave—the place in the lake where the water moves only back and forth—is called a *node*. In Lake Geneva the wavelength of the seiche is twice the length of the lake itself; the node lies at the center. The lake acts like a large version of the ice cube tray in the example earlier.

Seiches can be initiated by tides, storm surge, or seabed displacement, but damage from seiches along most ocean coasts is rare. The wavelength may be tremendous, but seiche wave height in the open ocean rarely exceeds a few inches. Larger tide-related seiches can occur in harbors: Coastal seiches at Nagasaki, on the southern coast of Japan, occasionally reach 3 meters (10 feet). Seiches may disturb shipping schedules by interfering with the predicted arrival times of tides; or they may cause currents in harbors, which could snap mooring lines.

© Cengage Learning

a A seiche is a long wave in a lake or ocean basin that sloshes back and forth from one end of the basin to another. The rocking frequency depends on the length of the basin. At the node, water moves sideways and does not rise or fall (see also Figure 10.24).

b A graph of a seiche in Lake Erie. Strong westerly winds in November 2003 caused a seiche with more than 4 meters (13 feet) of difference in water level from Toledo on the western end of the lake to Buffalo on the eastern end.

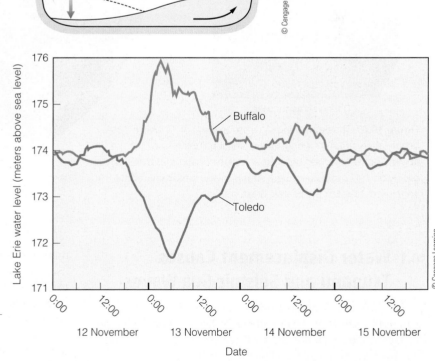

© Cengage Learning

Figure 10.28 Seiches.

💬 **THINKING BEYOND THE FIGURE**

Is a seiche dangerous?

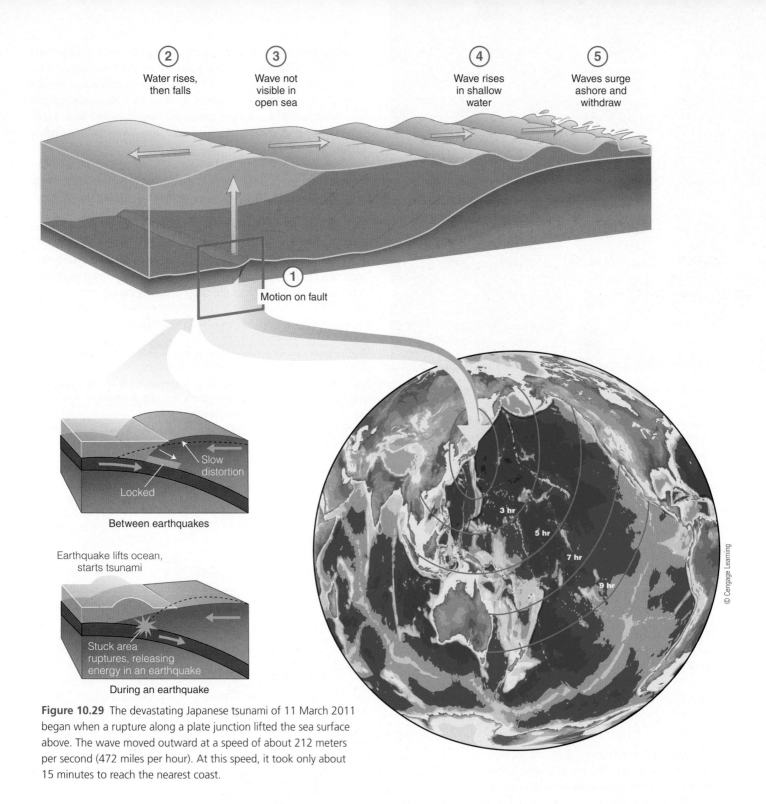

② **Water rises, then falls**

③ **Wave not visible in open sea**

④ **Wave rises in shallow water**

⑤ **Waves surge ashore and withdraw**

① **Motion on fault**

Slow distortion

Locked

Between earthquakes

Earthquake lifts ocean, starts tsunami

Stuck area ruptures, releasing energy in an earthquake

During an earthquake

3 hr

5 hr

7 hr

9 hr

© Cengage Learning

Figure 10.29 The devastating Japanese tsunami of 11 March 2011 began when a rupture along a plate junction lifted the sea surface above. The wave moved outward at a speed of about 212 meters per second (472 miles per hour). At this speed, it took only about 15 minutes to reach the nearest coast.

10.11 Water Displacement Causes Tsunami and Seismic Sea Waves

Long-wavelength, shallow-water progressive waves caused by the rapid displacement of ocean water are called **tsunami**, a descriptive Japanese term combining *tsu* ("harbor") with *nami* ("wave"). The word is both singular and plural. Tsunami caused by the sudden, vertical movement of Earth along faults (the same forces that cause earthquakes) are properly called **seismic sea waves**. Tsunami can also be caused by landslides, icebergs falling from glaciers, volcanic eruptions, asteroid impacts, and other direct displacements of the water surface. Note that all seismic sea waves are tsunami, but not all tsunami are seismic sea waves.

"Small" tsunami are caused by the displacement of surface water. Although less energy is released by landslides than by most seismic fractures, the resulting sea waves are still very destructive for people or structures near their point of origin. This is especially true if the wave is formed within a confined area.

Seismic sea waves originate on the seafloor when Earth movement along faults displaces seawater. **Figure 10.29** shows the birth of the 11 March 2011 seismic sea wave that struck northern Japan. Rupture along a submerged fault lifted the sea surface as much as 6 meters (20 feet) in places. Gravity pulled the crest downward, but the momentum of the water caused the crest to overshoot and become a trough. The oscillating ocean surface generated progressive waves that radiated from the epicenter in all directions **(Figure 10.30)**. Waves would also have formed if the fault movement were downward. In that case a depression in the water surface would propagate outward as a trough. The trough would be followed by smaller crests and troughs caused by surface oscillation.

An even larger tsunami struck near Sumatra in the Indian Ocean in December 2004. Again, a submerged fault ruptured and seawater was displaced—in this instance to about 10 meters (33 feet). The 2004 Sumatra quake and the 2011 event in Japan were respectively the third and fourth largest earthquakes ever recorded. These violent events are associated with subduction zones (about which you learned in Chapter 3). After the seabed stopped shaking in 2011, parts of northern Japan had moved about 4 meters (13 feet) closer to the United States![5]

It seems strange to refer to tsunami—waves with wavelengths of up to 200 kilometers (125 miles)—as *shallow-water*

---
[5]The violent shaking, by the way, lasted nearly 6 minutes!

> The 2004 Sumatra quake and the 2011 event in Japan were respectively the third and fourth largest earthquakes ever recorded. The two resulting tsunami took the lives of nearly a third of a million people.

waves. Yet half their wavelength would be 100 kilometers (62 miles), and even the deepest ocean trenches do not exceed 11 kilometers (7 miles) in depth. These immense waves therefore *never* find themselves in water deeper than half their wavelength.

Like any shallow-water wave, seismic sea waves are affected by the contour of the bottom and are commonly refracted, sometimes in unexpected ways. Detailed analysis of the 2004 and 2011 events showed that the midocean ridges act as topographic waveguides. These shallow-water waves were in constant contact with the seabed and appear to have followed the Southwest Indian Ridge below the southern tip of Africa to the Mid-Atlantic Ridge. **Figure 10.31** shows the speed of the dispersion of wave energy across the Pacific in the 2011 event.

Tsunami Move at High Speed

Remember, you learned at the beginning of the chapter that waves carry *energy*, not *mass*, across the ocean surface. To see the idea of speed, let's do some math:

The velocity (celerity) of tsunami waves depends on the water depth and gravity.

$$C = \sqrt{gd}$$

where

C = velocity in meters per second
d = depth in meters
g = gravitational acceleration (9.8 m/sec².)

Thus,

$$C = 3.13\sqrt{d}$$

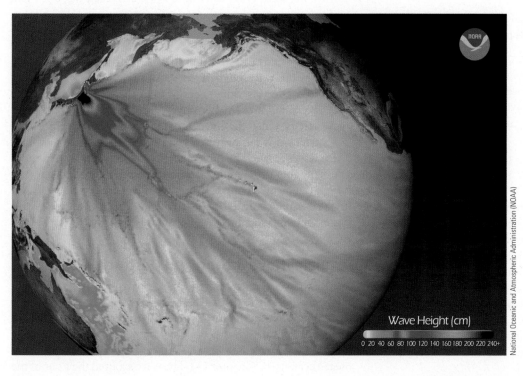

Figure 10.30 Maximum calculated open-ocean wave height for the 2011 tsunami off the coast of northern Japan. The scale indicates height in centimeters. Remember that the open-ocean height of a tsunami is much less than the nearshore or onshore height of the waves.

Wave Height (cm)

0 20 40 60 80 100 120 140 160 180 200 220 240+

National Oceanic and Atmospheric Administration (NOAA)

Tsunami Travel Times

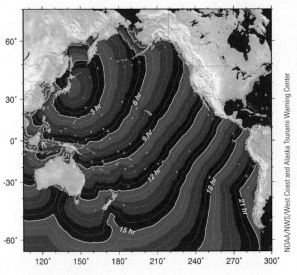

NOAA/NWS/West Coast and Alaska Tsunami Warning Center

Figure 10.31 Each faint concentric circle in this figure represents a travel time of 30 minutes for the 2011 tsunami. Waves 2 meters (6.6 feet) high struck parts of the coast of Chile 17,000 kilometers (11,000 miles away). The red dots indicate the positions of sensing buoys like that shown in Figure 10.37.

💬 **THINKING BEYOND THE FIGURE**

How long would it take for a tsunami to travel from the Aleutian Islands (west of Alaska) to the North Island of New Zealand?

For example, if $d = 4,600$ meters (deep ocean):

$C = 3.13\sqrt{4,600} = 3.13 \times 67.8$ meters per second or 763 kilometers per hour (473 miles per hour, the speed of some jet aircraft!)

If $d = 100$ meters (near shore)

$C = 3.13\sqrt{100} = 3.13 \times 10 = 31.3$ meters per second or 112.7 kilometers per hour (70 miles per hour, the speed of freeway traffic)

As the numbers show, the waves slow and bunch up as they approach shore. Their height increases. The waves come ashore as a wall of water, not a breaking wave (see **Figure 10.32**). Unless the locals have already started for high ground, they have no hope of survival.

What's It Like to Encounter a Tsunami?

We are familiar with the steepness of a wind wave and the short period of a few seconds between its crests. Tsunami are much different. Once a tsunami is generated, its steepness (ratio of height to wavelength) is extremely low. This lack of steepness, combined with the wave's very long period (5–20 minutes), enables it to pass unnoticed beneath ships at sea. A ship on the open ocean that encounters a tsunami with a 16-minute period would rise slowly and imperceptibly for about 8 minutes, to a crest only 0.3 to 0.6 meter (1 or 2 feet) above average sea level. It would then ease into the following trough 8 minutes later. With all the wind waves around, such a movement would not be noticed.

As the tsunami crest approaches shore, however, the situation changes rapidly and often dramatically. The period of the wave remains constant, its velocity drops, and the wave height greatly increases. As the crest arrives at the coast, observers would see water surge ashore in the same way as a very high, very fast tide (clearly seen in Figure 10.32). In confined coastal waters relatively close to their point of origin, tsunami can reach a height of perhaps 30 meters (100 feet). The wave is a fast onrushing flood of water, not the huge, plunging breaker of popular movies and folklore. Tsunami can be catastrophic even if the wave crests are not that high—imagine a smaller wave inundating a flat, low-lying coast. The *combination* of wave height and "run-up" (the distance the waves moves ashore) determines a tsunami's lethality.

The wave energy spreads through an enlarging circumference as a tsunami expands from its point of origin. People on-shore near the generating shock have reason to be concerned because the energy will not have dissipated very much. Because of its low elevation and proximity to the earthquake epicenter, parts of the Miyagi and Iwate prefectures of northern Japan were inundated by the March 2011 wave. The Sendai region, adjacent to the epicenter, suffered the greatest damage **(Figure 10.33)**. More than 20,000 people were killed or declared missing in the surrounding area. Nuclear power plants in the area were seriously damaged and leaked radioactive substances into air, land, and ocean. The economic loss will almost certainly amount to about 3% of a year's production by the world's third largest economy—more than US$310 billion.

In the more powerful December 2004 tsunami, the Indonesian city of Banda Aceh was essentially demolished when the Eurasian Plate was lifted about 5 meters (16 feet). The overlying water rose the same distance. Near the earthquake's epicenter the ocean is about 4,000 meters (13,000 feet) deep, and a column of water with an area the size of the state of Delaware was forced upward with an energy equivalent to 32,000 atomic bombs of Hiroshima size. This vast volume of water then began to fall, and the greatest tsunami in modern times was born. The human toll was vastly greater than that in Japan: more than a quarter of a million people died.

Note that the destruction in Japan and Indonesia was not caused by one wave, but by a *series* of waves following one another at regular intervals. Some energy from the main tsunami wave was distributed into smaller waves ahead of or behind the main wave as it moved. If the epicenter of the displacement responsible for a tsunami is far away, sea level at shore will rise and fall as these waves arrive. The interval between crests (the wave period) is usually about 15 minutes. Coastal residents far from a tsunami's origin can be lulled into thinking the waves are over; they return to the coastline only to be injured or killed by the next crest. This behavior contributed to the enormous loss of life.

Tsunami Have a Long and Destructive History

Researchers are uncovering evidence of astonishingly destructive tsunami in the distant past. Researchers have found signs of a huge wave, perhaps as high as 91 meters (300 feet), which crashed against the Texas coast 66 million years ago. It may have been caused by a comet or asteroid striking the Gulf of Mexico near Yucatán (see Chapter 13). The wave scoured the floor of the

Figure 10.32 The first wave of the December 2004 tsunami arrives at Hat Rai Lay Beach in southern Thailand. The wave was preceded by a trough—the ocean had receded, exposing rocks and attracting tourists farther away from shore. As noted in the text, the approaching wave is not only high (an estimated 7 meters, or 23 feet), it is also long. In these arresting views (recovered from a digital camera later found in the debris), we see a rapidly advancing *wall* of water that will continue ashore as if the ocean were spilling out of its basin, not a wind-wave-like hump. More than 8,000 people died in Thailand alone.

Gulf; picked up sand, gravel, and sharks' teeth; and deposited the material in what is now central Texas.

At the southern end of the island of Madagascar (east of southern Africa), marine sediments cover an area about twice the size of Manhattan Island to a depth of about 330 meters (1,000 feet). The chevrons of sediment point to a newly discovered crater 29 kilometers (18 miles) in diameter and 3,800 meters (12,500 feet) below the surface of the Indian Ocean. A large

a The Japanese town of Yamada lies in ruins on the afternoon of 11 March 2011. The waves had reached a maximum height of about 9 meters (30 feet). This part of the coast fell about 0.5 meters (2 feet) during the preceding earthquake, so the influx of water was even more extensive and devastating.

b The force of the onrushing water is clearly evident in this image of the city of Onagawa. In all, the tsunami inundated an area of about 470 square kilometers (181 square miles) of northeastern Japan.

Figure 10.33 Damage caused by the great earthquake and tsunami of March 2011.

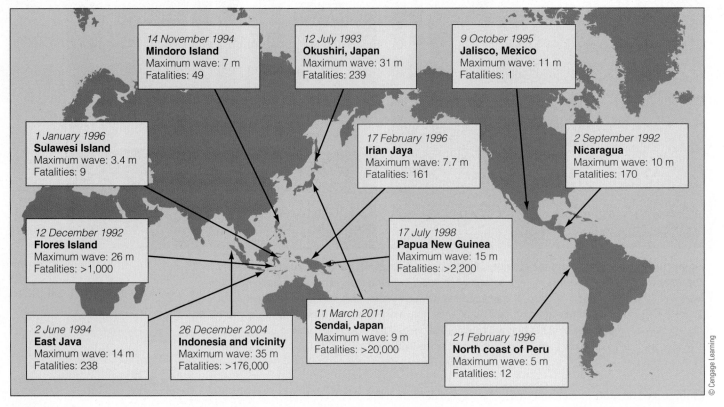

14 November 1994
Mindoro Island
Maximum wave: 7 m
Fatalities: 49

12 July 1993
Okushiri, Japan
Maximum wave: 31 m
Fatalities: 239

9 October 1995
Jalisco, Mexico
Maximum wave: 11 m
Fatalities: 1

1 January 1996
Sulawesi Island
Maximum wave: 3.4 m
Fatalities: 9

17 February 1996
Irian Jaya
Maximum wave: 7.7 m
Fatalities: 161

2 September 1992
Nicaragua
Maximum wave: 10 m
Fatalities: 170

12 December 1992
Flores Island
Maximum wave: 26 m
Fatalities: >1,000

17 July 1998
Papua New Guinea
Maximum wave: 15 m
Fatalities: >2,200

2 June 1994
East Java
Maximum wave: 14 m
Fatalities: 238

26 December 2004
Indonesia and vicinity
Maximum wave: 35 m
Fatalities: >176,000

11 March 2011
Sendai, Japan
Maximum wave: 9 m
Fatalities: >20,000

21 February 1996
North coast of Peru
Maximum wave: 5 m
Fatalities: 12

© Cengage Learning

Figure 10.34 Twelve destructive tsunami have claimed more than 200,000 lives since 1990.

asteroid or comet plunged into the seabed about 4,800 years ago creating waves at least 180 meters (600 feet) high, about 13 times as big as the waves that inundated Indonesia in 2004.

More than 300 tsunami have been recorded in the past 3,300 years in the Mediterranean alone. The most recent occurred in

December of 2002 when a landslide on Stromboli, a volcano north of the Italian island of Sicily, displaced seawater and generated a tsunami 10 meters (33 feet) high. The wave snapped moorings of ships in a harbor 100 kilometers (62 miles) away but did little other damage. Other events have been much more destructive. The collapse of the Stronghyle volcano on the Greek island of Thera (now Santorini) in about 1600 B.C.E. generated a tsunami that smashed into the advanced Bronze Age Minoan civilization. Cities on the island of Crete were shattered by waves more than 60 meters (200 feet) high; the Minoans never recovered. The Mycenaean Greeks came to dominate the area and the course of western civilization changed, but that's another story.

In the Pacific, on 27 August 1883, the enormous volcanic explosion of Krakatoa in Indonesia generated 35-meter (115-foot) waves that destroyed 163 villages and killed more than 36,000 people.

Destructive tsunami strike somewhere in the world an average of once each year. Some recent lethal tsunami are listed in **Figure 10.34**.

Tsunami Warning Networks Save Lives

Since 1948, an international tsunami warning network has been in operation around the seismically active Pacific to alert coastal residents of possible danger. Warnings must be issued rapidly because of the speed of these waves. Telephone books in coastal Hawai'ian towns contain maps and evacuation instructions for use when the warning siren sounds. This advice was crucial after the 2011 Japan earthquake when resorts and residences in

Figure 10.35 The harbor and shore at Crescent City, California, was rocked by the 11 March 2011 tsunami. Damage there exceeded US$30 million; one life was lost.

AP Photo/Bryant Anderson

a Failure of the edge of the continental shelf off the Oregon coast produced an underwater landslide that probably caused a large tsunami sometime around the end of the last ice age. The pocket from which the debris originated is about 6 kilometers (3.7 miles) wide.

b Tsunami hazard warning sign typical of central coastal Oregon. The configuration of the coast and the slope and instability of the bottom in this area make tsunami particularly dangerous.

Figure 10.36 Tsunami cause and consequence.

Hawai'i's Kona district were evacuated. Damage there exceeded US$17 million, but no lives were lost.

The tsunami warning system was also responsible for averting the loss of many lives after the great 27 March 1964 earthquake in Alaska (see Chapter 3). A 3.7-meter (12-foot) wave, probably the fourth crest to reach the coast, swept into Crescent City, California, about 6 hours after the quake. Though more than 300 buildings were destroyed or damaged, five gasoline storage tanks set ablaze, and 27 blocks of the city demolished, there were relatively few casualties because residents received a warning to evacuate the area. Crescent City suffered again in 2011 (**Figure 10.35**).

Some public safety experts suggest we have become complacent about the risks associated with these destructive waves. As can be seen in **Figure 10.36b**, some coastal communities remind residents and visitors of the danger. Residents of the United States northern Pacific coast should be especially wary. If you're curious, look again at Figure 3.25b—does that subduction zone west of Oregon and Washington coast look suspicious?

Modern tsunami warning systems depend on seabed seismometers and submerged devices and satellites that watch the shape of the sea surface (**Figure 10.37**). Sadly, no warning system was in place to provide an alarm in the Indian Ocean tragedy of December 2004. That has changed. Even when a tsunami rises from a seismic disturbance very close to shore, some warning may be possible (see Questions from Students #6).

How fast does a tsunami move?

Is a tsunami a shallow-water wave or a deep-water wave?

What is the wave height of a typical tsunami away from land? Are tsunami dangerous in the open sea?

Does a tsunami come ashore as a single wave? A series of waves? Does a tsunami wave break like a surfing wave?

What caused the most destructive tsunami in recent history? Where was the loss of life and property concentrated?

How might one detect and warn against tsunami?

Figure 10.37 A pressure sensor on the seabed detects subtle pressure changes of a rise and fall in sea level from the passage of a tsunami. The sensor transmits a signal to this floating buoy that relays the warning by satellite. The information is analyzed at the Pacific Tsunami Warning Center. Similar systems are now deployed in the Indian Ocean and central Atlantic.

CONCEPT CHECK

Before going on to the next section, check your understanding of some of the important ideas presented so far:

What causes tsunami? Do all geological displacements cause tsunami?

In this chapter you learned that waves transmit energy, not water mass, across the ocean's surface. The speed of ocean waves usually depends on their wavelength, with long waves moving fastest. Arranged from short to long wavelengths (and therefore from slowest to fastest), ocean waves are generated by very small disturbances (capillary waves), wind (wind waves), rocking of water in enclosed spaces (seiches), seismic and volcanic activity or other sudden displacements (tsunami), and gravitational attraction (tides). The behavior of waves depends largely on the relation between a wave's size and the depth of water through which it is moving. Waves can refract and reflect, break, and interfere with one another.

Wind waves can be deep-water waves if the water is more than half their wavelength deep. The waves of very long wavelengths are always in "shallow water" (water less than half their wavelength deep). These long waves travel at high speeds, and some may have great destructive power.

In the next chapter you will learn that even greater waves exist: the tides. You may be interested to know that two "tidal waves" move along most coasts each day. You don't read of daily mayhem and destruction from these passages because the crests are the day's high tides; and the troughs, the day's low tides. Tides are shallow-water waves no matter how deep the ocean they're moving through. Tides can be destructive, but among all waves their ability to cause damage is fortunately not proportional to their wavelengths.

QUESTIONS FROM STUDENTS

1. If so much energy is expended as wind waves break, why doesn't water in the surf zone get hot?

The amount of energy moving through the ocean in progressive waves is impressive. The energy of a wave is proportional to the square of its height. Each linear meter of a wave 2 meters (6.5 feet) above average sea level represents an energy flow of about 25 kilowatts (34 horsepower), enough to light 250 light bulbs (100 watts each); a wave twice as high contains four times as much energy. A single wave 1.2 meters high striking the west coast of the United States may release as much as 50 million horsepower!

The energy of wind waves is dissipated mostly as heat in the surf, but since water has such a high latent heat (discussed in Chapter 6), the injection of heat into the surf zone doesn't significantly increase water temperature. The surf zone is also an area of vigorous mixing, so any heat released is quickly distributed through a large volume of water.

A number of methods have been proposed to take advantage of this energy before it is dissipated. Among the more promising is the Pelamis System (appropriately named after a sea snake) pictured in Figure 17.11.

2. What about wind waves and the Coriolis effect? Do wind waves turn right in the Northern Hemisphere as they move across the ocean surface?

The Coriolis effect has no influence on waves with periods shorter than about 5 minutes. Large shallow-water waves such as tsunami, seiches, and tides involve the mass movement of water and are influenced by Earth's rotation. Wind waves, however, with a period rarely exceeding 20 seconds, are not.

3. Will people in California be in great danger from a seismic sea wave when a major earthquake occurs on the San Andreas Fault?

Probably not, for two reasons. First, San Andreas quakes are usually lateral-displacement quakes: the ground moves suddenly sideways rather than up or down. Tsunami are usually generated by vertical movement. Second, earthquake motion along the San Andreas Fault is parallel to the coast, not at right angles to it. If the ground movement were toward the coast (as could happen north of San Francisco), the "shove" might result in a wave. There may be some slopping about at the coastline during a large earthquake, but probably not a classic tsunami. One caveat: If massive offshore slumping occurs (see Figure 10.36a), all bets are off.

4. I love to surf. Where should I plan to live?

The Pacific has the largest potential fetch distances, so the best chances for large wind waves are there. The biggest wind waves are made in the West Wind Drift, but temperatures there are way too cold for comfortable surfing. Besides, the sea state in areas of continuous high wind is chaotic, and surfing requires orderly waves. It is best to let the wave trains sort out. Hawai'i is in the middle of the Pacific, so wind waves from polar storms in either hemisphere strike its shores only after much sorting by wavelength. Order is assured. The water is warm, too. We vote for Hawai'i.

5. I read that there was a second earthquake near Sumatra on 28 March 2005. It measured 8.7—nearly the same magnitude as the 26 December 2004 event—but no tsunami formed. Why not?

Because there was very little *vertical* movement of the seabed, and therefore very little displacement of seawater. Horizontal movement of the ocean floor tends not to cause tsunami unless it occurs adjacent to a coast.

6. Did the Japanese have any warning before the 2011 earthquake and tsunami struck?

Earthquakes (and resultant tsunami) cannot be reliably predicted. But in an interesting footnote to the March 2011 Japanese disaster, the Japan Meteorological Agency immediately detected P-waves from the quake. The slower S-waves—one of the type of earthquake waves capable of causing great damage— took about 90 seconds to reach Tokyo, about 373 kilometers (232 miles) southwest of the epicenter. About 1 minute before the shaking began, automatic alarms began to stop high-speed trains, open elevator doors, close water valves, shunt electrical power, and sound alarms. Many lives were saved.

7. Is there a reliable way to standardize the measurement of wave size or surface conditions at sea?

Yes, and it is a venerable method indeed. In the summer of 1805, when Rear Adm. Sir Francis Beaufort was captain of HMS *Woolwich*, a British warship, he invented a numbered scale to describe sea conditions. The lowest number, Beaufort 0, describes a "sea like a mirror." Beaufort 1 clearly describes capillary waves; 5 describes moderate waves and whitecaps; 9 describes dense streaks of foam and huge waves. Beaufort 12 (the highest number) conjures "... exceptionally high waves; sea covered with white foam; visibility greatly reduced." This method of describing **sea state** by number quickly spread through the fleet.

In 1912 the International Commission for Weather Telegraphy sought to correlate the Beaufort scale to specific wind velocities. A uniform set of equivalents was accepted in 1926, with revisions in 1946 and 1955. With new extension above Beaufort 12, the scale is still used today. A modification of the U.S. Navy's version is provided as Appendix 5.

TERMS AND CONCEPTS TO REMEMBER

C	gravity wave	seismic sea wave	wave diffraction
$C = \sqrt{gd}$	group velocity	shallow-water wave	wave frequency
$C = \dfrac{L}{T}$	interference	standing wave	wave height
capillary wave	internal wave	Stokes drift	wave period
constructive interference	orbit	storm surge	wave reflection
deep-water wave	orbital wave	surf	wave refraction
destructive interference	plunging wave	surf beat	wave steepness
dispersion	progressive wave	surf zone	wave train
disturbing force	restoring force	swell	wave trough
fetch	rogue wave	tsunami	wavelength
forced wave	sea	V	wind duration
free wave	sea state	wave	wind strength
fully developed sea	seiche	wave crest	wind wave

STUDY QUESTIONS

Thinking Critically

1. Though they move across the deepest ocean basins, seiches and tsunami are referred to as "shallow-water waves." How can this be?
2. How do particles move in an ocean wave? How is that movement similar to or different from the movement of particles in a wave in a spring or a rope? How does this relate to a *stadium wave*—a waveform made by sports fans in a circular arena?
3. What is the general relationship between wavelength and wave speed? How does water movement in a wave change with depth?
4. How can a rogue wave be larger than the theoretical maximum height of waves in a fully developed sea?
5. How is a progressive wave different from a standing wave? Must standing waves be orbital waves only, or can standing waves also form in shaken ropes or pushed-and-pulled springs?
6. How can large waves generated by a distant storm arrive at a shore first, to be followed later by small waves?

Thinking Analytically

1. What is the velocity of a deep-water wave that has a period of 10 seconds? What is its wavelength?
2. Assuming the alarm was raised immediately, how much tsunami warning would residents of southeastern Japan have if a strong earthquake struck the Peru–Chile Trench?
3. What is the wavelength of a typical storm surge?
4. Can deep-water waves and shallow-water waves exist at the *same* point offshore (that is, in the same depth of water)?

GLOBAL GEOSCIENCE WATCH

Visit www.cengagebrain.com to access MindTap, a complete digital course which includes access to Global Geoscience Watch and more. Search for "Japanese tsunami" and then "Banda Aceh tsunami" (separate searches) in the GREENR database. Find information about the earthquake and the resulting tsunami that occurred off of Sendai, Japan in 2011 and Banda Aceh, Indonesia in 2004, then summarize these two events and compare and contrast the characteristics and impacts from the tsunami in a two- to three-page report.

11 Tides

KEY CONCEPTS

Tides are periodic short-term changes in ocean surface height. Tides are forced waves formed by gravity and inertia. Tom Garrison

The *equilibrium* theory of tides explains tides by examining the balance and effects of forces that allow our planet to stay in orbit around the sun, or the moon to orbit Earth. © Cengage Learning

The *dynamic* theory of tides takes into account seabed contour, water's viscosity, and tide wave inertia. © Cengage Learning

Together, the equilibrium and dynamic theories allow tides to be predicted years in advance. Andrew Gunners/Getty Images

Power can be extracted from tidal flow. Marine Current Turbines Limited

Tourists rush away as a sudden tide wave (a tidal bore) strikes the bank of the Qiantang River in Hangzhou, China. The river has a wide, funnel-like mouth, and its tide is a famous (and dangerous) attraction.
AP Images/Color China Photo/Wang Chaoying

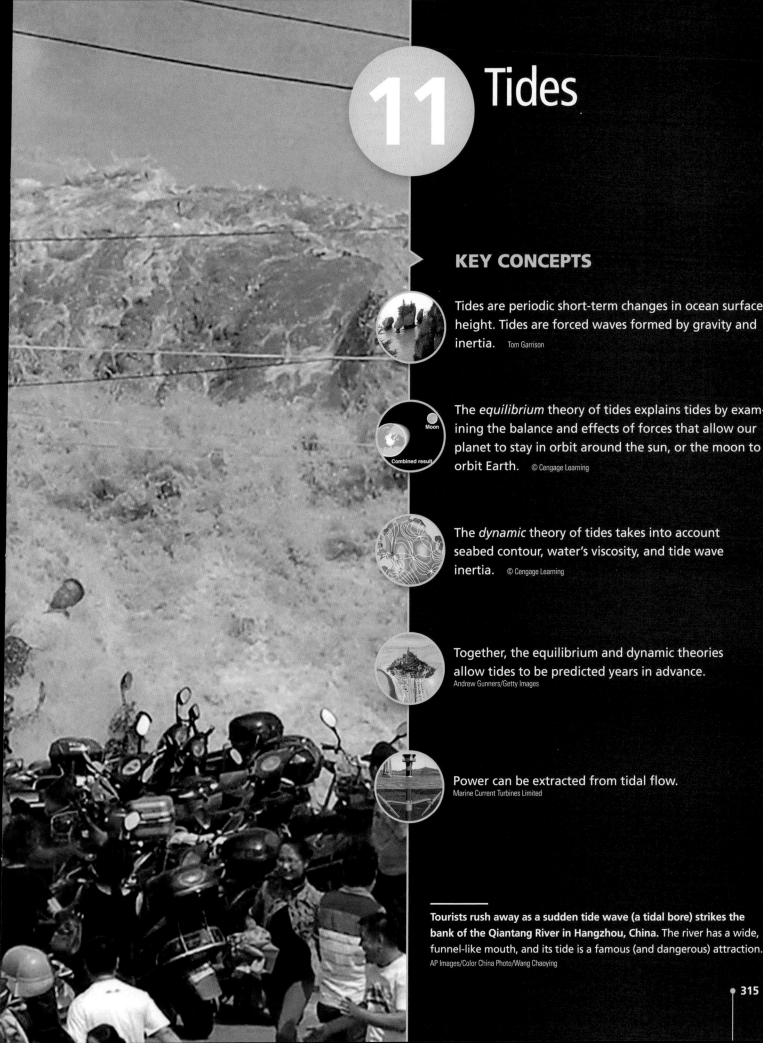

11.1 Tides Are the Longest of All Ocean Waves

Tides are periodic, short-term changes in the height of the ocean surface at a particular place, caused by a combination of the gravitational force of the moon and sun, the motion of Earth, and the inertia of water **(Figure 11.1)**. With a wavelength that can equal half of Earth's circumference, tides are the longest of all waves. Unlike the other waves we have met, these huge shallow-water waves are never free of the forces that cause them and thus are called *forced* waves. (After they are formed, wind waves, seiches, and tsunami are *free* waves, that is, they are no longer being acted upon by the force that created them and do not require a maintaining force to keep them in motion.)

The Greek navigator and explorer Pytheas first wrote of the connection between the position of the moon and the height of a tide around 300 B.C.E., but full understanding of tides had to await Newton's analysis of gravitation. Among many other things, Isaac Newton's brilliant 1687 book *Philosophiae Naturalis Principia Mathematica (Mathematical Principles of Natural Philosophy)* describes the motions of planets, moons, and all other bodies in gravitational fields. A central finding: The pull of gravity between two bodies is proportional to the masses of the bodies but inversely proportional to the square of the distance between them. This finding means that heavy bodies attract each other more strongly than light bodies do and that gravitational attraction quickly weakens as the distance grows larger. This may be expressed mathematically as

$$F = G\left(\frac{m_1 m_2}{r^2}\right)$$

where F is the gravitational attraction, G is the universal gravitational constant, m_1 and m_2 are the masses of the two bodies, and r is the distance between their centers. We can use this equation to calculate the gravitational attraction between the sun and Earth or between the moon and Earth.

Although the main cause of tides is the combined gravitational attraction of the moon and sun acting on the ocean, the forces that actually generate the tides vary inversely with the *cube* of the distance from Earth's center to the center of the tide-generating object (the moon or sun). Distance is thus even more important in this relationship, which may be expressed as

$$T = G\left(\frac{m_1 m_2}{r^3}\right)$$

where T is the tide-generating force, G is the universal gravitational constant, m_1 and m_2 are the masses of the two bodies, and

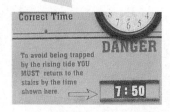

Figure 11.1 Prediction of the arrival times of high and low tides is possible because of our understanding of the physics involved. Tourists at this shore in New Brunswick, Canada—famous for its extreme tides—ignore these predictions at their peril. (Figure 11.17 shows where this photo was taken.)

Tom Garrison

r is the distance between their centers. The sun is about 27 million times as massive as the moon, but the sun is about 387 times as far away as the moon, so the sun's influence on the tides is only 46% that of the moon's.

As we will see, Newton's gravitational model of tides—the *equilibrium theory*—deals primarily with the position and attraction of Earth, moon, and sun and does not factor in the influence on tides of ocean depth or the positions of continental landmasses. The equilibrium theory would accurately describe tides on a planet uniformly covered by water. A modification proposed by Pierre-Simon Laplace about a century later—the *dynamic theory*—takes into account the speed of the long-wavelength tide wave in relatively shallow water, the presence of interfering continents, and the circular movement or rhythmic back-and-forth rocking of water in ocean basins. We will explore the idealized situation of the equilibrium theory before moving to the real-world dynamic view.

> **"The seas are the heart's blood of the earth. Plucked up and kneaded by the sun and the moon, the tides are systole and diastole of Earth's veins."**
> —HENRY BESTON

CONCEPT CHECK

Before going on to the next section, check your understanding of some of the important ideas presented so far:

How is a forced wave different from a free wave?

What celestial bodies are most important in determining tides?

11.2 Tides Are Forced Waves Formed by Gravity and Inertia

The **equilibrium theory of tides** explains many characteristics of ocean tides by examining the balance and effects of the forces that allow a planet to stay in a stable orbit around the sun, or the moon to orbit Earth. The equilibrium theory assumes that the seafloor does not influence the tides and that the ocean conforms instantly to the forces that affect the position of its surface; the ocean surface is presumed always to be in equilibrium (balance) with the forces acting on it.

The Movement of the Moon Generates Strong Tractive Forces

We begin our examination of these forces by looking at the moon's effect on the ocean surface. Gravity tends to pull Earth and the moon toward each other, but inertia—the tendency of moving objects to continue in a straight line—keeps them apart. Earth and the moon don't smash into each other (or fly apart) because they are in a stable orbit; their mutual gravitational attraction is exactly offset by their inertia **(Figure 11.2)**.

Contrary to what you might think, the moon does not revolve around the center of Earth. Rather, the Earth–moon *system* revolves once a month (27.3 days) around the system's center of mass. Because Earth's mass is 81 times that of the moon, this common center of mass is located not in space but 1,650 kilometers (1,023 miles) *inside* Earth. This center of mass is shown in **Figure 11.3**.

The moon's gravity attracts the ocean surface toward the moon. Earth's motion around the center of mass of the Earth–moon system throws up a bulge on the opposite side of Earth. Two tidal bulges result **(Figure 11.4)**.

Let's look more closely at the two bulges in the last image in Figure 11.4. In **Figure 11.5**, four places on Earth's surface are marked with numbers ① through ④. Each place has three arrows drawn to represent forces: the outward-flinging force of inertia is shown in blue, and the inward-pulling force of gravity is shown in brown. Combined, they are called *tractive forces*. Note that the inward pull of gravity and the outward-moving tendency of inertia don't always act in exactly the same balanced way on each particle of Earth and moon. The net strength and

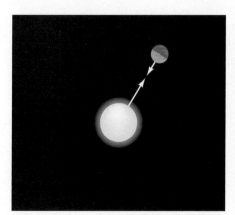

a If the planet is not moving, gravity will pull it into the sun.

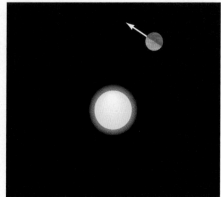

b If the planet is moving, the inertia of the planet will keep it moving in a straight line.

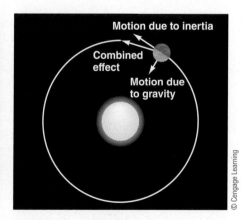

c In a stable orbit, gravity and inertia combine to cause the planet to travel in a fixed path around the sun.

© Cengage Learning

Figure 11.2 A planet orbits the sun in balance between gravity and inertia.

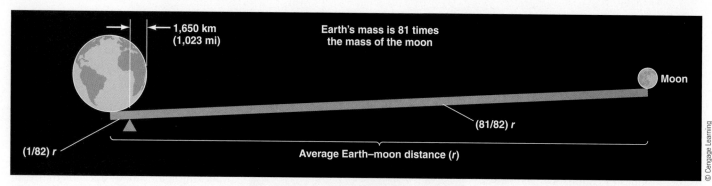

Figure 11.3 The moon does not revolve around the center of Earth. Earth and moon together—the Earth–moon system—revolve around a common center of mass about 1,650 kilometers (1,023 miles) beneath Earth's surface.

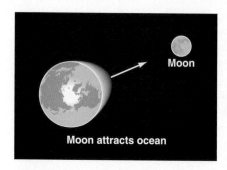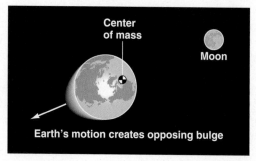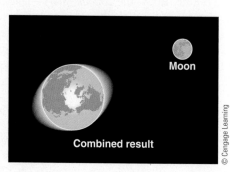

Figure 11.4 The moon's gravity attracts the ocean toward it. The motion of Earth around the center of mass of the Earth–moon system throws up a bulge on the side of Earth opposite the moon. The combination of the two effects creates two tidal bulges.

direction that result when the two forces are combined are shown as red arrows.

Points ① and ② are closer to the moon, so gravitational attraction at those points slightly exceeds the outward-moving tendency of inertia. Water there tends to be attracted toward the moon so is pulled along the ocean surface toward a spot beneath the moon. At points ③ and ④, slightly farther from the moon, inertia exceeds gravitational attraction. Water at those points

tends to be flung away from the moon so moves along the ocean surface toward a spot opposite the moon. Together, the tractive forces cause the two small bulges in the ocean, one in the direction of the moon, the other in the opposite direction.

Note that there is no point on Earth's surface where the force of the moon's gravity exactly equals the outward-moving tendency of inertia. Only at point **CE**—the center of Earth—are the inward pull of gravity and the outward-moving tendency

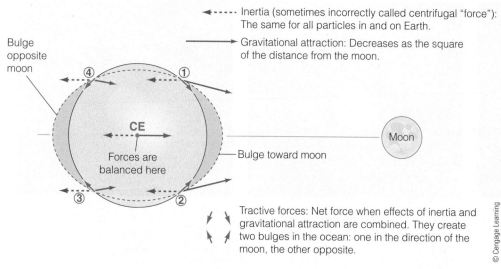

Figure 11.5 The actions of gravity and inertia on particles at five different locations on Earth. At points ① and ②, the gravitational attraction of the moon slightly exceeds the outward-moving tendency of inertia; the imbalance of forces causes water to move along Earth's surface, converging at a point toward the moon. At points ③ and ④, inertia exceeds gravitational force, so water moves along Earth's surface to converge at a point opposite the moon. Forces are balanced only at the center of Earth (point **CE**).

The two forces that can move the ocean—inertia and gravitational attraction—are precisely equal in strength but opposite in direction, and thus balanced, only at the center of Earth (point **CE**).

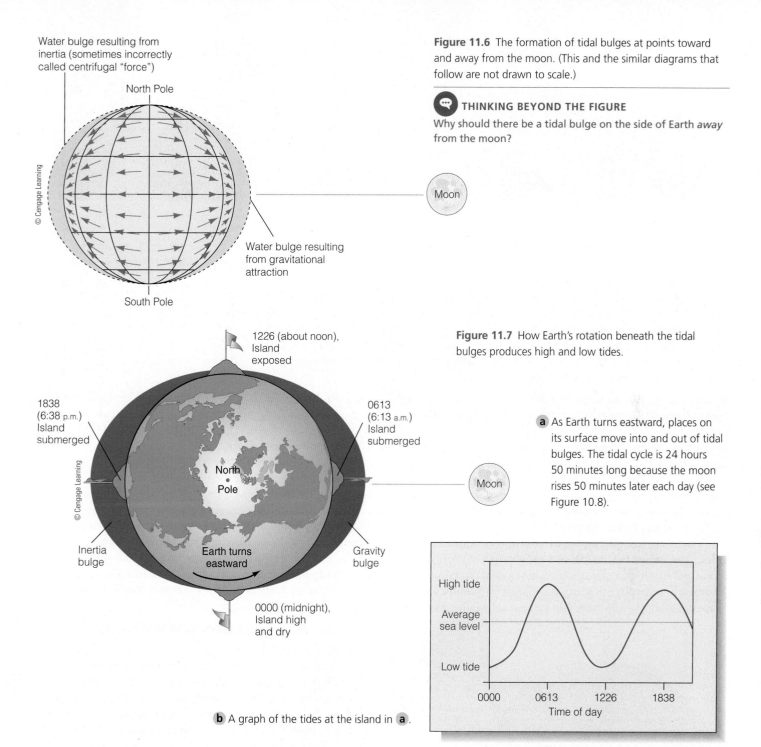

Water bulge resulting from inertia (sometimes incorrectly called centrifugal "force")

North Pole

© Cengage Learning

Water bulge resulting from gravitational attraction

South Pole

Figure 11.6 The formation of tidal bulges at points toward and away from the moon. (This and the similar diagrams that follow are not drawn to scale.)

💬 **THINKING BEYOND THE FIGURE**

Why should there be a tidal bulge on the side of Earth *away* from the moon?

Moon

1226 (about noon), Island exposed

1838 (6:38 p.m.) Island submerged

0613 (6:13 a.m.) Island submerged

© Cengage Learning

North Pole

Inertia bulge

Earth turns eastward

Gravity bulge

0000 (midnight), Island high and dry

Figure 11.7 How Earth's rotation beneath the tidal bulges produces high and low tides.

a As Earth turns eastward, places on its surface move into and out of tidal bulges. The tidal cycle is 24 hours 50 minutes long because the moon rises 50 minutes later each day (see Figure 10.8).

Moon

High tide

Average sea level

Low tide

0000   0613   1226   1838

Time of day

b A graph of the tides at the island in **a**.

of inertia exactly equal and opposite. The solid Earth cannot move much in response to these forces, but the fluid atmosphere and ocean can. We don't notice the changes in the height of the atmosphere, but changes in water level are visible to coastal observers. In **Figure 11.6**, tractive forces pull water toward a point beneath the moon and to a point opposite the moon.

How do these bulges cause the rhythmic rise and fall of the tides? In the idealized equilibrium model we are discussing, the bulges tend to stay aligned with the moon as Earth spins around its axis. **Figure 11.7** shows the situation in Figure 11.6 as it would look from above the North Pole. As Earth turns eastward, an island on the equator is seen to move in and out of these bulges

through one rotation (1 day). The bulges are the crests of the planet-sized waves that cause **high tides. Low tides** correspond to the troughs, the area between bulges. Starting at 0000 (midnight), we see the island in shallow water at low tide. Around 6 hours later, at 0613 (6:13 a.m.), the island is submerged in the lunar bulge at high tide. At 1226 (about noon), the island is within the tide-wave trough at low tide. At 1838 (6:38 p.m.) the island is again submerged, this time in the opposite crest caused by inertia. About an hour after midnight (0050) on the next day, the island is back in shallow water where it began.

The wave crests and troughs that cause high and low tides are actually very small: a 2-meter (7-foot) rise or fall in sea level

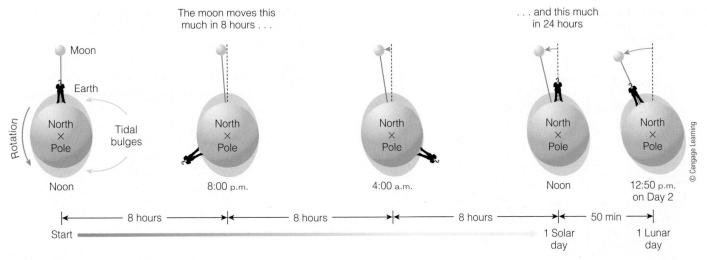

. . . and this much in 24 hours

Moon

Earth

Rotation

North × Pole   Tidal bulges

Noon

North × Pole

8:00 p.m.

North × Pole

4:00 a.m.

North × Pole

Noon

North × Pole

12:50 p.m. on Day 2

© Cengage Learning

|← 8 hours →|← 8 hours →|← 8 hours →|← 50 min →|

Start   1 Solar day   1 Lunar day

Figure 11.8 A lunar day is longer than a solar day. A lunar day is the time that elapses between the time the moon is highest in the sky and the *next* time it is highest in the sky. In a 24-hour solar day, the moon moves eastward about 12.2°. Earth must rotate another 12.2°—50 minutes—to again place the moon at the highest position overhead. A lunar day is therefore 24 hours 50 minutes long. Because Earth must turn an additional 50 minutes for the same tidal alignment, lunar tides usually arrive 50 minutes later each day.

is insignificant in comparison to the ocean's great size. Earth rotates beneath the bulges (tide wave crests) at about 1,600 kilometers (1,000 miles) per hour at the equator. The bulges appear to move across the ocean surface at this speed in an attempt to keep up with the moon. Theoretically, the wavelength of these tide waves is as long as 20,000 kilometers (12,500 miles)! The bulges tend to stay aligned with the moon as Earth spins around its axis. *The key to understanding the equilibrium theory of tides is to see Earth turning beneath these bulges.*

There are complications, of course. For example, **lunar tides**, tides caused by gravitational and inertial interaction of the moon and Earth, complete their cycle in a tidal day (also called a lunar day). A complete tidal day is 24 hours 50 minutes long, because the moon, which exerts the greatest tidal influence, moves through about $\frac{1}{27}$ of its orbit in a day. A point on Earth needs to move a bit more to be underneath it the next day, and this takes about 50 minutes **(Figure 11.8)**. Thus, the highest tide also arrives 50 minutes later each day.

Another complication arises from the fact that the moon does not stay right over the equator; each month, it moves from a position as high as 28½° above Earth's equator to 28½ below.[1] When the moon is above the equator, the bulges are offset accordingly **(Figure 11.9)**. When the moon is 28½° north of the equator, an island north of the equator will pass through the bulge on one side of Earth but miss the bulge on the other side. During 1 day, the island passes through a very high tide, a low tide, a lower high tide, and another low tide. This is shown in **Figure 11.10**.

The Sun Also Generates Tractive Forces

The sun's gravity also attracts particles on Earth. Remember that closeness counts for much in determining the strength of gravitational attraction. As we saw earlier, the sun is about 27 million times as massive as the moon but about 387 times as far from Earth as the moon, so the sun's influence on the tides is only 46% that of the moon's. The sun's tractive forces develop in the same way as the moon's, and the smaller solar bulges tend to follow the sun through the day. These are the **solar tides**, caused by the gravitational and inertial interaction of the sun and Earth.

Like the moon, the sun also appears to move above and below the equator (23½° north to 23½° south, as you may recall from Chapter 8), so the position of the solar bulges varies like that of the lunar bulges. Earth revolves around the sun only once a year, however, so the position of the solar bulges above or below the equator changes much more slowly than the position of the lunar bulges. (Figure 8.8, used to explain the cause of the seasons, shows this well.)

N

Moon

© Cengage Learning

S

Figure 11.9 Tidal bulges follow the moon. When the moon's position is north of the equator, the gravitational bulge toward the moon is also located north of the equator, and the opposite inertial bulge is below the equator. (Compare with Figure 11.6.)

---

[1]If Earth, moon, and sun were all moving in the same plane, lunar and solar eclipses would happen every 2 weeks.

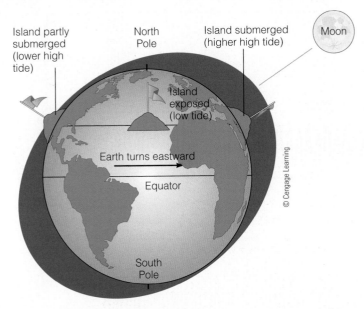

Figure 11.10 How the changing position of the moon relative to the Earth's equator produces higher and lower high tides. Sometimes the moon is below the equator; sometimes it is above.

💬 THINKING BEYOND THE FIGURE

Why are there *high* high tides and *low* high tides—that is, why are high tides sometimes very high, and not so high at other times?

Sun and Moon Influence the Tides Together

The ocean responds simultaneously to inertia and to the gravitational force of both the sun and moon. If Earth, moon, and sun are all in a line (as shown in **Figure 11.11a**), the lunar and solar tides will be additive, resulting in higher high tides and lower low tides.

But if the moon, Earth, and sun form a right angle (as shown in **Figure 11.11b**), the solar tide will tend to diminish the lunar tide. Because the moon's contribution is more than twice that of the sun, the solar tide will not completely cancel the lunar tide.

The large tides caused by the linear alignment of the sun, Earth, and moon are called **spring tides** (*springen*, "to move quickly"). During spring tides, high tides are very high and low tides very low. These tides occur at 2-week intervals corresponding to the new and full moons. (Please note that spring tides don't happen only in the spring of the year.) **Neap tides** (*næpa*, "hardly disturbed") occur when the moon, Earth, and sun form a right angle. During neap tides, high tides are not very high and low tides not very low. Neap tides also occur at 2-week intervals, with the neap tide arriving a week after the spring tide. **Figure 11.12** plots tides at two coastal sites through spring and neap cycles.

Because their orbits are ellipses, not perfect circles, the moon and the sun are closer to Earth at some times than at others. The difference between **apogee** (the moon's greatest distance from Earth) and **perigee** (its closest approach) is 30,600 kilometers (19,015 miles). Because the tidal force is inversely proportional to the cube of the distance between the bodies, the closer moon raises a noticeably higher tidal crest. The difference between **aphelion** (Earth's greatest distance from the sun) and **perihelion** (its closest approach) is 3.7 million kilometers (2.3 million miles). If the moon and sun are over nearly the same latitude, and if Earth is also close to the sun, extreme spring tides will result. Interestingly, spring tides will have greater ranges in the Northern Hemisphere winter than in the Northern Hemisphere summer because Earth is closest to the sun during the northern winter.

Tides caused by inertia and the gravitational force of the sun and moon are called **astronomical tides**. As explained in

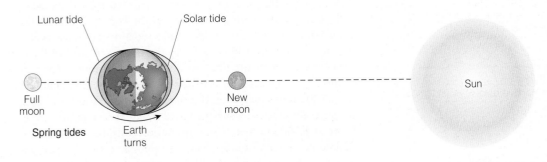

a At the new and full moons, the solar and lunar tides reinforce each other, making spring tides, the highest high and lowest low tides.

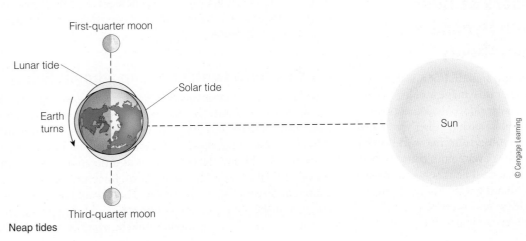

b At the first- and third-quarter moons, the sun, Earth, and moon form a right angle, creating neap tides, the lowest high and the highest low tides.

Figure 11.11 Relative positions of the sun, moon, and Earth during spring and neap tides.

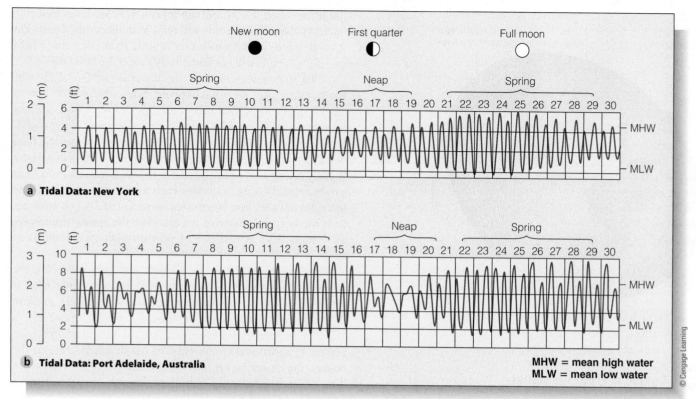

Figure 11.12 Tidal records for a typical month at (a) New York and (b) Port Adelaide, Australia. Note the relationship of spring and neap tides to the phases of the moon.

the next section, storms can affect tide height—a phenomenon known as a *meteorological tide*.

11.3 The Dynamic Theory of Tides Adds Fluid Motion Dynamics to the Equilibrium Theory

Newton knew his explanation was incomplete. For one thing, the maximum theoretical range of a lunar tidal bulge is only 55 centimeters (about 22 inches) and of a solar tide, only 24 centimeters (about 10 inches), both considerably smaller than the 2-meter (7-foot) average tidal range we observe in the world ocean. The reason is that the ocean surface never comes completely to the equilibrium position at any instant. The moon and the sun change their positions so rapidly that the water cannot keep up.

The **dynamic theory of tides**, first proposed in 1775 by Laplace, added a fundamental understanding of the problems of fluid motion to Newton's breakthrough in celestial mechanics. The dynamic theory explains the differences between predictions based on Newton's model and the observed behaviors of tides.

Remember that tides are a form of *wave*. The crests of these waves—the tidal bulges—are separated by a distance of half of Earth's circumference (see again Figure 11.7). In the equilibrium model, the crests would remain stationary, pointing steadily toward (or away from) the moon (or sun) as Earth turned beneath them. They would appear to move across the idealized water-covered Earth at a speed of about 1,600 kilometers (1,000 miles) per hour. But how deep would the ocean have to be to allow these waves to move freely? For a tidal crest to move at 1,600 kilometers per hour, the ocean would have to be 22 kilometers (13.7 miles) deep. As you may recall, the average depth of the ocean is only 3.8 kilometers (2.4 miles). So, tidal crests (tidal bulges) move as forced waves, their velocity determined by ocean depth.

Tidal Patterns Center on Amphidromic Points

This behavior of tides as *shallow-water waves* is only one variation from the ideal that the dynamic theory explains. The continents also get in the way. As Earth turns, landmasses obstruct the tidal crests, diverting, slowing, and otherwise complicating their movements. This interference produces different patterns in the arrival of tidal crests at different places. Imagine, for example, a continent directly facing the moon. There would be no oceanic bulge, and the shores of the continent would experience high tide. A few hours later the moon would be over the ocean. When the continent was not aligned with the moon, but the ocean was, the tidal bulge would reform and the continent's edges would experience low tide.

The shape of the basin itself has a strong influence on the patterns and heights of tides. As we have seen, water in large basins can rock rhythmically back and forth in seiches. Though they are small, tidal crests can stimulate this resonant oscillation, and the configuration of coasts around a basin can alter its rhythm.

For these and other reasons, some coastlines experience **semidiurnal** (twice daily) **tides**: two high tides and two low tides of nearly equal level each lunar day. Others have **diurnal** (daily) **tides**: one high and one low. The tidal pattern is called **mixed** (or **semidiurnal mixed**) if successive high tides or low tides are of significantly different heights throughout the cycle. This pattern is caused by blending diurnal and semidiurnal tides.

Figure 11.13 shows an example of each tidal pattern. The Pacific coast of the United States has a mixed tidal pattern: often a *higher* high tide, followed by a *lower* low tide, a *lower* high tide,

and a *higher* low tide each lunar day. (That's not as confusing as it seems—see Figure 11.13a.) The natural tendency of water in an enclosed ocean basin to rock at a specific frequency modifies the pattern in the Gulf of Mexico, so Mobile sees one crest per lunar day, a diurnal pattern (see Figure 11.13b). At Cape Cod, two tidal crests arrive per lunar day, a semidiurnal pattern (see Figure 11.13c). The Pacific Ocean has a unique pattern of diurnal, semidiurnal, and mixed tides. As Figure 11.13d shows, it has the most complex of all tidal patterns. The east coast of Australia, all of New Zealand, and much of the west coasts of Central and South America have a semidiurnal tidal pattern. The Aleutians have diurnal tides. The Pacific coasts of North America and some of South America have mixed tides. *Why the differences?*

Remember the surface of Lake Geneva in our discussion of seiches? The water level at the center of the lake remains at the same height while water at the ends rises and falls (see again Figure 10.28). The long axis of the lake stretches east and west. Because of the Coriolis effect, water moving east at the center of the lake is deflected slightly to the right (to the south). If the lake

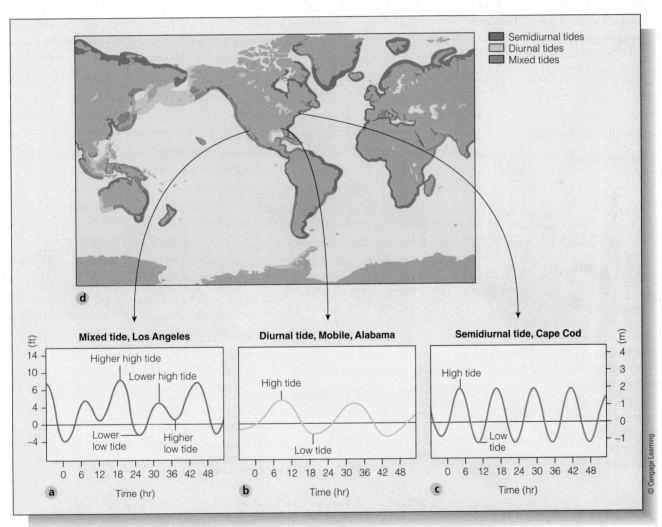

Figure 11.13 Tide curves for the three common types of tides. **a** A mixed tide pattern at Los Angeles, California. **b** A diurnal tide pattern at Mobile, Alabama. **c** A semidiurnal tide pattern at Cape Cod, Massachusetts. **d** The worldwide geographical distribution of the three tidal patterns. Most of the world's ocean coasts have semidiurnal tides.

were larger and the flow of water greater (and thus the Coriolis effect stronger), water would hug the southern shore as it traveled eastward. When the water began to rock the other way—back to the west—the Coriolis effect would move the water to the right toward (and along) the northern shore. Note that the overall movement of water would be counterclockwise.

Water moving in a tide wave tends to stay to the right of an ocean basin for the same reason. As water moves north in a Northern Hemisphere ocean, it moves toward the eastern boundary of the basin; as it moves south, it moves toward the western boundary. A wave crest moving counterclockwise will develop around a node if this motion continues to be stimulated by tidal forces. This rotary motion is shown in **Figure 11.14**.

The node (or nodes) near the center of an ocean basin is called an **amphidromic point** (*amphi*, "around"; *dromas*, "running"). An amphidromic point is a no-tide point in the ocean, around which the tidal crest rotates through one tidal cycle. Because of the shape and placement of landmasses around ocean basins, the tidal crests and troughs cancel each other at these points. The crests sweep around amphidromic points like wheel spokes from a rotating hub, radiating crests toward distant shores. Tide waves are influenced by the Coriolis effect because a large volume of water moves with the waves. They move counterclockwise around the amphidromic point in the Northern Hemisphere and clockwise in the Southern Hemisphere. The height of the tides increases with distance from an amphidromic point.

About a dozen amphidromic points exist in the world ocean; **Figure 11.15** shows their locations. Notice the complexity of the Pacific, which contains five. It's no wonder that the arrival of tide wave crests at the Pacific's edges produces such a complex mixture of tide patterns, depending on shoreline location.

The Tidal Reference Level Is Called the Tidal Datum

The reference level to which tidal height is compared is called the **tidal datum**. The tidal datum is the zero point (0.0) seen in tide graphs such as Figures 11.12 and 11.13. This reference plane is not always set at **mean sea level**, which is the height of the ocean surface averaged over a few years' time. On coasts with mixed tides, the zero tide level is the average level of the lower of the two daily low tides (*mean lower low water*, or MLLW). On coasts with diurnal and semidiurnal tides, the zero tide level is the average level of all low tides (*mean low water*, MLW).

Tidal Patterns Vary with Ocean Basin Shape and Size

The **tidal range** (high-water to low-water height difference) varies with basin configuration. In small areas such as lakes, the tidal range is small. In larger enclosed areas such as the Baltic and Mediterranean seas, the tidal range is also moderate. The tidal range is not the same over a whole ocean basin; it var-

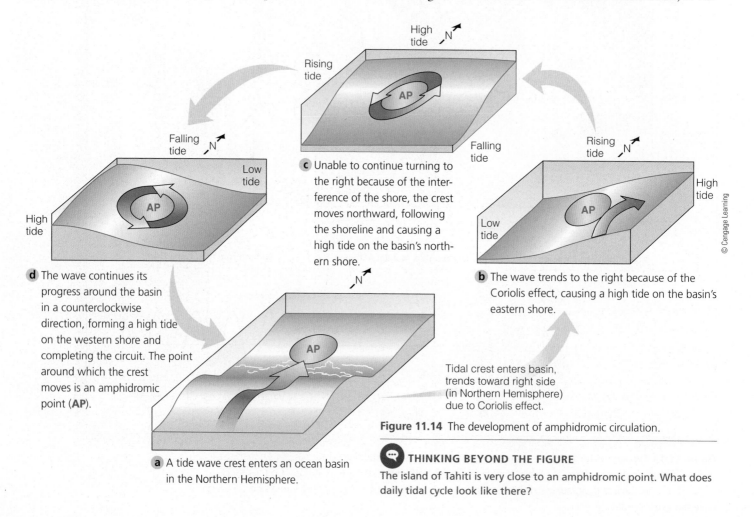

c Unable to continue turning to the right because of the interference of the shore, the crest moves northward, following the shoreline and causing a high tide on the basin's northern shore.

d The wave continues its progress around the basin in a counterclockwise direction, forming a high tide on the western shore and completing the circuit. The point around which the crest moves is an amphidromic point (**AP**).

a A tide wave crest enters an ocean basin in the Northern Hemisphere.

b The wave trends to the right because of the Coriolis effect, causing a high tide on the basin's eastern shore.

Tidal crest enters basin, trends toward right side (in Northern Hemisphere) due to Coriolis effect.

© Cengage Learning

Figure 11.14 The development of amphidromic circulation.

💬 **THINKING BEYOND THE FIGURE**
The island of Tahiti is very close to an amphidromic point. What does daily tidal cycle look like there?

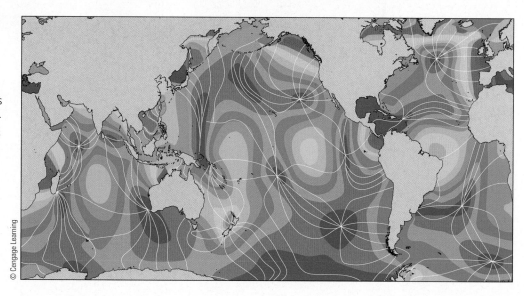

Figure 11.15 Amphidromic points in the world ocean. Tidal ranges generally increase with increasing distance from amphidromic points. The colors indicate where tides are most extreme (highest highs, lowest lows), with blues being least extreme. White lines radiating from the points indicate tide waves moving around these points. In almost a dozen places on this map, the lines converge.

Notice how at each of these places the surrounding color—the tidal force for that region—is blue, indicating little or no apparent tide. These convergent areas are called amphidromic points. Tide waves move around these points, counterclockwise in the Northern Hemisphere and clockwise in the Southern Hemisphere.

© Cengage Learning

ies from the coasts to the centers of oceans. The largest tidal ranges occur at the edges of the largest ocean basins, especially in bays or inlets that concentrate tidal energy because of their shape.

If a basin is wide and symmetrical, like the Gulf of St. Lawrence in eastern Canada, a miniature amphidromic system develops that resembles the large systems of the open ocean (**Figure 11.16**). If the basin is narrow and restricted, the tide wave crest cannot rotate around an amphidromic point and simply moves into and out of the bay (**Figure 11.17a**). Extreme tides occur in places where arriving tide crests stimulate natural oscillation periods of around 12 or 24 hours. In rare cases, water in the bay naturally resonates (seiches) at the same frequency as the lunar tide (12 hours 25 minutes). This rhythmic sloshing results in extreme tides. In the eastern reaches of the Bay of Fundy near Moncton, New Brunswick (Canada), the tidal range is especially great: up to 15 meters (50 feet) from highs to lows (**Figures 11.17b** and **c**). The northern reaches of the Sea of Cortez east of Baja California have a tidal range of about 9 meters (30 feet). Tide waves sweeping toward the narrow southern end of the North Sea can build to great heights along the southeastern coast of England and the northern coast of France.

If conditions are ideal, a **tidal bore** (*bara*, "wave") will form in some inlets (and their associated rivers) exposed to great tidal fluctuation. Here, at last, is a true **tidal wave**—a steep wave moving upstream generated by the action of the tide crest in the enclosed area of a river mouth (see this chapter's opener and **Figure 11.18**). The confining river mouth forces the tide wave to move

> The tidal range is not the same over a whole ocean basin; it varies from the coasts to the centers of oceans. The largest tidal ranges occur at the edges of the largest ocean basins, especially in bays or inlets that concentrate tidal energy because of their shape.

toward land at a speed that exceeds the theoretical shallow-water wave speed for that depth. The forced wave then breaks, forming a spilling wave front that moves upriver. Though most are less than 1 meter (3 feet) high, some bores may be up to 8 meters (26 feet) high and move 11 meters per second (25 miles per hour). Their potential danger is lessened by their predictability. Accurately predicting the arrival of tidal bores is essential to safe navigation. In addition to those in southwestern China and the Bay of Fundy, tidal bores are common in the Amazon, the Ganges Delta, and England's River Severn.

Tide Waves Generate Tidal Currents

The rise or fall in sea level as a tide crest approaches and passes will cause a **tidal current** of water to flow into or out of bays and harbors. Water rushing into an enclosed area because of the rise in sea level as a tide crest approaches is called a **flood current**. Water rushing out because of the fall in sea level as the tide trough approaches is called an **ebb current**. (The terms *ebb tide* and *flood tide* have no technical meaning.) Tidal currents reach maximum velocity midway between high tide and low tide. **Slack water**, a time of no currents, occurs at high and low tides when the current changes direction.

Anyone who has stood at the narrow mouth of a large bay or harbor cannot help being impressed with the speed and volume of the tidal current that occurs between tidal extremes. Midway between high and low spring tides, the ebb current rushing from San Francisco Bay strikes the base of the south tower of the Golden Gate Bridge with such force that a bow wave is formed,

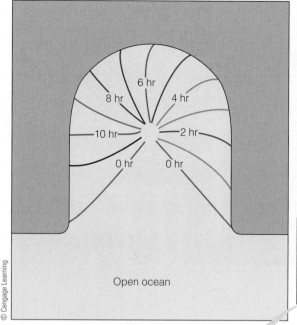

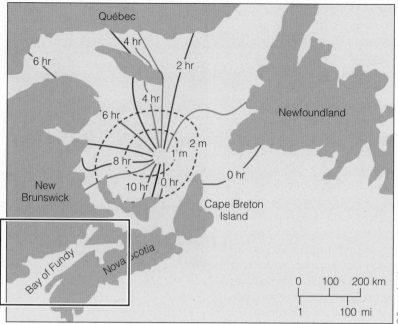

a An imaginary amphidromic system in a broad, shallow basin. The numbers indicate the hourly positions of tide crests as a cycle progresses.

Figure 11.16 Tides in broad confined basins.

b The amphidromic system for the Gulf of St. Lawrence between New Brunswick and Newfoundland, southeastern Canada. Dashed lines show the tide heights when the tide crest is passing.

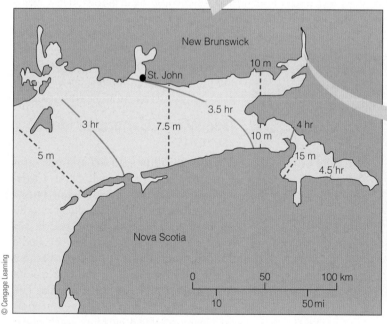

a True amphidromic systems do not develop in narrow basins because there is no space for rotation. Tides in the narrow Bay of Fundy, Nova Scotia, are extreme because water in the bay naturally resonates (seiches) at the same frequency as the lunar tide.

Tidal range there can be 15 meters (50 feet) (**b** and **c**). At the peak of the flood, water rises 1 meter (3.3 feet) in 23 minutes! As you saw in Figure 11.1, forewarned is forearmed!

Figure 11.17 Tides in narrow, restricted basins.

Figure 11.18 An onrushing tidal bore overwhelms the bank of China's Qiantang River. The time of the bore's arrival can be very accurately predicted. The phenomenon draws tourists and can be dangerous. Typhoon Trami intensified the bore's effects in 2013 (seen here).

giving the convincing illusion that the bridge itself is moving rapidly. Tidal currents at the Golden Gate can reach 3 meters per second (about 7 miles per hour) because of the volume of enclosed water and the narrowness of the channel through which it must escape. Navigators must know the times of tidal currents to safely negotiate any harbor entrance or other narrow strait—in some places, their lives may depend on this knowledge.

Tidal currents become more complex in the open sea. One's position relative to an amphidromic point, the shape of the basin, and the magnitude of gravitational forces and inertia must all be considered to calculate the speed and direction of tidal currents over a deep bottom. The velocity of tidal currents is less in the open sea because the water is not confined, as it is in a harbor. The speed of open-sea tidal currents has been measured at a few centimeters per second, and their velocity tends to decrease with depth.

Tidal Friction Slows Earth's Rotation

The daily rise and fall of the tides consumes a very large amount of energy, and this energy is ultimately dissipated as heat. Most of this energy comes directly from the rotation of Earth itself, and tidal friction is gradually slowing Earth's rotation by a few hundredths of a second per century. Even such a small change has long-term planetary effects, however. Geologists studying the daily growth rings of fossil corals and clams estimate that the days have grown longer, so the number of days in a year has decreased as planetary rotation has slowed.

Evidence suggests that 350 million years ago, a year contained between 400 and 410 days, with each day being about 22 hours long; and 280 million years ago, there were about 390 days in a year, each about 22 ½ hours long.

Tidal friction affects other bodies. Tidal forces have locked the rotation of the moon to that of Earth. As a result, the same side of the moon is always facing Earth, and a day on the moon is a month long.

CONCEPT CHECK

Before going on to the next section, check your understanding of some of the important ideas presented so far:

How does the equilibrium theory of tides differ from the dynamic theory?

Are tides always shallow-water waves? Are they ever in "deep" water?

What tidal patterns are observed on the world's coasts?

Are there tides in the open ocean?

How does basin shape influence tidal activity?

What's a tidal bore?

11.4 Most Tides Can Be Accurately Predicted

There are at least 140 tide-generating and tide-altering forces and factors in addition to the ones we have discussed. About seven important factors must be considered if we wish to predict tides mathematically. The interactions of all the forces and factors are so complex that if a previously unknown continent were discovered on Earth, the coastal tide times and ranges on its shores could not be predicted accurately, because only the study of past records allows tide tables to be projected into the future. Experience permits prediction of tidal height to an accuracy of about 3 centimeters (1.2 inches) for years in advance **(Figure 11.19)**.

Even so, extraneous factors can affect the estimates. For example, the arrival of a storm surge will greatly affect the height or timing of a tide—as will gentle, atmospherically induced seiching of the basin or excitement of large-scale resonances by a tsunami. Even a strong, steady wind on-shore or offshore will affect tidal height and the arrival time of the crest. Weather-related alterations are sometimes called **meteorological tides** after their origin.

Systems for studying the tides range from the simple (teams of students chasing oranges dropped into a harbor) to the complex (lines of pressure-sensing transducers arrayed on the ocean floor). As our discussion has shown, the complexities of tidal behavior are so great—the number of variables is so large—that any new method of tracking and predicting tidal movements would be welcome. Researchers in Germany and the United States have recently found a way to study tides through extremely subtle variations in Earth's main magnetic field sensed by orbiting satellites.

How can this be done? The ocean is an electrically conducting fluid. As it flows through Earth's magnetic field, it generates a small but distinct secondary magnetic field that interferes with

Figure 11.19 Accurate tidal prediction is important. The Benedictine abbey of Mont-Saint-Michel was built on a small, rocky, tidal island off the coast of Normandy, France. The Mount is connected to the mainland by a thin, natural land bridge that, until recently, was covered at high tide and exposed at low tide. Tides in the area vary greatly, sometimes reaching a difference of 14 meters (46 feet) between high and low water. Victor Hugo described high tides coming "as swiftly as a galloping horse." Even today, visitors are occasionally drowned trying to walk to the abbey across the tidal flats.

the main field. The trace of the lunar semidiurnal tide has been identified from satellite observations, and efforts are under way to increase the sensitivity of the readings.

CONCEPT CHECK

Before going on to the next section, check your understanding of some of the important ideas presented so far:

What is a meteorological tide?

Can astronomical and meteorological tides interact?

11.5 Tidal Patterns Can Affect Marine Organisms

Not surprisingly, as pervasive a phenomenon as the tides has a significant influence on coastal marine life. Organisms that live between the high-tide and low-tide marks experience very different conditions from those that reside below the low-tide line. Within the intertidal zone itself, organisms are exposed to varying amounts of emergence and submergence. Because some organisms can tolerate many hours of exposure and others are able to tolerate only a very few hours per week or month, the animals

Figure 11.20 During spring and summer months, these small fish (genus *Leuresthes*) swim ashore at night in large numbers just after the highest spring tides, deposit and fertilize their eggs below the sand surface, and return to the sea. Nearly 2 weeks later, when spring tides return, the eggs hatch. No one is certain how grunion time their reproductive behavior so precisely to the tidal cycle. The grunion are found only along the Pacific coast of North America and in the Gulf of California. Unlike the Pacific coast species, Gulf of California grunion spawn during daylight.

 THINKING BEYOND THE FIGURE
How would this unusual reproductive behavior benefit these small fish?

and plants sort themselves into three or more horizontal bands, or subzones, within the intertidal zone. Each distinct zone is an aggregation of animals and plants best adapted to the conditions within that particular narrow habitat. The zones are often strikingly different in appearance, even to a person unfamiliar with shoreline characteristics. (This zonation is clearly evident in the rocky shore of Figure 16.8b.)

Less obvious are periodic visitors that time their arrivals and departures to the rise and fall of the tide. At low tide, the tiny diatom (one-celled alga) *Hantzschia* migrates upward through wet sand to photosynthesize at the sunlit surface. At the first hint of the returning tide, these plantlike organisms descend to a relatively safe depth in the sand, where they are protected from wave action. Fiddler crabs (genus *Uca*) return to their burrows at high tide to avoid marine predators but emerge at low tide to look for any bits of food the ocean might have deposited at their doorsteps. Filter-feeding animals such as sand crabs (*Emerita*) and bean clams (*Donax*) migrate up and down the beach to stay in the surf zone where the chaos can provide both food and protection (see Figure 16.10).

Among the most famous tide-driven visitors are the grunion (*Leuresthes*), a small fish named after the Spanish word *gruñon*, which means "grunter," a reference to the squeaking noise they sometimes make during spawning. From late February through early September, these small fish (which reach a length of 15 centimeters, or 6 inches) swim ashore at night in large numbers just after the highest spring tides, deposit and fertilize their eggs below the sand surface, and return to the sea (Figure 11.20). Safe from marine predators, the eggs develop and are ready to

Figure 11.21
This tidal power installation at Lake Sihwa in South Korea is the world's largest. At high tide, water flows through turbines from the ocean into the lake, generating 254 megawatts of electricity. ANDRITZ

hatch 9 days after spawning. After a few more days, the spring tides return, soak and erode the beach, and stimulate the eggs to hatch. No one is certain how grunion time their reproductive behavior so precisely to the tidal cycle, but some research suggests they sense very small changes in hydrostatic pressure caused by tidal change or that they visually time their spawning 3 or 4 nights following each full and new moon. Not surprisingly, these accommodating little fish were an important food for Native Americans.

CONCEPT CHECK

Before going on to the next section, check your understanding of some of the important ideas presented so far:

Zones of marine organisms can usually be seen along rocky shores. How might tidal patterns result in this sort of differential growth?

11.6 Power Can Be Extracted from Tidal Motion

Humans have found ways to use the tides. Ships sail to sea and return to port with the tides. Intentional grounding of a ship with the fall of a tide can provide a convenient, if temporary,

dry-dock. To these traditional uses has been added a potential alternative to our growing dependence on fossil fuels: taking advantage of trapped high-tide water to generate electricity.

Tidal power is the only marine energy source that has been successfully exploited on a large scale. The first major tidal power station was opened in 1966 in France on the estuary of the Rance River, where tidal range reaches a maximum of 13.4 meters (44 feet). Built at a cost of US$75 million, this dam, which is 850 meters (2,800 feet) long, contains 24 turbo-alternators capable of generating 544 million kilowatt-hours of electricity annually. At high tide, seawater flows from the ocean through the generators into the estuary. At low tide, the seawater and river water from the estuary flow out through the same generators. Power is generated in both directions.

A similar though larger facility has been built at Lake Sihwa near Inchon in South Korea **(Figure 11.21)**. The project came online in 2010 and is producing 254 megawatts of power. The power plant connects the ocean to Sihwa Lake through a long seawall. At Sihwa, power is generated on tidal inflow only—outflow bypasses the turbines.

Another way to generate tidal power is through systems like those being developed by Marine Current Turbines of Bristol, England **(Figure 11.22)**. Unlike the Rance project, these turbines are submerged in open water. The first large generators were

b The turbine's size is evident in this photograph. The machinery is being built by Harlan and Wolff, the Belfast shipyard that built the RMS *Titanic*.

a An artist's impression of the tidal turbine installation in Northern Ireland. Pairs of turbines up to 20 meters (66 feet) in diameter rotate slowly in opposite directions as the tidal current passes them.

Figure 11.22 Tidal currents are being exploited to provide electrical power.

placed in the narrow entrance to Strangford Lough, Northern Ireland, in August 2007. Their vast size, seen in Figure 11.22b, will allow a set of turbines to generate 1.2 megawatts of clean electrical power as water flows from the Irish Sea through the gap, enough for about 1,000 homes. Larger facilities are planned around the United Kingdom and Scandinavia.

Tidal power has many advantages: operating costs are low, the source of power is free, and no carbon dioxide or other pollutants are added to the atmosphere. But even if tidal power sta-

tions were built at every appropriate site worldwide, the power generated would amount to less than 1% of current world needs.

CONCEPT CHECK

Before going on to the next section, check your understanding of some of the important ideas presented so far:

Where is electrical power being generated from tidal movement?

Why isn't tidal power being developed more aggressively?

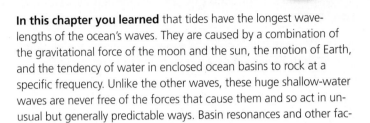

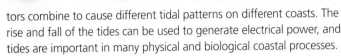

In this chapter you learned that tides have the longest wavelengths of the ocean's waves. They are caused by a combination of the gravitational force of the moon and the sun, the motion of Earth, and the tendency of water in enclosed ocean basins to rock at a specific frequency. Unlike the other waves, these huge shallow-water waves are never free of the forces that cause them and so act in unusual but generally predictable ways. Basin resonances and other factors combine to cause different tidal patterns on different coasts. The rise and fall of the tides can be used to generate electrical power, and tides are important in many physical and biological coastal processes.

In the next chapter you will learn how the interaction of wind, waves, and weather affects the edges of the land—the coasts. Coasts are complex, dynamic places where the only constant is change.

QUESTIONS FROM STUDENTS

1. Has anybody surfed a tidal bore?

Yes, indeed. Some are too large and unruly to be surfed—the Qiantang "silver dragon" (seen at its most violent in this chapter's opener and Figure 11.18) has never been ridden for more than 11 seconds, and the non-local surfer was pretty badly beaten up in the process. The Amazon bore (the pororoca, which means "great noise" in the local language) is a bit smaller but adds a new level of difficulty in the presence of small parasitic fishes (genus *Vandellia*) that can swim up a surfer's urethra, erect spines, and require surgical removal. (No, we're *not* making this up.) And then there are the alligators and bugs and bitey fish. The record Amazon ride was 37 *very* wary minutes **(Figure 11.23)**.

2. Are there tides in the solid Earth?

Yes. Even Earth isn't stiff enough to resist the tidal pulls of the moon and sun. Bulges occur in the solid Earth just as they appear in the ocean or the atmosphere. The crests (bulges) of Earth are, of course, much smaller; 25 to 30 centimeters (10–12 inches) is about average. They pass unnoticed beneath us twice a day.

Figure 11.23 Brazilian surfer Alex Salazar rides the turbulent pororoca, a tidal bore in the Amazon River. He set a record for the longest wave ever surfed: 37 minutes and 12.6 kilometers (7.8 miles).

J rgen Skarwan/Reuters/Corbis

The atmosphere also rises and falls in a tidal rhythm. Tidal variability in the height of the atmosphere has been measured in miles.

3. Are there tides in lakes?

Yes, but they are very, very small. In the case of Lake Michigan (one of the world's largest lakes), tidal variance is less than 1 centimeter (½ an inch). In ocean tides, the great mass of the ocean is spread across a relatively large span of the globe and can be acted upon by the combination of gravity and inertia you learned about in this chapter. The spread of water mass in a lake—even a large one—covers a much smaller area. The mass of water in the lake is small, so the effects of gravity and inertia are also small.

4. Are there tides in a glass of water?

Yes. They're too small to detect, but equilibrium tides do exist. Each molecule of water in the glass responds to the same planetary forces that affect molecules in the ocean.

5. In my newspaper, one of today's low tides is listed as 1.0, 1445. What does that mean?

In the United States, it means that the water will be 1.0 foot below tidal datum (that is, below mean lower low water [MLLW]—the long-term average position of the lower of the daily low tides) at 2:45 p.m. local time. This might be a good afternoon to spend at the shore digging for clams because a tide that low would expose intertidal organisms only rarely seen above water. See Figure 16.8a for more information on the relationship between tidal height and exposure.

6. A TV news reporter said to expect "astronomical high tides" tonight. Should we pack our stuff and head for the hills?

Not necessarily. The reporter is calling attention to an alignment of the sun and moon that produces a high spring tide. "Astronomical" here refers to an alignment of heavenly bodies; it is not synonymous with "gigantic" or "spectacular."

7. What is a "king tide"?

Spring tides have greater ranges in the northern hemisphere winter than in the northern hemisphere summer because Earth is closest to the sun during the northern winter. Australians and New Zealanders refer to these very high tides as *king tides*. This term is slowly migrating to North America.

TERMS AND CONCEPTS TO REMEMBER

amphidromic point	flood current	neap tide	$T = G\left(\dfrac{m_1 m_2}{r^3}\right)$
aphelion	high tide	perigee	
apogee	low tide	perihelion	tidal bore
astronomical tide	lunar tide	semidiurnal tide	tidal current
diurnal tide	mean sea level	slack water	tidal datum
dynamic theory of tides	meteorological tide	solar tide	tidal range
ebb current	mixed tide (or semidiurnal	spring tide	tidal wave
equilibrium theory of tides	mixed tide)		tide

STUDY QUESTIONS

Thinking Critically

1. Though they move through all the ocean, tides are referred to as shallow-water waves. How can that be?
2. What are the most important factors influencing the heights and times of tides? What tidal patterns are observed? Are there tides in the open ocean? If so, how do they behave?
3. How does the latitude of a coastal city affect the tides there—or does it?
4. From what you learned about tides in this chapter, where would you locate a plant that generated electricity from tidal power? What would be some advantages and disadvantages of using tides as an energy source?

Thinking Analytically

1. If you live near a coast, find a source of local tidal data (newspaper, Internet site, pamphlet from bait shop, and so on). Plot the rise and fall of the tides for 2 weeks. Note the cycle, and point out spring and neap tides. How does the tide correlate with the position of the moon and sun?
2. You know that tides always act as shallow-water waves. What is the speed of a tide wave in the open ocean? Assume the average depth of the ocean is 4,000 meters.
3. What would tides be like near an amphidromic node? In Tahiti, for instance.
4. What difficulties do you imagine intertidal organisms would need to overcome in order to be successful? (We will cover this in a later chapter, but the physics of the tides makes the question very interesting.)

GLOBAL GEOSCIENCE WATCH

Visit www.cengagebrain.com to access MindTap, a complete digital course which includes access to Global Geoscience Watch and more. Within the GREENR database, search for "tidal power" and research the potential and challenges associated with generating commercial energy from the tides. Develop a position statement detailing your opinion on whether we should expand the use of tidal power. Describe your position in a two- to three-page essay that uses information from your research to support your views.

CENGAGE brain.com Visit www.cengagebrain.com to access course materials for this text, including interactive learning tools, videos, and more.

TIDES 333

12 Coasts

KEY CONCEPTS

The *location* of a coast depends primarily on global tectonic activity and the ocean's water volume. Tom Garrison

The *shape* of a coast is a product of uplift and subsidence, the wearing down of land by erosion, and the redistribution of material by sediment transport and deposition. Sunset Avenue Productions/Digital Vision/Getty Images

Coasts may be classified as erosional coasts (on which erosion dominates) or depositional coasts (on which deposition dominates). University of Washington Libraries, Special Collections, John Shelton Collection, KC9589

Beaches change shape and volume as a function of wave energy and the balance of sediment input and removal. Tom Garrison

Human interference with coastal processes has generally accelerated the erosion of coasts near inhabited areas. ANAM Collection/Alamy

Coasts are places of rapid change, tenacious organisms, and sometimes great beauty. They can also be strange. Giant's Causeway in Northern Ireland qualifies as strange: It is an area of about 40,000 interlocking hexagonal basaltic columns, the result of slow cooling of lava from a volcanic eruption that occurred about 60 million years ago. The tallest columns are 12 meters (39 feet) high. The sounds of waves echoing off the basaltic shafts, the play of light and shadow among the formations, and the unusual regularity of a normally irregular place combine to give the visitor pause. . . .
Chris Hill/National Geographic Creative

12.1 Coasts Are Shaped by Marine and Terrestrial Processes

Coastal areas join land and sea. Our personal experience with the ocean usually begins at the coast. Have you ever wondered why a coast is in a particular location or why it is shaped as you see it? These temporary, often beautiful junctions of land and sea are subject to rearrangement by waves and tides, by gradual changes in sea level, by biological processes, and by tectonic activity.

The place where ocean meets land is usually called the **shore**, and the term **coast** refers to the larger zone affected by the processes that occur at this boundary. A sandy beach might form the shore in an area, but the coast (or coastal zone) includes the marshes, sand dunes, and cliffs just inland of the beach, as well as the **sandbars** and troughs immediately offshore. The world ocean is bounded by about 1,000,000 kilometers (620,000 miles) of shore, about one third of which are sandy beaches **(Figure 12.1)**.

Because of its proximity to both ocean and land, a coast is subject to natural events and processes common to both realms. A coast is an active place. Here is the battleground on which wind waves break and expend their energy. Tides sweep water on and off the rim of land, rivers drop most of their sediments at the coasts, and ocean storms pound the continents. The *location* of a coast depends primarily on global tectonic activity and the volume of water in the ocean. The *shape* of a coast is a product of many processes: uplift and subsidence, the wearing down of land by **erosion**, and the redistribution of material by sediment transport and deposition.

As we saw in Chapter 3, no area of geology has been left undisturbed by the revelations of plate tectonics. In the 1960s, geologists began to classify coasts according to their tectonic position. *Active* coasts, near the leading edge of moving continental plates, were found to be fundamentally different from the more *passive* coasts near trailing edges. The shapes, compositions, and ages of coasts are better understood by taking plate movements into account. But as we'll see, the slow forces of plate movement

are frequently obscured by the more rapid action of waves, by the erosion of land, and by the transport of sediments.

Another important consideration in understanding coasts is long-term change in sea level. Five factors can cause sea level to change. Three of these factors are responsible for **eustatic change**—variations in sea level that can be measured all over the world ocean:

- The amount of water in the world ocean can vary. Sea level is lower during periods of global glaciation (ice ages) because there is less water in the ocean. It is higher during warm periods, when the glaciers are smaller. Periods of abundant volcanic outgassing can also add water to the ocean and raise sea level.
- The volume of the ocean's "container" may vary. High rates of seafloor spreading are associated with the expansion in volume of the oceanic ridges. This expansion displaces the ocean's water, which climbs higher on the edges of the continents. Sediments shed by the continents during periods of rapid erosion can also decrease the volume of ocean basins and raise sea level.
- The water itself may occupy more or less volume as its temperature varies. During times of global warming, seawater expands and occupies more volume, raising sea level.

Of course, the continents rarely stay still as sea level rises and falls. Local changes are bound to occur, and two other factors produce variations in *local* sea level:

- Tectonic motions and isostatic adjustment can change the height and shape of a coast. Coasts can experience uplift as lithospheric plates converge or can be weighed down by masses of ice during a period of widespread glaciation. The continents slowly rise when the ice melts.
- Wind and currents, seiches, storm surges, an El Niño or La Niña event, and other effects of water in motion can force water against the shore or draw it away.

Sea level has been at its current elevation (give or take 0.5 meter, or 1.5 feet) for only about 2,500 years. Over the past 2 million years, worldwide sea level has varied from about 6 meters (20 feet) above to about 125 meters (410 feet) below its present position. The recent low point occurred about 18,000 years ago at the height of the most recent glaciation. Indeed, sea level has been at the modern "high" only rarely in the past 2 million years—the dominant state for Earth is a much lower eustatic sea-level position **(Figure 12.2a)**. It is important to realize that coastlines have not yet come into equilibrium with modern sea level and that an accelerating rate of sea-level rise probably lies ahead **(Figure 12.2b)**.

Figure 12.1 A high-energy shore on the American northwest coast. Violent forces of wind and waves, tidal range and currents, intruding roots, and surging rivers assure its features will be transitory.

Ocean/Corbis

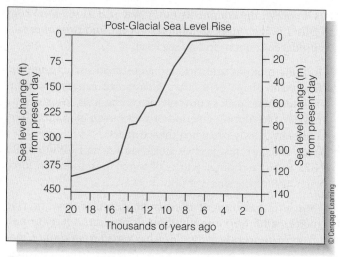

a Sea level rose rapidly at the end of the last ice age as glaciers and ice caps melted and water returned to the ocean. The rate of rise has slowed over the past 4,000 years and is now about 3 millimeters per year.

Figure 12.2 Sea levels past and future.

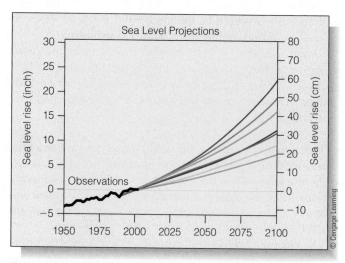

b Projections of sea level through the year 2100. Seven research groups (represented here by colored lines) have estimated future sea level based on historical observations and climate models. The most conservative of these predictions estimates a 20-centimeter (8-inch) rise.

Changes in sea level produce major differences in the position and nature of coastlines, especially in areas where the edge of the continent slopes gradually or where the coast is rising or sinking. **Figure 12.3** shows an estimate of previous shore positions along the southern coast of the United States in the geologically recent past and a prediction for the distant future should the present warming trend cause more polar ice to melt.

Because coasts are influenced by so many factors, perhaps the most useful scheme for classifying a coast is based on the predominant events that occur there: erosion and deposition. **Erosional coasts** are new coasts in which the dominant processes are those that *remove* coastal material. **Depositional coasts** are *steady or growing* because of their rate of sediment accumulation or the action of living organisms (such as corals).

The rocky shores of Maine are erosional because erosion exceeds deposition there; the sandy coastline from New Jersey to Florida is typically depositional because deposits of sediment tend to protect the shore from new erosion. The rocky central California coast is erosional, and the broad beaches of southern California are depositional. About 30% of the U.S. coastline is depositional, and 70% is erosional. We will use the erosional-depositional classification scheme in the rest of this chapter.

CONCEPT CHECK

Before going on to the next section, check your understanding of some of the important ideas presented so far:

How is a shore different from a coast?

What factors affect sea level and the location of a coast?

How is an erosional coast different from a depositional coast?

12.2 Erosional Processes Dominate Some Coasts

Land erosion and marine erosion both work to modify the nature of a rocky coast. Erosional coasts are shaped and attacked from the land by stream erosion, the abrasion of wind-driven grit, the alternate freezing and thawing of water in rock cracks, the probing of plant roots, glacial activity, rainfall, dissolution by acids from soil, and slumping.

From the sea, large storm surf routinely generates tremendous pressure. The crashing waves push air and water into tiny rock crevices. The repeated buildup and release of pressure within these crevices can weaken and fracture the rock. But it is not the hydraulic pressure of moving water alone that abrades the coasts. Tiny pieces of sand, bits of gravel, or stones hurled by the waves are even more effective at eroding the shore. Some indication of the violence of this activity may be inferred from **Figure 12.4**. Water dissolves minerals in the rocks, contributing to the erosion of easily soluble coastal rocks such as limestone. Even the digging and scraping of marine organisms have an effect.

The rate at which a shore erodes depends on the hardness and resistance of the rock, the violence of the wave shock to which it is exposed, and the local range of tides. Hard rock resists wear. Coasts made of granite or basalt may retreat an insignificant amount over a human lifetime; the granite coast of Maine erodes only a few centimeters per decade. Coasts of soft sandstone or other weak (or soluble) materials, however, may disappear at a rate of a few meters per year.

Marine erosion is usually most rapid on **high-energy coasts**, areas frequently battered by large waves. High-en-

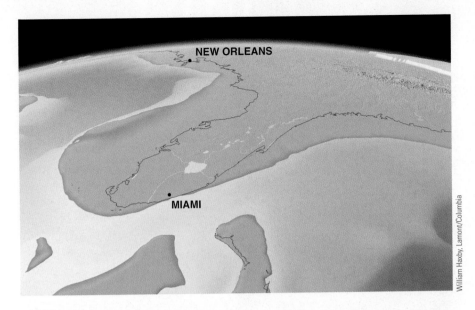

William Haxby, Lamont/Columbia

a About 18,000 years ago, during the last ice age, sea level was much lower. The position of the gently sloping southeastern coast was as much as 200 kilometers (125 miles) seaward from the present shoreline, leaving much of the continental shelf exposed.

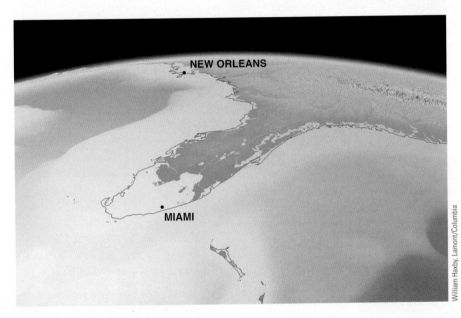

William Haxby, Lamont/Columbia

b In the future, if the ocean were to expand and some of the polar ice caps were to melt because of global warming, sea level could rise perhaps 5 meters (16.5 feet), driving the coast inland as much as 250 kilometers (160 miles).

Figure 12.3 The southeastern coast of the United States, past and future.

ergy coasts are most common adjacent to stormy ocean areas of great fetch and along the eastern edges of continents exposed to tropical storms. The coasts of Maine and British Columbia and the southern tips of South America and South Africa are typical high-energy coasts. **Low-energy coasts** are only infrequently attacked by large waves. Because of their generally protected location in the Gulf of Mexico, the U.S. Gulf states share a low-energy coast—at least between hurricanes!

Waves can affect the coast only where they strike, so erosion is concentrated near average sea level. A shore with little tidal variation can erode quickly because the wave action is concentrated near one level for longer times. Low-energy coasts protected by offshore islands usually erode slowly, as do areas below the low-tide line. Some erosion does occur below the sur-

face because of the orbital motion of water in waves, but even the largest waves have little erosive effect at depths greater than about 15 meters (50 feet) below average sea level. Cliffs above shore are subject to pounding either directly from waves or by rocks hurled by waves.

Erosional Coasts Often Have Complex Features

Erosive forces can produce a wave-cut shore that shows some or all of the features illustrated in **Figure 12.5**. Note the complex, small-scale irregularities of this rocky coastline. **Sea cliffs** slope abruptly from land into the ocean, their steepness usually resulting from the collapse of undercut notches. The position of the sea cliffs marks the shoreward limit of marine erosion on a coast. The parade of waves cuts **sea caves** into the cliffs at local zones

Figure 12.4 Attack from the sea is by waves and currents. On high-energy shores, the continuous onslaught of waves does most of the erosional work, with currents distributing the results of the waves' labor.

💬 **THINKING BEYOND THE FIGURE**

Where are the highest-energy shores in North America?

a Wave erosion of a rocky sea cliff produces a shelf-like, wave-cut platform visible at low tide. Remnants of the original cliff can protrude as sea stacks.

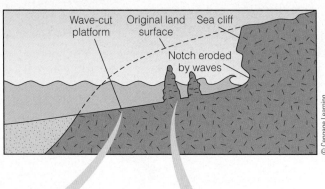

c Sea stacks at Port Campbell National Park, Australia. The large stack on the left fell in July 2005.

b A sea cliff and wave-cut platform off Tasmania's southeastern coast.

Figure 12.5 The results of wave action on a coast.

a The concentrated forces shape the headland into platforms and stacks. The accumulation of sediment from the headland in the tranquil bays eventually forms beaches and straightens the contours of the shore.

Figure 12.6 Wave energy converges on headlands and diverges in the adjoining bays.

b Waves approaching a shallow Caribbean reef refract around it. Energy is clearly concentrated on the headland.

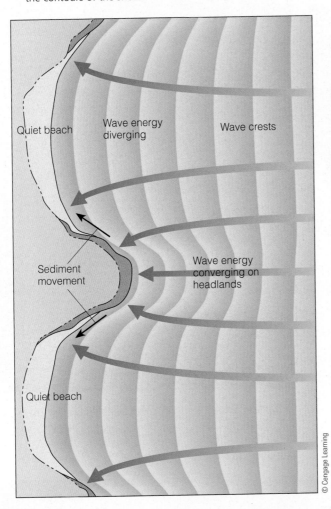

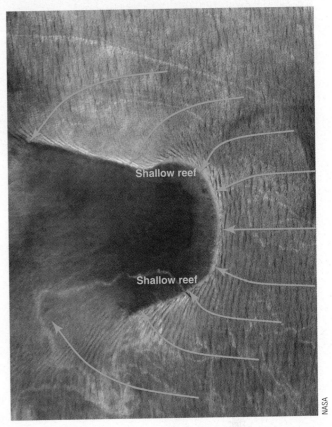

NASA

of weakness in the rocks. Most sea caves are accessible only at low tide. A blowhole can form if erosion follows a zone of weakness upward to the top of the cliff. When the tide is at just the right height, spray can blast from the fissure as waves crash into the cliff. Offshore features of rocky coasts can include natural arches, sea stacks, and a smooth, nearly level **wave-cut platform** just offshore, which marks the submerged limit of rapid marine erosion. Much of the debris removed from cliffs during the formation of these structures is deposited in the quieter water farther offshore, but some can rest at the bottom of the cliffs as exposed beaches. As we shall soon see, broad beaches are often features of depositional coasts.

Shorelines Can Be Straightened by Selective Erosion

The first effect that marine erosion has on a newly exposed coast is to intensify the irregularity of the coastline. This happens because coastal rocks are usually not uniform in composition over long horizontal distances. Some hard rocks will resist erosion

well, while softer rocks on the same coast may disappear almost overnight. (This explains the uneven character of the stacks, arches, and sea cliffs described earlier.)

Eventually, however, coastal erosion tends to produce a smooth shoreline. Because of wave refraction (see Figure 10.20), wave energy is focused onto headlands and away from bays (**Figure 12.6**). Sediment eroded from the headlands tends to collect as beaches in the relatively calm bays. As erosion continues, the deposits may eventually protect the base of the shore cliffs from the waves. Coastal irregularities are thus smoothed with the passage of time. As you might expect, straightening occurs most rapidly on high-energy coasts.

Coasts Are Also Shaped by Land Erosion and Sea-Level Change

When sea level was lower during the last glaciations, rivers cut across the land and eroded sediment to form coastal river valleys. When higher sea level returned, the valleys were flooded, or *drowned*, with seawater. Chesapeake Bay, the Hudson River

Figure 12.7 Drowned river valleys: submerging coasts that filled with water as the last ice age ended.

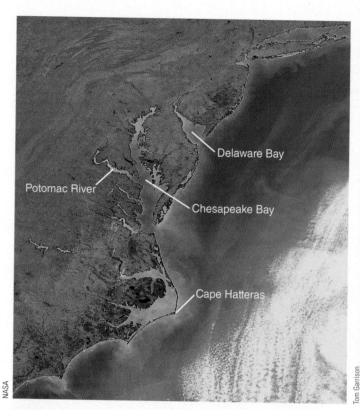

NASA

a Delaware and Chesapeake bays on the East Coast of the United States.

Tom Garrison

b Sydney Harbour, Australia, called by its discoverer "The finest harbour in the world."

and valley, and Sydney harbor (Australia) are examples of drowned river mouths (Figure 12.7).

Glaciers sometimes form in river valleys when rivers cut through the edges of continents at high latitudes. Deep, narrow bays known as **fjords** are often formed by tectonic forces and later modified by glaciers eroding valleys into deep, U-shaped troughs. Fjords are found in British Columbia, Greenland, Alaska, Norway, New Zealand, and other cold, mountainous places (Figure 12.8).

Volcanism and Earth Movements Affect Coasts

As we saw in Chapter 4, most islands that rise from the deep ocean are of volcanic origin. If the volcanism has been recent, the coasts of a volcanic island will consist of lobed lava flows extending seaward, common features in the Hawai'ian Islands (Figure 12.9a). Volcanic craters at a coast can also collapse and fill with seawater (Figure 12.9b).

Coasts can coincide with places where Earth's crust is being warped or faulted. When the seabed on the seaward side of a coastal fault moves *downward*, the result is a steep escarpment that continues to a greater depth than a wave-cut cliff. When the landward side of the fault moves *upward*, the part of the coast previously submerged during high tides can be left high and dry. The 1964 Alaska earthquake (described in Chapter 3) caused parts of the shore in Prince William Sound to rise as

> **" The seashore is a sort of neutral ground, a most advantageous point from which to contemplate this world. "** —HENRY DAVID THOREAU, 1864

much as 8 meters (26 feet). As can be seen in **Figure 12.10**, in some places a band of the old sea-cut bench nearly half a kilometer (one-fourth mile) wide was exposed, even at high tide.

Fault coasts sometimes occur along transform faults. The Pacific coast of North America provides two striking examples: the Gulf of California (located along the San Andreas Fault between Baja California and the mainland of Mexico) and the area around Point Reyes, north of San Francisco. Baja California was once a part of the North American continent, but movement at the boundary between the Pacific and North American plates has torn the finger of land on the west side of the fault from the land mass to the east. A young, narrow, straight gulf has formed as seawater intruded. As can be seen in **Figure 12.11**, a photo of Tomales Bay north of San Francisco, fault coasts can be startlingly straight.

CONCEPT CHECK

Before going on to the next section, check your understanding of some of the important ideas presented so far:

What wears down erosional coasts?

What are some features common to erosional coasts?

Over time, coastal erosion tends to produce a straight shoreline. Why?

How might volcanic activity shape a coast?

Figure 12.8 Sognefjord, Norway, one of the world's deepest fjords.

Ocean/Corbis

a Lava flowing seaward from an eruption on the island of Hawai'i forms a fresh coast exposed to erosion for the first time.

Figure 12.9 Volcanic alterations to coasts.

b Two volcanic cones on the southeastern coast of the Hawai'ian island of O'ahu. One of the volcanoes has collapsed, and its crater has filled with seawater.

Tom Garrison

George P. Lafaker, USGS

Figure 12.10 This former seafloor at Prince William Sound, Alaska, was raised 3.5 meters (12 feet) above sea level by tectonic uplift during the great earthquake of 27 March 1964. The exposed surface, which slopes gently from the base of the sea cliffs to the water, is about 400 meters (¼ mile) wide. The light-colored coating on the rocks consists mainly of the dried remains of small marine organisms. The photo was taken at about 0.0 tide, 30 May 1964.

💬 **THINKING BEYOND THE FIGURE**

Recalling Chapter 3's discussion of plate tectonics, can you suggest where shores might be most subject to tectonic uplift or subsidence?

12.3 Beaches Dominate Depositional Coasts

The features found on depositional coasts are usually composed of sediments rather than rock. Accumulation and distribution of a layer of protective sediments along a coast can insulate that coast from rapid erosion; the wave energy expended in churning overlying sediment particles cannot erode the underlying rock. So, with time, erosional shorelines can evolve into depositional ones. Unless the coast is rapidly rising or sinking, or unless other large-scale geological processes interfere, the inevitable process of erosion will tend to change the character of any coast from erosional to depositional.

Beaches Consist of Loose Particles

The most familiar feature of a depositional coast is the beach. A **beach** is a zone of loose particles that covers part or all of a shore **(Figure 12.12)**. The landward limit of a beach may be vegetation, a sea cliff, relatively permanent sand dunes, or construction such as a seawall. The seaward limit occurs where sediment movement onshore and offshore ceases—a depth of about 10 meters (33 feet) at low tide. The continental United States has 17,672 kilometers (10,983 miles) of beaches, about 30% of the total shoreline.

Beaches result when sediment, usually sand, is transported to places suitable for deposition. Such places include the calm spots between headlands, shores sheltered by offshore islands, and regions with moderate surf or broad stretches of high-energy coasts. Sometimes the sediment is transported a very short distance—particles may simply fall from the cliff above and accumulate at the shoreline—but more often the sediment on a beach has been moved for long distances by rivers or ocean currents to its present location.

Wherever they are found, beaches are in a constant state of change. As we will see, they may be thought of as rivers of sand—zones of continuous sediment transport.

Wave Action, Particle Size, and Beach Permeability Combine to Build Beaches

The material that makes up a beach can range from boulders through cobbles, pebbles, and gravel to very fine silt. The rare black sand beaches of Hawai'i are made of finely fragmented lava. Some beaches consist of shells and shell debris, or fragments of coral. Unfortunately, some also include large quantities of human junk: glass or plastic beaches are not unknown. Cobble beaches can be very steep (occasionally with slopes in excess of 20°), but wide beaches of fine sand are sometimes nearly as flat as a parking lot.

In general, the flatter the beach, the finer the material from which it is made. The relationship between particle size and beach slope depends on wave energy, particle shape, and the porosity of the packed sediments. Water from waves washing onto a beach—the **swash**—carries particles onshore, increasing the beach's slope. If water returning to the ocean—the **backwash**—carries back the same amount of material as it delivered, the beach slope will be in equilibrium; that is, the beach will not become larger or steeper.

Figure 12.11 A characteristically straight fault coast at Tomales Bay, California. Point Reyes is visible to the left (west); the city of San Francisco is just out of view to the south. The San Andreas Fault trace disappears below sea level in the Bay (arrow); the straight sides of the bay closely parallel the submerged fault.

Figure 12.12 A calm beach at the boundary between the eastern Australian states of New South Wales and Queensland. Bathers clustered on the raised berm appreciate the calm waves, and youngsters like the warm seawater pools of the backshore. Tom Garrison

a A steeply sloped beach consisting of coarse pebbles.

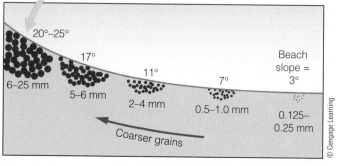

20°–25°

17°

11°

7°

Beach
slope =
3°

6–25 mm

5–6 mm

2–4 mm

0.5–1.0 mm

0.125–
0.25 mm

Coarser grains

© Cengage Learning

b The general relationship between grain size and beach slope.

Figure 12.13 A beach's slope is determined by the energy necessary to move the sand grains of which it is composed. Shallow water, smaller waves, and coarse grains combine to form steep beach slopes, but large waves very easily move small particles and smooth out the slope.

On fine-grain beaches, the ability of small, sharp-edged particles to interlock discourages water from percolating down into the beach itself, so water from waves runs quickly back down the beach, carrying surface particles toward the ocean. This process results in a very gradual slope. Broad, flat beaches also have a large area on which to dissipate wave energy, and they can provide a calm environment for the settling of fine sediment particles. In contrast, coarse particles (gravel, pebbles) do not fit together well and readily allow water to drain between them. Onrushing water disappears *into* a beach made of coarse particles, so little water is left to rush down the slope, thereby minimizing the transport of sediments back to the ocean. Larger particles tend to build up at the back of the beach where they are thrown by large waves, increasing the steepness of the beach **(Figure 12.13)**.

Beaches Often Have a Distinct Profile

Figure 12.14 shows a profile, or cross section, of a beach affected by small to moderate wave and tidal action. Most beaches have these key features:

- The **berm** (or berms), an accumulation of sediment that runs parallel to shore and marks the normal limit of sand deposition by wave action.
- The peaked top of the highest berm, called the **berm crest**, is usually the highest point on a beach. It corresponds to the shoreward limit of wave action during the most recent high tides.
- Inland of the berm crest, extending to the farthest point where beach sand has been deposited, is the **backshore**. The backshore is the relatively inactive portion of the beach, which may include windblown dunes and grasses.
- The **foreshore**, seaward of the berm crest, is the active zone of the beach, washed by waves during the daily rise and fall of the tides. It extends from the base of the berm—where a **beach scarp** (a vertical wall of variable height) is often carved by wave action at high tide—to the low-tide mark where the offshore zone begins.

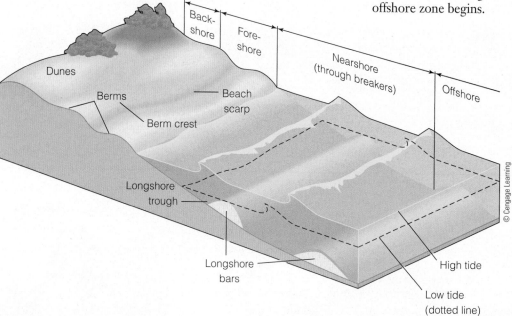

Back-
shore

Fore-
shore

Nearshore
(through breakers)

Offshore

Dunes

Berms

Beach
scarp

Berm crest

Longshore
trough

Longshore
bars

High tide

Low tide
(dotted line)

© Cengage Learning

Figure 12.14 A typical beach profile. The scale is exaggerated vertically to show detail.

a

b

Figure 12.15 As seasons change, sand moves on and off Boomer Beach near La Jolla, California. Gentle summer waves move sand onshore **a**, but larger winter waves remove the sand to offshore bars, exposing the basement rock **b**.

- Below the low-tide mark, wave action, turbulent backwash, and longshore currents excavate a **longshore trough** parallel to shore.
- Irregular **longshore bars** (submerged or exposed accumulations of sand) complete the seaward profile.

This beach profile is only temporary, generated by the interplay of sediments, waves, and tides. Great storm waves can rearrange a beach in a day, transporting thousands of tons of sediment from the beach to hidden sandbars offshore. Most temperate-climate beaches undergo a seasonal transformation. Beaches are cut to a lower level in winter than in summer because higher waves accompany winter storms. Changes from summer to winter on a beach are shown in **Figure 12.15**.

Waves Transport Sediment on Beaches

If the submerged slope of the seafloor is steep, eroded sediments will soon drain to deeper waters. If the slope is not too steep, sediments will be transported along the coast by wave and current action. The movement of sediment (usually sand) along the coast, driven by wave action, is referred to as **longshore drift**.

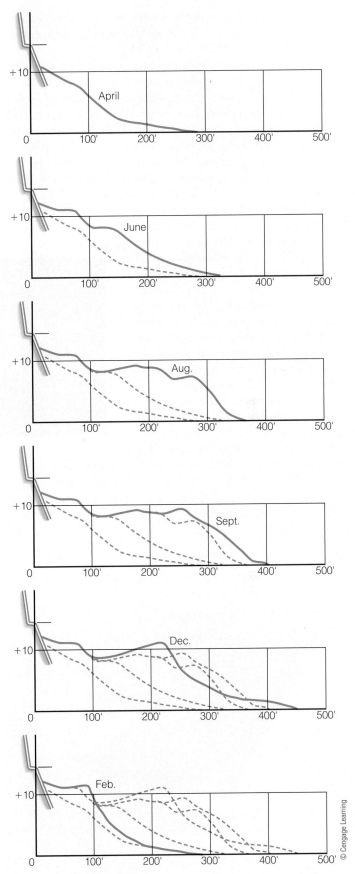

c The annual progression of sand on- and offshore at Boomer Beach for a typical year. East coast beaches are usually not as seasonally varied but can change dramatically with the advent of nor'easters or tropical cyclones.

Longshore drift occurs in two ways: the wave-driven movement of sand along the exposed beach and the current-driven movement of sand in the surf zone just offshore.

Most wind waves approach at an angle and then refract in shallow water to break almost parallel to shore. Refraction is usually incomplete, however, and some angle remains when the waves break. If sediments have accumulated to form a beach, water from the breaking wave will rush up the beach at a slight angle but return to the ocean by running straight downhill under the influence of gravity. The sand grains disturbed by the wave will follow the water's path, moving up the beach at an angle but retreating down the beach straight down the slope (Figure 12.16). Net transport of the grains is longshore, parallel to the coast, away from the direction of the approaching waves.

Sediments are also transported in the surf zone in a **longshore current**. The waves breaking at a slight angle distribute a portion of their energy away from their direction of approach. This energy propels a narrow current in which sediment already suspended by wave action can be transported downcoast. The speed of the longshore current can approach 4 kilometers (about 2½ miles) per hour.

Sand moving in the wash of waves along the beach and sediments propelled in the longshore current just offshore are often joined by much greater loads of sediment brought to the coast by rivers. Net southward transport of all this material along the central California coast exceeds 230,000 cubic meters (300,000 cubic yards) per year. Typical figures for the Atlantic coast of the United States are about two

> " The three great elemental sounds in nature are the sound of rain, the sound of wind in a primeval wood, and the sound of outer ocean on a beach. " —HENRY BESTON, 1928

thirds of this value. Though the direction may switch due to temporary conditions, net sand flow along both the Pacific and Atlantic coasts of the United States is usually to the south because the waves that drive the transport system usually approach from the north, where storms most commonly occur.

Sand Input and Outflow Are Balanced in Coastal Cells

Most new sand on a coast is brought in by rivers. The sand is moved parallel to the beach by longshore drift, and it is moved onshore and offshore at right angles to the beach as the seasons change. If a beach is stable in size, neither growing nor shrinking, the amount of new sand entering must be balanced by the amount of old sand being removed. Sand that drifts below the reach of wave action is lost from the coast and may migrate farther out on the continental shelf. Some sand is driven by longshore currents into the nearshore heads of submarine canyons. Sand moving away from shore in these canyons sometimes forms impressive sandfalls (see Figure 4.20 for an example) and is lost from the beaches above. The bulk of this material is transported by gravity down the axis of the canyon and ultimately deposited on a submarine fan at the base of the slope.

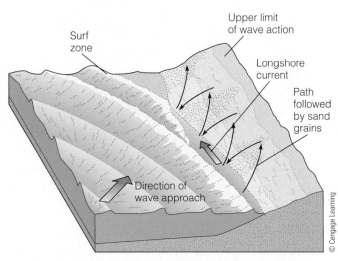

a A longshore current moves sediment along the shoreline between the surf zone and the upper limit of wave action.

Figure 12.16 Longshore transport.

💬 **THINKING BEYOND THE FIGURE**
Would longshore currents also trend north to south on the west coast of the United States?

b Groins built at right angles to the shore at Cape May, New Jersey, to slow the migration of sand. The groins interrupt the flow of longshore currents, so sand is trapped on their upcurrent sides. This view is toward the south, and south of the groins, on the downcurrent sides, sand is eroded.

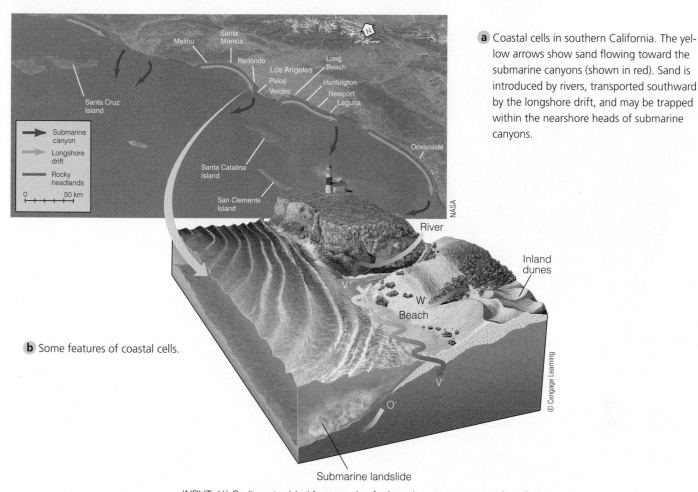

a Coastal cells in southern California. The yellow arrows show sand flowing toward the submarine canyons (shown in red). Sand is introduced by rivers, transported southward by the longshore drift, and may be trapped within the nearshore heads of submarine canyons.

b Some features of coastal cells.

INPUT	V^+	Sediment added from erosion for longshore transport onto beach
OUTPUTS	V^-	Sediment carried downcoast from the beach by longshore transport
	W^-	Sediment blown inland by wind
	O^-	Sediment cascading down the submarine slope
STABLE BEACH:		$(V^+) + (V^- + W^- + O^-) = 0$

Figure 12.17 Coastal sediment transport cells.

The natural sector of a coastline in which sand *input* and sand *outflow* are balanced may be thought of as a **coastal cell**. The main features of such a cell are illustrated in **Figure 12.17**. Coastal cells are usually bounded by submarine canyons that conduct sediments to the deep sea. Their size varies greatly. They are often very large along the relatively smooth, tectonically passive trailing edges of continents; coastal cells along the southeastern coast of the United States, for example, are hundreds of kilometers long. On the active leading edge of the continent, they are smaller. Four cells exist in the 360 kilometers (225 miles) between southern California's Point Conception and the Mexican border. Each terminates in a submarine canyon at the downcoast end.

CONCEPT CHECK

Before going on to the next section, check your understanding of some of the important ideas presented so far:

Do erosional coasts tend to evolve into depositional coasts, or is it the other way around?

What is the most common feature of a depositional coast?

What two marine factors are most important in shaping beaches?

How does sand move on a beach?

What is a coastal cell? Where does sand in a coastal cell come from? Where does it go?

12.4 Larger-Scale Features Accumulate on Depositional Coasts

Aside from beaches, depositional coasts exhibit some other large-scale features that result from the deposit of sediments. Some of these features are illustrated in **Figure 12.18**.

Sand Spits and Bay Mouth Bars Form When the Longshore Current Slows

Sand spits are among the most common of these features. A sand spit forms where the longshore current slows as it clears a headland and approaches a quiet bay. The slower current in the

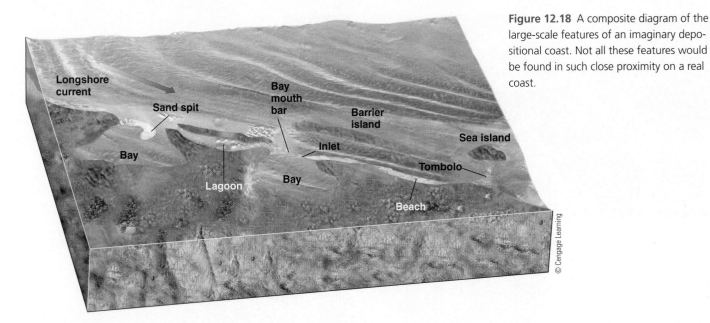

Figure 12.18 A composite diagram of the large-scale features of an imaginary depositional coast. Not all these features would be found in such close proximity on a real coast.

mouth of the bay is unable to carry as much sediment, so sand and gravel are deposited in a line downcurrent of the headland. As can be seen in Figure 12.18, sand spits often have a curl at the tip, which is caused by the current-generating waves being refracted around the tip of the spit.

A **bay mouth bar** forms when a sand spit closes off a bay by attaching to a headland adjacent to the bay. The bay mouth bar protects the bay from waves and turbulence and encourages the accumulation of sediments there. An **inlet**—a passage to the ocean—may be cut through a bay mouth bar by tidal action, by water flowing from a river emptying into the bay, or by heavy storm rains. **Figure 12.19** shows a bay mouth bar.

Barrier Islands and Sea Islands Are Separated from Land

Depositional coasts can also develop narrow, exposed sandbars that are parallel to but separated from land. These are known as **barrier islands (Figure 12.20)**. About 13% of the world's coasts are fringed with barrier islands.

Barrier islands can form when sediments accumulate on submerged rises parallel to the shoreline. Some islands off the Mississippi–Alabama coast developed in this way. Larger barrier islands are thought to form in a different way, however. Near the end of the last major rise in sea level, about 6,000 years ago, coastal plains near the edge of the continental shelf were fronted by lines of sand dunes. Rising sea level caused the ocean to break through the dunes and form a **lagoon** or a **sound**—a long, shallow body of seawater isolated from the ocean—behind these sand dunes. The high lines of coastal dunes became islands. As sea level continued to rise, wave action caused the islands and lagoons to migrate landward. Most of the barrier islands off the

Figure 12.19 A bay mouth bar. The inlet is now closed, but increased river flow (from inland rainfall) or large waves combined with very high tides could break the bar. For an indication of scale, note the freeway bridges at the top of the photograph.

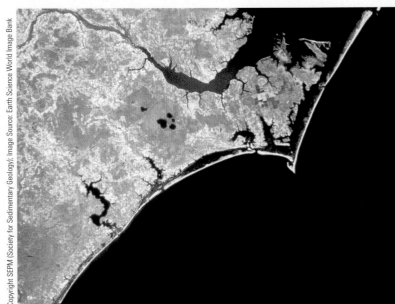

Figure 12.20 Barrier islands off the North Carolina Coast. (This photo, taken from space, is on a much larger scale than Figure 12.19.)

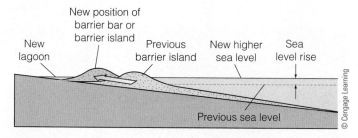

a Barrier islands migrate as sea level rises. A 30-centimeter (1-foot) rise through the next century would move barrier islands landward by 100 to 150 meters (330–500 feet).

Figure 12.21 The migration of barrier islands.

💬 **THINKING BEYOND THE FIGURE**

What steps were taken in 1998–1999 to preserve the Cape Hatteras lighthouse? Why were these efforts necessary?

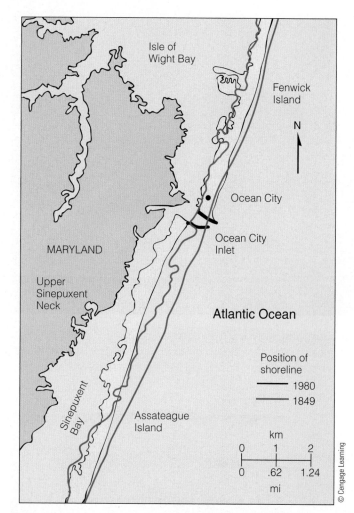

b The heavy black lines south of Ocean City represent jetties constructed in the 1930s to protect the inlet. The jetties disrupt the north-to-south longshore current. As a result, Assateague Island has been starved of sediment and has migrated about 500 meters (1,640 feet) westward.

southeastern coast of the United States probably originated in this way. They are still migrating slowly landward as sea level continues to rise. The process is accelerated if sediment inflow is restricted **(Figure 12.21)**.

Every year severe storms generate waves intense enough to erode barrier island beaches. The largest of these storms can generate waves that overwash the low islands. Runoff from rivers swollen by rains, coupled with water driven by wind waves and storm surge, can rapidly flood a lagoon and cut new inlets through barrier islands.

Despite these dangers, about 70 barrier islands off the U.S. coast have been commercially developed, and millions of people live on them. The most famous barrier islands include Atlantic City, New Jersey; Ocean City, Maryland; Miami Beach and Palm Beach, Florida; and Galveston, Texas. Roughly once every hundred years a winter storm has catastrophic effects on populated areas of Atlantic barrier islands, and the southeastern Atlantic and Gulf coasts must contend with occasional large hurricanes. The continuing subsidence of these passive coasts (combined with changes caused by commercial development and the ongoing rise in sea level) will undoubtedly cost lives and destroy property. **Figure 12.22** illustrates the extent of the threat.

Unlike barrier islands, **sea islands** are composite structures that contain a firm central core that was part of the mainland when sea level was lower. The rising ocean separated these high points from land, and sedimentary processes surrounded them with beaches. Hilton Head, South Carolina, and Cumberland Island, Georgia, are sea islands. If the island is close to shore, a bridge of sediments called a **tombolo** may accumulate to connect the island to the mainland. Tombolos can also connect offshore rocky outcrops or volcanoes to the mainland. A sea island and a tombolo are shown in Figure 12.18.

Deltas Can Form at River Mouths

In a few places, sediments washing off the land have built out the coasts extensively. The shoreline in such places is much different

from its configuration at the end of the last ice age. The most important of these coastal features are **deltas**.[1]

Deltas do not form at the mouth of every sediment-laden river. A broad continental shelf must be present to provide a platform on which sediment can accumulate. Tidal range is usually low, and waves and currents generally mild. There are no large deltas along the Atlantic coast of the United States because sediments that arrive at the coast are deposited in the sunken river mouths or dispersed by tides and currents. Also, there are no large deltas along the western margins of North and South America because these coasts are converging margins, where an oceanic plate is being subducted and the continental shelf is very narrow; sediment that would form a delta is swept down the continental slope or dispersed along the coast by waves. Deltas are most common on the low-energy shores of enclosed seas (where the tidal range is not extreme) and along the tectonically stable

---

[1]The term is derived from the triangular shape of the capital Greek letter *delta* (Δ).

a Ocean City, Maryland, a developed barrier island. Host to 8 million visitors a year, this city (and others similarly situated) has no effective protection against flooding and damage from severe storms.

b This breach of Hatteras Island on North Carolina's outer banks was caused by storm surge and flooding by Hurricane Isabel in September 2003. The ocean is to the right.

Figure 12.22 Barrier island modification: actual and potential.

c The breach severed the only road along the island's length and isolated many homes. The arrow in **b** and the yellow line in **c** mark the centerline of the road.

trailing edges of some continents. The largest deltas are those of the Gulf of Mexico (the Mississippi, **Figure 12.23a**), the Mediterranean Sea (the Nile), the Ganges–Brahmaputra river system in the Bay of Bengal **(Figure 12.23b)**, and the huge deltas formed by the rivers of China that empty into the South China Sea.

The shape of a delta represents a balance between the accumulation of sediments and their removal by the ocean. For a delta to maintain its size or grow, the river must carry enough sediment to keep marine processes in check. The combined effects of waves, tides, and river flow determine the shape of a delta. *River-dominated deltas* are fed by a strong flow of fresh water and continental sediments, and they form in protected marginal seas. They terminate in a well-developed set of *distributaries*—the split ends of the river—in a characteristic bird's-foot shape

(as shown in the Mississippi). In *tide-dominated deltas*, freshwater discharge is overpowered by tidal currents that mold sediments into long islands parallel to the river flow and perpendicular to the trend of the coast. The largest tide-dominated delta has formed at the mouths of the Ganges–Brahmaputra river system. *Wave-dominated deltas* are generally smaller than either tide- or river-dominated deltas and have a smooth shoreline punctuated by beaches and sand dunes. Instead of a bird's-foot pattern of distributaries, a wave-dominated delta will have one main exit channel.

Deltas are not the only types of coasts built out by the land. The glaciers that covered the poleward parts of the continents during the last ice age deposited great quantities of sediments and rocks near their outer margins. When the glaciers retreated, they left streamlined hills known as **drumlins** and hills and ridges of sediments called **moraines**—some of which still stand above sea level. Part of Long Island, New York, is a glacial moraine, and the oval-shaped hills of Boston and sections of the Puget Sound coast at Seattle were shaped in part by glaciers. Perhaps the most famous glacial moraine in the United States is the area around Cape Cod, Massachusetts. The cores of the islands of Martha's Vineyard and Nantucket represent the farthest advance of the glaciers around 18,000 years ago, and the spine of the Cape itself is a remnant of material dropped by the glacier as it stabilized and then retreated northward when the climate warmed **(Figure 12.24)**.

CONCEPT CHECK

Before going on to the next section, check your understanding of some of the important ideas presented so far:

What is the difference between sand spits and bay mouth bars?

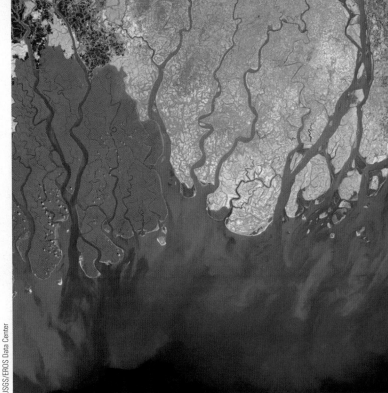

a The bird's-foot shape of the Mississippi Delta is seen clearly in this photograph. Lobed and bird's-foot deltas form where deposition overwhelms the processes of coastal erosion and sediment transportation. The sediment-laden water looks brown or tan in this photograph taken from low orbit.

Figure 12.23 River deltas form at places where sediment-laden rivers enter enclosed or semi-enclosed seas, where wave energy is limited.

b The mouths of the Ganges–Brahmaputra river system on 28 February 2000. This tide-dominated delta, home to about 120 million people, is routinely flooded during cyclones and monsoon rains. Note the sediment (milky blue color) flowing from the delta into the Bay of Bengal.

What is the difference between sea islands and barrier islands?

Why don't deltas form at every river mouth?

12.5 Biological Activity Forms and Modifies Coasts

Coasts can be extensively modified by the activities of animals and plants. Some kinds of marine algae and plants can build coasts, but the most dramatic biological modifications occur in the tropics, where coral organisms form reefs around volcanic islands or along the margin of a continent.

Reefs Can Be Built by Coral Animals

A **coral reef** is a linear mass of calcium carbonate assembled from and by multitudes of coral animals. Coral animals (about which you will learn more in Chapter 15) are related to the familiar sea anemones found along temperate coasts. Individual reef-building coral animals secrete a cup-shaped calcium carbonate skeleton, which remains behind after the animal dies; the accumulation of skeletons gradually forms the reef. Reef-building corals grow best in brightly lighted water about 5 to

Figure 12.24 Most of the islands of Martha's Vineyard and Nantucket were formed from debris deposited or shaped by an advancing glacier. The spine of Cape Cod is built of material dropped as the glaciers stabilized and then retreated.

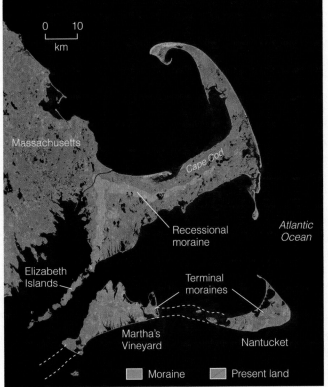

Tororo reaction/Shutterstock.com

Figure 12.25 A small section of the Great Barrier Reef, Queensland, Australia. This coast has been extensively modified by biological activity.

10 meters (16–33 feet) deep, and in ideal conditions they grow at a rate of about 1 centimeter (½ inch) per year.

The greatest of all reefs is the Australian Great Barrier Reef (Figure 12.25), which begins in the Torres Strait separating New Guinea and Australia and runs down the northeastern coast of Australia for 2,500 kilometers (1,500 miles).

Coral Reefs Are Classified into Three Types

In 1842, Charles Darwin classified tropical coral reef structures into three types: Fringing reefs, barrier reefs, and atolls (Figure 12.26a). We still use his classification today.

Fringing Reefs As their name implies, **fringing reefs** cling to the margin of land. As can be seen in Figure 12.26a, a fringing reef connects to shore near the water surface. Fringing reefs form in areas of low rainfall runoff primarily on the leeward (downwind side) of tropical islands. The greatest concentration of living material will be at the reef's seaward edge, where plankton and clear water of normal salinity are dependably available. Most new islands anywhere in the tropics have fringing reefs as their first reef form. Permanent fringing reefs are common in the Hawai'ian Islands and in similar areas near the boundaries of the tropics.

Barrier Reefs **Barrier reefs** are separated from land by a lagoon (Figure 12.26a). They tend to occur at lower latitudes than fringing reefs and can form around islands or in lines parallel to continental shores. The outer edge—the barrier—is raised because the seaward part of the reef is supplied with more food and is able to grow more rapidly than the shore side. The lagoon may be anywhere from a few meters to 60 meters (200 feet) deep, and it may separate the barrier from shore by only tens of meters or by as much as 300 kilometers (190 miles). Coral grows more slowly within the lagoon because fewer nutrients are available and because sediments and freshwater run off from shore. As you would expect, conditions and species within the lagoon are much different from those of the wave-swept barrier. The calm lagoon is often littered with eroded coral debris moved from the barrier by storms.

The Great Barrier Reef isn't a single reef, but a conglomeration of more than 3,000 interlinked segments covering 350,000 square kilometers (135,000 square miles)—collectively the largest structure made by living organisms on Earth. The segments present a steep outer wall to the prevailing currents and trade winds. At a growth rate of 1 centimeter (½ inch) per year, the structure is obviously of great age and astonishing volume. The huge reef is younger and thinner at its southern end; the slow northward movement of the Indian–Australian lithospheric plate in which Australia is embedded is thought to be responsible for this. The variety of organisms within the Australian Barrier Reef staggers the imagination.

Atolls An **atoll** (Figure 12.26b) is a ring-shaped island of coral reefs and coral debris enclosing, or almost enclosing, a shallow lagoon from which no land protrudes. Coral debris may be driven onto the reef by waves and wind to form an emergent arc on which coconut palms and other land plants take root. These plants stabilize the sand and lead to colonization by birds and other species. This is the tropical island of the travel posters.

Though an atoll's central lagoon connects to the deep water outside through a series of channels or grooves, coral does not usually thrive in the lagoon, both because the shallow water may become too hot from the sun or too fresh during rains and because feeding opportunities for the coral are limited there.

Some atolls are isolated, but most occur in loose groups in shallow continental-shelf areas or in the deep open ocean. More than 300 atolls exist—most in the Pacific. They range in size from a few kilometers in diameter to Kwajalein in the Marshall Islands, whose slender, 280-kilometer (176-mile) ring of coral encloses a lagoon of 2,850 square kilometers (1,100 square miles).

How do atolls form? Scientists began speculating on the cause of their ring shape soon after the first scientific voyages published their reports. Charles Darwin imagined a volcanic island growing from the sea, accumulating a skirt of coral around its shore, and then slowly subsiding at a rate equal to the growth rate of the coral. The central volcanic island would eventually sink from view, but the coral could grow continuously atop skeletons of past generations to maintain a living presence near the surface. Figure 12.26a shows the progression. Note that the island begins with a fringing reef, passes through a barrier-reef stage as it sinks, and eventually becomes an atoll as the peak disappears beneath the ocean surface. Should the island subside faster than about 1 centimeter (½ inch) per year (the growth rate of coral), all trace of both the island and the reefs would disappear. (The submerged island might become a guyot.)

This theory seemed reasonable, but Darwin couldn't explain what would cause volcanic islands to subside because he didn't know about plate tectonics. Now we know that volcanoes can form near spreading centers, ride outward and downward from their birthplaces, cease to be active as they leave their source of mantle heat, and sink as they are carried into deeper water—just slowly enough to permit coral growth to continue as they go.

Mangrove Coasts Are Dominated by Sediment-Trapping Root Systems

Other coasts have been formed by mangroves, trees that can grow in salt water. The coast of southwestern Florida has been extended and shaped by the activity of mangroves, whose root systems trap and hold sediments around the plant (**Figure 12.27**, see p. 358). The root complex forms an impenetrable barrier and safe haven for organisms around the base of the trees. You will learn more about mangroves in the section on marine plants in Chapter 14.

CONCEPT CHECK

Before going on to the next section, check your understanding of some of the important ideas presented so far:

What organisms can affect coastal configuration?

How are coral reefs classified? Who first proposed this classification scheme?

12.6 Freshwater Meets the Ocean in Estuaries

An **estuary** is a body of water partially surrounded by land, where freshwater from a river mixes with ocean water. Estuaries are areas of remarkable biological productivity and diversity. The coasts of the United States contain about 15,150 square kilometers (5,850 square miles) of estuarine waters. Chesapeake Bay, San Francisco Bay, and Puget Sound are all estuaries.

Estuaries Are Classified by Their Origins

Estuaries are classified into four types depending on their origins (**Figure 12.28**, see p. 358).

1. Drowned river mouths
2. Fjords
3. Bar-built
4. Tectonic

> **"Ever heard of a place... I think it's called Norway? That was one of mine. I got an award for it."**
> —SLARTIBARTFAST, IN *HITCHHIKER'S GUIDE TO THE GALAXY*, DOUGLAS ADAMS, 1979

Estuaries formed at drowned river mouths are common throughout the world, particularly along the Atlantic coast of the United States. Remember that sea level has risen about 125 meters (410 feet) in the 18,000 years since the end of the last major period of glaciation, and the result has been the incursion of seawater into river mouths. The mouths of the York, James, and Susquehanna rivers and Chesapeake Bay are examples of this type of estuary.

As Figure 12.8 suggests, fjords are steep, glacially eroded, U-shaped troughs. They are often about 300 to 400 meters (1,000–1,300 feet) deep but typically terminate in a shallow lip, or sill, of glacial deposits. In fjords with shallow sills, little vertical mixing occurs below the sill depth, and the bottom waters can become stagnant (look ahead to Figure 12.30d). In fjords with deeper sills, the bottom waters mix slowly with adjacent oceanic waters. Fjords are common in Norway, Greenland, New Zealand, Alaska, and western Canada. They are rare in the lower 48 states, but the Strait of Juan de Fuca in Washington is a good example.

Bar-built estuaries form when a barrier island or a barrier spit is built parallel to the coast above sea level. Since these estuaries are shallow and usually have only a narrow inlet connecting them to the ocean, tidal action is limited. Waters in bar-built estuaries are mainly mixed by the wind. Albemarle and Pamlico

The Development of Atolls and Guyots

a A fringing reef forms around an island in the tropics. The island sinks as the oceanic plate on which it rides moves away from a spreading center. In this case, the island does not sink at a rate faster than coral organisms can build upward. The island eventually disappears beneath the surface, but the coral remains at the surface as an atoll.

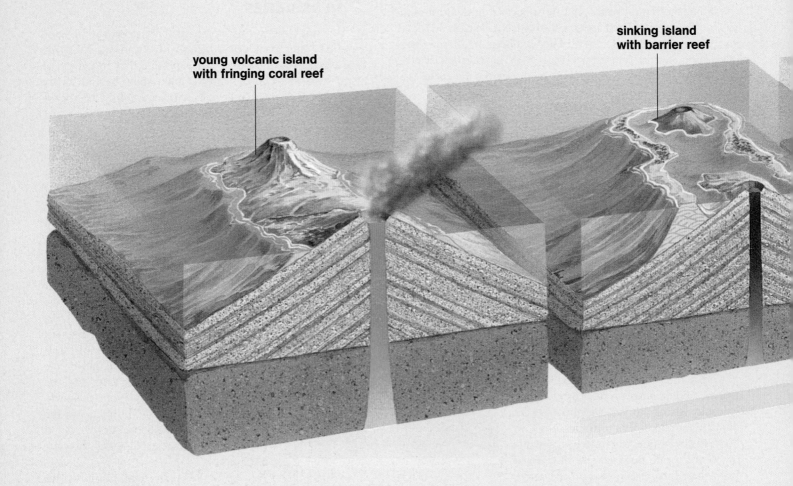

young volcanic island with fringing coral reef

sinking island with barrier reef

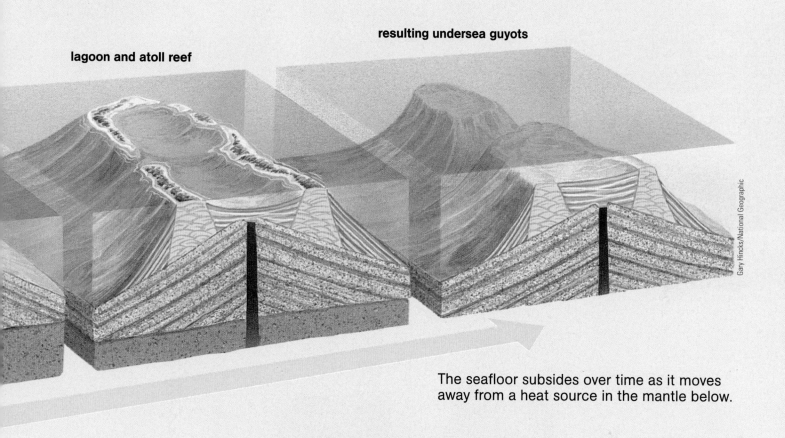

lagoon and atoll reef

resulting undersea guyots

Gary Hincks/National Geographic

The seafloor subsides over time as it moves away from a heat source in the mantle below.

Douglas Faulkner/Science Source

b The typical ring shape of an atoll—the alluring tropical island of travel posters.

💬 **THINKING BEYOND THE FIGURE**

Where are most atolls and guyots found?

a Mangrove trees trap sediments, building and stabilizing the coast. Their roots also provide safe havens for a great variety of marine life.

Figure 12.27 Mangrove coasts.

b Coastal restoration in progress in a damaged mangrove forest in Hong Kong's New Territories.   Tom Garrison

Brian Parker/Tom Stack and Associates

a Drowned river mouths: the mouths of the James, York, and Susquehanna rivers; Chesapeake Bay; Sydney Harbour, Australia.

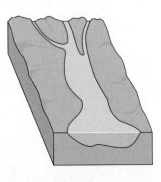

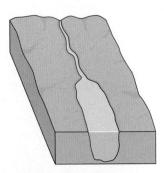

b Fjords: New Zealand's Milford Sound; the Strait of Juan de Fuca in Washington state.

c Bar-built: Albemarle and Pamlico sounds in North Carolina.

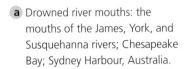
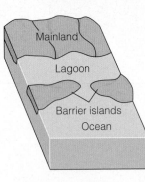

Mainland

Lagoon

Barrier islands
Ocean

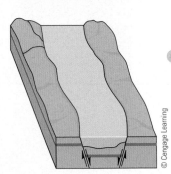

d Tectonic: San Francisco Bay; Tomales Bay (see again Figure 12.11).

© Cengage Learning

Figure 12.28 Estuaries classified by their origins.

sounds in North Carolina and Chincoteague Bay in Maryland are bar-built estuaries.

Estuaries produced by tectonic processes are coastal indentations formed by faulting and local subsidence. Freshwater and seawater both flow into the depression, and an estuary results. San Francisco Bay is, in part, a tectonic estuary.

Estuary Characteristics Are Influenced by Water Density and Flow

Three factors determine the characteristics of estuaries: the shape of the estuary, the volume of river flow at the head of the estuary, and the range of tides at the estuary's mouth. The mingling of waters of different densities, the rise and fall of the tide, and variations in river flow—along with the actions of wind, ice, and the Coriolis effect—guarantee that patterns of water circulation in an estuary will be complex.

Estuaries are categorized by their circulation patterns. The simplest circulation patterns are found in **salt-wedge estuaries**, which form where a rapidly flowing large river enters the ocean in an area where tidal range is low or moderate. The exiting freshwater holds back a wedge of intruding seawater (**Figure 12.29a**). Note that density differences cause freshwater to flow over salt water. The seawater wedge retreats seaward at times of low tide or strong river flow, and it returns landward as the tide rises or when river flow diminishes. Some seawater from the wedge joins the seaward-flowing freshwater at the steeply sloped upper boundary of the wedge, and new seawater from the ocean replaces it. Nutrients and sediments from the ocean can enter the estuary in this way. Examples of salt-wedge estuaries are the mouths of the Hudson and Mississippi rivers.

A different pattern occurs where the river flows more slowly and the tidal range is moderate to high. As their name implies, **well-mixed estuaries** contain differing mixtures of freshwater and salt water through most of their length. Tidal turbulence stirs the waters together as river runoff pushes the mixtures to sea. A well-mixed estuary is illustrated in **Figure 12.29b**. The mouth of the Columbia River is an example.

Deeper estuaries exposed to similar tidal conditions but greater river flow become **partially mixed estuaries**. Partially mixed estuaries share some of the properties of salt-wedge and well-mixed estuaries. Note in **Figure 12.29c** the influx of seawater beneath a surface layer of freshwater flowing seaward; mixing occurs along the junction. Energy for mixing comes from both tidal turbulence and river flow. England's River Thames, San Francisco Bay, and Chesapeake Bay are examples.

Fjord estuaries form where glaciers have gouged steep, U-shaped valleys below sea level. Typically, fjord estuaries have small surface areas, high river input, and little tidal mixing. River water tends to flow seaward at the surface with little contact with the seawater below (**Figure 12.29d**). In fjord estuaries with steep sills, a layer of stagnant water—cold water containing little oxygen and few nutrients—can form above the floor.

In well-mixed and partially mixed estuaries in the Northern Hemisphere, the incoming seawater presses against the right

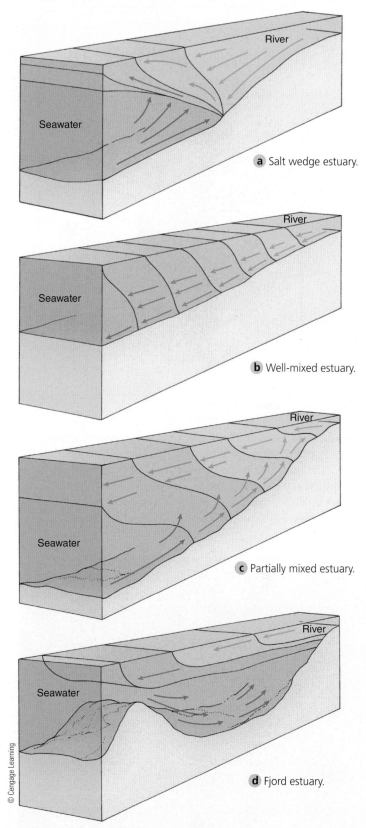

a Salt wedge estuary.

b Well-mixed estuary.

c Partially mixed estuary.

d Fjord estuary.

Figure 12.29 Types of estuaries in vertical cross sections. The salinity values show the amount of mixing between freshwater and seawater in the various types. Green color represents freshwater.

© Cengage Learning

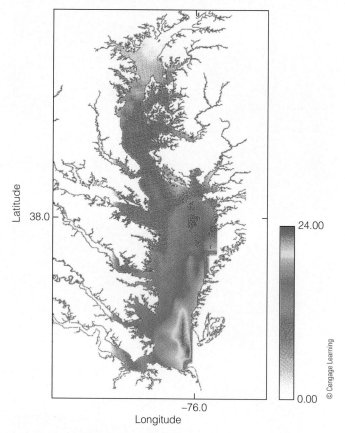

Figure 12.30 Range of salinity in Chesapeake Bay, an example of a partially mixed estuary. The colors indicate salinity in parts per thousand. The typical distribution of surface salinity in the estuary ranges from 28‰ at the mouth to 1‰ near the upper reaches. The Coriolis effect forces the inflowing salt water against the right (eastern) bank—notice how the 20‰ contour lines trend toward the right bank.

Figure 12.31 Commercial development of estuarine wetlands affects species diversity and biological productivity and leads to increased coastal erosion.

side of the estuary because of the Coriolis effect. Out-flowing river water also trends to the right of its direction of travel. This rightward drift can be seen in the contour of lines representing surface salinity in Chesapeake Bay (Figure 12.30).

Estuaries Support Complex Marine Communities

Some of the oldest continuous civilizations have flourished in estuarine environments. The lower regions of the Tigris and Euphrates rivers, the Po River Delta region of Italy, the Nile Delta, the mouths of the Ganges, and the lower Hwang Ho Valley have supported dense human habitation for thousands of years. Estuaries continue to be irresistibly attractive to developers. In areas of high population density, estuaries are routinely dredged to provide harbors, marinas, and recreational resources and filled to make space for homes and agricultural land.

Estuaries often support a tremendous number of living organisms. The easy availability of nutrients and sunlight, protection from wave shock, and the presence of various habitats permit the growth of many species and individuals. Biological productivity and diversity in estuaries is usually very high. Estuaries are frequently nurseries for marine animals; several species of perch, anchovy, and Pacific herring take advantage of the abundant food in estuaries during their first weeks of life. Unfortunately for their inhabitants, the high demand for development is incompatible with a healthy estuarine ecosystem.

Estuaries have also become the most polluted of all marine environments. Some of the plants growing in shallow, temperate estuaries have the ability to "scrub" polluted water—to remove inorganic nitrogen compounds and metals from seawater polluted by sources on land. Plants use electrostatic attraction and sticky surface layers to accumulate clay-sized particles from the water and deposit them on their surfaces. When the tide falls, this material is deposited at the base of the plant and helps protect it from erosion. Bacteria in the mud can decompose the nitrogen compounds and bind the metals, making the water cleaner. Ironically, development is destroying the very ecosystems capable of helping clean the water. More than one-half the nation's estuaries and other wetlands have been lost. Of the original 870,000 square kilometers (215 million acres) of wetlands that once existed in the lower 48 states, only about 360,000 square kilometers (90 million acres) remain. Figure 12.31 suggests the extent of the problem.

CONCEPT CHECK

Before going on to the next section, check your understanding of some of the important ideas presented so far:

What is an estuary?

Estuaries are classified by their origins. What types of estuaries exist?

Estuaries are also classified by the type of water they contain and the flow characteristics of that water. How are estuaries classified by water circulation patterns?

Of what value are estuaries?

Figure 12.32 Wave-cut terraces on San Clemente Island off the coast of southern California. Tectonic uplift and the erosive forces shown in Figure 12.5b explain their origin.

University of Washington Libraries, Special Collections, John Shelton Collection, KC8974

12.7 The Characteristics of U.S. Coasts

Plate tectonic forces have had immense influence on the margins of continents, and the edges of the United States are no exception. The results of plate movement on the Pacific coast differ greatly from those on the Atlantic and Gulf coasts, primarily because the Pacific coast is near an active plate margin and the Atlantic and Gulf coasts are not.

The Pacific Coast

The Pacific coast is an actively rising margin on which volcanoes, earthquakes, and other indications of recent tectonic activity are easily observed. Pacific coast beaches are typically interrupted by jagged rocky headlands, volcanic intrusions, and the effects of submarine canyons. Wave-cut terraces are found as much as 400 meters (1,300 feet) above sea level in a number of places, evidence that tectonic uplift has exceeded the general rise in sea level over the past million years (Figure 12.32).

Most of the sediments on the Pacific coast originated from erosion of relatively young granitic or volcanic rocks of nearby mountains. The particles of quartz and feldspar that constitute most of the sand were transported to the shore by flowing rivers. The volume of sedimentary material transported to Pacific coast beaches from inland areas activity greatly exceeds the amount originating at the coastal cliffs. Deltas tend not to form at Pacific coast river mouths because the continental shelf is narrow, river flow is generally low (except the Columbia River), and beaches are usually high in wave energy. The predominant direction of longshore drift is to the south because northern storms provide most of the wave energy.

The Atlantic Coast

The Atlantic coast is a passive margin, tectonically calm and subsiding because of its trailing position on the North American Plate. Subsidence along the coast has been considerable— 3,000 meters (10,000 feet) over the last 150 million years. A deep layer of sediment has built up offshore, material that helped produce today's barrier islands. Relatively recent subsidence has been more important in shaping the present coast, however. With the exception of the coast of Maine (which is still in isostatic rebound after the recent departure of the glaciers), coastal sinking and rising sea level have combined to submerge some parts of the Atlantic coast at a rate of about 0.3 meter (1 foot) per century. This process has formed the huge flooded valleys of Chesapeake and Delaware bays, the landward-migrating barrier islands, and the shrinking lowlands of Florida and Georgia.

Rocks to the north (in Maine, for example) are among the hardest and most resistant to erosion of any on the continent, so beaches are uncommon in Maine. But from New Jersey southward, the rocks are more easily fragmented and weathered, and beaches are much more common. As on the Pacific coast, sediments are transported coastward by rivers from eroding inland mountains, but the transported material is trapped in estuaries and therefore plays a less important role on beaches. Eastern beaches are typically formed of sediments from shores eroding nearby or from the shoreward movement of offshore deposits laid down when the sea level was lower. The amount of sand in an area thus depends in part on the resistance or susceptibility of nearby shores to erosion. Sand moves generally south on these beaches just as it does on the Pacific coast, but the volume of moving sand is less in the East.

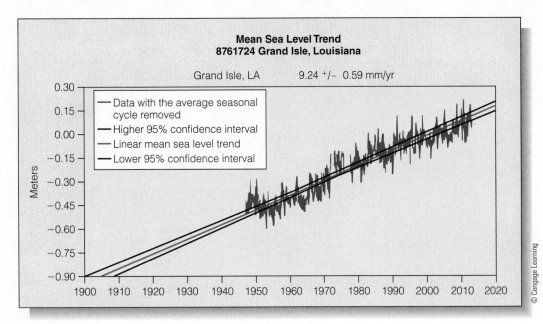

Mean Sea Level Trend
8761724 Grand Isle, Louisiana

Grand Isle, LA   9.24 +/− 0.59 mm/yr

Legend:
- Data with the average seasonal cycle removed
- Higher 95% confidence interval
- Linear mean sea level trend
- Lower 95% confidence interval

(Y-axis: Meters — 0.30, 0.15, 0.00, −0.15, −0.30, −0.45, −0.60, −0.75, −0.90)
(X-axis: 1900, 1910, 1920, 1930, 1940, 1950, 1960, 1970, 1980, 1990, 2000, 2010, 2020)

© Cengage Learning

Figure 12.33 Mean sea level trend at Grand Isle, Louisiana. This part of the Gulf coast is experiencing the most rapid rise in sea level of any U.S. coast. Accelerated subsidence of the land combines with rising sea level to cause a rise equivalent to 9.24 millimeters per year (3 feet in 100 years).

As we have seen, glaciers have also contributed to the shaping of the northern part of the Atlantic coast: Large portions of Long Island and all of Cape Cod are remnants of debris deposited by glaciers.

The Gulf Coast

The Gulf coast experiences a smaller tidal range and—between hurricanes—a smaller average wave size than either the Pacific or Atlantic coasts. Reduced longshore drift and an absence of interrupting submarine canyons allow the great volume of accumulated sediments from the Mississippi and other rivers to form large deltas, barrier islands, and a long raised "super berm" that prevents the ocean from inundating much of this sinking coast.

These are fortunate conditions because the rate of subsidence in the Gulf coast is considerably greater than that for nearly all of the Atlantic or Pacific coasts (Figure 12.33). Subsidence here is not the result of tectonic activity but of sediment compaction, de-watering, and the removal of oil and natural gas. Sediment starvation and dredging have made the situation worse around some large cities. At Galveston, Texas, for example, sea level appears nearly 64 centimeters (25 inches) higher than it was a century ago, and parts of New Orleans are now 2 meters (6.6 feet) below sea level. As we have seen, the results of hurricanes at such places can be tragic. The protective natural berm can easily be breached, and floodwaters can surge far inland.

CONCEPT CHECK

Before going on to the next section, check your understanding of some of the important ideas presented so far:

Briefly compare the U.S. Pacific, Atlantic, and Gulf coasts. What are the most important forces influencing these coasts?

12.8 Humans Interfere in Coastal Processes

Coasts are active areas where marine, terrestrial, atmospheric, and human factors converge. No single one of these factors dominates for long. We enjoy visiting and living near coasts, but human interference in coastal processes does not always produce the desired result. Steps taken to preserve or "improve" a stretch of rocky coast or a beach may have the opposite effect, and coastal residents do not always learn by example.

Beaches exist in a tenuous balance between accumulation and destruction. Human activity can tip the balance one way or the other. For example, consider the rocky **breakwater** shown in **Figure 12.34**. The breakwater interrupts the progress of waves to the beach, weakening the longshore current and allowing sand to accumulate there. Without dredging, the beach will eventually reach the breakwater and fill the small-boat anchorage the breakwater was built to provide. This is a minor example of human alteration of a beach, yet it serves to introduce the growing problem of human influences on coastal processes.

We often divert or dam rivers, build harbors, and develop property with surprisingly little understanding of the impact our actions will have on the adjacent coast. Our role then becomes that of powerless observers. Residents of eroding coasts can only accept the inevitable loss of their property to the attack of natural forces, but residents of coasts in which deposition exceeds erosion are sometimes presented with alternatives. The choices are almost never simple. For example, should rivers be dammed to control devastating floods? If the dams are built, they will trap sediments on their way from mountains to coast. Beaches within the coastal cell fed by the dammed river will shrink because the sand on which they depend (to replenish losses at the shore) is blocked. Alarmed coastal residents will then take

a The shoreline as it appeared in 1931.

b The same shoreline in 1949 after the breakwater was built. The boat anchorage formed by the breakwater is filling with sand deposited by disruption of the longshore current.

c The breakwater has deteriorated and can now be overtopped by waves. This 2007 image shows the beach has returned to its earlier shape.

Figure 12.34 Growth of a beach protected by a breakwater: Santa Monica, California.

steps to hang onto whatever sand remains. They may try to trap "their" beaches by erecting jetties or **groins**, short extensions of rock or other material placed at right angles to longshore drift, to stop the longshore transport of sediments. This temporary expedient usually accelerates erosion downcoast (**Figure 12.35a**; see also Figure 12.16b). Diminished beaches then expose shore cliffs to accelerated erosion. Wind-wave energy that would have harmlessly churned sand grains now speeds the destruction of natural and artificial structures. Seawalls don't help either. They increase beach erosion by deflecting wave energy onto the sand. Churning by this increased energy eventually undermines the seawall, causing it to collapse (**Figure 12.35b**). The importation of sand trapped behind dams (or from other sources) is also only

a temporary—and very expensive—expedient (**Figure 12.35c**). Scenes like that shown in **Figure 12.36** will be more common.

On high-energy coasts, harbor entrances can be protected by breakwaters made of large boulders transported to the site or by artificial structures called *dolos*. Dolos (**Figure 12.37**) interlock to form a permeable wall that dissipates wave energy and retards sand movement into harbors.

What are the implications of these unlooked-for sand movements? Douglas Inman, former director of the Center for Coastal Studies at Scripps Institution of Oceanography, believes that *at least 20% of the beach-bounded coastline of the United States is in danger of serious or catastrophic alteration.* On the West Coast, a long period of relatively mild weather may be ending. During

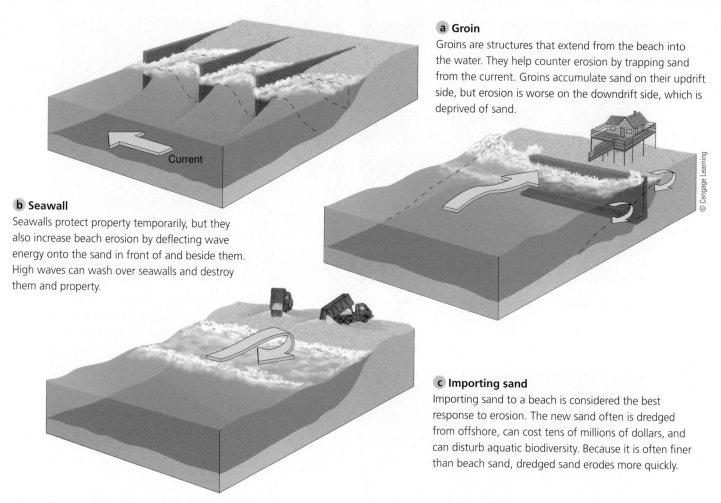

a Groin

Groins are structures that extend from the beach into the water. They help counter erosion by trapping sand from the current. Groins accumulate sand on their updrift side, but erosion is worse on the downdrift side, which is deprived of sand.

© Cengage Learning

b Seawall

Seawalls protect property temporarily, but they also increase beach erosion by deflecting wave energy onto the sand in front of and beside them. High waves can wash over seawalls and destroy them and property.

c Importing sand

Importing sand to a beach is considered the best response to erosion. The new sand often is dredged from offshore, can cost tens of millions of dollars, and can disturb aquatic biodiversity. Because it is often finer than beach sand, dredged sand erodes more quickly.

Figure 12.35 Some measures taken to slow beach erosion.

Figure 12.36 Hurricane Ivan battered Dauphin Island, at the southwestern entrance of Mobile Bay, in 2004; Hurricane Katrina finished the job 1 year later. The extent of beach erosion is dramatically clear.

Tyrone Turner/National Geographic Creative

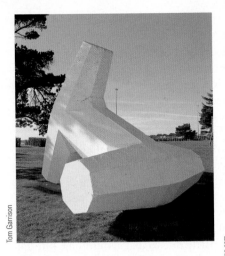

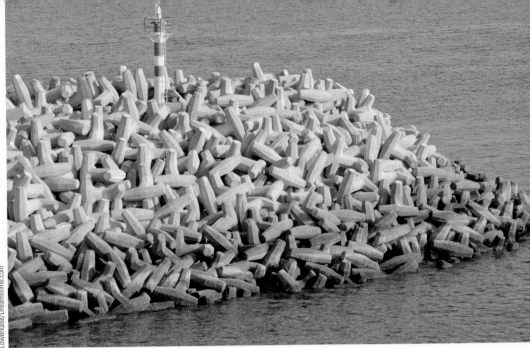

a Huge reinforced concrete structures called *dolos* are used to dissipate wave energy at high-energy shorelines, especially at harbor entrances and breakwaters.

Figure 12.37 Shore strengthening.

b Resembling a child's jacks, dolos interlock to reinforce a shore.

this time, people felt it was safe to build close to the shore. Increased dam building and breakwater, jetty, and groin construction have made southern California's beaches more vulnerable. In the 1997–1998 El Niño, coastal California alone suffered losses exceeding US$750 million. The barrier islands of the Atlantic and Gulf coasts are at least as vulnerable.

Shores that look permanent through the short perspective of a human lifetime are in fact among the most temporary of all marine structures. Let's enjoy them in their present stages.

CONCEPT CHECK

Before going on to the next section, check your understanding of some of the important ideas presented so far:

Generally speaking, would you say human intervention in coastal processes has been largely successful in achieving long-term goals of stabilization?

Again, generally speaking, would you say beaches on U.S. coasts are growing, shrinking, or staying about the same size?

Chapter in Perspective

In this chapter you learned that the *location* of a coast depends primarily on global tectonic activity and the ocean's water volume, while the *shape* of a coast is a product of many processes: uplift and subsidence, the wearing-down of land by erosion, and the redistribution of material by sediment transport and deposition. Coasts are classified as erosional coasts (on which erosion dominates) or depositional coasts (on which deposition dominates). Natural rock bridges, tall stacks, and sea caves are found on erosional coasts. Depositional coasts often support beaches, accumulations of loose particles. Generally, the finer the particles on the beach, the flatter its slope.

Beaches change shape and volume as a function of wave energy and the balance of sediment input and removal. Coral reefs and estuaries are among the most complex and biologically productive coasts. Human interference with coastal processes has generally accelerated the erosion of coasts near inhabited areas.

In the next chapter you'll learn that the study of oceanography includes a marvelous variety of living things. The next chapter's discussion of the general nature and characteristics of marine life will launch us into the biological part of our journey.

1. I have seen photos of the massive coastal modifications being made in Dubai. What happens when humans modify a coastline so extensively?

The Persian Gulf Emirate of Dubai has embarked on the most extensive human-made coastal developments in the world (**Figure 12.38**). Most of the additions were islands shaped like palm trees to maximize waterfront footage. Construction of the first island, Palm Jumeirah, began in June 2001. Shortly after, the Palm Jebel Ali was announced and reclamation work began. The Palm Deira, the third, is planned to have a surface area larger than that of Paris! If completed, it will become the world's largest man-made island, housing more than a million people. The first two islands will require about 100 million cubic meters of rock and sand. As planned, Palm Deira will be composed of more than 1 billion cubic meters of material! Between the three islands are planned more than 100 luxury hotels, hundreds of exclusive residential beachside villas, apartments, and marinas. Residents began moving into their Palm Jumeirah properties at the end of 2006.

There's more. Nearby lies The World, an archipelago of 300 artificial islands in the shape of the continents. The entire development is 9 kilometers (6 miles) in length by 6 kilometers (4 miles) wide and surrounded by an oval breakwater. Overall development cost of The World was estimated to be US$14 billion in 2005. And then there's The Universe... *If* present plans proceed, the coast of Dubai will exceed 440 miles (700 kilometers), an increase of 10 times!

It will come as no surprise that massive dredging and mining activities in a relatively confined and shallow area have greatly

Figure 12.38 The extensively modified coast of the Emirate of Dubai.

Motivate Publishing/Gallo Images/Getty Images

disrupted offshore waters. Rocks and sand have buried oyster beds and smothered coral reefs. Altered currents have eroded the mainland shore. Elsewhere currents are too weak: new island residents are appalled by the odorous and unsightly algae blooms in channels between the "fronds" of the palm-shaped constructions. Billions of additional dollars have been spent to raise the second and third palm islands by an additional meter (3.3 feet) against a predicted global rise in sea level.

And the original coastline? A stretch of Jumeirah Beach, a magnet for Western tourists and home to Burj Al Arab, arguably the world's most expensive and elaborate hotel, has been beset by a muddy brown tide of toilet paper, raw sewage and chemical waste. This inundation is partly due to illegal dumping, but the diminished natural circulation of seawater along the beach has greatly worsened the situation.

A combination of slowing world economies, high oil prices, currency revaluations, and political maturity are forcing more than a few residents of Dubai to rethink their priorities. Plans are being scaled back. As you have learned by reading this chapter, modifying any coast is an effort full of surprises.

2. My foot tends to sink whenever I stand on the beach and let water from a wave run over it. The sand moves away from the edges of my foot, and I sink in. Why?

This is a good example of water's ability to carry more sediment as its speed of flow increases. Your foot interrupts the flow of water up or down the beach after a wave breaks, and the water must speed up to get around your foot. Fast-running water moves sand more effectively than slow-running water, so the sand immediately next to your foot is removed. Could this be the "undertow" mentioned by inexperienced swimmers? This process, termed *scouring*, becomes a problem when structures are placed in shallow water.

3. What are those little white pellets I find on the beach along the high-tide line? Surfers call them "nurdles."

Those ubiquitous, insidious particles are the raw material for molded plastic goods. They are transported from producers to fabricators in containers loaded onto container ships. The pellets escape if a container is mishandled, breaks open during a storm, or is lost overboard. Virtually indestructible, "nurdles" float with the winds and currents until they encounter a shore. One researcher calculated that just 25 containers would carry enough plastic pellets to spread 100,000 "nurdles" per mile along all the seashores of the world!

You'll learn more about the problem of plastic in the ocean in this book's last chapter.

4. I hope someday to live near the ocean. What should I look for in buying property there?

Firm ground! Coastal Maine would be an ideal bet. The dense metamorphic rock of much of the Maine shore is stable (within the human time frame) and hard—ideal footings for a house. Make sure the site is far enough inland to avoid high surf, storm surge, and tides. If the winters in Maine don't appeal to you, coastal Florida might make a good choice—if your children aren't hoping to inherit the property. If you insist on building on a barrier island, make sure your home is on the mainland side of the southern end! Parts of the Pacific coast are all right, but local variability on that active margin makes some knowledge of the geological history of the area very valuable. For example, some parts of the San Diego shoreline are eroding about 3 meters (10 feet) per year—hardly a solid investment.

5. What's the world's longest uninterrupted beach?

Visitors to Long Beach, Washington, in the northwestern United States, are greeted by a sign that says: "World's Longest Beach." At best, the beach is 37 to 45 kilometers (23–28 miles) long—not even close to world-class. The top honor probably belongs to Coorong-Middleton Beach in South Australia, with some 194 kilometers (120 miles) of continuous sandy beach.

6. I've noticed that beaches composed of pebbles and cobbles tend to be fairly steep, but beaches made of fine sand are relatively flat. Why is that?

A beach's slope is determined by the energy necessary to move the sand grains of which it is composed. Shallow water, smaller waves, and coarse grains combine to form steep beach slopes, but large waves can very easily move small particles and smooth out the slope.

TERMS AND CONCEPTS TO REMEMBER

atoll	coastal cell	fringing reef	salt-wedge estuary
backshore	coral reef	groin	sand spit
backwash	delta	high-energy coast	sandbar
barrier island	depositional coast	inlet	sea cave
barrier reef	drumlin	lagoon	sea cliff
bay mouth bar	erosion	longshore bar	sea island
beach	erosional coast	longshore current	shore
beach scarp	estuary	longshore drift	sound
berm	eustatic change	longshore trough	swash
berm crest	fjord	low-energy coast	tombolo
breakwater	fjord estuary	moraine	wave-cut platform
coast	foreshore	partially mixed estuary	well-mixed estuary

STUDY QUESTIONS

Thinking Critically

1. How is an erosional coast different from a depositional coast?
2. What features would you expect to see along an erosional coast? A depositional coast? What determines how long the features will last?
3. What two processes contribute to longshore drift? What powers longshore drift? What is the predominant direction of drift on U.S. coasts? Why?
4. What are some of the features of a sandy beach? Are they temporary or permanent? What is the relationship between wave energy on a coast and the size (or slope, or grain size) of beaches found there?
5. How are deltas classified? Why are there deltas at the mouths of the Mississippi and Nile rivers, but not at the mouth of the Columbia River?
6. What is a coastal cell? Where does sand in a coastal cell come from? Where does it go?
7. How are estuaries classified? Upon what does the classification depend? Why are estuaries important?
8. Compare and contrast the U.S. West, East, and Gulf coasts.
9. How do human activities interfere with coastal processes? What steps can be taken to minimize loss of life and property along U.S. coasts?

Thinking Analytically

Internet sites like Zillow™ allow potential buyers to track home prices. Given the tenuous nature of some coastal real estate, use a trend-tracking feature to note price fluctuations. Do you suspect any of these might be linked to coastal erosion? Good places to start your search: the barrier islands of Maryland; Miami Beach; the town of Pacifica in northern California; the San Diego suburb of Ocean Beach.

GLOBAL GEOSCIENCE WATCH

Visit www.cengagebrain.com to access MindTap, a complete digital course which includes access to Global Geoscience Watch and more. Choose one of the following three topics to research: coastal armoring techniques like seawalls and revetments, soft engineered solutions like beach nourishment and vegetated buffers, or managed retreat. Within the GREENR database, search for your topic (for further information to support your topic, search "sea level rise"). Be prepared to discuss the characteristics, advantages, and disadvantages of your assigned strategy in an in-class debate that will be held to determine which strategy would work best for a particular coastline.

CENGAGE**brain**.com Visit www.cengagebrain.com to access course materials for this text, including interactive learning tools, videos, and more.

13 Life in the Ocean

KEY CONCEPTS

All Earth's life forms are related. All have apparently evolved from a single ancient instance of origin. Exploring Alaska's Seamounts 2002, NOAA/OER

Evolution happens. Organisms change as time passes, adapting by natural selection to their environments. Portrait of Charles Darwin (1809–1882) 1883 (oil on canvas), Collier, John (1850–1934) (after)/National Portrait Gallery, London, UK/The Bridgeman Art Library

All life activity is involved, directly or indirectly, in energy transformation and transfer. NASA

The atoms and small molecules that make up the biochemicals, and thus the bodies, of organisms move between the living and nonliving realms in biogeochemical cycles. © Cengage Learning

The success of marine organisms depends on their relation with the physical and biological factors influencing them. Rapid chance may result in mass extinction. Don Davis

Fishes swarm around a diver at Cabo Pulma, Baja California, Mexico.
Octavio Aburto

David Ashley/Shutterstock

life-forms are related; all apparently share a common origin. Earth and life have grown old together, generation by persistent generation, across some 4 billion years (**Figure 13.1**).

13.1 Life on Earth Is Notable for Unity and Its Diversity

Life on Earth exhibits unity and diversity: *diversity* because Earth may house as many as 100 million different species (kinds) of living organisms; *unity* because all species share the same underlying mechanisms for capturing and storing energy, manufacturing proteins, and transmitting information between generations. *This idea is especially important: In a sense, all life on Earth is fundamentally the same; it's just packaged in different ways.* All Earth's

> In a sense, all life on Earth is fundamentally the same; it's just packaged in thousands of different ways.

13.2 The Concept of Evolution Helps Explain the Nature of Life in the Ocean

Marine life has not accidentally capitalized on its rich fluid home. Earth's organisms did not arise in a few thousand years. They have changed, generation by generation, over more than 4 billion years. The ability of living things to change through time to fit the physical and chemical environment, to become ever more efficient at extracting energy from their surroundings, to colonize virtually every location capable of sustaining them, and finally to investigate themselves by scientific logic has come about through **evolution**.

Evolution means change. Clothing styles evolve, computer operating systems evolve, methods of government evolve, our perceptions of the world evolve; *things change*. The concept that animals and plants might be capable of change with the passage of time was not a popular idea in the 19th century. Indeed, the proposal that organisms *do* change with time began what has since been called a revolution in biology. The unlikely revolutionaries were Charles Darwin (**Figure 13.2**), a quiet and thoughtful English naturalist, and Alfred Wallace, a British biologist working in the Malay Archipelago.

Evolution Appears to Operate by Natural Selection

Why was Darwin and Wallace's theory of evolution controversial then, and why is it capable of stirring passions even today? By the mid-1850s, the two men had independently discovered a mechanism—now called **natural selection**—for *how* living things might evolve (change) with the passage of time. Here are Darwin's main points:

1. In any group of organisms, more offspring are produced than can survive to reproductive age.
2. Random variations occur in all organisms. Some of these traits are inheritable; they can be passed on to offspring.
3. Some inheritable traits make an organism better suited to its environment (most do not).

Figure 13.1 A tardigrade, a small aquatic animal with eight legs. Commonly called "water bears," these tiny (1 millimeter; 0.04 inch) animals are noted for their ability to survive extreme conditions of temperature, pressure, ionizing radiation, starvation, and even complete vacuum. About a million tardigrades exist for every one of us.

Eye of Science/Science Source

Figure 13.2 Charles Darwin, codiscoverer of the principle of natural selection, as an old man. Darwin visited the Galápagos Islands off the Pacific coast of South America during his voyage aboard HMS *Beagle* in the early 1830s; his landmark book *On the Origin of Species* was published in 1859.

④ Because bearers of favorable traits are more likely to survive, they are also more likely to reproduce successfully than bearers of unfavorable traits. Favorable traits tend to accumulate in the population; they are *selected*. Unfavorable traits are weeded out by competition.

⑤ The physical and biological (*natural*) environment itself does the selection. Favorable traits that contribute to the organism's success show up more often in succeeding generations (if the environment stays the same). If the environment changes, other traits become favorable, and the organisms with those traits live most effectively in the new environment.

It is easy to see how random variations are selected for or selected against by environmental pressures, but how do entirely new traits arise? They come about by spontaneous **mutation**, an inheritable change in an organism's genes (the structures that contain its assembly instruc-

tions). The vast majority of mutations are unfavorable, and the organisms possessing them are eliminated by other organisms or by the physical environment. For example, a tuna born with no eyes could not see to feed, so it would not live to reproductive age. But a tuna born with extraordinarily good eyesight might have more to eat than its cohorts, and being better nourished it might be especially effective in its reproductive efforts, spreading its genes far and wide. It is the accumulation of these favorable traits, either variations or mutations—and the elimination of unfavorable traits—that makes life possible in changing conditions on Earth.

Note that although mutations occur randomly, evolution by natural selection is anything but random. The natural environment winnows favorable mutations from unfavorable ones—hence the origin of the term *natural selection*. The process takes a great deal of time, but time is in abundant supply now that geologists have shown the Earth to be about 4.6 billion years old.

The evolutionary viewpoint has provided a new way of looking at life: An organism is a vessel holding a particular combination of traits, *testing* that combination. If the organism is successful it will reproduce, and the genes responsible for its good traits will continue in the population. If the organism is not successful, it will not reproduce and that combination of genes will be eliminated.

There is no biological predeterminism. Organisms don't *want* to evolve, and individual organisms *don't* evolve. Rather, generation after generation, groups of individuals respond to environmental pressures by change. The changes can be in shape, size, color, biochemistry, behavior, or any other aspect of the organism. Evolution by natural selection is the accumulation of these beneficial inheritable structural or behavioral traits, known as favorable **adaptations**. Again, organisms with favorable adaptations have more reproductive success than less well adapted organisms **(Figure 13.3)**.

A **species** is a group of actually (or potentially) interbreeding organisms that is reproductively isolated from all other forms of living things. How do new species arise? One way is by physical isolation, such as that created when land animals or birds are rafted or blown from a mainland shore to an isolated oceanic island. Because the number of breeding animals within a species on an island may be small, evolutionary change may be rapid; that is, favorable traits may accumulate quickly in the population, and, generation after generation, the species will change relatively rapidly to suit its new habitat. In general, the smaller the reproducing population, the more rapid the rate of evolutionary change.

Evolution is the maintenance of life under changing conditions by continuous adaptation of successive generations of a species to its environment.

For example, on the Galápagos Islands off the coast of Ecuador, Darwin observed finches and marine lizards, most of which closely resembled their mainland South American ancestors. They were not the same species of birds and reptiles that occurred in mainland South America, however. This suggested to Darwin that isolation was a driving force of evolution.

Evolution, then, is the maintenance of life under changing conditions by continu-

Figure 13.3 This small squid combines excellent eyesight, an aquatic form of jet propulsion, sensitive arms and tentacles, and a blending of camouflage with communication to cruise the reef looking for trouble. Its ancient body architecture has seen its ancestors through millions of years of challenges—this unique suite of evolutionary adaptations makes it ideal for its environment. Redbrickstock.com/Alamy

💬 **THINKING BEYOND THE FIGURE**
What would happen if the squid's environment gradually changed—would all squid die?

ous adaptation of successive generations of a species to its environment. It is a remarkable, beautiful, productive theory.[1]

Evolution "Fine-Tunes" Organisms to Their Environment

Evolutionary theory contains important implications for marine science. The ocean has a much larger inhabitable volume than dry land, and it is often easier to live there than in either the terrestrial or freshwater realms. There are more ways of "making a living" in the ocean than on land.

> **T**he environment isn't right for the organisms; rather, the organisms are right for the environment.

Indeed, some important oceanic lifestyles have no terrestrial counterparts. For example, filter feeding—practiced by clams, barnacles, and some whales—relies on a relatively dense fluid matrix and is therefore not seen on land. Also, marine food chains tend to be more complex than terrestrial ones and contain more trophic levels. Marine life has evolved in countless ways to take advantage of nearly every scrap of energy and nearly every available space.

Since physical conditions in the open ocean are relatively uniform, large marine animals with similar lifestyles but different evolutionary heritages eventually tend to look much the same. That is, similar conditions may result in coincidentally similar organisms. The shark (a fish), the ichthyosaur (an extinct marine reptile), the penguin (a bird), and the porpoise

---

[1]If you think "theory" is synonymous with "maybe," you may wish to review the distinctions between theories and laws in the discussion of the scientific method in Chapter 1.

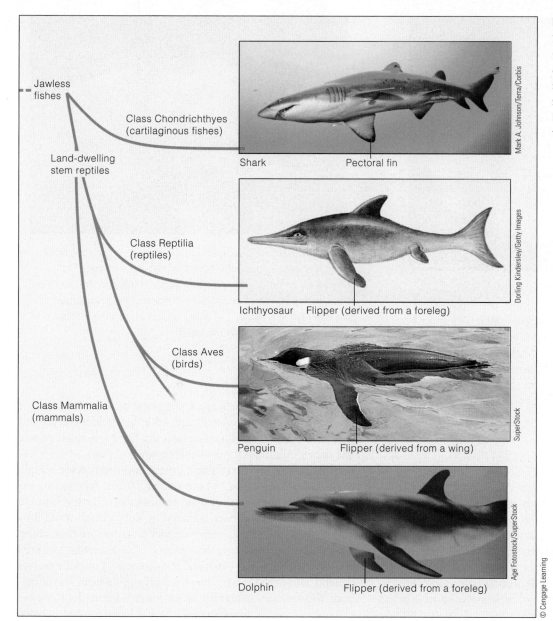

Figure 13.4 Convergent evolution in sharks, ichthyosaurs, penguins, and dolphins. Selection for adaptations that permitted rapid swimming resulted in superficially similar shapes among these four kinds of vertebrates, even though they are only remotely related.

Jawless fishes

Class Chondrichthyes (cartilaginous fishes)

Shark — Pectoral fin

Mark A. Johnson/Terra/Corbis

Land-dwelling stem reptiles

Class Reptilia (reptiles)

Ichthyosaur   Flipper (derived from a foreleg)

Dorling Kindersley/Getty Images

Class Aves (birds)

Penguin   Flipper (derived from a wing)

SuperStock

Class Mammalia (mammals)

Dolphin   Flipper (derived from a foreleg)

Age Fotostock/SuperStock

© Cengage Learning

(a mammal) are only remotely related, yet they resemble each other in shape, because the physics of rapid movement through water requires a similar streamlined shape (see **Figure 13.4**). Traits leading to this shape were independently selected by environmental conditions. These accumulated adaptations resulted in superficially similar animals, each derived from different and diverse stock. The process is known as **convergent evolution**.

Through these processes, life and Earth change together. Life is tenacious; it has survived catastrophe and calm. In every instance, however, *the environment isn't right for the organisms; rather, the organisms are right for the environment*. One is reminded of author Pår Lagerkvist's observation, "Things need be as they are." Living things have adapted to the physical conditions of their environment, and they continue to increase in numbers, complexity, and efficiency.

CONCEPT CHECK

Before going on to the next section, check your understanding of some of the important ideas presented so far:

Is evolution by natural selection a random process?

How is evolution by natural selection thought to operate?

How are new species thought to originate?

What's convergent evolution?

13.3 Rapid, Violent Change Causes Mass Extinctions

The physical and biological environment in which living things evolve changes as time passes, and the changes are not always gradual. The biological history of the Earth has been inter-

Figure 13.5 What an asteroid looks like, courtesy of the *Dawn* spacecraft. Asteroids are rocky, metallic bodies with diameters ranging from a few meters to 1,000 kilometers (600 miles). Most were swept into the planets during their formation, but about 6,000 large asteroids are still orbiting the sun in a belt between Mars and Jupiter. Unfortunately, the orbits of many dozens of others take them across Earth's orbit. This asteroid, named *Vesta,* would extend from Washington, D.C., to New York City. The image is in 3-D—get your old red-blue glasses out of the drawer and have a look.

Figure 13.6 Filled with snow and ice, northeastern Canada's Manicouagan crater is easily visible from the International Space Station. The crater is a multiple-ring structure about 100 kilometers (60 miles) across, with a 70-kilometer (40-mile) diameter inner ring its most prominent feature. About 214 million years old, its circular shape has been preserved in unusually hard rock.

rupted—catastrophically—at least six times in the past 450 million years. In these events, known as **mass extinctions**, a great many species died off simultaneously (in geological terms). Scientists are not certain of the causes of the mass extinctions, but leading candidates for a couple of these events include the collision of the Earth with an asteroid or comet.

Flocks of asteroids **(Figure 13.5)** orbit the sun along with the Earth and the other planets. Some of these asteroids have orbits that cross our own, and meetings are inevitable. The Earth and its neighbors are pocked with impact craters **(Figure 13.6)** as evidence of these meetings. The consequences to Earth of a collision with even a small asteroid are all but unimaginable. As you saw in Box 5.1, an asteroid only 10 kilometers (6 miles) in diameter would strike with an energy equivalent to the explosion of half a trillion tons of TNT. More than 100 million metric tons of the Earth's crust would be thrown into the atmosphere, obscuring the sun for decades and causing acid rain that would poison the planet's surface. Concussive shock waves would shatter structures, crush large organisms, and trigger earthquakes for a radius of hundreds of kilometers. If the impact occurred in the Atlantic Ocean, say 1,600 kilometers (1,000 miles) east of Bermuda, the resulting tsunami would wash away the resort islands and swamp most of Florida. Boston would be struck by a 100-meter (300-foot) surge of water. A hole more than 25 kilometers (16 miles) wide and perhaps 10 kilome-

> **" Civilization exists by geological consent, subject to change without notice. "** —WILL DURANT

ters (6 miles) deep would mark the point of impact. Clouds of fine particles, accelerated to escape velocity, would travel around the sun in orbits that would intersect the Earth's; a steady rain of fine debris might fall for tens of thousands of years.

These kinds of cataclysmic events have been disquietingly common in Earth's past. Where are the craters? As you may recall from the discussion of plate tectonics in Chapter 3, much of the ocean floor has been recycled by the movement of lithospheric plates, so any undersea impact craters more than about 100 million years old have disappeared. The distortion and erosion of continents has obscured the outlines of ancient craters on land, although they are more readily visible from space than from the surface—especially if we know what to look for (as in Figure 13.6). Evidence for one massive impact has been bolstered by the discovery of a thin, worldwide layer of iridium-rich continental rock dated at the boundary between Cretaceous and Tertiary periods (see Appendix 2). Iridium is rare on the Earth but common in asteroids. The thin iridium-rich layer may have formed from the dust settling after a collision some 65 million years ago.

Asteroids are not the only culprits. The greatest of all mass extinctions was caused by massive volcanic eruptions in what is now Siberia. The P-T event (so named because it marks the transition from the Permian to the Triassic periods in Earth history about 251 million years ago) drove up to 95% of all marine species and nearly 70% of terrestrial vertebrates to extinction! Death was caused by massive injections of carbon dioxide into the atmosphere. Atmospheric temperature rose by 10°C (18°F).

Billions of tons of sulfur dioxide gas also streamed from the volcanoes. Dissolved in water, sulfur dioxide makes sulfuric acid.

Rain in the Northern Hemisphere became as acidic as lemon juice (estimated at pH 2)—catastrophic to exposed vegetation. Oceanic pH also fell dramatically. Because so much biodiversity was lost, recovery from this mass extinction event took perhaps 10 million years, significantly longer than any other extinction event.

Clearly these tumultuous interruptions do not represent biological business as usual. The animals and plants, bacteria and single-celled organisms we see on the Earth have evolved from the survivors of these cataclysmic events.

CONCEPT CHECK

Before going on to the next section, check your understanding of some of the important ideas presented so far:

Can you think of any way to prevent a cataclysmic asteroid or comet impact once the object's path has been shown to be on a certain collision course?

Why do we see relatively few impact craters on Earth?

13.4 Oceanic Life Is Classified by Evolutionary Heritage

Long before Darwin's and Wallace's discovery of the mechanism of evolution, biologists realized the value of being able to classify living things into categories and give them universally understood names. The study of biological classification is called **taxonomy**.

Systems of Classification May Be Artificial or Natural

Classification schemes have been around for as long as people have looked at living things. Putting animals in one category and plants in another is an ancient distinction, for example.

The Greek philosopher Aristotle proposed a system of classifying animals based on their exterior similarities, but his results were not very useful. Using his system we would place airline pilots, gliding squirrels, flying fish, owls, and grasshoppers into the same group as birds because each can fly! Such an arrangement is an **artificial system of classification**. (Another artificial system of classification is the arrangement of books by jacket color or page size or typeface.) By contrast, the **natural system of classification** for living organisms that biologists use today relies on the evolutionary history and developmental characteristics of organisms. We place all insects together regardless of their flying ability just as we place all books by Melville together, all compositions by German baroque composers together, and all sea stars together because each group has a *common underlying natural origin*. The groups are arranged systematically—that is, in some order that makes structural and evolutionary sense.

One of the first persons to classify groups of organisms into natural categories was the 18th-century Swedish naturalist Carl von Linné, or as he called himself, **Linnaeus (Figure 13.7)**. In his zeal to classify every aspect of the natural world, Linnaeus invented three supreme categories, or **kingdoms**: animal, veg-

etable, and mineral. Linnaeus grouped organisms by external similarities and likenesses of developmental details. Linnaeus's great contribution was a system of classification based on **hierarchy**, a grouping of objects by degrees of complexity, grade, or class. In this boxes-within-boxes approach, sets of small categories are nested within larger categories. Linnaeus devised names for the categories, starting with kingdom (his largest category) and passing down through phylum, class, order, family, and genus, to species (the smallest category). In 1758, he published a catalog of all animals then known, his monumental *Systema Naturæ* (The System of Nature).

Much has changed since Linnaeus's time. Modern biologists have access to decoded sequences of nucleic acids, the molecules that determine an organism's genetic heritage. Analysis of fundamental similarities and differences in these molecules suggests three main kinds of living things above the Linnaean level of kingdom: *Bacteria*, *Archaea*, and *Eukarya*. These overarching groups have been labeled **domains**.

Michael Nicholson/Historical/Corbis

Figure 13.7 Carolus Linnaeus—the father of modern taxonomy—in Laplander costume. (He went on a scientific expedition to northern Finland in 1732.) This 1775 painting by Alexander Roslin is on view at the Royal Science Academy of Sweden.

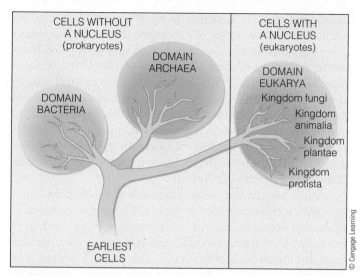

Figure 13.8 A family tree showing the relationship of the three domains of living things presumably evolved from a distant common ancestor. The Bacteria and Archaea contain single-celled organisms without nuclei or organelles; collectively, they are called *prokaryotes*. The fungi, protists, animals, and plants contain organisms with cells having nuclei and organelles; collectively, they are called *eukaryotes*.

Bacteria and Archaea are both domains of very small single-celled organisms that lack distinct compartments within their cells (there is no cell nucleus, for example). They have some important chemical and genetic differences, however, and the walls surrounding their cells are structurally different. The Bacteria have evolved diverse metabolic abilities—some are photosynthetic, others heterotrophic—and are familiar to us as decomposers and disease agents. Some Archaeans are called *extremophiles* because of their ability to withstand extremely hot or corrosive environments.

Cells of the domain Eukarya are larger than those of the Bacteria or Archaea, and each cell has a nucleus. Most Eukarya—including all animals and plants—are multicellular. Some fungi and some protists (protozoa and algae) are multicellular while others are single-celled Eukarya. Interestingly, the Eukarya and Archaea share so many biochemical characteristics that some researchers suggest they are more closely related to each other than either is to the Bacteria.

The names and characteristics of the domains and their largest subdivisions are listed in **Figure 13.8. Figure 13.9** shows the classification of a familiar seagull using the Linnaean method. Note the nested arrangement of category-within-category, each category becoming more specific with every downward step.

Figure 13.9 The modern system of biological classification using the California gull *(Larus californicus)* as an example. Note the boxes-within-boxes approach—a hierarchy.

💬 **THINKING BEYOND THE FIGURE**

Can you think of any organisms you refer to only by their scientific names?

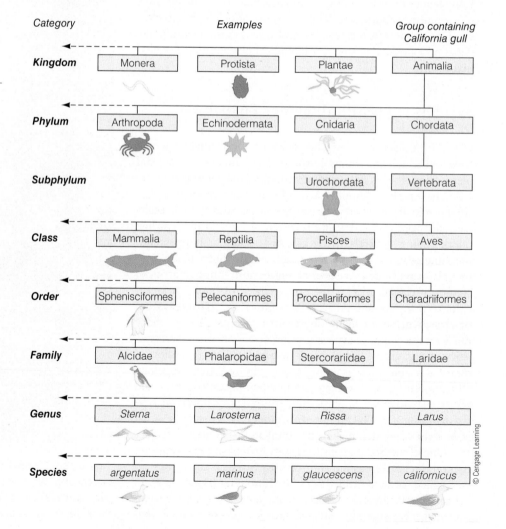

Scientific Names Describe Organisms

Linnaeus also perfected the technique of naming organisms. The *genus* and *species* names—the names of the last two nested categories—constitute an organism's **scientific name**. *Octopus bimaculatus* is the scientific name of a common west coast octopus: *Octopus* is the generic name, *bimaculatus* the specific name. A closely related species, *Octopus dofleini*, is a larger animal that ranges to Alaska. *Octopus bimaculatus* and *Octopus dofleini* are not interfertile (they're not the same species), but, as their shared generic name suggests, they *are* closely related.

The advantage of a scientific name over a common name is immediately apparent to anyone trying to identify a shell found on the beach. The same shell may have many different common names in many different languages, but it will have *only one scientific name*. When you discover that name in a good key to shells, you can use it to find references that will tell you what is known about the animal, its lifestyle, its range, and its evolutionary history.

CONCEPT CHECK

Before going on to the next section, check your understanding of some of the important ideas presented so far:

How is a natural system of classification different from an artificial system?

What are the three domains of living things?

How are organisms named?

What advantages do scientific names have over common names?

Linda E. Tway

Figure 13.10 Living or nonliving? The brightly colored "rock" in this picture is a colony of small marine photosynthesizers, primarily algae. The distinction between life and nonlife does not lie in composition or outward appearance but in the ability to manipulate energy.

 THINKING BEYOND THE FIGURE

We seem instinctively to know whether something is alive or not. How do you suppose we make that judgment?

13.5 The Flow of Energy Allows Living Things to Maintain Complex Organization

Biologists know there is nothing special about the atoms or energy of life, nothing to distinguish them from their nonliving counterparts. What *does* distinguish life from nonlife is the ability of living things to capture, store, and transmit energy—and the ability to reproduce.

It would seem easy to differentiate between the living and nonliving components of the marine environment, but, as **Figure 13.10** suggests, we cannot always tell the difference. The same atoms move continuously in and out of living and nonliving systems. With your last breath, you exhaled millions of carbon atoms that entered your body as food in your last meal. Before the day is done, some of those atoms may be incorporated into a nearby houseplant. The free exchange of identical components between life and nonlife complicates any attempts at formal definition. Yet intuitively, we all make this distinction when we observe the organization of matter in living things, and—as you will discover in a moment—*especially* when we consider the ways living things manipulate energy.

Living matter can't function without **energy**, the capacity to do work. Living organisms can't create new energy, but they can transform one kind of energy to a different kind. A plant can transform light energy into chemical energy; an animal can transform chemical energy into energy of movement by the muscles and can transform energy of movement into heat, and so on. In one way or another, all life activity is involved, directly or indirectly, in energy transformation and transfer. Energy must therefore be central to anyone's definition of life.

The main source of energy for living things on Earth is the sun. Life prospers, becomes ever more complex, and evolves into millions of forms by accepting sunlight and radiating waste heat to the cold of space. As we will see, with some striking exceptions, most organisms get their power directly or indirectly through the capture, storage, and transmission of energy from sunlight. Light energy is transformed into chemical energy and finally into heat as organisms grow, reproduce, and wear out.

How does this work?

Energy Can Be Stored through Photosynthesis

The sun produces enormous quantities of energy—some of it in the form of visible light, a tiny portion of which strikes Earth. Only about one part in 2,000 of the light that reaches Earth's surface is captured by organisms, but that "small" input of energy powers nearly all the growth and activity of living things on Earth's surface. Light energy from the sun is trapped by chlorophyll in organisms called *primary producers* (certain bacteria, algae, and green plants) and changed into chemical energy. The chemical energy is used to build simple carbohydrates and other

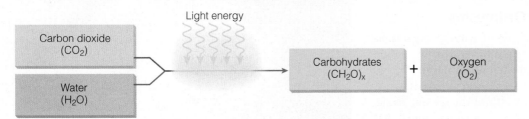

a In photosynthesis, energy from sunlight is used to bond six separate carbon atoms (derived from carbon dioxide) into a single energy-rich six-carbon molecule—the sugar glucose—here represented as $(CH_2O)_x$. The pigment chlorophyll absorbs and briefly stores the light energy needed to drive the reactions. Water is broken down in the process, and oxygen is released.

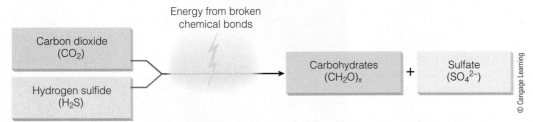

b A form of chemosynthesis. In this example, 6 molecules of carbon dioxide combine with 6 molecules of oxygen and 24 molecules of hydrogen sulfide to form glucose—$(CH_2O)_x$. (Other products include molecules of sulfate and water.) The energy to bond carbon atoms into glucose comes from breaking the chemical bonds holding the sulfur and hydrogen atoms together in hydrogen sulfide.

Figure 13.11 Photosynthesis and chemosynthesis.

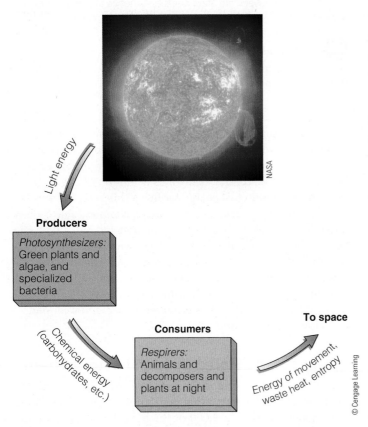

Figure 13.12 The flow of energy through living systems. At each step, energy is degraded (that is, transformed into a less useful form).

 **THINKING BEYOND THE FIGURE**

What happens to an organism if it has no access to energy? What happens if the organism cannot dump waste heat?

organic molecules—**food**—which is then used by the primary producers or eaten by animals (or other organisms) called *consumers*. Because light energy is used to synthesize molecules rich in stored energy, the process is called **photosynthesis**. A general formula for photosynthesis is shown in **Figure 13.11a**.

The stored energy is released when food is used for growth, repair, movement, reproduction, and the other functions of organisms. The breakdown of food eventually produces waste heat, which flows away from Earth into the coldness of space. This one-way flow of energy is shown in **Figure 13.12**.

Energy Can Also Be Stored through Chemosynthesis

Photosynthesis appears to be the dominant method of binding energy into carbohydrates on this planet (at least at Earth's surface), but there is another. **Chemosynthesis**, employed by some species of bacteria and archaea, is the biological conversion of simple carbon molecules (usually carbon dioxide or methane) into carbohydrate, using the oxidation of inorganic molecules (such as hydrogen gas or hydrogen sulfide or methane) as a source of energy. No sunlight is required. A general formula for a common type of chemosynthesis is shown in **Figure 13.11b**.

Some unusual forms of marine life depend on chemosynthesis, as we will see in the next two chapters. Until about 20 years ago, it was thought that chemosynthetic production of food in the ocean was relatively unimportant. Discoveries of extensive chemosynthetic communities at hydrothermal vents, deep in marine sediments, and in the seabed itself—and subsequent research in the microbiology of extreme environments—indicates that chemosynthesis may be far more extensive than previously recognized.

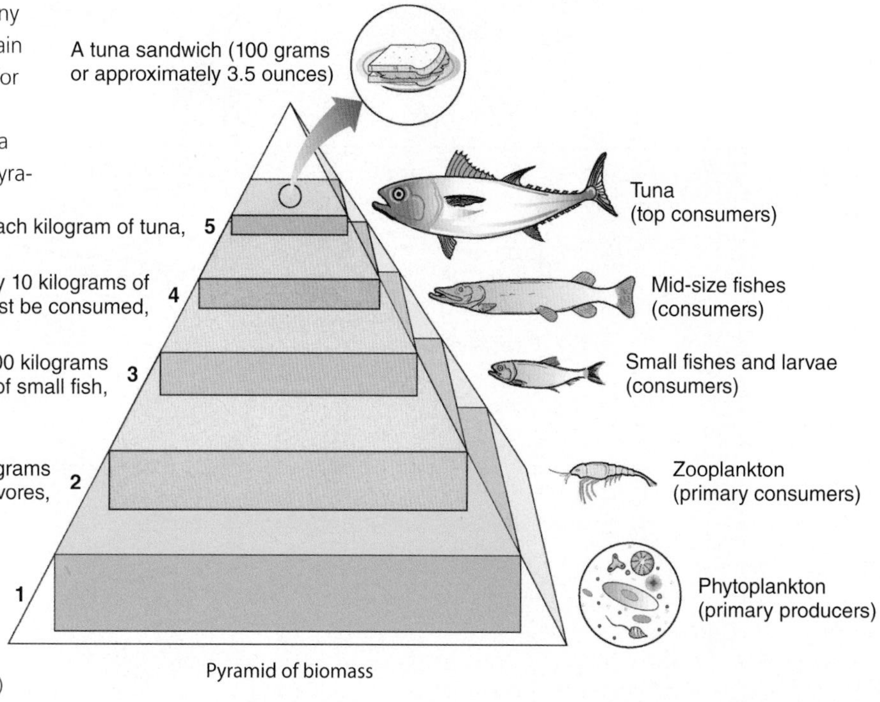

Figure 13.13 A generalized trophic pyramid. How many kilograms of primary producers are necessary to maintain 1 kilogram of tuna, a top carnivore? What is required for an average tuna sandwich? Using the trophic pyramid model shown here, you can see that 1 kilogram of tuna at the fifth trophic level (the fifth feeding step of the pyramid) is supported by 10 kilograms of midsize fish at the fourth, which in turn is supported by 100 kilograms of small fish at the third, which have fed on 1,000 kilograms of zooplankton (primary consumers) at the second, which have eaten 10,000 kilograms of phytoplankton (small autotrophs, primary producers) at the first. The quarter-pound tuna sandwich has a long and energetic history. (These figures have been rounded off to illustrate the general principle. The actual measurements are difficult to make and quite variable.)

A tuna sandwich (100 grams or approximately 3.5 ounces)

For each kilogram of tuna, 5 — Tuna (top consumers)

roughly 10 kilograms of mid-size fish must be consumed, 4 — Mid-size fishes (consumers)

and 100 kilograms of small fish, 3 — Small fishes and larvae (consumers)

and 1,000 kilograms of small herbivores, 2 — Zooplankton (primary consumers)

and 10,000 kilograms of primary producers. 1 — Phytoplankton (primary producers)

Pyramid of biomass

© Cengage Learning

Food Webs Disperse Energy through Communities

Photosynthetic and chemosynthetic organisms can be called either primary producers or **autotrophs** because they make their own food. The bodies of autotrophs are rich sources of chemical energy for any organisms capable of consuming them. **Heterotrophs** are organisms such as animals that must consume food from other organisms because they are unable to synthesize their own food molecules. Some heterotrophs consume autotrophs, and some consume other heterotrophs.

We can label organisms by their positions in a "who eats whom" feeding hierarchy called a **trophic pyramid**. The primary producers shown at the bottom of the pyramid in **Figure 13.13** are mostly chlorophyll-containing photosynthesizers. The animal heterotrophs that eat the primary producers are called **primary consumers** (or herbivores), the animals that eat the primary consumers are called secondary consumers, and so on to the **top consumer** (or top carnivore).

Note that the mass of consumers becomes smaller as energy flows toward the top of the pyramid. There are many small primary producers at the base, and a very few large top consumers at the apex. Only about 10% of the energy from the organisms consumed is stored in the consumers as flesh, so each level is about one-tenth the mass of the level directly below. The rest of the energy is lost as waste heat as organisms live and work to maintain themselves.

Pyramids such as the one in Figure 13.13 can lead to the misconception that one kind of fish eats only one other kind of fish, and so on. Real communities are more accurately described as food webs, an example of which is included as **Figure 13.14**. A **food web** is a group of organisms linked by complex feeding relationships in which the flow of energy can be followed from primary producers through consumers. Organisms in a food web almost always have some choices of food species.

So organisms interact with each other, feed on one another, and transfer energy as food from producing autotrophs through a web of consuming heterotrophs. Nearly all are ultimately dependent on sunlight and photosynthesis. What is the physical role of the ocean in this?

CONCEPT CHECK

Before going on to the next section, check your understanding of some of the important ideas presented so far:

What are the starting products for photosynthesis? The end products?

How is chemosynthesis different from photosynthesis?

What is an autotroph? A heterotroph? How are they similar? How are they different?

What is a trophic pyramid? What is the relationship of organisms in a trophic pyramid? Does this have anything to do with food webs? How?

13.6 Living Organisms Are Built from a Few Elements

We've been discussing energy and life. Let's turn to life's material substance. All of Earth's organisms are composed of about 23 of the 107 known chemical elements.[2] Four of these

---

[2]A periodic table of the elements—an arrangement of elements by their characteristics—is included as Appendix 7.

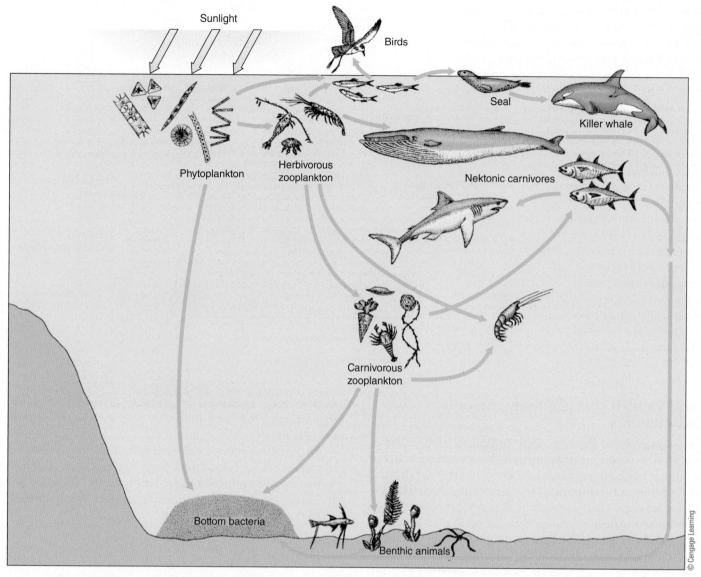

Figure 13.14 A simplified food web, illustrating the major trophic relationships leading to an adult killer whale. The arrows show the direction of energy flow. Note that feeding relationships are not as simple as one might assume from Figure 13.13.

elements—carbon, hydrogen, oxygen, and nitrogen—make up 99% of the mass of all living things, with nine additional elements composing nearly all of the remainder. Atoms of these elements combine to form the classes of biological chemicals common to all life. The main categories of these chemicals are probably familiar to you: carbohydrates, lipids (fats, waxes, oils), proteins, and nucleic acids (such as DNA, the primary molecule of heredity).

Before moving on, a central idea is worth repeating: Despite the astonishing variety of life-forms on Earth, biologists have come to appreciate that *unity, not diversity, is the central message of biology*. That is, the extraordinary sameness of biological chemicals as they are found in the structure of all living things strongly suggests all life is related by a common evolutionary history.

13.7 Elements Cycle between Living Organisms and Their Surroundings

The atoms and small molecules that make up the biochemicals, and thus the bodies, of organisms move between the living and nonliving realms in **biogeochemical cycles**. Living organisms are supported and sustained by huge, nonliving chemical re-

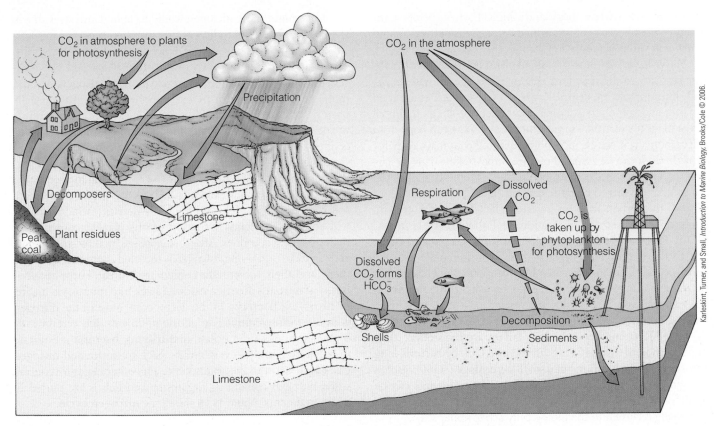

Figure 13.15 Carbon dioxide dissolved in seawater is the source of the carbon atoms assembled into food (initially glucose) by photosynthesizers and most chemosynthetic organisms. When this food is metabolized, the carbon dioxide is returned to the environment. Some carbon dioxide is converted into bicarbonate ions and incorporated into the shells of marine organisms. When these organisms die, their shells can sink to the bottom and be compressed to form limestone. Tectonic forces may eventually bring the limestone to the surface, where erosion will return the carbon to the ocean.

serves, and there is a large-scale transport of elements between the reserves and the organisms themselves. Sometimes the environment contains enough of a required element to sustain life; sometimes the element is in short supply. You learned in Chapter 6 about the ocean's density stratification. The tropical and temperate ocean is usually highly stratified, with a warm, less-dense layer of water—the *mixed zone*—separated from the cold, dense *deep zone* by a strong pycnocline (see Figure 6.18). In the surface mixed layer, the atoms and small molecules that make up the bodies of organisms may cycle rapidly for a time between predators, prey, scavengers, and decomposers. When these organisms die, their bodies can sink below the sunlit upper sea, beneath the pycnocline, where they are isolated from the rapid biological activity of the surface. Regions of upwelling are critical in returning these substances to the surface, if only for a short reprieve before their eventual incorporation into deep sediments from which only the very slow progress of tectonic cycles will liberate them.

Unity, not diversity, is the central message of biology. All life on Earth is fundamentally alike at the molecular level.

As you read about the biogeochemical cycles described below, remember that the elements and small molecules forming the tissues of an organism *are always on the move*. They may cycle rapidly in and out of living things, or they may be trapped in Earth for great spans of time, but the nature of the cycles dictates what will live where, which creatures will be successful, and ultimately, what the composition of the ocean and atmosphere itself will be.

The Carbon Cycle Is Earth's Largest Biogeochemical Cycle

The largest of all biogeochemical cycles is the global **carbon cycle** (Figure 13.15). Because of its ability to form long chains to which other atoms can attach, carbon is considered the basic building block of all life on Earth. Carbon enters the atmosphere by the respiration of living organisms (as carbon dioxide), volcanic eruptions that release carbon from rocks deep in Earth's crust, the burning of fossil fuels, and other sources. When levels

Karleskint, Turner, and Small, *Introduction to Marine Biology*, Brooks/Cole © 2006.

of atmospheric carbon dioxide are high, Earth's surface temperature rises because of intensified "greenhouse effect" (about which you will learn more in Chapter 18).

As we have seen, large and small plants (and plantlike organisms) capture sunlight and use this energy to incorporate, or *fix*, CO_2 into organic molecules. Some of these molecules are used as food, and some as structural components. When an animal eats a plant (or plantlike organism), three things can happen to the carbon: (1) it can be incorporated into the animal's body for growth, (2) it can be respired by the animal (taken apart to harvest the energy), or (3) it can be wasted, excreted back into the seawater as **dissolved organic carbon (DOC)**. Typically, about 45% of the carbon from an ingested plant is used for growth, about 45% is used for respiration, and about 10% is lost as DOC. The end product of respiration is CO_2, a gas eventually lost to the atmosphere. Most of the DOC is rapidly used by bacteria, which are in turn eaten by protozoans, which are eaten by zooplankton, which are then eaten by fish, in what is termed the *microbial loop*. Eventually, the organisms—or at least their hard parts containing calcium carbonate ($CaCO_3$)—sink below the mixed layer and begin the long fall toward the seabed. Most of the carbon in this $CaCO_3$ is turned into CO_2 by bacteria long before it hits the bottom, but a small percentage (<1%) reaches the sediments and is buried. The carbonate sediments can be uplifted over geologic time and weathered so that the carbon is eventually returned to the biologically active upper sea.

Because of the large amount of carbon dioxide available in the ocean, and because CO_2 from the atmosphere dissolves readily in seawater, marine organisms almost never suffer from a deficit of available carbon. For life in the sea, the critical bottlenecks lie elsewhere—mainly in the nitrogen, phosphorus, and iron cycles.

Nitrogen Must Be "Fixed" to Be Available to Organisms

Nitrogen is a critical component of proteins, chlorophyll, and nucleic acids. Like carbon, nitrogen may be found in the bodies of organisms as a dissolved gas (N_2) and as dissolved organic matter known as **dissolved organic nitrogen (DON)**.

One might think nitrogen would be abundantly available in the ocean—Table 7.4 indicates that nitrogen accounts for 48% of the dissolved gas in seawater, by volume. But most organisms cannot use the free nitrogen in the atmosphere and ocean directly. It must first be bound with oxygen or hydrogen, or *fixed*, into usable chemical forms by specialized organisms, usually bacteria or cyanobacteria. Thus, oceanic regions are frequently nitrogen limited; the growth of plants and plantlike organisms is often held back by a lack of available nitrogen.

The forms of nitrogen available for uptake by living things are ammonium (NH_4^+) and nitrate (NO_3^-), an ion formed by the oxidation of ammonium and nitrite (NO_2^-). Nitrate runoff from soil is an especially rich source of this nutrient, which explains why coastal water tends to support greater plankton populations than oceanic water does. After being assimilated by small plants and plantlike organisms, nitrogen is recycled as animals consume them and then excrete ammonium and urea. These reduced forms of nitrogen are then oxidized back into nitrate, via nitrite, by **nitrifying bacteria**. In the deep ocean, most of the nitrogen is in the form of nitrate. In anoxic sediments and certain low-oxygen regions of the ocean, **denitrifying bacteria** use nitrate in respiration and convert nitrate back to nitrite and nitrogen gas, which is lost to the atmosphere. The other major loss occurs when nitrogen-containing organisms and debris are buried in ocean sediments. **Figure 13.16** shows the **nitrogen cycle**.

Lack of Iron and Other Trace Metals May Restrict the Growth of Marine Life

Iron is used in minute quantities in the reactions of photosynthesis, in certain enzymes crucial to nitrogen fixation, and in the structure of proteins. Other essential trace metals, such as zinc, copper, and manganese, are also used by organisms in small quantities, primarily in enzymes. Although in absolute terms organisms require only very small quantities of iron, iron's concentration in seawater (relative to the concentration of nutrients such as nitrogen and phosphorus) can sometimes be so low that phytoplankton growth is limited by the availability of iron. This may seem strange—after all, iron is one of the most abundant elements in Earth's crust. But iron is nearly insoluble in

Figure 13.16 The nitrogen cycle. The atmosphere's vast reserve of nitrogen cannot be assimilated by living organisms until it is "fixed" by bacteria and cyanobacteria, usually in the form of ammonium and nitrite ions. Nitrogen is an essential element in the construction of proteins, nucleic acids, and a few other critical biochemicals. Upwelling and runoff from the land bring useful nitrogen into the photic zone, where producers can incorporate it into essential molecules.

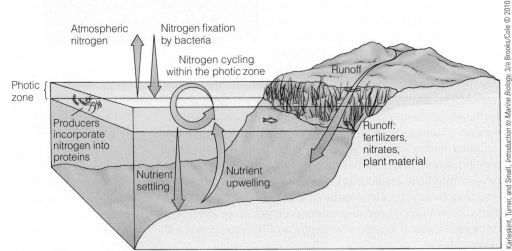

oxygenated seawater, and the little dissolved iron that is present is highly reactive, sticks to falling particles, and sinks to the bottom of the water column.

In general, the biogeochemical cycles of the trace metals follow the pattern we have seen above: uptake and recycling in the surface ocean and regeneration—sometimes over long periods of time—at depth. However, much remains to be learned about the interactions between living organisms and trace metals. Iron and other trace metals exist in many chemical forms in seawater. Discovering what these forms are, how transformation between forms occurs, and how available the different forms are to marine organisms is a major focus of current research in trace metal biochemistry.

CONCEPT CHECK

Before going on to the next section, check your understanding of some of the important ideas presented so far:

What does the ocean's density stratification have to do with biogeochemical cycles?

What is the source of the carbon made into carbohydrates by primary producers?

Why can't marine organisms use the nitrogen gas dissolved in seawater?

How does a "long biogeochemical loop" differ from a short one? (Think of the phosphorus and silicon cycles.)

13.8 Environmental Factors Influence the Success of Marine Organisms

Marine organisms depend on the ocean's chemical composition and physical characteristics for life support. Any aspect of the physical environment that affects living organisms is called a **physical factor**. Living in the ocean often has advantages over living on land—physical conditions in the sea are usually milder and less variable than physical conditions on land. The most important physical factors for marine organisms are light, temperature, dissolved nutrients, salinity, dissolved gases, acid-base balance, and hydrostatic pressure.

These physical factors work in concert to provide the physical environment for oceanic life. But some **biological factors**—biologically generated aspects of the environment—also affect living organisms. These biological factors include diffusion, osmosis, active transport, and surface-to-volume ratio. Let's examine all of these factors in more detail.

Often too much or too little of a single factor can adversely affect the function of an organism. We call that factor a **limiting factor**, a physical or biological necessity whose presence in inappropriate amounts limits the normal action of the organism. Imagine, for example, an ocean area in which everything is perfect for photosynthesis—warmth, nutrients, adequate CO_2—everything *except light*. In that circumstance, no photosynthesis would occur; light is the limiting factor. If light were present but nitrates were absent, nitrate nutrients would be the limiting factor. Sometimes *too much* of something—heat, for instance—can be limiting.

Photosynthesis Depends on Light

On land, most photosynthesis proceeds at or just above ground level. But seawater, unlike soil, is relatively transparent, which allows photosynthesis to proceed for some distance below the ocean surface. Incoming sunlight must run a gauntlet of difficulties, however, before the chlorophyll in marine autotrophs can absorb it.

Most sunlight approaching at a low angle (near sunrise or sunset, or in the polar regions) reflects off the water surface and doesn't enter the ocean. Light that does penetrate the surface is selectively absorbed—water is more transparent to some colors of light than others. In clear water, blue light penetrates to the greatest depth, while red light is absorbed near the surface. Light energy absorbed by water turns to heat.

The number and characteristics of particles in the water also limit the depth to which light penetrates. These particles, which may include suspended sediments, dustlike bits of once-living tissue, or the organisms themselves, scatter and absorb light. High concentrations of particles quickly absorb most blue and ultraviolet light. This absorption, combined with the reflection of green light by chlorophyll within the producers, changes the color of productive coastal waters to green.

How far down does light penetrate? Figure 6.23 and **Figure 13.17** show the depths reached by light of various wavelengths (colors) in the ocean. The **photic zone** (*photos*, "light") is the uppermost layer of seawater lit by the sun. Because of the abundant small organisms and light-scattering particles, the photic zone near the coasts usually extends to about 100 meters (330 feet), and in mid-latitude waters, it reaches down to about 150 meters (500 feet). In clear tropical waters in the open ocean, instruments much more sensitive than human eyes have detected light at much greater depths—the present record is 590 meters (1,935 feet) in the tropical Pacific! The **aphotic zone** (*a*, "without")—the permanently dark layer of seawater beneath the photic zone—extends below the sunlit surface to the seabed. The vast bulk of the ocean is never brightened by sunlight.

Photosynthesis proceeds slowly at low-light levels. Most of the biological productivity of the ocean occurs in the upper part of the photic zone called the **euphotic zone** (*eu*, "good"), shown in **Figure 13.18**. This is where marine autotrophs can capture enough sunlight energy for primary production by photosynthesis to exceed the loss of carbohydrates by respiration. Though it is difficult to generalize for the ocean as a whole, the euphotic zone typically extends to a depth of approximately 70 meters (230 feet) in mid-latitudes, averaged over the whole year. The upper productive layer of ocean is a very thin skin indeed—the water within this zone amounts to less than 1% of world ocean volume—and yet nearly all marine life depends on this fine illuminated band.

Below the euphotic zone (and still in the photic zone) lies the **disphotic zone** (*dys*, "difficult"), also seen in Figure 13.18. Though light is present in this zone, it is not bright enough to allow photosynthesis to generate as much carbohydrate as would be used by an autotroph through a day.

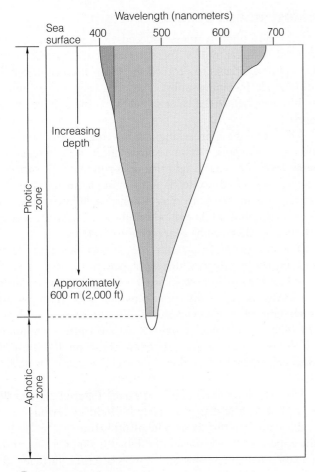

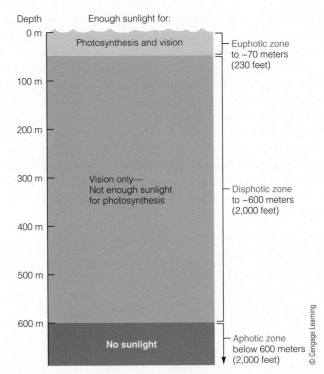

Figure 13.18 The upper, sunlit zone of the ocean is known as the photic zone. In the euphotic zone, there is enough light for photosynthesis. Below that depth, in the disphotic zone, light may be present but not in adequate quantity for the photosynthetic production of glucose to exceed its consumption. The aphotic zone lies in permanent darkness. Most of the ocean is without sunlight all of the time, and all of it is dark some of the time. (See also Figure 6.24.)

💬 **THINKING BEYOND THE FIGURE**

How can photographers take pictures showing red or yellow organisms below the photic zone?

a Clear, open ocean water.

In clear, open ocean water, sensitive instruments can detect light to a depth of 600 meters (2,000 feet).

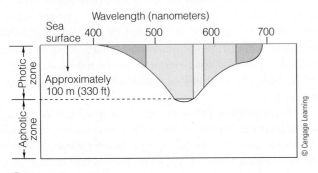

b Coastal ocean water.

Because of the suspended particles often present in coastal waters, light cannot penetrate so far—about 100 meters (330 feet) is typical. The sunlit upper zone is called the *photic zone*. The dark ocean beneath is called the *aphotic zone*.

Figure 13.17 Penetration of light into the ocean.

Temperature Influences Metabolic Rate

Ocean temperature varies with depth and latitude. The average temperature of the world ocean is only a few degrees above freezing; warmer water is found only in the lighted surface zones of the temperate and tropical ocean and in deep, warm chemosynthetic communities. Though temperature ranges of the ocean are considerable **(Figure 13.19)**, they are much narrower than comparable ranges on land.

What are the implications of the ocean's temperature for living things? The rate at which chemical reactions occur in an organism is largely dependent on the molecular vibration we call heat. Since agitation brings reactants together, warmer temperatures increase the rate at which chemical reactions occur. Thus, an organism's **metabolic rate**, the rate at which energy-releasing reactions proceed within an organism, increases with temperature. The metabolic rate approximately doubles with a 10°C (18°F) temperature rise. The interior temperature of an organism is directly related to the rate at which it moves, reacts, and lives.

The great majority of marine organisms are "cold blooded," or **ectothermic**, having an internal temperature that stays very close to that of their surroundings. A few complex animals—mammals and birds and some of the larger, faster fishes—are "warm blooded," or **endothermic**, meaning that they have a stable, high internal temperature.

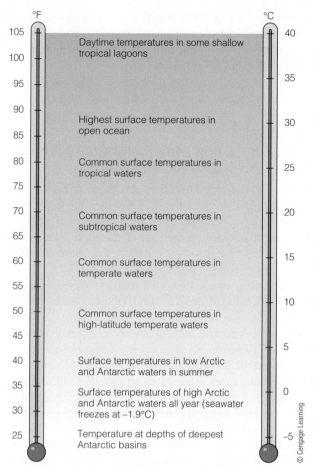

°F / °C scale

- 105°F / 40°C — Daytime temperatures in some shallow tropical lagoons
- Highest surface temperatures in open ocean
- Common surface temperatures in tropical waters
- Common surface temperatures in subtropical waters
- Common surface temperatures in temperate waters
- Common surface temperatures in high-latitude temperate waters
- Surface temperatures in low Arctic and Antarctic waters in summer
- Surface temperatures of high Arctic and Antarctic waters all year (seawater freezes at −1.9°C)
- Temperature at depths of deepest Antarctic basins

© Cengage Learning

Figure 13.19 Temperatures of marine waters capable of supporting life. Some isolated areas of the ocean, notably within and beneath hydrothermal vents, may support specialized living organisms at temperatures to 400°C (750°F)!

In general, the warmer the environment of an ectotherm within its tolerance range, the more rapidly its metabolic processes will proceed. Tropical fish in a heated aquarium will therefore eat more food and require more oxygen than goldfish of the same size living in an unheated but otherwise identical aquarium. The tropical fish will generally grow faster, have a faster heartbeat, reproduce more rapidly, swim more swiftly, and live shorter lives. But you can't just crank the heater up another notch for zippier fish—raising the temperature too much will kill the fish. The upper limit of temperature that an ectotherm can tolerate is often not much higher than its optimum temperature. The lower limit is usually more forgiving because molecules are merely slowed.

Do endotherms have narrow temperature requirements? Yes and no. Endotherms can tolerate a tremendous range of *external* temperature compared to ectotherms; think of a whale migrating from polar waters to the tropics, or an Emperor penguin incubating an egg at −51°C (−60°F). Their *internal* temperatures, however, vary only slightly. In our own case, consider the temperatures of places inhabited by humans in contrast to the narrow internal temperature range physicians consider nor-

mal. Sophisticated mechanisms for thermal regulation make it possible for endotherms to live in a variety of habitats, but they pay a price. Their high metabolic rates make proportionally high demands on food supply and gas transport, but the benefit of having a biochemistry fine-tuned to a single efficient temperature is worth the regulatory difficulties involved.

Dissolved Nutrients Are Required for the Production of Organic Matter

A **nutrient** is a compound required for the production of organic matter. Some nutrients help form the structural parts of organisms, some make up the chemicals that directly manipulate energy, and some have other functions. A few of these necessary nutrients are always present in seawater, but most are not readily available.

The main inorganic nutrients required in primary productivity include nitrogen (as nitrate, NO_3^-) and phosphorus (as phosphate, PO_4^{3-}). As any gardener knows, plants require fertilizer—mainly nitrates and phosphates—for success. Ocean gardeners would have more trouble raising crops than their terrestrial counterparts, though, because the most fertile ocean water contains only about 1/10,000 the available nitrogen of topsoil. Phosphorus is even more scarce in the ocean, but fortunately less of it is required by living things, which have about 1 atom of phosphorus for every 16 atoms of nitrogen.

Nitrogen and phosphorus are often depleted by autotrophs during times of high productivity and rapid reproduction. Also in short supply during rapid growth are dissolved silicates (used for shells and other hard parts) and trace elements such as iron and copper (used in enzymes, vitamins, and other large molecules). Marine plants have no choice but to recycle these nutrients.

Salinity Influences the Function of Cell Membranes

All the cells of every organism are enclosed by membranes, complex films through which a few selected substances can move. A cell's membranes are greatly affected by the salinity of the surrounding water.

The salinity of seawater (see Chapter 6) can vary in places because of rainfall, evaporation, runoff of water and salts from land, and other factors. Surface salinity varies most, with lows of 6‰ or less along the coast of the inner Baltic Sea in early summer, to year-around highs exceeding 40‰ in the Red Sea. Salinity is less variable with increasing depth, with the ocean typically becoming slightly saltier with depth.

A change in salinity can physically damage cell membranes, and concentrated salts can alter protein structure. Salinity can affect the specific gravity and density of seawater, and therefore the buoyancy of an organism. As we'll see in a moment, salinity is also important because it can cause water to enter or leave a cell through the membrane, changing the cell's overall water balance. Seawater is nearly identical in salinity to the interior of all but the most advanced forms of marine life; this means that maintaining salt balance, and therefore water balance, is easy for most marine species.

Dissolved Gas Concentrations Vary with Temperature

Nearly all marine organisms require dissolved gases—in particular carbon dioxide and oxygen—to stay alive. Oxygen does not easily dissolve in water, and as a result there is about 100 times more gaseous oxygen in the atmosphere than in the ocean. But CO_2, essential to primary productivity, is much more soluble and reactive in seawater than oxygen. Although as much as 1,000 times more carbon dioxide than oxygen can dissolve in water, normal values at the ocean surface average around 50 milliliters per liter for CO_2 and around 6 milliliters per liter for oxygen. At present, the ocean holds about 60 times as much carbon dioxide as the atmosphere. Because of this abundance, marine plants almost never run out of CO_2.

Deep water tends to contain more carbon dioxide than surface water. Why should this be? Table 13.1 shows the relationship between water temperature and its ability to dissolve gases. Note that colder water contains more gas at saturation. You may recall that the deepest and densest seawater masses are formed at the surface in the cold polar regions, and, as we have seen, more CO_2 can dissolve in that low temperature environment. The dense water sinks, taking its large load of CO_2 to the bottom, and the pressure at depth helps to keep it in solution. CO_2 also builds in deep water because only heterotrophs (animals) live and metabolize there and because CO_2 is produced as decomposers consume falling organic matter. No photosynthetic primary producers are present in the dark depths to use this excess CO_2 since there is not enough sunlight for photosynthesis to occur.

Rapid photosynthesis at the surface lowers CO_2 concentrations and increases the quantity of dissolved oxygen. Oxygen is least plentiful just below the limit of photosynthesis because of respiration by many small animals at middle depths. (These relationships were shown in Figure 7.8).

Low oxygen levels can sometimes be a problem at the ocean surface. Plants produce more oxygen than they use, but they produce it only during daylight hours. The continuing respiration of plants at night will sometimes remove much of the oxygen from the surrounding water. In extreme cases, this oxygen depletion may lead to the death of the plants and animals in the area, a phenomenon most noticeable in enclosed coastal waters during spring and fall plankton blooms.

The greatest variability in levels of dissolved gas is found at the surface near shore. Less dramatic changes occur in the open sea.

Dissolved Carbon Dioxide Influences the Ocean's Acid-Base Balance

Another physical condition that affects life in the ocean is the acid-base balance of seawater. The complex chemistry of Earth's life-forms depends on precisely shaped enzymes, large protein molecules that speed up the rate of chemical reactions. When strong acids or bases distort the shapes of these vital proteins, they lose their ability to function normally.

The acidity or alkalinity of a solution is expressed in terms of a *pH scale*, a logarithmic measure of the concentration of hydrogen ions in a solution. Recall (from Figure 7.9) that 7 on the pH scale is neutral, with smaller numbers indicating greater acidity and larger numbers indicating greater alkalinity.

Seawater is slightly alkaline; its average pH is about 8. The dissolved substances in seawater act to *buffer* pH changes, preventing broad swings of pH when acids or bases are introduced. The normal pH range of seawater is much less variable than that of soil—terrestrial organisms are sometimes limited by the presence of harsh alkali soils that damage cell components.

Though seawater remains slightly alkaline, it is subject to some variation. When dissolved in water, some CO_2 becomes carbonic acid. In areas of rapid plant growth, pH will rise because CO_2 is used by the plants for photosynthesis. And because temperatures are generally warmer at the surface, less CO_2 can dissolve in the first place. Surface pH in warm productive water is usually around 8.5.

At middle depths and in deep water, more CO_2 may be present. Its source is the respiration of animals and bacteria. With cold temperatures, high pressure, and no photosynthetic plants to remove it, this CO_2 will lower the pH of water, making it more acid with depth. Thus, deep, cold seawater below 4,500 meters (15,000 feet) has a pH of around 7.5. This lower pH can dissolve calcium-containing marine sediments. A drop to pH 7 can occur at the deep ocean floor when bottom bacteria consume oxygen and produce hydrogen sulfide.

As will see in Chapter 18, CO_2 concentrations are rising in the atmosphere. Much (perhaps most) of this rise is due to the burning of fossil fuels to support human industry and economic growth. Some researchers believe that this rapid increase in CO_2 could overwhelm the carbonate buffer system in surface waters and cause surface oceanic pH to drop.[3] Even slightly more acidic seawater would interfere with the formation of calcareous materials—coral skeletons, plankton tests, and some other hard parts of marine organisms (Figure 13.20).

Table 13.1 The Solubility of Gases in Seawater Decreases as Temperature Increases

Temperature	Solubility (mL/L at atmosphere pressure and salinity of 33‰)[a]		
	N₂	O₂	CO₂
0°C (32°F)	14.47	8.14	8,700.0
10°C (50°F)	11.59	6.42	8,030.0
20°C (68°F)	9.65	5.26	7,350.0
30°C (86°F)	8.26	4.41	6,600.0

[a]Figures are given at *saturation*, the maximum amount of gas held in solution before bubbling begins.

Source: F.G. Walton-Smith, *CRC Handbook of Marine Science* (Cleveland, OH: CRC Press, 1974). Table created by Tom Garrison.

---

[3]For a discussion of the carbonate buffer system, please see Section 7.4.

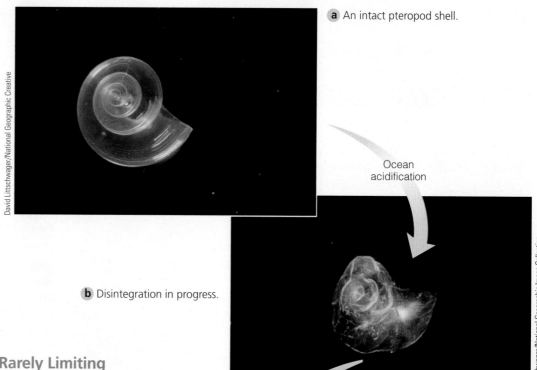

a An intact pteropod shell.

Ocean acidification

b Disintegration in progress.

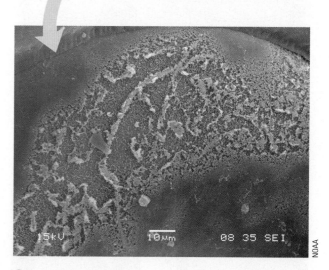

c A close view of surface etching.

Hydrostatic Pressure Is Rarely Limiting

Marine organisms are often subject to great pressure from the constant weight of water above them, but this so-called **hydrostatic pressure** presents very little difficulty. In fact, the situation in the ocean is parallel to that on land. Land animals live in air pressurized by the weight of the atmosphere above them (1 kilogram per square centimeter, or 14.7 pounds per square inch, at sea level) without experiencing any problems. Indeed, atmospheric pressure is necessary for breathing, flight, and some other physical necessities of life.

Pressures inside and outside an organism are virtually the same, both in the ocean and at the bottom of the atmosphere. Thus marine organisms do not need heavy shells to keep from being crushed by hydrostatic pressure. Great pressure does have some chemical effects—gases become more soluble at high pressure, some enzymes are inactivated, and metabolic rates for a given temperature tend to be slightly higher. These effects are felt only at great depth, though. Unless marine organisms have gas-filled spaces in their bodies (lungs, swim bladders), a moderate change in pressure has little effect.

Substances Move through Cells by Diffusion, Osmosis, and Active Transport

If a cube of dye is left undisturbed in a container of still water, the dye molecules will eventually be distributed evenly throughout the water. This process, known as **diffusion**, might take weeks to complete **(Figure 13.21)**. The energy to distribute the dye comes from heat, the random vibration of molecules—the warmer the water, the faster the diffusion. The net transfer of material in diffusion occurs from a region of high concentration to regions of lower concentration.

Diffusion, an important marine process, can be a physical factor. For example, minerals dissolve, and their components tend to diffuse randomly throughout a liquid environment. Liquids and gases can also diffuse through water from zones of high concentration to zones of low concentration. But mass transport (the movement of substances in currents, for example) is more important than diffusion in moving dissolved substances over large distances.

Diffusion is most important over small distances, particularly the distances within and between living cells. Selected substances can diffuse across cell membranes. They tend to diffuse

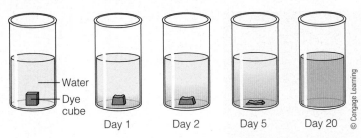

Figure 13.21 An example of diffusion. Molecules of dye gradually diffuse through the surrounding water as a dye cube dissolves. Random molecular movement spreads the dye away from the region of high concentration at the cube's surface. At the same time, water moves from its own region of high concentration (next to the dye cube) to areas of lower concentration (within the disintegrating dye cube itself). After many days, dye will be distributed evenly throughout the water. Stirring would greatly accelerate the mixing process, but this input of mechanical energy is not required if you don't mind waiting for diffusion alone to accomplish the task.

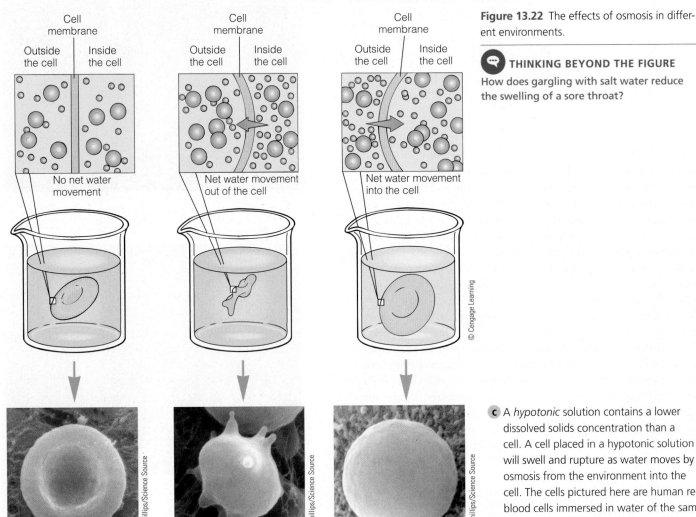

Figure 13.22 The effects of osmosis in different environments.

💬 **THINKING BEYOND THE FIGURE**
How does gargling with salt water reduce the swelling of a sore throat?

a An *isotonic* solution contains the same concentration of dissolved solids (green) and water molecules (blue) as a cell. Cells placed in isotonic solutions do not change size since there is no net movement of water.

b A *hypertonic* solution contains a higher concentration of dissolved solids than a cell. A cell placed in a hypertonic solution will shrink as water moves out of the cell to the surrounding solution by osmosis.

c A *hypotonic* solution contains a lower dissolved solids concentration than a cell. A cell placed in a hypotonic solution will swell and rupture as water moves by osmosis from the environment into the cell. The cells pictured here are human red blood cells immersed in water of the same tonicity as human blood, concentrated seawater, and distilled water.

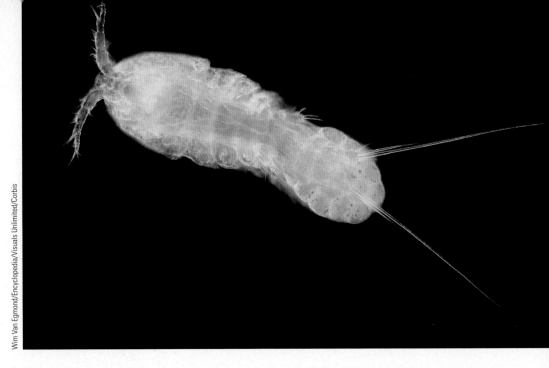

Figure 13.23 *Tigriopus,* a champion osmoregulator. This small copepod can survive in hot, highly saline "splash pools" at the top of the intertidal zone. As the name implies, splash pools are supplied by surf spray and are rarely overwashed. Evaporation and solar heating may raise the pools' salinity, which sometimes reaches saturation levels—salinity so high that crystals of salts form in the pool! No predator can withstand this hypersalinity, an advantage for the copepod. But maintaining an *interior* salinity suitable for normal metabolism is very expensive—scientists estimate that in extreme cases, as much as 90% of *Tigriopus's* food intake is dedicated to actively transporting water molecules into its body and excluding salt ions.

Wim Van Egmond/Encyclopedia/Visuals Unlimited/Corbis

from areas where they are highly concentrated to areas where they are less concentrated. For example, oxygen resulting from photosynthesis will diffuse through a membrane from inside a plant cell (a region of high oxygen concentration) to the outside (a region of lower oxygen concentration). Because biological membranes allow only certain kinds of small molecules to pass, they are considered *selectively permeable.*

Diffusion of water through a membrane is called **osmosis** (*osmos,* "a thrusting"). In osmosis, water moves between two solutions of different water concentrations through a membrane permeable to water but not to salts **(Figure 13.22)**. If the water outside a cell membrane contains less dissolved salt than the water inside, it will diffuse from the region of higher water concentration (outside the cell) to the region of lower water concentration (inside the cell), causing the cell to swell. Salt is prevented by the nature of the membrane from moving outside to balance the situation. This is why human swimmers avoid getting freshwater in nasal membranes—our body fluids are more saline than freshwater, so cells in our sinuses take up the water and swell painfully.

Most simple marine organisms have nearly the same concentration of dissolved substances in their body fluids as seawater does. They are almost **isotonic** to their fluid environment (*isos,* "equal"; *tonos,* "strength") and so experience little net flow of water through their outer membranes. In freshwater, a marine animal would be **hypertonic** (*hyper,* "over") to its surroundings; water would move *into* the animal through its cell membranes. If the animal had no way to eliminate the water, its cells would burst. The same kind of marine animal moved to Utah's highly saline Great Salt Lake would be **hypotonic** (*hypo,* "under"), and water would flow *out of* its cells, causing it to dehydrate and collapse.

Because they are nearly isotonic to their surroundings, simple marine organisms have a big advantage over their freshwater counterparts; freshwater animals must expend large amounts of energy in excretory systems designed to transport water from their tissues back to the outside. Because of these specialized excretory mechanisms, very few freshwater and marine organisms can successfully change places.

Active transport is the reverse of passive diffusion. In active transport, dissolved substances are "pumped" through a membrane "uphill" from a region of low concentration to a region of high concentration. Active transport requires energy because this "uphill" movement defies the normal "downhill" functioning of the second law of thermodynamics. When a cell exports a finished product (the sugars made in photosynthesis, for example) through a membrane to a storage area in which this product is concentrated, it uses active transport. Active transport is a common process in living things, and much of any organism's energy is expended on facilitated movement against the normal concentration gradient **(Figure 13.23)**.

CONCEPT CHECK

Before going on to the next section, check your understanding of some of the important ideas presented so far:

What's a limiting factor? Can you provide an example?

What characterizes the photic, euphotic, and disphotic zones?

How does metabolic rate vary with temperature?

How do dissolved gas concentrations vary with temperature? Do you see a potential problem for marine organisms?

Does the great hydrostatic pressure of the seabed crush organisms?

How is diffusion different from osmosis?

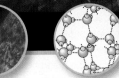

In this chapter you learned that life on Earth is notable for both unity and diversity: *diversity* because there are at least 50 million different species (kinds) of living things on Earth; *unity* because each species shares the same underlying mechanisms for basic life processes. The atoms in living things are no different from the atoms in nonliving things, and the energy that powers living things is the same energy found in inanimate objects.

Because of its watery nature and origin, all life on this planet is in a sense marine. The oceanic environment is a relatively easy place for cells to live. In part at least, life in the ocean is so successful—*total marine productivity is as high as it is*—because of the ocean's physical characteristics, but these characteristics may also limit the success of an organism.

Marine life has not accidentally capitalized on its rich fluid home. Earth's organisms did not arise in just a few thousand years. They have changed, generation by generation, over almost 4 billion years. The ability of living things to change through time to fit the physical and chemical environment, to become ever more efficient at extracting energy from their surroundings, to colonize virtually every loca-

tion capable of sustaining them, and finally to investigate themselves by scientific logic has come about through evolution.

Primary producers—autotrophs—are organisms that synthesize food from inorganic substances by photosynthesis and chemosynthesis. Autotrophic marine organisms transform energy from the sun (or from certain inorganic molecules) into chemical energy to power their own growth; they are in turn consumed by heterotrophic organisms. The feeding relationships in a community resemble complex webs.

A variety of physical factors affects the density, variety, and success of the life-forms in each marine habitat. These factors include water's transparency, temperature, dissolved nutrients, salinity, dissolved gases, hydrostatic pressure, acid-base balance, and others.

Marine organisms are naturally classified by their physical characteristics and by the degree to which they resemble other organisms. The various marine environments populated by marine life may be classified by physical characteristics.

In the next three chapters you will learn about the organisms—producers, then consumers—that are distributed through the marine environment in specific groups we call communities.

QUESTIONS FROM STUDENTS

1. How can evolution possibly produce complex organisms like fish or people? Isn't that like expecting a tornado blowing through a junkyard to result in the spontaneous assembly of a Boeing 747?

Imagine a state lottery in which the winning numbers stay the same for a whole year. You buy a ticket each week. Once in a while, your ticket contains one of the winning numbers. Now imagine you get to keep that number. As the weeks go by, you accumulate most of the numbers until—if things work out—the last of your numbers comes up and you claim your prize. If not, there's always next year.

Evolution by natural selection works in a similar way. Most of the numbers (genetic traits) are useless or even dangerous. But a few are useful, and you get to keep those.

The whole 747 (complex organism) is not assembled in one swoop—rather, it is assembled by accumulating favorable trait after favorable trait, one after another. Given 4.3 billion years or so, the process has resulted in the astonishing variety of living things you see around you.

2. Mass extinctions seem pretty scary. Is anybody watching for asteroids that might impact Earth in the future?

Yes, indeed. NASA's Jet Propulsion Laboratory (JPL) posts asteroids and comets that will make relatively close approaches to Earth. JPL provides a Widget displaying the date of closest approach, approximate object diameter, relative size, and distance from Earth for each encounter. It indicates the next five Earth approaches to within 4.6 million miles (7.5 million kilometers or 19.5 times the distance to the moon); an object larger than about 150 meters that can approach the Earth to within this distance is

termed a potentially hazardous object. The Widget is available at http://www.jpl.nasa.gov/asteroidwatch/.

But a different kind of mass extinction is under way, and it doesn't involve collisions with asteroids. Human-generated environmental change has accelerated the rate of species extinction. The exact rate is controversial, perhaps 100 to 1,000 times the normal background rate of extinction, but human impact is certainly modifying long-established patterns of biological distribution and diversity. We will have more to say on this equally scary topic in Chapter 18.

3. If the atoms that make up molecules in living things are identical to the atoms in rocks and water, why are "organic" molecules better and more nutritious?

They aren't. They're the same. A complex biological structure like a vitamin will have identical structure and properties whether synthesized in a chemical laboratory or occurring naturally in the skin of an orange. (If there was a difference, one of them wouldn't function as a vitamin.) "Organic" foods may be relatively free of pollutants like pesticides or synthetic fertilizers, but glucose is glucose whether from corn or petroleum.

4. If plants need CO_2 to construct glucose (by photosynthesis), why is the rising level of CO_2 in the atmosphere cause for concern? You'd think more CO_2 would accelerate plant growth, and that would be a good thing.

It's all about balance. Some terrestrial plant populations might benefit while others would suffer. In the ocean, you've read that a rise in dissolved CO_2 results in a drop in pH (increasing acidity), and that adversely affects calcium metabolism (the deposition of skeletons, shells, and other hard parts). The effect on coral-reef communities will probably be significant. Also, global surface temperature tends to rise as CO_2 accumulates in the atmosphere (a topic we will reserve for Chapter 18). Taken together, these deleterious effects will almost certainly overwhelm any increase in terrestrial productivity.

5. What's the difference between the photic zone and the euphotic zone?

The photic zone is the sunlit uppermost layer of the ocean. The euphotic zone is part of the photic zone. Within the euphotic zone, autotrophic organisms receive enough sunlight to make enough food (by photosynthesis) for their lives to continue. During daylight hours, light is available below the euphotic zone, but it is not bright enough to allow the photosynthetic machinery of autotrophs to produce enough food to sustain them indefinitely. Unless they rise into the euphotic zone, they will eventually die.

6. If humans have a fluid much like seawater bathing their cells, why can't we drink seawater and survive?

Human cells function in an environment hypotonic to seawater; that is, blood plasma is less saline than seawater. Drinking seawater therefore causes water to leave the intestinal walls, flood the intestine, and leave the body. There is a net loss via intestine or kidneys even if the seawater is diluted with freshwater before drinking. Moral: *Never* drink seawater in a survival situation at sea, and never dilute freshwater with seawater (in any proportion) to stretch your supply.

TERMS AND CONCEPTS TO REMEMBER

active transport	dissolved organic carbon (DOC)	hydrostatic pressure	nitrifying bacteria
adaptation	dissolved organic nitrogen (DON)	hypertonic	nitrogen cycle
aphotic zone		hypotonic	nutrient
artificial system of classification	domain	isotonic	osmosis
autotroph	ectothermic	kingdom	photic zone
biogeochemical cycle	endothermic	limiting factor	photosynthesis
biological factor	energy	Linnaeus (Carl von Linné)	physical factor
carbon cycle	euphotic zone	mass extinction	primary consumer
chemosynthesis	evolution	metabolic rate	scientific name
convergent evolution	food	mutation	species
denitrifying bacteria	food web	natural selection	taxonomy
diffusion	heterotroph	natural system of classification	top consumer
disphotic zone	hierarchy		trophic pyramid

STUDY QUESTIONS

Thinking Critically

1. The second law of thermodynamics states that entropy (disorganization) tends to increase with time. But living things tend to become *more* complex with time (embryos grow to adults; populations evolve). How can that be?
2. Can you suggest any ways humans might be altering biogeochemical cycles?
3. What is a limiting factor? Can you think of some examples not given in the text?
4. How is evolution by natural selection thought to work?
5. How would you define *biological success*? Does success depend on the size of an organism? Its beauty? The amount of space it controls? Its numbers?
6. How does a natural system of classification differ from an artificial system? Can you give an example of each? Was the hierarchy-based system invented by Linnaeus natural or artificial? What *is* a hierarchy-based system?
7. Are asteroid collisions the only causes of mass extinction events?

Thinking Analytically

1. Researchers believe there may be as many as 100 million species of living things on and in Earth. We know of about a million and a half species so far. Where do you think the rest of the species are hiding?
2. You'll find some statistics concerning marine fisheries in Chapter 17. Select a fishery, investigate the trophic level at which the organisms comprising that fishery are feeding, and then estimate the mass of primary producers needed to maintain the fishery at that level of harvest.

GLOBAL GEOSCIENCE WATCH

Visit www.cengagebrain.com to access MindTap, a complete digital course which includes access to Global Geoscience Watch and more. Search for "extremophiles" in the GREENR database and summarize some of the recent discoveries about the characteristics, amount of biomass, and diversity of microbial life found in extreme environments. Use this information to write a two- to three-page report on the primary production of these organisms and a general comparison of them to other marine microbial communities.

CENGAGE**brain**.com   Visit www.cengagebrain.com to access course materials for this text, including interactive learning tools, videos, and more.

14 Primary Producers

KEY CONCEPTS

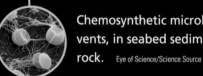

Primary productivity involves the synthesis of organic materials from inorganic substances by photosynthesis or chemosynthesis. SeaWiFS Project, NASA/Goddard Space Flight Center/GeoEye

Most primary producers are plankton—organisms that drift or swim weakly, unable to move consistently against waves or current flow. Lester V. Bergman/CORBIS

The ocean's most productive producers are very small cyanobacteria working in a "microbial loop." The Natural History Museum/Alamy

Not all producers are drifters. Attached seaweeds, sea grasses, and mangroves are also important contributors. MAURICIO HANDLER/National Geographic Creative

Chemosynthetic microbes can live near hydrothermal vents, in seabed sediments, and even in solid rock. Eye of Science/Science Source

Bioluminescent phytoplankton glow beneath the Milky Way as waves gently break on the shores of Gippsland Lakes, Victoria, Australia.

Philip Hart/Stocktrek Images/Getty Images

14.1 Primary Producers Synthesize Organic Material

The diversity of primary producers in the ocean ranges from microscopic bacteria to giant kelp that can grow to more than 60 meters (200 feet) in length! These organisms are able to

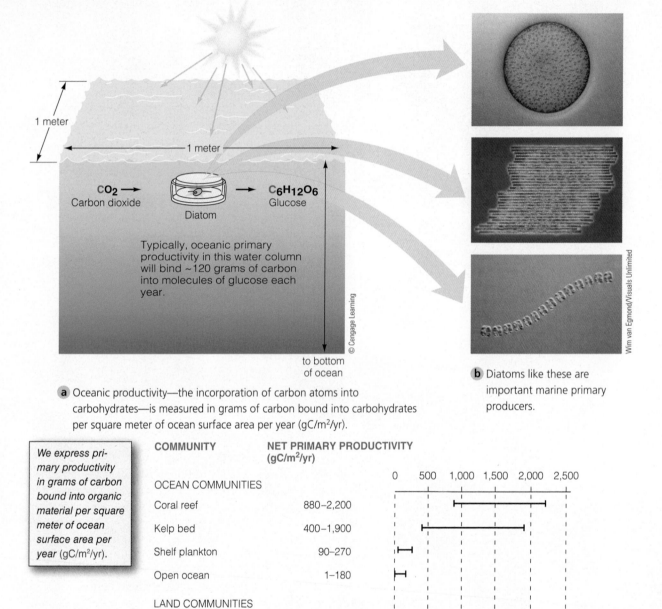

1 meter

1 meter

CO_2 → Carbon dioxide → Diatom → $C_6H_{12}O_6$ Glucose

Typically, oceanic primary productivity in this water column will bind ~120 grams of carbon into molecules of glucose each year.

to bottom of ocean

© Cengage Learning

a Oceanic productivity—the incorporation of carbon atoms into carbohydrates—is measured in grams of carbon bound into carbohydrates per square meter of ocean surface area per year (gC/m²/yr).

Wim van Egmond/Visuals Unlimited

b Diatoms like these are important marine primary producers.

We express primary productivity in grams of carbon bound into organic material per square meter of ocean surface area per year (gC/m²/yr).

COMMUNITY	NET PRIMARY PRODUCTIVITY (gC/m²/yr)
OCEAN COMMUNITIES	
Coral reef	880–2,200
Kelp bed	400–1,900
Shelf plankton	90–270
Open ocean	1–180
LAND COMMUNITIES	
Rain forest	460–1,600
Temperate forest	270–1,140
Freshwater swamp	360–1,820
Cropland	45–1,820

0   500   1,000 1,500 2,000 2,500

ALL OCEAN = 120 (average)   ALL LAND = 150 (average)

© Cengage Learning

c Net annual primary productivity in some marine and terrestrial communities.

Figure 14.1 Expressions of oceanic productivity.

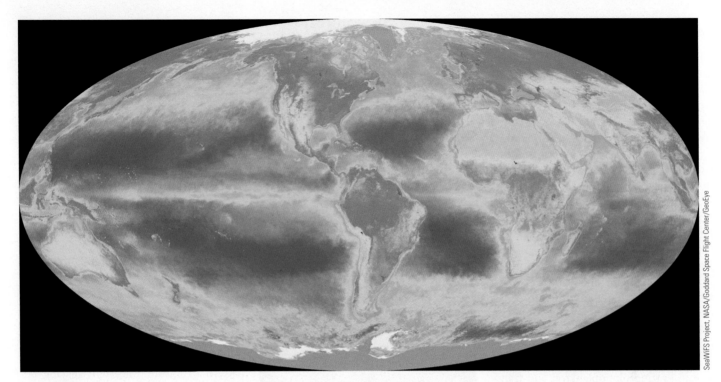

Figure 14.2 Oceanic productivity can be observed from space. NASA's *SeaWiFS* satellite, launched in 1997, can detect the amount of chlorophyll in ocean surface water. Chlorophyll content provides an estimate of productivity. Red, yellow, and green areas indicate high primary productivity; blue areas indicate low. This image was derived from measurements made between September 1997 and August 1998.

THINKING BEYOND THE FIGURE

Why are some ocean areas productive and some areas oceanic deserts?

synthesize organic materials from inorganic substances by photosynthesis or chemosynthesis. **Phytoplankton**—minute, drifting photosynthetic organisms—are responsible for producing between 90% and 96% of the surface ocean's carbohydrates. Seaweeds—larger marine photosynthesizers—contribute from 1% to 5% of the ocean's primary productivity. Chemosynthetic organisms probably account for between 2% and 5% of the total productivity in the water column.

Primary productivity is expressed in *grams of carbon bound into organic material per square meter of ocean surface area per year* (gC/m²/yr) **(Figure 14.1)**. The immediate organic material produced is the carbohydrate glucose. The source of carbon for glucose is dissolved carbon dioxide (CO_2). Though estimates vary widely, recent studies suggest that total ocean productivity ranges from 75 to 150 grams of carbon bound into carbohydrates per square meter of ocean surface per year (75–150 gC/m²/yr). (For comparison, a well-tended alfalfa field produces about 1,600 gC/m²/yr.)

How does marine productivity compare with terrestrial productivity? Recent research suggests the global net productivity in *marine* ecosystems is 35 to 50 billion metric tons of carbon bound into carbohydrates per year; global *terrestrial* productivity is roughly similar at 50 to 70 billion metric tons per year.[1] However, the total producer **biomass** (the mass of

living tissue) in the ocean is only 1 to 2 billion metric tons, compared with 600 to 1,000 billion metric tons of living biomass on land! As the rapid turnover time indicates, nutrients cycle from producer to consumer and back much more quickly in marine ecosystems.

The total mass of a **primary producer** is assumed to be about 10 times the mass of the carbon it has bound into carbohydrates. Thus, a primary productivity of 100 gC/m²/yr represents the yearly growth of about 1,000 grams of primary producers for each square meter of ocean surface (see **Figure 14.2**). Since 35 to 50 billion metric tons of carbon are bound into carbohydrates in the ocean each year, between 350 and 500 billion metric tons of marine plants and plantlike organisms are produced annually. Each year this vast bulk is consumed by the metabolic activity of the producers themselves and by the consumers that graze on them. The component atoms are then reassembled by photosynthesis into carbohydrates in a continuous solar-powered cycle.

CONCEPT CHECK

Before going on to the next section, check your understanding of some of the important ideas presented so far:

What are the most important kinds of primary producers?

What do primary producers produce? How is productivity expressed?

[1] 1 billion metric tons = 1.1 billion tons.

14.2 Plankton Drift with Ocean Currents

The organisms identified as **plankton** are as important as they are inconspicuous. The word is derived from the Greek word *planktos*, which means "wandering." Plankton drift or swim weakly, going where the ocean goes, unable to move consistently against waves or current flow. (In contrast, the word **nekton** (*nektos*, "swimming") describes organisms that actively swim.)

The term *plankton* or *nekton* is not a collective natural category like *molluscs* or *algae*, which implies an ancestral (evolutionary) relationship among the organisms; instead, it describes a common ecological connection—a lifestyle. Members of the plankton community, also referred to as **plankters**, can and do interact with one another. Some can swim weakly. Grazing, predation, parasitism, and competition occur among members of this dynamic group.

Plankton include many different photosynthetic and chemosynthetic species that are responsible for contributing to ocean primary productivity. There are also planktonic forms of every major group of animals, many of which play an important role in grazing phytoplankton and transferring energy to larger marine life. These heterotrophic plankton, or **zooplankton**, combine with the phytoplankton (discussed later) to constitute perhaps the most important of all marine communities. This diverse assemblage will be examined in additional detail in Chapter 16's discussion of marine communities.

> **P**lankton drift or swim weakly, going where the ocean goes, unable to move consistently against waves or current flow.

CONCEPT CHECK

Before going on to the next section, check your understanding of some of the important ideas presented so far:

Plankton are said to be drifters, yet some of them can swim. Which is it?

Is *plankton* a natural or artificial category?

14.3 Plankton Collection Methods Depend on the Organism's Size

The first large-scale systematic study of plankton was carried out by biologists aboard the research vessel *Meteor* during the German Atlantic Oceanographic Expedition of 1925. Many of the tools and techniques they pioneered are still in use today. **Plankton nets** of the type perfected for the *Meteor* expedition are essential to plankton studies (**Figure 14.3**). These conical nets are customarily made of nylon or Dacron cloth woven in a fine interlocking pattern to ensure consistent spacing between threads. The net is hauled slowly for a known distance behind a ship, or it is cast to a set depth and then reeled in. Trapped organisms are flushed to the pointed end of the net and carefully removed for analysis. Quantitative analysis of plankton requires both a count of the organisms and an estimate of the sampled volume of water.

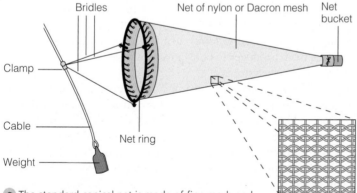

Plankton net fabric as seen under a microscope

© Cengage Learning

a The standard conical net is made of fine mesh and has a mouth up to 1 meter (3.3 feet) in diameter. The net is towed behind a ship for a set distance. The number of organisms present in the water can be estimated if the trapped organisms are counted and the volume of sampled water is known.

b The net shown here has a somewhat coarser mesh because its target organisms, small shrimplike crustaceans known as krill, are relatively large.

Dennis Kelly, Orange Coast College

Figure 14.3 Plankton nets.

Very small plankton can slip through a plankton net. Their capture and study requires concentration by centrifuge, or entrapment by a fine plankton filter through which water is drawn. The filter is later disassembled and the plankton studied in place. The smallest of plankton are trapped by specially made unglazed porcelain filters through which water is forced under very high pressure. As you'll see, organisms recently discovered in this way are some of the most intriguing members of the plankton community.

Measurements of physical ocean conditions such as dissolved carbon dioxide and oxygen content, pH, temperature, and light intensity at the time and place of sampling are of special importance in interpreting the samples. Simultaneous sampling at many locations can be useful in pinpointing the often subtle interplay between species and physical conditions that complicate our understanding of plankton biology.

Historically, plankton specialists have categorized plankton by their visibility or by the methods used to collect them. The recent discovery of astonishing numbers of submicroscopic phytoplankters, however, has stimulated the development of a more consistent system of classifying planktonic organisms by *size*. One evolving scheme of classification divides the full range of plankton into seven categories (**Figure 14.4**), each roughly an order of magnitude (10 times) larger or smaller than the next. You can refer to this guide for size relationships as you meet the important members of the phytoplankton community in the next few pages.

CONCEPT CHECK

Before going on to the next section, check your understanding of some of the important ideas presented so far:

Are fine nets sufficient to capture all the plankton in a water parcel?

What data are usually collected along with the plankton sample? Why is this information important?

14.4 Phytoplankton

Autotrophic plankton that generate glucose by photosynthesis—the primary producers—are generally called **phytoplankton** (*phyton*, "plant"). A huge, nearly invisible mass of phytoplankton drifts within the euphotic zone, the sunlit surface layer of the world ocean. This upper productive layer of ocean is a very thin skin indeed. The water within the euphotic zone amounts to less than 2% of world ocean volume, but most pelagic marine life depends on this fine illuminated band.

Phytoplankton are critical to all life on Earth because of their great contribution to food webs and their generation of large amounts of atmospheric oxygen through photosynthesis. Planktonic autotrophs are thought to bind *at least* 50 trillion kilograms of carbon into carbohy-

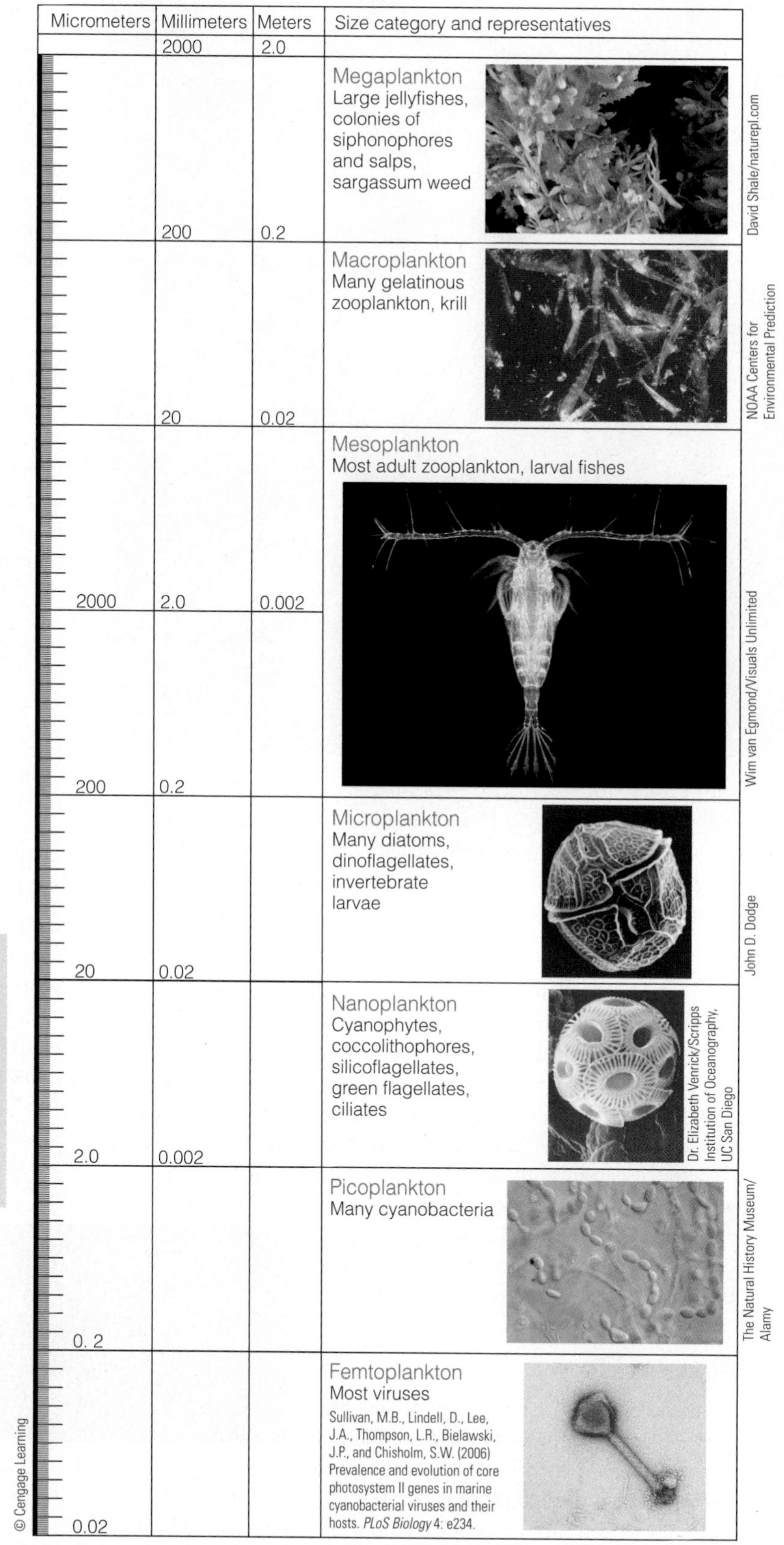

Micrometers	Millimeters	Meters	Size category and representatives
	2000	2.0	**Megaplankton** Large jellyfishes, colonies of siphonophores and salps, sargassum weed
	200	0.2	**Macroplankton** Many gelatinous zooplankton, krill
	20	0.02	**Mesoplankton** Most adult zooplankton, larval fishes
2000	2.0	0.002	
200	0.2		**Microplankton** Many diatoms, dinoflagellates, invertebrate larvae
20	0.02		**Nanoplankton** Cyanophytes, coccolithophores, silicoflagellates, green flagellates, ciliates
2.0	0.002		**Picoplankton** Many cyanobacteria
0.2			**Femtoplankton** Most viruses Sullivan, M.B., Lindell, D., Lee, J.A., Thompson, L.R., Bielawski, J.P., and Chisholm, S.W. (2006) Prevalence and evolution of core photosystem II genes in marine cyanobacterial viruses and their hosts. *PLoS Biology* 4: e234.
0.02			

Image credits (right margin): David Shale/naturepl.com; NOAA Centers for Environmental Prediction; Wim van Egmond/Visuals Unlimited; John D. Dodge; Dr. Elizabeth Venrick/Scripps Institution of Oceanography, UC San Diego; The Natural History Museum/Alamy

© Cengage Learning

Figure 14.4 A scheme of plankton classification that divides organisms into seven categories by size. Each category is roughly an order of magnitude (10 times) larger than the one below it.

 THINKING BEYOND THE FIGURE

Which of these classifications likely contains the most biomass?

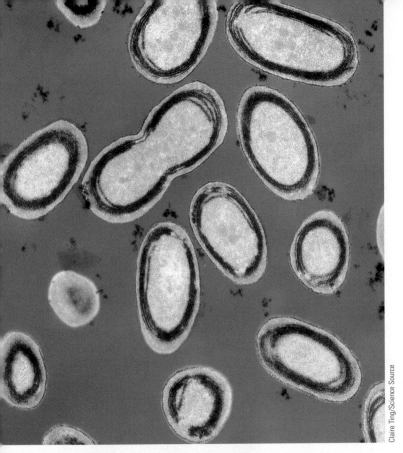

Figure 14.5 *Prochlorococcus*, a cyanobacterium. Along with *Synechococcus* (seen in Figure 14.6), this extraordinary tiny creature (not discovered until 1988) dominates the photosynthetic picoplankton of the world ocean. These autotrophs are able to absorb dim blue light in the deep euphotic zone.

Claire Ting/Science Source

The **cyanobacterium** *Prochlorococcus* (**Figure 14.5**) is typical of this newly recognized type of organism. It was discovered when individual cells—little more than naked photosynthetic machines—fluoresced a bright orange when struck by ultraviolet light. Living examples of *Prochlorococcus* are too small to study directly, but analysis of the fluorescence spectrum revealed the presence of an odd chlorophyll variant that permits the phytoplankter to absorb blue light at low light intensities in the deep euphotic zone.

Recent estimates suggest that picoplankton may account for up to 80% of all the photosynthetic activity in some parts of the open ocean, especially in the tropics, where surface nutrient concentrations are low. How could such a huge contribution to oceanic productivity have gone unnoticed for so long? It is in part because these autotrophs are extremely small, and in part because they are efficiently grazed by microflagellates and microciliates (tiny protistans). Additionally, the products of their photosynthetic activity are promptly used by even smaller heterotrophic bacteria in the immediate vicinity.

Here is a complete microecosystem—a community operating on the smallest possible scale—that manufactures and consumes particulate and dissolved carbon in amounts almost beyond comprehension. They function as a sort of ecological black market below the "official economy" of the relatively huge diatoms and dinoflagellates. As if they weren't busy enough, these heterotrophic bacteria also decompose organic material spilled into the water when phytoplankton are eaten by zooplankton, turn soluble organic materials released by zooplankton back into inorganic nutrients, and break down particulate organic matter into a dissolved form they can consume for their own growth. Biological oceanographers now believe that the greatest fraction of organic particles in the water column of the open ocean is composed of these metabolically active, heterotrophic bacterial cells operating in this **microbial loop** (**Figure 14.6**). The productivity associated with this "black market economy" is thought to be considerably larger than the "official economy." It is not available to fishes and other larger consumers because the small animals on which they prey are unable to separate these exceedingly small organisms from the surrounding water. Microconsumers simply utilize the carbon and shuttle the metabolic products back to the small cyanobacterial producers.

And what happens to the cyanobacteria? Those that aren't consumed by microflagellates may become infected by viruses that cause the cells to burst, adding to the supply of dissolved organic material. Viruses that infect bacteria are referred to as *bacteriophages;* those infecting phytoplankton are termed *phycoviruses.* Viruses, classified as femtoplankton, are extremely small (usually 20–250 nanometers in diameter) and are fundamentally different from other forms of life. Viruses have no metabolism of their own and must rely on a host organism for energy-requiring processes, including reproduction. Although we have known about viruses for some time, only in the late 1980s were their high abundances confirmed in a wide range of marine environments (**Figure 14.7**). How many are there? Between 10 million

drates each year, roughly 50% of the food made by photosynthesis on Earth! These easily overlooked, mostly single-celled, drifting photosynthesizers are much more important to marine productivity than the larger and more conspicuous seaweeds.

There are at least eight major types of phytoplankton, of which the most prominent are the diatoms and dinoflagellates. Recent research suggests that *very* small producers, most of which are forms of cyanobacteria and archaea, may be responsible for much more oceanic primary productivity than their larger and better-known counterparts!

Picoplankton

In the early 1980s, biological oceanographers began to appreciate the contribution to oceanic productivity of extremely small phytoplankton termed **picoplankton** (*pico,* "a trillionth part; very small"). These organisms are often too small to be resolved by light microscopes and slip undetected through all but the finest filters. Their size, typically about 0.2 to 2 micrometers (4–40 millionths of an inch) across, is made up for by their abundance: *an astonishing 100 million in every liter of seawater, at all depths and latitudes![2]*

> **P**hytoplankton make roughly 50% of the food made by photosynthesis on Earth.

---

[2]Someone with a bit of time on his hands calculated there are ~100 octillion picoplankters in the uppermost 200 meters (660 feet) of the world ocean!

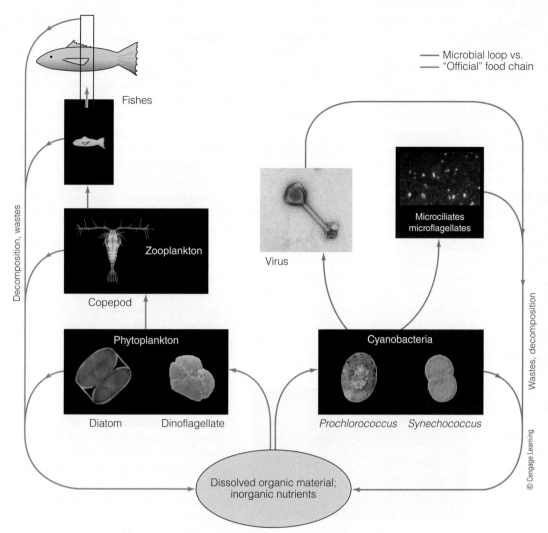

Fishes

Zooplankton

Copepod

Phytoplankton

Diatom   Dinoflagellate

Virus

Microciliates
microflagellates

Cyanobacteria

Prochlorococcus   *Synechococcus*

Dissolved organic material;
inorganic nutrients

Decomposition, wastes

Wastes, decomposition

—— Microbial loop vs.
—— "Official" food chain

© Cengage Learning

Figure 14.6 The "official" food chain of larger planktonic organisms (green) contrasts with the "black market economy" of the microbial loop (red). Larger planktonic organisms are unable to separate the astonishingly small cyanobacteria and microscopic consumers from the water and so cannot utilize them as food.

Photos: copepod: Wim van Egmond/Visuals Unlimited; diatom: D.P. Wilson/FLPA/Science Source; dinoflagellate: Florida Fish & Wildlife Conservation Commission, Florida Marine Research Institute, St. Petersburg; *Prochlorococcus*: Photograph courtesy of William Li and Frederic Partensky/(C) Fisheries and Oceans Canada, 2016; *Synechococcus*: blickwinkel/Alamy; microciliates: Gordon T. Taylor, Marine Sciences Research Center, SUNY—Stony Brook; virus: Sullivan, M.B., Lindell, D., Lee, J.A., Thompson, L.R., Bielawski, J.P., and Chisholm, S.W. (2006) Prevalence and evolution of core photosystem II genes in marine cyanobacterial viruses and their hosts. *PLoS Biology* 4: e234.

Figure 14.7 Marine viruses. Normally invisible in light microscopes, marine viruses glow as tiny dots when struck by ultraviolet light in this epifluorescence photograph. The number of viruses can be truly staggering—between 10 and 100 million *per milliliter* (12 drops) of surface seawater.

Courtesy Grieg Steward

and 100 million per milliliter of water—up to about 3 billion per ounce!

Diatoms

Apart from cyanobacteria, the most productive photosynthetic organisms in the plankton are the **diatoms**. Diatoms evolved comparatively recently and began to dominate phytoplanktonic productivity in the Cretaceous period about 100 million years ago. Their abundance and photosynthetic efficiency increased the proportion of free oxygen in Earth's atmosphere. More than 5,600 species of diatoms are known to exist. The larger species are barely visible to the unaided eye. Most are round, but some are elongated, branched, or triangular.

Typical diatoms are shown in **Figure 14.8**. The name means "to cut through" (*dia*, "through"; *tomos*, "to cut"), a reference to the patterns of perforations through the diatom's rigid cell wall, or **frustule**. As much as 95% of the mass of the frustule consists of silica (SiO_2), giving this heavy but beautiful covering the optical, physical, and chemical characteristics of glass—clearly an ideal protective window for a photosynthesizer. Magnification reveals that the frustule consists of two closely matched halves, or *valves*, which fit together like a well-made gift box, the top valve adhering tightly over the lip of the bottom one. The pattern of perforations, slits, striations, dots, and lines on the surface of the valves is different for each diatom species.

Inside the diatom's tailored valves lies a highly efficient photosynthetic machine. Fully 55% of the energy of sunlight absorbed by a diatom can be converted into the energy of carbohydrate chemical bonds, one of the most efficient energy conversion rates known. Excess oxygen not needed in the cell's respiration is released through the perforations in the frustule into the water. Some of this oxygen is absorbed by marine animals, some is incorporated into bottom sediments, and some diffuses into the atmosphere. Most of the oxygen we breathe has moved recently through the glistening pores of diatoms.

For more effective light absorption in diatoms, accessory pigments accompany chlorophyll, the main photosynthetic pigment. These yellow or brown pigments give most diatoms a yellow-green or tan appearance. Diatoms store energy as fatty acids and oils, compounds that are lighter than their equivalent volume of water and assist in flotation. As you might guess, flotation is a potential problem for diatoms because the weight of their heavy silica frustule seems at odds with their need to stay near the sunlit ocean surface. Oil floats, glass sinks, but a balanced amount of both reduce cell density and lighten the load. Diatom's high lipid content and abundance in many petroleum source rocks suggest that they are significant contributors to the petroleum we extract from oil fields.

Not all diatoms need to float, however. Many nonplanktonic species lie on shallow bottoms, where light and nutrients are able to support photosynthesis. These benthic species are nearly always elongated (or *pennate*) in shape.

Like most single-celled organisms, diatoms reproduce by dividing in half and drifting apart (or, in the species of diatoms that form chains, remaining linked in long lines of cells). The cells may divide as rapidly as once each day. In most species, the new valve is generated *within* the old one during division, so the average size of individuals in the population becomes smaller with time. Individuals reach a minimum of about one-fourth the size of the original cell. When the cells become too small—or more accurately, when the cells have too high a ratio of glass to living tissue—they become too heavy to float in spite of their buoyant oils. The problem is solved by sexual reproduction, which generates an *auxospore*, a naked cell without valves (**Figure 14.9**). If conditions are still favorable for growth, the auxospore will expand to the diatom's original size, form two thin new valves, and begin the cycle anew. If growing conditions are unsuitable, the auxospore will become dormant to await an opportunity for growth—which may be weeks or months away. When diatoms die, their valves fall to the seafloor to accumulate as layers of siliceous ooze (see Chapter 5).

> **T**otal productivity of the microbial loop is thought to be considerably larger than that of the "official" food chain.

Dinoflagellates

Most **dinoflagellates** are single-celled autotrophs (**Figure 14.10**). A few species live within the tissues of other organisms (the zooxanthellae of coral animals you will meet in Chapter 15, for example), but the great majority of dinoflagellates lives free in the water. Most have two whiplike projections called **flagella**, in channels grooved in their protective outer cell wall of cellulose. One flagellum drives the organism forward, while the other causes it to rotate in the water (hence the name: *dino*, "whirling"; *flagellum*, "whip"). Although unable to move them against currents, their flagella allow dinoflagellates to adjust their orientation and vertical position to make the best photosynthetic use of available light or to move vertically in the water column to obtain nutrients.

As you saw in this chapter's opener, some phytoplankton are strongly bioluminescent. Of the many types of planktonic organisms, dinoflagellates are the most common source of surface bioluminescence. **Bioluminescence** is the process by which energy from a chemical reaction is transformed into light energy. The imaginatively named compound *luciferin* is oxidized by action of the enzyme *luciferase;* in the process, a blue-green light is emitted. The release of light is very efficient and thus is not accompanied by a release of heat, so the organism does not overheat in the oxidation process. Bioluminescence may have evolved as a sort of "intrusion alarm" (see Questions from Students #4).

Some species of dinoflagellates can become so numerous that the water turns a rusty red as light reflects from the accessory pigments within each cell. These species are responsible

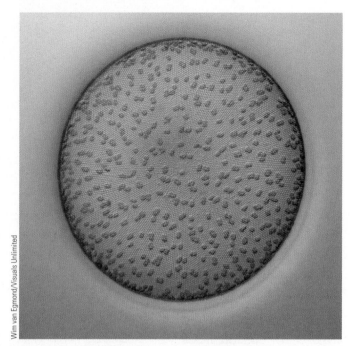

a Marine diatoms come in a variety of shapes and sizes. The largest of these is about the size of the period at the end of this sentence.

b The transparent frustrule of the diatom *Coscinodiscus* as shown with a light micrograph. The many small perforations that give diatoms their name are clearly visible. Note also the many green chloroplasts—cell organelles responsible for photosynthesis.

d An even closer view shows the perforations to be groups of still smaller holes.

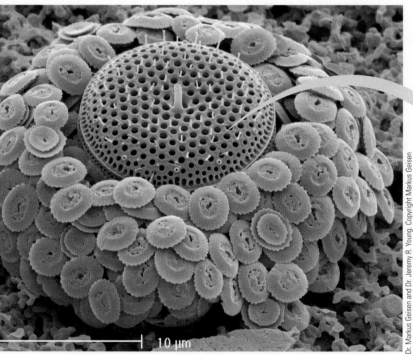

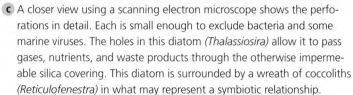

c A closer view using a scanning electron microscope shows the perforations in detail. Each is small enough to exclude bacteria and some marine viruses. The holes in this diatom *(Thalassiosira)* allow it to pass gases, nutrients, and waste products through the otherwise impermeable silica covering. This diatom is surrounded by a wreath of coccoliths *(Reticulofenestra)* in what may represent a symbiotic relationship.

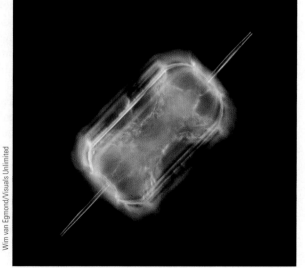

e *Ditylum*, a diatom, photographed in visible light. Note the junction between the valves.

Figure 14.8 Diatoms.

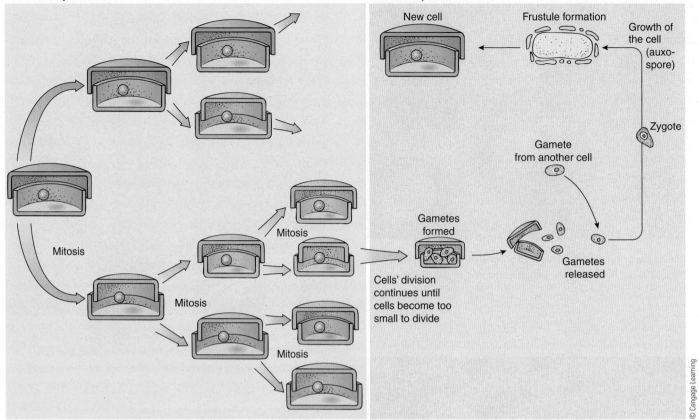

Asexual Reproduction

Mitosis

Mitosis

Mitosis

Mitosis

Sexual Reproduction

New cell

Frustule formation

Growth of the cell (auxospore)

Zygote

Gamete from another cell

Gametes formed

Gametes released

Cells' division continues until cells become too small to divide

© Cengage Learning

Figure 14.9 Diatom reproduction. Diatom cells can divide asexually into smaller daughter cells until those cells reach a critical size. They must then reproduce sexually to form an auxospore, which can eventually grow to the original cell's size.

for many harmful algal blooms—**HABs** (Figure 14.10b). During times of such rapid growth (usually in springtime), concentration of these microscopic organisms may briefly reach 6 million per liter (23 million per gallon)! At night, the huge numbers of bioluminescing dinoflagellates in an HAB (sometimes called a "red tide") can cause breaking waves to glow a bright blue (Figure 14.10c).

HABs can be dangerous because some dinoflagellate species synthesize potent toxins as by-products of metabolism. Among the most effective poisons known, these toxins may affect nearby marine life or even humans. Some of the toxins are similar in chemical structure to the muscle relaxant curare but are tens of times more powerful. Humans should avoid eating certain species of clams, mussels, and other filter feeders during summer months when toxin-producing dinoflagellates are abundant in the plankton. If shellfish from a particular area are unsafe, a state governmental agency will issue an advisory, which may remain in effect for 6 weeks or more until the danger is past.

HABs dominated by dinoflagellates of the genus *Karenia* can be especially pesky. Asthma sufferers are in danger when these dinoflagellates dry on the beach and are blown inland. Significant decreases in lung function were measured among Florida residents as far as a mile inland after an hour of exposure to brevetoxin, the poison produced by these organisms.

Coccolithophores

Coccolithophores are small, single-celled autotrophs covered with disks of calcium carbonate (coccoliths) fixed to the outside of their cell walls (Figure 14.11). Coccolithophores live near the ocean surface in brightly lit areas. Their translucent covering of coccoliths may act as a sunshade to prevent absorption of too much light. In areas of high coccolithophore productivity, most notably in the Mediterranean and Sargasso seas, their numbers occasionally become so great that the water appears milky or chalky. Coccoliths can also build seabed deposits of calcareous ooze. The famous White Cliffs of Dover in southeastern England consist largely of fossil coccolith deposits uplifted by geological forces (Figure 5.13).

CONCEPT CHECK

Before going on to the next section, check your understanding of some of the important ideas presented so far:

Can you provide an example of a nanoplankter? Microplankton? A femtoplanktonic organism?

What is an autotroph? Are all plankton autotrophic?

What is unique about picoplankton?

What is the "microbial loop"? What role does it play in the marine environment?

How are diatoms similar to and different from dinoflagellates?

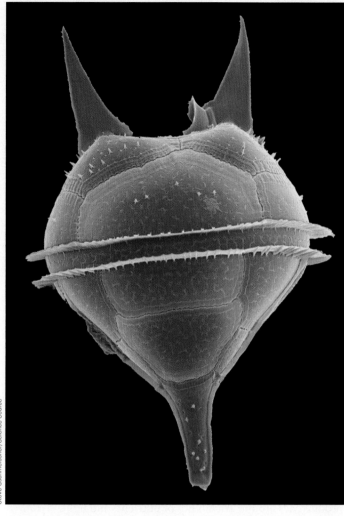

Steve Gschmeissner/Science Source

a *Ceratium*, a common photosynthetic dinoflagellate. As their name implies, dinoflagellates have two flagella—one flagellum beats within a central girdle and causes the cell to rotate so that all surfaces are exposed to sunlight; the other extends away from the organism and acts as a propeller. (Neither flagellum is visible in this scanning electron micrograph.) This specimen is about 0.5 millimeter (0.02 inch) across.

Pete Atkinson/Getty Images

b During a red tide, the presence of millions of dinoflagellates turns seawater brownish-red. The term *red tide* is misleading—they are not caused by the tide and they are not always red. *Harmful algal bloom* (HAB) is the preferred term. Some red tides may contain high abundances of different planktonic organisms like the diatom *Pseudonitzschia*.

North County Times/ZUMA Press/Corbis

c Visitors are drawn to the beach during a red tide in Southern California in 2011. The number of bioluminescent dinoflagellates in the water causes the waves to glow as they break.

Figure 14.10 Dinoflagellates.

💬 **THINKING BEYOND THE FIGURE**

Why do these dinoflagellates have a brownish-red color?

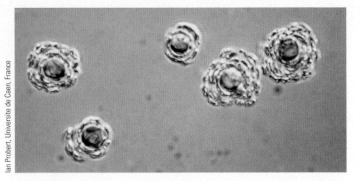

a A rare light microscope photograph of *Emiliania huxleyi,* a coccolithophore. The tiny calcified plates (coccoliths) covering the cells are 6 micrometers (120 millionths of an inch) across. Photosynthetic pigments give the cells a golden or golden-brown color.

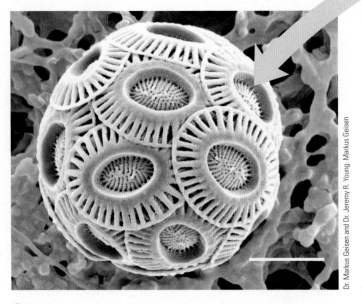

b This electron micrograph of the abundant coccolithophore *E. huxleyi* shows the small calcium carbonate plates (coccoliths) covering the cell's exterior. The coccoliths (not the organic cells themselves) act like mirrors suspended in the water and can reflect a significant amount of the incoming sunlight. The reflectance from the blooms can be picked up by satellites in space, allowing the extent of the blooms of this species to be distinguished in fine detail.

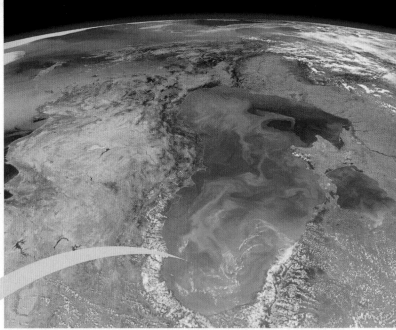

c A coccolithophore bloom is clearly visible in this natural color image of the Black Sea captured by NASA's *Aqua* satellite. The species responsible for this bloom is likely *E. huxleyi,* the same species in both **a** and **b** of this figure.

Figure 14.11 Coccolithophores.

💬 **THINKING BEYOND THE FIGURE**
Why are coccolithophores more easily seen from space than diatoms?

14.5 Lack of Nutrients and Light Can Limit Primary Productivity

As you may recall from Chapter 13, a *limiting factor* is a physical or biological necessity whose presence in inappropriate amounts limits the normal actions of an organism—in this case, production of carbohydrates. Photosynthetic autotrophs require four main ingredients to produce carbohydrates: water, carbon dioxide, inorganic nutrients, and sunlight. Obviously, water is not a limiting factor in the ocean. Carbon dioxide is almost never a limiting factor either because of its high solubility in water and because of the large quantity of carbon dioxide dissolved in the ocean. So the two potential limiting factors in marine primary productivity are the availability of nutrients and light.

Nutrient Availability Can Be a Limiting Factor

Autotrophs require inorganic nutrients for two purposes: to construct the large organic molecules that make primary productivity possible and to construct their skeletons or protective shells. **Nonconservative nutrients** are nutrients that change in concentration with biological activity. After a period of rapid phytoplankton growth—a phenomenon known as a **plankton bloom**—ocean surface waters are often depleted of nonconservative nutrients such as nitrate, phosphate, iron, and silicate. These nutrients become parts of the producers or of the animals that have eaten the producers. Unfortunately for the producers, valuable nutrients incorporated into the bodies of dead organisms tend to sink below the sunlit zone. The resulting low nutrient concentrations are the most important factors limiting the growth of marine producers. Photosynthetic productivity cannot continue unless the upwelling of deep water returns these nutrients to the surface (as in **Figure 14.12**).

The opportunity for the natural exchange of nutrient-depleted surface water and nutrient-rich deep water is relatively high where there is little or no thermocline. In the Antarctic, the return of sunlight in the southern summer triggers tremendous phytoplankton blooms; Antarctic water sometimes becomes a rich planktonic soup as millions of diatoms compete for nutri-

How Much Primary Productivity Occurs in the Ocean

NASA

Figure A Remote sensing of phytoplankton blooms. Scientists captured this image with NASA's Aqua satellite using seven different spectral bands to distinguish different plankton communities.

We've mentioned that the phytoplankton are believed to contribute at least 50% of the food made by photosynthesis on Earth, but the actual numbers are not easy to measure. At first glance the problem might seem simple to solve: Collect all the organisms—large and small, autotrophs and heterotrophs—in an area and analyze their organic content and weight. Since all organisms are dependent on primary productivity, the mass of living tissue (biomass) in the area would seem directly proportional to productivity.

This approach has drawbacks, though. A dense population of tiny drifting autotrophs (a high biomass) would interfere with light penetration; so the autotrophs would manufacture carbohydrates slowly, and productivity would be low. In contrast, a sparse population of drifting autotrophs (a low biomass) might encounter ideal conditions for photosynthesis and produce carbohydrates at a rapid rate. Small animals might immediately consume this production and keep the biomass at low levels, but productivity would be high.

What researchers need is a method of measuring the rate of productivity *directly*. Since we know the formula for photosynthesis (see again Figure 13.11), measuring any one component of photosynthesis will tell us about the others. For example, from a measurement of carbon taken up by primary producers we can calculate the rate of carbohydrate production.

Researchers can make this calculation using atoms of carbon "tagged" by radioactivity. Though it is radioactive, carbon-14 (^{14}C) behaves chemically in the same way as the much more common carbon-12 (^{12}C). And because ^{14}C is radioactive, its progress through photosynthesis can be monitored. One of the ions formed when carbon dioxide dissolves in seawater is bicarbonate (HCO_3^-).[3] The scientists tag the carbon in bicarbonate and add known amounts of radioactive bicarbonate to two bottles of seawater. One bottle is exposed to light, and photosynthesis and respiration take place. The other bottle is shielded from light, so only respiration occurs. The amount of radioactive carbon incorporated into carbohydrates is measured when the organisms are filtered out of the samples. The radioactivity is measured and productivity calculated:

$$\text{Rate of production} = (R_L - R_D) \times M/R \times t$$

where R is the total radioactivity added to the sample, t is the number of hours of incubation, R_L is the radioactive count in the "light" bottle sample, and R_D is the count in the "dark" sample. M is the total mass of all forms of carbon dioxide in the sample (in milligrams of carbon per cubic meter). Productivity is expressed as the amount of carbon (in milligrams) of carbon bound—or *fixed*—in new carbohydrate per volume of water (in cubic meters) per unit time (per hour). The rate of production varies from zero to as much as about 80 milligrams of carbon fixed into carbohydrates per cubic meter per hour. These data can be extrapolated to provide the amount of carbon fixed in the water column per square meter of surface per day ($gC/m^2/day$).

Ocean conditions are complex and variable, however—rarely controlled. A new and promising method of gauging productivity may be the most effective and useful of all. Recent advances in remote sensing have made it possible to estimate the chlorophyll content of ocean water from orbiting satellites (as in **Figure A** and Figures 14.2, 14.12, and 14.14). Because the amount of chlorophyll present is directly related to the rate of photosynthesis, chlorophyll content is a good indicator of productivity. As we have seen, however, the biomass of drifting autotrophs—the amount of phytoplankton in an area—is not always a true indicator of how rapidly substances are cycling through that biomass.

ents and sunlight. Indeed, because of the high nutrient levels from upwelling supplemented by iron suspended in glacial run-off from land, the summer waters between the Antarctic Convergence and the mainland of Antarctica are among the most productive on Earth. Phytoplankton blooms in the northern polar ocean are not usually as exuberant because the volume of upwelling water is much less there, and glaciers don't contribute as much rock dust to the ocean.

Because of the stability of the horizontal layers, upwelling is uncommon in the tropical ocean, except in the equatorial Pacific or in areas where currents impinge on interrupting islands or continents. The clear blue of the tropical ocean is the sure signature of an oceanic desert in which deep upwelling rarely occurs. When nutrients are available or are tightly recycled, as

[3]This is shown in Figure 7.11.

10 August 2003

SeaWiFS Project
NASA / GSFC
ORBIMAGE

Chlorophyll Concentration (mg /m²)

NASA/GGSFC and GeoEye

Figure 14.12 Upwelling along the California and Baja California coasts produced a large plankton bloom in August 2003. Red colors indicate chlorophyll-rich waters; blue and purple colors indicate little suspended chlorophyll.

in shallow coral reefs, the tropical ocean explodes into vigorous productivity.

Light May Also Be Limiting

If adequate nutrients are present, primary productivity depends on illumination. Too little light is obviously limiting for photosynthesizers; very little photosynthesis proceeds below 100 meters (330 feet), and no solar photosynthesizers are known to function below 268 meters (879 feet). Too much light can also be inhibiting, however. You might think that productivity would be greatest right at the ocean surface, where light is brightest. It isn't. Light there is often strong enough to overwhelm the photosynthetic chemistry of some photosynthesizers, especially diatoms.

Quantity is one important aspect of the light received by marine autotrophs. Quality, or color, is another. Chlorophyll is a green pigment and thus absorbs best in the red and violet wavelengths. Chlorophyll looks green because it reflects green light. But red and infrared light are effectively absorbed and converted into heat near the ocean surface; very little red light penetrates past 3 meters (10 feet). Phytoplankton, except photosynthetic cyanobacteria (which can accept blue light), stay near the surface to absorb red light, and primary productivity is thus highest in this top part of the euphotic zone.

CONCEPT CHECK

Before going on to the next section, check your understanding of some of the important ideas presented so far:

What factors most often limit phytoplankton productivity?

What element, when missing, most often limits planktonic productivity in the open ocean?

Which planktonic primary producers can succeed at the greatest depths?

14.6 Production Equals Consumption at the Compensation Depth

So far, we've emphasized the importance of phytoplankton as producers, but remember that autotrophs *respire* as they photosynthesize—they use some of the carbohydrates and oxygen they produce. Carbohydrate production usually exceeds consumption, but not always.

The deeper a phytoplankter's position, the less light it receives. At a certain depth, the production of carbohydrates and oxygen by photosynthesis in a day will exactly equal the consumption of carbohydrates and oxygen by respiration. This break-even depth is called the **compensation depth**. Compensation depth usually corresponds to the depth to which about 1% of surface light penetrates; it marks the bottom of the euphotic zone.

Figure 14.13 shows compensation depth graphically. Like the depth of greatest productivity, compensation depth changes with sun angle, turbidity, surface turbulence, and other factors. Remember that the compensation depth is always below the depth of greatest productivity and that producers can still "make a profit" between the depth of greatest productivity and the compensation depth. If a producer slips below its compensation depth for more than a few days, however, it will consume its carbohydrate reserves and die.

Because of their greater efficiency, diatoms have a deeper compensation depth than dinoflagellates. Open tropical seas have the deepest potential compensation depths, but as we've seen, these regions are not generally productive because of nutrient deficiencies.

14.7 Phytoplankton Productivity Varies with Local Conditions

Where is phytoplankton productivity the greatest? This question is among the most important in biological oceanography. Since phytoplankton form the base of nearly all oceanic food webs, the biological characteristics of any ocean area will depend heavily on the presence and success of phytoplankton.

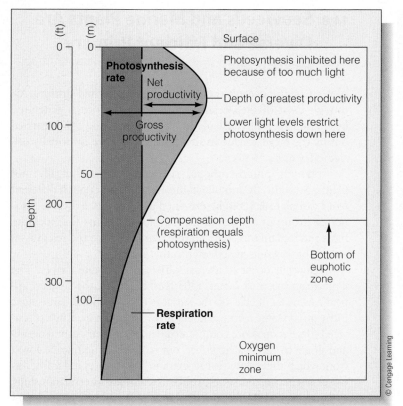

Figure 14.13 Compensation depth and its relationship to other aspects of productivity. Note the position of the bottom of the euphotic zone.

💬 **THINKING BEYOND THE FIGURE**

Which has a higher (higher in the water) compensation depth: diatom or dinoflagellate?

With some exceptions, the distribution of phytoplankton corresponds to the distribution of nutrients. Because of coastal upwelling and land runoff, nutrient levels are highest near the continents. Plankton are most abundant there, and productivity is highest. The water above some continental shelves sustains productivity in excess of 1 gC/m²/*day*! Whole-ocean productivity averages about one third that amount. But what of the open ocean? Where is productivity greatest away from land?

The open tropical oceans have abundant sunlight and CO_2 but are generally low in surface nutrients because the strong thermocline discourages the vertical mixing necessary to bring nutrients from the lower depths. The tropical oceans away from land are therefore oceanic deserts nearly devoid of visible (that is, non-cyanobacterial) plankton. The typical clarity of tropical water underscores this point. In most of the tropics, productivity rarely exceeds 30 g C/m²/yr, and seasonal fluctuation in productivity is low.

Tropical coral reefs are exceptions to this general rule. These are productive places because autotrophic dinoflagellates live *within* the tissues of coral animals and don't drift with the plankton. Nutrients are tightly cycled through the reef and not lost to sinking.

At very high latitudes, the low sun angle, reduced light penetration due to ice cover, and weeks or months of darkness in winter severely limit productivity. At the height of summer, however, 24-hour daylight, a lack of surface ice, and the presence of upwelled nutrients can lead to spectacular plankton blooms. The surface of some sheltered bays can look like tomato soup because dinoflagellates and other plankton are so abundant. This bloom cannot last because nutrients are not quickly recycled and because the sun is above the critical angle for a few weeks at best. The short-lived summer peak does not compensate for the long, unproductive winter months.

With the tropics generally out of the running for reasons of nutrient deficiency and the north polar ocean suffering from slow nutrient turnover and low illumination, the overall productivity prize is left to the temperate and southern subpolar zones. Thanks to the dependable light and the moderate nutrient supply, annual production over temperate continental shelves and in southern subpolar ocean areas is the greatest of any open ocean area.

Figure 14.14 shows the levels of productivity in tropical, temperate, and northern polar ocean areas. Note that *nearshore* productivity is nearly always higher than *open ocean* productivity, even in the relatively productive temperate and south subpolar zones. Zones of high and low productivity change as climate changes. Water runoff from land (which provides nearshore nutrients), increasing or decreasing levels of sunlight due to variations in cloud cover, and even changes in the strength of ultraviolet radiation at high latitudes due to ozone molecules high in the atmosphere (about which more will be found in Chapter 18) all affect primary productivity.

Curiously, the open ocean area with the greatest annual productivity is an exception to the general picture developed in this section. The slender, cold finger of high productivity pointing west from South America along the equator is a result of wind-propelled upwelling due to Ekman transport on either side of the geographic equator. Look for this area in Figure 9.19b.

Figure 14.15 shows the relationship of phytoplankton biomass to season and latitude. The low, flat line representing annual tropical productivity contrasts with the high, thin peak representing the arctic summer. The higher of the two peaks for the temperate zone indicates the plankton bloom of northern spring caused by increasing illumination, while the smaller peak representing the northern fall is caused by nutrients returning to the surface with increasing storm turbulence and less thermal stratification.

> **T**he tropical ocean—the paradise of travel posters—is an oceanic desert. Away from the reefs (where nutrients are quickly recycled), a lack of nutrients limits productivity. Clear blue water is dead water.

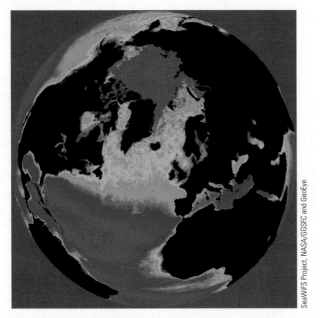

Figure 14.14 Measuring the concentration of chlorophyll in open ocean water by satellite. The scanner aboard the *Nimbus 7* satellite shows the concentration of chlorophyll in the upper layer of the ocean, with higher amounts indicated by green, orange, and red colors. Note the high phytoplankton concentrations induced by increased nutrient availability along the coasts. As shown by their purple hue, the centers of the oceanic gyres contain relatively few phytoplankters, Compare this image to Figure 14.12.

CONCEPT CHECK

Before going on to the next section, check your understanding of some of the important ideas presented so far:

Where is oceanic primary productivity highest?

Why is surface productivity generally low in the tropics?

At what time(s) of year does plankton productivity increase in different latitudes?

14.8 Seaweeds and Marine Plants Are Diverse and Efficient Primary Producers

Not all of the ocean's producers are single-celled drifters. Although most of the ocean's primary productivity is generated by single-celled phytoplankton, between 1% and 5% is carried out by the large marine **multicellular algae** we informally call *seaweeds* (Figure 14.16).

Though photosynthetic, seaweeds are technically not plants. Structurally and biochemically they are enough different from vascular plants to be classified as *protistans*, a diverse taxonomic group of comparatively simple organisms that includes most phytoplankton. True marine plants include the sea grasses and mangroves discussed later in this section.

Seaweeds occur in a great variety of sizes and shapes. The largest can reach 62 meters (205 feet) in length; the smallest appear as smears of cells on the surfaces of rocks. Some large algae form underwater forests; others grow in isolation. Lifeless seaweeds drying on shore communicate none of their natural beauty and grace to the beachcomber, but to a diver the marine forest from which they came reveals extraordinary grace and form, discloses sheltering nurseries, conceals complex interrelationships, and even provides nourishment. None of these algae grows below the euphotic zone because all depend on photosynthesis to produce the energy-rich compounds necessary for life. Nearly 7,000 species of multicellular marine algae have been identified.

Complex Adaptations Permit Seaweeds to Thrive in Shallow Waters

At first glance, marine algae and plants appear to have an easy life, but living near the ocean surface is not without hazards. Large intertidal autotrophs may find themselves exposed to the drying effects of air and sunlight when the tide is out and may

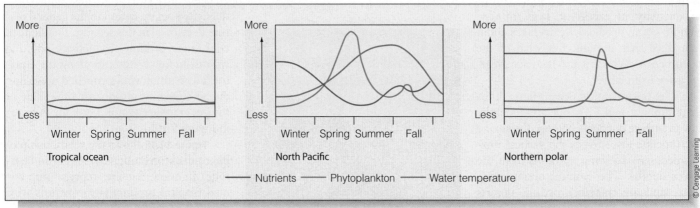

a In the tropics, an intense thermocline prevents nutrient-rich water from rising to the surface. Productivity is low throughout the year.

b In the northern temperate ocean, nutrients rising to the surface combine with spring and summer sunlight to stimulate a plankton bloom.

c In the northern polar ocean, a high and thin productivity spike occurs when the sun reaches high enough above the horizon to allow light to penetrate the ocean surface.

Figure 14.15 Variation in oceanic primary productivity by season and latitude. The area under each green curve (phytoplankton biomass) represents total productivity.

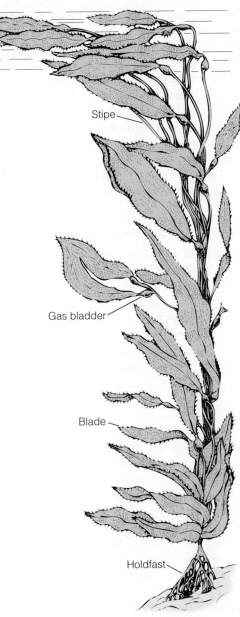

Stipe

Gas bladder

Blade

Holdfast

MAURICIO HANDLER/National Geographic Creative

© Cengage Learning

b *Macrocystis,* a fast-growing, very productive brown alga (phaeophyte), can form spectacular kelp forests off the coast of western North America. Gas bladders at the base of each blade help this alga create a dense canopy at the surface.

a The thallus (body) of a typical multicellular alga. These organisms can grow at a rate of 50 centimeters (20 inches) per day and reach a length of 40 meters (132 feet).

Figure 14.16 Seaweeds.

💬 **THINKING BEYOND THE FIGURE**
If the holdfast of this alga breaks free in a storm, will the organism die?

be lashed against rocks by waves when the tide is in. The physical nature of the organisms themselves provides some defense against these difficulties—their bodies are flexible, easily able to absorb shock, resistant to abrasion, streamlined to reduce water drag, and very strong.

Warmth and a lack of nutrients often limit the success of seaweeds. Higher temperatures lead to higher metabolic rates, and the oxygen level in warm seawater may not be high enough to support the respiratory needs of the algae at night. Warmer temperatures can also shatter the delicate accessory pigments and proteins required for photosynthesis and respiration. Large algae are rare in warm nutrient-poor waters, and divers visiting the tropics are usually surprised to find no sign of kelp forests. Chilly temperate and subpolar zones of nutrient upwelling often support thick algal mats and dense marine forests.

Yet another difficulty is the location of adequate anchorage or substrate. Attached seaweeds require a firm footing. A sandy

or muddy bottom is unsuitable for colonization by most large algae, and less than 2% of the ocean floor is shallow enough and solid enough to permit their growth.

The marine lifestyle also offers some advantages. Marine algae and plants suffer no droughts and nearly always have enough carbon dioxide for photosynthesis. Assuming suitable nutrient levels and a good foothold, only sunlight is required for productivity. Being submerged in seawater brings the additional advantage of lightweight construction. A seaweed doesn't require strong support structures because it has nearly the same density as the surrounding seawater. More of its bulk can thus be dedicated to photosynthesis. Indeed, productivity in some seaweed beds may be the highest of *any* autotrophic community on Earth.

The high productivity of seaweeds is underscored by the fact that many species of large seaweeds are surprisingly leaky—carbohydrates and other products of photosynthesis diffuse from their blades and stipes like tea from a teabag. Up to half of all the organic matter they produce can be lost in this way! The foam visible in surf near kelp beds is produced in part by these substances. Sea urchins and other heterotrophs can absorb these molecules directly through their skin or outer membranes, "feeding" on kelp plants that may be tens or even thousands of meters away.

Seaweeds Are Nonvascular Organisms

Photosynthetic organisms require CO_2, water, nutrients, and sunlight to make carbohydrates. Land plants lift water and nutrients into leaves exposed to light and air, construct food, and ship some of the food down to the roots to repay their efforts. Bundles of conductive vessels accomplish this task. Such vessels are not required in seaweeds because all four physical requirements are simultaneously present within their bodies.

Common terms like *leaf*, *stem*, and *root* are inappropriate for seaweeds because the definitions of those parts assume the presence of vessels. Instead, blades, stipes, and holdfasts comprise the body of the organism, the **thallus**. These parts are labeled in **Figure 14.16a**.

Some species of the largest of algae, which include the **kelps**, can reach lengths of 40 meters (132 feet)—the record length exceeds 60 meters (200 feet). To attain these dimensions, the plant can grow at the spectacular rate of 50 centimeters (20 inches) per day! Some brown algae are annuals; others live for up to 7 years. In ideal circumstances, kelp can grow in water up to about 35 meters (115 feet) deep.

Seaweeds Are Classified by Their Photosynthetic Pigments

Seaweeds are classified in part by the presence of colored compounds in their tissues. These **accessory pigments** (or masking pigments) are light-absorbing compounds closely associated with chlorophyll molecules. They don't resemble chlorophyll chemically, but they bind loosely with it. Their presence in plants greatly enhances photosynthesis because they absorb the dim blue light at depth and transfer its energy to the adjacent chlorophyll molecules. Accessory pigments may be brown, tan, olive green, or red; they are what give most marine autotrophs, especially seaweeds, their characteristic color.

Multicellular marine algae are segregated into three divisions based on their observable color. The green algae, with their unmasked chlorophyll, are the *Chlorophyta*, the brown algae *Phaeophyta*, and the red algae *Rhodophyta*. Phaeophytes are most familiar to beachcombers, and rhodophytes are the most numerous.

Seaweeds Are Commercially Important

Chances are good that you have had some recent contact with marine autotrophs even if you haven't been in the ocean. The mucilaginous material that is so effective in making algal blades slick, in lowering friction, and in deterring grazers is also harvested and made into an important commercial product called *algin*. When separated and purified, its long, intertwining molecules are used to stiffen fabrics; make adhesives; suspend water and oil together in salad dressings; prevent the formation of gritty crystals in ice cream; clarify beer and wine; and manufac-

ture shoe stains, soaps, and shaving cream. Fast-food restaurants now use carageenan, a similar seaweed extract, to replace some of the fat in healthier hamburgers. These substances also prevent fire-extinguishing foams from dispersing, they permit chocolate milk to remain on the refrigerator shelf without separating, and they keep the abrasives in liquid car waxes from settling to the bottom of the bottle. In biological laboratories, bacteria are cultured on agar made from seaweed extracts. Very likely, even the ink that forms the letters you are now reading has an algin or carageenan component!

You can find out more about the uses of marine plants in the discussion of marine resources in Chapter 17.

True Marine Plants Are Vascular Plants

Seaweeds may look like plants, but they are actually a form of *multicellular algae*. The single-celled diatoms and dinoflagellates discussed earlier are **unicellular algae**. As we have seen, **algae** lack the vessels and other structural and chemical features of true plants. Nearly all large land plants are vascular plants. A few species of vascular plants have recolonized the ocean. All have descended from land ancestors, and all live in shallow coastal water. The most conspicuous marine vascular plants are the sea grasses and the mangroves.

Sea Grasses

Many people lump sea grasses in the informal seaweed group, but their resemblance to large marine algae is only superficial. These plants are not true grasses, but they do have leaves and stems, as well as roots capable of extracting nutrients from the substrate. Extensive stands of sea grasses are found on the coasts of North America, on the Atlantic coast of Europe, in Eastern Asia, in temperate Australia, and in Southern Africa. They form broad gray or green submerged meadows, which support extraordinarily rich communities of heterotrophs. The life cycle of sea grasses is much like that of other flowering plants, but their stringy pollen is distributed by flowing water rather than by insects or wind. About 45 species of sea grasses are known.

Sea grasses can be remarkably productive. Some are capable of binding 1,000 $gC/m^2/yr$, three to five times the nearshore average for phytoplankton. One explanation for this efficiency is the advantage conferred by roots. Anaerobic bacteria in subtidal mud have been shown to bind dissolved nitrogen into nitrates easily available to these plants.

Perhaps the most beautiful sea grass is the vivid, emerald-green surf grass, genus *Phyllospadix*, with its seasonal flowers and fuzzy fruit. These hardy plants survive in the turbulent, wave-swept intertidal and subtidal zones of temperate East Asia and western North America (**Figure 14.17**).

Mangroves

Low, muddy coasts in tropical and some subtropical areas are often home to tangled masses of trees known as **mangroves**. These large, flowering plants are never completely submerged, but we consider them to be marine plants because of their intimate association with the ocean. They thrive in the sediment-

Figure 14.17 The bright green sea grass *Phyllospadix* in a tide pool. Sea grasses are vascular plants, not seaweeds.

rich lagoons, bays, and estuaries of the Indo-Pacific, tropical Africa, and the tropical Americas.

The root system traps and holds sediments around the plant by interfering with the transport of suspended particles by currents. Mangrove forests thus assist in the stabilization and expansion of deltas and other coastal wetlands. The root complex also forms an impenetrable barrier and safe haven for organisms around the base of the trees.

The mangrove communities of south Florida, among the world's most widespread, consist primarily of the red mangrove (genus *Rhizophora*). The spreading leaves of the tree protect the residents from the tropical sun, and the tangled roots keep out large predators and also harbor huge numbers of fiddler crabs, worms, marine and terrestrial snails, fish, oysters, and red algae (Figure 14.18). Birds and insects inhabit the treetops, their droppings enriching the sediments below.

Mangrove seeds germinate on the trees. If the tide is out when the seed drops, the force of the fall will plant the seed in the muck, where it will continue to grow. If the tide is in, the seed will drop into the water, suspend growth, float to a new location, and resume growth when it gets a foothold in the mud. The trees mature in 20 to 30 years and can attain a height of 8 to 10 meters (26–33 feet). New mangroves colonize mud and sandbars away from shore; they stabilize them and eventually create new land.

The distribution of kelp beds and mangrove communities is shown in Figure 14.19.

CONCEPT CHECK

Before going on to the next section, check your understanding of some of the important ideas presented so far:

What are algae? Are all algae seaweeds?

How are seaweeds classified? Which seaweeds live at the greatest depths? Why are accessory pigments important in algal physiology?

Give examples of marine flowering plants. Are they vascular plants or nonvascular algae?

Of what commercial importance are marine algae and plants?

Figure 14.18 A mangrove (genus *Rhizophora*) growing in saltwater in Everglades National Park, Florida. The tangled roots descending from the main branch are called prop roots or stilt roots. They provide anchorage for the mangrove, trap sediment, and provide shelter for small organisms. Mangroves are also pictured in Figure 12.27.

14.9 Primary Productivity Also Occurs Deep in the Water Column, at Hydrothermal Vents, in Seabed Sediments, and in Solid Rock

As you would expect, production of carbohydrates by *photosynthesis* dominates ocean surface productivity. Imagine researchers' surprise when they recovered emerald green photosynthesizers from the darkness 2,500 meters (8,200 feet) below the surface on the East Pacific Rise! The water streaming from hydrothermal vents can reach 400°C (750°F). The vents radiate a dark red light in the way hot electric-stove elements glow. The newly discovered organisms use this dim light to power photosynthesis. Could photosynthetic organisms have originated in the deep ocean and then drifted upward to colonize the sunlit surface?

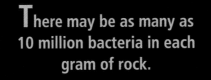

There may be as many as 10 million bacteria in each gram of rock.

While some photosynthetic organisms can survive on the dim light from hydrothermal vents, most of the deep sea primary productivity results from *chemosynthesis*. Specialized archaea and bacteria that live near hydrothermal vents are able to use hydrogen sulfide, carbon dioxide, and oxygen to produce carbohydrates. These organisms form the base of an entire food web. In other areas of the sea floor, mats of chemosynthetic bacteria metabolize sulfur-containing compounds or methane at cold seeps. You'll learn more about these chemosynthetic communities in Chapter 16.

The extent of primary productivity by chemosynthesis *within* the seabed itself has been another surprise. Recent research has shown that seabed communities are not confined to the uppermost layer of sediments but are also found deep beneath the seafloor. What may prove to be the world's largest communities are only now being discovered. In the late 1970s, researchers studying the quality of groundwater discovered that microorganisms could live in deep sediments and water-yielding rock formations. Because water coming from great depths can easily be contaminated by surface bacteria, a special drilling device was invented that could take uncontaminated samples of solid rock, even below the ocean bottom. Biologists were astonished to find microbial ecosystems existing in the pores between interlocking mineral grains of many rocks at drilling depths to 1,220 meters (4,000 feet) and temperatures to 400°C (750°F). Studies of cores from the Ocean Drilling Program show bacterial and archaean communities as deep as 842 meters (2,800 feet) below the seabed in sediments 14 million years old, and a Princeton geologist has recently extracted bacteria from water collected more than 3.2 kilometers (2 miles) beneath the South African coast! These bacteria are thriving in extreme conditions at these depths; they have high diversity and are well adapted to life in the subsurface.

What a vast habitat!

There may be as few as 100 or as many as 10 million bacteria in each gram of rock. What are they doing? What do they live on? Because their habitat receives no light, the autotrophs must be capable of chemosynthesis. These primary producers are consumed by equally tiny primary consumers. Although there are a great many organisms present, their metabolic rates appear to be very slow—some of these organisms divide once every 100 to 2,000 years! As depth and pressure increase, the already microscopic pores in the rocks become smaller and the availability of chemosynthetic raw materials dwindles, but some

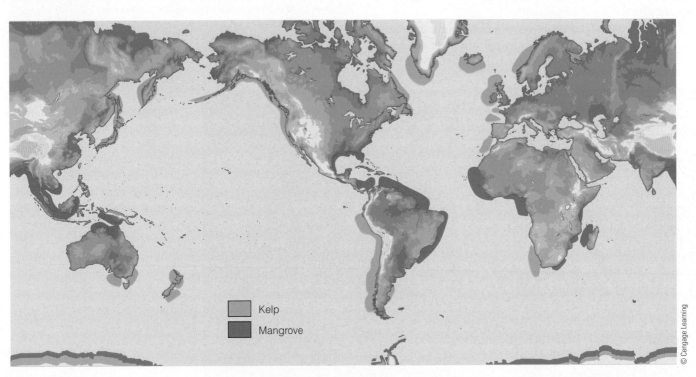

Kelp

Mangrove

© Cengage Learning

Figure 14.19 Distribution of kelp beds and mangrove communities worldwide.

researchers believe these well-hidden communities comprise about one third of Earth's total biomass!

These aggregations of producers and consumers have been called SLIMES (for *subsurface lithoautotrophic microbial ecosystems*). New categories of bacteria have been found in their midst, including "ultramicrobacteria," dwarf bacteria apparently adapted to exceedingly small rock pores. Tantalizingly, these simple cells may be the remnants of Earth's first life-forms. Conditions on Earth at the time of the origin of life were hot and oxygen free, and the genetic make-up of these bacteria and archaeans suggests they've evolved more slowly and in different directions than other forms of life here.

These specialized organisms are usually called **extremophiles** because they are capable of life under extreme conditions **(Figure 14.20)**. Bacteria and similar organisms known as archaea have been found in fractured rocks more than 3 kilometers (1.9 miles) below the surface of Africa. Specialized organisms have been seen in hot oil reservoirs below the North Sea and the North Slope of Alaska (where they cause oil to "sour") and in volcanic rock below the surface of the island of Hawai'i.

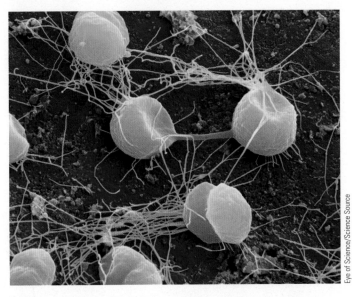

Eye of Science/Science Source

Figure 14.20 The deep-living (and appropriately named) archaean *Pyrococcus furiosus* lives in the sediments around hydrothermal vents. Although it functions most efficiently at around 100°C (212°F), this chemosynthesizer can tolerate much higher temperatures.

💬 THINKING BEYOND THE FIGURE
Why can't larger multicellular organisms live in the same places as these archaeans?

CONCEPT CHECK

Before going on to the next section, check your understanding of some of the important ideas presented so far:

What is an extremophile?

How are extremophiles thought to produce energy?

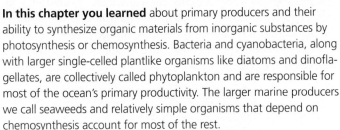

Chapter in Perspective

In this chapter you learned about primary producers and their ability to synthesize organic materials from inorganic substances by photosynthesis or chemosynthesis. Bacteria and cyanobacteria, along with larger single-celled plantlike organisms like diatoms and dinoflagellates, are collectively called phytoplankton and are responsible for most of the ocean's primary productivity. The larger marine producers we call seaweeds and relatively simple organisms that depend on chemosynthesis account for most of the rest.

Phytoplankton are most common along the coasts, in the upper sunlit layers of the temperate zone, in areas of equatorial upwelling, and in the southern subpolar ocean. Planktonic cyanobacteria are often present in astonishingly high numbers, especially in areas such as the tropics that lack adequate nutrients for the larger phytoplankters.

Many physical and biological factors influence marine primary productivity, the most important being the availability of light (or energy-rich chemicals) and inorganic nutrients. Worldwide oceanic productivity almost certainly exceeds land productivity, but a *much* smaller mass of producers is responsible for productivity in the ocean

than on land—marine producers are considerably more efficient in assembling glucose molecules.

The larger producers informally known as seaweeds are classified by color (that is, pigment composition) into three large groups: green, brown, and red algae. Some forms of brown algae, which we call kelp, grow in great underwater forests. Note that not all large marine autotrophs are algae; some are plants—sea grasses and mangroves.

Scientists are still learning about the extent of primary productivity attributable to chemosynthesis in the ocean. Researchers have found chemosynthetic organisms thriving deep in the water column, near hydrothermal vents, in seabed sediments, and even in solid rock. What may prove to be the world's largest communities are only now being discovered beneath the seafloor.

In the next chapter you will learn more about the world of marine heterotrophs—animals. Freed from the need to make their own food, animals have evolved astonishing adaptations for grazing, predation, and parasitism.

QUESTIONS FROM STUDENTS

1. How abundant are marine viruses? Should I be concerned about a virus infection if I go swimming in the ocean?

Although we have known there were viruses in the ocean for some time, only in the late 1980s were their astonishingly high abundances confirmed in a wide range of marine environments. How many are there? Between 10 million and 100 million per milliliter of water—up to about 3 billion per ounce!

Are they dangerous to humans? No. These hyperabundant viruses are bacteriophages—host-specific viruses that have evolved to infect only small cyanobacteria.

2. If I read correctly between the lines, it seems that recent discoveries of very small microorganisms (cyanobacteria, viruses) in the ocean are a *really* big deal. Where do you think this line of research is leading?

We've only scratched the surface. Before the middle 1980s, these organisms literally fell through the cracks—the traditional sampling techniques were incapable of capturing them. Mitchell Sogin of Woods Hole suggested that the number of different kinds of bacteria in the ocean could exceed 5 to 10 million![4] That's different *kinds*, not individual organisms per milliliter. Sogin wrote, "This implies there is a whole world of unexplored habitat that oceanographers are only just becoming aware of." Indeed!

3. Why are marine cyanobacteria so amazingly successful?

They're very small, so their surface-to-volume ratio is very large. They can take up and metabolize materials rapidly. Their small size also prevents rapid sinking, so substances stay longer in the photic zone's microbial loop production–metabolism cycle. Strains of *Prochlorococcus* are thought to contribute half of the photosynthesis in the ocean!

---

[4]Published in Sogin, M, et al.: Microbial diversity in the deep sea and the underexplored "rare biosphere" (2006), *Proceedings of the National Academy of Sciences* vol. 103 no. 32, pp. 12115–12120.

4. What's this about bioluminescence in dinoflagellates having evolved, possibly, as an "intrusion alarm"?

Some dinoflagellates flash when being attacked by zooplankton. The glow may startle the zooplankter and distract it from feeding. It may also alert larger organisms to the presence of that zooplankter and attract a predator toward the offending grazer. Behavioral studies suggest that dinoflagellate grazers react negatively to a dinoflagellate's blue flash, and perhaps this is why.

5. Total terrestrial primary productivity appears to be roughly the same as total marine primary productivity. But the total *biomass* of producers in the ocean is at least 300 times smaller (and maybe 1,000 times smaller) than the total *biomass* of producers on land! How can that be? How can terrestrial and marine productivity be so similar?

Because of the astonishing efficiency of small marine autotrophs (phytoplankton), nutrient and carbohydrate molecules are cycled with great speed and efficiency. There may not be nearly as great a biomass of producers in the ocean, but they appear to be *very* busy indeed!

6. What proportion of total world productivity is achieved by chemosynthesis?

An interesting and controversial question! Some biological oceanographers have suggested that vent communities are abundant on ridges and that bacteria (and archaea) can grow at much hotter temperatures than previously thought. Chemosynthesis may therefore account for a much larger proportion of total oceanic productivity than was previously thought. And, as you read, recent discoveries have shown vast chemosynthetic communities deep *within* the seabed itself!

We're reminded of a quote from Andrew Koll that eukaryotic food webs (that is, ecosystems composed of cells possessing nuclei and organelles) ". . . form a crown—intricate and unnecessary—atop ecosystems fundamentally maintained by prokaryotic [bacterial, archaean] metabolism." All the animals and plants that we see are just frosting on the cake—the dominant life form on Earth is simple, ancient, and deeply buried. We have *much* to learn.

TERMS AND CONCEPTS TO REMEMBER

accessory pigment
algae
autotroph
bioluminescence
biomass
chemosynthesis
coccolithophore
compensation depth
cyanobacterium

diatom
dinoflagellate
extremophile
flagellum
frustule
HABs
heterotroph
kelp

mangrove
microbial loop
multicellular algae
nekton
nonconservative nutrient
phytoplankton
picoplankton
plankter

plankton
plankton bloom
plankton net
primary producer
primary productivity
thallus
unicellular algae
zooplankton

STUDY QUESTIONS

Thinking Critically

1. What factors limit productivity? What methods have marine producers evolved to cope with the lack of red light needed by chlorophyll for photosynthesis?
2. What is compensation depth? What happens to phytoplankton below that depth? To zooplankton?
3. Why was the microbial loop overlooked for so long?
4. Where in the ocean is plankton productivity the greatest? Why?
5. How does a nonvascular alga differ from a vascular plant? Why are most marine autotrophs nonvascular?

Thinking Analytically

1. Is phytoplankton productivity highest at the ocean surface? What advantage would optimum productivity at a depth *below* the surface provide to phytoplankton?
2. What implications do you think the discovery of living things in deep rocks might have for discovering life on other planets?
3. What percent of the total biosphere do you think deep extremophiles comprise? Which is the dominant mechanism of primary productivity on Earth—photosynthesis or chemosynthesis?

GLOBAL GEOSCIENCE WATCH

Visit www.cengagebrain.com to access MindTap, a complete digital course which includes access to Global Geoscience Watch and more. Picoplankton (also known as bacterioplankton) are part of an entire microecosystem that represents a large component of ocean photosynthetic activity. Although they are not readily visible, recent research demonstrates that they have a number of important functions in the marine environment. Within the GREENR database search for "bacterioplankton" and use the information you find to create a two- to three-page report on the important role they play within the ocean.

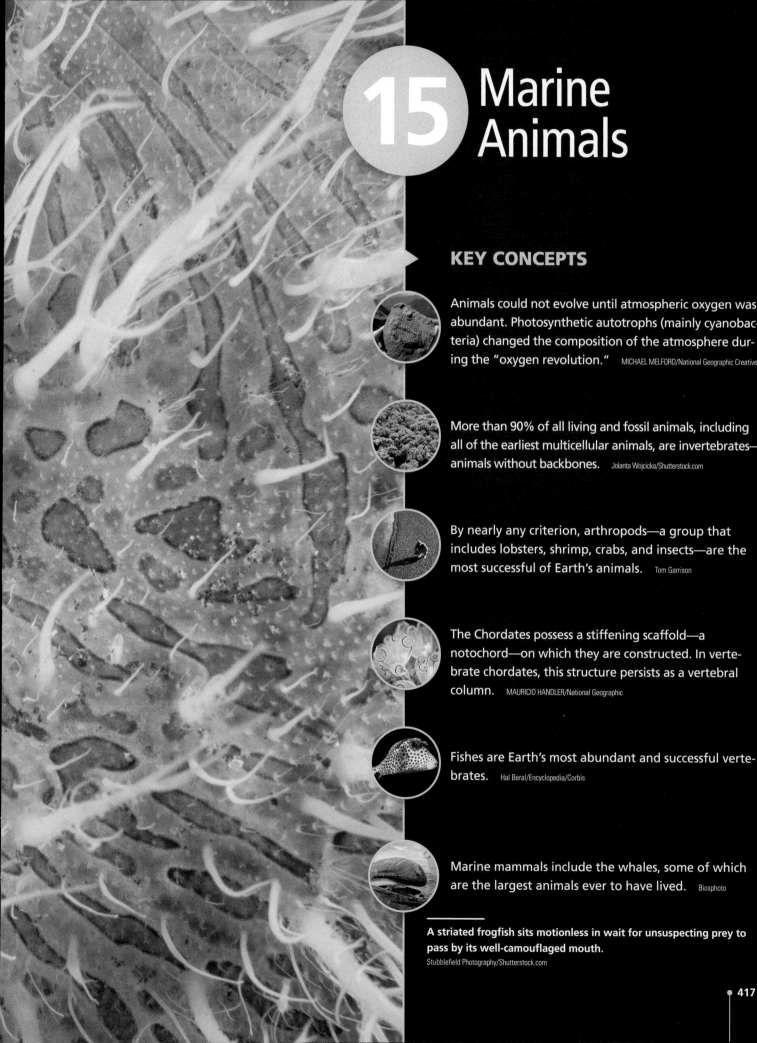

15 Marine Animals

KEY CONCEPTS

Animals could not evolve until atmospheric oxygen was abundant. Photosynthetic autotrophs (mainly cyanobacteria) changed the composition of the atmosphere during the "oxygen revolution." MICHAEL MELFORD/National Geographic Creative

More than 90% of all living and fossil animals, including all of the earliest multicellular animals, are invertebrates—animals without backbones. Jolanta Wojcicka/Shutterstock.com

By nearly any criterion, arthropods—a group that includes lobsters, shrimp, crabs, and insects—are the most successful of Earth's animals. Tom Garrison

The Chordates possess a stiffening scaffold—a notochord—on which they are constructed. In vertebrate chordates, this structure persists as a vertebral column. MAURICIO HANDLER/National Geographic

Fishes are Earth's most abundant and successful vertebrates. Hal Beral/Encyclopedia/Corbis

Marine mammals include the whales, some of which are the largest animals ever to have lived. Biosphoto

---

A striated frogfish sits motionless in wait for unsuspecting prey to pass by its well-camouflaged mouth.
Stubblefield Photography/Shutterstock.com

Media Connection
Start off this chapter by listening to a podcast with Explorer Sylvia Earle, as she talks about swimming among jellyfish in Palau. Visit www.cengagebrain.com to access MindTap, a complete digital course that includes this podcast and other resources.

15.1 Animals Evolved When Food and Oxygen Became Plentiful

Successful Animals Blend Effective Form and Function

The striated frogfish pictured in the chapter opener exhibits a unique combination of physical traits and adaptive strategies. Although not a particularly strong swimmer, it has survived because it is well suited to its specific environment. This small fish can camouflage itself in sandy habitats near sponges, as well as on rocky or coral reefs. Its striped coloration pattern and hairlike extensions help to break up its shape, making it difficult to identify. The first dorsal spine has been modified into a wormlike fishing lure that the frogfish dangles in front of its upturned mouth. When an unsuspecting crustacean or benthic fish gets close to the lure, the frogfish quickly opens its mouth and sucks in its prey with one of the fastest strike speeds of any animal on Earth. Its highly expandable stomach allows it to eat animals nearly twice its own size. If it needs to move into a different position, the frogfish can use its modified pectoral and pelvic fins to "walk" to an area better suited for ambushing its prey.

Animals like the striated frogfish teach us an important lesson: *No matter how improbable a structure or behavior seems, it must benefit the species in some way—or it would not be present.* In the long run, every shape and movement helps an organism succeed in its surroundings. When conditions change, either the organisms adapt to the change or they are replaced by creatures better suited to the new environment. Adaptations that seem jaw-droppingly amazing to us are simply demonstrations of the countless ways animals solve the problems of survival and reproduction in a specific environment. That is, after all, what successful animals *do.*

The Oxygen Revolution

The history of animals begins far back in the history of the ocean. The path leading to today's formidable array of animals began with the availability of food and oxygen.

As you may recall from Chapter 1, the first organisms to evolve on Earth were probably tiny creatures adept at absorbing organic molecules that formed spontaneously in the ocean. Primitive life-forms could use the energy stored in these food molecules for growth and reproduction. Competition for food increased as these organisms grew more numerous. Had photosynthesis not evolved, life would probably have died out when the supply of usable energy-rich molecules in the environment

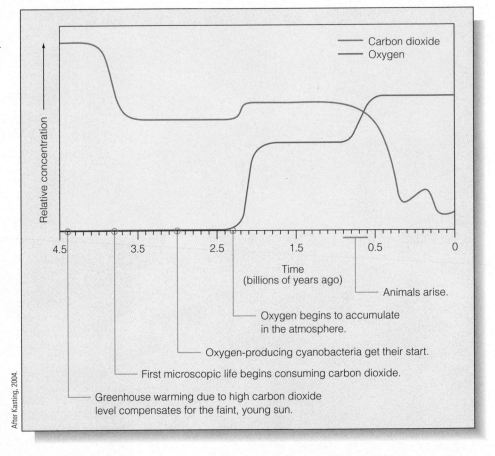

Figure 15.1 During the oxygen revolution between about 2 billion and 400 million years ago, the activity of photosynthetic autotrophs—mostly cyanobacteria—caused a rapid rise in the amount of oxygen in the air, making possible the evolution of animals. Animals are thought to have arisen between 900 and 600 million years ago.

was exhausted. The first primary producers—probably an early form of cyanobacteria—assembled their own food from inorganic molecules and then broke the food down to release energy. With the success and proliferation of these simple autotrophs, and with the abundance of oxygen they eventually provided, the way was clear for the evolution of animals.

The first animal-like creatures were single-celled organisms. They began to prosper in the ocean during the oxygen revolution, a time of radical change in the Earth's atmosphere. During the **oxygen revolution**, between about 2 billion and 400 million years ago, the activity of photosynthetic autotrophs changed the composition of the atmosphere from less than 1% free oxygen to its present oxygen-rich mixture of more than 20% **(Figure 15.1)**. The growing abundance of free oxygen made aerobic respiration practical, speeding the disassembly of food molecules that early animals obtained by eating the autotrophs. Ozone derived from this oxygen blocked most of the sun's dangerous ultraviolet radiation from reaching Earth's surface, permitting life to survive at the surface of the ocean and, later, on land.

An **animal** is a multicellular organism unable to synthesize its own food and often capable of movement. Animals grew in complexity as they became more abundant. Instead of drifting apart after reproduction, some dividing cells stuck together and formed colonies. True animals evolved as these colonies distributed labor among specialized cells, eventually increasing the degree of interdependence among cells within the colony. The colonies ceased to be simple aggregations of individuals and began to take on specific architectures for specific tasks.

A group of animals that shares similar architecture, level of complexity, and evolutionary history is known as a **phylum** (*phy-lon*, "tribe"). The plural form is *phyla*. No one knows how many phyla of animals may have developed during the time of rapid animal proliferation that occurred near the end of the oxygen revolution. Small but fascinating "shapshots" of early marine life are preserved in fossils found in the Ediacara Hills of Australia, in the Burgess shales of British Columbia, and in the Chengjiang beds of southwestern China. These sites were once parts of the warm, sediment-covered continental shelves of equatorial landmasses. When animals inhabiting these places were abruptly buried—perhaps by turbidity currents—their delicate features were preserved. The animal in **Figure 15.2a** is about 560 million years old and shows evidence of a segmented body plan. The 510-million-year-old proto-arthropod in **Figure 15.2b** is similar to species alive today.

Not all groups of early marine animals survived, however. Once abundant, trilobites like the one shown in **Figure 15.2c** have all perished. Indeed, most of the animals in these ancient fossil beds are extinct and may represent failed branches in animal evolution, unique designs that were not suited to later environmental conditions.

CONCEPT CHECK

Before going on to the next section, check your understanding of some of the important ideas presented so far:

What is an animal? How is an animal different from an autotroph?

What was the oxygen revolution? What caused it?

Why was the oxygen revolution necessary for the evolution of animals?

What is a phylum?

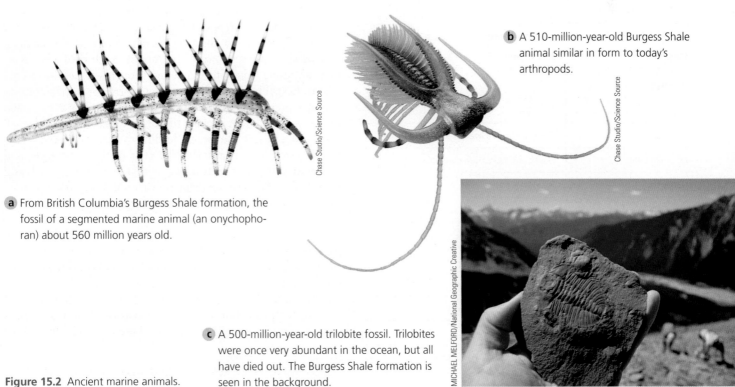

a From British Columbia's Burgess Shale formation, the fossil of a segmented marine animal (an onychophoran) about 560 million years old.

Chase Studio/Science Source

b A 510-million-year-old Burgess Shale animal similar in form to today's arthropods.

Chase Studio/Science Source

c A 500-million-year-old trilobite fossil. Trilobites were once very abundant in the ocean, but all have died out. The Burgess Shale formation is seen in the background.

MICHAEL MELFORD/National Geographic Creative

Figure 15.2 Ancient marine animals.

15.2 Invertebrates Are the Most Successful and Abundant Animals

More than 90% of all living and fossil animals, including all of the earliest multicellular animals, are categorized as invertebrates. **Invertebrates** are generally soft-bodied animals that lack a rigid internal skeleton for the attachment of muscles, but many invertebrates possess some sort of hard, protective outer covering, which can be continuous (like a snail shell) or segmented (like a lobster shell). The category is convenient though somewhat artificial, containing as it does a vast variety of different organisms in many diverse phyla that have little in common. Invertebrates range in size from microscopic worms to giant squids. Biologists currently recognize at least 33 invertebrate phyla, and almost every one of them has marine representatives. The nine most conspicuous invertebrate phyla are presented in this chapter in order of increasing complexity (for an overview, see **Table 15.1**).

> **M**ore than 90% of all living and fossil animals, including all of the earliest multicellular animals, are categorized as invertebrates.

Phylum Porifera Contains the Sponges

Sponges belong to the phylum **Porifera** (*porus*, "holes"; *ferre*, "to bear"), the most primitive true animals. Nearly all of the 10,000 species of these simple attached animals are marine. Sponges are widely distributed from intertidal zone to abyss and are found at all latitudes and in most benthic habitats. They range from the size of a bean to the size of a small automobile and come in a few basic shapes: branching, vaselike, and encrusting.

All sponges are **suspension feeders**—they strain plankton and tiny organic food particles from the surrounding water. A large sponge may filter more than 1,500 liters (400 gallons) of water each day. **Figure 15.3** shows a cutaway diagram of a simple upright sponge. Water carrying food and oxygen enters the sponge through pores on its surface and is swept toward the exit opening by flagellated collar cells. The sticky collars snare food particles drifting past, and digestion begins. The captured nutrients are distributed to other cells of the organism by wandering amoeboid cells in the body of the sponge. Sponges have no digestive system, only individual digestive cells; they have no circulatory, respiratory, or nervous systems. Excretion and the movement of gases into and out of the animal occur by simple diffusion.

The flagellated cells of more elaborate sponges are concentrated in hundreds of tiny chambers. Their interiors resemble Swiss cheese. A skeletal network of *spicules* (needles) of calcium carbonate or glassy silica prevents the internal chambers and canals from collapsing; a fibrous protein called *spongin* often serves the same purpose. Commercial natural bath sponges (not to be confused with the brightly colored synthetic sponges encrusting supermarket shelves) have been treated to retain only this spongin matrix; all the cells are gone.

Stinging Cells Define the Phylum Cnidaria

Jellyfish,[1] sea anemones, and corals belong to the phylum **Cnidaria** (*knide*, "nettle"), which contains about 9,000 mostly marine species. (You may be more familiar with this phylum's old name, Coelenterata.) This group of carnivorous animals takes its name from the large, stinging cells called **cnidoblasts** (*knide*, "nettle"; *blastikos*, "to shoot upward"), deployed on tentacles that bend or retract toward the mouth. Each cnidoblast contains a capsule from which a coiled thread is forcefully ejected **(Figure 15.4)**. The thread can repel an aggressor or penetrate or entangle prey, often immobilizing the victim with a toxin. Then the prey—which may include the larger zooplankters and small fish—is drawn into the mouth, leading to a saclike digestive cavity. Digested food is absorbed by cells of the inner layer and transported to other parts of the animal by migratory cells and by diffusion. Because the digestive cavity has only one opening, indigestible bones and other wastes are eliminated through the mouth.

Cnidarians exhibit **radial symmetry**; that is, their body parts radiate from a central axis like the spokes of a wheel. Some, such as sea anemones and corals, attach to rocks or other objects; others, such as jellies, swim freely in the water.

No cnidarian possesses a definite head or concentration of sensory receptors, but a primitive network of nerves permits some species to respond to stimuli. These relatively simple ani-

Table 15.1 Major Animal Phyla with Marine Examples

Phylum	Examples
Invertebrates	
Porifera	Sponges
Cnidaria	Coral, jellies, sea anemones, siphonophores
Platyhelminthes	Flatworms, flukes, tapeworms
Nematoda	Roundworms
Annelida	Segmented worms
Mollusca	Chitons, snails, bivalves, squid, octopuses
Arthropoda	Crabs, shrimp, barnacles, copepods, krill
Echinodermata	Sea stars, sea urchins, sea cucumbers
Chordata[a]	Tunicates, salps, Amphioxus
Vertebrates	
Chordata	Fishes, reptiles, birds, mammals

[a]Phylum Chordata includes both vertebrate and invertebrate classes.

---

[1]Because they're not fish, we'll refer to jellyfish as *jellies* in the rest of this section.

Figure 15.3 Sponges, phylum Porifera.

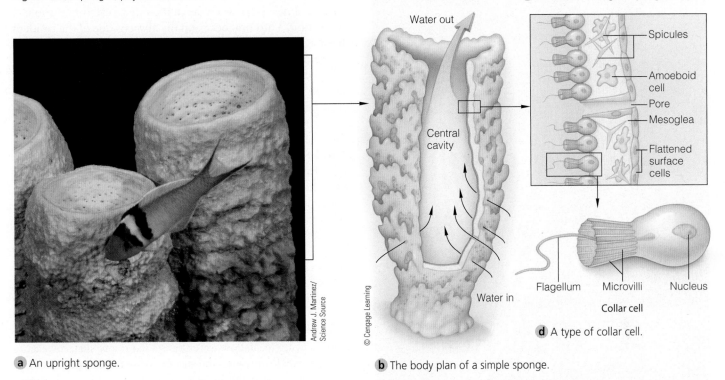

c A section through a sponge body wall.

Labels: Spicules, Amoeboid cell, Pore, Mesoglea, Flattened surface cells, Water out, Central cavity, Water in

d A type of collar cell.

Labels: Flagellum, Microvilli, Nucleus, Collar cell

Andrew J. Martinez/ Science Source

© Cengage Learning

a An upright sponge.

b The body plan of a simple sponge.

mals depend on diffusion to move wastes and gases; they have no excretory or circulatory systems.

Cnidarians occur in two forms: medusae and polyps. Jellies are examples of the **medusa** body plan, named after a Greek mythological monster with a woman's face and hair that was a mass of writhing snakes. Medusae are predatory animals that swim by the rhythmic contraction of their bell-shaped bodies. They catch their prey with trailing tentacles armed with cnido-blasts. Some medusae are microscopically small, but the genus *Cyanea* can reach a bell diameter of 3.5 meters (12 feet), with tentacles 18 meters (60 feet) long! A smaller, more dangerous species is shown in **Figure 15.5**.

Sea anemones and corals are examples of the sedentary **polyp** (*polypous,* "many-footed") body plan. Sea anemones have no skeleton and attach firmly to the substrate or burrow into it with a sticky basal disc on which they can slide slowly, like a

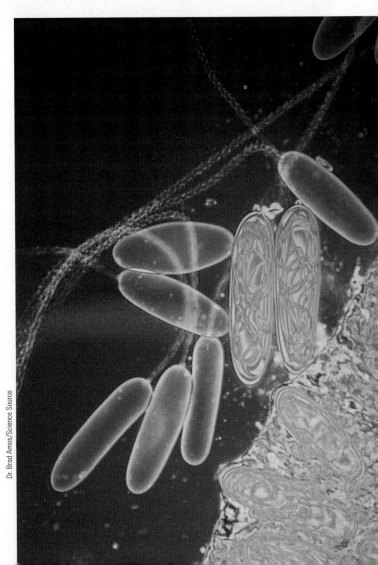

Dr. Brad Amos/Science Source

Figure 15.4 Cnidoblasts contain capsules that can forcibly eject coiled threads. Some threads entangle prey, but each cnidoblast of the sea anemone *Rhodactis,* shown here, consists of a penetrating barb with hollow tubing connecting to a poison sac. Batteries of such cells form the armament of jellies, sea anemones, and other cnidarians.

💬 **THINKING BEYOND THE FIGURE**

All cnidarians have cnidoblasts, but not all cnidoblasts sting. What else could an ejecting thread do to capture a prey organism?

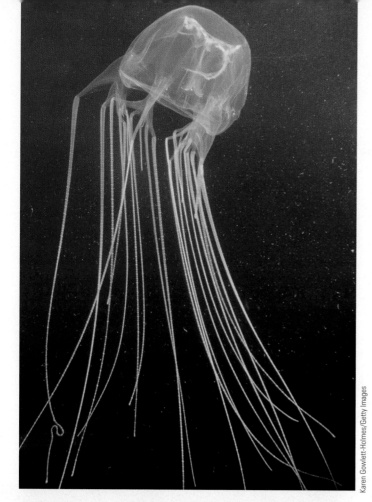

Karen Gowlett-Holmes/Getty Images

Figure 15.5 A sea wasp (*Chironix*), one of the most dangerous jellies. An inhabitant of tropical waters from Africa to northeastern Australia, it can kill a human within 3 minutes. The tentacles of a large specimen can be 15 meters (50 feet) long. *Chironix* has probably been responsible for more human deaths than sharks.

snail. Corals contain a calcareous skeleton covered by living tissue and are permanently cemented in place. The anatomy of a coral polyp is shown in **Figure 15.6**.

Some corals are solitary animals with bodies up to 30 centimeters (12 inches) in diameter, but most of the more than 500 species are ant-sized organisms crowded into colonies called coral reefs (see Figure 12.25). The coral animals themselves construct the reefs by secreting hard skeletons of aragonite (a fibrous, crystalline form of calcium carbonate). The matrix of cup-shaped individual skeletons secreted by coral animals gives the colony its characteristic shape.

Tropical, reef-building corals are **hermatypic**, a term derived from the Greek word *hermatos*, meaning "mound-builder." Their bodies contain masses of single-celled symbiotic dinoflagellates, which carry on photosynthesis, absorb waste products, grow, and divide within their coral host. The coral animals provide a safe and stable environment and a source of carbon dioxide and nutrients; the dinoflagellates reciprocate by providing oxygen, carbohydrates, and the alkaline pH necessary to enhance the rate of calcium carbonate deposition. The coral occasionally absorbs one of the resident dinoflagellates, "harvesting" the organic compounds for its own use.

Because of the needs of their symbiotic dinoflagellates, hermatypic corals depend on light and warmth. Reef corals grow best in brightly lighted water about 5 to 10 meters (16–33 feet) deep. Coral reefs can form to depths of 90 meters (300 feet), but growth rates decline rapidly past the optimum 5- to 10-meter depth. In ideal conditions, coral animals grow at a rate of about 1 centimeter (½ inch) per year. They prefer clear water because turbidity prevents light penetration and because suspended inorganic particles interfere with feeding. The animals are protected from the harmful effects of bright sunlight by a mucous coating that contains an ultraviolet-blocking "suntan lotion" and by their habit of feeding at night; the polyps retract into their skeletal cups during the day to escape strong sunlight and predators.

Hermatypic corals also prefer water of normal or slightly elevated salinity. Coral animals are highly susceptible to osmotic shock, and exposure to freshwater is rapidly fatal; thus reefs growing in shallow water have a flat upper surface because rain is lethal. Freshwater and suspended sediments prevent reefs from forming near the mouths of rivers or in areas adjacent to islands or continents where rainfall is abundant.

Nearly all of the reef-building corals are found in warm water. But some individual coral organisms are found in some cold, high-latitude waters as well. These corals lack interior dinoflagellates, so they deposit calcium carbonate much more slowly, and the structures they build do not resemble those found in the tropics. Instead, these deepwater corals, known as **ahermatypic** corals, build smooth banks on the cold, dark, outer edges of temperate continental shelves from Norway to the Cape Verde Islands, and off New Zealand and Japan. The rarest corals of all are large solitary organisms living on the abyssal floors and outer continental shelves of the Antarctic.

The Cnidaria are a very successful group, but their simple architecture would not serve more advanced organisms. Their blind-sac digestive system allows only one batch of food to be processed at a time; feeding opportunities arising during digestion cannot easily be accommodated. The lack of a distinct head with a concentration of sense organs is a drawback, as is the absence of circulatory, respiratory, and excretory systems. Having only two structural cell layers limits the complexity of systems and structures that can form within the organisms.

CONCEPT CHECK

Before going on to the next section, check your understanding of some of the important ideas presented so far:

What's an invertebrate?

What's a suspension feeder?

What feature is unique to cnidarians? Name some cnidarians.

Pat Mason

a Close-up of hermatypic coral, showing expanded polyps.

Figure 15.6 Coral animals, a form of cnidarian.

💬 **THINKING BEYOND THE FIGURE**

What would happen to a coral animal if all its resident zooxanthellae died, but the animal itself remained relatively healthy?

15.3 The Worm Phyla Are the Link to Advanced Animals

A transition from relatively simple to more advanced organisms is made in the worm phyla, three of which are discussed here. The worm body plan exhibits **bilateral symmetry**, not radial symmetry; that is, the body has a left side and a right side that are mirror images of each another. Nearly all worms have some concentration of sensory tissue in what may be termed a head, and many have flow-through digestive systems and systems to circulate fluids and eliminate waste. Some are efficient parasites, but most are free-living. A few burrow in cavities; others roam the seabed or lurk under rocks.

The simplest worms are the well-named flatworms of phylum **Platyhelminthes** (*platys*, "flat"; *helmins*, "worm"). Some are parasitic of vertebrates, such as the tapeworms of fish and

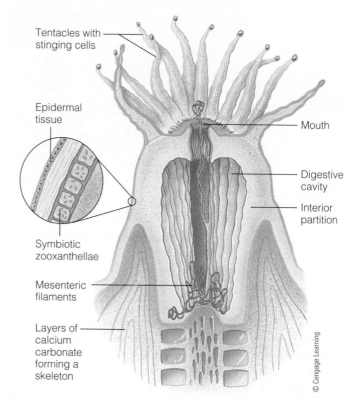

© Cengage Learning

b Anatomy of a reef coral polyp, with enlarged detail showing a cross section of the outer covering and tissue. The symbiotic photosynthetic zooxanthellae are crucial to the survival of this type of coral.

Jolanta Wojcicka/Shutterstock.com

c Coral colonies can be composed of hundreds or even thousands of individual polyps, which may eventually build entire reefs.

marine mammals. However, most marine flatworms are free-living predators and scavengers; they can be found on the shady underside of intertidal rocks or sharing colonies or burrows with other animals **(Figure 15.7)**. Few examples exceed 3 centimeters (1¼ inches) in length.

Flatworms are the most primitive organisms with a central nervous system. In some species, a complex of nerve cells serving as a rudimentary brain connects the animal's simple nervous sys-

Figure 15.7 An intertidal marine flatworm navigates the surface of an overturned rock. This primitive worm is very thin to allow quick diffusion of gases, nutrients, and waste.

Figure 15.8 This larval form of a polychaete (genus *Polydora*), a form of annelid worm. The division of the body into similar segments is clearly visible.

tem to a pair of light-sensitive eyespots. The eyespots are small pigmented cups that can sense only the presence or absence of light—certainly a critical factor for an animal only a few cells thick that must avoid light to prevent overheating or detection.

Larger flatworms are necessarily thin because they lack a true respiratory or excretory system. Gases must be exchanged and wastes eliminated by diffusion through the animal's surface, so no cell can be more than a few cell diameters from the outside.

The first animals in our phylum survey that possess a flow-through digestive system (a digestive tract with mouth and anus rather than a digestive cavity) are members of the plentiful phylum **Nematoda** (*nematos*, "thread"). Also called *roundworms*, these most successful of the worm phyla are present in nearly every imaginable terrestrial, aquatic, and marine habitat. Most of the 12,000 known species are free-living and microscopic, thriving in garden soil and marine sediments. Some make perfect parasites, however, and nearly all vertebrates and many invertebrates are parasitized by species of these long, thin worms. Readers who enjoy eating raw fish—sashimi—should be aware of the rare presence of fish parasites of the genus *Anasakis* that occasionally infect people.

Members of the phylum **Annelida** (*annelus*, "ring") are the most evolutionarily advanced worms. Their bodies are divided into a number of similar rings or segments. **Metamerism**, as this segmentation is called, is a convenient strategy for increasing the size of an animal simply by adding nearly identical units. We see evidence of metamerism in most higher animals (look at your fingers). Each segment of an annelid can have its own circulatory, excretory, nervous, muscular, and reproductive systems, but some segments (such as those forming the head) are specialized for specific tasks. The familiar garden earthworm is an annelid.

The 5,400 species of class **Polychaeta** (*poly*, "many"; *chaetae*, "bristles"), the largest and most diverse class of annelids, are also the most important marine annelids. Polychaetes are often brightly colored or iridescent worms with pairs of bristly projections extending from each segment **(Figure 15.8)**. They range in length from about 1 to 15 centimeters (½–6 inches). Some polychaetes burrow through and devour sediments or move freely over the bottom in search of food; others construct fixed parchment-like or calcareous tubes from which only parts of their heads emerge. Mobile polychaetes have well-developed heads with prominent sense organs, and they can function as efficient predators. Some skewer prey with their sharp mouthparts.

CONCEPT CHECK

Before going on to the next section, check your understanding of some of the important ideas presented so far:

How is bilateral symmetry different from radial symmetry?

What evolutionary advances characteristic of higher organisms are first seen in the worms?

15.4 Advanced Invertebrates Have Complex Bodies and Internal Systems

The Phylum Mollusca Is Exceptionally Diverse

The conspicuous phylum **Mollusca** (*molluscus*, "soft bodied") contains 80,000 species, second in size only to the vast phylum Arthropoda. The phylum Mollusca includes such diverse members as clams, snails, octopuses, and squid. Most molluscs are marine, and most have an external or internal shell. A few molluscan species possess acute sight and considerable intelligence.

Molluscs and annelids probably shared a common origin—possibly a distantly ancestral segmented worm—and therefore share a few basic characteristics. Like annelids, molluscs are bilaterally symmetrical and generally have obvious heads, flow-through digestive tracts, and well-developed nervous systems. A few molluscs are segmented. Unlike annelids, however, some molluscs achieve great size, secrete beautifully fitted shells in which to take refuge, and exhibit great structural diversity.

We will briefly discuss three molluscan classes here—the class **Gastropoda**, the snails; the class **Bivalvia**, the clams, oysters, and mussels; and the class **Cephalopoda**, the nautiluses,

Figure 15.9 Gastropod shells are often aesthetically pleasing and have been prized by collectors for centuries.

💬 **THINKING BEYOND THE FIGURE**

What are the most valuable shells? What is your rough estimate of the highest price ever paid for a gastropod shell?

octopuses, cuttlefish, and squid. Though greatly different in shape and habits, each class shows its link to a common ancient ancestor by sharing similar underlying parts.

Gastropods (*gaster*, "stomach"; *pod*, "foot") include the abalone, conch, limpet, and garden snail. Members of this largest class of molluscs usually inhabit relatively large shells, where they can seek refuge in case of danger. Some gastropods are grazers, some are suspension feeders, and some are predators. A few marine species—the pteropods and heteropods—are planktonic, but most marine gastropods are found wandering on rocky bottoms or other firm substrates.

Gastropod shells are often structures of great beauty **(Figure 15.9)**. The animal adds to and enlarges the opening as it grows. The shell is frequently coiled to compress its mass and allow for easier maneuverability. A foot and head protrude from the shell while the snail moves about. The shell itself is secreted in three principal layers: a fibrous outer covering that may serve to distribute shock; a strong, crystalline layer of calcium carbonate ($CaCO_3$) to provide strength; and an inner layer of smooth $CaCO_3$ to provide nonabrasive surroundings for the resident. Not all gastropods have shells. Nudibranchs (*nudus*, "naked"; *branchia*, "gills"), also called *sea slugs*, are lovely shell-less gastropods **(Figure 15.10)**.

Although some snails use their foot to burrow, a gastropod foot cannot attach to sand or mud. The shell-and-foot configuration of gastropods is therefore not well suited to life on most of the ocean floor, which is covered by sediments. Evolution of the bivalves (*bi*, "two"; *valv*, "door") admitted molluscs to this rich sedimentary habitat. Animals enclosed in twin shells (clams, oysters, mussels, and scallops), bivalves surrender mobility for protection, and usually gather food by suspension feeding rather than pursuit (see **Figure 15.11**). Burrowing species dig with a strong muscular foot and extend their siphons to the surface to obtain water and eject wastes. In other species, the foot has other functions: in mussels it secretes the tough threads that attach

Figure 15.10 A brightly colored reef nudibranch (genus *Flabellina*) glides over a diver's mask. The brilliant gill-like structures on its back assist in gas exchange. Although nudibranchs usually live in plain sight, their terrible taste seems to discourage animals from eating them.

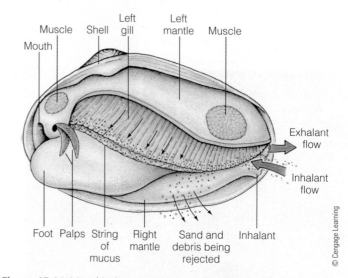

Figure 15.11 Most bivalves are suspension feeders that make their living by filtering the water for edible particles. In this diagram (showing a bivalve with its left shell removed), water and tiny bits of food are swept into the animal by the movement of tracts of cilia on the gills. Food settles into the gills and is then driven toward the mouth and swallowed.

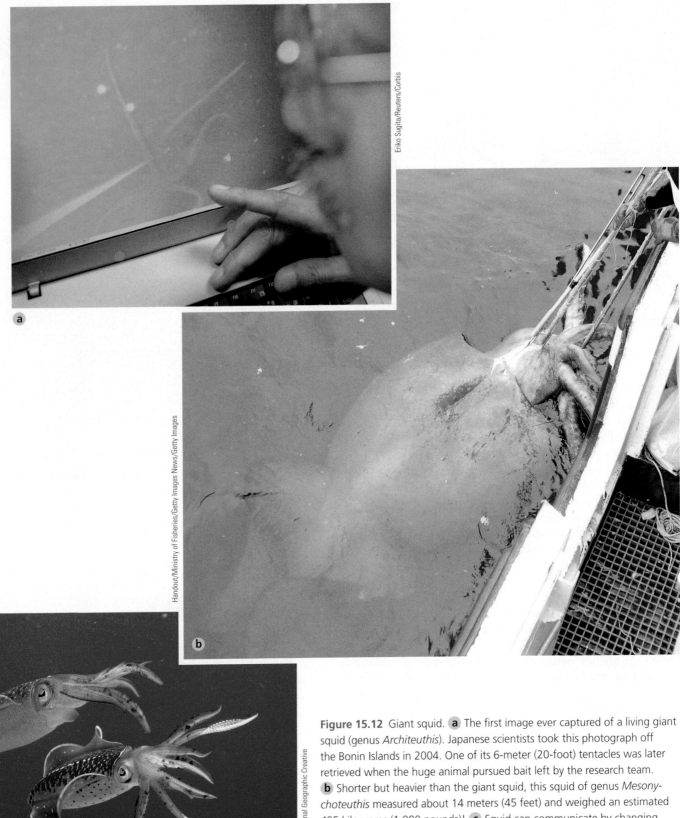

Figure 15.12 Giant squid. **a** The first image ever captured of a living giant squid (genus *Architeuthis*). Japanese scientists took this photograph off the Bonin Islands in 2004. One of its 6-meter (20-foot) tentacles was later retrieved when the huge animal pursued bait left by the research team. **b** Shorter but heavier than the giant squid, this squid of genus *Mesonychoteuthis* measured about 14 meters (45 feet) and weighed an estimated 495 kilograms (1,090 pounds)! **c** Squid can communicate by changing their color patterns and "body language." These two small Caribbean reef squid are seen communicating in what may be a mating ritual or some form of threat display.

DR. SYLVIA EARLE encounters cephalopods.

From my perch in the depths, watching the silvery red creature swim toward me, I realized that I was being inspected not by a squid but rather a large, intensely curious octopus. Eight, not ten arms reached out to touch the mechanical arms of the sub, revealing a sheltered cluster of pale eggs. Officially, my dive was over, but I begged and got permission to remain nearly an hour, staying close as the octopus glided upward in the water column, stopping when she stopped, moving when she moved, keeping pace in a slow-motion dance until, finally, I had to pull away.

Seeing a creature unlike any I had ever seen before, knowing that she had never seen a creature such as I, inspired in me a sense of wonder and hope. Humans are newcomers on this ancient planet, the first and only species to so radically dominate Earth's natural systems—land, water, air, and biosphere—and the first and only to provoke changes that not only threaten our survival but could initiate Earth's sixth planetwide wave of extinction.

Cephalopods may have an advantage over us with their resilience, their half-a-billion-year lineage that has endured several waves of mass extinction, including the cataclysmic events 65 million years ago when dinosaurs faded into oblivion and mammals began a new era of prosperity. Some variation on the cephalopod theme will likely survive, one way or another, to swim in future seas.

Source: National Geographic Ocean Atlas, p. 325
Intertwined octopuses.
Background photo: PULSE/Corbis

---

the organisms to wave-swept rocks—a neat turnabout trick by which bivalves can invade a gastropod habitat.

The most highly evolved molluscs are the magnificent cephalopods (*cephalon*, "head"; *pod*, "foot"), a group of marine predators containing nautiluses, octopuses, cuttlefish, and squid. These well-named animals have a head surrounded by a foot divided into tentacles. The nautiluses retain a large coiled external shell, but squid have only a thin vestige of the shell within their bodies—and octopuses have none at all. Cephalopods can move by creeping across the bottom, by swimming with special fins, or by squirting jets of water from an interior cavity.

Most cephalopods catch prey with stiff adhesive discs on their tentacles that function as suction cups, and they tear or bite the flesh with horny beaks. Nautiluses are an open-ocean group that hunts at considerable depths, their strong shells buoyed with gas-filled chambers. Little is known of their natural history. Squid and octopuses are more advanced cephalopods. Squid and octopuses can confuse predators with clouds of ink. Some kinds of squid eject a kind of "dummy" of coagulated ink that's a rough duplicate of their size and shape—the squid is long gone by the time the attacker discovers the deception! At least one species of squid living below the euphotic zone produces sparkling luminous ink instead of black ink (which would, of course, be ineffec-

tive in the darkness). Squid can grow to surprising sizes (**Figure 15.12**). The record length, including tentacles, is 18 meters (59 feet)! Most are much smaller.

Squid may be the largest invertebrates, but octopuses are the most intelligent. About as smart as puppies and with even better eyesight, some nearshore species of octopuses kept in captivity soon learn to recognize their keepers and forage at night through adjacent aquariums for tidbits. Octopuses use their visual acuity, ability to survive out of water for a short time, and intelligence to good advantage in their intertidal or subtidal homes, memorizing the positions of hiding places, escape holes, and good hunting locations.

Only about 450 species of cephalopods live today, a small percentage of the number of species known from fossils. The small number of species suggests that sophistication is no guarantee of biological success.

The Phylum Arthropoda Is the Most Successful Animal Group

The phylum **Arthropoda** (*arthron*, "joint"; *pod*, "foot")—a group that includes the lobsters, shrimp, crabs, krill, and barnacles—is a phylum of superlatives. Over a million species of

Figure 15.13 Arthropods define success! There are more species of arthropods, and more individuals, than of all other animals combined. A visitor wonders at their countless numbers on a deserted beach in Baja California, Mexico.

They do, however, exhibit three remarkable evolutionary advances that have led to their great success:

- An *exoskeleton:* a strong, lightweight, form-fitted external covering and support.
- *Striated muscle:* a quick, strong, lightweight form of muscle that makes rapid movement and flight possible.
- *Articulation:* the ability to bend appendages at specific points. The appendages of more primitive phyla can usually bend anywhere (like wet spaghetti), but arthropod appendages bend at a joint. There are no ball-and-socket joints in arthropods; instead, each joint along an appendage moves through a different plane to ensure a full range of motion.

Most important of these advances is the **exoskeleton**. Unlike the often cumbersome shell of a gastropod, the exoskeleton of an arthropod fits and articulates like a finely tailored suit of armor. It is made in part of a tough, nitrogen-rich carbohydrate called **chitin**, which may be strengthened by calcium carbonate. Its three layers serve to waterproof the covering, tint it a protective color, and make it resilient and strong. Muscles within the animal are attached to the exoskeleton to move the appendages.

Such an arrangement sounds ideal, but the difficulties encountered by an organism with an exoskeleton are profound. How can muscle leverage be obtained? How can encrusting organisms be discouraged? How can ducts and feeding passages remain unblocked? And, perhaps most critically, how can the organism grow? That each of these problems has been solved is obvious by the group's overwhelming success, but the growth issue deserves a closer look.

We vertebrates grow by steadily adding length to the bones of our *internal* skeleton and bulk to our bodies. An *external* skeleton obviously limits growth and must be shed, or **molted**, at regular intervals. Arthropods do not have a steady growth pattern; instead, their external growth progresses in a series of steplike jumps **(Figure 15.14)** as the animal molts and replaces its exoskeleton. The arthropod grows without getting bigger between these jumps in size. An aquatic arthropod slowly substitutes body mass for water held in the tissues between molts. When molting, it suddenly takes on water from outside the body, expanding its tissues without growing in muscle mass. The shell splits and falls away, and, through a magnificently orchestrated sequence of glandular secretions, the animal quickly regenerates a new exoskeleton one size larger **(Figure 15.15)**.

The largest class of arthropods, the class Insecta, is poorly represented in the sea. There is only one marine genus and five

arthropods are now known! Arthropods are by far the most successful of Earth's animal phyla, occupying the greatest variety of habitats, consuming the greatest quantities of food, and existing in almost unimaginable numbers **(Figure 15.13)**. As you will read in the next chapter, one type of planktonic arthropod—the krill—comprises the greatest biomass of any single species on Earth. The arthropod body plan is a variation on the basic annelid theme; bodies show clear segmentation with a pair (or pairs) of appendages per segment. All are bilaterally symmetrical.

Arthropods have not achieved the nervous system development of cephalopod molluscs, nor do they have the advantages of intelligence or extraordinary eyesight.

By nearly any criterion, arthropods are the most successful of Earth's animals.

known open-ocean species, all of which are water striders. The class **Crustacea** (*crustaceus,* "having a shell or rind"), however, includes 30,000 species of primarily marine, gill-breathing lobsters, crayfish, shrimp, crabs, water fleas, copepods, krill, amphipods, barnacles, and others. Their bodies usually have between 16 and 20 segments; the appendages may be specialized for sensing, food handling, walking, fighting, defense, and so forth. Copepods, minute crustaceans that graze on diatoms

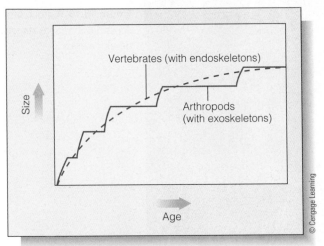

Figure 15.14 Generalized arthropod and vertebrate growth curves compared. The vertical segments of the arthropod curve represent molting periods.

💬 **THINKING BEYOND THE FIGURE**
How does an arthropod growth curve compare with yours?

a A Dungeness crab *(Cancer magister)* backing out of the exoskeleton (right) that it is abandoning.

b Clear of the old exoskeleton, the soft-bodied crab takes in water and expands. It immediately begins to secrete a new exoskeleton. Note the obvious increase in the animal's size.

Figure 15.15 A molting arthropod.

and dinoflagellates, are among the most abundant multicellular animals on the planet (**Figures 15.16** and **16.17**). More familiar to most of us are the lobsters and crabs, including those we esteem for food. The largest crustacean is the king crab, which can reach a leg span of 3.6 meters (12 feet). The heaviest individual, however, was a lobster caught off Chatham, Massachusetts, in 1949. The beast weighed 22 kilograms, or 48 pounds, and reportedly served 10!

The activity and importance of billions of crustaceans as scavengers, predators, parasites, grazers, and general participants in oceanic biology are hard to overestimate. Large or small, obvious or retiring, crustaceans dominate the world of marine animals.

A Water Vascular System Is Unique to the Phylum Echinodermata

The exclusively marine phylum **Echinodermata** (*echinos*, "hedgehog"; *derma*, "skin") is an odd group, sharply different from other members of the animal kingdom. The 6,000 species of echinoderms lack eyes or brains, have a radially symmetrical body plan based on five sections or projections, move slowly using a unique water vascular system, and include only two known parasitic representatives.

Living echinoderms are divided into five classes: The four most familiar ones are the class *Asteroidea*, the sea stars; the class *Ophiuroidea*, the brittle stars; the class *Echinoidea*, the sea urchins; and the class *Holothuroidea*, the sea cucumbers.

Nearly all asteroids (*aster*, "star"; *oidea*, "resembling"), or sea stars, are star-shaped echinoderms with five or more arms that are not completely delineated from the central disc. The arms usually have spiny projections on top and delicate tube feet beneath. The tube feet work like suction cups and can grip

objects; they also participate in gas exchange. Tube feet are part of a sea star's most striking feature—its unique **water-vascular system** (**Figure 15.17**), a complex of water-filled canals, valves, and projections used for locomotion and feeding. Operating like a hydraulic power system, the water-vascular system's plumbing can transmit forces generated by muscles at one side of the sea star to arms on the other side. Using this system, the tube feet of a sea star can grip a clam or mussel and exert a continuous pull (sometimes for hours) to force open its valves, even though one or more of the star's arms may tire. When the mussel finally opens, the sea star expels its stomach from its mouth, slips it between the mussel's shells, and digests the victim in place.

Delicate ophiuroids (*ophidion*, "a snake") have long, slender arms. They are called *brittle stars* because of their unusual

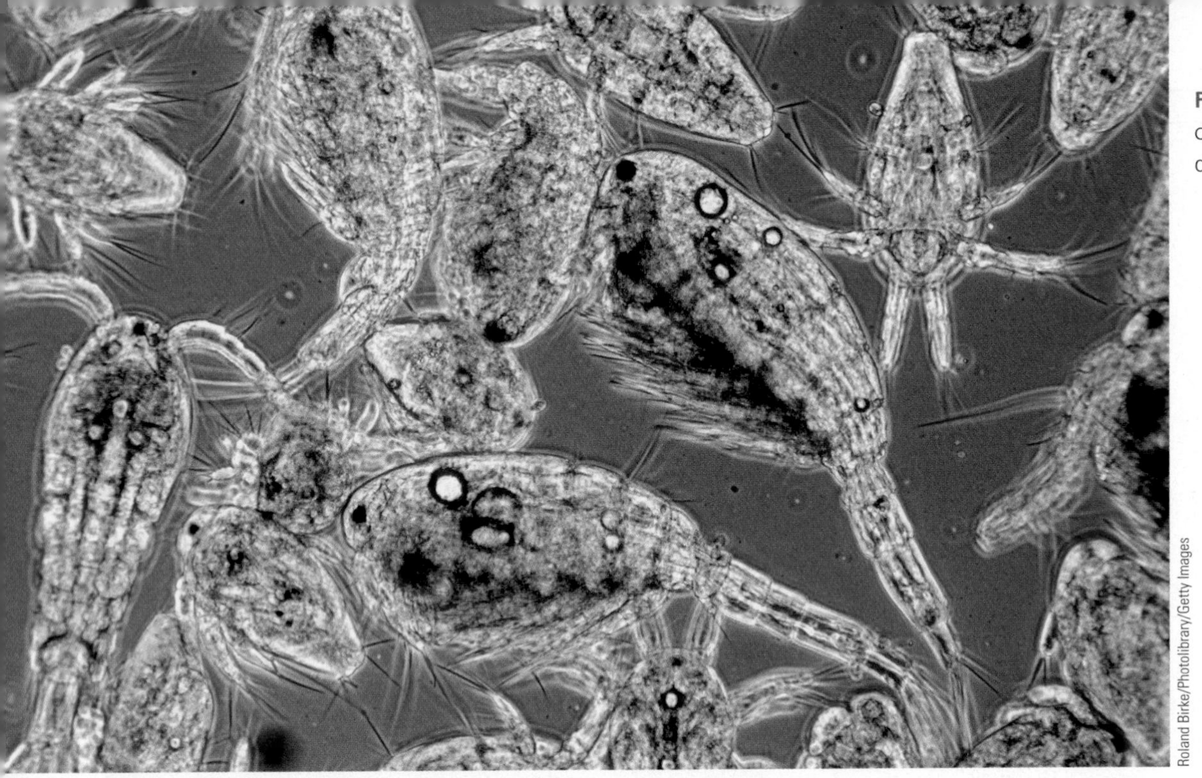

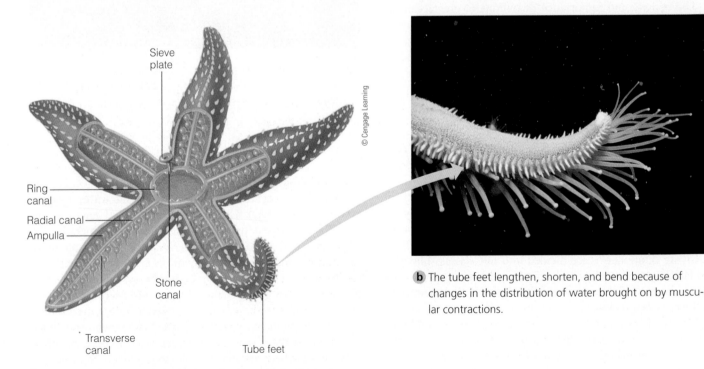

Roland Birke/Photolibrary/Getty Images

Figure 15.16 A group of copepods in various stages of development.

strategy for evading capture. If grasped by a predator, a brittle star will often detach its arm and escape. (The brittle star will later regenerate the arm.) Ophiuroids are perhaps the most widely distributed benthic marine animals; a few species are found in great numbers on the deep sedimented seabeds of the world ocean **(Figure 15.18a)**. Many ophiuroid species also live beneath intertidal and subtidal rocks. Longitudinal grooves on the underside of each arm enable some species of brittle stars to locate food particles and transfer them by cilia to the mouth. Other species wave their arms through the water to capture plankton on sticky strands of mucus strung between adjacent arm spines.

Sieve plate

Ring canal

Radial canal

Ampulla

Stone canal

Transverse canal

Tube feet

© Cengage Learning

Jeff Rotman/Science Source

b The tube feet lengthen, shorten, and bend because of changes in the distribution of water brought on by muscular contractions.

a Water enters the animal's body through a sieve plate, which excludes material that might clog the tubes and valves, and circulates through canals.

Figure 15.17 The water-vascular system in a sea star (shown in blue).

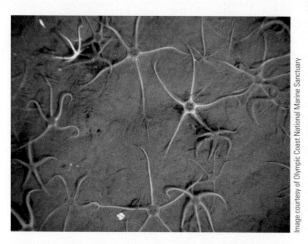

a Ophiuroids (brittle stars) feed on edible particles in the surface layer of sediments on the continental slope off New England. Ophiuroids are among the most widely distributed of all benthic animals.

b A blood star (class Asteroidea) is surrounded by a group of purple sea urchins (class Echinoidea).

c A pineapple sea cucumber *(Thelenota ananas)* moves slowly along the seafloor near the reef's edge.

Figure 15.18 Echinoderms.

💬 **THINKING BEYOND THE FIGURE**

Why do you think brittle stars are found in high densities in so many marine environments?

Elongated holothuroids (*holothurian*, "tube-shaped"), or sea cucumbers, have a unique appearance among echinoderms. They usually lie on their side, which makes them appear to lack radial symmetry **(Figure 15.18c)**. Many species are detritus feeders that filter organic particles from seafloor sediments. Lacking the hard bodies or spines that other echinoderms possess, sea cucumbers have developed a unique defense mechanism, which can include excreting toxic substances or even expelling some of their internal organs as a distraction to potential predators. Like other classes of echinoderms, sea cucumbers are able to regenerate some of their lost body parts.

Echinoids (*echinos*, "hedgehog") are familiar to most coastal residents **(Figure 15.18b)**. The prickly appearance of a sea urchin or the smooth velvety surface of a sand dollar do not at first suggest the phylum's five-sided symmetry. Close observation, however, reveals just such an arrangement, overlain by a few bilaterally symmetrical features. Urchins can feed either by taking bits of food into the mouth with complex grasping, chewing jaws, or simply by absorbing food molecules into the mucus layer that covers their bodies and flows toward the mouth.

CONCEPT CHECK

Before going on to the next section, check your understanding of some of the important ideas presented so far:

Which group of animals is Earth's most successful (by nearly any definition of success)?

Clams and squid are in the same phylum. How can that be?

Provide an example of each of the major groups of molluscs.

It has been said that arthropods "grow without getting bigger" and "get bigger without growing." What's meant by that?

What structural system is unique to the phylum Echinodermata?

15.5 Construction of Complex Chordate Bodies Begins on a Stiffening Scaffold

All members of the phylum **Chordata**, the most structurally complex animal phylum, possess a stiffening **notochord** (*notus*, "back"; *chorda*, cord), a tubular dorsal nervous system, and gill slits behind the oral opening at some time in their development. The notochord was critical in chordate evolution. It permitted a more complex embryonic development by providing a rigid "scaffold" on which the developing embryo could be constructed, and it provided an internal mechanical foundation for skeletal and muscle development. About 5% of the 58,000 species of chordates lose their notochord as they develop; these are called *invertebrate chordates*. The other 95% of chordates retain their notochord (or the vertebral column that forms from it) into adulthood; these are the familiar *vertebrate chordates* (such as fish, reptiles, birds, and mammals).

Not All Chordates Have Backbones

Two invertebrate chordates are of interest here. The **tunicates**, or sea squirts, are suspension feeders that superficially look and function like sponges. Their common name comes from an extraordinarily strong and flexible tunic (outer covering). Close investigation reveals a body plan much different from the primitive sponges, however (see **Figure 15.19**). Solitary or colonial, attached as adults or free-swimming, tunicates filter water with a special mucus plankton net capable of trapping a wide variety of microscopic food particles. The mucus is generated by a long glandular seam on one side of a basket-like interior structure (the pharynx); it is then driven by tiny cilia around to the other side and is collected and swallowed by an esophagus leading to a small stomach. Salps, related zooplanktonic forms, act almost like miniature jet engines—taking in water at one end, filtering it, and ejecting it from the other end to force themselves ahead. It stretches the imagination to consider these animals within the same phylum as seagulls or dolphins or people, but all chordate embryos share the same fundamental architecture.

Vertebrate Chordates Have Backbones

Vertebrates are the members of the phylum Chordata that possess backbones. The word is derived from *vertebratus*, meaning "jointed," a reference to the segments of the backbone. About 95% of all chordates are vertebrates—nearly 55,000 species. Here we find the familiar creatures drawn by generations of children—the fishes, frogs, lizards, chickens, cats, and dogs most of us first think of when we hear the word *animal*.

15.6 Vertebrate Evolution Traces a Long and Diverse History

Like other chordates, vertebrates had a remote marine ancestor. The first chordates evolved about 500 million years ago and had the stiffening notochord that gives the phylum its name, but they lacked the backbones of true vertebrates. Besides the backbone, vertebrate chordates differ from the invertebrate chordates by having an internal skeleton of calcified bone or cartilage (or both). This scaffold allows uninterrupted support during growth; it also protects vital organs and provides a foundation to which muscles may attach to permit the strength and rapid responses characteristic of active animals. The vertebrate skull, a special unit of the skeleton, provides secure housing for the brain, eyes, and other sense organs that have made the evolution of intelligence possible. Only the simplest vertebrates lack jaws. The central nervous system is partially enclosed within the backbone, which extends from the skull, and the pairs of nerves passing between the vertebral segments allow rapid and efficient communication between brain and body. The most abundant and successful vertebrates are the fishes, which exist in an extraordinary variety of form and habitat. Least successful in the marine environment are the amphibians.

As is the case with any natural system of organization, vertebrate classification reflects our understanding of vertebrate evolution. **Figure 15.20** indicates the likely evolutionary relationship among vertebrate species. Note that all higher vertebrates appear to be derived from fishlike ancestors. In an architectural sense,

Reinhard Dirscherl/Visuals Unlimited, Inc.

Figure 15.19 An adult tunicate.

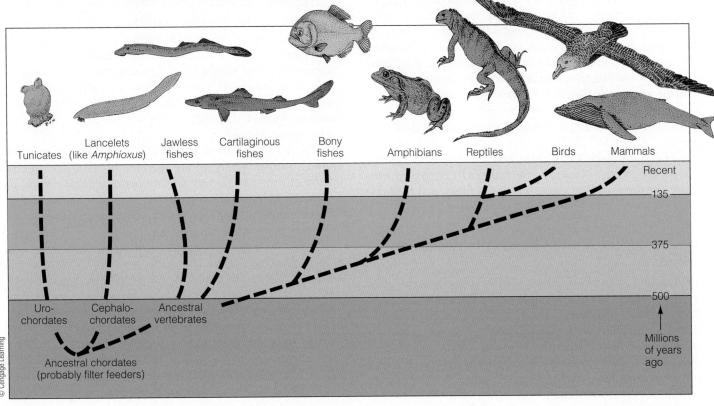

Tunicates | Lancelets (like *Amphioxus*) | Jawless fishes | Cartilaginous fishes | Bony fishes | Amphibians | Reptiles | Birds | Mammals

Recent
135
375
500

Millions of years ago

Uro-chordates
Cephalo-chordates
Ancestral vertebrates

Ancestral chordates (probably filter feeders)

© Cengage Learning

Figure 15.20 One proposed family tree for the vertebrates and their relatives, the invertebrate chordates.

THINKING BEYOND THE FIGURE
Why is this referred to as just one "proposed" family tree of the chordates?

all higher vertebrates (including amphibians, reptiles, birds, and mammals) are highly modified, four-limbed, air-breathing fish!

CONCEPT CHECK

Before going on to the next section, check your understanding of some of the important ideas presented so far:

How are the vertebrates related?

What are the evolutionary advantages of possessing vertebrae?

15.7 Fishes Are Earth's Most Abundant and Successful Vertebrates

Fishes are vertebrates that usually live in water and possess gills for breathing and fins for swimming. There are more species of fishes, and more individuals, than species and individuals of all other vertebrates *combined*. This is not a surprising fact considering the vast oceanic habitat the planet provides. Fishes range in adult length from less than 10 millimeters to over 20 meters (0.4 inches–60 feet) and weigh

from about 0.1 gram to about 41,000 kilograms (0.004 ounce–45 tons). Some fishes are capable of short bursts of speed in excess of 113 kilometers per hour (70 miles per hour); some species hardly ever move.

Fishes live near the surface and at great depth, in warm water and cold, even frozen within ice or dried in balls of mud. Like other ectothermic (cold-blooded) organisms, the great majority of fishes are incapable of generating and maintaining a steady internal temperature from metabolic heat, so the internal body temperature of a fish is usually the same as that of the surrounding environment. About 40% of the 30,000-plus fish species live at least part of their lives in freshwater; 60% live exclusively in seawater. Fishes have evolved to fit almost every conceivable watery habitat but are most numerous on the bottom or in productive seawater over the continental shelves. Some species have a "sixth sense," an ability to detect small changes in the electrical field surrounding their bodies, which assists in the detection of prey or avoidance of predators. Some electric eels, catfish, and rays can use internally generated electricity for defense and offense.

The first fishes probably evolved in the ocean around 500 million years ago.

> There are more species of fishes, and more individuals, than species and individuals of all other vertebrates *combined*.

These jawless animals were little more than motile sucking digestive tubes, but they did have the structural advantages of the chordate body plan. Only a few jawless forms, such as lampreys and hagfish, have persisted to the present day. The earliest jawed fishes, with their grasping and crushing mouths, were far more successful at feeding on invertebrates with shells or exoskeletons than their jawless predecessors had been. Early jawed fishes were also equipped with paired fins, which stabilized their movements and minimized pitching and rolling when they attacked their prey.

Fishes are divided into three major groups based on the presence or absence of jaws and the material forming their skeletons: the jawless fishes, the cartilaginous fishes, and the bony fishes.

Jawless Fishes Are the Most Primitive Living Fishes

Hagfish and lampreys—members of the class Agnatha—lack jaws and have no paired appendages to aid in locomotion. Their thick, snakelike bodies are pierced by gill slits and (in some species) the openings of slime glands. Their round, sucking mouths are surrounded by organs sensitive to touch and smell. Hagfish often eat soft-bodied benthic invertebrates or scavenge for large vertebrates that have died and fallen to the seafloor. They are capable of tying their flexible bodies into a loose knot to help tear away flesh using two dental plates. When disturbed, hagfish can also produce copious amounts of slime to deter predators. Lampreys possess a toothed, funnel-shaped mouth, which they use to attach to and feed off other vertebrates **(Figure 15.21)**. Most marine species of lampreys are **anadromous**, traveling to freshwater rivers or lakes to spawn.

Sharks Are Cartilagenous Fishes

All members of the class **Chondrichthyes**—the group that includes sharks, skates, rays, and chimaeras—have a skeleton made of a tough, elastic tissue called **cartilage**. Though there is some calcification in the cartilaginous skeleton, true bone is entirely absent from this group. These fish have jaws with teeth, paired fins, and often active lifestyles. Sharks and rays tend to be larger than bony fishes, and except for some whales, sharks are the largest living vertebrates.

About 350 species of sharks and 320 species of rays are known to exist. Nearly all are marine, though a few species inhabit estuaries, and a very few are permanent inhabitants of freshwater. Although there are many exceptions, sharks tend to favor swimming through open water, while rays tend to be found on or near the bottom.

Sharks have an undeservedly bad reputation. More than 80% of shark species are less than 2 meters (6.6 feet) long as adults, and only a few of the remaining 20% are aggressive toward humans. Like other cartilaginous fishes, sharks are not very intelligent and certainly don't hold grudges or behave in the malignant ways so vividly portrayed in popular novels and movies. Still, some sharks are indeed dangerous to humans, and the great white sharks in the genus *Carcharodon* (*karcharos*, "sharp"; *odontos*, "tooth") **(Figure 15.22)** are perhaps the most dangerous of all.[2] These swimmer's nightmares can attain lengths of up to

---

[2]Some perspective: Worldwide, sharks are responsible for about six known human fatalities each year. Each year, more people are killed in the United States by dogs than have been killed by sharks in the past century. For every human killed by a shark, humans kill more than 16 million sharks, mostly for food and medicines. And no, shark cartilage pills don't prevent human cancers.

Blickwinkel/Alamy

Figure 15.21 The abrasive mouth of the sea lamprey often leaves circular scars on larger invertebrates that it attacks. This species is found in the Atlantic Ocean and the North American Great Lakes.

Figure 15.22 The great white shark (genus *Carcharodon*), a predator of seals, sea lions, and large fish. The great white shark is one of the most dangerous sharks encountered by human swimmers. In this extraordinary photo taken off the coast of South Africa, a large great white shark rockets from the ocean with a freshly caught sea lion in its jaws.

Richard Packwood/Oxford Scientific/Getty Images

7 meters (23 feet) and weigh up to 1,400 kilograms (3,000 pounds) but average about 4.5 to 5 meters (about 15 feet). Great whites are not white, but grayish brown or blue above, and creamy on the lower half. A dangerous relative, the mako shark, reaches lengths of 4 meters (13 feet) and is known also to attack small boats. These and other predatory species, such as tiger sharks and hammerhead sharks, are attracted to prey by vibrations in the water, which they detect with sensitive organs arrayed in lines beneath the surface of their skin. Smell also plays an important role in hunting their prey, which is usually composed of fish and marine mammals.

Though most famous, the man-eaters are not the largest of shark species. This honor goes to the immense warm-water whale sharks in the genus *Rhincodon* (*rhincos*, "a rasp"), which reach sizes in excess of 18 meters (60 feet) and 41,000 kilograms (90,000 pounds). Whale sharks and their somewhat smaller relatives, the basking sharks, are docile and present little threat to people. These greatest of fishes swim slowly near the surface with their huge mouths open, feeding on plankton. They may filter as much as 2,200 cubic meters (2,000 tons) of water per hour through a fine mesh of gill rakers. Accumulated plankton is periodically back-flushed into the mouth, where it is concentrated for swallowing.

Bony Fishes Are the Most Abundant and Successful Fishes

The 27,000-plus species of bony fishes, members of the class **Osteichthyes**, owe much of their great success to the hard, strong, lightweight skeleton that supports them. These most numerous of fish—and most abundant, most diverse, and successful of all vertebrates—are found in almost every marine habitat from tidepools to the abyssal depths. Their numbers include the air-breathing lungfishes and lobe-finned coelacanths, whose ancient relatives broke from the path of fish evolution to establish the dynasties of land vertebrates.

About 90% of all living fishes are contained within the Osteichthyan order **Teleostei**, which contains the cod, tuna, halibut, goldfish, and other familiar species. Within this large category (see **Figure 15.23**) are varieties of fishes with numerous advanced features, including independently movable fins for well-controlled swimming, great speed for pursuit or avoidance of predators, highly effective camouflage, social organization, and orderly patterns of migration. Their economic importance is great—some 101 million metric tons (111 million tons) of bony fishes are taken annually from the ocean to help satisfy the human demand for protein.

CONCEPT CHECK

Before going on to the next section, check your understanding of some of the important ideas presented so far:

What are the classes of living fishes? Which class of living fishes is considered most primitive? The most advanced?

Which class has the largest individuals? Which is the most economically important?

Are fishes warm-blooded (endothermic) or cold-blooded (ectothermic)?

Are most sharks dangerous to human swimmers?

15.8 Fishes Are Successful Because of Unique Adaptations

Seawater may seem to be an ideal habitat, but living in it does present difficulties. Water is about 800 times denser than air and 100 times more viscous, and it impedes motion effectively at low speeds. How can a fish best move through it? How can a fish maintain its vertical position in the water column? Must a fish swim constantly to offset the weight of muscle and bone? And how about breathing? Can oxygen and carbon dioxide be exchanged efficiently under water? How can predators be

Rolex Awards/Kurt Amsler

Dr. Brad Norman, an expert in whale shark biology, answers questions about his experiences.

What did you want to be when you were growing up?
An adventurer.

How did you get started in your field?
Volunteering for a friend who was studying fish at Ningaloo Reef, Western Australia. I got excited about the virtually pristine nature of the reef—and especially its mysterious inhabitants, the whale sharks.

What is a typical day like for you?
A typical day centers around hard work and a need to be very focused (on whale sharks). There's a lot of research, plenty of education and community outreach, and (ultimately) continual efforts toward whale shark conservation. I often have extended field trips and at these times, the ocean is my office.

What inspires you to dedicate your life to the ocean?
It's such a fantastic environment, filled with amazing and interesting creatures. I love to go diving. Sometimes I'll stay focused on an area no greater than one meter by one meter for 30 minutes, just looking at the tiny animals that inhabit and interact with the reef. Other times, I'll swim with schools of so many different types of fish. And then there's the whale shark—a truly majestic creature to see in its natural environment.

To borrow a phrase from another National Geographic explorer, Dr. Sylvia Earle, the ocean is the "world bank" and the life support system of our planet. We can't survive without it. I fully subscribe to this belief. We owe it to future generations—and to those that came before us—to ensure we respect and protect this wonderful underwater world.

What has been your favorite experience in the field?
Apart from the obvious—swimming with a powerful yet graceful creature on a daily basis (during whale shark season)—there was a time a few years ago where I had the opportunity to swim (by myself) only a few meters above a tiger shark. This was in the shallow lagoon of Ningaloo Reef, an experience that lasted for probably 20 minutes. Amazing!

Do you have a hero?
I have a great respect for Sir David Attenborough. A brilliant man and one that has brought the wonder of nature to so many people throughout the world.

If you could have people do one thing to help the ocean, what would it be?
Consider it like your own home and respect and protect it. It is a living organism.

Source: http://ocean.nationalgeographic.com/ocean/take-action/ocean-hero-brad-norman/
A whale shark drifts quietly in search of food.

Background photo: Ethan Daniels/Shutterstock

thwarted? Despite many problems, these most successful vertebrates have evolved structures and behaviors to cope.

Movement, Shape, and Propulsion

Active fish usually have streamlined shapes that make their propulsive efforts more effective. A fish's resistance to movement, or **drag**, is determined by frontal area, body contour, and surface texture. Drag increases geometrically with increasing speed.

Faster-swimming fish are therefore more highly modified to minimize the slowing effects of the dense, relatively sticky medium in which they live. The most effective anti-drag shape is the tapering torpedo-like body plan **(Figure 15.23a)**.

A fish's forward thrust comes from the combined effort of body and fins. Muscles within slender flexible fish (such as eels) cause the body to undulate in S-shaped waves that pass down the body from head to tail in a snake-like motion. The eel pushes forward against the water much as a snake pushes against the

Figure 15.23 Some of the diversity exhibited by teleost (bony) fishes. Not all fish are streamlined like the yellowfin tuna **a**. Some, like the spotted trunkfish **b**, are tough and rigid to withstand the pummeling of reef surf, while others, like the leafy seadragon **c**, blend exquisitely with their surroundings. The colorful blue ribbon eel **d** is marked to attract a mate. And nobody is quite sure why the huge ocean sunfish **d** is mostly head!

ground. This type of movement is not very efficient, however. More advanced fishes have a relatively inflexible body, which undulates rapidly through a shorter distance, and a hinged tail to couple muscular energy to the water. The fish's body can be shorter and can face more squarely in the direction of travel; the drag losses are lower.

Maintenance of Level

The density of fish tissue is typically greater than that of the surrounding water, so fishes will sink unless their weight is offset by propulsive forces or by buoyant gas- or fat-filled bladders. Cartilaginous fishes have no swim bladders and must work continuously to maintain their position in the water column. Sharks generate lift with a high tail and fins that act like airplane wings. Their lighter cartilaginous skeletons and large, oily livers also help to reduce their overall density. In contrast, bony fishes

that appear to hover motionless in the water usually have well-developed swim bladders just below their spinal columns. The volume of gas in these structures provides enough buoyancy to offset the animal's weight. The quantity of gas is controlled by secretion and absorption of gas from the blood, and by muscular contraction of the swim bladder to compensate for temporary changes in depth.

Gas Exchange

How can fish breathe underwater? **Gas exchange**, the process of bringing oxygen into the body and eliminating carbon dioxide (CO_2), is essential to all animals. At first glance, the task may seem more difficult for water breathers than for air breathers, but air-breathing animals add an extra step. We air breathers must first dissolve gases in a thin film of water in our lungs before the gases can diffuse across a membrane.

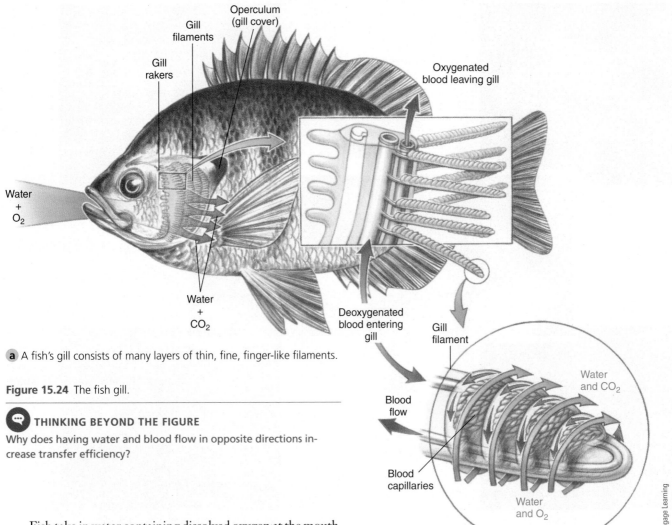

Operculum
(gill cover)

Gill
filaments

Gill
rakers

Oxygenated
blood leaving gill

Water
+
O₂

Water
+
CO₂

Deoxygenated
blood entering
gill

Gill
filament

Water
and CO₂

Blood
flow

Blood
capillaries

Water
and O₂

© Cengage Learning

a A fish's gill consists of many layers of thin, fine, finger-like filaments.

Figure 15.24 The fish gill.

💬 **THINKING BEYOND THE FIGURE**
Why does having water and blood flow in opposite directions increase transfer efficiency?

b The direction of blood flow through the capillaries of the gill filament is opposite to that of the incoming water. This mechanism, called *counter current flow*, allows efficient gas exchange between blood and water to occur along the entire length of each blood vessel.

Fish take in water containing dissolved oxygen at the mouth, pump it past fine **gill membranes**, and exhaust it through rear-facing slots. The higher concentration of free oxygen dissolved in the water causes oxygen to diffuse through the gill membranes into the animal; the higher concentration of CO_2 dissolved in the blood causes CO_2 to diffuse through the gill membranes to the outside. The gill membranes themselves are arranged in thin filaments and plates efficiently packaged into a very small space **(Figure 15.24)**. Water and blood circulate in opposite directions—in a countercurrent flow—which increases transfer efficiency.

An active fish like a mackerel requires so much oxygen and generates so much waste CO_2 that its gill surface area must be 10 times its body surface area! (Sedentary fish have proportionally less gill area.) With their large gill area and countercurrent flow, active fish extract about 85% of the dissolved oxygen in water flowing past their gills. Air-breathing vertebrates, by contrast, extract only about 25% of the oxygen from air entering their lungs.

Feeding and Defense

Competitive pressure among the large number of fish species has given rise to a wonderful variety of feeding and defense tactics. Sight is very important to most fishes, enabling them to see

their prey or avoid being eaten. Even some deepwater fishes that live below the photic zone have excellent eyesight for seeing luminous cues from potential mates or meals. Hearing is also well developed, as is the ability to detect low-frequency vibrations. The bizarre flattened crossbar that gives the hammerhead shark its name may provide a kind of stereo smell to sense differing amounts of interesting substances in the water. Salmon smell their way to their home streams after years at sea by detecting faint chemical traces characteristic of the water from the stream in which they hatched.

About a quarter of all bony fish species exhibit **schooling** behavior at some time during their life cycle. A fish school is a massed group of individuals of a single species and size class packed closely together and moving as a unit. There is no leadership in fish schools, and the movement of fish within them seems

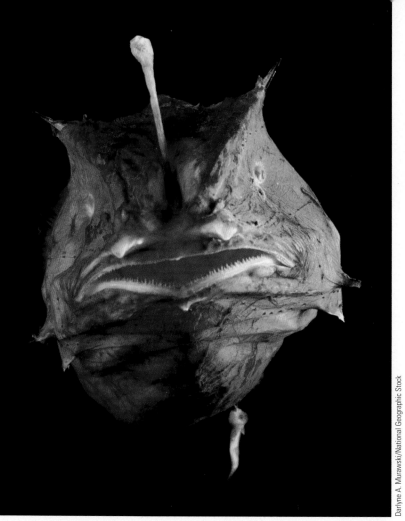

Figure 15.25 Some species of deep sea anglerfish have bioluminescent lures. Victims curious about the lure are quickly eaten by the predator. This female of genus *Melanocoetus* is about 10 centimeters (4 inches) long and carries a permanently attached parasitic male.

💬 **THINKING BEYOND THE FIGURE**
Why do you suppose a male attaches permanently to the female in this species?

to be controlled automatically by direct interaction between lateral-line sensors and the locomotor muscles themselves. We can personally attest to the effectiveness of schooling as a means of defense. On a few diving trips, we've noticed a large moving mass just beyond the limit of clear visibility. Is it a fish school, or is it a single large animal? Many predators might not stay around long enough to find out! Schools have the added benefits of reducing chance detection by a predator, providing ready mates at the appropriate time, and increasing feeding efficiency.

Fishes living in the twilight world at the bottom of the photic zone use bioluminescence in feeding, avoiding being eaten, and mate attraction. Some members of this sparsely populated community have built-in luminescent organs that cast dim blue light downward; this light masks their own shadows, so they have less chance of being detected and eaten. Other deep-swimming organisms attract their infrequent meals with a luminous lure

(Figure 15.25). These animals also use patterns of glowing spots or lines to identify themselves to members of the same species, a necessary first step in mating. Some use flashes of light to dazzle or frighten potential predators.

CONCEPT CHECK

Before going on to the next section, check your understanding of some of the important ideas presented so far:

How do fishes move effectively through such a dense medium as water?

How do fishes breathe?

How does schooling behavior benefit fishes?

15.9 Sea Turtles and Marine Crocodiles Are Ocean-Going Reptiles

Each of the three main groups of reptiles has marine representatives: turtles, sea snakes and marine lizards (iguanas), and marine crocodiles **(Figure 15.26)**. Like all reptiles, marine reptiles are ectothermic, breathe air with lungs, are covered with scales and a relatively impermeable skin, and are equipped with special **salt glands** to concentrate and excrete excess salts from body fluids.

Figure 15.26 Compare the size of this marine crocodile to the late Steve Irwin, its trainer. Extremely aggressive, marine crocs are the world's largest, reaching a length of 7 meters (23 feet). They hunt in packs and can swim long distances between tropical islands.

Figure 15.27 A leatherback turtle (genus *Dermochelys*) is the largest living turtle. Instead of a bony shell, its carapace is covered by skin and oily flesh. It can dive to a depth of 1,280 meters (4,200 feet).

Joel Sartore/National Geographic Creative

Except for one widely ranging species of turtle, all marine reptiles require the warmth of tropical or subtropical waters.

The best-known and most successful living marine reptiles are the seven species of sea turtles. Unlike land turtles, sea turtles have relatively small streamlined shells without enough interior space to retract head or limbs. The shell provides an effective passive defense, and adult sea turtles have few predators other than humans. Their forelimbs are modified as flippers and provide propulsive power; their hind limbs act as rudders.

Green sea turtles are the most abundant and widespread species. They are the only fully herbivorous species as an adult, and they range over great distances looking for marine algae, turtle grass, and other plants. The largest living turtles are the massive leatherbacks, which feed primarily on jellies (Figure 15.27). The other five species have adapted to eating various types of benthic invertebrates.

Sea turtles are justly famous for their remarkable feats of navigation. Sea turtles return at 2-, 3-, or 4-year intervals to lay eggs on the beaches at which they were hatched. Homing behavior can be a great advantage to any animal. If the parent survived its earliest childhood at this location, it will probably be a suitable place for hatching the next generation. The navigation of green turtles to tiny Ascension Island, an emergent point of the Mid-Atlantic Ridge between Brazil and Africa, has been extensively studied. Researchers have found that the turtles use the Earth's magnetic field, the solar angle (to derive latitude), wave direction, smell, and visual cues first to find the island, then to discover the spot on the beach where they hatched perhaps 20 years before!

While the females of most species prefer to come ashore individually to nest at night, the two species of ridley sea turtles (genus *Lepidochelys*) can storm the shores of specific beaches by the hundreds or even thousands. These mass nestings are known as *arribadas* (Figure 15.28). In contrast to many animal species that have a genetic sex determination, sea-turtle gender is actually based on the temperature of the nest. Higher temperatures cause embryos to develop as females, while lower temperatures lead to a greater relative proportion of male hatchlings. As hatchlings and eventually juveniles, sea turtles are vulnerable to a variety of natural and human-made threats—only about 0.1% will survive to sexual maturity.

In the distant past, marine reptiles ruled the seas. Among the more spectacular were plesiosaurs (Figure 15.29). During the Me-

Figure 15.28 Arribada (or mass nesting). Olive ridley turtles arrive in great numbers to a nesting beach in Costa Rica. In some cases, the beach becomes so dense with nesting turtles that they must climb over each other to find an open patch of sand.

 **THINKING BEYOND THE FIGURE**

Why do you think ridley turtles nest in such large numbers on the same beach?

Roland Seitre/Minden Pictures/Getty Images

Figure 15.29
A plesiosaur of genus *Thalassomedon* ambushes a school of fish, using its 6-meter (20-foot) long neck to hide its great bulk in the dim waters behind. MATTE FX, MATTE FX INC./National Geographic

sozoic Era (250–65 million years ago), while dinosaurs roamed the continents, marine reptiles like these were the ocean's top predators. The largest plesiosaurs were about 20 meters (66 feet) in length. They gave birth to living young and never came onto land. Their powerful four-flippered propulsion, long mobile neck, and presumably stealthy hunting habits led to their diversification into many species and habitats. All are extinct.

CONCEPT CHECK

Before going on to the next section, check your understanding of some of the important ideas presented so far:

How are sea turtles able to find their way back to the beach where they originally hatched?

What is an arribada?

15.10 Some Marine Birds Are the World's Most Efficient Flyers

Birds probably evolved from small, fast-running dinosaurs about 160 million years ago. Their reptilian heritage is clearly visible in their scaly legs and claws, and in the configuration of their internal organs and skeletons. The success of the 8,600 living species of birds is due in large part to the evolution of feathers (derivatives of reptilian scales) used to insulate the body and to provide aerodynamic surfaces for flight. Birds (and mammals) are *endothermic;* they generate and regulate metabolic heat to maintain a constant internal temperature that is generally higher than their surroundings.

Flying birds have light, thin, hollow bones without fatty insulation; they have forsaken the heavy teeth and jaws of reptiles for a lightweight beak. Their highly efficient respiratory system can accept great quantities of oxygen, and their large four-chambered heart circulates blood under high pressure. All birds lay eggs on land, and most incubate them and provide care for the young. Some seabirds may stay at sea for years, but all must eventually return to land to breed.

Only about 270 kinds of birds, about 3% of known bird species, qualify as seabirds. Most seabirds live in the southern hemisphere. Like the marine reptiles, seabirds have special salt-excreting glands in their heads to eliminate the excess salt taken in with their food. Salty brine from these glands may sometimes be seen dripping from the tip of their beaks. Marine birds are voracious feeders, and the ocean will usually be teeming with life wherever they are found. True seabirds generally avoid land unless they are breeding, obtain nearly all their food from the sea, and seek isolated areas for reproduction.

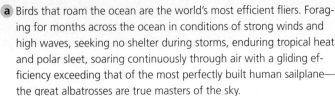

a Birds that roam the ocean are the world's most efficient fliers. Foraging for months across the ocean in conditions of strong winds and high waves, seeking no shelter during storms, enduring tropical heat and polar sleet, soaring continuously through air with a gliding efficiency exceeding that of the most perfectly built human sailplane—the great albatrosses are true masters of the sky.

Figure 15.30 Seabirds.

b The magnificent wandering albatrosses (genus *Diomedea*) are the largest of these birds, with a wingspan reaching 3.6 meters (12 feet) and a weight of 10 kilograms (22 pounds). The key to the albatross's success is its aerodynamically efficient wing—a very long, thin, narrow, cupped, and pointed structure ideal for high-speed soaring and gliding with very little expense of energy. Flying is its natural state—the heart of an albatross beats more slowly in flight than while it is sitting calmly on the ocean surface.

Of the four general groups of seabirds, the gulls and the pelicans may be most familiar to us because there are many species and because they spend much of their time near shore. But the groups best adapted to the pelagic world are the tubenoses (albatrosses, petrels) and the penguins.

The 100 species of tubenoses are the world's most oceanic birds. Their prosaic common name does not convey any sense of their beauty and grace; it refers to the plumbing in their beak responsible for sensing air speed, detecting smells, and ducting saline water from the salt glands. The supremely competent albatrosses—and their relatives the petrels and shearwaters—are masters of the ocean sky **(Figure 15.30)**. Their long, thin, narrow, cupped wings are wonders of aerodynamic efficiency.

Penguins have completely lost the ability to fly, but they use their reduced wings to swim for long distances and with great maneuverability. Their flightlessness makes it practical to have fatty insulation, greasy peg-like feathers, stubby appendages, and large size and weight; indeed such heat-conserving adaptations are critical to survival of aquatic organisms in very cold climates.

Their neutral buoyancy is an advantage as they forage for food underwater. Emperor penguins, the largest of the living penguin species, may dive to depths of 265 meters (875 feet) and stay submerged for 10 minutes or more. Penguins feed on fish, large zooplankters, bottom-dwelling molluscs or crustaceans, and squid. A few of the 18 species spend 2 uninterrupted years at sea between breedings.

Penguins are native only to the Southern Hemisphere and range from the size of a large duck to a height of more than a meter (3.3 feet) and weight exceeding 36 kilograms (80 pounds). They are thought to consume about 86% of all food taken by birds in the southern ocean—about 34 million metric tons (37 million tons) per year, mostly larger zooplanktonic crustaceans. The small Galápagos penguin lives a comparatively easy life fishing the cold, nutrient-rich Humboldt Current at the equator, but its Antarctic relatives lead what must surely be the most rigorous existence of any seabird. For example, emperor penguins breed and incubate during the bitterly cold Antarctic winter **(Figure 15.31)**.

Figure 15.31 Emperor penguins and a maturing chick. One of the larger penguin species, Emperors subsist on a diet mainly of fish and squid.

CONCEPT CHECK

Before going on to the next section, check your understanding of some of the important ideas presented so far:

From what group are birds thought to have evolved?

How are seabirds different from land birds?

Which marine birds are the world's most efficient flyers?

How do penguins differ from other marine birds?

15.11 Marine Mammals Include the Largest Animals Ever to Have Lived

The class **Mammalia**, to which humans belong, is the most advanced vertebrate group. About 4,300 species of mammals are known. The three living groups of marine mammals are the porpoises, dolphins, and whales of order **Cetacea**; the pinnipeds (seals, sea lions, and walruses), sea otters, and polar bears of order **Carnivora**; and the manatees and dugongs of order **Sirenia**.

> **A** gentle joyfulness—the mighty mildness of repose in swiftness, invested the gliding whale. —MELVILLE, MOBY DICK

Each of these orders arose independently from land ancestors. They exhibit the mammalian traits of being endothermic, breathing air, giving birth to living young that they suckle with milk from mammary glands, and having hair at some time in their lives. Unlike other mammals, however, these extraordinary creatures have become adapted to life in the ocean.

All marine mammals share four common features:

- Their *streamlined body shape* with limbs adapted for swimming makes an aquatic lifestyle possible. Drag is reduced by a slippery skin or hair covering.
- They *generate internal body heat* from a high metabolic rate and *conserve* this heat with layers of insulating fat and, in some cases, fur. Their large size gives them a favorable surface-to-volume ratio; with less surface area per unit of volume, they lose less heat through the skin.
- The *respiratory system is modified* to collect and retain large quantities of oxygen. The biochemistry of blood and muscle is optimized for the retention of oxygen during deep, prolonged dives.
- A number of *osmotic adaptations* free marine mammals from any requirement for freshwater. Minimal intake of seawater, coupled with the ability of their kidneys to excrete a concentrated and highly saline urine, permit them to meet their water needs with the metabolic water derived from the oxidation of food.

Order Cetacea—the Whales

The 79 living species of cetaceans are thought to have evolved from an early line of ungulates—hooved land mammals related to today's horses and sheep—whose descendants spent more and more time in productive shallow waters searching for food. Modern whales range in size from 1.8 meters (6 feet) to 33 meters (110 feet) in length and weigh up to 100,000 kilograms (110 tons). Their paddle-shaped forelimbs are used primarily for steering, and their hind limbs are reduced to vestigial bones that do not protrude from the body. They are propelled mainly by horizontal tail flukes moved up and down by powerful muscles at the animal's posterior end. A thick layer of oily blubber provides insulation, buoyancy, and energy storage. One or two nostrils are located at the top of the head and have special valves to prevent intake of water when submerged. Whales have large, deeply convoluted brains and are thought to form complex family and social groupings.

Modern cetaceans are further divided into two suborders. **Figure 15.32** shows representative whales in each division. Suborder **Odontoceti**, the toothed whales, are active predators and possess teeth to subdue their prey. Toothed whales have a high brain-weight-to-body-weight ratio, and though much of their brain tissue is involved in formulating and receiving the sounds on which they depend for feeding and socializing, many researchers believe them to be quite intelligent. Smaller whales in this group include the killer whale (the largest of all dolphins) and the familiar dolphins and porpoises of

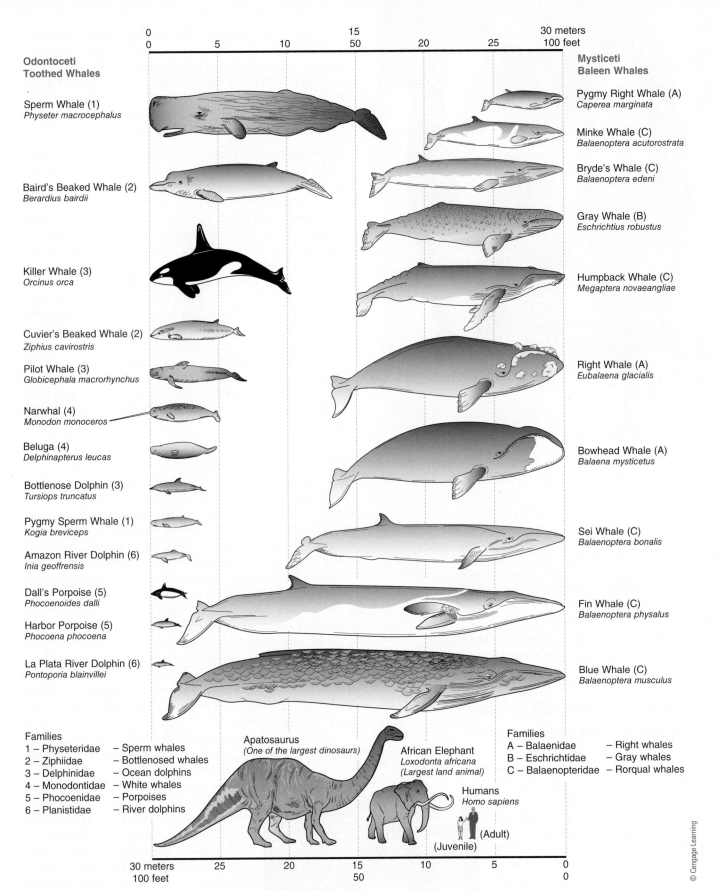

Figure 15.32 Some representatives of the order Cetacea. Note the differences between Mysticetes and Odontocetes.

Odontoceti
Toothed Whales

Sperm Whale (1)
Physeter macrocephalus

Baird's Beaked Whale (2)
Berardius bairdii

Killer Whale (3)
Orcinus orca

Cuvier's Beaked Whale (2)
Ziphius cavirostris

Pilot Whale (3)
Globicephala macrorhynchus

Narwhal (4)
Monodon monoceros

Beluga (4)
Delphinapterus leucas

Bottlenose Dolphin (3)
Tursiops truncatus

Pygmy Sperm Whale (1)
Kogia breviceps

Amazon River Dolphin (6)
Inia geoffrensis

Dall's Porpoise (5)
Phocoenoides dalli

Harbor Porpoise (5)
Phocoena phocoena

La Plata River Dolphin (6)
Pontoporia blainvillei

Mysticeti
Baleen Whales

Pygmy Right Whale (A)
Caperea marginata

Minke Whale (C)
Balaenoptera acutorostrata

Bryde's Whale (C)
Balaenoptera edeni

Gray Whale (B)
Eschrichtius robustus

Humpback Whale (C)
Megaptera novaeangliae

Right Whale (A)
Eubalaena glacialis

Bowhead Whale (A)
Balaena mysticetus

Sei Whale (C)
Balaenoptera bonalis

Fin Whale (C)
Balaenoptera physalus

Blue Whale (C)
Balaenoptera musculus

Families
1 – Physeteridae – Sperm whales
2 – Ziphiidae – Bottlenosed whales
3 – Delphinidae – Ocean dolphins
4 – Monodontidae – White whales
5 – Phocoenidae – Porpoises
6 – Planistidae – River dolphins

Apatosaurus
(One of the largest dinosaurs)

African Elephant
Loxodonta africana
(Largest land animal)

Humans
Homo sapiens
(Juvenile) (Adult)

Families
A – Balaenidae – Right whales
B – Eschrichtidae – Gray whales
C – Balaenopteridae – Rorqual whales

© Cengage Learning

THINKING BEYOND THE FIGURE
Why do you think baleen whales are generally larger than toothed whales? Why are they capable of growing larger than even the largest dinosaurs?

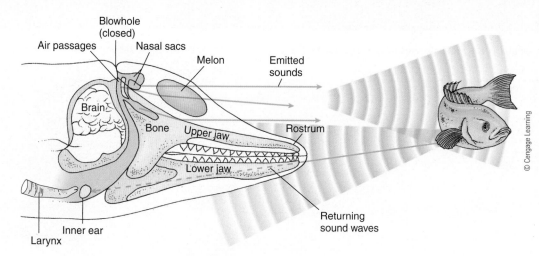

Figure 15.33 Echolocation. Toothed whales are able to generate different frequencies of sounds by passing air through sacs found just inside the blowhole. These *clicks* are focused by directing them through the fat-filled melon that is characteristic of many toothed whales. The echolocation clicks travel through the water and reflect back off of any surrounding solid objects. The returning sound waves are primarily received through the animal's fat-filled lower jaw and then transmitted to its inner ear.

© Cengage Learning

oceanarium shows. The largest toothed whale is the 18-meter (60-foot) sperm whale, which can dive to at least 1,140 meters (3,740 feet) in search of the large squids that provide much of its diet.

Toothed whales search for prey using **echolocation**, the biological equivalent of sonar. They generate sharp clicks and other sounds that bounce off prey species and return to be recognized **(Figure 15.33)**. Odontocete whales are now thought to use sound offensively as well. Recent research indicates that some odontocetes can generate sounds loud enough to stun, debilitate, or even kill their prey. In one experiment, dolphins produced clicks as loud as 229 decibels, equivalent to a blasting cap exploding close to the target organism. Sperm whales, it has been calculated, may generate sounds exceeding 260 decibels! (The decibel scale is logarithmic; compare this figure to the 130-decibel noise of a military jet engine at full power 20 feet away!) How this prodigious noise is generated is not yet known, nor do we know how such energy is radiated from the whale without damaging the organs that produce and focus it.

Suborder **Mysticeti**, the whalebone or baleen whales, have no teeth. Filter feeders rather than active predators, these whales subsist primarily on krill, a relatively large shrimp-like crustacean zooplankter obtained in productive polar or subpolar waters. With the exception of the gray whales, which primarily eat benthic organisms living in the sediments, baleen whales commonly feed only a few meters below the surface. Their mouths contain interleaving triangular plates of bristly hornlike **baleen** (Figure **15.34**), used to filter the zooplankton from great mouthfuls of water. The plankton is concentrated as water is expelled, swept from the baleen plates by the whale's tongue, compressed to wring out as much seawater as possible, and swallowed through a throat not much larger in diameter than a grapefruit. A great blue whale, largest of all animals, requires about 3 metric tons (6,600 pounds) of krill each day during the feeding season. The short, efficient food chain from phytoplankton to zooplankton to whale provides the huge quantity of food required for their survival.

Mysticeti is an excellent name for these odd and wonderful animals. We know comparatively little of their social structure, intelligence, sound-producing abilities, navigational skills, or physiology. We do know that humpback whales use complex songs in group communication and that blue whales may use very low-frequency sound to communicate over tremendous distances. Some species migrate annually from polar to tropical waters and back. Until recently, our primary response to all whales has been to slaughter them in countless numbers for meat and oil, with little thought for their extraordinary abilities and assets. More of this depressing history may be found in the discussion of marine resources in Chapter 17.

Order Carnivora—Oceanic Carnivores

The order Carnivora includes land predators ranging from dogs and cats to bears and weasels, but the members of the carnivoran suborder **Pinnipedia**—the seals, sea lions, and walruses (see **Figure 15.35**, p. 448)—are almost exclusively marine. Unlike the cetaceans, the gregarious pinnipeds leave the ocean for varying periods of time to mate and raise their young.

True seals have a smooth head with no external ear flaps, the external part of the ear having been sacrificed to further streamline the body. They are covered with a short coarse hair without soft underfur. Seals are graceful swimmers that pursue small fish, their usual prey, with powerful side-to-side strokes of their hind limbs. These rear appendages are partially fused and always point back from the hind end of the body, so they are of very little use for locomotion on land. The elephant seal, named for its long snout and large size, holds the diving depth record for all air-breathing vertebrates: 1,560 meters (5,120 feet). Sea lions, familiar to many as the performers in "seal" shows, have hind limbs with a greater range of motion, and thus

Plates of baleen line the jaw of a gray whale. The gray whale uses the stiff, coarse plates to sieve crustaceans from shallow bottom mud.

Tom Garrison

Figure 15.36 A female sea otter nurtures her nearly mature offspring (genus *Enhydra*) in a California kelp bed. Sea otters have the densest fur of any mammal. *Courtesy of Friends of the Sea Otter/Richard Bucich*

Figure 15.35 Female elephant seals relax and molt on a central California beach.

💬 **THINKING BEYOND THE FIGURE**

Can you imagine the smell?

are more mobile on land. They have a streamlined head with small external ears and a pelt with soft underfur; unlike seals, they use their front flippers for propulsion. Walruses are much larger than either seals or sea lions and may reach weights of 1,800 kilograms (2 tons). Their characteristic large tusks help to guide their sensitive "whiskers" just above the sediments looking for clams. Walrus tusks are also useful for hauling their heavy bodies onto ice floes.

The suborder Fissipedia has many members (including cats, dogs, raccoons, and bears) but only one truly marine representative, the sea otter **(Figure 15.36)**, a relative newcomer to the marine environment. Sea otters do not have a layer of blubber like most other marine mammals; they instead have a thick fur, which holds in a layer of insulating air. Human demand for this fur, the densest and warmest of any animal, caused its near extermination. The modern population of the Pacific sea otter descends from a very few individuals accidentally overlooked by fur hunters of the late 19th and early 20th centuries. Playful

and intelligent, otters rarely exceed 120 centimeters (4 feet) in length; they eat voraciously, consuming up to 20% of their body weight in molluscs, crustaceans, and echinoderms each day. A sea otter will sometimes lie on its back in the water, balance a rock on its chest, and hammer the shell of the prey against the rock until the shell cracks. Morsels of food are extracted with small nimble fingers, and rolling over in the water cleans away the debris. A morning spent watching sea otters in their coastal habitat is a morning well spent!

Polar bears (*Ursus maritimus*) spend most of their lives stalking seals and stranded whales on the frozen northern polar ocean **(Figure 15.37)**. They swim between ice floes with large, oarlike forepaws and can cross 100 kilometers (62 miles) of open water. The world's largest bears, male polar bears may grow to 2.5 meters (8.2 feet) and weigh 800 kilograms (1,800 pounds). Polar bears are dependent on sea ice for finding food, so as the Arctic sea ice declines with changing climate, many polar bears may have a more difficult time finding food.

Order Sirenia—Manatees and Their Kin

The bulky, lethargic, small-brained dugongs and manatees, collectively called sirenians **(Figure 15.38)**, are the only herbivorous marine mammals. Like the cetaceans, they appear to have evolved from the same ancestors as modern ungulates. They make their living grazing on sea grasses, marine algae, and estuarine plants in coastal temperate and tropical waters of the Americas, Asia, Africa, and Australia. Some species live in freshwater. The largest sirenians reach 4.5 meters (15 feet) in length and weigh 680 kilograms (1,500 pounds). They were first compared to mermaids by early Greeks, who noted the manatee's habit of resting in an upright position in the water and holding a suckling calf to her breast. Sirenians

Figure 15.37 A female polar bear with two cubs surveys the pack ice. Note the large size of the front paws, which help make polar bears efficient swimmers.

have been hunted extensively. All four living species—three manatee species and one dugong species—are currently classified as internationally threatened. Even though protected now, many are killed or wounded each year in Florida by the propellers of powerboats.

The animals we have discussed in this chapter do not exist in isolation but interact with each other and with plants and other autotrophs in complex marine communities. In the next chapter we will turn our attention to these groupings.

CONCEPT CHECK

Before going on to the next section, check your understanding of some of the important ideas presented so far:

What characteristics are unique to marine mammals?

What are the largest animals ever to have lived on Earth? From what animals are they thought to have evolved?

What is echolocation? How do whales use it?

How is a seal different from a sea lion? Which animal is featured in "seal shows" at oceanaria?

Figure 15.38 A West Indian or Florida manatee (genus *Trichechus*). In spite of their appearance, manatees have a relatively low fat content and are vulnerable to cold shock.

💬 **THINKING BEYOND THE FIGURE**
Why do you think manatees evolved to be so much slower than other marine mammals?

The Age of Marine Organisms

The lifespan of marine organisms varies greatly. Some animals like the pygmy goby may only live for 2 months. The longest-lived species may survive for many decades or even centuries. Accurately determining the age of an animal can give scientists an insight into many aspects of their biology, including how long they are capable of living, how fast they grow, and when they reach sexual maturity. This is also important to resource managers, fishermen, and conservationists who are concerned with how much of the population can sustainably be harvested without dangerously depleting it.

Probably the most well-known technique for aging organisms is to count their annual growth rings. The most reliable site for growth-ring deposition is different for individual species and can range from otoliths (ear bones), scales, vertebrae, opercula, fin rays, pectoral spines, or teeth for fish to shells or other calcareous parts of some invertebrate species like squid statoliths or coral skeletons. The width of these rings is often dependent on the physical conditions of the animal's environment. In more favorable seasons, material is deposited at a greater rate than when conditions are not as favorable. Therefore, scientists can often observe a darker ring associated with a slower growing season and can use the number of rings present to estimate the animal's age **(Figure A)**.

Radiocarbon dating is another technique that provides reliable age estimates for certain long-lived animals. Deep-sea corals are often aged using this method, which relies upon the decay of ^{14}C, which has a half-life of 5,730 years.[3] While this information is useful to scientists, caution is needed since there are a number of complicating factors that can lead to imprecise results. For example, since both coral growth rates and the ^{14}C levels are affected by the surrounding environment, scientists need to ensure that they also have accurate environmental records to be confident in the numbers. This complexity is compounded if the coral's diet includes a different source of carbonate.

The designation of longest-living animal (as far as we know) is currently held by Ming the clam, at 507 years old **(Figure B)**. Ming is a quahog clam that was dredged off the coast of Iceland in 2007 and later named for the Chinese dynasty that was in power when the clam was very young. Unfortunately, in the harvest, transfer and subsequent examination process, Ming did not survive. Scientists were unaware

[3]For a review of radiometric dating, please see Box 3.1 on page 69.

of its great age until annual growth rings were counted and its age was estimated at over 400 years. A later reanalysis focused on the narrow bands that were initially too close to distinguish, as well as radiocarbon dating, and determined that the clam was over 500 years old. Since the number of clams sampled was so small, the odds of another quahog clam in the North Atlantic being even older than Ming are highly likely.

Although we have the ability to decipher the age of many marine organisms, it is difficult for scientists to accurately estimate the age of some soft-bodied species. Other species seem

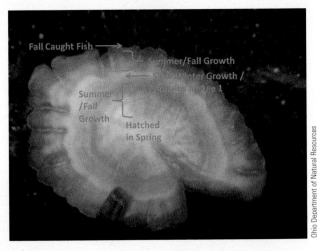

Figure A Cross section of a fish otolith used to age fish.

Figure B Ming, the clam.

to confound the concept of age altogether. One notable marine species has been dubbed the "immortal jelly" **(Figure C)**. *Turritopsis dohrnii* is a species of hydrozoan that has the unique ability to revert back to a younger version of itself. More specifically, when an adult medusa is stressed by starvation or physical damage, it can convert its cells back into immature polyp tissue. The polyp can subsequently form a colony that buds off genetically identical medusa and in essence restart its life cycle. This allows it to theoretically achieve a form of biological immortality. However, each individual medusa is subject to the same challenges as other planktonic organisms and may be eaten or die of disease.

Figure C *Turritopsis dohrnii,* the "immortal jelly."

Images & Stories/Alamy

Chapter in Perspective

In this chapter you learned that animals must ultimately depend on primary producers (autotrophs) for nutrition. Animals could not exist on Earth until increasing levels of free oxygen in the atmosphere permitted them to metabolize food obtained from autotrophs. And remember, it was the photosynthetic autotrophs themselves that contributed huge quantities of oxygen to the environment. True multicellular animals arose between 900 and 700 million years ago, near the end of this "oxygen revolution." Their variety is astonishing—a tribute to millions of years of complex interplay between environment, producer, and consumer.

Our survey of marine animals followed the course of their evolution. The complexity of animals increased as we moved from groups (phyla) whose basic structure seems to have solidified relatively early in the history of animals to groups that evolved more recently. Every marine animal has evolved effective adaptations for capturing prey, avoiding danger, maintaining thermal and fluid balance with their surroundings, and competing for space. Our survey of marine animals stressed these adaptations.

In the next chapter you'll learn how these animals interact with one another and with their environment. The organisms you met in the previous two chapters don't live alone. They are distributed throughout the marine environment in specific communities: groups of interacting producers, consumers, and decomposers that share a common living space. The types and variety of organisms found in a particular community depend on the physical and biological characteristics of that living space.

QUESTIONS FROM STUDENTS

1. **Representatives of the phylum Mollusca are both the largest and the most intelligent invertebrates. Yet molluscs are *not* considered the most successful nor the most highly evolved invertebrates. Why is that?**

The most often used measure of success in biology is the *number* of species and individuals within a group. The number of arthropod species and individuals greatly exceeds the number of molluscan species and individuals; thus, arthropods are more successful.

The question of evolutionary position—that is, which group is most highly evolved—is a matter of some controversy. Consider this automotive analogy: Modern cars possess a number of sophisticated technological features, such as electronic fuel injection, turbochargers, independent suspension, high-speed radial tires, antilock brakes, aerodynamic enhancements, nifty stereo systems, and so forth. All of these innovations *could* be fitted to a 1959 Cadillac, but the older car's primitive chassis design would limit the performance of the total package (especially when encountering the first high-speed turn!). Cephalopod molluscs are an *old* chassis design. Excellent eyes and relatively high intelligence have been added to the ancient chassis, but the physical limitations of musculature and the lack of a jointed skeleton do not allow the class to exploit the full potential of these "options." Though well represented in the fossil record, fewer than 500 species of cephalopods exist today. They may represent an evolutionary dead-end.

2. **If copepods and krill are the most successful marine *invertebrates,* what is the most widely distributed and successful *vertebrate*?**

Probably small pelagic bony fishes of genus *Cyclothone.* These *bristlemouths* are found just below the euphotic zone in virtually all the world ocean away from shore. Bristlemouths feed on plankton and on organisms swimming into their visual field at the limit of light penetration. Some ichthyologists believe there are more living individuals of *Cyclothone microdon* than of any other single vertebrate species on Earth.

3. **I've heard that sharks don't get cancer. Could we find out what prevents cancer in sharks and synthesize it for human use?**

In fact sharks do develop cancer. The mistaken impression that shark cartilage may prevent or cure cancer in humans has contributed to sales of shark-derived supplements (derived mainly from hammerheads and spiny dogfish) worth more than $45 million in 2009. A large percentage of those 100 million sharks killed each year are taken for these ineffective medicines.

4. **How do birds like the arctic tern and wandering albatross navigate across the trackless ocean?**

No one is certain. Experiments done at Cornell University with homing pigeons suggest that homing birds use a combination of magnetic and optical cues to return to their starting points. Even polarized light and the positions of certain stars might be involved. However it works, the behavior is not learned but is instinctive to the bird.

5. **What's the difference between a dolphin *fish* and a dolphin *mammal*?**

The name confusion began in Aristotle's time and has not abated. Dolphin mammals are small-toothed whales; dolphin fish are teleosts (bony fish). The dolphin fish appears on restaurant menus and in seafood markets as mahi-mahi, dorado, and other names. Dolphin mammals are slaughtered in dismaying numbers in association with tuna fishing by some non-U.S. fishing fleets, but none of their meat appears on U.S. plates.

6. **OK, how is a *porpoise* different from a *dolphin*?**

Dolphin and *porpoise* are common names of two subtly different groups of odontocetes. *Porpoises* are the smaller members of the group; they have spade-shaped teeth, a triangular dorsal fin, and a smooth front end tapering to a point. *Dolphins* are usually larger and have an extended bottle-like jaw filled with sharp, round teeth. The small jumping whales in most oceanarium shows are dolphins. Indeed, a killer whale is technically a dolphin, and by far the largest member of that group. To make matters even more complicated, the common dolphin seen in ocean-themed amusement parks, *Tursiops truncatus,* is often referred to as a porpoise, even by show announcers. This confusion between common names points out how useful scientific names can be; the real name of the animal, *Tursiops,* is clear and unambiguous.

7. **Could plesiosaurs like the one shown in Figure 15.29 still be alive somewhere—maybe in a place like Scotland's Loch Ness?**

Almost certainly not, sadly. This fascinating topic is covered in Question #3 in Questions from Students, Chapter 16.

TERMS AND CONCEPTS TO REMEMBER

ahermatypic	Chordata	Mammalia	Platyhelminthes
anadromous	Cnidaria	medusa	Polychaeta
animal	cnidoblast	metamerism	polyp
Annelida	Crustacea	Mollusca	Porifera
Arthropoda	drag	molt	radial symmetry
baleen	Echinodermata	Mysticeti	salt glands
bilateral symmetry	echolocation	Nematoda	schooling
Bivalvia	exoskeleton	notochord	Sirenia
Carnivora	gas exchange	Odontoceti	suspension feeder
cartilage	Gastropoda	Osteichthyes	Teleostei
Cephalopoda	gill membranes	oxygen revolution	tunicate
Cetacea	hermatypic	phylum	vertebrate
chitin	invertebrate	Pinnipedia	water-vascular system
Chondrichthyes			

STUDY QUESTIONS

Thinking Critically

1. When did the first true animals evolve? What atmospheric changes had to happen before animal life was possible? Are descendants of most of the early forms of animal life represented in the ocean today? Explain why.
2. How can an arthropod grow within a "tailored" shell? How can an animal grow without getting bigger, or get bigger without growing?
3. Are all chordates vertebrates? Are all vertebrates chordates? What distinguishes a vertebrate?
4. Of the living classes of vertebrates, only one has no permanent marine representatives. Which is it? Why can't these animals live in the ocean?

5. How are odontocete (toothed) whales different from mysticete (baleen) whales? Which are the better known and studied?

Thinking Analytically

1. Do you think there are more species of animals on land or in the ocean? What about absolute numbers of animals—are there more animals on land or in the ocean?
2. There are more parasitic species of animals than all other sorts of animals combined. (If this sounds impossible, consider how many animals parasitize any animal you can think about, including ourselves.) Why do you think parasitism is such a runaway success? What are the drawbacks of being a parasite?

GLOBAL GEOSCIENCE WATCH

Visit www.cengagebrain.com to access MindTap, a complete digital course which includes access to Global Geoscience Watch and more. Within the GREENR database, search for "new marine species" or other similar search terms. Utilize as many different types of sources as possible. Research information about new marine species that have been discovered some time in the past few years. Use this information to write a short outline that addresses the following: 1) The characteristics of the new species. 2) How and where we discovered the species? 3) Any interesting facts that you learned about the organism. Write an outline that contains this information and be prepared to report back to the class on this new species' characteristics.

16 Marine Communities

KEY CONCEPTS

A community is composed of the many populations of organisms that interact at a particular location. A population is a group of organisms of the same species occupying a specific area. Tom Garrison

A habitat is an organism's "address" within its community, its physical location. Each habitat has a degree of environmental uniformity. An organism's niche is its "occupation" within that habitat, its relationship to food and enemies, an expression of what the organism is doing. Georgie Holland/Age Fotostock

Physical and biological factors in the environment determine the location and composition of a community. Tom Garrison

A stable, long-established community is known as a climax community. This self-perpetuating aggregation of species tends not to change unless disrupted by severe external forces. Reinhard Dirscherl/Glow Images

More than half of the animal species known to science are not free-living. Most are actively involved in close symbiotic relationships with at least one other life-form in their community. Amos Nachoum/Encyclopedia/Corbis

Despite being located in warm, nutrient-poor waters, coral reefs house some of Earth's most diverse and densely populated communities.
Specta/Shutterstock.com

Media Connection

Start off this chapter by listening to a podcast with Explorer Enric Sala, as he talks about his exploration of Salas y Gómez Island in the Pacific Ocean. Visit www.cengagebrain.com to access MindTap, a complete digital course that includes this podcast and other resources.

CONCEPT CHECK

Before going on to the next section, check your understanding of some of the important ideas presented so far:

What is a community?

How does a population differ from a community?

Which is the largest marine community?

16.1 Marine Organisms Live in Communities

Organisms are distributed throughout the marine environment in specific groups of interacting producers, consumers, and re-cyclers that share a common living space. These groups are called *communities*. A **community** is composed of the many populations of organisms that interact at a particular location. A **population** is a group of organisms of the same species occupying a specific area. The location of a community and the populations that comprise it depend on the physical and biological characteristics of that living space (Figure 16.1).

> **A**n interacting set of populations can exist on a single grain of sand or on one decomposing fish scale.

The largest marine community—also the most sparsely populated—lies within the uniform mass of permanently dark water between the sunlit surface and the deep bottom. Few animals live there because so little food is available, but those organisms that survive are among the strangest in the ocean. Opportunities for feeding in the deep open-ocean community are few and far between, and some animals are able to consume prey larger than themselves should the occasion arise. Because so few animals are present, mating is also a rare event; in a few species, males and females become permanently bonded during their first encounter, the male burrowing into the female's body for a lifelong free ride (see again Figure 15.25).

In contrast, the smallest obvious marine communities may be those established against solitary rocks on an otherwise flat, featureless seabed. Drifting larvae will colonize the place; the es-tablished community can seem an oasis of life and activity in an otherwise static sedimentary desert. Seaweeds will grow, worms will burrow, snails will scrape food from the hard surfaces, and small fishes will nestle among crevices. Hundreds of small plants and animals can live their lives within a meter of each other, interacting in a compact solitary community with no similar en-vironment available for thousands of meters. The larvae of the next generation drift away, with little chance of finding a suitable place to carry on their lives. Microscopic communities also ex-ist; in fact, an interacting set of populations can exist on a single grain of sand or on one decomposing fish scale.

16.2 Communities Consist of Interacting Producers, Consumers, and Decomposers

Communities are dependent on the avail-ability of energy. As you learned in Chapter 13, living things cannot create new energy, but they can transform one kind of energy to a different kind. Using energy from the sun and the reactions of photosynthesis (or using energy from the chemical reduc-tion of iron, sulfate, and manganese ions—chemosynthesis), primary producers as-semble molecules of food (typically glu-

Figure 16.1 An unusual marine community on the edge of the Irish Sea. Brightly colored lichens compete for space in the high intertidal zone.

Tom Garrison

cose) from atoms of carbon, hydrogen, and oxygen. The energy is passed from organism to organism in a food web (see Figure 13.14). These food webs and their related interactions usually define communities—the organisms in the web often share a common location and similar tolerances to the physical factors to which they are exposed.

There are many different places to live and many different "jobs" for organisms within even a simple community. A **habitat** is an organism's "address" within its community, its physical *location*. Each habitat has a degree of environmental uniformity. An organism's **niche** (*nidus*, "nest") is its "occupation" within that habitat, its relationship to food and enemies, an expression of what the organism is *doing*. For example, the small fishes living among the coral heads in a coral reef community share the same habitat, but each species has a slightly different niche. Each population in the community has a different "job" for which its shape, size, color, behavior, feeding habits, and other characteristics particularly suit it.

Biologists also survey the *range* of species in a community. The variety of species in a given area is an expression of **biodiversity** (biological diversity). Communities with high biodiversity—a coral reef, for example—are characterized by complex interactions between and among residents.

Physical and Biological Environmental Factors Affect Communities

Physical and biological factors in the environment determine the location and composition of a community. Physical factors such as temperature, pressure, salinity, and the degree of force exerted by waves or currents affect the success of an organism. Seaweeds adapted to cold water usually cannot survive prolonged exposure to abnormally warm water, and lower-than-normal salinity can dangerously disrupt the fluid balance of most marine invertebrates. Biological factors are influences on an organism by members of its own population or other populations, in its own or other communities. Biological factors include crowding, predation, grazing, parasitism, shading from light, generation of waste substances, and competition for limited oxygen.

Figure 16.2 provides an idealized look at the tolerance of organisms to a varying physical factor such as temperature. When an organism's optimal temperature range is wide (represented by

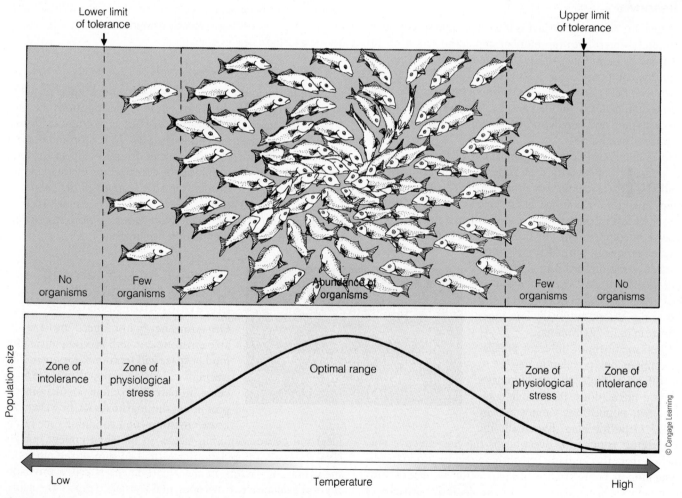

Figure 16.2 Range of tolerance to a physical factor—in this case, temperature—for a population of organisms.

THINKING BEYOND THE FIGURE
What kinds of organisms have a wide optimal temperature range? What kinds have a narrow optimal temperature range?

this broad curve), it can survive wide swings of warm to cold. If the curve were narrow (represented by a steeper, tighter curve), any small temperature deviation might be lethal. The same reasoning applies to salinity tolerance or to pressure tolerance.

An animal or plant is almost never exposed to fluctuations of only one physical or biological factor in its environment. An organism may die if subjected to changes in several environmental factors at once, even though each individual change is within its range of tolerance. Even small changes in temperature or salinity, survivable in themselves, might prove lethal if a particular species of tropical fish were exposed simultaneously to both.

A favorable balance of physical and biological factors is critical to each individual organism's success—and therefore to community success and longevity. The study of this balance and of the relationships of organisms and interactions within communities is called **ecology** (*oikos*, "house"; *logos*, "study of"). Marine ecologists are concerned with the types of organisms within marine communities, as well as their habitats, niches, distribution, numbers, and reactions to variations in their environment.

Organisms within a Community Compete for Resources

The availability of resources such as food, light, and space within a community determines the number and composition of the populations of organisms within that community. Within the community, competition for the necessities of life may occur between members of the *same* population or between members of *different* populations. Subtle swings in physical or biological factors may give one population the advantage for a time but then shift to favor another.

When members of the same population (all members of the same species) compete with each other, some individuals will naturally be larger, stronger, or more adept at gathering food, avoiding enemies, or mating. These animals tend to prosper, forcing their less successful relatives to emigrate, fight, or die in the course of competition. The most successful organisms have the most surviving offspring; so useful inheritable variations are passed along in greater quantity to the next generation. As we saw in our discussion of evolution, this kind of competition continually fine-tunes a population to its environment.

When members of different populations compete, one population may be so successful in its "job" that it eliminates competing populations. In a stable community, two populations cannot occupy the same niche for long. Eventually the more effective competitor overwhelms the less effective one. Extinction from this kind of head-to-head competition is probably uncommon, but restriction of a population because of competition between species is not. For example, little barnacles of genus *Chthamalus* live on the uppermost rocks in many intertidal communities; larger limpets (*Collisella*) live lower on the rocks (see **Figure 16.3**). Planktonic larvae of both species can attach themselves to rocks anywhere in the intertidal zone and begin

Figure 16.3 Competition between two species prevents either from occupying as much of the intertidal zone as might otherwise be possible. Small encrusting barnacles dominate most of this rock, but limpets at the bottom of the rock have probably prevented larval barnacles from gaining a foothold near its bottom.

to grow. In the lower zone, the faster-growing limpets push the weaker barnacles off the rocks, but at higher positions the limpets cannot survive because they are not as resistant to drying and exposure as the tough little barnacles. At the top and bottom of their distribution, the two species do not compete for food or space. The competition at the intersection of their ranges prevents each species from occupying as much of the habitat as might otherwise be possible. See "How Do We Know? 16.1" in this chapter for a more detailed discussion on how researchers are able to determine population dynamics.

> **S**ubtle swings in physical or biological factors may give one population the advantage for a time but then shift to favor another.

Growth Rate and Carrying Capacity Are Limited by Environmental Resistance

Organisms newly introduced into a favorable environment with no competitors for food or space will reproduce exponentially, tracing a J-shaped population growth curve. In nature, very few populations reproduce at this maximal rate, however, because environmental conditions are rarely ideal and because limiting factors in the environment quickly slow the rate of population growth. The sum of the effects of these limiting factors in the environment is called **environmental resistance**. Environmental resistance causes the actual population growth curve to be lower than the maximum potential growth curve. **Figure 16.4** shows the growth rate in number of individuals over time for both potential and actual situations. When limiting factors intrude, note that the curve is S-shaped;

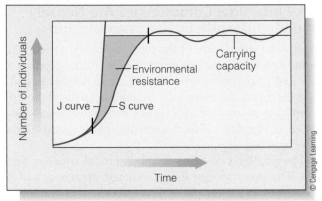

Figure 16.4 The J-shaped curve of population growth of a species is converted to an S-shaped curve when the population encounters environmental resistance. The physical or biological conditions responsible for the cessation of growth are called limiting factors.

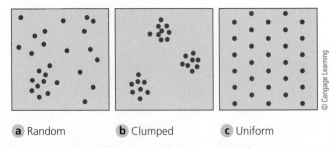

a Random   **b** Clumped   **c** Uniform

Figure 16.5 Random **a**, clumped **b**, and uniform **c** population distribution patterns.

it gradually flattens toward an upper limit of the number of individuals in the population. The final number of organisms oscillates around the **carrying capacity** of the environment for that species—the population size of each species that a community can support indefinitely under a stable set of environmental conditions.

The carrying capacity changes if environmental conditions change. A marine population could *crash* if upwelling ceased, if new predators were introduced, if climate varied, if food supplies dwindled, or if a new parasite infiltrated the population.

Population Density and Distribution Depend on Community Conditions

As we have seen, physical and biological factors affect the number and positions of organisms in a community. In general, the level of disturbance resulting from these factors determines the number of niches available. In an environment where the level of disturbance is high, such as the surface of a polar ocean, few organisms are capable of surviving, which often leads to lower species diversity. In contrast, where physical and biological factors result in many different niches, such as a tropical coral reef, the biodiversity is usually relatively high. In regions with high biodiversity, often many populations coexist in a relatively small area. The number of individuals per unit area (or volume) is known as the **population density**. Rare individuals have a much lower population density than dominant ones.

Individual organisms are almost never distributed randomly throughout their habitat. A **random distribution** implies that the position of one organism in a community *in no way influences* the position of other organisms in the same community (**Figure 16.5a**). Further, a truly random distribution indicates that conditions are precisely the same throughout the habitat, an extremely unlikely situation, except possibly in the unvarying benthic communities of abyssal plains.

The most common pattern for distribution of organisms is small patchy aggregations, or clumps. **Clumped distribution** occurs when conditions for growth are optimal in small areas because of physical protection (in cracks in an intertidal rock), nutrient concentration (near a dead body on the bottom), initial dispersal (near the position of a parent), or social interaction (**Figure 16.5b**).

Uniform distribution, with equal space between individuals (**Figure 16.5c**), such as the arrangement of trees planted in orchards, is the rarest natural pattern of all. The distribution of some garden eels through their territories becomes almost uniform because each eel can extend from its burrow just far enough to hassle neighbor eels spaced equally distant. There is eventually a break in the order of position—they don't line up row-upon-row like apple trees.

16.3 Marine Communities Change as Time Passes

Like the organisms that constitute them, communities change through time, but marine communities generally do not evolve as rapidly as terrestrial communities. The slow changes associated with seafloor spreading, climate cycles, atmospheric composition, or newly evolved species have shaped this generally slow evolution. As on land, the species, community composition, and location of a marine community are changed by the environmental factors to which members of the community are exposed. Communities themselves can gradually modify the physical aspects of their environment. A coral reef is an extreme example. The massive accumulation of coral and sediments within the reef can alter current patterns, influence ocean temperature, and change the proportions of dissolved gases.

Figure 16.6 Bluestripe snappers (genus *Lutjanus*) school over a reef in the Maldives. This reef has been stable for a long time and so represents a climax community.

But rapid changes can occur in marine communities. A natural catastrophe—a volcano erupting, a landslide that blocks a river, the collision of an asteroid with Earth, for example—can disrupt a community. Similarly, human activities—such as altering an estuary by damming a river, dumping excess nutrients into a nearshore area, or stressing organisms with toxic wastes—can cause rapid, disruptive changes. Offshore communities change abruptly near new sewage outfalls, for example.

A stable, long-established community is known as a **climax community (Figure 16.6)**. This self-perpetuating aggregation of species tends not to change unless disrupted by severe external forces, such as violent storms, significant changes in current patterns, epidemic diseases, or influx of great amounts of freshwater or pollutants. A disrupted climax community can be reestablished through the process of **succession**, the orderly changes of a community's species composition from temporary inhabitants to long-term inhabitants. Disruption makes the environment more hostile to the original species, but destruction of species in the original community leaves open habitats and niches. A few highly tolerant species will move into the area, eventually drawing in other species that depend on them. If the environment is permanently changed by the disruption, a different climax community will be established than was previously present.

CONCEPT CHECK

Before going on to the next section, check your understanding of some of the important ideas presented so far:

What is a climax community?

16.4 Examples of Shoreline Marine Communities

There are many distinct marine communities. In the next three sections, we survey a few of the most interesting examples.

Rocky Intertidal Communities Are Densely Populated Despite Environmental Rigors

Anyone who spends time at the shore, especially a rocky shore, is soon struck by a curious contradiction. Although the rocky shore looks like a very difficult place for organisms to make a living, the **intertidal zone**—the band between the highest high tide and lowest low tide—is one of Earth's most densely populated areas. Hundreds of species and individuals crowd this junction of land and sea.

The problems of living in the intertidal zone are formidable. The tide rises and falls, alternately drenching and drying out the animals and plants. **Wave shock**, the powerful force of crashing waves, tears at the structures and underpinnings of the residents. For intertidal areas exposed to the open sea, wave shock is a formidable physical factor. Fist-sized rocks have been thrown 100 meters (330 feet) into the air by the force of breaking waves. **Motile** animals, like crabs, move to protective overhangs and crevices, where they cower during intense wave activity. Attached, or **sessile**, animals hang on tightly, often gaining assistance from rounded or very low-profile shells, which deflect the violent forces of rushing water around their bodies. Some sessile animals have a flexible foot that wedges into small cracks to provide a good hold; others, like mussels, form shock-absorbing cables that attach to something solid. Large intertidal plants must be immensely strong, elastic, and slippery to avoid being shredded by wave energy.

Temperature can change rapidly as cold water hits warm shells or as the sun shines directly on newly exposed organisms. In high latitudes, ice grinds against the shoreline; in the tropics, intense sunlight bakes the rocks. Predators and grazers from the ocean visit the area at high tide, and those from land have access at low tide. Too much freshwater can osmotically shock the occupants during storms. Annual movement of sediment onshore and offshore can cover and uncover habitats. Yet, astonishingly, the richness, productivity, and diversity of the intertidal rocky community—especially in the world's temperate zones—are matched by very few other places. There is intense competition for space. Life abounds.

One reason for the great diversity and success of organisms in the rocky intertidal zone is the large quantity of food available. The junction between land and ocean is a natural sink for living and once-living material. Minerals dissolved in water running off the land serve as nutrients for the inhabitants of the intertidal zone, as well as for plankton in the area. The crashing surf and strong tidal currents keep nutrients stirred and ensure a high concentration of dissolved gases to support a rich population of autotrophs. Many of the larval forms and adult organisms of the intertidal community depend on plankton as their primary food source.

Another reason for the success of organisms here is the large number of habitats and niches to be occupied (see **Figure 16.7**). The habitats of intertidal animals and plants vary from hot, high, salty splash pools to cool, dark crevices. These spaces provide hiding places, quiet places to rest, attachment sites, jumping-off spots, cracks from which to peer to obtain a surprise meal, footing from which to launch a sneak attack, secluded mating nooks, or darkness to shield a retreat. The niches of the creatures

Figure 16.7 Students lure an octopus from its hiding place in a Pacific Coast tide pool. Shiny car keys work well—this octopus has a large collection.

in this community are varied and numerous—encrusting algae produce carbohydrates, snails scrape algae from rocks, hermit crabs scavenge for tidbits, and octopuses wait in ambush for likely meals. Sea stars pry open mussels, and barnacles sweep bits of food from the water.

The most obvious and important physical factor in intertidal communities is the rise and fall of the tides. Organisms living between the high- and low-tide marks experience very different conditions from those residing below the low-tide line. Within the intertidal zone itself, organisms are exposed to varying amounts of emergence and submergence. For example, **Figure 16.8a** plots the number of hours of exposure to air in a California intertidal zone over 6 months versus the tidal height. Because some organisms can tolerate many hours of exposure while others can tolerate only a very few hours per week or month, the animals and plants sort themselves into three or more horizontal bands, or subzones, within the intertidal zone.

a A graph showing intertidal height versus hours of exposure. The 0.0 point on the graph, the tidal datum, is the height of mean lower low water.

b Vertical zonation, showing four distinct zones. The uppermost zone (I) is darkened by lichens and cyanobacteria; the middle zone (II) is dominated by a dark band of the red alga *Endocladia;* the low zone (III) contains mussels and gooseneck barnacles; and the bottom zone (IV) is home to sea stars *(Pisaster)* and anemones *(Anthopleura).* The bands in the photograph correspond approximately to the heights shown in the graph.

Figure 16.8 The relationship between amount of exposure and vertical zonation in a rocky intertidal community.

From Between Pacific Tides 5/e by Edward F. Ricketts et al. Revised by David W Phillip. Copyright © 1985 by the Board of Trustees of the Leland Stanford Jr. University. All Rights Reserved. Used with permission of Stanford University Press, www.sup.org.

What Influences Intertidal Community Structure

Figure A Experimental manipulations on a rocky shore in a Hong Kong marine preserve examine the effects of shading on the settlement of barnacle larvae at different tidal heights.

Intertidal communities are composed of many different organisms living together in a relatively narrow band of coastline. Differences in physical and biological environmental factors can significantly influence the diversity and abundance of species found in adjacent zones. This diversity, coupled with our ability to accurately predict the tides and keep track of slow-moving intertidal organisms, has made intertidal zones important ecosystems for studying community structure and population dynamics.

One of the most common methods that research ecologists use for studying intertidal community structure is experimental manipulation **(Figure A)**. This allows scientists to add or remove a specific variable or species to determine how important it is to the current community structure. Back in the 1960s, J. H. Connell performed a series of famous experiments that showed some intertidal organisms were not able to occupy their entire **fundamental niche**. This represents the theoretical maximum niche that the population can occupy under ideal conditions. Due to the interactions of other organisms, populations are often only able to survive in a smaller portion of their fundamental niche, otherwise known as their **realized niche**.

Connell focused his research on two species of intertidal barnacles, *Chthamalus* and *Semibalanus* **(Figure B)**. He observed their normal distribution patterns as a **control**, or standard, to compare the results from his experimental

manipulation. Each species was then removed from particular sections of the intertidal zone to see how the other population would respond. In the areas where *Chthamalus,* the species living higher in the intertidal zone, was removed, *Semibalanus* did not expand its range farther up the rocks. This suggested that *Semibalanus* was not capable of living higher in the intertidal zone as a result of physical factors, not competition. In contrast, when *Semibalanus* was experimentally removed, *Chthamalus* colonized the newly opened space below. This suggested that its fundamental niche extended much farther than its realized niche, which was limited due to competition with *Semibalanus.*

This relatively simple experiment set the stage for future experimental manipulations of increasing complexity. Modern researchers now commonly use intertidal cages bolted down over small populations designed to exclude predators to see what effect that will have on community structure. Settlement plates are often employed to collect larval organisms as they colonize the rocky shores. This allows researchers to compare what species initially settle in different regions of the intertidal zone with what species survive into adulthood. Statistical models, digital imagery, and Geographic Information System (GIS) maps are also used to help marine ecologists better understand why particular species inhabit the areas of the intertidal zone that they do and what specific factors limit their distribution.

Each distinct zone is an aggregation of animals and plants best adapted to the conditions within that particular narrow habitat. The zones are often strikingly different in appearance, even to a person unfamiliar with shoreline characteristics. This zonation is clearly evident in the rocky shore of **Figure 16.8b.**

Desiccation (drying) by exposure to air and sunlight is another source of intertidal stress. Again, motile organisms have an advantage because they can move toward water left in tidal pools or muddy depressions by the retreating ocean. Attached animals and plants must await the water's return, huddled in low spots, moist pockets, or cracks in the rocks—or in tightly closed shells. Water trapped within a shell can keep gills moist for the needed exchange of gases. A protective mucous coating can retard evaporative water loss from exposed soft animal body parts

or blades of seaweed. When the weather is too warm or when the tide is out for an unusually long time, a deceptive calm settles over the zone, to be relieved only by the returning ocean.

Sand and Cobble Beach Communities Exist in One of Earth's Most Rigorous Habitats

Some intertidal areas are sandy, some are muddy, and others consist of gravel or cobbles. (A few shores combine all of these elements within a small area.) The usual rigors of the intertidal zone are intensified for organisms surviving on loose substrates. Indeed, it may surprise you to learn that in spite of its generally benign appearance, the ocean contains what may well be the most hostile, rigorous, and dangerous

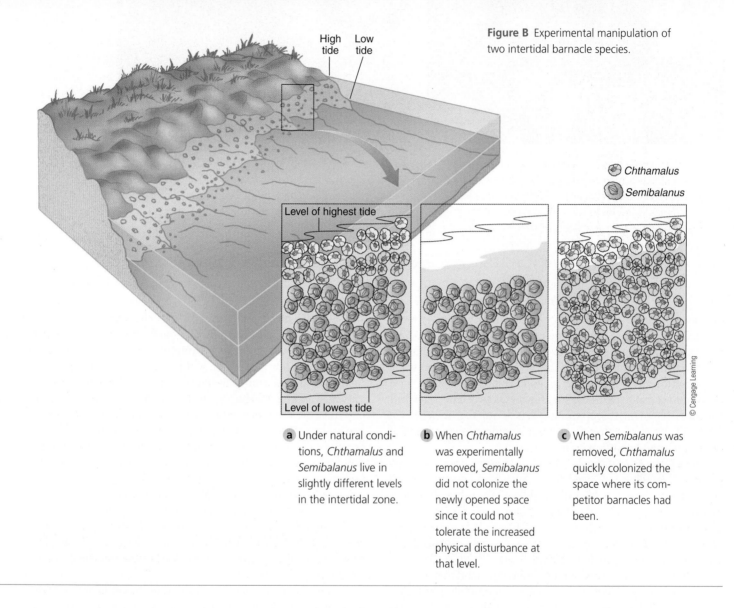

Figure B Experimental manipulation of two intertidal barnacle species.

High tide   Low tide

Chthamalus
Semibalanus

Level of highest tide

Level of lowest tide

© Cengage Learning

a Under natural conditions, *Chthamalus* and *Semibalanus* live in slightly different levels in the intertidal zone.

b When *Chthamalus* was experimentally removed, *Semibalanus* did not colonize the newly opened space since it could not tolerate the increased physical disturbance at that level.

c When *Semibalanus* was removed, *Chthamalus* quickly colonized the space where its competitor barnacles had been.

environments for small living things on Earth: high-energy sand and cobble beaches.

As environments go, sand beaches don't seem particularly nasty places to us; many people consider the beach to be about the finest habitat around. Seals and sea lions spend a lot of time at the beach and seem to enjoy the experience as much as people do. In short, for organisms of about our size, the problems of living on a beach are manageable. But for smaller organisms, a beach is a forbidding place. Sand itself is the key problem. Many sand grains have sharp, pointed edges, so rushing water turns the beach surface into a blizzard of abrasive particles. Jagged grit works its way into soft tissues and wears away protective shells. A small organism's only real protection is to burrow below the surface, but burrowing is difficult without a firm footing. When

the grain size of the beach is small, capillary forces can pin down small animals and prevent them from moving at all. If these organisms are trapped near the sand surface, they may be exposed to predation, to overheating or freezing, to osmotic shock from rain, or to crushing as heavy animals walk or slide on the beach.

As if this weren't enough, those that survive must contend with the difficulty of separating food from swirling sand and the dangers of leaving telltale signs of their position for predators or being excavated by crashing waves. A few can run for their lives; some larger beach-dwelling crabs depend on their good eyesight and sprinting ability to outrace onrushing waves!

To these horrors must be added the usual problems of intertidal life discussed earlier. Not surprisingly, very few species have adapted to wave-swept sandy beaches! Some of the successful

a Dime-sized bean clams *(Donax)* lie at the surface awaiting a ride up the beach on an incoming wave. They will bury themselves in the loose sediment, push up their siphons, and filter the water for food. When the tide retreats, they will again pop to the surface and allow the waves to take them back down the beach.

b A sand crab *(Emerita),* beloved of all beach-going children and beginning lab students, attempts to bury itself in anticipation of an onrushing wave. It gleans food from passing water with its feathery antennae.

Figure 16.9 Sand beach organisms.

ones are shown in **Figure 16.9.** The few that have done so—mostly small, fast-burrowing clams, sand crabs, sturdy polychaetes, and other minute worms—consume a rich harvest of plankton and organic particles washed onto the beach and filtered from the water by the uppermost layer of sand. Other organisms have evolved to actually live in the small spaces between the sediment particles. These microscopic animals, or *meiofauna,* are a diverse group of organisms that make up a significant component of sandy and muddy bottom communities.

Surely the most difficult of all sandy beaches are those composed of black sand, derived from pulverized lava on tropical volcanic islands such as Hawai'i. Besides all the other difficulties mentioned is the ability of pulverized lava to store solar heat until temperatures approach 71°C (160°F) just below the surface of the sand. Almost nothing can tolerate these beaches for more than a few minutes—including human feet!

Cobble beaches are even more uninviting (and they're murder on bare feet). The rounded rocks clack and bump together as waves pound the shore; most small animals are crushed. Except for nimble insect-like "beach hoppers" and a few species of scavenging terrestrial insects, most loose rock-strewn shores are understandably devoid of anything much larger than microscopic organisms.

Salt Marshes and Estuaries Often Act as Marine Nurseries

Muddy-bottomed salt marshes are among the most interesting intertidal shores. Much of the high primary productivity of a salt marsh comes from sea grasses, mangroves, and other vascular plants that can prosper in a marine (or partly marine) environment.

As you may recall from Chapter 12, salt marshes often form in an **estuary,** a broad, shallow, river mouth where freshwater

and saltwater mix (see Figure 12.29). A characteristic of estuaries is the reduction of wave shock—surf is blocked from estuaries by longshore bars or by twisting passages connecting to the ocean. The salinity of water within an estuary may vary with tidal fluctuations, from seawater through **brackish** water (mixed saltwater and freshwater) to freshwater. In areas near the river entrance, the water may be almost fresh, whereas near the outlet it may be of oceanic salinity. Many of the organisms living in estuaries are necessarily tolerant of a range of salinities, but the different salinities in an estuary often lead to a distinct horizontal zonation of organisms. Temperature range is also potentially extreme, especially in the tropics or during the temperate-zone summer when a receding tide abandons residents to the heat of the sun. Strong currents may move in estuaries as the tide rises and falls and the river flows. Flowing water takes the place of waves in mixing nutrients and gases in the intertidal estuary community.

Estuarine marshes (such as the one shown in **Figure 16.10**) are richer and exhibit greater species diversity than marshes exposed only to seawater. Primary productivity in estuaries is often extraordinarily high because of the availability of nutrients, the great variety of organisms present, strong sunlight, and the large number of niches. Decomposition of fast-growing, salt-tolerant plants provides the raw material for the large and complex food webs and rapid nutrient turnover characteristic of these communities. The standing biomass (mass of living matter per unit area or volume) in a typical estuary is among the highest per unit of surface area of any marine community.

Estuarine organisms show unique adaptations to their rich and variable environment. Some estuarine plants trap fine silt particles at their roots, thus countering the erosive action of current flow. Small plants are often filamentous, bristling with tiny projections used to anchor themselves to the substrate. Larger plants have extensive root systems to hold themselves in place

a Urban developers often destroy coastal marshes to build marinas or homes, but citizens near this marsh in Orange County, California, have recognized its natural value and have set it aside as a marine preserve.

b Snails (genus *Cerithidea*) in an estuary search the surface mud for food.

Figure 16.10 An estuarine marsh.

 THINKING BEYOND THE FIGURE

What kind of food is abundant enough in the mud to support such large estuarine snail populations?

and to colonize new areas. Most of the resident animals burrow into the muck, scurry rapidly across the surface, or hide in the vegetation. Clams and snails work their way through the substrate, obtaining food and shelter at the same time. Polychaete worms dig for targets of opportunity, and crabs dart for any interesting morsels. Since planktonic larvae would be washed out to sea, most estuarine organisms produce nonplanktonic larvae, lay eggs on firm objects, or carry eggs on their bodies.

Estuaries are sometimes called marine nurseries because so many juvenile organisms are found there. This is especially true for fish. Many pelagic species spend their larval lives in the protective confines of an estuary, taking advantage of the lavish feeding opportunities available. Most of the commercially exploited fish species on the North American Atlantic coast utilize estuaries as juvenile feeding grounds. The human pressures of development and pollution are thus doubly stressful in estuaries, affecting both permanent residents and the sensitive larval stages of open-water animals.

You may also recall that estuaries are not permanent features—they are very sensitive to changes in sea level. Most East Coast estuaries probably formed during the rise in sea level of the last 3,000 to 10,000 years. Estuaries are probably more common today than they were at the height of the last glaciation, when sea level was lower.

CONCEPT CHECK

Before going on to the next section, check your understanding of some of the important ideas presented so far:

With conditions so violent, how can the rocky intertidal community be so populous and diverse?

Wave-swept beaches are subjected to the same conditions as the intertidal community but support vastly fewer organisms. Why is that?

What might be some consequences of widespread commercial development of estuaries?

16.5 Examples of Shallow Benthic and Open-Ocean Marine Communities

Seaweed Communities Shelter Organisms

The shelter and high productivity of a kelp forest can help provide a near-ideal environment for animals. When light and nutrient conditions are optimal, large algae can make carbohydrate molecules so rapidly and in such quantity that sugars leak from their tissues like tea from a tea bag. Resident animals like sea

Dan Sullivan/Alamy

Figure 16.11 Sea urchins in a kelp bed. The urchins can absorb carbohydrates that leak from the algae or gnaw the stipes and holdfasts with their teeth. Too many urchins can destroy a kelp forest by releasing the kelp from its holdfasts.

urchins (**Figure 16.11**) are able to grow rapidly by collecting these molecules on their surfaces and transporting them directly into their bodies. As the algae weaken with age and productivity declines, the urchins' sharp teeth can gnaw at the thalluses. Other animals graze on the blades, nestle within the holdfasts, and consume kelp flakes and debris. Sea otters may eventually move in to eat the urchins. As with all other communities, the kelp forest changes as its inhabitants come and go and as time passes.

Coral Reef Are Earth's Most Densely Populated and Diverse Communities

Tropical coral reefs typically form in areas of high wave energy; indeed, reef organisms prefer to build into high-energy environments in an attempt to be first to obtain dissolved and suspended material in the water. In most reefs, there is an approximate balance between construction and destruction. The reef consists of actively growing coral colonies and fragments of material of different sizes, from coral boulders worn down to fine sand.

Corals are by no means the only participants in reef life; however, they may account for only about half of the biomass in these areas. Other reef residents include calcareous algae, whose secretions help "cement" the reef together, along with a bewildering array of encrusting, burrowing, producing, and consuming creatures ranging upward in size from the microscopic. Some tunnel into the coral or shatter it in search of food, contributing to the erosion of the reef. Fierce competition exists among reef organisms for food, living space, mates, and protection from predators. The bright colors, protective camouflage, spines, and various toxins and venoms common to tropical organisms are probably related to the intense struggle for existence that goes on in these beautiful but deceptively calm looking places. A colorful reef scene is depicted in **Figure 16.12**. More than 1 million species are thought to inhabit the ocean's coral reef ecosystems!

Figure 16.12 A coral reef habitat in the Red Sea.

Planktonic Communities Are Common throughout the Ocean

Entire communities of planktonic organisms drift throughout every region of the world ocean, making them the most common marine communities on Earth. The diversity of planktonic organisms is surprising. There are giant, drifting jellyfish with ten-

Georgie Holland/Age Fotostock

Rebecca Hale/National Geographic Creative

DR. ENRIC SALA shares his thoughts upon completion of an expedition to the Southern Line Islands, a pristine area of coral reefs in the South Pacific Ocean.

All good things come to an end. After an extraordinary six weeks studying the southern Line Islands, we're back home. It's time to reflect on what we learned, and what are the implications.

Our expedition was strenuous but it yielded amazing insights, and your support—right up to the hundreds of wonderful welcome back notes you sent our way—kept us going through it all.

We set out on this journey to discover and to share with you what coral reefs were like before the impact of humans. What we found—abundant top predators and spectacular coral formations—exceeded our expectations. Moreover, all the scientific data confirm that humans are the most important factor in determining the health of coral reefs.

What kills reef ecosystems is not natural events or oceanography, but a combination of the local impact of human activities such as fishing and pollution with the global impact of human-induced climate change. We also learned that reefs need all of their parts, including sharks and other top predators, to be functional and resilient—that is, so that they can recover from warming events and overfishing.

How can we ensure that all parts of the ecosystem are present and functioning? Simply by taking out less and throwing in less. Measures that have proven effective include marine reserves that allow for the recovery of marine life within and beyond their boundaries.

Although the expedition is over and we're back on terra firma now, Ocean Now is just beginning. It won't be long before we set out again in pursuit of our mission: Documenting the seas' last pristine places, increasing awareness, and fighting to preserve as much of the ocean as possible. As we move ahead to the next step, we're counting on your continued engagement and support.

http://ocean.nationalgeographic.com/ocean/explore/pristine-seas-southern-line-blog-archives/
A pristine reef in the central South Pacific.
Background photo: Frank Burek/Corbis

tacles 8 meters (25 feet) long; small but voracious arrowworms; molluscs with slowly beating flaps that resemble butterfly wings; crustaceans that look like microscopic shrimp; miniature, jet-propelled animals that live in jelly-like houses and filter food from water; and shimmering, crystal-shelled algae. There is also a recently discovered hidden component to the plankton—important organisms smaller than a wavelength of light! While the majority of planktonic organisms are often found closer to the surface, many are also found at depth. The only feature common to all plankton is their inability to move consistently against the currents. (Many can and do move vertically through the water column.) The organisms within the white ovals of **Figure 16.13** are plankton, shown here in the context of the larger open-ocean community (see Figure 14.4 for a review of major categories of plankton).

Phytoplankton—the autotrophic plankton responsible for most of the ocean's production—were discussed in Chapter 14. These organisms, while a significant component of plank-

ters, are not the only principle members of this community. Heterotrophic plankton—the planktonic organisms that eat the primary producers—are collectively called **zooplankton** (*zoion*, "animal"). Zooplankters are the most numerous primary consumers of the ocean. They graze on larger cyanobacteria, diatoms, dinoflagellates, and other phytoplankton at the bottom of the trophic pyramid the way cows graze on grass. The mass of zooplankton is typically about 10% that of phytoplankton, which is reasonable because of the harvesting relationship that exists between them.

The variety of zooplankton is surprising; nearly every major animal group is represented. Each is an expert at meticulously concentrating food from the water. The most abundant zoo-plankters are the vanishingly small microflagellates and micro-ciliates of the microbial loop. Of the larger consumers, about 70% of individuals are tiny shrimplike animals called **copepods** **(Figure 16.14)**. Copepods are crustaceans, a group that also includes crabs, lobsters, and shrimp. Their typical size is about half

a A stylized representation of some of the plankton and nekton in the pelagic zone of the subtropical Atlantic Ocean. Note the relative magnification of organisms in the plankton community (in the white ovals).

Figure 16.13 Plankton and nekton.

1 dolphins, *Delphinus*
2 tropic birds, *Phaethon*
3 paper nautilus, *Argonauta*
4 anchovies, *Engraulis*
5 mackerel, *Pneumatophorus*, and sardines, *Sardinops*
6 squid, *Onykia*
7 *Sargassum*
8 sargassum fish (*Histro*)
9 pilot fish (*Naucrates*)
10 white-tipped shark, *Carcharhinus*
11 pompano, *Palometa*
12 ocean sunfish, *Mola mola*
13 squids, *Loligo*
14 rabbitfish, *Chimaera*
15 eel larva, *Leptocephalus*
16 deep sea fish
17 deep sea angler, *Melanocetus*
18 lantern fish, *Diaphus*
19 hatchetfish, *Polyipnus*
20 "widemouth," *Malacosteus*
21 euphausid shrimp, *Nematoscelis*
22 arrowworm, *Sagitta*
23 amphipod, *Hyperoche*
24 sole larva, *Solea*
25 sunfish larva, *Mola mola*

26 mullet larva, *Mullus*
27 sea butterfly, *Clione*
28 copepods, *Calanus*
29 assorted fish eggs
30 stomatopod larva
31 hydromedusa, *Hybocodon*
32 hydromedusa, *Bougainvilli*
33 salp (pelagic tunicate), *Doliolum*
34 brittle star larva
35 copepod, *Calocalaus*
36 cladoceran, *Podon*
37 foraminifer, *Hastigerina*
38 luminescent dinoflagellates, *Noctiluca*
39 dinoflagellates, *Ceratium*
40 diatom, *Coscinodiscus*
41 diatoms, *Chaetoceras*
42 diatoms, *Ceraulutus*
43 diatom, *Fragilaria*
44 diatom, *Melosira*
45 dinoflagellate, *Dinophysis*
46 diatoms, *Biddulphia regia*
47 diatoms, *B. arctica*
48 dinoflagellate, *Lingulodinium*
49 diatom, *Thalassiosira*
50 diatom, *Eucampia*
51 diatom, *B. vesiculosa*

b Key. This drawing shows organisms to be much more crowded than they are in real life.

a millimeter (0.02 inch). Not all zooplankton are small. Many range from 1 to 2 centimeters (½ to 1 inch) in size. The largest drifters are giant jellyfish of genus *Cyanea*; their bells may be more than 3.5 meters (12 feet) in diameter! We have a special term for plankton larger than about 1 centimeter (½ inch) across: **macroplankton**.

Most zooplankton spend their whole lives in the plankton community, so we call them **holoplankton**. But some planktonic animals are the juvenile stages of crabs, barnacles, clams, sea stars, and other organisms that will later adopt a benthic or nektonic lifestyle. These temporary visitors are **meroplankton (Figure 16.15)**. Most animal groups are represented in the meroplankton; even the powerful tuna serves a brief planktonic apprenticeship. These useful categories can be applied to phytoplankton as well as zooplankton. Holoplanktonic organisms are by far the most numerous forms of both phytoplankton and zooplankton.

One of the ocean's most important zooplankters is the pelagic arthropod known as **krill** (genus *Euphausia*; **Figure 16.16**), the keystone of the Antarctic ecosystem. This thumb-sized, shrimplike crustacean mostly grazes on the abundant diatoms of the southern polar ocean. In turn, krill are eaten in tremendous numbers by seabirds, squids, fishes, and whales. Some 500 million to 750 million metric tons (550 million to 825 million tons) of krill inhabit the Antarctic Ocean, with the greatest concentrations in the productive upwelling currents of the Weddell Sea. Krill travel in great schools that can extend over several square miles and collectively exceed the biomass of Earth's entire human population! They behave more like schooling fish than planktonic crustaceans.

Though their primary swimming mode is horizontal, not vertical, krill (and some other relatively large plankters like jellies) migrate up and down through the water column about 100 meters (330 feet) every day. Recent research has shown that this daily vertical migration mixes nutrient-rich deeper water with nutrient-poor surface water. This "calm turbulence" is a previously unrecognized way for nutrients to be delivered to phytoplankton in the sunlit surface layer.

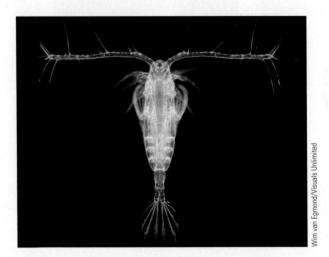

Figure 16.14 A zooplanktonic copepod. Copepods of the genus *Calanus*—named after a mythical East Indian wanderer—probably are the most abundant and widely distributed animal in the world. The species shown here reaches a maximum size of about 0.5 millimeter (about 0.02 inch).

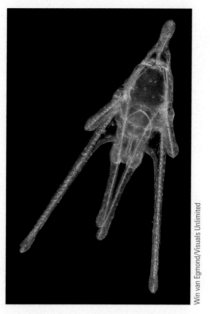

Figure 16.15 This larval sea urchin is a part-time member of the plankton community, so we call it *meroplanktonic*. As it matures, it will leave the plankton and settle to the seabed.

Win van Egmond/Visuals Unlimited

THINKING BEYOND THE FIGURE

What is the advantage of having a planktonic larval stage for benthic animals such as the sea urchin?

Hiroya Minakuchi/Minden Pictures/Getty Images

Figure 16.16 Krill *(Euphausia superba)*. These shrimplike crustaceans, shown here about twice actual size, occur throughout the world ocean. They are particularly numerous in Antarctic seas, where they are the main prey of the largest whales.

The great diversity of members of the plankton community is perhaps best illustrated by the informally named "jellies." Common to all oceans, these diaphanous animals come from several taxonomic categories, including jellyfishes, pteropods,

Wim van Egmond/Visuals Unlimited/Corbis

Figure 16.17 A foraminiferan (genus *Globigerina*) rests on the head of a match. The word means "bearers of windows." Light streams into the organism through thin parts of the shell.

salps, and ctenophores (see Chapter 15). They range in size from microscopic to immense and employ a wonderful range of adaptation to reduce weight, retard sinking, and snare prey.

Small but important, planktonic **foraminifera (Figure 16.17)** are related to amoebas. Like amoebas, they extend long protoplasmic filaments to snare food. Most foraminifera have calcium carbonate shells. As we saw in Chapter 5, extensive white deposits of calcareous ooze have been built on the seabed from their skeletons. As was the case with some phytoplankton sediments, some of these layers have been uplifted and can be found on land.

It is interesting to note that the largest marine animals, such as whale sharks (fish) and baleen whales (mammals), do not expend their energy tracking down and attacking big animals. Instead, these largest of all feeders concentrate zooplankton from the water and consume it in vast quantity. The zooplankton they eat are not usually the primary consumers but the somewhat larger secondary consumers, usually crustaceans such as krill, which have themselves fed on the microscopic primary consumers. In this way whales and other large filter feeders can harvest energy closer to the source, gaining the advantage of efficiency and quantity.

The Open-Ocean Community Is Concentrated at the Surface

About 83% of the ocean's total biomass is concentrated in its uppermost 200 meters (660 feet). Here we find most of the fishes and plankton. Of the ocean's biomass, only 0.8% is found below 3,000 meters (10,000 feet). Nearly all deep-ocean habitats are sparsely populated, but the few species of animals that have adapted to this impoverished place range through virtually all oceanic latitudes. There are no photosynthetic autotrophs in the deep ocean because there is no light. With the exception of rift communities (about which more in a moment), consumers at great depths must depend on the productivity of the water column above.

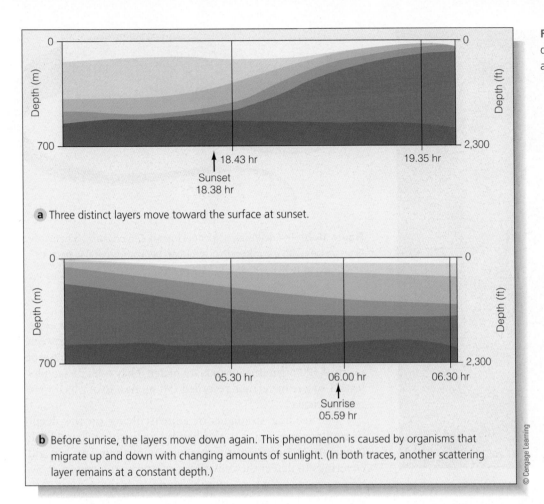

Figure 16.18 The movements of deep scattering layers, as recorded by an echo sounder.

a Three distinct layers move toward the surface at sunset.

b Before sunrise, the layers move down again. This phenomenon is caused by organisms that migrate up and down with changing amounts of sunlight. (In both traces, another scattering layer remains at a constant depth.)

© Cengage Learning

A peculiar pelagic community lives at the uppermost limits of the permanent darkness. Named after its ability to reflect sound pulses and appear to echo sounders as a false bottom, the **deep scattering layer (DSL)** is a relatively dense aggregate of fishes, squid, and other animals that usually migrate up and down in synchrony with daylight **(Figure 16.18)**. Deep scattering layers—there is often more than one—are found in all ocean areas except the Arctic, and they are best developed in regions with high surface productivity. The DSL is most pronounced during daylight hours, when members of the community congregate at the lowest limit of light penetration. At nightfall, many of the organisms migrate to the surface to feed on plankton. Most residents of the deep scattering layer have large, sensitive eyes, which permit them to feed by detecting the faint shadows of prey above. Some members of the community have built-in luminescent organs that cast dim blue light downward; this light masks their own shadows, so they have less chance of being detected and eaten (see **Figure 16.19**).

Between the deep scattering layer and the bottom, in the bathypelagic zone, the ocean is nearly devoid of life. Usually little food is available in this zone. Tiny crumbs of organic material are often broken down by microbes before reaching mid-depths, and the bodies of large organisms continue their fall to the seafloor. But recent research has shown that the deep-water environment is "patchy"—nutrient-rich zones caused by sinking remnants of phytoplankton blooms or the falling excretions of large fish populations can temporarily enrich a water parcel to the benefit of deep residents.

The few animals in this vast middle volume of ocean below the DSL are among the Earth's most bizarre. Gulper eels **(Figure 16.20)** have extendable jaws and a stomach capable of consuming prey larger than the eels themselves, an adaptation of great importance when one considers that a gulper eel may not encounter a feeding opportunity more often than once or twice a year! Bioluminescence, the biological production of light, is important here in both feeding and mate attraction. Some deep-swimming organisms attract their infrequent meals with a luminous lure (you met one of these in Figure 15.25). These animals also use patterns of glowing spots or lines to identify themselves to members of the same species, a

> **E**ntire communities of planktonic organisms drift throughout every region of the world ocean, making them the most common marine communities on Earth.

Figure 16.19 A bioluminescent mesopelagic lantern fish. Large light-producing organs on the body mask the fish's shadow and may identify it to potential mates. This fish is 8 centimeters (3.5 inches) long.

necessary first step in mating. Some use flashes of light to dazzle or frighten potential predators.

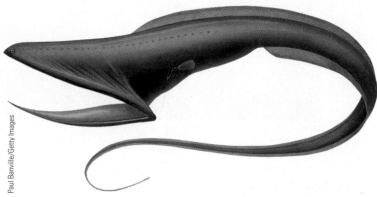

Figure 16.20 The deep-sea gulper eel (genus *Eurypharynx*), a bathypelagic species with a worldwide distribution beneath tropical waters. Its length is about 60 centimeters (24 inches).

CONCEPT CHECK

Before going on to the next section, check your understanding of some of the important ideas presented so far:

How are zooplankton different from phytoplankton? Heterotroph from autotroph?

How are holoplankton different from meroplankton?

What is the deep scattering layer?

16.6 Examples of Deep-Sea Marine Communities

The Deep-Sea Floor Is Earth's Most Uniform Community

Most of the deep-ocean floor is an area of endless sameness. It is eternally dark, almost always very cold, slightly hypersaline (to 36‰), and highly pressurized. Scientists once thought that such

rigors would limit the extent of communities there. Not so. In the 1980s, researchers investigating bottoms at depths between 1,500 and 2,500 meters (5,000 to 8,000 feet) found an average of nearly 4,500 organisms per square meter. They took 21 samples (each 1 square meter) and recorded 798 species, 46 of which were new to science!

The feeding strategies of animals living on the deep-ocean floor are unique to this harsh environment. Tripod fish **(Figure 16.21c)** use sensitive extensions of their fins and gill coverings to detect the movement of prey many meters away. Some organisms whose mouths blend with the natural contours of the ooze act as living caves into which small creatures crawl for protection. The predator need not even swallow to get the prey into its gut—back-pointing spines direct the victim along a one-way path to the stomach! Other species are capable of smelling large, sunken dead animals for many kilometers downcurrent and then spending weeks or months slowly following the scent to its source. The metabolic rate of organisms in cold water tends to be low, so most deep animals require relatively little food, move slowly, and live very long lives. Some may feed less than once in a year and may live to be hundreds of years old. Common deep benthic representatives are seen in **Figure 16.21**.

The organisms in deep pelagic and benthic communities share some curious adaptations. Gigantism is a common characteristic. Individuals of representative families tend to be much larger in deep water than related individuals in the shallow ocean. Fragility is also common in the depths. Not only are heavy support structures unnecessary in the calm deep environment, but the relatively low water pH and high pressure discourage the deposition of calcium—and thus skeletal development. Some animals have slender legs or stalks to raise themselves above the sediment, and some come apart like warm gelatin at the slightest touch. Except for its influence on enzyme activity, hydrostatic pressure is not a problem for these animals. They live in balance with the great pressure; their internal pressure is precisely the same as that outside their bodies.

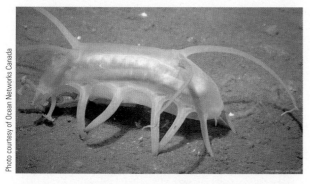

Photo courtesy of Ocean Networks Canada

a An abyssal holothuran *(Scotoplanes),* a relative of the sea star, cruises the deep seabed for nutritious bits. (They're informally called "sea pigs.")

Alan Jamieson, University of Aberdeen

b Deep-sea gigantism in amphipods. Most amphipods are the size of large ants. These, recovered from the Kermadec Trench, measured 28 centimeters (11 inches).

Bahamas Deep-Sea Coral Expedition Science Party, NOAA-OER

c A blind tripod fish. The very long, curved projections on the fish's fins and gills are thought to aid in sensing the distant vibrations of prospective prey.

Image Courtesy of IFE, URI-JAO, Lost City science party, and NOAA

d A xenophyophore, the largest single-celled animal known, rests on the seabed absorbing nutrients.

Figure 16.21 Some representative deep benthic animals.

 THINKING BEYOND THE FIGURE

What benefits, if any, do abyssal benthic organisms have over their shallow-dwelling counterparts?

Hydrothermal Vents and Cold Seeps Support Diverse Communities

Other organisms that can withstand extremes of temperature and pressure have also been found. The oceanographic world was excited in 1976 when scientists from Scripps Institution of Oceanography discovered an entirely new type of marine community over 3,000 meters (10,000 feet) below the surface. Using a towed camera platform, they were searching the seafloor along a spreading center 350 kilometers (220 miles) north and east of the Galápagos Islands. They found jets of water superheated to 350°C (660°F) blasting from rift vents in the young oceanic ridge. Seawater had percolated into fractures in the active crust and had been warmed by heat from nearby magma chambers at a depth of 1 to 3 kilometers (0.6 to 1.9 miles). Convection moved this water back to the seafloor, where it emerged as hot springs. The heated water dissolved minerals from the surrounding basaltic rock. As this water emerged and cooled, some inorganic sulfides precipitated, turning the water black. The term "black smoker" was quickly applied to these active vents **(Figure 16.22)**.

Clustered around the vents were dense aggregations of large, previously unknown animals. Bottomwater in the area was laden with hydrogen sulfide (H_2S), carbon dioxide, and oxygen, on which specialized archaea and bacteria were found to live. These chemosynthesizers form the base of a food chain that extends to the animals. Large crabs, clams, sea anemones, shrimp, and unusual worms contained in long parchment-like tubes were found in this warm oasis. These were termed hydrothermal vent communities.

Some of the so-called tube worms measured 3 to 4 meters (10 to 13 feet) in length and were the diameter of a human arm. They have been identified as pogonophorans, members of a small phylum of invertebrates also found in fairly shallow water. Three species of the new genus *Riftia* **(Figure 16.23a)** have been found so far. The tubes of these pogonophorans are flexible and capable of housing the length of the animal when it retracts. The animals extend tufts of tentacles from the openings of their

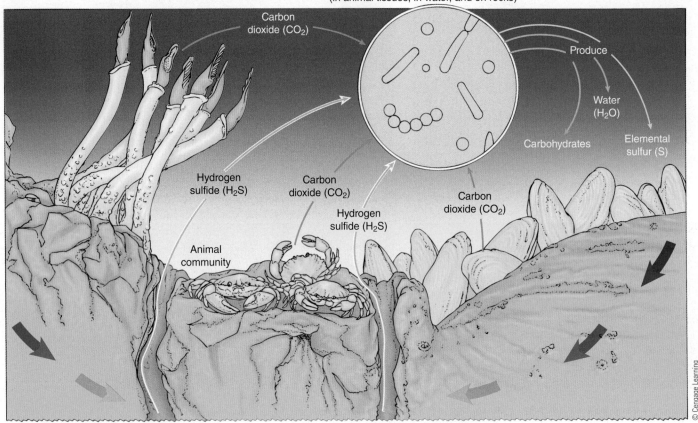

Chemosynthetic bacteria
(in animal tissues, in water, and on rocks)

Carbon
dioxide (CO₂)

Produce

Water
(H₂O)

Carbohydrates

Elemental
sulfur (S)

Hydrogen
sulfide (H₂S)

Carbon
dioxide (CO₂)

Carbon
dioxide (CO₂)

Hydrogen
sulfide (H₂S)

Animal
community

© Cengage Learning

Figure 16.22 The path of water associated with a hydrothermal vent. Seawater enters the fractured seabed near an active spreading center and percolates downward, where it comes into contact with rocks heated by a nearby magma chamber. The warmed water expands and rises in a convection current. As it rises, the hot water dissolves minerals from the surrounding fresh basalt. When the water shoots from a weak spot in the seabed, some of these minerals condense to form a "chimney" up to 20 meters (66 feet) high and 1 meter (3.3 feet) in diameter. As the vented water cools, metal sulfides precipitate out and form a sedimentary layer downcurrent from the vent. Bacteria in the sediment, in the surrounding water, and within specialized organisms make use of the hydrogen sulfide (H₂S) in the water to bind carbon into glucose by chemosynthesis. This chemosynthesis forms the base of the food chains of vent organisms. (This illustration is not to scale.)

tubes, but feeding was something of a puzzle because these animals have no mouth, digestive tract, or anus. The trunks of the worms were found to contain large feeding bodies tightly packed with bacteria similar to those seen in the water and on the bottom near the hydrothermal vents. The worms' tentacles absorb hydrogen sulfide from the water and transport it to the bacteria, which then use the hydrogen sulfide as an energy source to convert carbon dioxide to organic molecules. The ultimate source of the worms' energy (and the energy of most other residents in this community) is this energy-binding process—chemosynthesis—which replaces photosynthesis in the world of darkness.

The clams and shrimp of these vent communities are equally unusual. For example, the large, white clam *Calyptogena* grows among uneven basaltic mounds **(Figure 16.23b)**. Each the size of a shoe, the clams shelter the same kinds of bacteria as *Riftia*. Though the clam retains its filter-feeding structures, it derives nutrition from the specialized bacteria embedded in the

cells of its gill filaments. Small shrimp discovered at the vents in 1985 have been found to possess special organs that may allow them to sense heat from the vents. Such an adaptation would permit them to range away from the vents for food yet return to the warmth and richness of the home community. Hydrothermal vent communities have now been found off Florida and Louisiana, off California and Oregon, in the North Sea, east of Japan, and at least 30 other locations.

Not all vent-type communities are on oceanic ridges, and not all are in areas spouting hot water, however. Cold-seep communities are less dramatic but probably more widespread. These areas are not always associated with the edges of tectonic plates, and about 25 large fields have been discovered so far. At cold seeps, hypersaline water rich in minerals, hydrogen sulfide, and sometimes methane percolates upward from beneath the seabed to emerge in broad fields at near-ambient temperatures. The slow upward seeping of cool, mineral-rich water

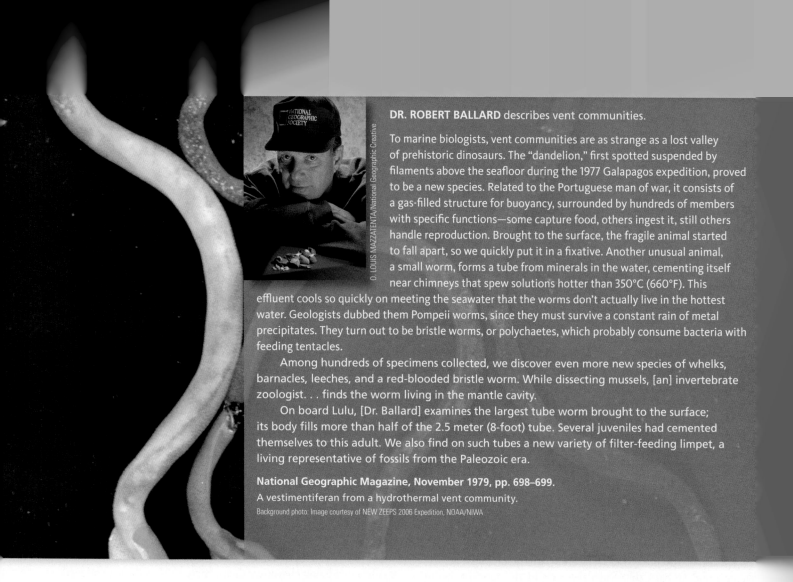

DR. ROBERT BALLARD describes vent communities.

To marine biologists, vent communities are as strange as a lost valley of prehistoric dinosaurs. The "dandelion," first spotted suspended by filaments above the seafloor during the 1977 Galapagos expedition, proved to be a new species. Related to the Portuguese man of war, it consists of a gas-filled structure for buoyancy, surrounded by hundreds of members with specific functions—some capture food, others ingest it, still others handle reproduction. Brought to the surface, the fragile animal started to fall apart, so we quickly put it in a fixative. Another unusual animal, a small worm, forms a tube from minerals in the water, cementing itself near chimneys that spew solutions hotter than 350°C (660°F). This effluent cools so quickly on meeting the seawater that the worms don't actually live in the hottest water. Geologists dubbed them Pompeii worms, since they must survive a constant rain of metal precipitates. They turn out to be bristle worms, or polychaetes, which probably consume bacteria with feeding tentacles.

Among hundreds of specimens collected, we discover even more new species of whelks, barnacles, leeches, and a red-blooded bristle worm. While dissecting mussels, [an] invertebrate zoologist. . . finds the worm living in the mantle cavity.

On board Lulu, [Dr. Ballard] examines the largest tube worm brought to the surface; its body fills more than half of the 2.5 meter (8-foot) tube. Several juveniles had cemented themselves to this adult. We also find on such tubes a new variety of filter-feeding limpet, a living representative of fossils from the Paleozoic era.

National Geographic Magazine, November 1979, pp. 698–699.

A vestimentiferan from a hydrothermal vent community.

Background photo: Image courtesy of NEW ZEEPS 2006 Expedition, NOAA/NIWA

a *Riftia,* large tube worms (pogonophorans) that contain masses of chemosynthetic bacteria in special interior pouches.

Figure 16.23 Some large organisms dependent on hydrothermal vents.

b A vent field dominated by the giant white clam *Calyptogena*. Each clam is about the size of a man's shoe and contains chemosynthetic bacteria within its gill filaments.

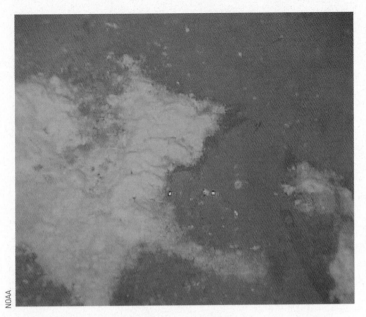

Figure 16.24 A bacterial mat coats the seabed at a cold seep. These chemosynthetic organisms form the base of a limited food chain in the area. The two red dots, 10 centimeters apart, are used to measure the size of objects at a distance from the robot submersible.

Figure 16.25 A whale-fall community off the California coast. These communities may act as "stepping stones" for the specialized organisms that inhabit vent communities.

💬 **THINKING BEYOND THE FIGURE**

If whale carcasses act as stepping stones for hydrothermal vent organisms to get from one place to another, how could unregulated whaling influence vent communities?

encourages the growth of mats of chemosynthetic bacteria that are able to metabolize sulfur-containing compounds or methane (Figure 16.24). The methane, where present, appears to come from the decomposition of organic material in the sediment or underlying sedimentary rocks. The dominant large organisms at cold seeps are bivalve mollusks, pogonophoran worms, and a few species of sponges—but the heroes are the chemosynthetic bacteria and archaea that form the base of the food chains in these deep, mysterious places.

Studies of hydrothermal-vent and cold-seep communities suggest *many* questions. Do such communities occupy the active central rift valleys of a significant percentage of the 65,000 kilometers (41,000 miles) of Earth's oceanic ridges? Did the hot or cold vents—or even seemingly solid rock—serve as the birthplace of life on Earth? Perhaps the deep vent and seep communities will prove to be more important to overall marine productivity than has been previously supposed. Marine biologists are eager to continue their explorations.

Whale-Fall Communities Represent Unique Opportunities

You may wonder how the specialized organisms that inhabit vent and seep communities disperse over the great distances between vent systems. The problem becomes more acute with the recent knowledge that the lifetimes of the vents themselves may be measured in the tens of years at most. How are such unique organisms recruited? How are they dispersed?

The answer may lie in the "stepping stones" provided by the fallen bodies of whales (Figure 16.25). Even though humans have greatly diminished the numbers of living whales, researchers estimate that whale carcasses may be spaced at roughly 25-kilometer (16-mile) intervals across areas like the North Pacific. Studies of fallen whale skeletons have shown the presence of sulfur-oxidizing chemosynthetic bacteria. As sulfide produced by these bacteria diffuses out of the bone, planktonic larvae of vent organisms might sense its presence, settle, grow, and reproduce. With luck, their offspring might drift to another whale fall and repeat the process. After many steps, a new or newly active vent would be reached.

CONCEPT CHECK

Before going on to the next section, check your understanding of some of the important ideas presented so far:

How is primary production accomplished in hydrothermal vent communities?

How might whale-fall communities assist the spread of vent organisms?

16.7 Organisms in Communities Can Live in Symbiosis

Newcomers to biology are often surprised to learn that more than half of the animal species known to science are not free-living. Most are actively involved in close symbiotic relation-

Amos Nachoum/Encyclopedia/Corbis

Figure 16.26 Mutualism. Some species of sea anemones have a symbiotic relationship with anemone fish, in which the fish receives protection from predators and the anemone receives scraps of food from the fish.

ships with at least one other life-form in their community. These relationships are often intricate and sometimes quite strange.

Symbiosis (*sym*, "with"; *bios*, "life") is the biologist's term for the co-occurrence of two species in which the life of one is closely interwoven with the life of the other. The symbiotic bond is often so strong that one organism (the symbiont) is totally dependent on the other (the host). There are three general types of symbioses—mutualism, commensalism, and parasitism.

In **mutualism**, as the name implies, both the symbiont and the host—the larger organism with which the symbiont lives—benefit from the relationship. True mutualism is rare among marine organisms, but a few examples have been observed. One is the relationship between an anemone fish and its sea anemone. In this symbiosis, a small, brightly colored anemone fish nestles within the stinging tentacles of a sea anemone **(Figure 16.26)**. The mechanism that permits the little fish to do this without being stung is not well understood, but biologists believe that the fish gradually desensitizes the stinging cells of the anemone by using mucous secretions from its own skin. In return for the anemone's protection, the fish feeds the anemone scraps of food and may even lure prey within the anemone's reach.

Another example of mutualism is the relationship between certain cnidarians, such as reef-building coral animals, and the specialized dinoflagellates known collectively as *zooxanthellae* that live within their tissues. Both organisms benefit. The autotrophic dinoflagellates receive a safe home, a ready source of carbon dioxide, and nutrients that are otherwise scarce in warm tropical waters. In exchange, the animals have a handy, built-in

Newcomers to biology are often surprised to learn that more than half of the animal species known to science are parasites.

source of carbohydrates, which can provide up to 90% of the coral's nutritional needs. Without their resident dinoflagellates, the reef corals would be unable to deposit their calcium carbonate skeletons, and the rich tropical reef communities we know today would not exist.

In **commensalism**, the symbiont benefits from the association while its host neither benefits nor is harmed. For example, biologists once believed that the association between pilot fish and shark was mutualistic; the pilot fish was thought to guide the shark to a meal and in turn be permitted to dine on the scraps. We now know the pilot fish is only an opportunistic commensal, taking what scraps it can from the shark's meal. (On the other hand, the mutualistic anemone fish/anemone partnership was thought to be commensal before the anemone fish was observed feeding the anemone.)

Some of the most curious commensals are those that enter their partner's habitat and, after a period of growth, cannot escape. Small pea crabs of genus *Fabia*, for example, live inside mussel shells, eating food particles that are brought inside by the normal feeding and respiratory actions of the mussel. After a time, the crab grows too large to fit through the gap between the mussel's shells.

Parasitism is the most highly evolved and by far the most common symbiotic relationship. The parasite lives in (or on) the host for at least part of its life cycle and obtains food at the host's expense. For obvious reasons, parasites do not usually kill their hosts, but they can seriously affect the host organism by reducing its feeding efficiency, depleting its food reserves, reducing its reproductive potential, lowering its resistance to disease, or

Norman Cole

Figure 16.27 Whale barnacles. These arthropods can be buried up to 3 centimeters (1¼ inches) deep in the skin of certain whales. They derive part of their nutrition from the flesh and circulating fluids of the whale. Some individuals are up to 7.6 centimeters (3 inches) across.

otherwise sapping its energy. The host–parasite relationship is finely balanced and extraordinarily delicate. The parasite must in some way be aware of the host's physical condition in order to avoid weakening the host so much that it dies. On the other hand, the parasite must take as much energy from the host as possible in order to ensure its own success.

All major phyla have parasitic marine representatives. However, the most widely distributed and successful parasitic marine animals are the roundworms of the phylum Nematoda. Like nearly all parasites, nematodes have a **species-specific relationship** with a host. A species-specific relationship is an exclusive relationship between two species; parasites can usually parasitize only one species of host. The reason for this interdependency is the delicacy of the biochemical feedback mechanisms informing the parasite that its activity may be overstressing the host. The feedback responses are, by necessity, tailored to specific host–parasite pairs. The parasite will not usually survive if it settles in or on a host for which it is not specifically "programmed."

More than one species of parasite *can* infect a single host, however. A significant percentage of the weight of many fishes may be in nematode worms and other parasites. The parasitic

burden of a normal, apparently healthy sea lion may exceed 2.3 kilograms (5 pounds) and 20 species! Parasites are usually small, but one species of nematode parasite of the fin whale reaches 7 meters (23 feet) in length and is the diameter of a pencil. And, by the way, parasites can have their own parasites!

These three categories (mutualism, commensalism, parasitism) form a continuum in nature. There are few clearly mutualistic relationships that do not suggest at least a touch of commensalism, and few commensalistic relationships that do not hint of parasitism. The whale barnacles pictured in **Figure 16.27** glean most of their food directly from the ocean, but some of their nutrition is derived from the flesh and circulating fluids of their host. It's sometimes hard to tell where one kind of relationship stops and another begins.

CONCEPT CHECK

Before going on to the next section, check your understanding of some of the important ideas presented so far:

What's symbiosis?

What types of symbiosis exist?

Chapter in Perspective

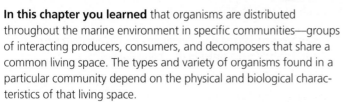

In this chapter you learned that organisms are distributed throughout the marine environment in specific communities—groups of interacting producers, consumers, and decomposers that share a common living space. The types and variety of organisms found in a particular community depend on the physical and biological characteristics of that living space.

Any community is a dynamic place, growing and shrinking and changing its composition as residents respond to environmental fluctuations. The growth and distribution of organisms within communities depend on the often subtle interplay of physical and biological factors. The relative numbers of species and individuals in a community depend in part on whether its environment is relatively easy and free of stressors, or relatively hard and full of potential limiting factors. A few prominent marine communities were compared and contrasted, and some representative adaptations of their residents were discussed. The main principle explored in this chapter was why organisms live where they do.

In the next chapter you'll learn about the range of marine resources, including physical resources such as oil, natural gas, building materials, and chemicals; marine energy; biological resources such as seafood, kelp, and pharmaceuticals; and nonextractive resources like transportation and recreation. World economies are now dependent on oceanic resources, but we find that we cannot exploit those resources without damaging their source.

QUESTIONS FROM STUDENTS

1. **If the rocky, sandy, or muddy intertidal zones represent such a challenging mix of environmental factors, why do so many organisms live there?**

Difficulty in biology is a relative term. It may seem a circular argument, but wherever organisms live, conditions for life at that place are biologically tolerable, food is available, and environmental conditions are not so extreme as to preclude success. Organisms live in abundance where energy is available. Where there is food, or sunlight, or biodegradable compounds, there is life. Natural selection has sorted out the ways that work in this zone from the ways that do not, and the adaptations that work give the organisms living in the intertidal zone's many niches access to a rich harvest of nutrients.

2. **Is the species-specificity rule of parasitism ever broken?**

Yes, sometimes with catastrophic results. As an example, the lung flukes that inhabit the respiratory tract of most sea lions will "abandon ship" if the sea lion is weakened or dying. A sea lion in this condition sometimes comes out of the water onto a beach to rest. If your pet dog should discover the animal there and sniff at its nose, some of the parasites could transfer from sea lion to dog and establish themselves in the dog's lungs. Because the dog is not the species-specific host for the lung parasites, the biochemical machinery that tells the parasite that it is weakening its host is missing. The dog may die a painful death in a few weeks from an uncontrolled infestation of lung flukes. The parasites will, of course, also die. The interaction will have been a failure in all respects.

3. **Could there be any huge undiscovered Godzilla-type sea monsters in the deep ocean?**

Probably not, unless they can extract energy directly from water molecules! The deep pelagic feeding situation is simply not rich enough to support the energy needs of an active population of violent, aggressive, city-eating (metrophagous?) reptiles. Scientists never say never, but classic science fiction films aside, it doesn't look promising.

4. **Thinking again about the richness of the intertidal environment, who first studied the interactions of organisms there?**

This is an interesting story, and one with important implications in the development of the science of ecology.

In 1939, Ed Ricketts, a friend of Nobel-prize–winning novelist John Steinbeck published a landmark book, *Between Pacific Tides.* Unlike previous guides to seashore life, *Between Pacific Tides* was organized by community and not by organism type. "The treatment," he wrote in the preface to the first edition, "is ecological and inductive; that is, the animals are treated according to their most characteristic habitat, and in the order of their commonness, conspicuousness, and interest." The graceful writing and accurate observations quickly made it a classic, a book that has deeply influenced generations of marine scientists. *Between Pacific Tides* is now in its fifth edition.

Ed Ricketts died in an automobile accident in May 1948. Steinbeck wrote a foreword to the second edition, which Ricketts was preparing. The foreword concludes: "There are good things to see in the tide pools and there are exciting and interesting thoughts to be generated from the seeing. Every new eye applied to the peep hole which looks out at the world may fish in some new beauty and some new pattern, and the world of the human mind must be enriched by such fishing."

5. Do zooplankton have a compensation depth? If so, would it be above, below, or at the same level as the compensation depth of most phytoplankters?

The concept of compensation depth is meaningless for zooplankton. Compensation depth applies only to *autotrophs*, organisms capable of both photosynthesis and respiration. Since animals don't make their own food—aren't autotrophic—productivity in them can *never* equal consumption.

6. You mentioned that krill can swim horizontally over considerable distances. Doesn't that disqualify them for inclusion in the plankton?

Japanese researchers, using two ships equipped with side-scan sonar, tracked a large school of swimming krill for 14 days across 278 kilometers (172 miles). This new finding does indeed threaten krill's usual classification as zooplankton (animals unable to move consistently in one direction against waves or current flow). Traditions die hard, though, and we don't anticipate this keystone of the Antarctic ecosystem will lose its status as the premier zooplankter anytime soon.

7. Can the specialized dinoflagellates that live within hermatypic coral polyps exist outside of a coral animal?

In laboratory cultures, yes. They change from the spherical shape seen within the coral to the typical biflagellate form characteristic of motile dinoflagellates. Researchers are uncertain whether all corals are host to the same species of dinoflagellates or whether several species exist. As far as we know, they do not normally live free in the ocean.

TERMS AND CONCEPTS TO REMEMBER

biodiversity	desiccation	krill	random distribution
brackish	ecology	macroplankton	realized niche
carrying capacity	environmental resistance	meroplankton	sessile
climax community	estuary	motile	species-specific relationship
clumped distribution	foraminiferan	mutualism	succession
commensalism	fundamental niche	niche	symbiosis
community	habitat	parasitism	uniform distribution
control	holoplankton	population	wave shock
copepod	intertidal zone	population density	zooplankton
deep scattering layer (DSL)			

STUDY QUESTIONS

Thinking Critically

1. How is a population different from a community? A niche from a habitat?
2. What is a limiting factor? Give a few examples.
3. In what ways can members of the same population compete with one another? How might members of different populations compete? Contrast the results of these kinds of competition.
4. What factors influence the distribution of organisms within a community? How are these distributions described? Why is random distribution so rare?
5. What problems confront the inhabitants of the intertidal zone? How do you explain the richness of the intertidal zone in spite of these rigors?

6. Why must the host–parasite relationship be so finely balanced? What would be the result of an imbalance?

Thinking Analytically

1. Why would larvae dispersing from hydrothermal vent communities need the assistance of whale-fall "stepping stones" to colonize new vents? (Hint: Do you think hydrothermal vents are steady, relatively stable phenomena?)
2. Considering the fact that we have already found more than 20,000 deep-sea marine species and we have only explored a few percent of the deep ocean, how many total species do you think there may be in the deep sea?

GLOBAL GEOSCIENCE WATCH

Visit www.cengagebrain.com to access MindTap, a complete digital course which includes access to Global Geoscience Watch and more. In the GREENR database, go to the "Coral Reef" topic portal. On this page, look for "Case Studies" in the right-hand column and click "VIEW ALL." Select a case study and create an outline of the specific characteristics and challenges that the coral reefs in your case study face. Prepare a brief presentation and be prepared to discuss your topic with the class. What do you believe is the future outlook for the corals discussed in your case study, and what do you think should be done to help protect them?

CENGAGE**brain**.com Visit www.cengagebrain.com to access course materials for this text, including interactive learning tools, videos, and more.

MARINE COMMUNITIES • **481**

17 Marine Resources

KEY CONCEPTS

By most calculations we have used more natural resources since 1955 than in all of recorded human history up to that time. Alaska Stock/Alamy

Petroleum and natural gas are the ocean's most valuable resources. Shell U.K. Limited

Fish provide more than 2.9 billion people with nearly 20% of their average per capita animal protein intake. Brian J. Skerry/National Geographic Creative

The International Law of the Sea nominally governs marine resource allocation. National Geographic Maps

With very few exceptions, our present level of growth and exploitation of marine resources is unsustainable. Brian J. Skerry/National Geographic Creative

---

A commercial fishing boat targets salmon in Prince William Sound. Fishing is an important industry in Alaska, which harvests 5.35 billion pounds of fish annually and generates up to 60,000 jobs and over US$3 billion to the regional economy.

Alaska Stock Images/National Geographic Creative

17.1 Marine Resources Are Subject to the Economic Laws of Supply and Demand

The human population grew by 400% during the 20th century. This growth, coupled with a 4.5-fold increase in economic activity per person, resulted in accelerating exploitation of Earth's resources. *By most calculations, we have used more natural resources since 1955 than in all of recorded human history up to that time.*

Our increasingly anxious efforts at resource extraction and utilization have affected the ocean and atmosphere on a global scale. World economies are now dependent on oceanic materials—nations fight each other for access to them. We are unwilling to abandon or diminish the use of marine resources until we see clear signs of severe environmental damage. *With few exceptions, our present level of growth and exploitation of marine resources is unsustainable.*

Resources are allocated by systems of economic checks and balances. An economy is a system of production, distribution, and consumption of goods and services that satisfies people's wants or needs. In marine economics, individuals, businesses, and governments make economic decisions about what ocean-related goods and services to produce, how to produce them, how much to produce, and how to distribute and consume them.

In a free-market economic system, buying and selling are based on pure competition, and no seller or buyer can control or manipulate the market. Economic decisions are governed solely by supply, demand, and price; sellers and buyers have full access to information about the beneficial and harmful effects of goods and services to make informed decisions. Ideally, prices reflect all harmful costs of goods and services to the environment.

But ours is not a pure free-market economic system. Often, prices do not reflect all harmful costs of goods and services to the environment. For example, seafloor habitat degradation and the catch of nontarget species that can result from some fishing methods are not factored into the cost we pay for the fish caught. Nor is the continued catch of marine organisms from lost or discarded nets like the one illustrated in **Figure 17.1**. These unincorporated costs are known as *externalities*. In addition, consumers rarely have full access to information about the beneficial and harmful effects of goods and services to make informed decisions. If consumers had to pay higher prices to account for these externalities or had full access to information about the impacts of particular fishing methods, the fishing industry would be less likely to harvest in unsustainable or wasteful ways.

As you read this chapter, remember the nature of supply and demand of economic markets and think about the long-term implications of our growing dependence on oceanic resources. What comes next? Although the answer is uncertain, as you will see, this story will probably not have a happy ending.

Figure 17.1 This abandoned trawl net, often referred to as a *ghost net*, continues to catch fish and disturb the seafloor along a productive seamount in Mexico.

Brian J. Skerry/National Geographic Creative

We will discuss four groups of marine resources in this chapter:

- **Physical resources** result from the deposition, precipitation, or accumulation of useful substances in the ocean or seabed. Most physical resources are mineral deposits, but petroleum and natural gas, mostly remnants of once-living organisms, are included in this category. Freshwater obtained from the ocean is also a physical resource.
- **Marine energy resources** result from the extraction of energy directly from the heat or motion of ocean water.
- **Biological resources** are living animals and plants collected for human use and animal feed.
- **Nonextractive resources** are uses of the ocean in place. Transportation of people and commodities by sea, recreation, and waste disposal are examples.

Marine resources can be classified as either renewable or nonrenewable:

- **Renewable resources** are naturally replaced by the growth of marine organisms or by natural physical processes.
- **Nonrenewable resources** such as oil, gas, and solid mineral deposits are present in the ocean in fixed amounts and cannot be replenished over time spans as short as human lifetimes.

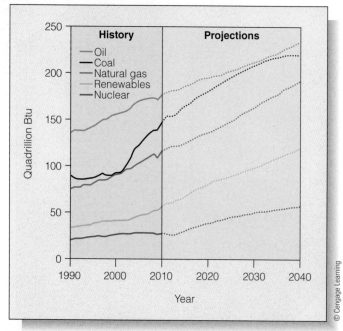

Figure 17.2 World energy consumption from 1990 to 2040 (as projected by the U.S. Department of Energy). Renewable energy includes hydroelectric power and electricity from wind and geothermal sources.

CONCEPT CHECK

Before going on to the next section, check your understanding of some of the important ideas presented so far:

Human population grew explosively in the past century. Is the number of humans itself the main driver of resource demand?

What are the differences between physical and biological resources?

What are the differences between renewable and nonrenewable resources?

17.2 Physical Resources

Physical resources from the ocean include hydrocarbon deposits (petroleum, natural gas, and methane hydrate); mineral deposits (sand and gravel, magnesium and its compounds, salts of various kinds, manganese nodules, phosphorites, and metallic sulfides); and freshwater.

Petroleum and Natural Gas Are the Ocean's Most Valuable Resources

Global demand for oil grows each year **(Figure 17.2)**. The world's accelerating thirst for oil is currently running at about 1,000 gallons per second (over 32 billion barrels a year[1]), a demand enhanced by the increasingly robust Chinese, Indian, and Brazilian economies and by the lack of a coherent energy policy in the United States. The United States alone consumes over 20% of the global oil supply—nearly 18.5 million barrels each day in 2012, approximately equal to the next three nations combined. The United States imports about 57% of its petroleum,

mostly from Canada. Although rapidly rising prices have weakened demand slightly, world oil consumption is forecasted to grow from 87 million barrels per day in 2010 to 115 million barrels per day in 2040. By that same year, China is projected to consume more than twice as much energy as the United States, much of it coming from petroleum and natural gas.[2]

Proven oil reserves stand at around 1,500 billion barrels, and estimates of undiscovered reserves vary from 198 to 1,208 billion barrels (565 billion barrels mean estimate). There is a growing deficit between consumption and the discovery of new reserves. Huge, easily exploitable conventional oil fields are almost certainly a thing of the past. However, accurate estimates are becoming more difficult to forecast due to shifting technologies like hydraulic fracturing and development from unconventional petroleum sources like oil sands.

Offshore petroleum and natural gas generated nearly US$1 trillion in worldwide revenues in 2012. Here the ocean makes a significant contribution to present world needs: almost 33% of the crude oil and 30% of the natural gas produced in 2009 came from the seabed. About a third of known world reserves of oil and natural gas lie along continental margins. Major U.S. marine reserves are located on the continental shelf of southern California, off the Texas and Louisiana Gulf Coast, and along the North Slope of Alaska. The deep-sea floors probably contain little or no oil or natural gas.

---

[1]One petroleum barrel = 159 liters = 42 U.S. gallons.

[2]Most oil is consumed in automobiles. In 2009, there were 1,201 automobiles for every 1,000 eligible drivers in the United States. In 2010, we consumed 1 billion gallons of gasoline every 30 hours and travelled more than 6.5 billion miles each day. In China, there were 9 automobiles per 1,000 people in 2007, but in 2009, China surpassed the United States as the world's largest automobile market.

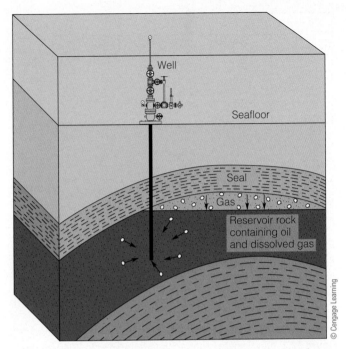

Figure 17.3 Oil and gas are not found in vast hollow reservoirs but within pore spaces in rock. The pressure of natural gas and compression by the weight of overlying strata drive oil through the porous rock and toward the drill pipe.

💬 **THINKING BEYOND THE FIGURE**
What might be done to "rejuvenate" oil and gas fields that have been pumped nearly dry?

Figure 17.4 Platform *Brent Charlie* braces against a North Atlantic storm. About a third of crude oil comes from the seabed.

Oil is a complex chemical soup containing perhaps a thousand compounds, mostly hydrocarbons. Petroleum is almost always associated with marine sediments, suggesting that the organic substances from which it was formed were once marine. Over millions of years, large quantities of planktonic organisms and masses of bacteria accumulated in quiet basins where the supply of oxygen was low and there were few bottom scavengers. The action of anaerobic bacteria converted the original tissues into simpler, relatively insoluble organic compounds that were buried— possibly first by turbidity currents, then later by the continuous fall of sediments from the ocean above. Further conversion of the hydrocarbons by high temperatures and pressures must have taken place at considerable depth, probably 2 kilometers (1.2 miles) or more beneath the surface of the ocean floor. Slow cooking under this thick sedimentary blanket for millions of years completed the chemical changes that produce oil.[3] It should be noted that it

> **I**t took about 3 million years to make one year's worth of oil at the present rate of consumption.

took about 3 million years to make one year's worth of oil at the present rate of consumption!

Oil is less dense than the surrounding sediments, so it can migrate toward the surface from its source rock through porous overlying formations. It collects in the pore spaces of reservoir rocks when an impermeable overlying layer prevents further upward migration of the oil **(Figure 17.3)**.

Drilling for oil offshore is far more costly than drilling on land because special drilling equipment and transport systems are required. Most marine oil deposits are tapped from offshore platforms resting in water less than 100 meters (330 feet) deep. As oil demand (and therefore price) continues to rise, however, deeper deposits farther offshore will be exploited from larger platforms **(Figure 17.4)**. Currently, the tallest and heaviest platform is *Troll-A*, in position since 1996. The base of *Troll-A* stands on the seafloor 303 meters (994 feet) below the surface of the North Sea.

Hydraulic fracturing or "fracking" has become an increasingly common practice in many parts of the world. Fracking uses a pressurized injection of water, sand, and chemicals into ex-

---

[3]How much marine life was needed to make a gallon of gasoline? Make a guess and then read student question #1 on page 508.

Where to Find New Oil Reserves

Oil is not found everywhere. Relatively few locations exhibit the specific characteristics necessary to form and accumulate oil in economically recoverable reserves.

When attempting to find new oil reserves, petroleum geologists and engineers must initially survey potential sites and map areas where oil could be found based on their geologic history. Since oil is primarily formed from ancient plankton and marine bacteria, the most likely place to find oil reserves is in plankton-rich, sedimentary *source rocks*. These rocks are often found beneath the seafloor of continental shelves but can also be found onshore in areas where historically productive seas once existed. This is exemplified by the great petroleum reserves of the Middle East; even though much of the oil is currently extracted from onshore environments in Saudi Arabia, Iraq, Iran, and Kuwait, the oil itself originated from millions of years of plankton deposition on the seafloor of the ancient Tethys Sea.

Potential sites must also contain different rock types that allow the oil to collect in areas where it is practical to recover. *Reservoir rocks* like sandstone have a relatively high amount of pore space where the oil can migrate and accumulate. Nonporous *cap rocks* keep the oil from rising so that it is trapped within the reservoir rocks. *Structural traps* like folds or faults help oil and natural gas to concentrate in areas where it can be extracted.

Technology can then help to further isolate areas that may be worth drilling. Oil exploration ships commonly tow seismic air guns that emit blasts of compressed air at regular intervals that penetrate the seafloor. These waves are reflected by subsurface rock layers and detected by an array of sensors towed behind the vessels **(Figure A)**. This allows geophysicists to look for the signature combination

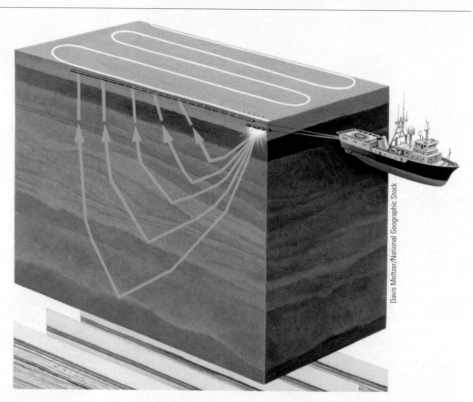

Davis Meltzer/National Geographic Stock

Figure A Oil exploration vessels often follow a predetermined grid pattern to search for potential new oil reserves. The patterns of reflected waves help engineers to identify seafloor sediment layers where oil is likely to be found.

of layered sediments, depth, and reservoir structures that are commonly exhibited by petroleum reserves.

Gravity meters and magnetometers can also be used to indicate changes in sediment density or magnetic field caused by porous reservoir rocks, dense cap rocks, and even flowing oil. This information is collectively used to create three-dimensional maps that can be analyzed to identify areas that are likely to contain significant petroleum reserves. Satellite imaging and

even chemical "sniffers" can help isolate the location of these potential reserves.

If the collective evidence suggests the presence of an economically recoverable amount of oil, an exploratory well is drilled. This well allows engineers to estimate the size and characteristics of the reserve by examining the materials brought to the surface. Based on this direct evidence, it can then be determined whether the site is productive enough to support a commercial production well.

---

isting oil wells to fracture the rock and enhance the productivity of the well by increasing the amount of oil and natural gas that can be recovered. The recent expansion of hydraulic fracturing into California offshore wells has led to concern about coastal water pollution and a potential increase in seismic events.

Large Methane Hydrate Deposits Exist in Shallow Sediments

The largest known reservoir of hydrocarbons on Earth is not coal or oil but methane-laced ice crystals—methane hydrate—in the sediments of some continental slopes. Methane is formed

Figure 17.5 A sample of methane hydrate burns after ignition at normal surface pressure.

a Normally associated with banks and beaches, oolites are concentric sand-sized concretions of mineral matter, usually calcium carbonate. Water in the north Atlantic gyre is forced to shallow depths where the calcium carbonate comes out of solution and forms smooth grains.

b Large oolite deposits are mined in the Bahamas, an island group east of Florida. This nearly pure form of calcium carbonate is used in the production of Portland cement, animal feed, and soil amendments.

Figure 17.6 Oolite sands.

from the decomposition of organic matter by microorganisms living in the seafloor sediments. Due to high pressures and cold temperatures, the methane is trapped within a crystal lattice of interlocking water molecules. Sediment rich in methane hydrate looks like green Play-Doh. Methane hydrates are often found in marine environments in waters at least 250 meters (833 feet) in depth where they are stable and long-lived. They are not found in shallower waters along the continental shelf since the lower pressure is not sufficient to keep them stable, and they are rare in the deep ocean due to the lack of organic material that reaches the deep ocean floor. When brought to the warm, low-pressure conditions at the ocean surface, the sediment fizzes as the methane escapes. The gas burns vigorously if ignited (Figure 17.5).

Although methane hydrate is abundant, exploitation of this resource would be very costly and quite dangerous. In 2013, Japanese engineers using a deep-sea drilling rig aboard the R/V *Chikyu* bored through 330 meters (1,000 feet) of overlying sediment into a 60-meter-thick layer rich in methane hydrate. They were able to extract methane from the formation and estimate the deposit contains 11 years' worth of natural gas at Japan's current rate of importation. Exploitation is planned.

The prospect of widespread methane hydrate mining worries environmentalists. The escape of methane from marine sediments may have played a role in ancient climate change.

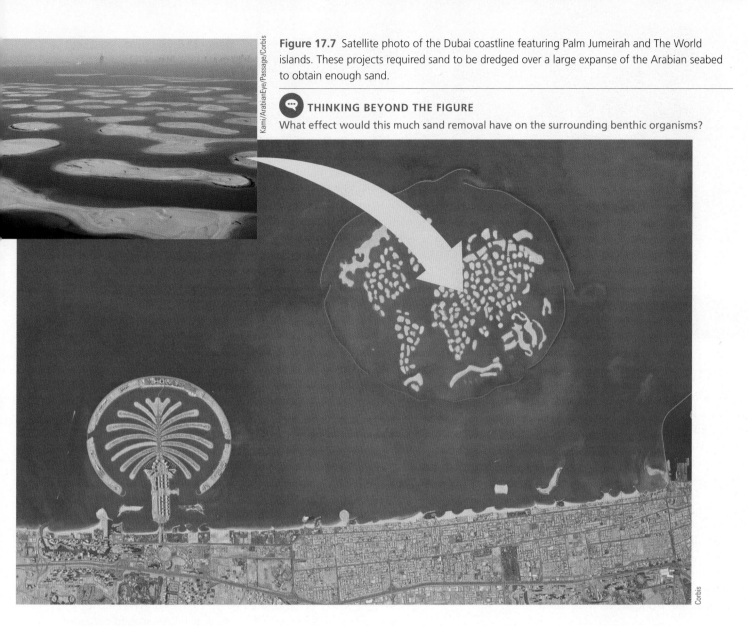

Figure 17.7 Satellite photo of the Dubai coastline featuring Palm Jumeirah and The World islands. These projects required sand to be dredged over a large expanse of the Arabian seabed to obtain enough sand.

💬 **THINKING BEYOND THE FIGURE**
What effect would this much sand removal have on the surrounding benthic organisms?

Kami/ArabianEye/Passage/Corbis

Corbis

Methane is a powerful greenhouse gas (explained in Chapter 18), and changes in ocean circulation that result in deep-ocean warming could release large quantities of methane. About 55 million years ago, the deep ocean warmed by at least 4°C (7°F). The large-scale escape of methane from the seabed may have raised surface temperatures abruptly, melted surface ice, decreased the pH, and lowered oxygen levels in the deep sea. Will methane freed from deep sediments exacerbate the problem of climate change? Would this massive new source of available energy increase our dependency on carbon-based fossil fuels? Would it reduce the incentive to pursue clean, renewable energy sources?

Marine Sand and Gravel Are Used in Construction

Sand and gravel are not very glamorous marine resources, but they are second in dollar value only to oil and natural gas. More than 1.6 billion metric tons (1.7 billion tons) of

sand and gravel, valued at more than half a billion dollars, were mined offshore in 2010. Only about 1% of the world's total sand and gravel production is scraped and dredged from continental shelves each year, but the seafloor supplies about 20% of the sand and gravel used in the island nations of Japan and the United Kingdom. The world's largest single mining operation is the extraction of aragonite sands at Ocean Cay in the Bahamas (**Figure 17.6**). Sand is suction-dredged onto an artificial island and then shipped on specially designed vessels. This sand, about 97% calcium carbonate, is used in Portland cement, glass, animal feed supplements, and in the reduction of soil acidity.

Two of the largest and most recognizable projects constructed with sand mined from the seafloor are Palm Jumeirah and The World in Dubai, United Arab Emirates (**Figure 17.7** and Figure 12.38). Construction of these islands required a massive 430 million cubic meters of sand that was dredged from the nearby waters of the Persian Gulf.

Figure 17.8 Salt evaporation ponds at the southern end of San Francisco Bay in California. Operators can segregate the various salts from one another by shifting the residual brine from pond to pond at just the right time during the evaporation process. The colors in the ponds are imparted by algae and other microorganisms that thrive at varying levels of salinity. In general, the highest-salinity ponds have a reddish cast.

Most of the exploitable U.S. deposits of marine sand and gravel are found off the coasts of Alaska, California, Washington, the East Coast states from Virginia to Maine, and Louisiana and Mississippi. Nearshore deposits are widespread, easily accessible, and used extensively in buildings and highways and in the supplementation of eroded offshore islands. Offshore oil wells in Alaska are built on huge human-made gravel platforms; the large quantities of gravel available at those locations make offshore drilling practical there.

Not all gravel is dull. Diamonds have been found in offshore gravel deposits in Australia and Africa. In 2009, offshore mining vessels dredged 1 million carats of diamonds from the Namibian coast, which accounted for nearly 60% of the total diamond output from all Namibia-based mines.

Deep-Sea Mining

Technological advances in robotic control are making abyssal mining practical. Nautilus Minerals Corporation discovered high-grade copper and gold ore at a depth of 1,600 meters (5,200 feet) in the southwestern Pacific's Bismarck Sea. Nautilus proposed the Solwara-1 project to extract this valuable material, but disputes with the nearby state of Papua New Guinea have put this project on hold. Other private companies are eagerly exploring seeps and deep-sea rift zones for sulfites, lead, and zinc.

Reports of a 2013 Japanese discovery of high concentrations of rare-earth elements in sediments near the Pacific island of Minami Torishima are particularly interesting. Rare-earth elements like neodymium, terbium, and yttrium are essential to the manufacture of cell phones, solar panels, LCD displays, and other modern necessities. China controls more than 95% of global rare-earth production, and other sources would be welcome. The rich concentration of these newly discovered rare earth deposits—far higher than in deposits found elsewhere in the Pacific or on land in China—will stimulate ways to recover them from their 5,800-meter (19,000-foot) depth.

Manganese nodules, which were discussed in Chapter 5, also present an abundant source of a variety of metals. For now, the low market value of the minerals in manganese and other nodules makes them too expensive to recover. As techniques for deep-sea mining become more advanced and raw material prices increase, however, the concentration of valuable materials in these nodules will almost certainly be exploited.

Powdery deposits of metal sulfides have been found in the vicinity of hydrothermal vents at oceanic ridges. Hot, metal-rich brines blasting from the vents meet cold water, cool rapidly, and lose the heavy metal sulfides by precipitation. Iron sulfides and manganese precipitates fall in thick blankets around the vents. The cobalt crusts of rift zones also seem to be associated with this phenomenon. These areas may also one day be mined for their metal content.

Salts Are Harvested from Evaporation Basins

As you may remember from Chapter 7, the ocean's salinity varies from about 3.3% to 3.7% by weight. When seawater evaporates, the remaining major constituent ions (see Table 7.1) combine to form various salts, including calcium carbonate ($CaCO_3$), gypsum ($CaSO_4$), table salt (NaCl), and a complex mixture of magnesium and potassium salts. Table salt makes up slightly more than 78% of the total salt residue. "Sea salt," an increasingly popular alternative to table salt, contains all the residual solids.

Seawater is evaporated in large salt ponds in arid parts of the world to recover the salts (**Figure 17.8**). Operators can segregate the various salts from one another by shifting the residual brine from pond to pond at just the right time during the evaporation process.

Figure 17.9 This desalination plant in Dubai, the world's largest, is capable of producing about 822,000 cubic meters (220 million gallons) of pure water daily, about 90% of the Emirate's domestic consumer and industrial demand.

The magnesium salts are used as a source of magnesium metal and magnesium compounds; the potassium salts are processed into chemicals and fertilizers. Bromine (a useful component of certain medicines, chemical processes, and antiknock gasoline) is also extracted from the residue. Gypsum is an important component of wallboard and other building materials.

About a third of the world's table salt is currently produced from seawater by evaporation. In North America, some of this salt is used for snow and ice removal. Salt is also used in water softeners, agriculture, and food processing. In 2013, the United States produced 40.1 metric tons of salt with a value of US $1.6 billion; about 46% of this was produced by evaporation of saltwater.

Freshwater Is Obtained by Desalination

Only 0.017% of Earth's water is liquid, fresh, and available at the surface for easy use by people. Another 0.6% is available as groundwater within half a mile of the surface. Unfortunately, much of this water is polluted or otherwise unfit for human consumption. The fact that fresh, pure water can cost more per gallon than gasoline emphasizes its scarcity and importance. More than any other factor in nature, the availability of **potable water** (water suitable for drinking) determines the number of people who can inhabit any geographic area, their use of other natural resources, and their lifestyle.

Freshwater is becoming an important marine resource. Exploitation of that resource by **desalination**, the separation of pure water from seawater, is already under way, mainly in the Middle East, West Africa, Peru, Florida, Texas, and California. More than 15,000 desalination plants are currently operating in 125 countries, producing a total of about 32.4 million cubic meters (8.5 billion gallons) of freshwater per day **(Figure 17.9)**. A large desalination plant in Ashkelon, Israel, produces about 330,000 cubic meters (87 million gallons) of pure water daily, about 13% of the country's domestic consumer demand.

Several desalination methods are currently in use. *Distillation* by boiling is the most familiar. Distillation uses a great deal of energy, making it a very expensive process. *Freezing* is another effective but costly method of desalination; ice crystals exclude salt as they form, and the ice can be "harvested" and melted for use. Solar or geothermal power may bring down the cost of distillation or freezing, but more efficient, less energy-intensive mechanisms are being developed. Among these is *reverse osmosis desalination*. In this process, seawater is forced against a semipermeable membrane at high pressure. Freshwater seeps through the membrane's pores while the salts stay behind. About 60% of desalinated water is produced in this way (including the Israeli plant mentioned earlier). Reverse osmosis uses less energy per unit of freshwater produced than distillation or freezing, but the necessary membranes are fragile and costly.

Desalination, water conservation, and perhaps even iceberg harvesting will become more common as water becomes more polluted, scarcer, and more valuable.

CONCEPT CHECK

Before going on to the next section, check your understanding of some of the important ideas presented so far:

What are the three most valuable physical resources? How does the contribution of each to the world economy compare to the contribution of that resource derived from land?

Is the discovery of new sources of oil keeping up with oil use? Is oil being made (by natural processes) as fast as it is being extracted?

What's the largest known reservoir of hydrocarbons on Earth? Why is this resource not being utilized?

What are the sources of metals mined or extracted (or potentially mined or extracted) from the sea?

Is recovery of fresh water from seawater economically viable?

17.3 Renewable Sources of Marine Energy

The energy crises of 1973 and 1979, and the precipitous rise in the cost of crude oil in 2007, focused public attention on the need for unconventional sources of power. Sources of energy that are not consumed in use—solar power or wind power, for example—are preferable to nonrenewable sources such as fossil fuels. Anyone who has watched the ocean knows that so restless a place must surely be rich in energy. The energy is certainly there, but extracting it in useful form is not easy.

Windmills Are Effective Energy Producers

The fastest-growing alternative to oil as an energy source is wind power. The world's largest offshore "wind farm" is currently the London Array. It is located 20 kilometers (12 miles) from the Essex coast in the outer Thames Estuary of England and covers an area of around 245 square kilometers (94.5 square miles). Its 175

Figure 17.10 This installation at Copenhagen's airport is one of the world's largest "wind farms." On some windy days, Denmark is said to have a 100% supply of electricity from wind power. Generation of electricity from wind is the fastest-growing source of energy in the world.

 THINKING BEYOND THE FIGURE
Why do you suppose the windmills are arrayed in an arc rather than in a straight line?

turbines were officially opened in July 2013 and are capable of generating electricity for up to half a million United Kingdom homes. Wind is the world's fastest-growing power source (**Figure 17.10**). Unlike oil and natural gas, wind can't be used up. A recent Department of Energy forecast predicts that electricity generation from wind power will be cost-competitive without subsidies by 2025 and will remain the largest source of non-hydropower renewable capacity through 2040.

Extraction of wind power over or near the ocean is especially effective. Winds tend to be more steady (less gusty) as they move over water, lessening stress on the blades and gears of the windmills. Also, average wind speeds tend to be higher over the ocean than on land. Estimates suggest that the near-coast wind-power potential off the U.S. coasts is 900,000 megawatts, the combined capacity of all conventional power plants in the country.

Waves, Currents, and Tides Can Be Harnessed to Generate Power

Waves are the most obvious manifestation of oceanic energy—ask any surfer about the energy in a wave. As you learned in Chapter 10, wind waves store wind energy and transport it toward shore.

Many devices have been proposed to harness this energy; Japan, Norway, Britain, Sweden, the United States, and Russia

Figure 17.11 Large tubes flexed by ocean waves are used to generate electricity in Portugal and on the Scottish coast. Pistons inside the tubes pressurize hydraulic fluid to turn a generator.

have built small experimental plants to evaluate their effectiveness. One of these devices uses the rush of air trapped by waves entering breakwater caissons to power a generator. Another, shown in **Figure 17.11**, uses long moored tubes flexed by passing waves to pressurize hydraulic fluid and generate power. This system, named Pelamis after the genus of sea snakes it resembles, was first used to generate power at Aguçadoura off the coast of northern Portugal. The project was discontinued in 2009 due to financial problems associated with the global recession, but the designers are now testing the second-generation Pelamis and plan to install a wave farm of up to 26 machines, which could generate 20 megawatts of power at Aguçadoura. The Edinburgh, Scotland-based company is planning multiple additional installations off the Scottish coast. The World Energy Council has estimated that the cost of wave energy could eventually be brought down to the current cost of offshore wind energy once the technologies mature.

Energy from ocean currents can also be harnessed. Huge, slowly turning turbines immersed in the Gulf Stream have been proposed, but their necessary size and complexity make them prohibitively expensive. Smaller versions operating in constricted places where tidal currents flow rapidly have proven successful. The first successful commercial marine power plant began operation in Northern Ireland's Strangford Lough in August, 2008. Its 1.2-megawatt generator provides clean electricity for about 1,000 homes. Similar but larger systems are being planned for Nova Scotia's Bay of Fundy on Canada's eastern seaboard, and also in Western Canada in the waters off Vancouver, British Columbia.

The potential of tidal power was discussed in Chapter 11. The world's largest tidal hydroelectric plant came online in 2011 at Lake Sihwa on the South Korean peninsula. Its 10 submerged turbines are capable of producing 254 megawatts of electricity. Estimates suggest that about 15% of the electrical power needs

Only 0.017% of Earth's water is liquid, fresh, and available at the surface for easy use by people.

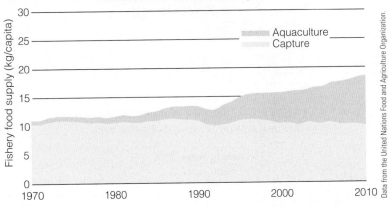

Figure 17.13 The relative contribution of aquaculture and capture fisheries to fish for human consumption. Note that aquaculture production has more than made up for a decline in capture fisheries.

💬 **THINKING BEYOND THE FIGURE**
What do you think the relative contribution of aquaculture and capture fisheries to food supply will be in 2050? In 2100?

Figure 17.12 A sidewalk vendor displays his lunch recommendations in the northern Chinese coastal city of Qingdao.

of the United States could be supplied by tidal generation by 2030, but the initial infrastructure investment is high and there are environmental concerns about how tidal plants would affect the local marine life.

CONCEPT CHECK

Before going on to the next section, check your understanding of some of the important ideas presented so far:

What renewable marine energy source is presently making a significant contribution to the world economy?

What special difficulties would engineers expect to encounter when designing energy-extracting devices suitable for the oceanic environment?

17.4 Biological Resources

Ancient kitchen middens (garbage dumps of bones and shells) found in many coastal regions demonstrate that people have used the sea for thousands of years as a source of food and medicines. Now the human population threatens to outgrow its food supply. Contemporary food production and distribution practices are unable to satisfy the nutritional needs of all the world's 7+ billion people, and starvation and malnutrition are major problems in many nations. Can the ocean help?

Compared to the production from land-based agriculture, the direct contribution of marine animals and plants to the human intake of all protein is small, probably around 6.5% (**Figure 17.12**). Marine sources account for only about 17% of the to-

tal *animal* protein consumed by humans. Fish, crustaceans, and molluscs contribute about 14.5% of that total; fish meal and by-products included in the diets of animals raised for food account for another 2.5%. About 88% of the annual catch of fish, crustaceans, and molluscs comes from the ocean; the rest comes from freshwater. Overall, in 2011, fish provided more than 2.9 billion people with almost 20% of their average per capita animal protein intake. In 2011, we caught or grew 428 million kilograms (943 million pounds) of fish each day!

The sea will not be able to provide substantially more food to help alleviate future problems of malnutrition and starvation caused by human overpopulation; indeed, population growth will likely absorb any resource increase. Nevertheless, these resources currently sustain a great many people.

Fish, Crustaceans, and Molluscs Are the Ocean's Most Valuable Biological Resources

Fish, crustaceans, and molluscs are the most valuable living marine resources. Fishers caught or grew 156 million metric tons (172 million tons) of these animals in 2011 (**Figure 17.13**). The recent distribution of the catch is shown in **Figure 17.14**.

Of the thousands of species of marine fish, crustaceans, and molluscs, fewer than 500 species are regularly caught and processed. The 10 species listed in **Figure 17.15** supply about a quarter of the live weight of all living marine fish caught each year.

Fishing is a big business, employing more than 38 million people worldwide. More than 2.2 million mechanized (that is, powered by an engine) fishing vessels were active in 2010. It is also the most dangerous job in the United States—commercial fishers suffer 124 deaths per 100,000 workers each year (**Figure**

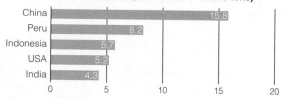

Top harvesters, 2011 (live catch of fishes, crustaceans, and molluscs, in millions of metric tons)

a Top harvesters in 2011 (live catch of marine fishes, crustaceans, and molluscs, in millions of metric tons).

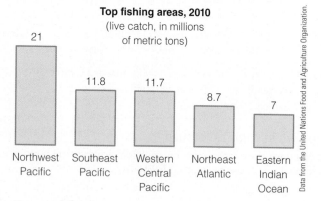

Top fishing areas, 2010
(live catch, in millions of metric tons)

b Top fishing areas in 2010 (live marine catch, in millions of metric tons).

Figure 17.14 Fisheries production, harvesters and capture locations.[4]

Data from the United Nations Food and Agriculture Organization

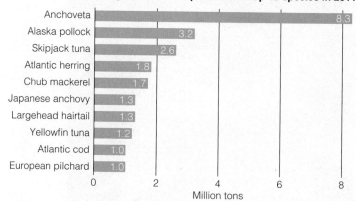

Marine capture fisheries production: top 10 species in 2011

Million tons

Figure 17.15 Marine capture fisheries production: top 10 species in 2011. Data from the United Nations Food and Agriculture Organization.

💬 **THINKING BEYOND THE FIGURE**

From the statistics you've seen, do you think marine food sources will keep up with human population growth?

Almost half the world's commercial marine catch is taken by the five countries shown in Figure 17.14a, with China accounting for the most rapid growth of both caught and farmed fish. About 75% of the annual harvest is taken by commercial fishers who operate vast fleets working year-round, using satellite sensors, aerial photography, scouting vessels, and sonar to pinpoint the location of fish schools. Huge factory ships often follow the largest fleets to process, can, or freeze the animals on the run. Catching methods no longer depend on hooks and lines but on large trawl nets **(Figure 17.17)**, purse seines, or gill nets. Between 1950 and 1997, the commercial marine fish catch increased more than fivefold.

The cost to obtain each unit of seafood has risen dramatically in spite of all this high-tech assistance. The increasing expense of fuel for the fishing fleets and processing plants,

17.16, see p. 495).[5] Though estimates vary widely, the first-sale value of the worldwide marine catch in 2006 was thought to be around US$100 billion. The United States recently displaced Japan as the world's #1 importer of seafood.

---

[4]1 metric ton = 1,000 kilograms = 1.1 tons. Metric tons are sometimes written "tonnes."

[5]Sebastian Junger's extraordinary 1997 book (and the 2000 movie) *The Perfect Storm* chillingly recalls these dangers, as does the Discovery Channel's program *Deadliest Catch*.

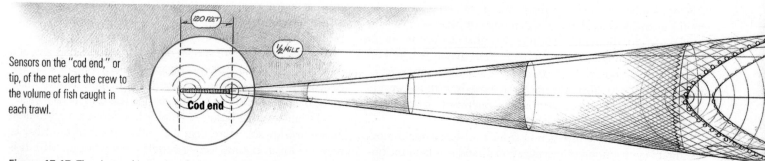

Sensors on the "cod end," or tip, of the net alert the crew to the volume of fish caught in each trawl.

Figure 17.17 The size and intensity of commercial fishing is evident from this diagram of the factory ship *Alaska Ocean* trawling for pollock. The opening in this net could engulf the Statue of Liberty. This 376-foot-long factory ship can process more than 600 metric tons (660 tons) of fish each day. If you have ever eaten a McDonald's Filet-o-fish sandwich, it probably came from the *Alaska Ocean*. Once thought inexhaustible, pollock stocks are plunging; about 25% of the pollock in the Bering Sea is caught each year, an unsustainable quantity. © Cengage Learning

Figure 17.18 As this 5-cent coin from Singapore suggests, the ocean is under furious assault. Tom Garrison

Alaska Stock/Alamy

Figure 17.16 A dangerous way to make a living: Crab fishers work on deck in bad weather in the Bering Sea, Alaska. The 120-foot crab boats and crews are often buffeted by winter storms—winds stronger than 100 miles per hour and seas to 50 feet are not unheard of. Large iron crab pots, some weighing more than 750 pounds, are moved by hand as rolling waves toss the boat like a cork in a bathtub. A crab fisher can make US$20,000 per catch—for a successful trip lasting about 6 weeks. Commercial fishing is the most dangerous profession in the United States.

the cost of wages for the crews, and the greater distances that boats must cover to catch each ton of fish have all helped drive up the cost of seafood. In spite of greater efforts, the total marine catch leveled off in about 1970 and remained surprisingly stable until 1980, when greater demand and increasing prices began to drive the tonnage upward again. Harvests are now declining in spite of increasingly desper-

After 50 years of increasingly aggressive industrial fishing, only 10% of all large fish remain in the sea.

ate attempts to increase yields (Figure 17.18). Since 1972, the world population of humans has grown, so the average per capita world fish catch has fallen significantly.

Most of Today's Fisheries Are Not Sustainable

Marine fish are the only wildlife still hunted on a large scale. About 90% of worldwide stocks of tuna, cod, and other large ocean fish have disappeared in the last 50 years. Can we continue to take huge amounts of food from the ocean? The **maximum sustainable yield**, the maximum amount of each type of fish, crustacean, and mollusc that can be caught without impairing future populations, probably lies between 100 million and 135 million metric tons (110 million and 150 million tons) annually. Current yield exceeds the top figure. Fleets are obtaining fewer tons per unit of effort and are ranging farther afield in their urgent search for food. We may be perilously close to the catastrophic collapse of more fisheries.

As of 2011, 30% of recognized marine fisheries were designated as overexploited, and over 57% more were at their presumed limit of exploitation. In 2012, the United States National Marine Fisheries Service estimated that 19% of the fish stocks whose status is known are now

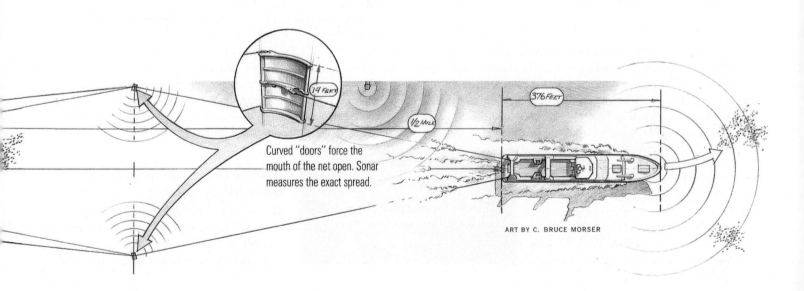

Curved "doors" force the mouth of the net open. Sonar measures the exact spread.

14 FEET

1/2 MILE

376 FEET

ART BY C. BRUCE MORSER

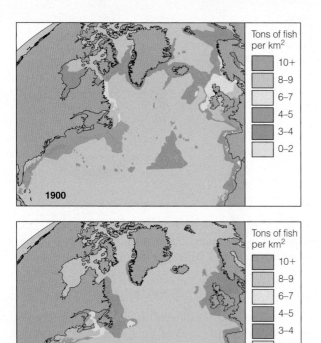

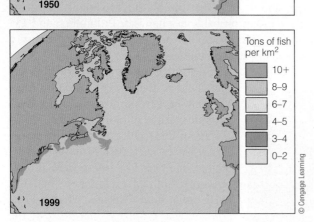

Figure 17.19 A century of dramatic decline in the fisheries of the North Atlantic. These data are for table fish (to be eaten directly by people), not for fish collected for oil or animal feed.

suffering from **overfishing**—so many fish have been harvested that there is not enough breeding stock left to replenish the species. However, we do not have enough scientific information to make a determination on the stock health for more than half of the species managed in federal fisheries management plans. Recent trends may be inferred from **Figure 17.19**. At one point, fishing pressure on Atlantic cod was so great that less than 1% of 1-year-old cod remained in the ocean long enough to spawn. Each year 60% of all the cod present were swept into nets. In 1992, the northwestern cod fishery collapsed, resulting in the loss of 35,000 jobs (18,000 of them in Newfoundland alone).

Aquaculture production currently accounts for almost half of all fish consumed by humans.

Even though substantial reductions in fishing were mandated, the stock has still not recovered.

As you might expect, removal of so many fish seriously impacts the balance of marine communities. Marine biologists have recently reported an alarming increase in the number of jellies in the world ocean, presumably due to lack of predation in their juvenile states by fish. Jellies can consume vast numbers of fish eggs and larvae, making the problem worse.

Can something be done? Maybe. Research has shown that fish stocks are much less likely to collapse if fishers own rights to them. These rights are called "catch shares." Like owning shares in a company, it gives fishers an incentive not to overharvest. Halibut fishing in Alaska, for example, became more efficient and much safer after a catch-share system was established in 1995. What was once a dangerous 2- or 3-day free-for-all (no matter what the weather) that flooded the market with fish has become a well-planned and executed extended mutual harvest. Prices for Alaskan halibut have climbed and fresh fish is available year round. A similar consortium has been proposed for the rapidly expanding fisheries of the thawing arctic. A 2012 study suggests that perhaps 80% of stressed fisheries can be rebuilt by these sorts of cooperative efforts.

Much of the Commercial Catch Is Discarded as "Bycatch"

The intended organism is not the only victim. In some fisheries, **bycatch**—animals unintentionally killed while collecting desirable organisms—sometimes greatly exceeds target catch. About 80% of the total catch, or 4 pounds of bycatch, is discarded for every pound of shrimp caught by Gulf Coast shrimpers **(Figure 17.20a)**. In 1995, the governor of Alaska said, "Last year's [Alaska bottom fishing] discards would have provided about 50 million meals." Worldwide bycatch reached 27 million metric tons (30 million tons) in 1995, *a quantity nearly one-third of total landings*! According to a 2011 National Marine Fisheries Service report, U.S. fishers discarded over 550 billion kilograms (1.2 billion pounds) of nontarget fish, which amounts to 17% of the commercial catch in 2005. Some of the fisheries assessed had greater than a 90% bycatch rate!

Some progress has been made to minimize bycatch. Turtle exclusion devices (TEDs), chutes through which sea turtles are ejected from nets, have been mandated for shrimp fishing in U.S. territorial waters **(Figure 17.20b)**.

Bottom trawling has more problems than bycatch—it is devastating to bottom communities. The habitat itself is disturbed; slow-growing organisms and complex ecosystems are ransacked **(Figure 17.21)**.

Marine Botanical Resources Have Many Uses

Marine algae are also commercially exploited. The most important commercial product is **algin**, made from the mucus that slickens seaweeds. When separated

a All of the marine life pictured here was caught in a shrimping trawl net. The small number of shrimp in the fisher's hands is the only thing that will be sold; everything else will be discarded.

Figure 17.20 Bycatch.

and purified, algin's long, intertwining molecules are used to stiffen fabrics; to form emulsions such as salad dressings, paint, and printer's ink; to prevent the formation of large crystals in ice cream; to clarify beer and wine; and to suspend abrasives. The U.S. seaweed gel industry produces more than US$220 million worth of algin each year, and the annual worldwide value of products containing algin (and other seaweed substances) was estimated to be around US$50 billion in 2007.

Seaweeds are also eaten directly, and some species are cultivated. People in Japan consume 150,000 metric tons (165,000 tons) of nori each year; seaweed and seaweed extracts are also eaten in the United States, Britain, Ireland, New Zealand, and Australia. Their mineral content and fiber are useful in human nutrition.

b An unintentional consequence of trawl net fishing.

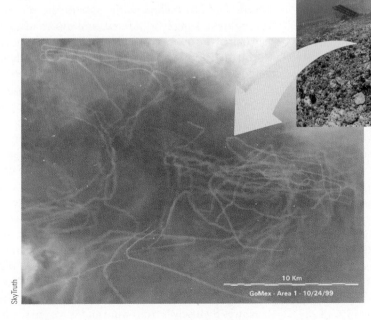

Figure 17.21 Bottom trawling in the Gulf of Mexico can be visible from space. Trails of bulldozed sediments thrown up by bottom-trawling nets (inset) can be traced for 27 kilometers (17 miles). Bright spots in this image are individual vessels and fixed oil and gas drilling platforms.

Figure 17.22 A deep-water pen off the coast of Puerto Rico utilizes the stronger, open ocean currents to farm cobia in a more environmentally friendly way. Brian J. Skerry/National Geographic Creative

Organisms Can Be Grown in Controlled Environments

Aquaculture is the growing or farming of plants and animals in any water environment under controlled conditions. Aquaculture production currently accounts for almost half of all fish consumed by humans. Most aquaculture production occurs in China and the other countries of Asia. In 2010, more than 60 million metric tons (66 million tons) of fish—mostly freshwater fish and shrimp—was produced worldwide, with a total value estimated at US$119 billion. In that same year, the total production of fish aquaculture overtook the production of cattle ranching.

Mariculture is the farming of *marine* organisms, usually in estuaries, bays, or nearshore environments or in specially designed structures using circulated seawater. Mariculture facilities are sometimes placed near power plants to take advantage of the warm seawater flowing from their cooling condensers.

Worldwide mariculture production is thought to be about one-eighth that of freshwater aquaculture. Several species of fish, including plaice and salmon, have been grown commercially, and marine and brackish-water fish account for two-thirds of the total production. Shrimp mariculture is the fastest-growing and most profitable segment, accounting for about 15% of the total international trade in fish products in 2011. The value of

the white-leg shrimp alone was estimated at more than US$12 billion that same year.

In the United States, annual revenue from mariculture now exceeds US$210 million, with most of the revenue generated from salmon, kampichi (a tuna), and oyster mariculture. More than half of the oysters consumed in North America are cultured. Not all mariculture produces food; cultured pearls are an important industry in Japan, French Polynesia, and northern Australia.

Marine fish farming involves production of fish in the hatchery and then rearing them in large moored enclosures, or cages. Some concerns with fish farming include the consumption of fish meal and fish oil by carnivorous species and the potential for escapes, which may reduce the genetic diversity of the wild fish or compete with them for food. As with any farmed animals, fish in close quarters are also more susceptible to disease, and their waste products (and feed) can contaminate areas around their holding pens.

Recent improvements in farm-management strategies have helped significantly alleviate many of these concerns. Raising fish in submerged pens in more exposed, open ocean waters helps to address some of the coastal pollution issues by allowing stronger currents to dilute the waste while also reducing damage to the structures themselves from wind and wave energy **(Figure 17.22)**. The greater flow rate coupled with moderating the number of

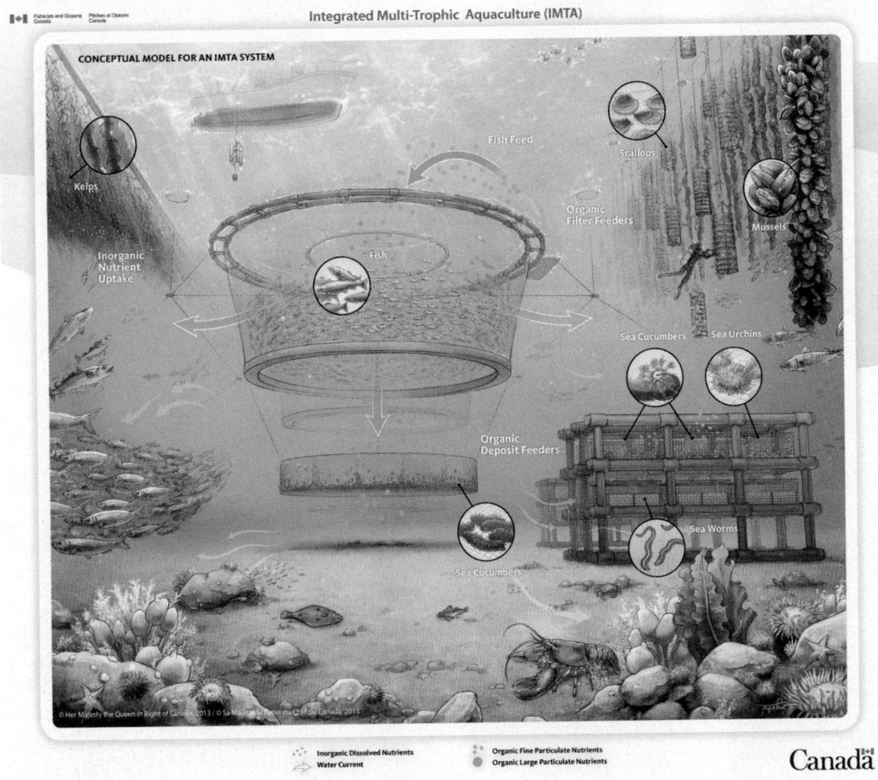

CONCEPTUAL MODEL FOR AN IMTA SYSTEM

Kelps

Inorganic Nutrient Uptake

Fish

Fish Feed

Scallops

Organic Filter Feeders

Mussels

Sea Cucumbers   Sea Urchins

Organic Deposit Feeders

Sea Worms

Sea Cucumbers

Inorganic Dissolved Nutrients
Water Current

Organic Fine Particulate Nutrients
Organic Large Particulate Nutrients

Canada

Figure 17.23 Integrated Multi-Trophic Aquaculture uses the wastes of higher trophic level species as input for additional cultivated organisms like mussels, sea cucumbers, and algae to increase the efficiency and minimize the amount of waste produced from the entire system.

THINKING BEYOND THE FIGURE
Why do you think this type of approach has not been used more in the past?

Reproduced with the permission of Her Majesty the Queen in Right of Canada, 2014.

fish in each pen can further reduce the amount of wasted feed and the need for as many antibiotics. More advanced systems that employ novel designs, such as Integrated Multi-Trophic Aquaculture (IMTA), can increase the system productivity and offer even greater opportunity for environmentally sound fish farming (**Figure 17.23**).

Mariculture may also include stock enhancement—the hatchery production of fish that are released into the wild as juveniles. The natural homing instinct of salmon makes salmon stock enhancement quite successful—a large proportion of wild salmon is originally reared in hatcheries, which imparts greater productivity to the fishery. Only about 1 in 50 of the hatchery-raised juveniles returns to the area of release.

Mariculture is an expanding industry; it is growing at about 8% annually, whereas the world marine fishery as a whole is declining. Mariculture produces mostly high-value seafoods, such as oysters, abalone, shrimp, and salmon. As world population increases and the demand for protein grows among the world's millions of undernourished people, however, there will be increasing demand for culture of lower-value species, primarily omnivores and herbivores. Alternative sources of proteins and oils may also alleviate pressure on the fisheries that supply most of the world's fish meal and fish oil. Of course there are trade-offs. About a third of Ecuador's mangroves have been converted to ponds used in shrimp mariculture.

Whaling Continues

Since the 1880s, whales have been hunted to provide meat for human and animal consumption; oil for lubrication, illumination, industrial products, cosmetics, and margarine; bones for fertilizers and food supplements; and baleen for corset stays. An estimated 4.4 million large whales existed in 1900; today those populations have been significantly reduced. Eight of the 11 species of large whales once hunted by the whaling industry are commercially extinct. The industry pursued immediate profits despite obvious signs that most of the "fishery" was exhausted.

Substitutes exist for all whale products, but the harvest of most commercial species did not stop until whaling became uneconomical. In 1986, the International Whaling Commission, an organization of whaling countries established to manage whale stocks, placed a moratorium on the slaughter of large whales. Except for a suspiciously large harvest of whales taken by Japanese fishers for "scientific purposes," commercial whaling ceased in 1987. Fewer than 700 large whales were taken in 1988.

After that, however, the numbers began to rise. What protection there is may have come too late to save some species from extinction—and protection may be only temporary. Under intense pressure from its major fishing industry, Norway resumed whaling in 1993. Japan never stopped. Their main target, the minke whale, is the smallest and most numerous of the great whale species (see **Figure 17.24** and Figure 15.32). The meat and blubber of this whale are prized in Japan as an expensive delicacy to be eaten on special occasions. The minke whale population has been estimated at 1,200,000, a number that may withstand the present level of harvest. Japan has killed more than 12,300 whales since starting its scientific whaling program in 1988—the great majority were minkes. Iceland took 148 fin whales in 2010, and Norway's quota for 2011 was set at 1,286 minke whales. Whalers in Chile, Peru, and North Korea have joined the hunt.

Figure 17.24 The crew of a Japanese whaling ship waits for a minke whale to bleed to death after being harpooned.

Figure 17.25 Medicines in a Chinese pharmacy in San Francisco, California. Much of the stock is derived from marine sources.

There is a glimmer of hope. In 1994, the International Whaling Commission voted overwhelmingly to ban whaling in about 21 million square kilometers (8 million square miles) around Antarctica, thus protecting most of the remaining large whales, which feed in those waters. The sanctuary is often ignored. In their quest for minke (and other whales), Japanese and Norwegian whalers have entered the area and harpooned animals.

Pirate whalers based in other countries can also catch whales in the Antarctic and sell the flesh on the Japanese and Korean markets. Conservation efforts can do some good, however. Although it was hunted nearly to extinction, the California gray whale has long been off limits to all but aboriginal hunters. Its numbers have grown, and it was removed from the endangered list in 1993.

New Drugs and Bioproducts of Oceanic Origin Are Being Discovered

The earliest recorded use of medicines derived from marine organisms appears in the *Materia Medica* of the emperor Shen Nung of China, 2700 B.C.E. **(Figure 17.25)**. Modern medical researchers estimate that perhaps 10% of all marine organisms are likely to yield clinically useful compounds. One such medicine is derived from a Caribbean sponge and is already in use: Acyclovir, the first antiviral compound approved for humans, has been fighting herpes infections of the skin and nervous system since 1982. A class of anti-inflammatory drugs known as pseudopterosins, developed by researchers at the University of California, has also been successful and is now incorporated in a popular commercial line of "cosmeceuticals."

Newly discovered compounds are also being tested. A common bryozoan—a small encrusting invertebrate—has been found to produce a potent anticancer chemical. Extracts from 30% of all tunicate species investigated show antiviral and antitumor activity; some of these extracts, known collectively as didemnins, shows promise as a treatment for malignant melanoma, the deadliest form of skin cancer. Another tunicate derivative, Ecteinascidin 743 (approved for use in Europe and Asia under the brand name *Yondelis*), has been found useful in the treatment of skin, breast, lung, and ovarian cancers. A related compound found in the same organism, aplidine, shows promise in shrinking tumors in pancreatic, stomach, bladder, and prostate cancer and may be useful in the treatment of leukemia.

Cancer is not the only target. A compound derived from cyanobacteria stimulated the immune system of test animals by 225% and cells in culture by 2,000%; related compounds exhibit pharmacological promise for treating AIDS and arthritis, as well as displaying anticancer and antiviral properties. Synthetically made cone-shell toxins are currently used to treat chronic pain in people with cancer, AIDS, or certain neurological disorders, and they show great promise to help patients with epilepsy, schizophrenia, Parkinson's, and Alzheimer's disease. Autoimmune diseases such as multiple sclerosis, rheumatoid arthritis, and type 1 diabetes may soon be treated with a sea anemone toxin to block destructive cells. Even marine algae harbor active compounds: A red alga of genus *Callophycus* produces a host of newly discovered molecules with potential for the treatment of malaria and even exhibits the ability to kill drug-resistant malaria parasites.

Compounds derived from intertidal sandcastle worms are currently being tested to help repair shattered bones and even patch together heart tissue and torn arteries. Sandcastle worms (genus *Phragmatopoma*) naturally produce compounds that can securely glue together individual sand grains while underwater. These compounds have helped researchers to develop an adhesive that is nontoxic, biodegradable, and capable of hardening while inside the body.

At least 25 drugs derived from marine life—such as bacteria, sponges, and tunicates—are either in clinical trials or have been approved for medical use. Drugs from marine sources may be promising, but commercial materials from extremophiles are a reality. Biotech companies have isolated and slightly modified enzymes from primitive organisms that thrive in deep-ocean sediments and near hydrothermal vents. The most widely used products are enzymes that function as cleaning agents in the detergents used in washing clothes and dishes. These agents remove protein stains and grease more effectively and at lower temperatures and lower concentrations than the phosphate-based chemicals they replace.

> **T**he global whale-watching industry is estimated to be worth more than US$2 billion annually and employs 13,000 people in 119 countries.

CONCEPT CHECK

Before going on to the next section, check your understanding of some of the important ideas presented so far:

What's the most valuable biological resource?

Fishing effort has increased greatly over the past decade. Has the *per capita* harvest also increased?

What is meant by "overfishing"? Are most of the world's marine fishes overfished?

What is "bycatch"?

Does anyone still hunt whales? Why?

Can mariculture make a significant contribution to marine economics?

Are any drugs derived from the ocean presently approved for use by humans?

17.5 Nonextractive Resources Use the Ocean in Place

Transportation and recreation are the main nonextractive resources the oceans provide. People have been using the ocean for transportation for thousands of years, and our capacity to move goods across ocean basins has dramatically increased. The growth of the global transportation fleet was unprecedented during the first decade of the 21st century, doubling its size since 2001. In 2013, the world fleet had a capacity of over 1.6 billion metric tons (1.76 billion tons). Over 80% of global trade is transported by maritime vessels, which accounts for over 70% of its value. If current trends continue, global maritime trade may double again by 2033.

Through most of this time, the transport of cargo has produced far more revenue than the movement of passengers. At present, oil tankers ship the greatest gross tonnage of any type of cargo—over 2.8 billion metric tons (3.1 billion tons) in 2012. Oil alone accounts for about 30% of the global trade, followed by various bulk materials like iron, coal, and grain, which account for over 40% of the rest.

More than half of the world's crude-oil production is transported to market by ships. Tankers are needed because very few of the major oil-drilling sites are close to areas where the demand for refined oil products is highest. The largest tankers are more than 430 meters (1,300 feet) long and 66 meters (206 feet) wide, and they can carry more than 500,000 metric tons (3.5 million barrels) of oil.

Modern harbors are essential to transportation. Cargoes are no longer loaded and off-loaded piece by piece by teams of longshoremen. Today's harbors bristle with automated bulk terminals, high-volume tanker terminals (both offshore and dockside), containership facilities (see **Figures 17.26**), roll-on–roll-off ports for automobiles and trucks, and passenger facilities required by the growing popularity of cruising. Most of this specialized construction has occurred since 1960. The world's busiest container terminal is in Shanghai—in 2011, more than 30 million containers passed through the port! Six of the top eight container ports are located in China.

Global trade is significantly affected by where container ships can travel. As a result of the increase in the size of cargo

b The gloves are carefully sewn to match the agreed-on specifications.

Tom Garrison

Tom Garrison

c Sealed securely in a standard container, the finished goods are transported by truck from factory to port through specialized economic zones in China. The gloves will depart from the port of Hong Kong, one of the busiest container ports in the world (seen in the background). The largest ships can transport 15,000 containers!

Tom Garrison

a A representative of a U.S. manufacturer of high-tech cases and apparel inspects materials and negotiates the production of gloves to be made in a factory in Guangdong, China.

Tom Garrison

f This containership rides high in the water as it returns to Asia with America's #1 containerized export: air. The container cycle begins anew.

d After a 16-day journey across the Pacific in the containership *Hyundai Kingdom,* the gloves arrive in the largest U.S. container port complex near Los Angeles, California.

Port of Long Beach

Port of Long Beach

e Still sealed, the containers are placed on trucks (and trains) for movement inland. Total time from the gloves' manufacture until they are available in a Midwestern city averages about 8 weeks.

AP images/Arnulfo Franco

a One of the gates for a new lock that is part of Panama Canal expansion project. Each of the gates weighs an estimated 3,100 metric tons (3,410 tons). Note the relative size of the people on either side of the gate.

b Rapid melting of arctic ice makes transit of the Northwest Passage possible. The saving in time and fuel is evident in these charts.

London to Tokyo via Northeast Passage
13,000 kilometers

London to Tokyo via Suez Canal
20,900 kilometers

New York City to Tokyo via Northwest Passage
14,000 kilometers

New York City to Tokyo via Panama Canal
18,200 kilometers

© Cengage Learning

Figure 17.27 Changes that can influence global shipping patterns.

ships, many of the bigger ships can no longer pass through the Panama Canal. To address this issue, the Panama Canal is currently constructing a new, larger set of locks, as well as widening and deepening many of the channels to allow for larger vessels **(Figure 17.27a)**. When complete, this project is expected to double the capacity of the Canal and allow many of the new generation of large cargo ships to reduce their travel distance, resulting in an average cost savings of around 10%. The melting of Arctic ice—a by-product of the global climate change you'll learn about in the next chapter—is also making it possible for ships to take different trade routes. As the ice-free areas continue to expand, ships are able to move from the north Atlantic to the north Pacific without transiting either the Suez or Panama canals. A German shipping corporation had six vessels making the Northwest Passage voyage across Canada in 2010, at a savings of US$800,000 **(Figure 17.27b)**.

Transportation is sometimes combined with recreation. In the past decade, the cruise industry has experienced spectacular growth. Passengers on luxurious ocean liners and cruise ships can enjoy a few relaxing days on the ocean crossing the North Atlantic, visiting tropical islands, or touring places accessible to the public only by ship. (Indeed, tourism is now the world's largest industry.) Ocean-related leisure pursuits, including sport fishing, surfing, diving, day cruising, sunbathing, dining in seaside restaurants, and just plain relaxing contribute to the economy. In addition to being important producers of revenue, public aquariums and marine parks (like Sea World) are centers of education, research, and captive breeding programs **(Figure 17.28)**. Public interest and curiosity about whales is another source of recreational revenue. The global whale-watching industry is estimated to be worth more than US$2 billion annually and employs 13,000 people in 119 countries.

Tom Garrison

Figure 17.28 Public aquariums and marine parks are important resources. Monterey Bay Aquarium, California, is seen here. Its associated Research Institute conducts marine research.

And don't forget about real estate. As coastal land values will attest, people enjoy living near the ocean. Over 123 million people lived in coastal communities in the United States in 2010. This represented over 39% of the nation's population; it is estimated that this number will grow by more than 10 million new residents by 2020.

> **CONCEPT CHECK**
>
> Before going on to the next section, check your understanding of some of the important ideas presented so far:
>
> What use of the ocean in place (a nonextractive resource) is most valuable?
>
> How has the advent of containerized shipping changed world economics?
>
> Where are most of the world's busiest ports currently located? Which is the biggest?
>
> What is the world's largest industry? Is it ocean related?

17.6 The Law of the Sea Governs Marine Resource Allocation

Prehistoric peoples living near the shore were the earliest users of marine resources. With the rise of nation-states and military establishments, the fight for control of marine resources began. Some nations assumed that the ocean belonged to all and endeavored to guarantee free right to passage and resources. Others decided that it belonged to none and tried to control access to ports and resources by force. In 1604, Hugo Grotius, a learned Dutch jurist, wrote *De Jure Praedae (On the Law of Prize and Booty)*, a treatise justifying the action of a Dutch admiral who had successfully defended Dutch trading rights in a dispute with Portugal. One chapter of this work, in which Grotius defended free ocean access for all nations, was reprinted in 1609 under the title *Mare Liberum (A Free Ocean)*. *Mare Liberum* formed the basis for all modern international laws of the sea.

About a century later, in 1703, the concept of territorial seas adjacent to land was recognized. A country's seaward boundary was set at about 5 kilometers (3 miles)—the distance a cannonball could be fired from shore. This 5-kilometer (3-mile) limit stood until 1945.

The United Nations Formulated the International Law of the Sea

After World War II, the technology became available to search for oil and natural gas on continental shelves. After U.S. oil companies found rich deposits beyond the 5-kilometer (3-mile) limit off Louisiana, President Harry Truman issued a proclamation annexing the physical and biological resources of the continental shelf contiguous to the United States. Other nations rushed to make similar claims.

The United Nations then became involved. A committee of the General Assembly began to formulate policy, which was later presented at the First United Nations Conference on **Law of the Sea** in 1958 in New York. Twenty-four years of effort by delegates from many interested nations resulted in the 1982 Draft Convention on the Law of the Sea. In April 1982, the United Nations adopted the convention by a vote of 130 to 4, with 17

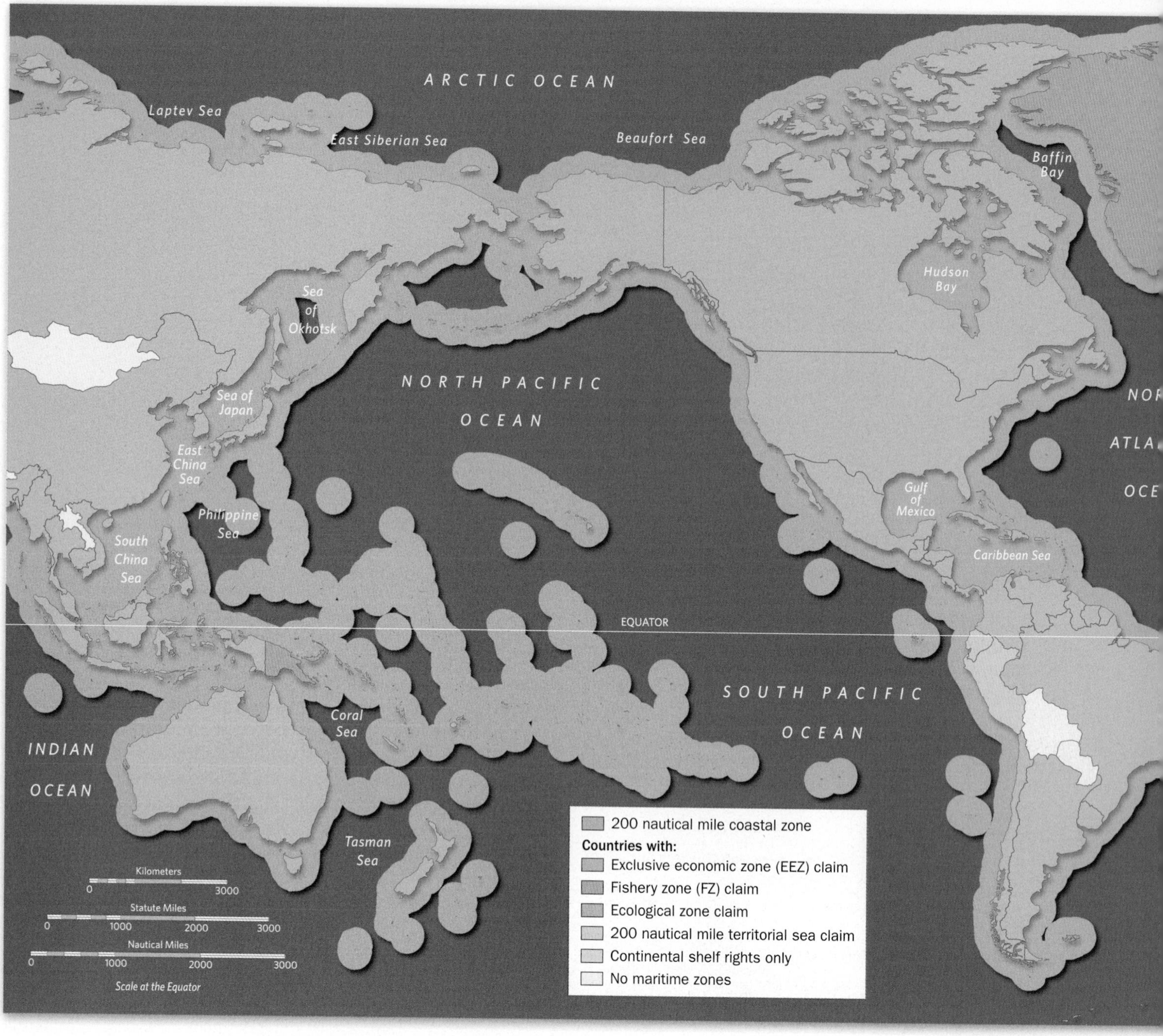
ARCTIC OCEAN
Laptev Sea
East Siberian Sea
Beaufort Sea
Baffin Bay
Hudson Bay
Sea of Okhotsk
Sea of Japan
East China Sea
NORTH PACIFIC OCEAN
Philippine Sea
South China Sea
Gulf of Mexico
Caribbean Sea
NORTH ATLANTIC OCEAN
EQUATOR
SOUTH PACIFIC OCEAN
Coral Sea
INDIAN OCEAN
Tasman Sea
200 nautical mile coastal zone
Countries with:
Exclusive economic zone (EEZ) claim
Fishery zone (FZ) claim
Ecological zone claim
200 nautical mile territorial sea claim
Continental shelf rights only
No maritime zones
Kilometers
0
3000
Statute Miles
0
1000
2000
3000
Nautical Miles
0
1000
2000
3000
Scale at the Equator

abstentions. (The United States, Turkey, Venezuela, and Israel voted against the convention.) By 1988, more than 140 countries had signed all or most parts of the treaty. It is now legally binding, but signatories have selectively chosen to respect or ignore its individual provisions.

Here are some important features of the 1982 Draft Convention:

- **Territorial waters** are defined as extending 18.2 kilometers (12 nautical miles) from shore. A nation has the right to jurisdiction within its territorial waters. Straits used for international navigation are excluded from a nation's territorial waters in that any vessel has the right to innocent passage.
- The **exclusive economic zone (EEZ)** extends 370 kilometers (200 nautical miles) from a nation's shoreline. If that country is able to prove that its continental shelf extends beyond that distance, it can extend that boundary out to nearly 650 kilometers (350 nautical miles) from shore. Nations hold sovereignty over resources, economic activity, and environmental protection within their EEZs.

Labels on map: ARCTIC OCEAN, Barents Sea, Kara Sea, Norwegian Sea, North Sea, Baltic Sea, Black Sea, Mediterranean Sea, Red Sea, Arabian Sea, EQUATOR, INDIAN OCEAN, SOUTH ATLANTIC OCEAN

National Geographic Maps

Figure 17.29 The world's Exclusive Economic Zones (EEZs) are shown in blue. Note how the EEZs are often associated with countries' distant possessions.

THINKING BEYOND THE FIGURE
Which country do you suppose has the largest EEZ?

The convention places about 40% of the world ocean under the control of the coastal countries, within the EEZs. The resources of the remaining 60%, the high seas, are to be shared by the citizens of the world.

The U.S. Exclusive Economic Zone Extends 200 Nautical Miles from Shore

The United States did not sign the 1982 Draft Convention for a variety of reasons. Among these was concern that private enterprise would be deprived of profits if it were made to share high-seas resources with other countries. Instead, the United States unilaterally claimed sovereign rights and jurisdiction over all marine resources within its own 200-nautical-mile region, which it called the **U.S. Exclusive Economic Zone**. The proclamation—similar in most ways to the 1982 United Nations Treaty but lacking the provision of shared high-seas resources—was signed by President Ronald Reagan on 10 March 1983. The U.S. EEZ brings within national domain more than 10.3 million square kilometers (4 million square miles) of continental margins, an area 30% larger than the contiguous 48 U.S. states and a region of diverse geological and oceanographic settings **(Figure 17.29)**.

- All ocean areas outside the EEZs are considered the **high seas**. In tradition, the high seas are common property to be shared by the citizens of the world. The International Seabed Authority was established to oversee the extraction of mineral resources from the floors of the high seas.
- The values of protecting the ocean and preventing marine pollution were endorsed.
- Subject to some conditions, the freedom of scientific research in the ocean was encouraged.

CONCEPT CHECK

Before going on to the next section, check your understanding of some of the important ideas presented so far:

When was the first set of laws governing the allocation of ocean resources proposed?

What is an "exclusive economic zone"?

Which body nominally governs the Law of the Sea?

In this chapter you saw two sides of humanity's use of the ocean. On the one hand, we find the ocean's resources useful, convenient, and essential. On the other, we find we cannot exploit those resources without damaging their source. World economies are now dependent on oceanic materials, and we are unwilling to abandon or diminish their use until we see unmistakable signs of severe environmental damage. By then, mitigation is usually too late.

Marine resources include physical resources such as oil, natural gas, building materials, and chemicals; marine energy; biological resources such as seafood, kelp, and pharmaceuticals; and nonextractive resources like transportation and recreation. The contribution of marine resources to the world economy has become so large that international laws now govern their allocation.

In spite of their abundance, marine resources provide only a fraction of the worldwide demand for raw materials, human food, and energy. Similar resources on land can usually be obtained more safely and at lower cost. The management of marine resources—especially biological resources—for long-term benefit has been largely unsuccessful.

In the last chapter you will learn that humanity has embarked on an unintentional global experiment in marine resource exploitation and waste management. We hesitate to adjust course as we rush into the unknown.

QUESTIONS FROM STUDENTS

1. **How much marine life had to die to provide my car with a gallon of gasoline?**

Jeffrey Dukes, an ecologist at the University of Massachusetts, recently wondered the same thing. He estimates that about 2% of phytoplankton fall to the seabed to be buried in sediments.

Heat transforms about 75% of this mass into oil, but only a small fraction of the oil accumulates where humans can get at it. Depending on the quality of the oil pumped from the ground, about half can be refined into gasoline. Back-of-the-envelope calculations show that about 90 metric tons (99 tons) of dying phytoplankton went into each gallon of gasoline. Your tank can hold the remains of more than 1,000 metric tons (1,100 tons) of ancient diatoms, dinoflagellates, coccolithophores, and other planktonic producers.

Suddenly the cost of a gallon of premium unleaded sounds like a bargain!

2. **When will we run out of oil?**

This is a question that divides many experts. The total amount of conventionally recoverable oil that can be extracted is expected to range from 1.6 trillion to 2.4 trillion barrels. If this estimate is correct and only conventional drilling methods are used, consumption could continue at the present level until sometime around 2025, at which time it would drop quite rapidly. But in fact, we will never run *completely* out of oil—there will always be some oil within Earth to reward great effort at extraction. As technology advances, new techniques are developed that make it more economical to extract a greater percentage of the remaining oil. Since it is hard to predict technological innovation, it is difficult to determine exactly when we will run out of oil. In addition, growing populations, the development of affordable alternatives, and more efficient uses of energy all play a role in the amount and rate of oil ultimately extracted. Whatever the actual date, future civilizations will surely look back in horror at the fact that their ancestors actually burned something as valuable as lubricating oil.

3. **How does the sea salt I see in the health-food stores differ from regular table salt?**

Unlike the producers of regular table salt, the producers of commercial sea salt do not move the evaporating brine from pond to pond to isolate the different precipitates. Sea salt therefore contains the ocean's salts in their natural proportions, with NaCl making up 78% of the mix. Sea salt has a slightly bitter taste from the potassium and magnesium salts present. Some people believe that the variety of minerals in sea salt is beneficial, but most people obtain adequate amounts of these minerals from a normal diet.

4. **What seafoods are the favorites in the United States?**

In 2002, shrimp replaced canned tuna as the nation's favorite seafood, and it has remained that way since. We ate approximately 1.8 kilograms (4 pounds) of shrimp per person in 2011. Canned tuna, long the reigning champ, is currently second, with Americans consuming about 1.2 kilograms (2.6 pounds) per capita. Other common marine species include salmon, pollock, crab, cod, and clams.

5. **I've heard that seafood contains mercury, a toxin. Can I still eat fish?**

That depends. Mercury occurs naturally in the ocean, but most arrives as a by-product of coal burning and mining and other industries. Once in the water, mercury takes the form of methyl mercury, which is concentrated in the tissues of fishes high in the food chain—fishes like swordfish, large mackerel, and (in some cases) tuna.

Methyl mercury is especially damaging to the developing brains of fetuses and children. The FDA advises pregnant women to avoid eating swordfish, shark, tilefish, and king mackerel and to limit consumption of other fishes to 341 grams

(12 ounces) a week. For information about an extreme case of mercury poisoning, see Figure 18.13.

6. The problems of overfishing and bycatch concern me. What seafood should I eat, and what should I avoid?

This list **(Figure 17.30)** prepared by the Monterey Bay Aquarium is an excellent guide.

For the most recent version, go to: http://www.seafoodwatch.org/.

7. Now that I know what to look for, I see shipping containers everywhere. How many containers are there, anyway? And how many are at sea at any given time?

At the height of the last cycle of growth, in mid-2007, about 20 million containers were making 220 million trips per year. More than a quarter of these trips originated or ended in China. In 2007, about 90% of non-bulk cargo was being moved by containers stacked on specially designed transport ships.

If a container spends half its time at sea, an astonishing 10 million containers are transferring products across the ocean as we write!

Figure 17.30 List of seafood recommended to be used or avoided. Visit http://www.seafood-watch.org/ for updates, regional information, and phone app downloads.

Seafood Watch

TERMS AND CONCEPTS TO REMEMBER

algin	exclusive economic zone (EEZ)	maximum sustainable yield	potable water
aquaculture	high seas	nonextractive resources	renewable resources
biological resources	Law of the Sea	nonrenewable resources	territorial waters
bycatch	mariculture	overfishing	U.S. Exclusive Economic
desalination	marine energy resources	physical resources	Zone

STUDY QUESTIONS

Thinking Critically

1. How are oil and natural gas thought to be formed? How can these substances be extracted from the seabed? Why are the physical characteristics of the surrounding rock important?
2. What methods of ocean energy extraction have been shown to be practical?
3. Does the ocean provide a substantial percentage of all protein needed in human nutrition? Of all *animal* protein? What is the most valuable biological resource? The fastest-growing fishery?
4. What are the signs of overfishing? How does the fishing industry often respond to these signs? What is the usual result? What is bycatch?
5. What are the advantages and disadvantages of a proclaimed EEZ? Do you think the United States was justified in proclaiming its own EEZ separate from the provisions of the 1982 United Nations Draft Convention?

Thinking Analytically

1. Imagine a conversation between the owner of a fishing fleet and a governmental official responsible for managing the fishery. List five talking points that each person would bring to a conference table. What would be the likely outcome of the resulting discussion?
2. Review the information about primary productivity in Chapters 13 and 14. Estimate the surface area of a wheat or alfalfa field that would produce energy equivalent to the 90 metric tons of phytoplankton required to produce one gallon of gasoline.

GLOBAL GEOSCIENCE WATCH

Visit www.cengagebrain.com to access MindTap, a complete digital course which includes access to Global Geoscience Watch and more. In the GREENR database, go to the "Fisheries" topic portal. From this page, you can search "overfishing" within the portal using the search feature in the left-hand column. Research some of the potential regulations and innovative strategies that have been used to realistically deal with this problem. Develop a position statement detailing your opinion on what you think should be done to reduce the problem of overfishing. Support this position in a two- to three-page essay that uses information from your research to support your opinion.

CENGAGE **brain** Visit www.cengagebrain.com to access course materials for this text, including interactive learning tools, videos, and more.

18 The Ocean and the Environment

KEY CONCEPTS

Since 1961, human demand on Earth's organisms and raw materials has more than doubled and now exceeds Earth's natural replacement capacity by at least 20%. U.S. Coast Guard photo

A pollutant causes damage by interfering directly or indirectly with the biochemical processes of an organism. About three quarters of the pollution entering the ocean comes from human activities on land. Tom Garrison

The ocean's increasing acidity is a serious threat to calcium-deposition organisms. Coral reefs are especially endangered by acidification. David Littschwager/National Geographic Image Collection

The global temperature trend has been generally upward in the 18,000 years since the last ice age, but the rate of increase has recently accelerated. We are entering a period of rapid climate warming, much of which is almost certainly caused by human activity. National Geographic Visual Atlas

Our present rates of resource consumption and environmental degradation are unsustainable. AP Images/Dave Martin

8000 B.C.E.: After retreating inland during a storm, a group of hunter-gatherers in Doggerland return to find their camp flooded.[1] Sea level is rising. Eventually there would be no land left to come home to. We face the same situation.
ALEXANDER MALEEV/National Geographic

[1]Doggerland is now part of the seabed between present-day Britain and France. It was exposed by low sea level during the last ice age.

Media Connection

Start off this chapter by watching news coverage of the 2010 BP oil spill in the Gulf of Mexico. Visit www.cengagebrain.com to access MindTap, a complete digital course that includes this video and other resources.

18.1 An Introduction to Marine Environmental Issues

Humanity has embarked on an unintentional global experiment in marine resource overexploitation and waste management. We only hesitatingly adjust course as we rush into the unknown. A 2014 study funded by NASA's Goddard Space Flight Center reminds us that civilizations are not permanent. Cases of severe civilizational disruption due to "precipitous collapse—often lasting for centuries—have been quite common." The study suggests that ". . . collapse can be avoided and population can reach equilibrium if the per capita depletion of nature is reduced to a sustainable level and if resources are distributed in a reasonably equitable fashion."

Human demand has exceeded Earth's ability to regenerate resources since at least the early 1980s. Since 1961, human demand on Earth's organisms and raw materials has more than doubled and now exceeds Earth's natural replacement capacity by at least 20%. Our present rate of growth is clearly unsustainable (see final endpaper).

This chapter outlines some of the ways our species has exercised its capacity to consume marine resources and pollute its surroundings. What comes next? The answer is uncertain, but the story will almost certainly have an unhappy ending.

18.2 Marine Pollutants May Be Natural or Human Generated

The ocean's great volume and relentless motion dissipate and distribute natural and synthetic substances. For this reason, we humans have long used the sea as a dump for our wastes. The ocean's ability to absorb is not inexhaustible.

Marine pollution is the introduction into the ocean of substances or energy that changes the quality of the water or affects the physical, chemical, or biological environment (Figure 18.1).

It is not always easy to identify a pollutant; some materials labeled as pollutants are produced in large quantities by natural processes. For example, a volcanic eruption can produce immense quantities of carbon dioxide, methane, sulfur compounds, and oxides of nitrogen. Excess amounts of these substances produced by human activity may alter global climate and generate acid rain. For this reason we need to distinguish between *natural pollutants* and *human-generated* ones.

No one knows to what extent we have contaminated the ocean. By the time the first oceanographers began widespread testing, the Industrial Revolution was well under way, and changes had already occurred. Traces of synthetic compounds have now found their way into every oceanic corner.

Pollutants Interfere with an Organism's Biochemical Processes

A **pollutant** causes damage by interfering directly or indirectly with the mechanical or biochemical processes of an organism. Many pollutants are harmful to human health. Some pollution-induced changes may be instantly lethal; other changes may weaken an organism over weeks or months, or alter the dynamics of the population of which it is a part, or gradually unbalance the entire community.

Figure 18.1 A sewage spill closes beaches near San Diego, California.

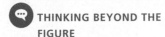 **THINKING BEYOND THE FIGURE**

How long do such closures usually last? What happens to the viruses and bacteria in the spill?

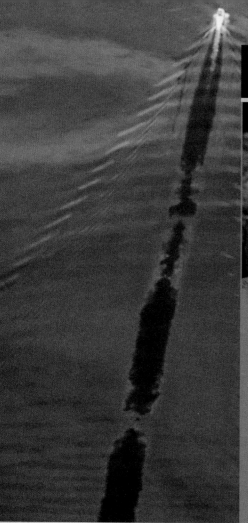

Becky Hale/National Geographic Creative

DR. JARED DIAMOND warns of overexploitation of resources and future turmoil.

Because we are rapidly advancing along this non-sustainable course, the world's environmental problems *will* get resolved, one way or another, within the lifetimes of the children and young adults alive today. The only question is whether they will become resolved in pleasant ways of our own choice, or in unpleasant ways not of our choice, such as warfare, genocide, starvation, disease epidemics, and collapses of societies. While all of those grim phenomena have been endemic to humanity throughout our history, their frequency increases with environmental degradation, population pressure, and the resulting poverty and political instability.

Globalization makes it impossible for modern societies to collapse in isolation, as did Easter Island and the Greenland Norse in the past. Any society in turmoil today, no matter how remote. . . can cause trouble for prosperous societies on other continents and is also subject to their influence (whether helpful or destabilizing). For the first time in history, we face the risk of a global decline. But we also are the first to enjoy the opportunity of learning quickly from developments in societies anywhere else in the world today, and from what has unfolded in societies at any time in the past.

Excerpts from Collapse: How Societies Choose to Fail or Succeed, Jared Diamond, Viking, 2005.

Boats leave wakes of clean water through the oil the day after the Exxon Valdez oil spill. Alaska, USA, 1989.

Background photo: Natalie Fobes/RGB Ventures/SuperStock/Alamy

In most cases, an organism's response to a particular pollutant will depend on its sensitivity to the combination of *quantity* and *toxicity* of that pollutant. Some pollutants are toxic to organisms in tiny concentrations. For example, the photosynthetic ability of some species of diatoms is diminished when chlorinated hydrocarbon compounds are present in parts-per-trillion quantities. Other pollutants may seem harmless, as when fertilizers flowing from agricultural land stimulate plant growth in estuaries. Still other pollutants may be hazardous to some organisms but not to others. For example, crude oil interferes with the delicate feeding structures of zooplankton and coats the feathers of birds, but it simultaneously serves as a feast for certain bacteria.

Pollutants also vary in their *persistence*; some reside in the environment for thousands of years, while others last only minutes. Some pollutants break down into harmless substances spontaneously or through physical processes (like the shattering of large molecules by sunlight). Sometimes pollutants are removed from the environment through biological activity. For example, some marine organisms escape permanent damage by metabolizing hazardous substances to harmless ones. Indeed, many pollutants are ultimately **biodegradable**—that is, they can be broken down by natural processes into simpler compounds. Many pollutants resist attack by water, air, sunlight, or living organisms, however, because the synthetic compounds of which they are composed resemble nothing in nature.

The ways in which pollutants are changing the ocean and the atmosphere are often difficult for researchers to determine. Environmental impact cannot always be predicted or explained. As a result, marine scientists vary widely in their opinions about what pollutants are doing to the ocean and atmosphere and what to do about it. Environmental issues are frequently emotional, and media reports tend to sensationalize short-term incidents (like oil spills) rather than more serious, long-term problems (like climate change or the effects of long-lived chlorinated hydrocarbon compounds).

The sources of marine pollution are summarized in **Figure 18.2.**

Oil Enters the Ocean from Many Sources

Oil is a natural part of the marine environment. Oil seeps have been leaking large quantities of oil into the sea for millions of years—indeed, natural seeps are the largest source of oil in the ocean. The amount of oil entering the sea has increased in recent years, however, because of our growing dependence on

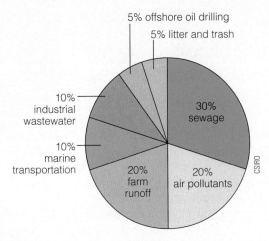

Figure 18.2 Sources of marine pollution.

North American Marine Waters

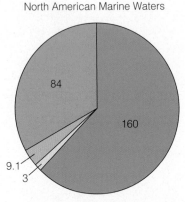

Worldwide Marine Waters

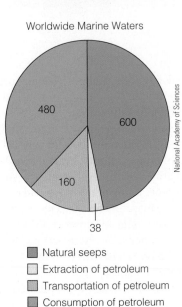

- ■ Natural seeps
- □ Extraction of petroleum
- ■ Transportation of petroleum
- ■ Consumption of petroleum

Figure 18.3 Average Worldwide Annual Releases of Petroleum by Source (1990–1999), in thousands of metric tons per year.

marine transportation for petroleum products, offshore drilling, nearshore refining, and street runoff carrying waste oil from automobiles **(Figure 18.3)**.

The world's accelerating thirst for oil is currently running at about 3,800 liters (1,000 gallons) *per second*, slightly more than

Figure 18.4 Storm drains empty directly into rivers, bays, and ultimately the ocean. Each year, more than 240 million gallons of used motor oil—about 8 times the volume of the 2010 Gulf of Mexico spill—is dumped down drains, poured into dirt, or concealed in trash headed for landfills. Much of this finds its way to the ocean.

💬 THINKING BEYOND THE FIGURE

Which is greater—the amount of oil dumped down drains or the amount of oil entering the ocean from natural seeps?

half of which is transported to market in large tankers. In the decade of the 1990s, about 1,300 thousand metric tons (1,430 thousand tons) of oil entered the world ocean each year. Natural seeps accounted for about half of this annual input—600,000 metric tons annually. About 8% of the total was associated with marine transportation. Some of this oil was not spilled in well-publicized tanker accidents but was released during the loading, discharging, and flushing of tanker ships. Between 150,000 and 450,000 marine birds are believed to be killed each year by oil released from tankers.

Much more oil reaches the ocean in runoff from city streets or as waste oil dumped down drains, poured into dirt, or hidden in trash destined for a landfill. Every year more than 900 million liters (about 240 million gallons) of used motor oil—about 8 times the volume of the 2010 Gulf of Mexico spill—finds its way to the sea **(Figure 18.4)**. This used oil is much more toxic than crude oil because it has developed carcinogenic and metallic components from the heat and pressure within internal combustion engines. And not all hydrocarbon pollution is "wet"—aromatic compounds released when crude oil evaporates will eventually find their way back into the ocean.

It is difficult to generalize about the effects that a concentrated release of oil—an oil spill from a tanker, coastal storage

Jeff Schmaltz/MODIS Rapid Response Team/NASA

Figure 18.5 The oil slick extending from the *Deepwater Horizon* accident on 24 May 2010, about a month after the spill began. Oil smooths the sea surface, making the sun's reflection brighter. Tendrils of oil extend to the north and east of the main body of the slick. A small dark plume along the edge of the slick indicates a possible controlled burn of oil on the ocean surface.

AP Images/Dave Martin

Figure 18.6 Congealed crude oil washes up on Orange Beach, Alabama, about 145 kilometers (90 miles) from the site of the *Deepwater Horizon* spill.

facility, or drilling platform—will have in the marine environment **(Figure 18.5)**. The consequences of a spill vary according to several factors: its location and proximity to shore; the quantity and composition of the oil; the season of the year, currents, and weather conditions at the time of release; and the composition and diversity of the affected communities. Intertidal and shallow-water subtidal communities are most sensitive to the effects of an oil spill.

Spills of *crude* oil are generally larger in volume and more frequent than spills of refined oil. Most components of crude oil do not dissolve easily in water, but those that do can harm the delicate juvenile forms of marine organisms even in minute concentrations. The remaining insoluble components form sticky layers on the surface that prevent free diffusion of gases, clog feeding structures in adult organisms, kill larvae, and decrease the sunlight available for photosynthesis. Crude oil is ultimately biodegradable. Though crude oil spills look terrible and generate great media attention, most forms of marine life in an area recover from the effects of a moderate spill within about 5 years.

Spills of *refined* oil, especially near shore where marine life is abundant, can be more disruptive for longer periods of time. The refining process removes and breaks up the heavier components of crude oil and concentrates the remaining lighter, more biologically active ones. Components added to oil during the refining process also make it more deadly. Spills of refined oil are of growing concern because the amount of refined oil transported to the United States rose dramatically through the 1980s and 1990s.

The volatile components of any oil spill eventually evaporate into the air, leaving the heavier tars behind. Wave action causes the tar to form into balls of varying sizes **(Figure 18.6)**. Some of the tar balls fall to the bottom, where they may be as-

similated by bottom organisms or incorporated into sediments. Bacteria will eventually decompose these spheres, but the process may take years to complete, especially in cold polar waters. This oil residue—especially if derived from refined oil—can have long-lasting effects on seafloor communities. The fate of spilled oil is summarized in **Figure 18.7**.

Cleaning a Spill Always Involves Trade-offs

The methods used to contain and clean up an oil spill sometimes cause more damage than the oil itself. Detergents used to disperse oil are especially harmful to living things. Cleanup of the 1969 *Torrey Canyon* accident off the southern coast of England, one of the first large tanker accidents, did much more environmental damage than the 100,000 metric tons (110,000 tons) of crude oil released. Some resort beaches in the south of England were closed for two seasons, not because of oil residue but because of the stench of decaying marine life killed by the chemicals used to make the shore look clean.

Even the more sophisticated compounds and methods that were used in dealing with the 2010 Gulf of Mexico oil spill—the worst oil spill in U.S. history (and the 19th worst spill ever—**Figure 18.8**) seem to have done more harm. An accident on the *Deepwater Horizon* oil platform off the Louisiana coast on 20 April 2010 resulted in the largest accidental release of oil in the history of the U.S petroleum industry. Eleven workers were killed, 17 injured, and the rig itself failed and sank **(Figure 18.9)**. Oil under high geological pressure shot from the stump of the broken drill pipe. The best estimate of the total release of oil during the 85 days required to plug the well is around 4.9 million barrels (206 million gallons) of crude oil, of which 800,000 barrels (33 million gallons, about 16%) were captured by direct recovery from the drillhead. Skimming and burning at the surface were also employed (recovering an additional 8%). The rest of the oil dispersed into the surrounding ocean and atmosphere. About 26% of it formed tar balls, washed ashore, was buried in sediments, or remained as surface sheen as seen in "Insight from

Figure 18.7 The fate of oil spilled at sea. Smaller molecules evaporate or dissolve into the water beneath the slick. Within a few days, water motion coalesces the oil into tar balls and semisolid emulsions of water-in-oil and oil-in-water. Tar balls and emulsions may persist for months after formation. If crude oil is left undisturbed, bacterial activity will eventually consume it. Refined oil, however, can be more toxic, and natural cleaning processes take a proportionally longer time to complete.

Evaporation
Photooxidation
Thin film on water
OIL SLICK ON SURFACE
Oil-in-water emulsion
Dissolution
Tar balls
Dissolution
Particles
Water-in-oil emulsion
Resuspension
Sedimentation
© Cengage Learning

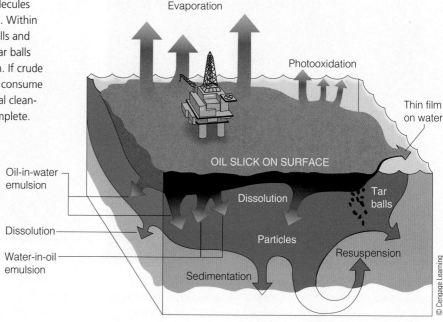

a National Geographic Explorer” on page 513. Lighter components of the oil (about 25%) evaporated or dissolved in seawater.

About 8% of the oil was dispersed by chemicals injected at the wellhead or sprayed on the ocean surface. The dispersants were toxic to the very organisms that would metabolize the oil naturally. A vast amount of dispersant, mostly the chemical Corexit, was deployed—estimates suggest about 2 million gallons were sprayed over the water or injected directly into the gushing wellhead on the seafloor (**Figure 18.10**). When mixed with the dispersant, oil that would normally float was able to linger far below the surface and affect fishes and bottom-dwelling organisms. Research continues, but one toxicologist at the University of Southern Mississippi suggested the dispersed oil was more likely to be toxic than the crude oil by itself.

What to do? It appears that an overambitious cleanup program can be counterproductive. Sylvia Earle, a scientist at the National Oceanic and Atmospheric Administration (NOAA), has said, “Sometimes the best, and ironically the most difficult, thing to do in the face of an ecological disaster is to do nothing.”[2]

Of course the best way to deal with oil pollution is to prevent it from happening in the first place. Tanker design is being modified to limit the amount of oil intentionally released in transport. Oil companies limit new tanker construction to stronger, double-hull designs, and platforms must contain redundant fail-safe components. Perhaps most important, crew testing and training has been upgraded. These moves are paying off. During the past decade, improved production technology and safety

[2]The public usually demands a cleanup. At the peak of the 2010 Gulf spill, more than 23,000 cleanup workers were on duty. By the time the spill is resolved, British Petroleum, the company leasing *Deepwater Horizon* at the time of the accident, will have paid an estimated US$60 billion in penalties, equipment expenditures, and cleanup costs.

Figure 18.8 The world's largest oil spills. For comparison, the 1989 *Exxon Valdez* spill in Alaska's Prince William Sound released about 11 million gallons of crude oil.

Data from Oil Spill Intelligence Report: *Los Angeles Times,* 28 May 2010.

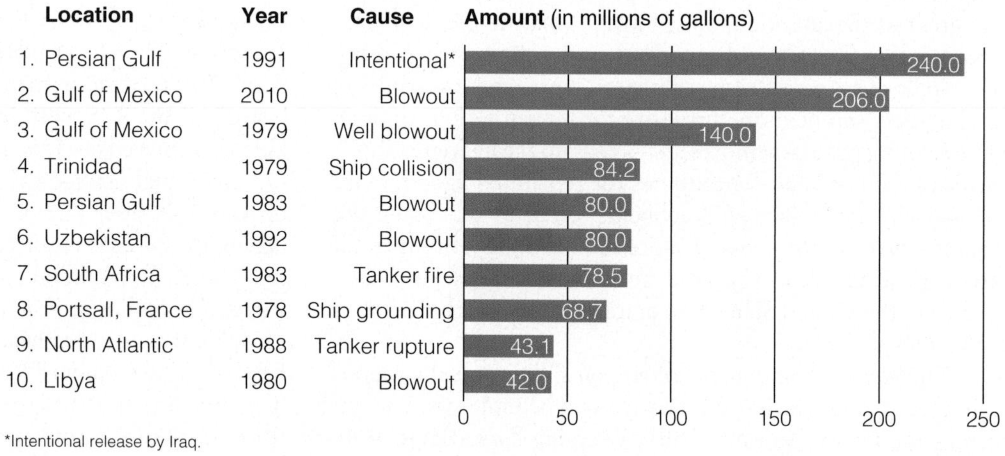

Location
Year
Cause
Amount (in millions of gallons)
1. Persian Gulf 1991 Intentional* 240.0
2. Gulf of Mexico 2010 Blowout 206.0
3. Gulf of Mexico 1979 Well blowout 140.0
4. Trinidad 1979 Ship collision 84.2
5. Persian Gulf 1983 Blowout 80.0
6. Uzbekistan 1992 Blowout 80.0
7. South Africa 1983 Tanker fire 78.5
8. Portsall, France 1978 Ship grounding 68.7
9. North Atlantic 1988 Tanker rupture 43.1
10. Libya 1980 Blowout 42.0
0
50
100
150
200
250
*Intentional release by Iraq.

Figure 18.9 *Deepwater Horizon* moments before sinking.

training of personnel have significantly reduced both blowouts and daily operational spills. Today, accidental spills from platforms represent about 1% of petroleum discharged in North American waters and about 3% worldwide.

Figure 18.10 A robot photographs oil gushing from the *Deepwater Horizon* wellhead. During the 85 days required to plug the well, around 4.9 million barrels (206 million gallons) of crude oil flowed into the Gulf of Mexico.

Toxic Synthetic Organic Chemicals May Be Biologically Amplified

Many different synthetic organic chemicals enter the ocean and become incorporated into its organisms. Ingestion of even small amounts of these compounds can cause illness or even death. Rachel Carson's alarming discussion of the effects of these compounds in her 1962 book *Silent Spring* began the environmental movement (Figure 18.11).

Halogenated hydrocarbons—a class of synthetic hydrocarbon compounds that contain chlorine, bromine, or iodine—are used in pesticides, flame retardants, industrial solvents, and cleaning fluids. The concentration of **chlorinated hydrocarbons**—the most abundant and dangerous halogenated hydrocarbons—is so high in the water off New York State that officials have warned women of childbearing age and children younger than 15 not to consume more than half a pound of local bluefish a week. (They are told *never* to eat striped bass caught in the area.)

The level of synthetic organic chemicals in seawater is usually very low, but some organisms at higher levels in the food chain can concentrate these toxic substances in their flesh. This **biomagnification** is especially hazardous to top carnivores in a food web (Figure 18.12). The concentration in marine organisms of polychlorinated biphenyls (PCBs), fluids once widely used to cool and insulate electrical devices and to strengthen wood or concrete, may be responsible for the behavior changes and declining fertility of some populations of seals and sea lions on islands off the California coast. Even more alarming to biologists is the recent discovery that the nearshore dolphins off U.S. coasts are intensely contaminated. The concentration of chlorinated hydrocarbons in these animals exceeds 6,900 parts

Figure 18.11 Rachel Carson, author of the influential 1962 book *Silent Spring* on the misuse of synthetic pesticides. This work is generally credited with beginning the environmental movement in the United States.

per million (ppm), concentrations high enough to disrupt the dolphins' immune systems, hormone production, reproductive success, neural function, and ability to stave off cancers.[3] These levels vastly exceed the limit of 50 ppm the U.S. government considers hazardous for animals and the 5 ppm considered the maximum acceptable level for humans!

Other synthetic organic poisons are accumulating in the oceanic sink. Two groups of widely used brominated flame retardants (seen in textiles, electronic equipment, upholstery, mattresses, and plastics), polybrominated biphenyls (PBBs) and polybrominated diphenyl ethers (PBDEs), are present in sperm whales, which normally feed in deep water. These compounds have reached the depths.

---

[3]If you drag a beached porpoise into the ocean, you could theoretically receive a $10,000 fine for improper disposal of polluted materials!

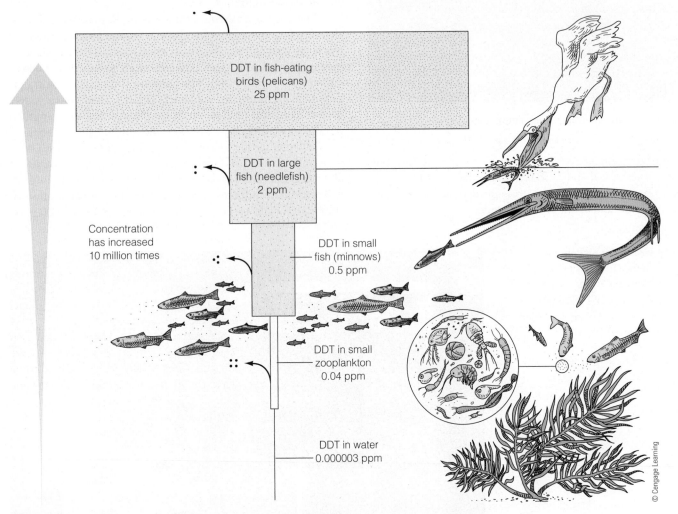

Figure 18.12 The concentration of the pesticide DDT in the fatty tissues of organisms was biologically amplified approximately 10 million times in this food chain of an estuary adjacent to Long Island Sound, near New York City. Dots represent DDT, and arrows show small losses of DDT through respiration and excretion. Concentrations of DDT interfere with, among other functions, calcium metabolism. Pelicans off the California coast were similarly affected and unable to form proper eggshells. Their population plummeted.

 THINKING BEYOND THE FIGURE

Is DDT still approved for use in the United States?

Marine animals are even being killed by illicit drug trafficking. In 1997, 42 dolphins and at least three large whales were found dead off Mexico's west coast, victims of the cyanide-based chemicals used by drug merchants to mark ocean drop-off sites. Ships approach the shore at night, drop bales of illegal drugs overboard, mark them with toxic luminescent compounds, and retreat into international waters. Pick-up crews in small boats retrieve the drugs. Squid and fishes are also drawn to the light, encounter the poisonous chemicals, and die. They are in turn eaten by larger animals, with predictably disastrous results.

Heavy Metals Can Be Toxic in Very Small Quantities

Synthetic organic chemicals are not the only poisons that contaminate marine life. Small quantities of heavy metals are capable of causing damage to organisms by interfering with normal cell metabolism. Among the dangerous heavy metals being introduced into the ocean are mercury, lead, copper, and tin.

Human activity releases about 5 times as much mercury and 17 times as much lead into the ocean as is derived from natural sources, and the incidences of mercury and lead poisoning—major causes of brain damage and behavioral disturbances in children—have increased dramatically over the past 2 decades. Lead particles from industrial wastes, landfills, and gasoline residue reach the ocean through runoff from land during rains, and the lead concentration in some shallow-water, bottom-feeding species is increasing at an alarming rate.

Mercury is an especially toxic pollutant. Exposure in the womb or in infancy to a small amount of mercury can result in severe neurological consequences (**Figure 18.13**). Mercury enters the ocean from mining operations, coal smoke, and as a by-product of industrial production. Larger species of fish, such as tuna or swordfish, are usually of greater concern than smaller species, since the mercury accumulates up the food chain. The U.S. Food and Drug Administration (FDA) advises women of child-bearing age and children to completely avoid swordfish, shark, king mackerel, and tilefish and to limit consumption of king crab, snow crab, albacore tuna, and tuna steaks to 170 grams (6 ounces) or less per week.

Wellness-conscious consumers see fish as a safe and healthful food. But with the ocean still receiving heavy metal–contaminated runoff from the land, a rain of pollutants from the air, and the fallout from shipwrecks, we can only wonder how much longer most seafood will be safe to eat. Consumers should be especially wary of seafood taken near shore in industrialized regions.[4]

> **"We have a long way to go to make peace with this planet, and with each other . . . It is not too late for us to come around, without losing the quality of life already gained."** —EDWARD O. WILSON

Figure 18.13 For the people of Minamata, Japan, who were poisoned by mercury released into the ocean from a nearby factory, heavy metal pollution is a continuing horror story. Between 1953 and 1960, more than 100 people who ate shellfish taken from Minamata Bay were afflicted by a form of mercury poisoning now called Minamata disease. Their symptoms included kidney damage, neuromuscular deterioration, birth defects, insanity, and eventually death.

Eutrophication Stimulates the Growth of Some Species to the Detriment of Others

Not all pollutants kill organisms. Some dissolved organic substances act as nutrients or fertilizers that speed the growth of marine autotrophs, causing eutrophication. **Eutrophication** (*eu*, "good, well"; *trophos*, "feeding") is a set of physical, chemical, and biological changes that take place when excessive nutrients are released into the water. Too much fertility can be as destructive as too little. Eutrophication stimulates the growth of some species to the detriment of others, destroying the natural biological balance of an ocean area. The extra nutrients come from wastewater treatment plants, factory effluent, accelerated soil erosion, or fertilizers spread on land. They usually enter the ocean from river runoff and are particularly prevalent in estuaries. Eutrophication is occurring at the mouths of almost all the world's rivers.

The most visible manifestations of eutrophication are the red tides, yellow foams, and thick green slimes of vigorous plankton blooms (**Figure 18.14a**). These blooms typically consist of one dominant phytoplanktonic species that grows explosively and overwhelms other organisms. Huge numbers of algal cells can choke the gills of some animals and (at night, when sunlight is unavailable for photosynthetic oxygen production) deplete the free oxygen content of surface water, a condition known as **hypoxia** (*hypo*, "low"; *oxia*, "oxygen"). In nearshore waters, hypoxia is now thought to cause more mass fish deaths than any other single agent, including oil spills. It is the leading threat to commercial shellfisheries. **Figure 18.14b** indicates how pervasive hypoxic incidents have become.

---

[4]As noted in Chapter 17, an up-to-date list of safe seafood can be obtained from the Monterey Bay Aquarium's website: http://www.mbayaq.org/cr/seafoodwatch.asp

Oxygen-starved waters are not limited to coastal zones. As ocean temperatures increase with global warming, deep, low-oxygen layers are beginning to surface off the coast of Central America and northern South America. These hypoxic areas drive fishes away from commercially important fishing grounds.

Plastic and Other Forms of Solid Waste Can Be Especially Hazardous to Marine Life

Not all pollutants enter the ocean in a dissolved state; much of the burden arrives in solid form. About 134 million metric tons (147 million tons) of plastic are produced each year, and about 10% ends up in the ocean **(Figure 18.15)**. Americans use more plastic per person than any other group **(Figure 18.16)**. We now generate about 31 million metric tons (34 million tons) of plastic waste, about 120 kilograms (240 pounds) per person, each year. We consume an average of 167 bottles of water per person per year—some 25 million per hour! Slightly more than 4% of world oil production goes to the manufacture of plastics.

The attributes that make plastic items useful to consumers—their durability and stability—also make them a problem in marine environments. Scientists estimate that some kinds of synthetic materials—plastic six-pack holders, for example—will not completely decompose for about 400 years! While oil spills

Figure 18.14 The consequences of eutrophication.

Paul Schmidt/Charlotte Sun

a A toxic algal bloom chokes the waters off Little Gasparilla Island, Florida.

b "Dead" hypoxic zones worldwide. Unless nutrient runoff is limited, the zones will grow in size and number.

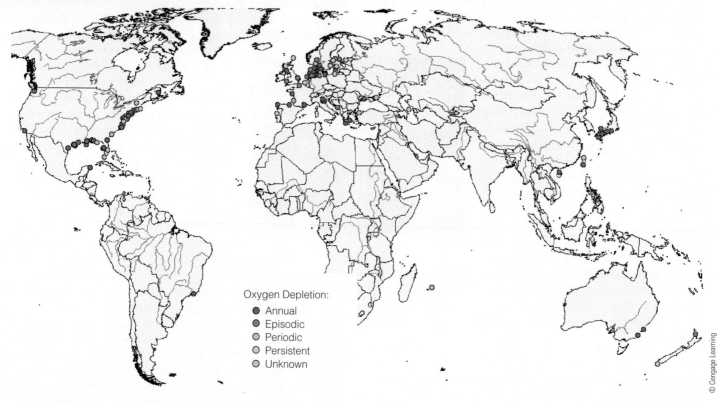

Oxygen Depletion:
- Annual
- Episodic
- Periodic
- Persistent
- Unknown

© Cengage Learning

Figure 18.15 A visitor marvels at a 3-meter (10-foot) tall puffin sculpture fashioned entirely out of debris washed ashore near Bandon, Oregon, USA.

Connie Mack Kostka

get more attention as a potential environmental threat, plastic is a far more serious danger. Oil is harmful but, unlike plastic, it eventually biodegrades. Dumped overboard from ships or swept to sea in flooding rivers, plastic debris is everywhere.

The problem is not confined to the coasts. The North Pacific sub-tropical gyre covers a large area of the Pacific in which the water circulates clockwise in a slow spiral (see again Figure 9.8). Winds there are light. The currents tend to move any floating material into the low-energy center of the gyre. There are few islands on which the floating material can beach, so it stays in the gyre. This area, about the size of Texas, has been dubbed "the Asian Trash Trail" or the "Eastern Pacific Garbage Patch." (A smaller western Pacific equivalent has formed midway between San Francisco and Hawai'i; another lies off the east coast of the United States—**Figure 18.17**.)[5] One researcher estimates the weight of the debris trapped in gyres to be about 3 million metric tons, comparable to a year's deposition at Los Angeles's largest landfill.

Hundreds of marine mammals and thousands of seabirds die *each year* after ingesting or being caught in plastic debris. Sea turtles mistake plastic bags for their jellyfish prey and die from intestinal blockages. Seals and sea lions starve after becoming entangled in nets (**Figure 18.18a**) or muzzled by six-pack rings. The same kinds of rings strangle fish and seabirds. About a quarter of a million Laysan albatross chicks die each year when their parents feed them bits of plastic instead of food (**Figure 18.18b**).

Not all plastic floats. Around 70% of discarded plastic sinks to the bottom. In the North Sea, Dutch scientists have counted around 110 pieces of litter for every square kilometer of the seabed, about 600,000 metric tons (660,000 tons) in the North Sea alone. These plastics can smother benthic life forms.

What should we do with plastic and other solid wastes such as glass and paper, disposable diapers, scrap metal, building debris, and all the rest? Dumping it into the ocean is clearly unacceptable, yet places to deposit this material are becoming scarce. California's Los Angeles and Orange Counties generate enough

---

[5]Contrary to popular belief, this plastic debris does not form vast mats that one could easily see or even walk across. The particles are small—sometimes too small to be seen from the deck of a ship—and widely spaced.

Figure 18.16 Generation and recovery (recycling) of plastics in the United States since 1960. Plastics are not usually biodegradable and accumulate in the marine environment.

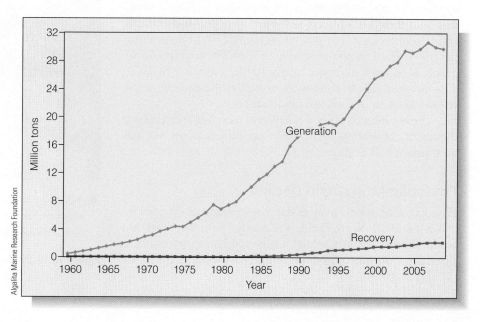

Algalita Marine Research Foundation

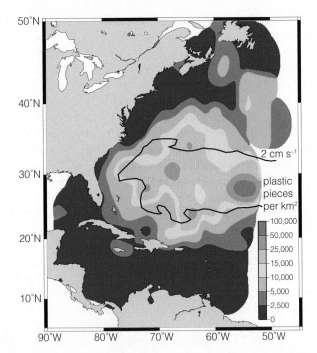

Figure 18.17 Plastic pieces in the North Atlantic gyre. The estimated total amount of plastic in the area shown (most of which consists of particles about 1 millimeter—¹⁄₁₆ of an inch—in size) is 1,100 metric tons (1,210 tons).

2010 by The American Association for the Advancement of Science. Reprinted by permission.

solid waste to fill Dodger Stadium every 8 days. Transportation of waste to sanitary landfills becomes more expensive as nearby landfills reach capacity.

Is recycling the answer? The United States generated 33.6 million metric tons (37.0 million tons) of post-consumer plastic waste in 2008. Of this, 6.5% was recycled and 7.7% burned for energy. Nearly all of the rest was discarded in landfills or abandoned at sea. The Japanese currently recycle about 50% of their solid waste and are importing even more; scrap metal and waste paper headed for Asia are the two biggest exports from the Port of New York. Americans are buying back their own refuse in the form of appliances, automobiles, and the cardboard boxes that hold their MP3 players and smart phones. Massachusetts and California have set a goal of recycling 25% of their waste; the city of Seattle is now approaching 30%. The country of Singapore is on track to recycle nearly 100% of its waste by 2020. The direct savings to consumers, as well as the environmental rewards to ocean and air, will be significant.

The *best* solution is a combination of recycling and reducing the amount of debris we generate by our daily activities. We will soon have no other choice.

Phytoplankton Are in Decline

Phytoplankton productivity in the past 25 years has dropped by about 9% in the North Pacific and nearly 7% in the North Atlantic. Researchers at Canada's Dalhousie University estimate that since 1950, phytoplankton biomass has shrunk an estimated

a Sea lions and seals die by the hundreds each year after becoming entangled in plastic debris, especially discarded and broken fishing nets.

b A decaying albatross reveals the cause of its death. Plastic trash (mistaken for food) has blocked its digestive system. On Midway Atoll, about 40% of albatrosses die in this way.

Figure 18.18 Discarded plastic can pose direct or indirect dangers.

Figure 18.19 As plastic degrades, it breaks into ever smaller pieces but does not disappear altogether. Even laundry lint containing synthetic fibers makes its way to sea. Broad areas of the ocean now contain more plastic than living plankton, as illustrated in this sample.

40% in 8 out of 10 major ocean basins. What could cause such a drastic change?

One possibility is increased ocean stratification. As the ocean surface grows warmer (a subject we will address in a moment), it becomes more difficult for nutrient-rich bottom water to mix with the sunlit upper layers where most primary productivity occurs. Warmer water may also diminish the midlatitude winds that provide the light dusting of terrestrial iron needed for phytoplankton metabolism. Another possibility is exposure of phytoplankton to higher levels of solar ultraviolet radiation because of a thinning ozone layer.

Yet another possibility is the chemical disruption caused by microfine plastic debris **(Figure 18.19)**. Sunlight, wave action, and mechanical abrasion break plastic into ever smaller particles. These tiny bits tend to attract oily toxic residues like PCBs, dioxin, brominated flame retardants, and other noxious organic chemicals. In the middle of the Pacific Ocean, 1,000,000 times more toxins are concentrated on the plastic debris and plastic particles (such as microbeads used in mildly abrasive skin cleaners) than in ambient seawater. The microscopic plastic particles outweighed zooplankton by 6 times in water taken from the North Pacific subtropical gyre (the "East Pacific Garbage Patch"). Might these synthetic chemicals be interfering with some aspect of phytoplankton physiology?

Less phytoplankton means less carbon dioxide uptake and significant changes in oceanic ecosystems.

Pollution Is Costly

In 2004, government and industry in the United States spent about US$310 billion on the control of atmospheric, terrestrial, and marine pollution—an average of almost US$900 for each American. This figure was equivalent to about 1.6% of the gross national product, or 2.8% of capital expenditures by U.S. business; over one-third of the United States Gross National Product originates in coastal areas. China expended an estimated 3.5% of its gross national product on pollution liabilities in 2011. Clearly the financial costs of pollution will continue to increase.

But there are other costs. Failure to control pollution will eventually threaten our food supply (marine and terrestrial), destroy whole industries, produce a greater disparity between have and have-not nations, and cause a decline in the health of all the planet's inhabitants.

To these costs must be added the aesthetic costs of an ocean despoiled by pollution; few of us look forward to sharing the beach with oiled birds, jettisoned diapers, or clumps of medical waste.

CONCEPT CHECK

Before going on to the next section, check your understanding of some of the important ideas presented so far:

What is pollution? What factors determine how dangerous a pollutant is?

How does oil enter the marine environment? Which source accounts for the greatest amount of introduced oil?

What is biomagnification? Why is it dangerous?

How do heavy metals enter the food chain? What can be the results?

Why is plastic so dangerous to marine organisms?

18.3 Organisms Cannot Prosper if Their Habitats Are Disturbed

The pollution processes we have discussed don't affect individual organisms alone. They influence whole habitats, especially the most complex and biologically sensitive shallow-water habitats.

Bays and Estuaries Are Especially Sensitive to the Effects of Pollution

The hardest-hit habitats are estuaries, the hugely productive coastal areas at the mouths of rivers where freshwater and seawater meet. Pollutants washing down rivers enter the ocean at estuaries, and estuaries often contain harbors, with their potentials for oil spills. As little as 1 part of oil for every 10 million parts of water is enough to seriously affect the reproduction and growth of the most sensitive bay and estuarine species. Some of the estuaries along Alaska's Prince William Sound, site of the 1989 *Exxon Valdez* accident, were covered with oil to a depth of 1 meter (3.3 feet) in places. The spill's effects on the $150-million-a-year salmon, herring, and shrimp fishery will be felt for years to come.

Estuaries and bays along the U.S. Gulf Coast, one of the most polluted bodies of water on Earth, are also being severely stressed. About 40% of the nation's most productive fishing grounds, including its most valuable shrimp beds, are found in the Gulf. Nearly 60% of the Gulf's oyster- and shrimp-harvesting areas—about 13,800 square kilometers (5,300 square miles)—are either permanently closed or have restrictions

Environmental Images/Universal Images Group/Age Fotostock

Figure 18.20 Plastic debris mixes with natural material in this southern California estuary.

placed on them because of rising concentrations of toxic chemicals and sewage. Half of Galveston Bay, once classed as the second most productive estuary in the United States, is off limits to oyster fishers because of sewage discharges.

Estuaries along the East Coast are also threatened. From southern Florida to central Georgia, more than 325 square kilometers (125 square miles) of sea grasses (which act as nurseries for a great variety of marine life) have been killed by a virus. Scientists speculate that pollutants from urban and agricultural runoff have weakened the plants' resistance. Fishers to the north in Chesapeake Bay have been mystified by a sudden decline in the abundance of fish and crustaceans, a change marine scientists attribute to increasing pollution in the area. Could these changes also be caused by toxic wastes **(Figure 18.20)**? The beluga whale population of Canada's St. Lawrence estuary collapsed in the early 1980s. High levels of PCBs, DDT, and heavy metals—substances biologically amplified in the whales' food—were blamed for the tumors, ulcers, respiratory ailments, and failed immune systems discovered during the autopsies of 72 dead whales.

People also "develop" estuaries into harbors and marinas (see again Figure 12.31). In the 1960s and 1970s, California led the world in the acreage of bays and estuaries filled for recreational marinas. Harbors grew smaller as more of their area was filled for docks and storage facilities. One hundred and fifty years ago, San Francisco Bay covered 1,131 square kilometers (437 square miles). Today only 463 square kilometers (179 square miles) remain—the rest has been filled in. Filling of estuaries is just as threatening to the natural reproductive cycles of shrimp and fish as poisoning by toxic wastes.

Some states control the development of coastal regions. A citizens' initiative passed by Californians in 1972 limited development of that state's coastal zone. Massachusetts laws make it illegal to fill any marsh or estuarine region, even areas that are privately owned. Similar legislation is pending in a few other coastal states.

Introduced Species Can Disrupt Established Ecosystems

Several thousand species are in transit every day in the ballast water of tankers and other ships. Juvenile forms of marine organisms can easily hitch rides across otherwise insurmountable oceanic barriers and set up housekeeping at distant shores. These foreign organisms—called **introduced species** or exotic species—sometimes outcompete native species and reduce biological diversity in their new habitats. New marine diseases can also be introduced in this way. Even canals and fishery enhancement projects can introduce potentially destabilizing new species.

The Chinese mitten crab (genus *Eriocheir*), a common decorative aquarium seaweed (genus *Caulerpa*), and the brown kelp *Undaria* (used in miso soup and now invading San Francisco Bay) are good examples of exotic marine species that are proving destructive in ecosystems not adapted to them.

The disastrous 2011 tsunami in Japan set adrift vast amounts of debris (some of which is radioactive), and the North Pacific Current is bringing organism-laden chunks ashore in the Pacific Northwest. At least 30 alien species accompanied a large segment of a pier and parts of a fishing boat to a beach in central Oregon **(Figure 18.21)**. Volunteers scraped away as much of the material as possible, and blowtorches were used to sterilize the rest. Some larvae doubtless escaped.

Coral Reefs Are Stressed by Environmental Change

Marine biologists have been baffled by recent incidents of coral bleaching—corals expelling their symbiotic dinoflagellates (zooxanthellae)—in the Caribbean and tropical Pacific **(Figure 18.22)**. As noted earlier, hermatypic corals depend on these photosynthesizing dinoflagellates for a portion of their carbohydrate and

Allen Pleus/Washington Department of Fish and Wildlife

Figure 18.21 As a result of the disastrous Japanese tsunami of 2011, hundreds of marine organisms alien to North America are being transported to our shores. Organisms riding debris on ocean currents can enter ecosystems in which they have no predators. They may overwhelm local inhabitants. This striped beakfish was found in a submerged compartment of a derelict boat washed up on the Washington coast. Tens of other foreign species were also along for the ride.

oxygen requirements. For reasons that are not well understood, when water temperature exceeds a normal summer high by 1°C (1.8°F) or more for a few weeks, coral polyps eject their dinoflagellates, turn pale, and begin to starve. If the water temperature returns to normal in a few weeks, the coral can regain their algae populations and survive the bleaching event. If not, filamentous algae or other decomposers overtake the polyps. A coral reef's ability to survive bleaching depends on the level of stress that it endures before and during such events. The warm El Niño year of 1998 saw the death of about 16% of living corals worldwide. As the ocean warms, bleaching events will probably be more widespread.

Climate change isn't the only problem. Especially damaging to tropical reefs has been the practice of using cyanide to collect tropical fishes. Fishers squirt a solution of sodium cyanide over the reef to stun valuable species. Many fish die; those that survive are sent to collectors all over the world. At the same time the invertebrate populations of the sensitive coral reef communities are decimated.[6]

Rising Ocean Acidity Is Jeopardizing Habitats and Food Webs

The ocean's increasing acidity is a serious threat to coral reefs and other organisms that construct rigid structures from calcium compounds. As you may remember from Chapter 7, as the ocean absorbs more of the carbon dioxide that results from the increased burning of fossil fuels, carbonic acid forms and the pH of seawater falls. Over the past 200 years, the ocean has taken up about 35% of the excess carbon dioxide generated by the burning of fossil fuels—ocean surface acidity has

[6]For more on this troubling practice, see Questions from Students #6, page 539.

increased by nearly 30% since the 17th century. Average oceanic pH has fallen by 0.025 units since the early 1990s and is expected to drop to pH 7.8 by 2100, lower than any time in the past 420,000 years (**Figure 18.23**). Fewer carbonate ions will be available for shell-building organisms. Eventually, corals, plankton, and other organisms will fail to form strong skeletons.

Some corals can survive acidic conditions by reverting to sea-anemone form and then when pH conditions become more normal, they resume skeleton building. Most would not. The effect of marine acidification on marine food webs could be severe—whole groups of calcium-forming phytoplankters might decline in numbers and the consumers dependent on them would be forced to find other sources of food.

Sound Is Also a Pollutant

The human contribution to ocean noise has increased substantially in the past few decades—human noise has become the dominant component of marine sound in some regions, and noise is directly correlated with increasing industrialization of the ocean and coastal zone (**Figure 18.24**, see pp. 528–529).

Do these unnatural sounds harm marine life? Any animal unlucky enough to be very near the intense sound pulses generated by the air guns deployed by ships prospecting for oil or by powerful military sonar arrays will suffer ill effects, but the intensity of sound falls off rapidly with distance, so most marine

NOAA

Figure 18.22 Coral bleaching. Environmental damage can cause coral animals to expel their symbiotic dinoflagellates (zooxanthellae). If conditions do not improve, the colony can weaken, be overtaken by algae, and die.

The ocean is becoming more acidic as it absorbs additional carbon dioxide from the atmosphere. A less alkaline environment will make it more difficult for organisms to build hard structures containing calcium (shells, coral skeletons, among others) from dissolved carbonates. The charts show changes in sea surface pH from the late 1800s to the year 2100. Tropical coral reefs may be unable to survive.

THE UNMAKING OF SHELLS AND SKELETONS

Snails, barnacles, sea urchins, corals—there's a long list of marine organisms that make their hard parts by combining calcium and carbonate ions they get from the water. When atmospheric carbon dioxide levels go up, the organisms' supply of essential carbonate goes down. Here's how.

❶ Increasing CO_2 in the air forces more CO_2 into surface waters. Slowly, it spreads into the deep.

❷ CO_2 reacts with water, releasing hydrogen ions, which acidify the water. (They lower its pH.)

❸ Hydrogen ties up carbonate ions, converting them into bicarbonate ions.

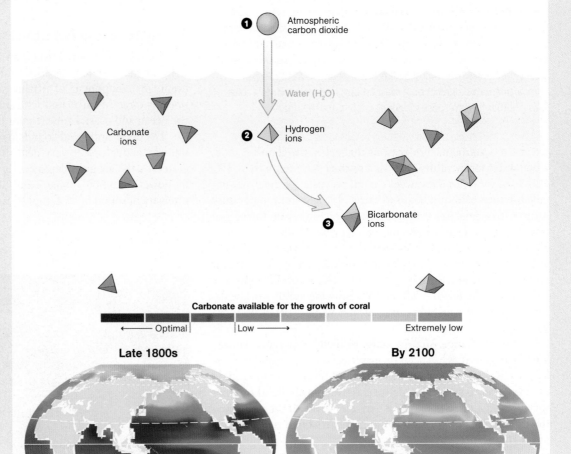

Carbonate available for the growth of coral

← Optimal | Low → Extremely low

Late 1800s   By 2100

A GROWING PROBLEM FOR CORAL REEFS

In the late 1800s, when fossil-fuel carbon dioxide began to pile up rapidly in the atmosphere and acidify the ocean, tropical corals weren't yet affected. But today carbonate levels have dropped substantially near the Poles; by 2100 they may be too low even in the tropics for reefs to survive.

GRAPHIC: JASON LEE; MARIEL FURLONG, NGM STAFF. MAPS: TED SICKLEY; NGM MAPS
SOURCES: ANDREW G. DICKSON, SCRIPPS INSTITUTION OF OCEANOGRAPHY, UC SAN DIEGO (GRAPHIC); SARAH COOLEY, WOODS HOLE OCEANOGRAPHIC INSTITUTION (MAPS)

organisms will be exposed only to a continuous and growing background hum.[7]

Many marine animals use sound to orient themselves to their surroundings, to search for prey (or avoid being eaten), to advertise availability for mating, and even as a form of camouflage. Active use of sound by an animal is easy to detect, but passive sensing—silent listening—is not. Would it make a difference if some populations of fishes were unable to hear rain falling on the sea surface? The distance to a surf zone? The subtle low-frequency pressure changes of a school of fish beyond the limit of sight? We know very little about the impact of anthropogenic (human-generated) noise on the fitness of individuals within populations.

A study has been proposed to predict how increasing anthropogenic sound will affect marine communities. The International Quiet Ocean Experiment (IQOE) proposes to quantify the ocean soundscape and examine the possible effects of increasing sound on marine organisms.

CONCEPT CHECK

Before going on to the next section, check your understanding of some of the important ideas presented so far:

Which areas are most at risk for disruption by invasive species?

What benefits do estuaries provide? What are some threats to marine estuaries?

What dangers threaten coral reef communities?

The ocean is becoming more acidic as it absorbs more carbon dioxide. What effects could this increasing acidity have on marine organisms?

How might unnatural sounds affect the behavior of marine organisms or the health of communities?

18.4 Marine Protected Areas Are Refuges

Marine protected areas (MPAs) are regions in which human activity is restricted to protect the natural environment. These areas are intended as safe havens for marine life. Despite their name, protected areas are not always off-limits to commercial fishing, trawling, or dredging.

Beginning in 1972, the U.S. federal government has established more than 1,600 MPAs covering 36% of U.S. marine waters. These vary in size but include coral reefs, whale migration corridors, undersea archaeological sites, deep canyons, and zones of extraordinary beauty and biodiversity. The MPAs vary widely in purpose, legal authorities, managing agencies, management approaches, level of protection, and restrictions on human uses.

Other countries are actively involved. By the end of 2010, more than 5,880 MPAs had been set aside, encompassing 1.17% of the world ocean. On 1 April 2010, the British government established the Chagos Archipelago in the Indian Ocean as the world's largest marine reserve. At 640,000 square kilometers (247,000 square miles), it is larger than France or the state of California. Its designation essentially doubled the total area of environmental "no take" zones worldwide.

Listed for reasons of species diversity, area, and unique habitats, other important MPAs are:

- The Bowie Seamount on the Coast of British Columbia, Canada.
- The Great Barrier Reef in Queensland, Australia.
- The Ligurian Sea Cetacean Sanctuary in the seas of Italy, Monaco, and France.
- The Dry Tortugas in the Florida Keys, United States.
- Papahānaumokuākea Marine National Monument in Hawai'i.
- The Phoenix Islands Protected Area, Kiribati.
- The Channel Islands Marine Protected Areas in California, United States.
- The Chagos Archipelago in the Indian Ocean.
- The Antarctic Whaling Preserve.[8]

CONCEPT CHECK

Before going on to the next section, check your understanding of some of the important ideas presented so far:

Have areas set aside for marine conservation areas and sanctuaries grown in overall size or become smaller in the past decade?

18.5 Earth's Climate Is Changing

Our planet's variable climate is shaped by hundreds of interlocked physical and biological factors. Of these, ocean and atmospheric interaction is the most important. The ocean and the atmosphere are extensions of each other, and natural processes (along with human activity) have changed the atmosphere as they have changed the ocean. Substances injected into the air can have global consequences for the ocean and for all of Earth's inhabitants. Among the most troublesome recent ocean–atmosphere changes are climate change and global warming.

Earth's Surface Temperature Is Rising

The surface temperature of Earth varies slowly over time. The global temperature trend has been generally upward in the 18,000 years since the last ice age, but the *rate* of increase has recently accelerated. This rapid warming is probably the result

---

[7]Ships' propellers are a major source of oceanic noise. The volume of cargo transported by sea has been doubling approximately every 20 years—the background noise level is relentlessly rising, especially along shipping lanes of the North Pacific and around the British Isles.

[8]In 1994 the International Whaling Commission voted to ban whaling in about 21 million square kilometers (8 million square miles) around Antarctica, thus protecting most of the remaining large whales, which feed in those waters. The killing ban is not enforced.

Human-made sound is affecting marine organisms. Animals use sound to find food and mates, to avoid predators, to orient to their surroundings, and to communicate. Anthropogenic noise has multiplied a hundredfold since 1960. The consequences are as yet unknown.

 THINKING BEYOND THE FIGURE

What are the most powerful sources of anthropogenic sound in the ocean?

Stefan Fichtel, Sources: C.W.Clark, Cornell Lab of Ornothoglogy; Brandon Southall, University of California, Santa Cruz; Kathleen Vigness-Raposa, Marine Acoustics, Inc.

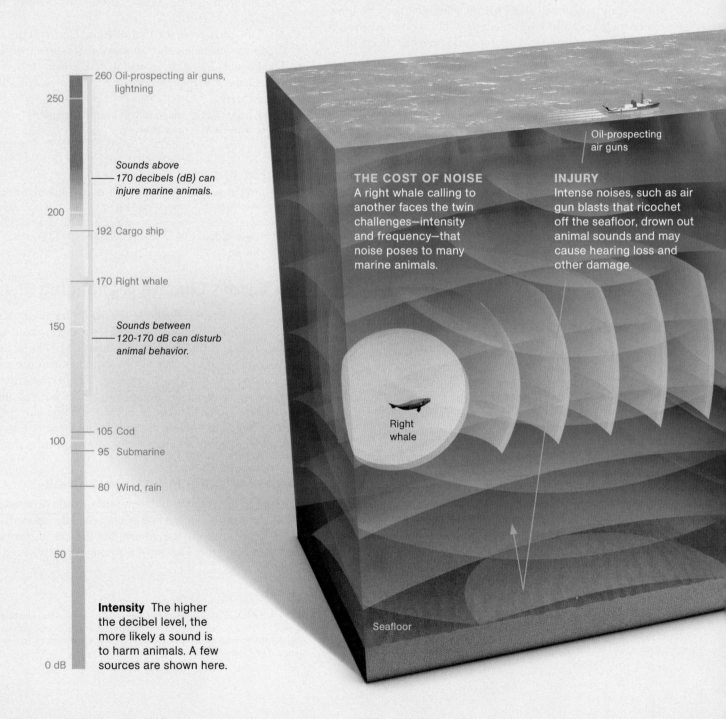

260 Oil-prospecting air guns, lightning

250

Sounds above 170 decibels (dB) can injure marine animals.

200

192 Cargo ship

170 Right whale

150

Sounds between 120-170 dB can disturb animal behavior.

105 Cod

100

95 Submarine

80 Wind, rain

50

Intensity The higher the decibel level, the more likely a sound is to harm animals. A few sources are shown here.

0 dB

Oil-prospecting air guns

THE COST OF NOISE
A right whale calling to another faces the twin challenges—intensity and frequency—that noise poses to many marine animals.

INJURY
Intense noises, such as air gun blasts that ricochet off the seafloor, drown out animal sounds and may cause hearing loss and other damage.

Right whale

Seafloor

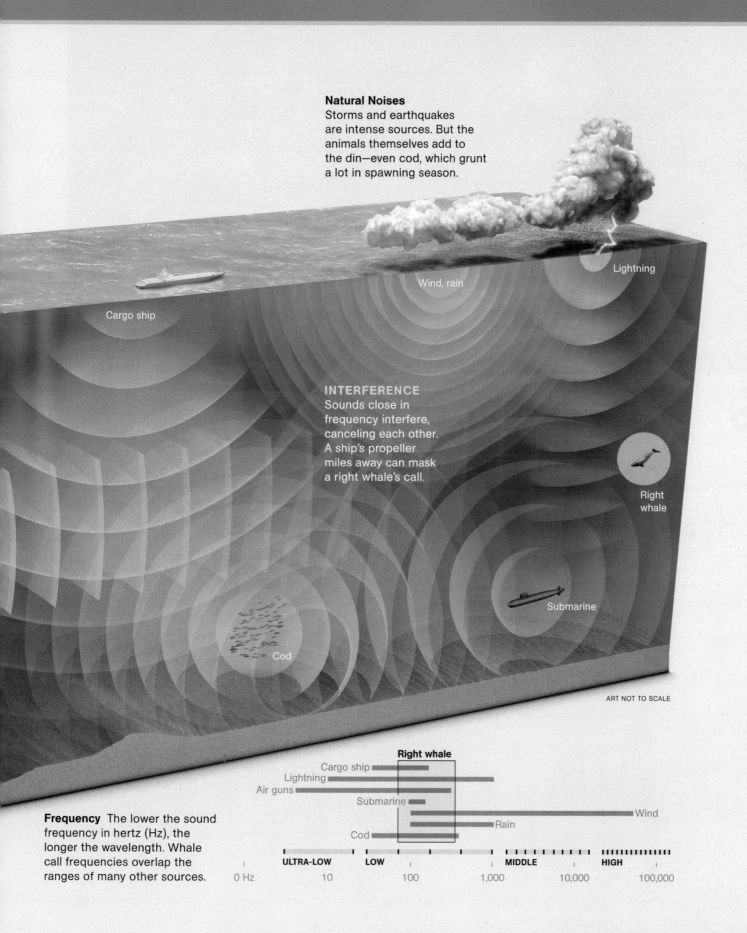

Natural Noises
Storms and earthquakes are intense sources. But the animals themselves add to the din—even cod, which grunt a lot in spawning season.

Lightning

Wind, rain

Cargo ship

INTERFERENCE
Sounds close in frequency interfere, canceling each other. A ship's propeller miles away can mask a right whale's call.

Right whale

Submarine

Cod

ART NOT TO SCALE

Frequency The lower the sound frequency in hertz (Hz), the longer the wavelength. Whale call frequencies overlap the ranges of many other sources.

Right whale

Cargo ship
Lightning
Air guns
Submarine
Wind
Rain
Cod

ULTRA-LOW   LOW   MIDDLE   HIGH

0 Hz   10   100   1,000   10,000   100,000

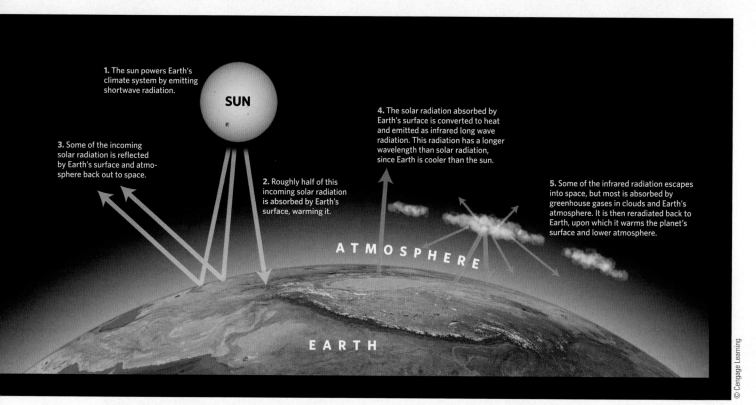

Figure 18.25 How the greenhouse effect works.

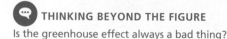 **THINKING BEYOND THE FIGURE**

Is the greenhouse effect always a bad thing?

of an enhanced **greenhouse effect**, the trapping of heat by the atmosphere.

Glass in a greenhouse is transparent to light but not to heat. The light is absorbed by objects inside the greenhouse, and its energy is converted into heat. The temperature inside a greenhouse rises because the heated air is unable to escape. On Earth **greenhouse gases**—water vapor, carbon dioxide, methane, CFCs, and others—take the place of glass. Heat that would

otherwise radiate away from the planet is absorbed and trapped by these gases, causing surface temperature to rise. **Figure 18.25** shows this mechanism.

The greenhouse effect is necessary for life; without it, Earth's average atmospheric temperature would be about −18°C (0°F). Earth has been kept warm by natural greenhouse gases. The sources of these gases are volcanic and geothermal processes, the decay and burning of organic matter, and respiration and other

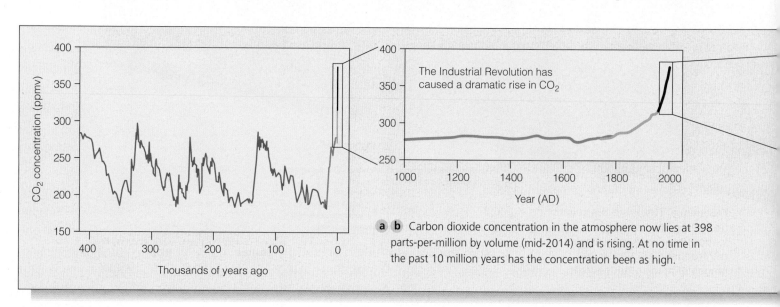

a b Carbon dioxide concentration in the atmosphere now lies at 398 parts-per-million by volume (mid-2014) and is rising. At no time in the past 10 million years has the concentration been as high.

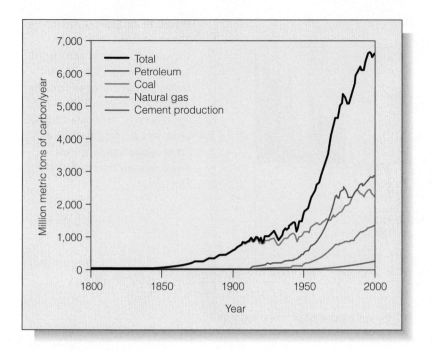

Figure 18.26 Global carbon emission by source, 1800–2000.

Source: Figure was prepared by Robert A. Rohde from publicly available data and is part of the Global Warming Art Project. Original data from G. Marland, T. A. Boden, and R. J. Andres. 2003. "Global, regional, and national CO_2 emissions." In *Trends: A Compendium of Data on Global Change.* Carbon Dioxide Information Analysis Center, Oak Ridge National Laboratory, U.S. Department of Energy, Oak Ridge, TN. http://cdiac.esd.ornl.gov/trends/emis/tre_glob.htm.

biological sources. The removal of these gases by photosynthesis and absorption by seawater appears to prevent the planet from overheating.

But the human demand for quick energy to fuel industrial growth, especially since the beginning of the Industrial Revolution, has injected unnatural amounts of new carbon dioxide into the atmosphere from the combustion of fossil fuels. The concentration of CO_2 in the atmosphere, second only to water vapor as the most effective greenhouse gas, has risen by 38% since that time. In 2009, the world burned through nearly 31 billion barrels of oil, 6 billion tons of coal, and 100 trillion cubic feet of natural gas. This added about 30 billion tons of excess carbon dioxide to the atmosphere. *If* the present rate of economic growth is sustained, these figures will rise by about 2% each year.

Carbon dioxide is now being produced at a greater rate than it can be absorbed by the ocean **(Figure 18.26). Figure 18.27** shows how much carbon dioxide concentrations have increased in the atmosphere over the past 400,000 years (note especially the spike beginning about 1850). The atmosphere's carbon dioxide content now rises at the rate of 0.4% each year. As we write (in the early summer of 2014), the atmospheric concentration of CO_2 is about 398 parts per million by volume. At no time in the past 10 million years has the concentration been so high.[9]

Earth is now absorbing about 0.85 watts per square meter more energy from the Sun than it is emitting into space. There has been a 5°C (9°F) rise in global temperature from the end of the last ice age until today; most of this increase has occurred only in the past 200 years. Carbon dioxide and other human-generated greenhouse gases produced since 1880 are thought

---

[9]For a daily report of CO_2 concentration, go to: http://keelingcurve.ucsd.edu/

Figure 18.27 Atmospheric carbon dioxide variations through time.

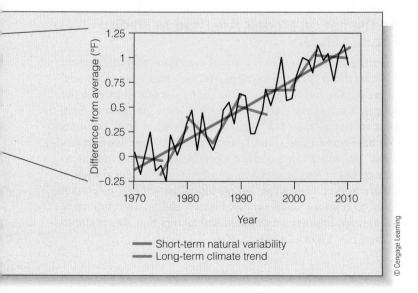

c Short-term variations versus long-term warming trend. Note that the most recent short-term trend (uppermost blue line) seems to indicate a pause in global warming. The long-term trend continues upward, however.

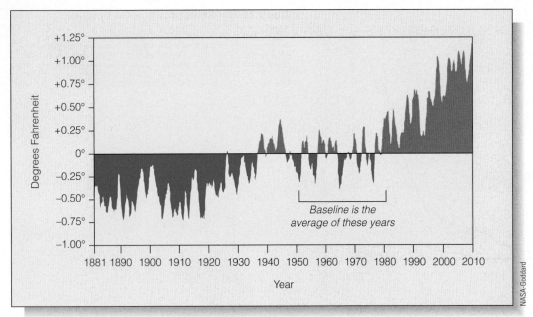

NASA-Goddard

Figure 18.28 Mean global surface air temperatures, 1881–2010, shown as a 12-month moving average. The chart shows how temperatures have varied from their average for the years 1951–1980 (the baseline). These data are based on surface air measurements at meteorological stations and satellite measurements of surface temperature.

to be responsible for much of that sudden increase. (As we'll see shortly, other factors also contribute to climate change.)

Figure 18.28 shows changes observed in global atmospheric temperature since 1881. The year 2010 was the warmest on record; the hottest recorded years have all occurred since 1998. The rise has not been steady—some periods actually show cooling, but the *overall* trend is up.

As **Figure 18.29** indicates, the ocean is warming, too. Observations suggest that about 90% of the excess heating of Earth's surface since the 1950s is stored in the ocean. Though the measured temperature increase is small (0.1°C over the past 50 years), remember water has a *very* high heat capacity (discussed in Chapter 6)—it takes a large amount of heat to raise water's temperature. Floating polar ice is also melting rapidly—when reflective ice is replaced by dark ocean water, more sunlight can be absorbed (**Figure 18.30**). Increasing warmth has caused the ocean to expand and sea level to rise.

The accelerated melting of the land-based Greenland and Antarctic ice caps is also adding water to the ocean. **Figure 18.31** indicates the nature of the problem. Imagine the effect of a significant rise in sea level on the harbors, coastal cities, river deltas, and wetlands, where one third of the world's people now live. Sea level is projected to rise another 60 centimeters (2 feet) by the year 2100.

These and other problems associated with climate warming are shown in **Figure 18.32**.

How are we to make decisions about how to respond to climate change in the face of such uncertainty? We can experiment with mathematical models.

Mathematical Models Are Used to Predict Future Climates

A **mathematical model** is a set of equations that attempts to describe the behavior of a system. You use a mathematical model to predict the future value of your savings account, based on the type of account you have, the length of time you plan to keep your money in the bank, experience with other bank accounts you have owned, and other factors. Based on their assumptions about the relative importance of individual factors that contribute to climate, researchers construct mathematical models of atmosphere and ocean interaction in an attempt to predict future conditions based on past history, and on assumptions of the relative importance of individual factors that contribute to climate (see Figure 18.32).

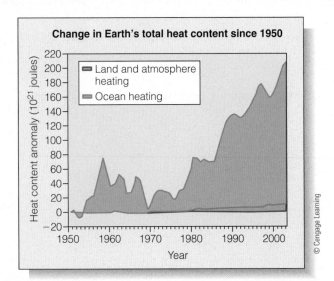

© Cengage Learning

Figure 18.29 Earth's total heat content is growing. About 90% of excess heating of Earth's surface since 1950 is stored in the upper 700 meters (2,330 feet) of the ocean.

September 1997

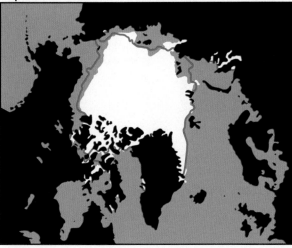

September 2007

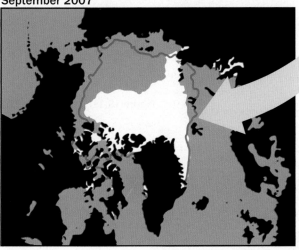

a The change in arctic ice from September 1997 to September 2007.

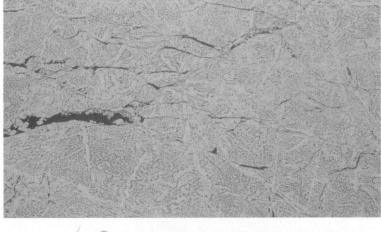

b A photo taken 3° south of the North Pole in the summer of 2011. Open water leads are clearly visible.

Figure 18.30 A melting arctic.

Consider, for example, some of the many factors thought to affect Earth's climate that are included in climate models such as that shown in **Figure 18.33**:

- How much greenhouse gas is being produced?
- What are the relative contributions of methane, water vapor, and carbon dioxide to warming?
- How have volcanic eruptions affected climate?
- How do variations in solar output affect warming?
- How fast is carbon dioxide being absorbed by the ocean?

The relative importance assigned by a climate model to each of these factors (and a great many others) will influence the predictions made by that model. Add to that the astonishing inherent complexity of atmosphere–ocean interaction, and one begins to appreciate the daunting task faced by climate researchers. Politicians, policy-makers, and the public look to specialists to make definitive statements about future climate change, but our present understanding of the weight of each variable makes such definition remarkably difficult.

Can Global Warming Be Curtailed?

Most climate researchers are confident that human activity is a major contributor, accounting for most of the observed heating. In 2010, the United States National Research Council stated: "...there is a strong, credible body of evidence, based on multiple lines of research, documenting that climate is changing and that these changes are in large part caused by human activities. While much remains to be learned, the core phenomenon, scientific questions, and hypotheses have been examined thoroughly and have stood firm in the face of serious scientific debate and careful evaluation of alternative explanations."[10]

By 2013 models and data showed an even stronger correlation. The Intergovernmental Panel on Climate Change (IPCC), an intergovernmental body under the auspices of the United Nations, issued a report saying: "... it is *extremely likely* that human influence has been the dominant cause of the observed warming since the mid-20th century."[11] (Emphasis in original)

---

[10]National Research Council. 2010. *Advancing the Science of Climate Change.* Washington, DC: The National Academies Press.

[11]IPCC. 2013. Summary for Policymakers. In: *Climate Change 2013: The Physical Science Basis. Contribution of Working Group I to the Fifth Assessment Report of the Intergovernmental Panel on Climate Change.* Stocker, T. F., D. Qin, G.-K. Plattner, M. Tignor, S. K. Allen, J. Boschung, A. Nauels, Y. Xia, V. Bex and P. M. Midgley (eds.). Cambridge University Press, Cambridge, United Kingdom and New York, NY, USA.

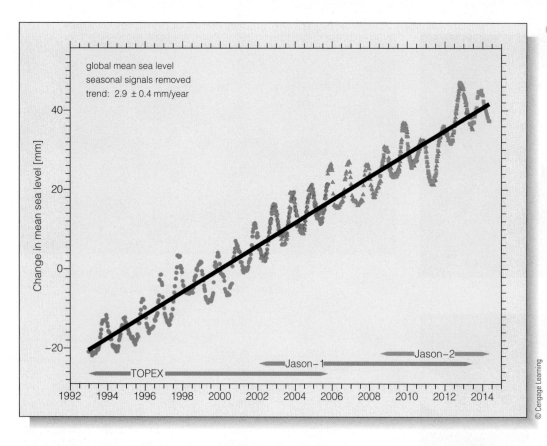

© Cengage Learning

a Sea level rise since 1992 as measured by the *TOPEX/Poseidon*, Jason-1 and Jason-2 satellites. Doggerlanders would feel right at home when reading this chart.

Figure 18.31 Sea level is rising.

b For island nations such as the Maldives, even a small rise in sea level could spell disaster. This lowest, flattest country on Earth is strung out across 880 kilometers (550 miles) of the Indian Ocean. Nearly all of its inhabitants live less than a meter (3.3 feet) above sea level. Most of the population lives in Maale, where the effects of this century's sea level rise of 10 to 25 centimeters (4 to 10 inches) have already been felt. Of the country's 1,180 islands, only a handful would survive the median estimate of sea level rise by 2100. Yann Arthus-Bertrand/Corbis

c Venice, perhaps the world's most beautiful city, is sinking at a rate of about 3 millimeters (0.12 inches) a year—about 30 centimeters (one foot) a century. The seabed on which it rests is also tilting eastward a millimeter or two each year. During times of high tides and storms, the ocean creeps into rooms on the "ground" floor, rendering them uninhabitable. The population of the city is falling as long-time residents abandon this great city for higher ground. A costly mitigation system is being planned. Carlo Morucchio/Robert Harding World Imagery/Corbis

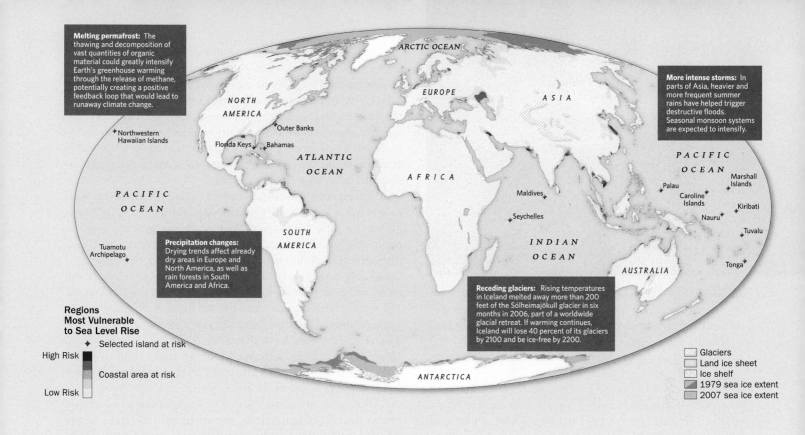

Figure 18.32 Some of the possible effects of climate warming. © Cengage Learning

Melting permafrost: The thawing and decomposition of vast quantities of organic material could greatly intensify Earth's greenhouse warming through the release of methane, potentially creating a positive feedback loop that would lead to runaway climate change.

More intense storms: In parts of Asia, heavier and more frequent summer rains have helped trigger destructive floods. Seasonal monsoon systems are expected to intensify.

Precipitation changes: Drying trends affect already dry areas in Europe and North America, as well as rain forests in South America and Africa.

Receding glaciers: Rising temperatures in Iceland melted away more than 200 feet of the Sólheimajökull glacier in six months in 2006, part of a worldwide glacial retreat. If warming continues, Iceland will lose 40 percent of its glaciers by 2100 and be ice-free by 2200.

Regions Most Vulnerable to Sea Level Rise

✦ Selected island at risk

High Risk
Coastal area at risk
Low Risk

Glaciers
Land ice sheet
Ice shelf
1979 sea ice extent
2007 sea ice extent

The impact of climate change is being felt. Writing in 2014 in the third official National Climate Assessment (NCA), the U. S. Global Research Program says that Americans face wide-ranging impacts in every region of our country and throughout our economy. The report notes that "[P]lanning is occurring in the public and private sectors at all levels of government; however, few measures have been implemented."[12]

It will be exceedingly difficult to limit global warming by decreasing the production of carbon dioxide. In the past hundred years, industrial production has increased 50-fold; we have burned truly astonishing quantities of fossil fuel. Carbon dioxide is a major product of combustion for all of these hydrocarbon compounds. We generate more than 71 million metric tons (78 million tons) of carbon dioxide *per day* (**Figure 18.34**)! The world's energy demand is projected to increase 3.5 times between now and 2025, with carbon dioxide emissions 65% higher than today. China alone plans to add 18,000 megawatts of coal-fired electric generating capacity each year—equal to Louisiana's entire power grid. For each kilowatt-hour of electrical energy produced, about 631 grams (1.39 pounds) of CO_2 is released into the air.

Climates change, and life adapts. But we may have no choice but to err on the side of safety and roll back the production of

[12]2014 U.S. National Climate Assessment (http://nca2014.globalchange.gov/)

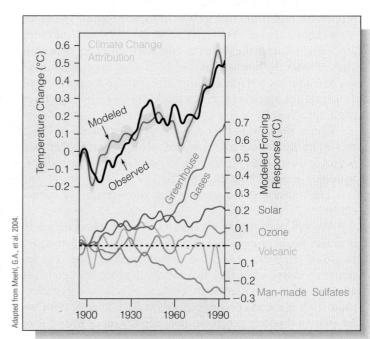

Adapted from Meehl, G.A., et al. 2004.

Figure 18.33 A mathematical model of climate change. Five factors capable of altering climate are summed to provide an estimate of change (brown line). The model is tested against observations (black line). The closer the agreement, the more accurately the model is said to predict future conditions.

Figure 18.34 Thilafushi, an island in the Indian Ocean nation of Maldives, is devoted almost exclusively to waste disposal by accumulation and incineration. The plume contains toxic heavy metals such as mercury, lead, and cadmium.

heat-retaining gases. Do we have the political and economic will to proceed? Attempts have been made. At a meeting in Kyoto, Japan, in 1997, leaders and representatives of 160 countries established carbon dioxide emission targets for each developed country. The United States, for example, would have reduced its carbon dioxide emissions to 7% less than 1990 levels by 2012. This goal was thought to be economically untenable, and the Kyoto Protocol was not ratified by the U.S. Senate. In May of 2014 the United States Environmental Protection Agency (EPA) proposed a regulation that would cut carbon dioxide emissions from existing coal plants by up to 30% by 2030 compared with 2005 levels. Under the draft rule, the EPA would let states and utilities meet the new standard with different approaches mixing four options including energy efficiency, shifting from coal to natural gas, investing in renewable energy, and making power plant upgrades.

China, the world's largest emitter, plans to set an absolute cap on its carbon dioxide emissions when its next 5-year plan begins in 2016. India is planning to make solar power available to every electrified household by 2019.

These measures would only slow the warming of Earth's climate.

We will need to find alternatives to fossil fuel if we are to maintain world economies and prevent an accelerating increase in global temperature with all its uncertainties. The only alternative source of energy that currently produces significant amounts of power is **nuclear energy**, which now generates about 19% of the electricity produced in the United States. (Solar and wind power generated only slightly more than 1% of total U.S. energy consumption in 2011.)

The problem with nuclear power lies not so much in the everyday operation of the reactors but in disposing of the nuclear wastes they produce. By 2009, there were 62,683 metric tons of commercial spent fuel stored in the United States. Of that total, about 48,818 metric tons—or about 78%—were in temporary storage in deep pools of cooled water, and another 13,856 metric tons—or about 22%—were stored in dry casks; they must be stored for 10,000 years before their levels of radioactivity will be low enough to pose no environmental hazard. Radioactive substances emit **ionizing radiation**, a form of energy that is able to penetrate and permanently damage cells. It is mandatory that artificial sources of ionizing radiation be isolated from the environment. Even functioning reactors are subject to catastrophic accident, as the March 2011, tsunami-induced incident at Fukushima in eastern Japan demonstrated.

Will citizens of industrial countries (and countries that wish to become industrialized) agree to lessen the danger of increased global warming by slowing their economic growth, limiting population growth, decreasing their dependence on fossil fuel combustion, and developing safe alternate sources of energy? Some insight may be gained from the behavior of ranchers and industrialists in the rain forests of New Guinea, the Philippines, and Brazil. The Amazon rain forest of Brazil is being burned at a rate of about 12 square kilometers (almost 5 square miles) *per hour*; acreage equivalent to the area of West Virginia every year. Huge stands of trees that should be nurtured to absorb excess carbon dioxide are being destroyed. The cleared land is used for farms, cattle ranches, roads, and cities.

Ultimately, of course, the cost of doing nothing will greatly exceed the cost of maintaining the status quo, so erring on the side of caution is not without important human, financial, and environmental costs. In the end, we will have to balance the enormous costs of cutting fossil fuel use against the benefits of reducing greenhouse gases.

CONCEPT CHECK

Before going on to the next section, check your understanding of some of the important ideas presented so far:

What is "greenhouse effect?" What gases are most responsible for it?

Is greenhouse effect always bad?

What causes global warming?

What effects might be caused by global warming?

What alternatives exist for burning hydrocarbon fuels for energy?

18.6 What Can Be Done?

In a pivotal paper published in 1968, biologist Garrett Hardin examined what he termed "The Tragedy of the Commons." Hardin's title was suggested by his study of societies in which some agricultural areas were held *in common*—that is, were jointly owned by all residents. Citizens of these societies owned small homes, plots of land, and perhaps a cow that was put to pasture on the commons. Each farmer *kept* the milk and cheese given by his cow but *distributed* the costs of cow ownership—overgrazing of the commons, cow excrement, fouled drinking water, and so on—among all the citizens. This arrangement worked well for centuries because wars, diseases, and poaching kept the numbers of people and cows well below the carrying capacity of the land. But eventually political stability and relative freedom from disease allowed the human (and cow) population to increase. Farmers pastured more cows on the commons and

Ibrahim Muneez/National Geographic Image Collection

Figure 18.35 Our Earth at night. Human-made light beams from populated areas on the surface, a striking display of the present extent of industrialization and resource use. NGDC/NOAA/U.S. Air Force Weather Agency

gained more benefits. Soon the overstressed commons could no longer sustain the growing numbers of cows, and the area held in common was ruined. Eventually no cows could survive there.

The lesson applies to our present situation. Hardin noted that in our social system, each individual tends to act in ways that maximize his or her material gain. Each of us gladly keeps the *positive* benefit of work but willingly distributes the *costs* among all. For example, this morning we drove to our college offices; the benefit to us was one trip to our office. A cost of this short drive was the air pollution generated by the fuel combustion in our car's engine. Did we route the exhaust fumes through a hose to a mask held tightly over our nose and mouth? (That is, did we reserve the environmental costs of our actions for our own use, just as we had reserved for ourselves the benefit of our ride to work?) No. we shared those fumes with our fellow Californians, just as you shared your morning's sewage with your fellow citizens, or just as the factory down the road shared its carbon dioxide with all the world. Indeed, *the world itself* is our commons. The modern tragedy of the commons rests on these kinds of actions.

The carrying capacity of the whole Earth as commons may already have been exceeded. Births now exceed deaths by about 3 people per second, 10,400 per hour. *Each year* there are 95 million more of us, a total equal to nearly one third of the population of the United States. The number of people tripled in the 20th century and is expected to double again before reaching a plateau sometime in this century. Another billion humans will join the world's present population of 7+ billion in the next 10 years, 92% of them in developing countries.[13] One fifth of the world's people already suffer in abject poverty and hunger.

This exploding population is not content with using the same proportion of resources used today. Citizens of the world's least developed countries are influenced by education and advertising to demand a developed-world standard of living. They look with misguided envy at the United States, a country with 5% of the planet's population that consumes 32% of its raw material resources and 24% of its energy, while generating 22% of industry- and transport-related carbon dioxide. As long as our economic systems require constant growth, we will continue to degrade the world ecosystem.

Human demand has exceeded Earth's ability to regenerate resources since at least the early 1980s. Since 1961, human demand on Earth's organisms and raw materials has more than doubled and now exceeds Earth's replacement capacity by about 20% **(Figure 18.35)**.

> The solution to environmental problems, if one exists, lies in education and action.

Can the world support a population whose expectations are rising as rapidly as their numbers? In Garrett Hardin's words, "We can maximize the number of humans living at the lowest possible level of comfort, or we can try to optimize the quality of life for a smaller population." The burgeoning human population is the greatest environmental problem of all.

---

[13]In the United States alone, the population is growing by the equivalent of four Washington D.C.s every year, another New Jersey every 3 years, another California every 12.

We cannot expect science to solve the problem for us. Most of the decisions and necessary actions fall outside pure science in the areas of values, ethics, morality, and philosophy. *The solution to environmental problems, if one exists, lies in education and action.* Each of us is obliged to become informed on issues that affect Earth, its ocean, and its air—to learn the arguments and weigh the evidence. Once informed, we would hope to act in rational ways. Chaining yourself to an oil tanker is not rational, but selecting well-designed, long-lasting, recyclable products made by responsible companies with minimal environmental impact (and encouraging others to do so) certainly is.

Obvious answers and quick solutions are often misleading; a great deal of research and work are needed to give reliable insight into the many difficult questions that confront us. The present trade-off between financial and ecological considerations is often strongly tilted in favor of immediate gain, short-term profit,

and instant convenience. Garrett Hardin suggests that absolute freedom in a commons brings ruin to all. Education may be the only way to modify these destructive behaviors.

Our cities are crowded and our tempers are short. As Jared Diamond wrote in this chapter's "Insight" piece, ". . . the world's environmental problems will get resolved, one way or another, within the lifetimes of the children and young adults alive today. The only question is whether they will become resolved in pleasant ways of our own choice, or in unpleasant ways not of our choice. . ." Times of turbulent change lie before us. The trials ahead will be severe.

What to do? Each of us, individually, needs to take a stand. We must preserve the sunsets and fog; the waves to ride; the cold, clean windblown spray on our faces—our one world ocean (**Figure 18.36**). Margaret Mead summarized our potential for making a difference: "Never doubt that a small group of thoughtful, committed citizens can change the world. Indeed it is the only thing that ever has." *We need to start now.*

Robert Ellis

Figure 18.36

"What would the world be, once bereft
Of wet and wildness? Let them be left,
O let them be left, wildness and wet;
Long live the weeds and the wilderness yet."

—Gerard Manley Hopkins

both photos: Tom Garrison

In this chapter you learned that our species has always exercised its capacity to consume resources and pollute its surroundings, but only in the past few generations have our efforts affected the ocean and atmosphere on a planetary scale. The introduction into the biosphere of unnatural compounds (or natural compounds in unnatural quantities) has had—and will continue to have—unexpected detrimental effects. The destruction of marine habitats and the uncontrolled harvesting of the ocean's living resources have also disturbed delicate ecological balances. We have embarked on a time of inadvertent global experimentation and find ourselves in difficult situations for which solutions do not come easily. We hesitate to adjust course as we rush into the unknown.

This is an unpleasant chapter—the only one we don't enjoy working on. Nonetheless, we urge you think carefully about what you've seen, then read the Afterword that follows. The problems are yours to solve. You'll find some encouragement there.

The next chapter is up to you.

QUESTIONS FROM STUDENTS

1. **What can an individual do to minimize his or her impact on the ocean and atmosphere?**

Remember that Earth and all its millions of life-forms are interconnected. There are no true consumers, only users. Nothing can truly be thrown away (there is no "away"). We must abandon the pollute-and-move-on ethic that has guided the actions of most humans for thousands of years. *Our task is not to multiply and subdue Earth;* we must work toward a society more in harmony with the fundamental rhythms of life that sustain us. It may not be too late to change our ways. We need to act individually to effect change. We should think globally and act locally.

2. **What happens to all those plastic water bottles?**

The United States has the safest water supply in the world. Despite this fact, American consumption of bottled water has doubled over 10 years. We now consume about 167 million bottles of water each day—nearly all of it packaged in disposable plastic. Filtered and bottled tap water, brilliantly marketed as an alternative to plain old boring tap water, can cost more than 4,000 times as much as the stuff flowing from the kitchen faucet. It often costs more than unleaded premium gasoline! Only about one in six plastic water bottles sold in the United States is recycled. Now add the cost of distribution (some water arrives on store shelves from great distances) to the price of the petroleum products needed to manufacture the bottle. Perhaps it's time to save one of those bottles and refill it at the sink.

3. **I don't really *see* any effects of climate change. Yes, it's hot in summer and cold in the winter, but the ocean doesn't seem any higher this year than it did a few years ago. And a lot of people deny anything important is happening—they say global warming seems to have stopped about 10 years ago.**

Take another look at Figure 18.27c in which short-term variations are superimposed on the long-term trend. It takes a while for natural systems to react—there is quite a bit of inertia in these large and complex systems. The historical record is disarmingly clear, however. Over the decades, we think the general public will gain direct experience with climate change.

4. **Do you think daylight-saving time could be contributing to global warming? The longer we have sunlight, the more it heats the atmosphere.**

What?

5. **Wait a minute—I remember the winter of 2013–2014. The U.S. east coast suffered the worst blizzards on record, and many cold-temperature records were set. How could 2013 have been one of the hottest years on record?**

Here's an excerpt from a 2010 paper by James Hansen of NASA's Goddard Space Flight Center. No one could put it better:

"Communication of the status of global warming to the public has always been hampered by weather variability. Laypeople's perception tends to be strongly influenced by the latest local fluctuation. This difficulty can be alleviated by stressing the need to focus on the frequency and magnitude of warm and cold anomalies, which change noticeably on decadal time scales as global warming increases."

"Other obstacles to public communication include the Media's difficulty in framing long-term problems as 'news,' a preference for sensationalism, a generally low level of familiarity with basic science, and a preference for 'balance' in every story. The difficulties are compounded by the politicization of reporting of global warming, a perhaps inevitable consequence of economic and social implications of efforts required to alter the course of human-made climate change."[14]

6. **How about marine aquarium fishes? Are we wrecking reef ecosystems by buying them?**

In the recent past, most reef fishes were collected indiscriminately by people with no interest in preserving the species or environment for the future. Fishes were stunned by squirting a solution of cyanide into reef crevices. The reef would be poisoned, and the fish would often die of liver failure shortly after arriving in their new homes. That's changing in many places, in part because aquarium fish are now big business,

---

[14]Hansen, J., R. Ruedy, Mki. Sato, and K. Lo, 2010: Global surface temperature change. Rev. Geophys., 48, RG4004, doi:10.1029/2010RG000345.

and consumers are becoming aware of the methods used in their collection.

Over 20 million fishes from 1,471 species, and another 10 million invertebrates, are caught every year for the trade, now worth about US$330 million annually. For countries like Sri Lanka, protecting and preserving the "fishery" provides steady employment to 50,000 low-income people. The Marine Aquarium Council, a trade association, is beginning a certification process to provide information to consumers about the methods used to collect any fishes or other reef animals they may want to purchase. This could force countries like the Philippines and Indonesia, whose collectors rely heavily on cyanide, to abide by sustainable collection methods or convert to mariculture.

7. What are the *most* dangerous threats to the environment, overall?

The underlying causes of the problems discussed in this chapter are human population growth and a growth-dependent economy. Stanford professor Paul Ehrlich said, "Arresting global population growth should be second in importance on humanity's agenda only to avoiding nuclear war." The present world population, now more than 6.9 billion, seems doomed to reach at least 10 billion before leveling off. And what if everybody wants to live in comfort associated with developed countries?

TERMS AND CONCEPTS TO REMEMBER

biodegradable	eutrophication	hypoxia	mathematical model
biomagnification	greenhouse effect	introduced species	nuclear energy
chlorinated hydrocarbons	greenhouse gases	ionizing radiation	pollutant
		marine pollution	

STUDY QUESTIONS

Thinking Critically

1. Why is refined oil more hazardous to the marine environment than crude oil? Which is spilled more often? What happens to oil after it enters the marine environment?

2. What heavy metals are most toxic? How do these substances enter the ocean? How do they move from the ocean to marine organisms and people?

3. Few synthetic organic chemicals are dangerous in the very low concentrations in which they enter the ocean. How are these concentrations increased? What can be the outcome when these substances are ingested by organisms in a marine food chain?

4. How can a decrease in stratospheric ozone affect planktonic productivity?

5. What is the greenhouse effect? Is it always detrimental? What gases contribute to the greenhouse effect? Why do most scientists believe that the Earth's average surface temperature will increase over the next few decades? What may result?

6. How might global warming *directly* affect the ocean?

7. What is the tragedy of the commons? Do you think Garrett Hardin was right in applying the old idea to modern times? What will you do to minimize your negative impact on the ocean and atmosphere?

Thinking Analytically

1. The cost of pollution and habitat mismanagement, over time, will be higher than the cost of doing nothing. But the cost *now* is cheaper. Arguing only from practical standpoints (that is, avoiding an appeal to emotion), how could you convince the executive board of an industrial corporation in a developed country dependent on an ocean resource to reduce or eliminate the negative effects of its activities?

2. Considering the same question, how would you convince the board of a corporation in a developing country (say, China or India)?

3. If a pollutant has effects at a dilution of one part per billion by weight, how much seawater is contaminated by the release of one metric ton of the material into a bay? Consider either Chesapeake Bay or San Francisco Bay in your calculations.

4. You and your wife spend a 2-week vacation in Hawai'i during which you eat top marine carnivores (bluefin and albacore tuna, swordfish, marlin) for most lunches and each evening for dinner. A month after your return, you and your wife are delighted to discover you're pregnant. How concerned should you be about the health of your child?

GLOBAL GEOSCIENCE WATCH

Visit www.cengagebrain.com to access MindTap, a complete digital course which includes access to Global Geoscience Watch and more. Within the GREENR database, search for "marine protected areas." You may wish to further narrow your results by searching for fully protected "marine reserves" within your results. Utilize as many different types of sources as possible. Research marine protected areas within the database. Use this information to write a short report which addresses the following: 1) What are the expected benefits of marine protected areas? 2) What are some of the challenges associated with their implementation? 3) Identify at least three locations where marine protected areas have been established and describe how the areas have changed since they were instituted.

CENGAGEbrain.com Visit www.cengagebrain.com to access course materials for this text, including interactive learning tools, videos, and more.

Afterword

The marine sciences are at the threshold of a new age. The 20th century's revolutions in biology and geology are being assimilated, and the road ahead seems clearer. A revolution in the design of sampling devices, robot submersible vehicles, and data processing has brought new vigor to oceanography. Satellite-borne sensors can provide in an instant data that would have taken decades to collect using surface ships. Shipboard technology has become so sophisticated that Wyville Thomson or Fridtjof Nansen would hardly recognize our sensors or sampling devices.

The tools may be different, but the spirits of those who use them remain the same. Today's marine scientists are like all the men and women who have gone before: *We want to know about the ocean.* We eagerly search our mailboxes for journals bearing the latest research news, scan Internet Web sites daily for discoveries, inspect new samples with the enthusiasm of little kids, and share our insights with anyone at the drop of a hat. We are personally delighted that you have traveled with us this far. Those of us who enjoy an oceanographic background (and this now includes you) look at Earth with greater understanding than we did before we began. The whole concept of an ocean world appeals to us, gives us profound pleasure, and sobers us with a deep sense of responsibility. In no other field of science do so many ideas interweave to form so rich a tapestry.

Our journey together is over, but before we go our separate ways, we have four last ideas to share:

- *Change* has been a recurrent theme of this book. Earth's climate has changed with time, as has its atmospheric composition, its ocean chemistry, the size and positions of its continents, and its life-forms. Our Earth may seem a calm and stable home, but it is really a violent place for inhabitation by such seemingly delicate objects as living things. Even so, life and the ocean have grown old together. The story of Earth is the story of change and chance; its history is written in the rocks, the water, and the genes of the millions of organisms that have evolved here. We are survivors.

  That survival may now be in question. Change is now progressing at an unnatural rate, and these human-induced changes are imposing stress on natural systems. What we do with and to the ocean is literally of planetary consequence. In the past century, we have developed the physical, chemical, and biological machinery to destroy or rejuvenate the world ocean and all of its life. A painful time of inadvertent global experimentation lies just ahead.

  All of us who love the colors and textures of this small wet world need to act to moderate the negative effects of the looming environmental crisis. In Chinese, the written character for the word *crisis* has two components: danger and opportunity. Informed citizens will express their concern, discuss this concern with others, and act whenever possible to minimize the threats and take advantage of new opportunities. Intelligence and beauty must triumph; we have no other rational alternative.

- Appreciation of the ocean doesn't come exclusively from the realm of science. Philosophers, artists, composers, and poets have had much to say about the sea. Read Homer's description of the ocean in *The Odyssey* (try Books IV, X, and XI). See how Lord Byron's feeling for the ocean colors his poetry (see, for instance, *Childe Harold's Pilgrimage*, stanzas 183 and 184). Read modern poet Robinson Jeffers's powerful *Continent's End.* Share Prospero's marine magic in Shakespeare's *The Tempest.* See what Chinese philosopher Lao Tsu has to say about our need to find solace in the natural world. Enjoy some of the evocative woodcuts of Rockwell Kent and the impressionistic ocean paintings of English artist J. M. W. Turner. Listen to Benjamin Britten's *Four Sea Interludes* from *Peter Grimes* and Ralph Vaughan-Williams's *Sea Symphony* and *Sinfonia Antarctica* (but take care not to blow out your sound system). Read the ocean novels of Herman Melville and Jack London, and try reading the journals and accounts of the famous explorers and scientists you have met in this book. Sit on a quiet beach at night with the stars of the Milky Way shining softly overhead. The pervasive inspiration of the wave-breathing ocean is never far away.

- Don't let your involvement stop here. Lifelong learning is the truest joy, a pleasure that does not diminish with age, a source of wisdom and calm. We can learn much about patience, hope, and optimism from the ocean. We can learn much about the world—and about ourselves—by looking for the oceanic connections among things. The authors hope your interest in learning about the ocean has been kindled. There is much good in the world. Go and add to it.

- Share your insight with family and friends. *Use* your new knowledge—make it your own. You don't need to be a college professor to talk to people about the beauty, history, and future of the ocean. Our families have always been patient with and receptive to our oceanic tilt. Our children and grandchildren are our tolerant built-in students. Some family members are participants in the preface and Figure 18.36. Along with your children, they will inherit the world.

In the Preface we wrote, "Readers are invited to see the connections between astronomy, economics, physics, chemistry, history, meteorology, geology, and ecology—areas of study they may once have considered separate." Thomas Traherne, an English poet and naturalist, encapsulated this unifying idea vividly in his long poem *Centuries of Meditation* (c. 1670).

You never enjoy the world aright,
till the sea floweth in your veins,
till you are clothed with the heavens,
and crowned with the stars.
Thomas Traherne, c.1670

Enjoy your newfound knowledge of oceanography. As you know, this is only a small planet, but how *very* beautiful it is!

Measurements and Conversions

Other than the United States, only two countries in the world—Liberia and Myanmar—do not use metric measurements. The metric system, a contribution of the French Revolution, conquered Europe along with Napoleon. It is based on a decimal system, a system familiar to Americans because of our decimal money system: 10 cents to a dime, 10 dimes to a dollar.

The first move toward a rational system of measurement was made in 1670 by Gabriel Mouton, the vicar of St. Paul's Church in Lyon, France. Instead of the then-prevalent measurement system based on the width of the king's hand, or the length of his outstretched right arm, or the weight of a particular basket of stones kept in the palace, Mouton suggested a length measure based on the arc of 1 minute of longitude, to be subdivided decimally. Other measurements would follow from this unit of length. His proposal contained the three major characteristics of the metric system: using Earth itself as a basis for measurement, subdividing decimally (by 10s), and using standard prefixes (*kilo, centi, milli,* and so on). These ideas were debated for 125 years before being implemented by a commission appointed by Louis XVI in one of his last official acts before being imprisoned during the French Revolution. One ten-millionth of the distance from the North Pole to the equator (on the line of longitude passing through Paris) was selected as the standard unit of length, the meter. A new unit of weight was derived from the weight of 1 cubic meter of pure water. Temperature was to be based on pure water's boiling and freezing points. A list of prefixes for decimal multiples and submultiples was proposed. In 1795, a firm decision was made to establish the system throughout France, and in 1799, the metric system was implemented "for all people, for all time."

At first, people objected to the changes, but the government insisted that old measurements be included side by side with the equivalent new (metric) ones. In everyday competition, the advantages of the metric system proved decisive; in 1840, it was declared a legal monopoly in France. The French public had been won over to the new, simple, rational system of measurement. All of Europe—and, eventually, virtually all other countries—followed.

Not the United States, however. Though Ben Franklin proposed that the country convert in the 18th century, the people of the United States have continued to insist that the metric system—now known as the Système International (SI)—is too difficult to learn and work with. The federal government has urged conversion to metric units to increase opportunities for international trade. In August 1988, President Ronald Reagan signed the Omnibus Trade and Competitiveness Act. This act amended the 1975 Metric Conversion Act, stating that by 1992 all federal agencies must, wherever feasible, use the metric (SI) system in their purchases, grants, and other business.[1] (It should be noted that Canada began to convert in 1970 and has been metric since 1980.)

The government may be making the change, but the public clings tenaciously to inches, pints, and pounds. Why? Is it really simpler to add $\frac{1}{16}$ of an inch, $\frac{1}{12}$ of an inch, and $\frac{3}{8}$ of an inch to cut a bookshelf to length? Can you remember how many cups to a quart? How many pints to a gallon? How many miles to a league? The reason we continue to use the old English Imperial system (which, of course, the English have long since abandoned) is that it is familiar to us. We know how long 5 inches is, and how much a quart is, and what 72° Fahrenheit represents. Perhaps by following the French example—by having measurements expressed everywhere in English *and* metric measurements—we may be able to make a complete conversion within a generation or two. That's why American and metric measurements are used together throughout this book. The process has already begun, of course: You use 35mm film, 2-liter soft drink containers, 750-milliliter wine bottles, 100-watt light bulbs—and you might run a 10-K (10-kilometer) race on Saturday.

The conversion factors listed here will give you an idea of how American and metric (SI) units are equivalent. Don't panic—the system is as rational and logical as it has always been. Note that 1 meter equals 100 centimeters and that 1 centimeter equals 10 millimeters. Note that 2.54 centimeters equals 1 inch. (See the table of conversion factors if you wish to convert from one system to the other.) Some numerical oceanographic data are included in supplemental tables.

---

[1]Sadly, Congress cancelled most of the provisions on the Metric Conversion Act in 2000.

Scientific Notation

Multiples and Submultiples	Name	Common Prefixes
$10^{18} = 1,000,000,000,000,000,000$		exa
$10^{15} = 1,000,000,000,000,000$		peta
$10^{12} = 1,000,000,000,000$	trillion	tera
$10^{9} = 1,000,000,000$	billion	giga
$10^{6} = 1,000,000$	million	mega
$10^{3} = 1,000$	thousand	kilo
$10^{2} = 100$	hundred	hecto
$10^{1} = 10$	ten	deka
$10^{-1} = 0.1$	tenth	deci
$10^{-2} = 0.01$	hundredth	centi
$10^{-3} = 0.001$	thousandth	milli
$10^{-6} = 0.000001$	millionth	micro
$10^{-9} = 0.000000001$	billionth	nano
$10^{-12} = 0.000000000001$	trillionth	pico

© Cengage Learning

Conversion Factors

Area

1 square inch (in.²)	6.45 square centimeters
1 square foot (ft²)	144 square inches
1 square centimeter (cm²)	0.155 square inch 100 square millimeters
1 square meter (m²)	10^4 square centimeters 10.8 square feet
1 square kilometer (km²)	247.1 acres 0.386 square mile 0.292 square nautical mile

© Cengage Learning

Mass

1 kilogram (kg)	2.2 pounds 1,000 grams
1 metric ton	2,205 pounds 1,000 kilograms 1.1 tons
1 pound	16 ounces 454 grams 0.45 kilogram
1 ton	2,000 pounds 907.2 kilograms 0.91 metric ton

© Cengage Learning

Pressure

1 atmosphere (sea level)	760 millimeters of mercury at 0°C 14.7 pounds per square inch
	33.9 feet of water (fresh)
	29.9 inches of mercury
	33 feet of seawater

© Cengage Learning

Length

1 micrometer (μm)	0.001 millimeter 0.0000349 inch
1 millimeter (mm)	1,000 micrometers 0.1 centimeter 0.001 meter
1 centimeter (cm)	10 millimeters 0.394 inch 10,000 micrometers
1 meter (m)	100 centimeters 39.4 inches 3.28 feet 1.09 yards
1 kilometer (km)	1,000 meters 1,093 yards 3,281 feet 0.62 statute mile 0.54 nautical mile
1 inch (in.)	25.4 millimeters 2.54 centimeters
1 foot (ft)	30.5 centimeters 0.305 meter
1 yard	3 feet 0.91 meter
1 fathom	6 feet 2 yards 1.83 meters
1 statute mile	5,280 feet 1,760 yards 1,609 meters 1.609 kilometers 0.87 nautical mile
1 nautical mile	6,076 feet 2,025 yards 1,852 meters 1.15 statute miles
1 league	15,840 feet 5,280 yards 4,804.8 meters 3 statute miles 2.61 nautical miles

© Cengage Learning

Volume

1 cubic inch (in.³)	16.4 cubic centimeters
1 cubic foot (ft³)	1,728 cubic inches 28.32 liters 7.48 gallons
1 cubic centimeter (cc; cm³)	1,000 cubic millimeters 0.061 cubic inch
1 liter	1,000 cubic centimeters 61 cubic inches 1.06 quarts 0.264 gallon
1 cubic meter (m³)	106 cubic centimeters 264.2 gallons 1,000 liters
1 cubic kilometer (km³)	10^9 cubic meters 10^{15} cubic centimeters 0.24 cubic mile

© Cengage Learning

Some Familiar Metric Approximations

Measurement	Metric Unit	Approximate Size of Unit
Length	millimeter	diameter of a paper clip wire
	centimeter	a little more than the width of a paper clip (about 0.4 inch)
	meter	a little longer than a yard (about 1.1 yards)
	kilometer	somewhat farther than $\frac{1}{2}$ mile (about 0.6 mile)
Mass (Weight)	gram	a little more than the mass (weight) of a paper clip
	kilogram	a little more than 2 pounds (about 2.2 pounds)
	metric ton	a little more than a ton (about 2,200 pounds)
Volume	milliliter	five of them make a teaspoon
	liter	a little larger than a quart (about 1.06 quarts)
Pressure	kilopascal	atmospheric pressure is about 100 kilopascals

Source: U.S. Metric Board Report.

Temperature

$$°C = \frac{(°F - 32)}{1.8}$$

$$°F = (1.8 \times °C) + 32$$

$$K = °C + 273.2$$

100°C = 212°F
(boiling point of water)

40°C = 104°F
(heat-wave conditions)

37°C = 98.6°F
(normal body temperature)

30°C = 86°F
(very warm—almost hot)

20°C = 68°F
(a mild spring day)

10°C = 50°F
(a warm winter day)

0°C = 32°F
(freezing point of water)

	°F	°C
Water boils	212	100
	160	80
		60
Body temperature	98.6	37
	80	20
Water freezes	32	0
	0	−20
	−40	−40

© Cengage Learning

Time

1 hour	3,600 seconds
1 day	86,400 seconds 1,440 minutes 24 hours
1 calendar year	31,536,000 seconds 525,600 minutes 8,760 hours 365 days

© Cengage Learning

Speed

1 statute mile per hour	1.61 kilometers per hour 0.87 knot
1 knot (nautical mile per hour)	51.5 centimeters per second 1.15 miles per hour 1.85 kilometers per hour
1 kilometer per hour	27.8 centimeters per second 0.62 mile per hour 0.54 knot

© Cengage Learning

Numerical Oceanographic Data

Equivalence in Concentration of Seawater

Seawater with 35 grams of salt per kilogram of seawater	3.5 percent 35 parts per thousand (‰) 35,000 parts per million (ppm)

Speed of Sound

Velocity of sound in seawater at 34.85 parts per thousand (‰)	4,945 feet per second 1,507 meters per second 824 fathoms per second

Area, Volume, and Depth of the World Ocean

Body of Water	Area (10^6 km^2)	Volume (10^6 km^3)	Mean Depth (m)
Atlantic Ocean	82.4	323.6	3,926
Pacific Ocean	165.2	707.6	4,282
Indian Ocean	73.4	291.0	3,963
All oceans and seas	361	1,370	3,796

APPENDIX 2

Geologic Time

As we saw in Chapter 1, astronomers and geologists have determined that Earth originated about 4.6 billion years ago. They have divided Earth's age into eras, roughly corresponding to major geologic and evolutionary changes that have taken place, as shown in the figure below. Note that the time spans of the different eras are not shown to scale; if they were, the chart would run off the page.

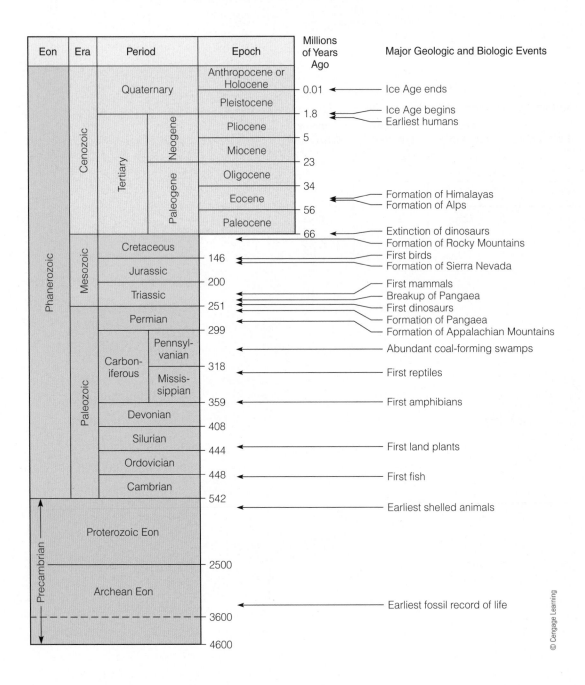

Eon	Era	Period		Epoch	Millions of Years Ago	Major Geologic and Biologic Events
Phanerozoic	Cenozoic	Quaternary		Anthropocene or Holocene	0.01	Ice Age ends
				Pleistocene	1.8	Ice Age begins / Earliest humans
		Tertiary	Neogene	Pliocene	5	
				Miocene	23	
			Paleogene	Oligocene	34	
				Eocene	56	Formation of Himalayas / Formation of Alps
				Paleocene	66	Extinction of dinosaurs / Formation of Rocky Mountains
	Mesozoic	Cretaceous			146	First birds / Formation of Sierra Nevada
		Jurassic			200	First mammals / Breakup of Pangaea
		Triassic			251	First dinosaurs / Formation of Pangaea
	Paleozoic	Permian			299	Formation of Appalachian Mountains / Abundant coal-forming swamps
		Carboniferous	Pennsylvanian		318	First reptiles
			Mississippian		359	First amphibians
		Devonian			408	
		Silurian			444	First land plants
		Ordovician			448	First fish
		Cambrian			542	Earliest shelled animals
Precambrian		Proterozoic Eon			2500	
		Archean Eon			3600	Earliest fossil record of life
					4600	

© Cengage Learning

Latitude and Longitude, Time, and Navigation

The ocean is large and easy to get lost in. A backyard, like that shown in **Figure 1**, is smaller, but we can still be lost in it if we don't have a frame of reference. Note that the yard is framed by a fence. We can refer to this frame to establish our position—in this case, at the intersection of perpendicular lines drawn from fence posts 2 and C. Many towns are arranged in this way: Fourth and D Streets intersect at a precise spot based on the municipal frame of reference.

But the World Is Round: Spherical Coordinates

If the world were flat, a simple scheme of rectangular coordinates would serve all mapping purposes—a rectangle, like the yard in Figure 1, has four sides from which to measure. A sphere has no edges, no beginnings or ends, so what shall we use as a frame of reference for Earth? Because Earth turns, the poles—the axis of rotation—are the only absolute points of reference. We can draw an imaginary line equidistant from the North and South Poles, a line that *equates* the globe into northern and southern halves: the equator. Other lines, drawn parallel to the equator, further divide the sphere north and south of the equator. These lines, or parallels, are lines of **latitude (Figure 2)**.

We can further subdivide Earth by drawing lines at regular intervals through both poles. Note that unlike the parallels, these lines, called meridians, are all equally long. Meridians are lines of **longitude (Figure 3)**.

If you travel north from the equator, you can count the parallels (lines of latitude) that cross your path to find out how far you have gone. Likewise, if you travel east from a reference meridian, you can count the meridians (lines of longitude) that cross your path to find out how far you have gone. Just as a football player on the field knows his distance from the goal line by the yard lines that cross his run, so you know how far north or east you have gone by the lines that have crossed your path.

Because there are no continuous lines of fence posts on the spherical Earth, our reference frame for latitude and longitude must be marked from the equator and poles by some other means. This is done by degrees.

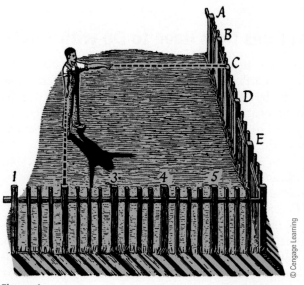

Figure 1

Figure 2

© Cengage Learning

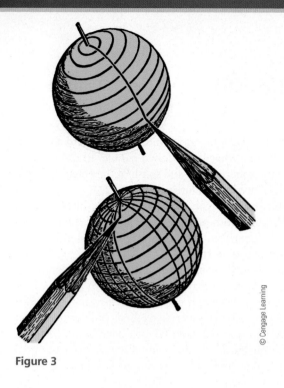

Figure 3

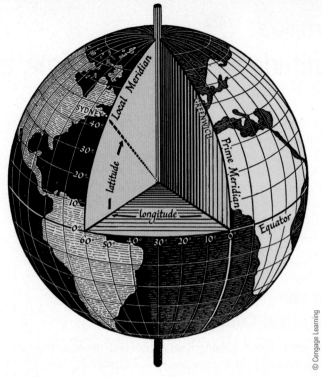

Figure 4

Why Degrees?

Degrees measure fractions of a circle. We need to know what fraction of Earth's circumference separates us from the equator and from the reference meridian to have a definite idea of our location.

Babylonian astronomers first divided the circle into 360 degrees (°). Why 360? The moon cycles around Earth every 30 days. It takes about 12 months ("moonths") to make a year. Thus, $30 \times 12 = 360$, the number of days they supposed was in a year. Circles were divided the same way. As we saw in Chapter 2, the Greek librarian Hipparchus applied this division to the surface of Earth.

In **Figure 4** we have marked the position of Sydney, Canada. A line drawn from Sydney to the center of Earth intersects the plane of the equator at an angle of 46° to the north. That is its latitude.

The reference meridian, the meridian from which all others are marked, is known as the prime meridian. Unlike the equator, there is no earthly reason the prime meridian should pass through any particular place. It passes through Greenwich, England, because an international agreement signed in 1884 decreed it so. The meridian on which Sydney, Canada, lies intersects the plane of the prime meridian at an angle of 60°. The angular distance of Sydney from the prime meridian is 60° to the west. That is its longitude. So its position is 46°N 60°W.

We can do this for each hemisphere. A line drawn to the center of Earth from Sydney, Australia, intersects the plane of the equator at an angle of 34° south latitude. Sydney, Australia, lies 151° east of the prime meridian. Thus, its position is 34°S

151°E. (Note that the greatest possible longitude is 180°; once you pass 180°, the line opposite the prime meridian, you begin to come around the other side of Earth, and the angle to Greenwich decreases.)

What Does Time Have to Do with This?

Meridians are often numbered from the prime meridian in 15° increments. Earth takes 24 hours to complete a 360° rotation. Divide 360° by 24 hours and you get 15, the number of degrees the sun moves across the sky in 1 hour. Meridians on a globe are often spaced to represent 1 hour's turning of Earth toward or away from the sun, toward or away from the moon.

You can use this fact to find your east-west position, your longitude. Imagine that you have a radio that can tell you the precise time of noon at Greenwich.[1] If your local noon comes *before* Greenwich noon, you are east of Greenwich. For instance, if the sun is highest in your sky at 10:00 a.m. Greenwich time, you are 2 hours before—30° east—of Greenwich. Earth must turn 2 more hours before the sun will shine directly above the Greenwich meridian. If your local noon is *after*

---

[1] Any shortwave radio will do. Tune it to 2.5, 5, 10, 15, or 20 mHz for radio stations WWV (Colorado) or WWVH (Hawai'i). These stations broadcast time signals giving a measure of coordinated universal time, an international time standard based on the time at Greenwich. For a telephone report of coordinated universal time, call WWV at (303) 499-7111 or go to www.time.gov on the Internet.

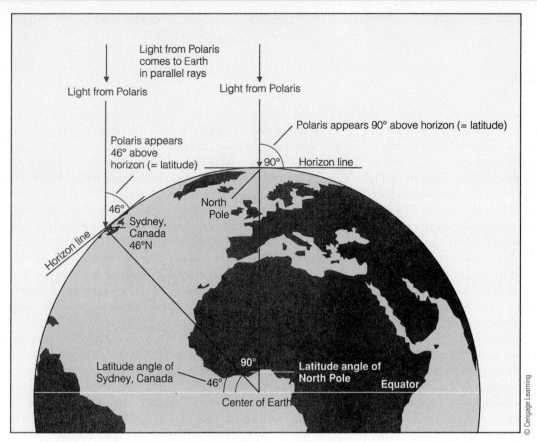

Figure 5 The estimation of latitude at Sydney, Canada, using the North Star.

Greenwich noon, you are west of Greenwich. Suppose that the sun is at high noon and your chronometer, set at Greenwich time, says 6:00 p.m. That means that Earth has been turning 6 hours since noon at Greenwich, and 6 hours times 15° per hour is 90°. That's your longitude relative to Greenwich: 90°W.

Navigation

Longitude is half the problem. To find latitude and obtain a position, we need to measure the angle north or south of the equator. But we can't use the time difference between local noon and Greenwich noon to determine latitude because the sun moves from east to west and we want to measure north-south position. Instead, we use the angle of the North Star above the horizon. Polaris, the current North Star, lies almost exactly above the North Pole. If we were standing at the North Pole, the North Star would appear almost directly overhead; ideally, the angle from the horizon to the star would be 90°, the same as the latitude of the North Pole (**Figure 5**). At Sydney, Canada, the angle from the horizon to the star would be about 46°—again, the same as the latitude. What would the angle of Polaris be at the equator, 0° latitude? (If you enjoyed your high school geometry course, you might try to prove that the angle from the horizon to Polaris is equal to the latitude at any position in the Northern Hemisphere.)

Polaris is not visible in the Southern Hemisphere, so how can we find south latitude? By finding the angle above the horizon of other stars. In practice, navigators in both hemispheres use a sextant to measure angles from the horizon to selected stars, planets, the moon, and the sun. The time of the observation is carefully noted. The navigator takes these readings to his or her stateroom, consults a series of mathematical tables, does some relatively simple calculations to compensate for observational errors, and comes up to the pilothouse with the vessel's latitude and longitude, accurate (in the best of circumstances) to within $\frac{1}{2}$ mile, marked on a small slip of paper. The daily results are always entered into the ship's log.

New Tricks

Discovering position by measuring the angular positions of heavenly bodies—celestial navigation—is a dying art. Global positioning satellites, Loran-C, inertial platforms, radar, and other electronic wonders have largely replaced the romance of a navigator standing on the bridge squinting through a sextant. The slip of paper has been supplanted by the glow of backlit liquid-crystal readouts or a chart with an X marking the ship's position, accurate to within about 2 meters (6 feet), feeding out of a slot. Still, when the power fails, the human navigator becomes the most popular person on board.

APPENDIX 4
Maps and Charts

It is easier to draw a diagram to show someone how to get to a place than to describe the process in words. For centuries, travelers have made special diagrams—maps and charts—to jog their own memories and to show others how to reach distant destinations. A **map** is a representation of some part of Earth's surface, showing political boundaries, physical features, cities and towns, and other geographical information. A **chart** is also a representation of Earth's surface, but it has been specially designed for convenient use in navigation. It is intended to be worked on, not merely looked at. A **nautical chart** is primarily concerned with navigable water areas. It includes information such as coastlines and harbors, channels, obstructions, currents, depths of water, and the positions of aids to navigation.

Any flat map or chart is necessarily a distortion of the spherical Earth. If we roll a flat sheet of paper around a globe to form a cylinder, the paper will contact the globe only along one curve. Let's assume that it's the equator. If the lines of latitude and longitude on the globe are covered with ink, only the equator will contact the paper and print an exact replica of itself. Unroll the cylinder, and that part of the new map will be a perfect representation of Earth. To include areas north and south of the equator, we will have to "throw them forward" onto the paper; we need to *project* them in some way.

Now imagine our globe to be a translucent sphere. If we place a bright light at its center, we can project the lines of latitude and longitude onto the rolled paper cylinder **(Figure 1)**. Careful tracing of these lines will result in a map, but the areas away from the equator will be distorted: The farther from the equator, the greater the distortion. A useful modification of this projection—one that does not distort high latitudes as dramatically—was devised by Gerardus Mercator, a Flemish cartographer who published a map of the world in 1569. Though landmasses and ocean areas are not depicted as accurately in a Mercator projection as they would be on a globe, such a map is still useful because it enables mariners to steer a course over long distances by plotting straight lines.

The distortion in Mercator projections has led generations of schoolchildren to believe that Greenland is the same size as South America **(Figure 2)**. Mercator charts can distort our perceptions of the ocean as well: The area of the continental shelves at high latitudes, the amount of primary productivity in the polar regions, and the importance of ocean currents at the northerly and southerly extremes of an ocean basin may be exaggerated if presented in Mercator projection. The projection used in this book—a further modification of the Mercator projection known as the Miller projection—was chosen for its more accurate representation of surface area at high latitudes.

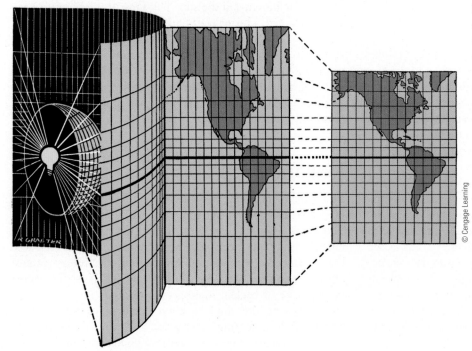

Figure 1 Central projection of a globe upon a cylinder, and a modified map structure, the Mercator, made to the same scale along the equator.

© Cengage Learning

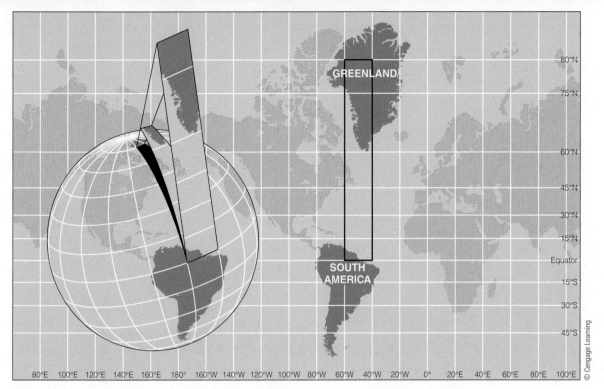

Figure 2 A gore of the globe peeled and projected according to the scheme devised by Gerardus Mercator. This is the projection used on modern sailing charts. Note that this projection's distortion at high latitudes makes Greenland and South America appear about the same size. Next time you're near a globe, check their real sizes.

Mapmakers have invented other projections, each with advantages and disadvantages for particular uses. Some are conical projections: a flat sheet of paper wrapped into a cone with its edge touching the globe at a line of latitude north (or south) of the equator and the point of the cone above the North (or South) Pole. Conical projections do not distort high-latitude areas in the same way a Mercator projection does and, if drawn for the ocean area in which a mariner is sailing, can be used to draw great circle routes as straight lines. However, the distortions inherent in a conical projection prevent it from being used to represent more than about one third of the globe on a single sheet of paper. Other projections, like the point-contact projection shown in **Figure 3**, try to minimize distortion around a specific location. All map and chart projections are distorted in some way; a sphere cannot be flattened onto a plane without deformation. Marine scientists necessarily become familiar with various chart projections and are careful to use the proper chart for the intended purpose.

Figure 4 is a Mercator projection of the world. On it are indicated areas of interest discussed in this book.

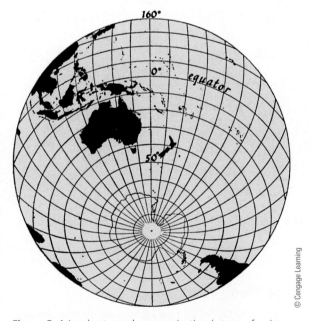

Figure 3 A Lambert equal-area projection (a type of point-contact projection), centered at 50°S 160°E.

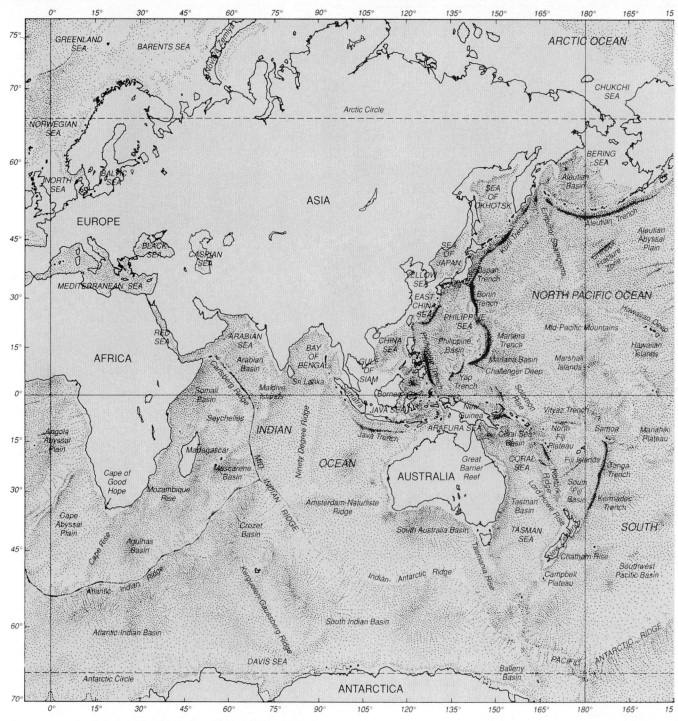

Figure 4 A map of the world, with many oceanic features labeled.

For Further Study

Herring, T. 1996. "The Global Positioning System." *Scientific American* 274 (no. 2): 44–50.

Krause, G., and M. Tomczak. 1995. "Do Marine Scientists Have a Scientific View of the Earth?" *Oceanography* 8 (no. 1): 11–16. Chart distortions often distort our interpretation of data, as this well-illustrated paper demonstrates.

Wilford, J. N. 1998. "Revolution in Mapping." *National Geographic* 193 (no. 2): 6–39. The usual thorough treatment of a rapidly changing topic.

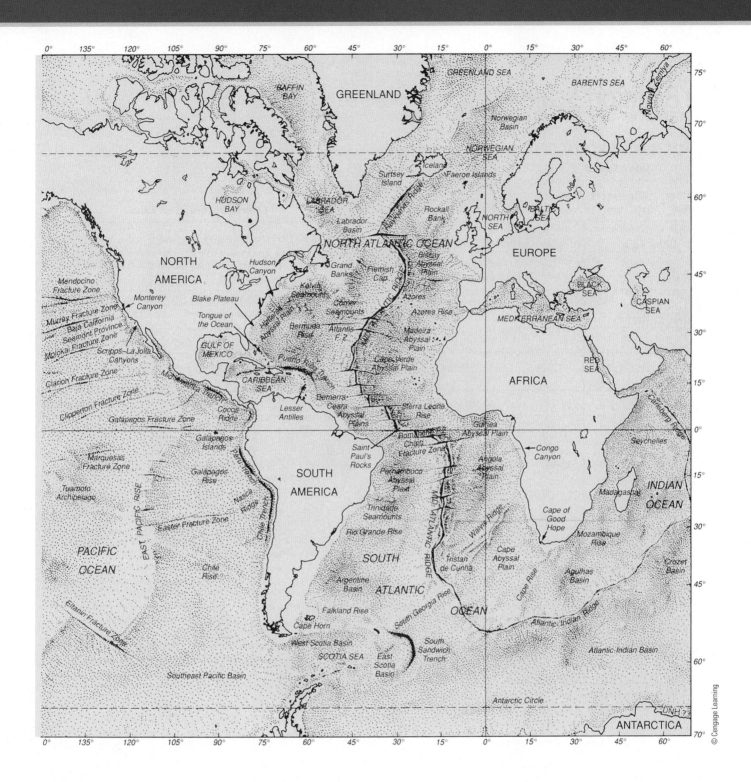

0°   135° 120° 105° 90° 75° 60° 45° 30° 15° 0° 15° 30° 45° 60°

75°

GREENLAND SEA

BARENTS SEA

Novaya Zemlya

BAFFIN
BAY

GREENLAND

70°

*Norwegian
Basin*

NORWEGIAN
SEA

*Surtsey
Island*   Iceland   *Faeroe Islands*

60°

HUDSON
BAY

LABRADOR
SEA

*Labrador
Basin*

Rockall
Bank

NORTH
SEA

BALTIC
SEA

NORTH

AMERICA

Hudson
Canyon

*Grand
Banks*

NORTH ATLANTIC OCEAN

*Flemish
Cap*

*Biscay
Abyssal
Plain*

EUROPE

45°

*Mendocino
Fracture Zone*

Monterey
Canyon

Blake Plateau

*Kelvin
Seamounts*

*Corner
Seamounts*

Azores

Azores Rise

BLACK
SEA

CASPIAN
SEA

Murray Fracture Zone

*Baja California
Seamont Province*

Molokai Fracture Zone

Tongue of
the Ocean

*Hatteras
Abyssal Plain*

*Bermuda
Rise*

*Atlantis
F.Z.*

*Madeira
Abyssal
Plain*

MEDITERRANEAN SEA

RED
SEA

30°

*Scripps–La Jolla
Canyons*

GULF OF
MEXICO

Puerto Rico Trench

CARIBBEAN
SEA

*Cape Verde
Abyssal Plain*

AFRICA

15°

Clarion Fracture Zone

MID-ATLANTIC RIDGE

Clipperton Fracture Zone

Mid-Atlantic Trench

Cocos
Ridge

Lesser
Antilles

*Demerara–
Ceara
Abyssal
Plains*

*Sierra Leone
Rise*

*Guinea
Abyssal Plain*

Carlsberg Ridge

0°

Galápagos Fracture Zone

*Marquesas
Fracture Zone*

Galápagos
Islands

Galápagos
Rise

SOUTH

AMERICA

Saint
Paul's
Rocks

Romanche F.Z.

*Chain
Fracture Zone*

*Pernambuco
Abyssal
Plain*

*Angola
Abyssal
Plain*

Congo
Canyon

Seychelles

INDIAN
OCEAN

15°

*Tuamoto
Archipelago*

Nasca Ridge

Peru-Chile Trench

EAST PACIFIC RISE

*Trinidade
Seamounts*

MID-ATLANTIC RIDGE

Walvis Ridge

Cape of
Good
Hope

Madagascar

*Mozambique
Rise*

30°

PACIFIC
OCEAN

Easter Fracture Zone

Rio Grande Rise

*Chile
Rise*

SOUTH

*Tristan
de Cunha*

*Cape
Abyssal
Plain*

Cape Rise

*Agulhas
Basin*

*Crozet
Basin*

45°

*Argentine
Basin*

ATLANTIC

South Georgia Rise

Atlantic-Indian Ridge

Falkland Rise

OCEAN

Elfanin Fracture Zone

Cape Horn

West Scotia Basin

SCOTIA SEA

*East
Scotia
Basin*

*South
Sandwich
Trench*

Atlantic-Indian Basin

60°

Southeast Pacific Basin

Antarctic Circle

UNH 7

70°

ANTARCTICA

0°   135° 120° 105° 90° 75° 60° 45° 30° 15° 0° 15° 30° 45° 60°

© Cengage Learning

The Beaufort Scale

| Beaufort Number | Wind Speed | | Effects Observed at Sea | Effects Observed on Land |
	Kilometers per Hour	Miles per Hour		
0	<1	<1	Sea like a mirror	Calm; smoke rises vertically
1	1–5	1–3	Ripples with appearance of scales (capillary waves), no foam	Smoke drifts to indicate wind direction; weather vanes don't move
2	6–11	4–7	Small wavelets, crest glassy and not breaking	Wind felt on face; leaves rustle; weather vanes begin to move
3	12–19	8–12	Wave crests begin to break; scattered whitecaps	Leaves in constant motion; lightweight flags extended
4	20–28	13–18	Small waves of longer wavelengths; numerous whitecaps	Dust, leaves, loose paper raised up; small branches move
5	29–38	19–24	Moderate waves, many whitecaps, some spray	Small trees begin to sway
6	39–49	25–31	Larger waves forming whitecaps everywhere, more spray	Large trees in motion; whistling heard in wires
7	50–61	32–38	Sea heaps up; white foam from breaking waves begins to be blown in streaks	Large trees in motion; resistance felt when walking against wind
8	62–74	39–46	Moderately high waves of long wavelength; foam is blown from crests in well-marked streaks	Small branches break off trees; walking progress impeded
9	75–88	47–54	High waves, dense streaks of foam, spray may reduce visibility	Slight structural damage; shingles are blown from roofs
10	89–102	55–63	High waves with overhanging crests; sea takes on white appearance from foam; visibility reduced	Trees broken or uprooted; considerable structural damage
11	103–117	64–72	Exceptionally high waves; sea covered with white foam; visibility sporadic	Very rarely experienced on land; usually accompanied by widespread damage
12	>118	>73	Hurricane conditions; air filled with foam; sea white with driven spray; visibility greatly reduced	Catastrophic damage to structures

Source: U.S. Navy, with modifications.

APPENDIX 6
Taxonomic Classification of Marine Organisms

Exclusively nonmarine phyla generally have been omitted, along with most extinct phyla and classes.

DOMAIN BACTERIA: Single-celled prokaryotes with a single chromosome that reproduce asexually and exhibit high metabolic diversity.

DOMAIN ARCHAEA: Superficially similar to bacteria, but with genes capable of producing different kinds of enzymes. Often live in extreme environments.

DOMAIN EUKARYA

KINGDOM PROTISTA: Eukaryotic single-celled, colonial, and multicellular autotrophs and heterotrophs.

PHYLUM CHRYSOPHYTA. Diatoms, coccolithophores, silicoflagellates.

PHYLUM PYRROPHYTA. Dinoflagellates, zooxanthellae.

PHYLUM CRYPTOPHYTA. Some "microflagellates"; cryptomonads.

PHYLUM EUGLENOPHYTA. A few "microflagellates"; mostly freshwater.

PHYLUM ZOOMASTIGINA. Nonphotosynthesizing flagellated protozoa.

PHYLUM SARCODINA. Amoebas and their relatives.

Class Rhizopodea. Foraminiferans.

Class Actinopodea. Radiolarians.

PHYLUM CILIOPHORA. Ciliated protozoa.

PHYLUM CHLOROPHYTA. Multicellular green algae.

PHYLUM PHAEOPHYTA. Brown algae, kelps.

PHYLUM RHODOPHYTA. Red algae, encrusting and coralline forms.

KINGDOM FUNGI: Fungi, mushrooms, molds, lichens; mostly land, freshwater, or highest supratidal organisms; heterotrophic.

KINGDOM PLANTAE: Photosynthetic autotrophs.

DIVISION ANTHOPHYTA. Flowering plants (angiosperms). Most species are freshwater or terrestrial. Marine eelgrass, manatee grass, surfgrass, turtle grass, salt marsh grasses, mangroves.

KINGDOM ANIMALIA: Multicellular heterotrophs.

PHYLUM PLACOZOA. Amoeba-like multicellular animals.

PHYLUM MESOZOA. Wormlike parasites of cephalopods.

PHYLUM PORIFERA. Sponges.

PHYLUM CNIDARIA. Jellyfish and their kin; all are equipped with stinging cells.

Class Hydrozoa. Polyp-like animals that often have a medusa-like stage in their life cycle, such as Portuguese man-of-war.

Class Scyphozoa. Jellyfish with no (or reduced) polyp stage in life cycle.

Class Cubozoa. Sea wasps.

Class Anthozoa. Sea anemones, coral.

PHYLUM CTENOPHORA. "Sea gooseberries," comb jellies; round, gelatinous, predatory, common.

PHYLUM PLATYHELMINTHES. Flatworms, tapeworms, flukes; many free-living predatory forms, many parasites.

PHYLUM NEMERTEA. Ribbon worms.

PHYLUM GNATHOSTOMULIDA. Microscopic, wormlike; live between grains in marine sediments.

PHYLUM GASTROTICHA. Microscopic, ciliated; live between grains in marine sediments.

PHYLUM ROTIFERA. Ciliated; common in freshwater, in plankton, and attached to benthic objects.

PHYLUM KINORYNCHA. Small, spiny, segmented, wormlike; live between grains in marine sediments; all marine.

PHYLUM ACANTHOCEPHALA. Spiny-headed worms; all parasitic in vertebrate intestines.

PHYLUM ENTOPROCTA. Polyp-like, small, benthic suspension feeders.

PHYLUM NEMATODA. Roundworms. Common, free-living, parasitic.

PHYLUM BRYOZOA. Common, small, encrusting colonial marine forms.

PHYLUM PHORONIDA. Shallow-water tube worms; suspension feeders; a few centimeters long; all marine.

PHYLUM BRACHIOPODA. Lampshells; bivalve animals, superficially like clams; scarce, mainly in deep water.

PHYLUM MOLLUSCA. Molluscs.

Class Monoplacophora. Rare deep-water forms with limpet-like shells.

Class Polyplacophora. Chitons.

Class Aplacophora. Shell-less, sand burrowing.

Class Gastropoda. Snails, limpets, abalones, sea slugs, pteropods.

Class Bivalvia. Clams, oysters, scallops, mussels, shipworms.

Class Cephalopoda. Squid, octopuses, nautiluses.

Class Scaphopoda. Tooth shells.

PHYLUM ARTHROPODA.

Subphylum Crustacea. Copepods, barnacles, krill, isopods, amphipods, shrimp, lobsters, crabs.

Subphylum Chelicerata. Horseshoe crabs, sea spiders.

Subphylum Uniramia. Insects, centipedes, millipedes; one genus and five species in the ocean.

PHYLUM PRIAPULIDA. Small, rare, wormlike, subtidal.

PHYLUM SIPUNCULA. Peanut worms; all marine.

PHYLUM ECHIURA. Spoon worms.

PHYLUM ANNELIDA. Segmented worms; includes polychaetes such as feather-duster worms and some oligochaete deep-sea bristle worms.

PHYLUM TARDIGRADA. "Water bears"; tiny, eight-legged animals with the ability to survive long periods of hibernation.

PHYLUM PENTASTOMA. Tongue worms; parasites of vertebrates.

PHYLUM POGONOPHORA. Beard worms; no digestive system; deep-water tube worms; all marine.

PHYLUM ECHINODERMATA. Spiny-skinned, benthic, radially symmetrical, most with a water-vascular system.

Class Asteroidea. Sea stars.

Class Ophiuroidea. Brittle stars, basket stars.

Class Echinoidea. Sea urchins, sand dollars, sea biscuits.

Class Holothuroidea. Sea cucumbers.

Class Crinoidea. Sea lilies, feather stars.

Class Concentricycloidea. Sea daisies.

PHYLUM CHAETOGNATHA. Arrowworms; stiff-bodied, planktonic, predaceous, common.

PHYLUM HEMICHORDATA. Acorn worms; unsegmented burrowers.

PHYLUM CHORDATA.

Subphylum Urochordata. Sea squirts, tunicates, salps.

Subphylum Cephalochordata. Lancelets, *Amphioxus*.

Subphylum Vertebrata.

Class Agnatha: Jawless fishes: lampreys, hagfishes; cartilaginous skeleton.

Class Chondrichthyes. Sharks, skates, rays, sawfish, chimaeras; cartilaginous skeleton.

Class Osteichthyes. Bony fishes.

Class Amphibia. Frogs, toads, salamanders; no marine species.

Class Reptilia. Sea snakes, turtles, one species of crocodile.

Class Aves. The birds.

Order Sphenisciformes. Penguins.

Order Procellariiformes. Albatrosses, petrels.

Order Charadriiformes. The gulls.

Order Pelecaniformes. The pelicans.

Class Mammalia. Warm-blooded, with hair and mammary glands.

Order Cetacea. Whales, porpoises, dolphins.

Order Sirenia. Manatees.

Order Carnivora. Two marine families.

Suborder Pinnipedia. Seals, sea lions, walruses.

Suborder Fissipedia. Sea otters.

Order Primates. One family that regularly enters the ocean.

Family Hominidae. Humans.

Periodic Table of the Elements

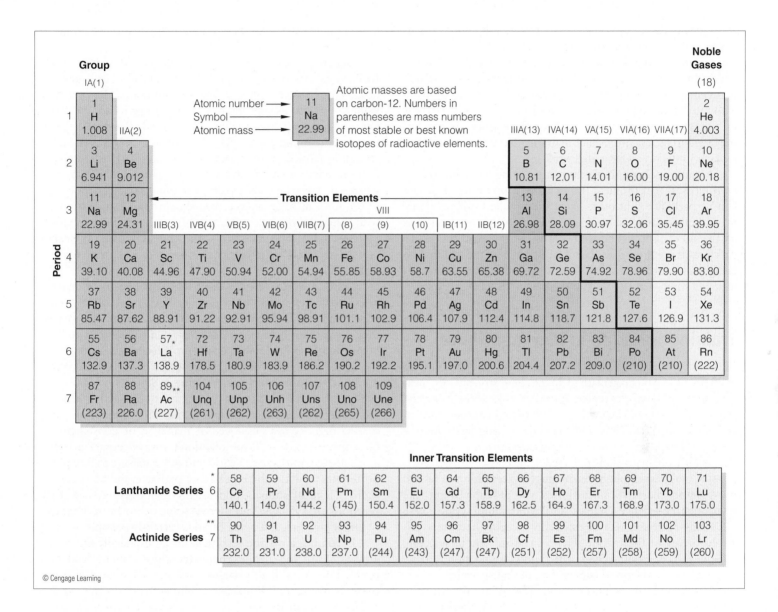

Group

Atomic number ⟶ 11
Symbol ⟶ Na
Atomic mass ⟶ 22.99

Atomic masses are based on carbon-12. Numbers in parentheses are mass numbers of most stable or best known isotopes of radioactive elements.

Noble Gases

Transition Elements

VIII

Inner Transition Elements

	IA(1)	IIA(2)	IIIB(3)	IVB(4)	VB(5)	VIB(6)	VIIB(7)	(8)	(9)	(10)	IB(11)	IIB(12)	IIIA(13)	IVA(14)	VA(15)	VIA(16)	VIIA(17)	(18)
1	1 H 1.008																	2 He 4.003
2	3 Li 6.941	4 Be 9.012											5 B 10.81	6 C 12.01	7 N 14.01	8 O 16.00	9 F 19.00	10 Ne 20.18
3	11 Na 22.99	12 Mg 24.31											13 Al 26.98	14 Si 28.09	15 P 30.97	16 S 32.06	17 Cl 35.45	18 Ar 39.95
4	19 K 39.10	20 Ca 40.08	21 Sc 44.96	22 Ti 47.90	23 V 50.94	24 Cr 52.00	25 Mn 54.94	26 Fe 55.85	27 Co 58.93	28 Ni 58.7	29 Cu 63.55	30 Zn 65.38	31 Ga 69.72	32 Ge 72.59	33 As 74.92	34 Se 78.96	35 Br 79.90	36 Kr 83.80
5	37 Rb 85.47	38 Sr 87.62	39 Y 88.91	40 Zr 91.22	41 Nb 92.91	42 Mo 95.94	43 Tc 98.91	44 Ru 101.1	45 Rh 102.9	46 Pd 106.4	47 Ag 107.9	48 Cd 112.4	49 In 114.8	50 Sn 118.7	51 Sb 121.8	52 Te 127.6	53 I 126.9	54 Xe 131.3
6	55 Cs 132.9	56 Ba 137.3	57* La 138.9	72 Hf 178.5	73 Ta 180.9	74 W 183.9	75 Re 186.2	76 Os 190.2	77 Ir 192.2	78 Pt 195.1	79 Au 197.0	80 Hg 200.6	81 Tl 204.4	82 Pb 207.2	83 Bi 209.0	84 Po (210)	85 At (210)	86 Rn (222)
7	87 Fr (223)	88 Ra 226.0	89** Ac (227)	104 Unq (261)	105 Unp (262)	106 Unh (263)	107 Uns (262)	108 Uno (265)	109 Une (266)									

Period

Inner Transition Elements

Lanthanide Series 6 *	58 Ce 140.1	59 Pr 140.9	60 Nd 144.2	61 Pm (145)	62 Sm 150.4	63 Eu 152.0	64 Gd 157.3	65 Tb 158.9	66 Dy 162.5	67 Ho 164.9	68 Er 167.3	69 Tm 168.9	70 Yb 173.0	71 Lu 175.0
Actinide Series 7 **	90 Th 232.0	91 Pa 231.0	92 U 238.0	93 Np 237.0	94 Pu (244)	95 Am (243)	96 Cm (247)	97 Bk (247)	98 Cf (251)	99 Es (252)	100 Fm (257)	101 Md (258)	102 No (259)	103 Lr (260)

© Cengage Learning

Working in Marine Science

Working in the marine sciences is wonderfully appealing to many people. They sometimes envision a life of diving in warm, clear water surrounded by tropical fish, or descending to the seabed in an exotic submersible outfitted like Captain Nemo's fictional submarine in *20,000 Leagues Under the Sea*, or living with intelligent dolphins in a marine-life park. Then reality sets in. There are rewards from working in the marine sciences, but they tend to be less spectacular than the first dreams of students looking to the ocean for a life's work.

A marine science worker is paid to bring a specific skill to a problem. If that problem lies in warm, tropical water or in a marine park, fine. But more likely, the problem will yield only to prolonged study in an uncomfortable, cold, or dangerous environment. The intangible rewards can be great; the physical rewards are often slim. Having said that, let me add that no endeavor is more interesting or exciting, and few are more intellectually stimulating. Doing marine science is its own reward.

Training for a Job in Marine Science

Marine science is, of course, science. And science requires mathematics—you need math to do the chemistry, physics, measurements, and statistics that lie at the heart of science. Your first step in college should be to take a math placement test, enroll in an appropriate math class, and spend time doing math. *Math is the key to further progress in any area of marine science.*

With your math skills polished, start classes in chemistry, physics, and basic biology. Surprisingly, except for one or two introductory marine-science classes, you probably won't take many marine-science courses until your junior year. These introductory classes will be especially valuable because a balanced survey of the marine sciences can aid you in selecting an appealing specialty. Then, with a good foundation in basic science, you can begin to concentrate in that specialty.

Other skills are important, too. The ability to write and speak well is crucial in any science job. Also critical is computer literacy. Expertise in photography or foreign languages or the ability to field-strip and rebuild a diesel engine or hydraulic winch will put you a step above the competition at hiring time. Certification as a scuba diver is almost mandatory; you can never have too much diving experience. (Remember, though, diving is only a tool, a way to deliver an informed set of eyes and an educated brain to a work site.) You should be in good health. Indeed, good aerobic fitness is essential in most marine-science jobs; stamina is often a crucial factor in long experiments under difficult conditions at sea. It is also desirable to be physically strong—marine equipment is heavy and

often bunglesome. And it helps greatly if you are not prone to seasickness.

Deciding what school to attend will depend on your skills. Readers of this book will probably be enrolled in a general oceanography course in a college or university. The first step would be to discuss your interests with your professor (or his or her teaching assistants). You'll need to attend a 4-year college or university to complete the first phase of your training. If you're attending a 2-year institution, picking a specific transfer institution can come later, but keep a few things in mind: No matter where you take your first 2 years of training, you need thorough preparation in basic science. You should attend an institution with strengths in the area of your specialty (such as geology, biology, and marine chemistry). And you should be reasonable in your expectations of acceptance if you're a transfer student (that is, don't try for Stanford or Yale with a B average).

Another thing: Most marine scientists have completed a graduate degree (a master's degree or doctorate). Most graduate students hold teaching or research assistantships (that is, they get paid for being grad students). In all, progress to a final degree is a long road, but the journey is itself a pleasure.

If the thought of four or more years of higher education does not appeal to you, does that mean there's no hope? Not at all. Many students begin a program with the goal of becoming a marine technician, animal trainer at a marine-life park, marina or boatyard employee or manager, or crew member on a private yacht. Those jobs don't always require a bachelor's degree. Jobs at Sea World and other marine theme parks do require athletic ability, extreme patience, public-speaking skills, a love of animals, and usually diving experience. Few positions are available, but there is some turnover in the ranks of junior trainers, and being hired is certainly possible.

Becoming a marine technician is an especially attractive alternative to the all-out chemistry-physics-math academic route. For every highly trained marine scientist, there are perhaps five technical assistants who actually do the experiments, maintain the equipment, work daily with organisms, and build special apparatus. Marine technicians tend to spend more time at hands-on tasks than marine scientists. Most of these folks (including the author of the letter that ends this appendix) have the equivalent of a 2-year technical degree, usually from a community college.

Don't quit your job, burn your bridges, leave your family, sell your possessions, and dedicate yourself monklike to marine science. Do some investigation. Nothing is as valuable as *actually going out and talking to people who do things that you'd like to do*. Ask them if they enjoy their work. Is the pay OK? Would they start down the same road if they had it to do all over again? You may decide to expand your involvement in marine science in a more informal way by becoming a volunteer; joining the Sierra Club, Audubon Society, Greenpeace, or other environmental group; working for your state's fish and game office as a seasonal aide; or attending lectures at local colleges and universities.

If you decide to continue your education, don't be discouraged by the time it will take. Have a general view of the big picture, but proceed one semester at a time. Again, remember that the educational journey is itself a great pleasure. Don Quixote reminds us of the joys of the road, not the inn.

The Job Market

Marine science is very attractive to the general public. People are naturally drawn to thoughts of working in the field. Unfortunately, there aren't a great many jobs in the marine sciences. But there will always be some jobs, and people will fill them. Those people will be the best prepared, most versatile, and most highly motivated of those who apply. Perhaps not surprisingly, marine biology is the most popular marine science specialty. Unfortunately, it is also the area with the smallest number of nonacademic jobs. Museums, aquariums, and marine theme parks employ biologists to care for animals and oversee interpretive programs for the public. A few marine biologists are employed as monitoring specialists by water-management agencies like sanitation districts, which discharge waste into the ocean. Electrical utilities that use seawater to cool the condensers in power-generating plants almost always have a handful of marine biologists on staff to watch the effects of discharged heat on local marine life and to write the reports required by watchdog agencies. State and federal agencies employ marine biologists to read and interpret those documents and to set standards. Relatively small businesses, like private shipyards, agricultural concerns, and chemical plants, can't afford their own staff biologists, so private consulting firms staffed by marine biologists and other specialists have arisen to assist in the preparation of the environmental impact reports required of businesses under various legislation.

There are more jobs in physical oceanography: marine geology, ocean engineering, and marine chemistry and physics. Thousands of marine geologists work for oil and mineral companies; indeed, with the increasing emphasis on offshore resources, the market for these people may be increasing. Marine engineers are needed to design, construct, and maintain offshore oil rigs, ships, and harbor structures. Marine chemists are hard at work figuring ways to stop corrosion and to extract chemicals from seawater. Physicists are vitally interested in the transmission of underwater sound and light, in the movement of the ocean, and in the role the ocean plays in global weather and climate. Economists, lawyers, writers, and mathematicians also work in the marine-science field.

Many biological and physical oceanographers are teachers and professors. Indeed, there are nearly as many marine scientists employed in the academic world as there are in private industry and government. If you like the idea of teaching, you might consider this avenue. The demand for science teachers at all educational levels is already great and is expected to increase.

Four factors will be significant in influencing your employability:

1. *Experience.* Employers are favorably impressed by experience, especially work experience related to the duties of the position for which you are applying. Volunteer work counts.
2. *Grades.* Good grades are important, especially for positions in government agencies. A grade point average of 3.0 or higher in all college work increases your chances of employment and should give you a higher starting salary.
3. *Geographical availability.* Don't restrict yourself geographically. Not everyone can work in Hawai'i or California, but 4 out of 10 marine scientists work in just three states: California, Maryland, and Virginia.
4. *Diversification.* Again, mastery of more than one specialty gives you an employment edge. Being a plankton connoisseur and also able to repair a balky computer while ordering in-port supplies over a radiotelephone in Spanish makes a lasting impression.

Report from a Student

Students in marine-science programs graduate, get jobs, and move on. One of the pleasures of being a professor is hearing from them. One of our former students, an employee of the Marine Science Institute at the University of California, Santa Barbara, reported his activities as part of a team using the submersible *Alvin* to investigate plumes of warm water issuing from hydrothermal vents along the southern Juan de Fuca Ridge. The nature of his work—and his enthusiasm for it—is clearly evident in this excerpt. Dan Dion writes:

> The buoyant plume experiment wasn't going very well. The chemistry dives were pushed back because of technical difficulties and poor weather (rough seas cut two dives). The first two buoyant plume dives ended in failure. The first one because of mechanical/electrical problems, the second because of a computer crash. Everyone worked around the clock to get things in order for dive 2440. I was scheduled to go down with John Trefrey, from the Florida Institute of Technology. Cindy Van Dover was our pilot. We launched *Alvin* right on schedule at 0800 and descended from the glacier blue water into the bioluminescent snowstorm of the euphotic zone. During the hour and a half descent, we listened to the music of Enya in the soft light of the sub as we busily prepared ourselves for the experiment: booting up the computers, loading film in the cameras, tapes in the recorders, etc. We had three laptop computers to deal with in the cramped spaces of the sub. I was in charge of two of them, one that plotted our in-sub navigation (from transponders), and the other that controlled and recorded data from the [continuous temperature-depth-conductivity probe]. The third laptop was connected to

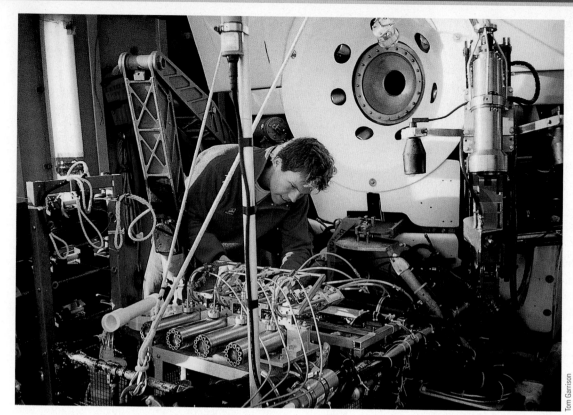

Tom Garrison

Figure 1 Dan Dion attaching hydraulic actuators to water-sampling bottles, in preparation for a dive by *Alvin*. The bottles were part of a sampling program that included measurements of water conductivity, transmissivity, temperature, and iron and manganese ion content near a hydrothermal vent. This information was later merged with data from transponder navigation to obtain a three-dimensional map of plume structure and chemistry.

Tom Garrison

Figure 2 Marine scientists are not always engaged in exciting research. Sometimes stacks of equipment need unpacking and tedious calibrating, messes need cleaning, supplies need ordering, and engines need coaxing back to life.

the chemical analyzer, and John was in control of that. All of the instruments were operating perfectly. I periodically saved the computer file in the event of another crash. We reached the bottom right on target; Monolith Vent was in sight, 2,261 meters below the surface. We did a video survey of the vent, especially a chimney that was rapidly growing back after the geologists had decapitated it just a few days earlier. We ascended to 55 meters off-bottom and began our drive-throughs. To me, the navigator, it was the ultimate video game. From the computer screen I would guide *Alvin* through a dark abyss, calling out headings that would maneuver us into a "lawn mower" pattern crisscrossing the plume. It was quite visible, and even beautiful; wispy, intricate patterns of "smoke" which seemed to dance like graceful ghosts. We completed passes at 35, 20, 10, and 5 meters above the bottom, then one last one at 45 meters. Eight hours of sub time went by so quickly! Our dive was a huge success; in addition to all the samples we obtained, we generated over 25 megabytes of data. I used everything I learned . . . from computer skills to navigation and marlinspike seamanship (and, of course, chemistry!).

Marine science is equipment training sessions, long cruises, seminars and lectures, visiting experts, hot sand volleyball games, and chilly labs with classical music. Marine science is a long and demanding road, but it is, quite honestly, great fun. Captain Nemo and his sub never had it this good!

For More Information

Useful Internet sites

The Centers for Ocean Sciences Education Excellence (COSEE) hosts this site, which contains information on ocean-related careers, colleges, jobs, and internships:

http://www.oceancareers.com/

The Monterey Bay Aquarium Research Institute (MBARI) site lists a number of links that provide information on ocean science careers:

http://www.mbari.org/education/careers.html

The National Oceanic and Atmospheric Administration's (NOAA) *OceanAGE Careers* page includes the profiles of professionals in various ocean-related disciplines and offers resources to help students to learn more about career paths:

http://oceanexplorer.noaa.gov/edu/oceanage/

This Web site, hosted by Joe Wible, Hopkins Marine Station of Stanford University, contains links to many career-related sites:

http://hopkins.stanford.edu/careers.htm

This Web site lists the member institutions of the Joint Oceanographic Institutions, universities at the forefront of oceanographic research:

www.oceanleadership.org

Organizations

American Society of Limnology and Oceanography A nonprofit, professional scientific society that seeks to promote the interests of limnology and oceanography and related sciences and to further the exchange of information across the range of aquatic science disciplines.

www.aslo.org

Association for Women in Science A nonprofit association dedicated to achieving equity and full participation for women in science, mathematics, engineering, and technology.

www.awis.org

The Ocean Conservancy A nonprofit membership organization dedicated to protecting marine wildlife and its habitats and to conserving coastal and ocean resources.

www.oceanconservancy.org

Marine Advanced Technology Training Center For training of marine technicians—specialists in the deployment and maintenance of tools used in marine science.

www.marinetech.org

Marine Technology Society An international, interdisciplinary society devoted to ocean and marine engineering science and policy.

www.mtsociety.org

National Marine Educators Association An organization that brings together those interested in the study and enjoyment of the world of water—both fresh and salt.

www.marine-ed.org

National Oceanic and Atmospheric Administration A government agency that guides the use and protection of our oceans and coastal resources, warns of dangerous weather, charts the seas and skies, and conducts research to improve our understanding and stewardship of the environment.

www.noaa.gov

Oceanic Engineering Society An organization that promotes the use of electronic and electrical engineers for instrumentation and measurement work in the ocean environment and the ocean/atmosphere interface.

www.oceanicengineering.org

The Oceanography Society A professional society for scientists in the field of oceanography.

www.tos.org

The Society for Marine Mammalogy A professional organization that supports the conservation of marine mammals and the educational, scientific, and managerial advancement of marine mammal science.

www.marinemammalogy.org

Joint Oceanographic Institutions The Joint Oceanographic Institutions (JOI) is a consortium of U.S. academic institutions that brings to bear the collective capabilities of the individual oceanographic institutions on research planning and management of the ocean sciences.

www.oceanleadership.org

Note: The organizations listed here are but a sampling of the resources that are available to help you learn more about careers in the marine sciences. Each contact you make in your search for information will lead you to more contacts and more information.

The World Ocean Seafloor

Pacific Ocean Seafloor

Atlantic Ocean Seafloor

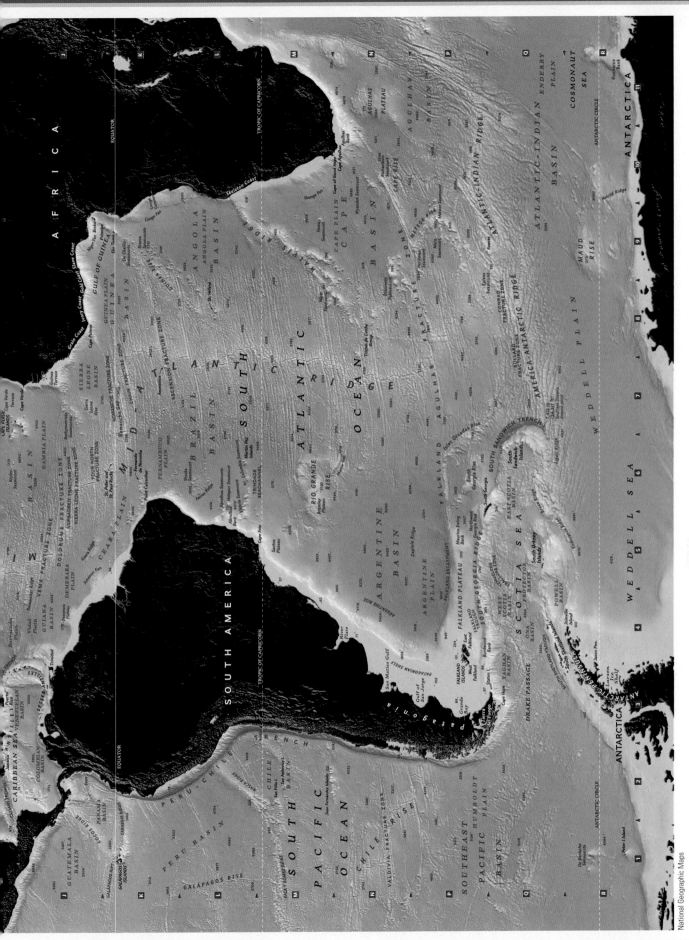

National Geographic Maps

The Ocean Seafloor

The ice-free ocean
showing floor details.
Compare with
Appendix 4, Figure 4.

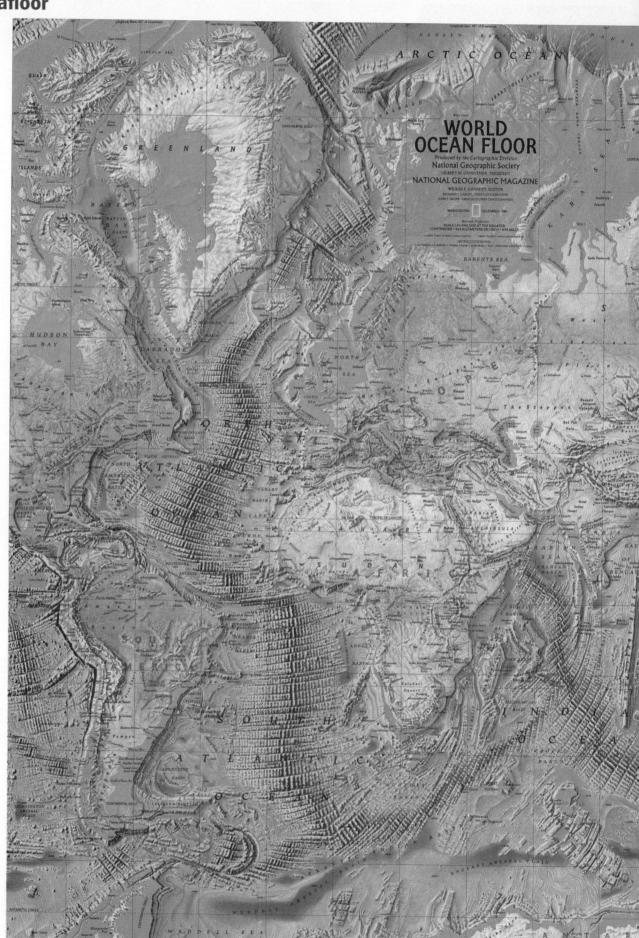

The Ocean Basics Map

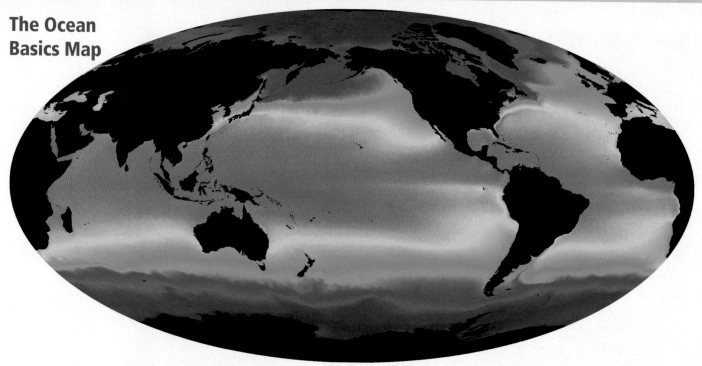

Average Sea-surface Temperature

Above: *Warmest surface waters occur in the tropics, which receive the most sun. Near the Poles, sunlight hits the Earth less directly, so surface waters are cooler. Stretched and pulled by ocean currents, temperature patterns reflect more than just warming from the sun. For instance, two of the great gyres of the Pacific Ocean can be seen bringing cooler waters toward the Equator along the west coast of the Americas. The Gulf Stream is also visible as an orange jet north of Florida carrying warm waters eastward across the Atlantic.*

Change in Sea-surface Height

Below: *Since the early 1990s, satellites have monitored the ever-changing height of the ocean surface. Areas of the ocean highlighted in red have risen by as much as 20 centimeters during the satellite record. In the blue regions, such as the eastern Pacific, sea levels have dropped, despite the addition of water from melting land-ice and a global rise in sea level. These patterns are caused by decade-long changes in ocean currents.*

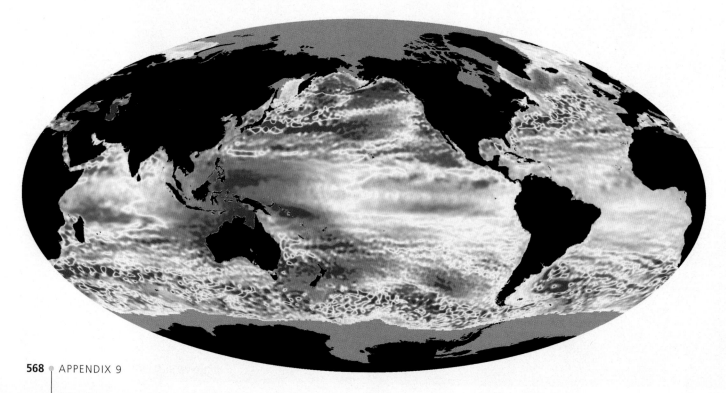

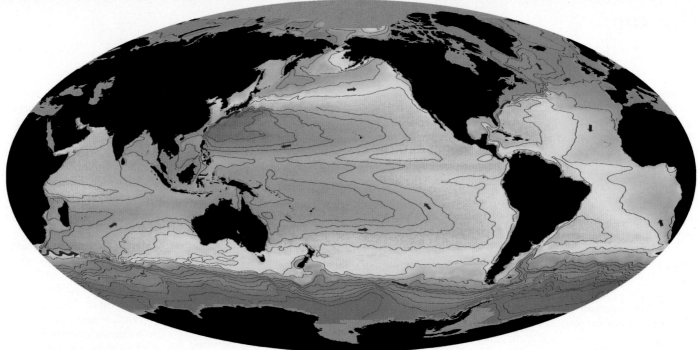

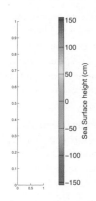 Sea Surface height (cm)

Average Sea-surface Height

Above: The height of the sea surface shows the ocean's major current systems. Water flows along the lines of constant sea-surface height (drawn in gray) and in the direction of the arrows. The red patches show the great subtropical gyres of the Pacific Ocean, spinning counterclockwise in the Northern Hemisphere and clockwise in the Southern. Tightly bunched lines show fast-moving currents such as the Gulf Stream, along the east coast of the United States, and the Antarctic Circumpolar Current, which circles the world around Antarctica.

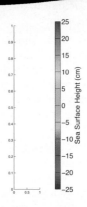 Sea Surface Height (cm)

Ocean Eddies

Below: Swirling currents called eddies fill the ocean. Shown here in a map of sea-surface height with other effects filtered out, eddies help to stir and mix the ocean, bringing nutrients to the surface and allowing phytoplankton to bloom. The fastest-spinning eddies, in red or purple, are associated with the fastest-flowing ocean currents such as the Gulf Stream, the Kuroshio (off the coast of Japan), and the Antarctic Circumpolar Current.

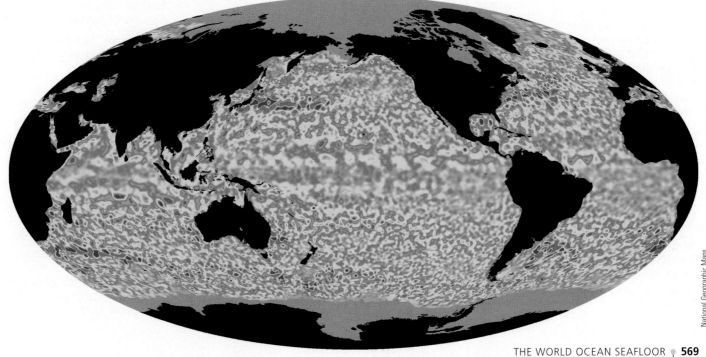

Glossary

absolute dating Determining the age of a geological sample by calculating radioactive decay and/or its position in relation to other samples.

absorption Conversion of sound or light energy into heat.

abyssal hill Small, sediment-covered, inactive volcano or intrusion of molten rock less than 200 meters (650 feet) high, thought to be associated with seafloor spreading. Abyssal hills punctuate the otherwise flat abyssal plain.

abyssal plain Flat, cold, sediment-covered ocean floor between the continental rise and the oceanic ridge at a depth of 3,700 to 5,500 meters (12,000–18,000 feet). Abyssal plains are more extensive in the Atlantic and Indian oceans than in the Pacific.

abyssal zone The ocean between about 4,000 and 5,000 meters (13,000 and 16,500 feet) deep.

accessory pigment One of a class of pigments (such as fucoxanthin, phycobilin, and xanthophyll) that are present in various photosynthetic plants and that assist in the absorption of light and the transfer of its energy to chlorophyll; also called *masking pigment.*

accretion An increase in the mass of a body by accumulation or clumping of smaller particles.

acid A substance that releases a hydrogen ion (H^+) in solution.

acid rain Rain containing acids and acid-forming compounds such as sulfur dioxide and oxides of nitrogen.

acoustical tomography A technique for studying ocean structure that depends on pulses of low-frequency sound to sense differences in water temperature, salinity, and movement beneath the surface.

active margin The continental margin near an area of lithospheric plate convergence; also called *Pacific-type margin.*

active sonar A device that generates underwater sound from special transducers and analyzes the returning echoes to gain information of geological, biological, or military importance.

active transport The movement of molecules from a region of low concentration to a region of high concentration through a semipermeable membrane at the expense of energy.

adaptation An inheritable structural or behavioral modification. A favorable adaptation gives a species an advantage in survival and reproduction. An unfavorable adaptation lessens a species' ability to survive and reproduce.

adhesion Attachment of water molecules to other substances by hydrogen bonds; wetting.

Agnatha The class of jawless fishes: hagfishes and lampreys.

ahermatypic Describing coral species that lack symbiotic zooxanthellae and are incapable of secreting calcium carbonate at a rate suitable for reef production.

air mass A large mass of air with nearly uniform temperature, humidity, and density throughout.

algae Collective term for nonvascular plants possessing chlorophyll and capable of photosynthesis. (Singular, *alga.*)

algin A mucilaginous commercial product of multicellular marine algae; widely used as a thickening and emulsifying agent.

alkaline Basic. See also *base.*

amphidromic point A "no-tide" point in an ocean caused by basin resonances, friction, and other factors around which tide crests rotate. About a dozen amphidromic points exist in the world ocean. Sometimes called a *node.*

anadromous A term applied to fishes that reproduce in freshwater and spend their adult lives in the marine environment.

angiosperm A flowering vascular plant that reproduces by means of a seed-bearing fruit. Examples are sea grasses and mangroves.

angle of incidence In meteorology, the angle of the sun above the horizon.

animal A multicellular organism unable to synthesize its own food and often capable of movement.

Animalia The kingdom to which multicellular heterotrophs belong.

Annelida The phylum of animals to which segmented worms belong.

Antarctic Bottom Water The densest ocean water (1.0279 g/cm^3), formed primarily in Antarctica's Weddell Sea during Southern Hemisphere winters.

Antarctic Circle The imaginary line around Earth parallel to the equator at 66°33'S, marking the southernmost limit of sunlight at the June solstice. The Antarctic Circle marks the northern limit of the area within which, for one day or more each year, the sun does not set (around 21 December) or rise (around 21 June).

Antarctic Circumpolar Current (West Wind Drift) The current driven by powerful westerly winds north of Antarctica. The largest of all ocean currents, it continues permanently eastward without changing direction.

Antarctic Convergence Convergence zone encircling Antarctica between about 50° and 60°S, marking the boundary between Antarctic Circumpolar Water and Subantarctic Surface Water.

Antarctic Ocean An ocean in the Southern Hemisphere bounded to the north by the Antarctic Convergence and to the south by Antarctica.

aphelion The point in the orbit of a satellite where it is farthest from the sun; opposite of *perihelion.*

aphotic zone The dark ocean below the depth to which light can penetrate.

apogee The point in the orbit of a satellite where it is farthest from the main body; opposite of *perigee.*

AQUA A NASA satellite designed to obtain data on Earth's water cycle.

aquaculture The growing or farming of plants and animals in a water environment under controlled conditions. Compare *mariculture.*

Arctic Circle The imaginary line around Earth parallel to the equator at 66°33'N, marking the northernmost limit of sunlight at the December solstice. The Arctic Circle marks the southern limit of the area within which, for one day or more each year, the sun does not set (around 21 June) or rise (around 21 December).

Arctic Convergence Convergence zone between Arctic Water and Subarctic Surface Water.

Arctic Ocean An ice-covered ocean north of the continents of North America and Eurasia.

Argo system A set of more than 3,000 drifting devices that relay ocean conditions to researchers.

Arthropoda The phylum of animals that includes shrimp, lobsters, krill, barnacles, and insects. The phylum Arthropoda is the world's most successful.

artificial system of classification A method of classifying an object based on attributes other than its reason for existence, its ancestry, or its origin. Compare *natural system of classification*.

Asteroidea The class of the phylum Echinodermata to which sea stars belong.

asthenosphere The hot, plastic layer of the upper mantle below the lithosphere, extending some 350 to 650 kilometers (220–400 miles) below the surface. Convection currents within the asthenosphere power plate tectonics.

astronomical tide Tides caused by inertia and the gravitational force of the sun and moon.

atmosphere The envelope of gases that surround a planet and are held to it by the planet's gravitational attraction.

atmospheric circulation cell Large circuit of air driven by uneven solar heating and the Coriolis effect. Three circulation cells form in each hemisphere. See also *Ferrel cell; Hadley cell; polar cell*.

atoll A ring-shaped island of coral reefs and coral debris enclosing, or almost enclosing, a shallow lagoon from which no land protrudes. Atolls often form over sinking, inactive volcanoes.

atom The smallest particle of an element that exhibits the characteristics of that element.

authigenic sediment Sediment formed directly by precipitation from seawater; also called *hydrogenous sediment*.

autotroph An organism that makes its own food by photosynthesis or chemosynthesis.

auxospore A naked diatom cell without valves; often a dormant stage in the life cycle following sexual reproduction.

Aves The class of birds.

backshore Sand on the shoreward side of the berm crest, sloping away from the ocean.

backwash Water returning to the ocean from waves washing onto a beach.

bacteria Single-celled prokaryotes, organisms that lack membrane-bound organelles.

baleen The interleaved, hard, fibrous, hornlike filters within the mouth of baleen whales.

barrier island A long, narrow, wave-built island lying parallel to the mainland and separated from it by a lagoon or bay. Compare *sea island*.

barrier reef A coral reef surrounding an island or lying parallel to the shore of a continent, separated from land by a deep lagoon. Coral debris islands may form along the reef.

basalt The relatively heavy crustal rock that forms the seabeds, composed mostly of oxygen, silicon, magnesium, and iron. Its density is about 2.9 g/cm^3.

base A substance that combines with a hydrogen ion (H^1) in solution.

bathyal zone The ocean between about 200 and 4,000 meters (700 and 13,000 feet) deep.

bathybius Thomas Henry Huxley's name for an artifact of marine specimen preservation he thought was a remnant of the "primeval living slime."

bathymetry The discovery and study of submerged contours.

bathyscaphe Deep-diving submersible designed like a blimp, which uses gasoline for buoyancy and can reach the bottom of the deepest ocean trenches. From the Greek *batheos* ("depth") and *skaphidion* ("a small ship").

bay mouth bar An exposed sandbar attached to a headland adjacent to a bay and extending across the mouth of the bay.

beach A zone of unconsolidated (loose) particles extending from below the water level to the edge of the coastal zone.

beach scarp A vertical wall of variable height marking the landward limit of the most recent high tides; corresponds with the berm at extreme high tides.

benthic zone The zone of the ocean bottom. See also *pelagic zone*.

berm A nearly horizontal accumulation of sediment parallel to shore; marks the normal limit of sand deposition by wave action.

berm crest The top of the berm; the highest point on most beaches; corresponds to the shoreward limit of wave action during most high tides.

big bang The hypothetical event that started the expansion of the universe from a geometric point; the beginning of time.

bilateral symmetry Body structure having left and right sides that are approximate mirror images of each other. Examples are crabs and humans. Compare *radial symmetry*.

biodegradable Able to be broken down by natural processes into simpler compounds.

biodiversity The variety of different species within a habitat.

biogenous sediment Sediment of biological origin. Organisms can deposit calcareous (calcium-containing) or siliceous (silicon-containing) residue.

biogeochemical cycle Natural processes that recycle nutrients in various chemical forms from the nonliving environment to living organisms and then back to the nonliving environment.

biological factor A biologically generated aspect of the environment, such as predation or metabolic waste products, that affects living organisms. Biological factors usually operate in association with purely physical factors such as light and temperature.

biological resource A living animal or plant collected for human use; also called a *living resource*.

bioluminescence Biologically produced light.

biomagnification Increase in the concentration of certain fat-soluble chemicals such as DDT or heavy-metal compounds in successively higher trophic levels within a food web.

biomass The mass of living material in a given area or volume of habitat.

biosynthesis The initial formation of life on Earth.

Bivalvia The class of the phylum Mollusca that includes clams, oysters, and mussels.

Bjerknes, Vilhelm (1862–1951) Pioneering Norwegian physicist and discoverer of the nature and formation of extratropical cyclones, which cause most mid-latitude weather.

blade Algal equivalent of a vascular plant's leaf; also called a *frond*.

body wave Seismic waves that travel through Earth's interior. See *P wave* and *S wave*.

bond See *chemical bond*.

brackish Describing water intermediate in salinity between seawater and freshwater.

breakwater An artificial structure of durable material that interrupts the progress of waves to shore. Harbors are often shielded by a breakwater.

buffer A group of substances that tends to resist change in the pH of a solution by combining with free ions.

buoyancy The ability of an object to float in a fluid by displacement of a volume of fluid equal to it in mass.

bycatch Animals unintentionally killed when desirable organisms are collected.

C [celerity] Velocity

$C = \sqrt{gd}$ Relationship of velocity (C), the acceleration due to gravity (g), and water depth (d) for shallow-water waves.

$C = \frac{L}{T}$ Relationship of velocity (C), wavelength (L), and period (T) for deep-water waves; velocity increases as wavelength increases.

Typically measured in meters per second.

caballing Mixing of two water masses of identical densities but different temperatures and salinities, such that the resulting mixture is denser than its components.

calcareous ooze Ooze composed mostly of the hard remains of organisms containing calcium carbonate.

calcium carbonate compensation depth (CCD) The depth at which the rate of accumulation of calcareous sediments equals the rate of dissolution of those sediments. Below this depth, sediment contains little or no calcium carbonate.

calorie The amount of heat needed to raise the temperature of 1 gram (0.035 ounce) of pure water by 1°C (1.8°F).

capillary wave A tiny wave with a wavelength of less than 1.73 centimeters (0.68 inch), whose restoring force is surface tension; the first type of wave to form when the wind blows.

carbon cycle The movement of carbon from reservoirs (sediment, rock, ocean) through the atmosphere (as carbon dioxide), through food webs, and back to the reservoirs.

Carnivora The order of mammals that includes seals, sea lions, walruses, and sea otters.

carrying capacity The size at which a particular population in a particular environment will stabilize when its supply of resources—including nutrients, energy, and living space—remains constant.

cartilage A tough, elastic tissue that stiffens or supports.

cartographer A person who makes maps and charts.

catastrophism The theory that Earth's surface features are formed by catastrophic forces such as the biblical flood. Catastrophists believe in a young Earth and a literal interpretation of the biblical account of Creation.

celestial navigation The technique of finding one's position on Earth by reference to the apparent positions of stars, planets, the moon, and the sun.

cell The basic organizational unit of life on this planet.

Cephalopoda The class of the phylum Mollusca that includes squid, octopuses, and nautiluses.

Cetacea The order of mammals that includes porpoises, dolphins, and whales.

CFCs See *chlorofluorocarbons*.

***Challenger* expedition** The first wholly scientific oceanographic expedition, 1872–1876; named for the steam corvette used in the voyage.

chart A map that depicts mostly water and the adjoining land areas.

chemical bond An energy relationship that holds two atoms together as a result of changes in their electron distribution.

chemical equilibrium In seawater, the condition in which the proportion and amounts of dissolved salts per unit volume of ocean are nearly constant.

chemosynthesis The synthesis of organic compounds from inorganic compounds using energy stored in inorganic substances such as sulfur, ammonia, and hydrogen. Energy is released when these substances are oxidized by certain organisms.

Chinese navigators Explorers led by Zheng He into the Indian Ocean and around the tip of Africa.

chitin A complex, nitrogen-rich carbohydrate from which parts of arthropod exoskeletons are constructed.

chiton A marine mollusk of the class Polyplacophora.

chlorinated hydrocarbons The most abundant and dangerous class of halogenated hydrocarbons, synthetic organic chemicals hazardous to the marine environment.

chlorinity A measure of the content of chloride, bromine, and iodide ions in seawater. We derive salinity from chlorinity by multiplying by 1.80655.

chlorofluorocarbons (CFCs) A class of halogenated hydrocarbons thought to be depleting Earth's atmospheric ozone. CFCs are used as cleaning agents, refrigerants, fire-extinguishing fluids, spray-can propellants, and insulating foams.

chlorophyll A pigment responsible for trapping sunlight and transferring its energy to electrons, thus initiating photosynthesis.

Chlorophyta Green algae.

Chondrichthyes The class of fishes with cartilaginous skeletons: the sharks, skates, rays, and chimaeras.

Chordata The phylum of animals to which tunicates, Amphioxus, fishes, amphibians, reptiles, birds, and mammals belong.

chromatophore A pigmented skin cell that expands or contracts to affect color change.

chronometer A very consistent clock. It doesn't need to tell accurate time, but its rate of gain or loss must be constant and known exactly so that accurate time may be calculated.

clamshell sampler A sampling device used to take shallow samples of the ocean bottom.

classification A way of grouping objects according to some stated criteria.

clay Sediment particle smaller than 0.004 millimeter in diameter; the smallest sediment size category.

climate The long-term average of weather in an area.

climax community A stable, long-established community of self-perpetuating organisms that tends not to change with time.

clockwise Rotation around a point in the direction that clock hands move.

clumped distribution Distribution of organisms within a community in small, patchy aggregations, or clumps; the most common distribution pattern.

Cnidaria The phylum of animals to which corals, jellyfish, and sea anemones belong.

cnidoblast Type of cell found in members of the phylum Cnidaria that contains a stinging capsule. The threads that evert from the capsules assist in capturing prey and repelling aggressors.

coast The zone extending from the ocean inland as far as the environment is immediately affected by marine processes.

coastal cell The natural sector of a coastline in which sand input and sand outflow are balanced.

coastal upwelling Upwelling adjacent to a coast, usually induced by wind.

coccolithophore A very small planktonic alga carrying discs of calcium carbonate, which contributes to biogenous sediments.

cohesion Attachment of water molecules to each other by hydrogen bonds.

colligative properties Those characteristics of a solution that differ from those of pure water because of material held in solution.

Columbus, Christopher (1451–1506) Italian explorer in the service of Spain who discovered islands in the Caribbean in 1492. Although traditionally credited as the discoverer of America, he never actually sighted the North American continent.

commensalism A symbiotic interaction between two species in which only one species benefits and neither is harmed.

commercial extinction Depletion of a resource species to a point where it is no longer profitable to harvest the species.

community The populations of all species that occupy a particular habitat and interact within that habitat.

compass An instrument for showing direction by means of a magnetic needle swinging freely on a pivot and pointing to magnetic north.

compensation depth The depth in the water column at which the production of carbohydrates and oxygen by photosynthesis exactly equals the consumption of carbohydrates and oxygen by respiration. The break-even point for autotrophs. Generally a function of light level.

compound A substance composed of two or more elements in a fixed proportion.

condensation theory Premise that stars and planets accumulate from contracting, accreting clouds of galactic gas, dust, and debris.

conduction The transfer of heat through matter by the collision of one atom with another.

conservative constituent An element that occurs in constant proportion in seawater; for example, chlorine, sodium, and magnesium.

constructive interference The addition of wave energy as waves interact, producing larger waves.

consumer A heterotrophic organism.

continental crust The solid masses of the continents, composed primarily of granite.

continental drift The theory that the continents move slowly across the surface of Earth.

continental margin The submerged outer edge of a continent, made of granitic crust; includes the continental shelf and continental slope. Compare *ocean basin*.

continental rise The wedge of sediment forming the gentle transition from the outer (lower) edge of the continental slope to the abyssal plain; usually associated with passive margins.

continental shelf The gradually sloping submerged extension of a continent, composed of granitic rock overlain by sediments; has features similar to the edge of the nearby continent.

continental slope The sloping transition between the granite of the continent and the basalt of the seabed; the true edge of a continent.

contour current A bottom current made up of dense water that flows around (rather than over) seabed projections.

control The trial group in an experiment that does not contain the experimental variable.

convection Movement within a fluid resulting from differential heating and cooling of the fluid. Convection produces mass transport or mixing of the fluid.

convection current A single closed-flow circuit of rising warm material and falling cool material.

convergence zone The line along which waters of different density converge. Convergence zones form the boundaries of tropical, subtropical, temperate, and polar areas.

convergent evolution The evolution of similar characteristics in organisms of different ancestry; the body shape of a porpoise and a shark, for instance.

convergent plate boundary A region where plates are pushing together and where a mountain range, island arc, and/or trench will eventually form; often a site of much seismic and volcanic activity.

Cook, James (1728–1779) Officer in the British Royal Navy who led the first European voyages of scientific discovery.

copepod A small planktonic arthropod, a major marine primary consumer.

coral Any of more than 6,000 species of small cnidarians, many of which are capable of generating hard calcareous (aragonite, $CaCO_3$) skeletons.

coral reef A linear mass of calcium carbonate (aragonite and calcite) assembled from coral organisms, algae, mollusks, worms, and so on. Coral may contribute less than half of the reef material.

core The innermost layer of Earth, composed primarily of iron, with nickel and heavy elements. The inner core is thought to be a solid 6,000°C (11,000°F) sphere, the outer core a 5,000°C (9,000°F) liquid mass. The average density of the outer core is about 11.8 g/cm³, and that of the inner core is about 16 g/cm³.

Coriolis, Gaspard-Gustave de (1792–1843) The French scientist who in 1835 worked out the mathematics of the motion of bodies on a rotating surface. See *Coriolis effect*.

Coriolis effect The apparent deflection of a moving object from its initial course when its speed and direction are measured in reference to the surface of the rotating Earth.

The object is deflected to the right of its anticipated course in the Northern Hemisphere and to the left in the Southern Hemisphere. The deflection occurs for any horizontal movement of objects with mass and has no effect at the equator.

cosmogenous sediment Sediment of extraterrestrial origin.

counterclockwise Rotation around a point in the direction opposite to that in which clock hands move; also called *anticlockwise*.

countercurrent A surface current flowing in the opposite direction from an adjacent surface current.

countershading A camouflage pattern featuring a dark upper surface and a lighter bottom surface.

covalent bond A chemical bond formed between two atoms by electron sharing.

crest See *wave crest*.

crust The outermost solid layer of Earth, composed mostly of granite and basalt; the top of the lithosphere. The crust has a density of 2.7 to 2.9 g/cm^3 and accounts for 0.4% of Earth's mass.

Crustacea The class of phylum Arthropoda to which lobsters, shrimp, crabs, barnacles, and copepods belong.

cryptic coloration Camouflage; may be active (under control of the animal) or passive (an unalterable color or shape).

Curie point The temperature above which a material loses its magnetism.

current Mass flow of water. (The term is usually reserved for horizontal movement.)

cyanobacterium A phylum of bacteria, often blue-green in color, that obtain energy through photosynthesis.

cyclone A weather system with a low-pressure area in the center around which winds blow counterclockwise in the Northern Hemisphere and clockwise in the Southern Hemisphere. Not to be confused with a tornado, a much smaller weather phenomenon associated with severe thunderstorms. See also *extratropical cyclone; tropical cyclone*.

Darwin, Charles (1809–1882) An English biologist and the codiscoverer (with Alfred Russell Wallace) of evolution by natural selection.

deep scattering layer (DSL) A relatively dense aggregation of fishes, squid, and other mesopelagic organisms capable of reflecting a sonar pulse that resembles a false bottom in the ocean. Its position varies with the time of day.

deep-water wave A wave in water deeper than half its wavelength.

deep zone The zone of the ocean below the pycnocline, in which there is little additional change of density with increasing depth; contains about 80% of the world's water.

degree An arbitrary measure of temperature. One degree Celsius (°C) = 1.8 degrees Fahrenheit (°F).

delta The deposit of sediments found at a river mouth, sometimes triangular in shape (named after the Greek letter D [Δ]).

denitrifying bacteria Bacteria capable of converting nitrite or nitrate to gaseous nitrogen.

density The mass per unit volume of a substance, usually expressed in grams per cubic centimeter (g/cm^3).

density curve A graph showing the relationship between a fluid's temperature or salinity and its density.

density stratification The formation of layers in a material, with each deeper layer being denser (weighing more per unit of volume) than the layer above.

dependency A feeding relationship in which an organism is limited to feeding on one species or, in extreme cases, on one size phase of one species.

deposition Accumulation, usually of sediments.

depositional coast A coast in which processes that deposit sediment exceed erosive processes.

desalination The process of removing salt from seawater or brackish water.

desiccation Drying.

destructive interference The subtraction of wave energy as waves interact, producing smaller waves.

diatom Earth's most abundant, successful, and efficient single-celled phytoplankton. Diatoms possess two interlocking valves made primarily of silica. The valves contribute to biogenous sediments.

diffusion The movement—driven by heat—of molecules from a region of high concentration to a region of low concentration.

dinoflagellate One of a class of microscopic single-celled flagellates, not all of which are autotrophic. The outer covering is often of stiff cellulose. Planktonic dinoflagellates are responsible for "red tides."

dispersion Separation of wind waves by wavelength (and therefore wave speed) as they move away from the fetch (the place of their formation). Dispersion occurs because waves with long wavelengths move more rapidly than waves with short wavelengths.

disphotic zone The lower part of the photic zone, where there is insufficient light for photosynthesis.

dissolution The dissolving by water of minerals in rocks.

dissolved organic carbon (DOC) Organic (carbon-containing) molecules dissolved in water.

dissolved organic nitrogen (DON) Nitrogen-containing organic molecules dissolved in seawater. Most DON is in the form of protein.

disturbing force The energy that causes a wave to form.

diurnal tide A tidal cycle of one high tide and one low tide per day.

divergent evolution Evolutionary radiation of different species from a common ancestor.

divergent plate boundary A region where plates are moving apart and where new ocean or rift valley will eventually form. A spreading center forms the junction.

doldrums The zone of rising air near the equator known for sultry air and variable breezes. See also *intertropical convergence zone (ITCZ)*.

domain The three main kinds of living things above the Linnaean level of kingdom: *Bacteria, Archaea,* and *Eukarya*.

downwelling Circulation pattern in which surface water moves vertically downward.

drag The resistance to movement of an organism induced by the fluid through which it swims.

drift net Fine, vertically suspended net that may be 7 meters (25 feet) high and 80 kilometers (50 miles) long.

drumlin A streamlined hill formed by a glacier.

DSL See *deep scattering layer*.

dynamic theory of tides Model of tides that takes into account the effects of finite ocean depth, basin resonances, and the interference of continents on tide waves.

earthquake A sudden motion of Earth's crust resulting from waves in Earth caused by faulting of the rocks or by volcanic activity.

eastern boundary current Weak, cold, diffuse, slow-moving current at the eastern boundary of an ocean (off the west coast of a continent). Examples include the Canary Current and the Humboldt Current.

ebb current Water rushing out of an enclosed harbor or bay because of the fall in sea level as a tide trough approaches.

Echinodermata The phylum of exclusively marine animals to which sea stars, brittle stars, sea urchins, and sea cucumbers belong.

Echinoidea The class of the phylum Echinodermata to which sea urchins and sand dollars belong.

echo sounder A device that reflects sound off the ocean bottom to sense water depth. Its accuracy is affected by the variability of the speed of sound through water.

echolocation The use of reflected sound to detect environmental objects. Cetaceans use echolocation to detect prey and avoid obstacles.

ecology Study of the interactions of organisms with one another and with their environment.

ectotherm An organism incapable of generating and maintaining steady internal temperature from metabolic heat and therefore whose internal body temperature is approximately the same as that of the surrounding environment; a cold-blooded organism.

eddy A circular movement of water usually formed where currents pass obstructions, or between two adjacent currents flowing in opposite directions, or along the edge of a permanent current.

EEZ See *exclusive economic zone.*

Ekman spiral A theoretical model of the effect on water of wind blowing over the ocean. Because of the Coriolis effect, the surface layer is expected to drift at an angle 45° to the right of the wind in the Northern Hemisphere and 45° to the left in the Southern Hemisphere. Water at successively lower layers drifts progressively to the right (N) or left (S), though not as swiftly as the surface flow.

Ekman transport Net water transport, the sum of layer movement due to the Ekman spiral. Theoretical Ekman transport in the Northern Hemisphere is 90° to the right of the wind direction.

El Niño A southward-flowing, nutrient-poor current of warm water off the coast of western South America, caused by a breakdown of trade-wind circulation.

electron A tiny, negatively charged particle in an atom responsible for chemical bonding.

element A substance composed of identical atoms that cannot be broken down into simpler substances by chemical means.

endotherm An organism capable of generating and regulating metabolic heat to maintain a steady internal temperature. Birds and mammals are the only animals capable of true endothermy. A warm-blooded organism.

energy The capacity to do work.

ENSO Acronym for the coupled phenomena of El Niño and the Southern

Oscillation. See also *El Niño; Southern Oscillation.*

entropy A measure of the disorder in a system.

environmental resistance All the limiting factors that act together to regulate the maximum allowable size, or carrying capacity, of a population.

epicenter The point on Earth's surface directly above the focus of an earthquake.

epipelagic zone The lighted, or photic, zone in the ocean.

equator See *geographical equator; meteorological equator.*

equatorial upwelling Upwelling in which water moving westward on either side of the geographical equator tends to be deflected slightly poleward and replaced by deep water often rich in nutrients. See also *upwelling.*

equilibrium theory of tides Idealized model of tides that considers Earth to be covered by an ocean of great and uniform depth capable of instantaneous response to the gravitational and inertial forces of the sun and the moon.

Eratosthenes of Cyrene (276–192 B.C.E.) Greek scholar and librarian at Alexandria who first calculated the circumference of Earth about 230 B.C.E.

erosion A process of being gradually worn away.

erosional coast A coast in which erosive processes exceed depositional ones.

estuary A body of water partially surrounded by land where freshwater from a river mixes with ocean water, creating an area of remarkable biological productivity.

euphotic zone The upper layer of the photic zone in which net photosynthetic gain occurs. Compare *photic zone.*

euryhaline Describing an organism able to tolerate a wide range in salinity.

eurythermal Describing an organism able to tolerate a wide variance in temperature.

eurythermal zone The upper layer of water, where temperature changes with the seasons.

eustatic change A worldwide change in sea level, as distinct from local changes.

eutrophication A set of physical, chemical, and biological changes brought about when excessive nutrients are released into water.

evaporite Deposit formed by the evaporation of ocean water.

evolution Change; the maintenance of life under constantly changing conditions by continuous adaptation of successive generations of a species to its environment.

excess volatiles A compound found in the ocean and atmosphere in quantities greater than can be accounted for by the weathering of surface rock. Such compounds probably entered the atmosphere and ocean from deep crustal and upper mantle sources through volcanism.

exclusive economic zone (EEZ) The offshore zone claimed by signatories to the 1982 United Nations Draft Convention on the law of the sea. The EEZ extends 200 nautical miles (370 kilometers) from a contiguous shoreline. See also *United States Exclusive Economic Zone.*

exoskeleton A strong, lightweight, form-fitted external covering and support common to animals of the phylum Arthropoda. The exoskeleton is made partly of chitin and may be strengthened by calcium carbonate.

experiments Tests that simplify observation in nature or in the laboratory by manipulating or controlling the conditions under which observations are made.

extratropical cyclone A low-pressure, mid-latitude weather system characterized by converging winds and ascending air rotating counterclockwise in the Northern Hemisphere and clockwise in the Southern Hemisphere. An extratropical cyclone forms at the front between the polar and Ferrel cells.

extremophile An organism capable of tolerating extreme environmental conditions, especially temperature or pH level.

fault A fracture in a rock mass along which movement has occurred.

Ferrel, William (1817–1891) The American scientist who discovered the mid-latitude circulation cells of each hemisphere.

Ferrel cell The middle atmospheric circulation cell in each hemisphere. Air in these cells rises at 60° latitude and falls at 30° latitude. See also *westerlies*.

fetch The uninterrupted distance over which the wind blows without a significant change in direction, a factor in wind-wave development.

Fissipedia The carnivoran suborder that includes sea otters.

fjord A deep, narrow estuary in a valley originally cut by a glacier.

fjord estuary An estuary in a fjord, a steep, submerged, U-shaped valley.

flagellum A whiplike structure used by some small organisms and gametes to move through the environment. (Plural, *flagella*.)

float method A method of current study that depends on the movement of a drift bottle or other free-floating object.

flood current Water rushing into an enclosed harbor or bay because of the rise in sea level as a tide crest approaches.

flow method A method of current study that measures the current as it flows past a fixed object.

food General term for organic molecules capable of providing energy to heterotrophs when combined with oxygen during biochemical respiration.

food web A group of organisms associated by a complex set of feeding relationships in which the flow of food energy can be followed from primary producers through consumers.

foraminiferan One of a group of planktonic amoeba-like animals with a calcareous shell, which contributes to biogenous sediments.

forced wave A progressive wave under the continuing influence of the forces that formed it.

Forchhammer's principle See *principle of constant proportions*.

foreshore Sand on the seaward side of the berm, sloping toward the ocean, to the low-tide mark.

fracture zone Area of irregular, seismically inactive topography marking the position of a once-active transform fault.

Franklin, Benjamin (1706–1790) Published the first chart of an ocean current in 1769.

free wave A progressive wave free of the forces that formed it.

freezing point The temperature at which a solid can begin to form as a liquid is cooled.

fringing reef A reef attached to the shore of a continent or island.

front The boundary between two air masses of different density. The density difference can be caused by differences in temperature and/or humidity.

frontal storm Precipitation and wind caused by the meeting of two air masses, associated with an extratropical cyclone. Generally, one air mass will slide over or under the other, and the resulting expansion of air will cause cooling and consequently, rain or snow.

frustule The siliceous external cell wall of a diatom consisting of two interlocking valves fitted together like the halves of a box.

fucoxanthin A brown or tan accessory pigment found in many species of brown algae and some species of diatoms.

fully developed sea The theoretical maximum height attainable by ocean waves given wind of a specific strength, duration, and fetch. Longer exposure to wind will not increase the size of the waves.

fundamental niche The broadest of all possible niches that an organism theoretically can occupy.

galaxy A large, rotating aggregation of stars, dust, gas, and other debris held together by gravity. There are perhaps 50 billion galaxies in the universe and 50 billion stars in each galaxy.

gas bladder In multicellular algae, an air-filled structure that assists in flotation.

gas exchange Simultaneous passage, through a semipermeable membrane, of oxygen into an animal and carbon dioxide out of it.

Gastropoda The class of the phylum Mollusca that includes snails and sea slugs.

geographical equator 0° latitude, an imaginary line equidistant from the geographical poles.

geostrophic Describing a gyre or current in balance between the Coriolis effect and gravity; literally, "turned by Earth."

geostrophic gyre Circular current around the periphery of an ocean basin in balance between the Coriolis effect and the pressure gradient.

gill membrane The thin boundary of living cells separating blood from water in a fish's (or other aquatic animal's) gills.

Global Positioning System (GPS) Satellite-based navigation system that provides a geographical position—longitude and latitude—accurate to less than 1 meter (3.3 feet).

GPS See *Global Positioning System*.

granite The relatively light crustal rock—composed mainly of oxygen, silicon, and aluminum—that forms the continents. Its density is about 2.7 g/cm^3.

gravimeter A sensitive device that measures variations in the pull of gravity at different places on Earth's surface.

gravity wave A wave with wavelength greater than 1.73 centimeters (0.68 inch), whose restoring forces are gravity and momentum.

greenhouse effect Trapping of heat in the atmosphere. Incoming short-wavelength solar radiation penetrates the atmosphere, but the outgoing longer-wavelength radiation is absorbed by greenhouse gases and reradiated to Earth, causing a rise in surface temperature.

greenhouse gases Gases in Earth's atmosphere that cause the greenhouse effect; include carbon dioxide, methane, and CFCs.

groin A short, artificial projection of durable material placed at a right angle to shore in an attempt to slow longshore transport of sand from a beach; usually deployed in repeating units.

group velocity Speed of advance of a wave train; for deep-water waves, half the speed of individual waves within the group.

Gulf Stream The strong western boundary current of the North Atlantic, off the Atlantic coast of the United States.

guyot A flat-topped, submerged, inactive volcano.

gyre Circuit of mid-latitude currents around the periphery of an ocean basin. Most oceanographers recognize five gyres plus the Antarctic Circumpolar Current.

habitat The place where an individual or population of a given species lives; its "mailing address."

HABs Harmful Algal Blooms.

hadal zone The deepest zone of the ocean, below a depth of 5,000 meters (16,500 feet).

Hadley, George (1685–1768) A London lawyer and philosopher who worked out the overall scheme of wind circulation in an effort to explain the trade winds.

Hadley cell The atmospheric circulation cell nearest the equator in each hemisphere. Air in these cells rises near the equator because of strong solar heating there and falls because of cooling at about 30° latitude. See also *trade winds*.

half-life Time required for half of all the unstable radioactive nuclei in a sample to decay.

halocline The zone of the ocean in which salinity increases rapidly with depth. See also *pycnocline*.

Harrison, John (1693–1776) British clockmaker who invented the modern chronometer in 1760.

heat A form of energy produced by the random vibration of atoms or molecules.

heat budget An expression of the total solar energy received on Earth during some period of time and the total heat lost from Earth by reflection and radiation into space through the same period.

heat capacity The heat, measured in calories, required to raise 1 gram of a substance 1°C (1.8°F). The input of 1 calorie of heat energy raises the temperature of 1 gram of pure water by 1°C (1.8°F).

heavy metal Heavy metals are capable of causing damage to organisms by interfering with normal cell metabolism. Among the dangerous heavy metals being introduced into the ocean are mercury, lead, copper, and tin.

Henry the Navigator (1394–1460) Prince of Portugal who established a school for the study of geography, seamanship, shipbuilding, and navigation.

hermatypic Describing coral species possessing symbiotic zooxanthellae within their tissues and capable of secreting calcium carbonate at a rate suitable for reef production.

heterotroph An organism that derives nourishment from other organisms because it is unable to synthesize its own food molecules.

hierarchy Grouping of objects by degrees of complexity, grade, or class. A hierarchical system of nomenclature is based on distinctions within groups and between groups.

high-energy coast A coast exposed to large waves.

high seas That part of the ocean past the exclusive economic zone that is considered common property to be shared by the citizens of the world; about 60% of the ocean area.

high tide The high-water position corresponding to a tidal crest.

holdfast A complex branching structure that anchors many kinds of multicellular algae to the substrate.

holoplankton Permanent members of the plankton community. Examples are diatoms and copepods. Compare *meroplankton*.

Holothuroidea The class of the phylum Echinodermata to which sea cucumbers belong.

horse latitudes Zones of erratic horizontal surface air circulation near 30°N and 30°S latitudes. Over land, dry air falling from high altitudes produces deserts at these latitudes (e.g., the Sahara).

hot spot A surface expression of a plume of magma rising from a stationary source of heat in the mantle.

hurricane A large tropical cyclone in the North Atlantic or eastern Pacific, whose winds exceed 118 kilometers (74 miles) per hour.

hydrogen bond A relatively weak bond formed between a partially positive hydrogen atom and a partially negative oxygen, fluorine, or nitrogen atom of an adjacent molecule.

hydrogenous sediment A sediment formed directly by precipitation from seawater; also called *authigenic sediment*.

hydrologic cycle The continuous cycle of water between ocean, atmosphere, and freshwater reservoirs. Powered by the sun.

hydrostatic pressure The constant pressure of water around a submerged organism.

hydrothermal vent A spring of hot, mineral- and gas-rich seawater found on some oceanic ridges in zones of active seafloor spreading.

hypertonic Referring to a solution having a higher concentration of dissolved substances than the solution that surrounds it.

hypothesis A speculation about the natural world that may be verified or disproved by observation and experiment.

hypotonic Referring to a solution having a lower concentration of dissolved substances than the solution that surrounds it.

hypoxia A state of low oxygen saturation.

ice age One of several periods (lasting several thousand years each) of low temperature during the past million years. Glaciers and polar ice were derived from ocean water, lowering sea level at least 100 meters (328 feet). (See Appendix 2, "Geologic Time.")

ice cap Permanent cover of ice; formally limited to ice atop land, but informally applied also to floating ice in the Arctic Ocean.

iceberg A large mass of ice floating in the ocean that was formed on or adjacent to land. Tabular icebergs are table-like or flat; pinnacled icebergs are castellated, or jagged. Southern icebergs are often tabular; northern icebergs are often pinnacled.

inlet A passage giving the ocean access to an enclosed lagoon, harbor, or bay.

insolation rate The amount of solar energy reaching Earth's surface per unit time.

interference Addition or subtraction of wave energy as waves interact; also called *resonance*. See also *constructive interference; destructive interference*.

internal wave A progressive wave occurring at the boundary between liquids of different densities.

intertidal zone The marine zone between the highest high-tide point on a shoreline and the lowest low-tide point. The intertidal zone is sometimes subdivided into four separate habitats by height above tidal datum, typically numbered 1 to 4, land to sea.

intertropical convergence zone (ITCZ) The equatorial area at which the trade winds converge. The ITCZ usually lies at or near the meteorological equator; also called the *doldrums*.

introduced species A species removed from its home range and established in a new and foreign location; also called *exotic species*.

invertebrate Animal lacking a backbone.

ion An atom (or small group of atoms) that becomes electrically charged by gaining or losing one or more electrons.

ionic bond A chemical bond resulting from attraction between oppositely charged ions. These forces are said to be "electrostatic" in nature.

ionizing radiation Fast-moving particles or high-energy electromagnetic radiation emitted as unstable atomic nuclei disintegrate. The radiation has enough energy to dislodge one or more electrons from atoms it

hits to form charged ions, which can react with and damage living tissue.

island arc Curving chain of volcanic islands and seamounts almost always found paralleling the concave edge of a trench.

isostatic equilibrium Balanced support of lighter material in a heavier, displaced supporting matrix; analogous to buoyancy in a liquid.

isotonic Referring to a solution having the same concentration of dissolved substances as the solution that surrounds it.

ITCZ See *intertropical convergence zone*.

Jason-1 Successor to the *TOPEX/Poseidon* satellite mission.

kelp Informal name for any species of large phaeophyte.

kingdom The largest category of biological classification. Five kingdoms are presently recognized.

knot A speed of 1 nautical mile per hour. See also *nautical mile*.

krill *Euphausia superba*, a thumb-size crustacean common in Antarctic waters.

La Niña An event during which normal tropical Pacific atmospheric and oceanic circulation strengthens and the surface temperature of the eastern South Pacific drops below average values; usually occurs at the end of an ENSO event. See also *ENSO*.

lagoon A shallow body of seawater generally isolated from the ocean by a barrier island. Also, the body of water enclosed within an atoll, or the water within a reverse estuary.

land breeze Movement of air offshore as marine air heats and rises.

Langmuir circulation Shallow, wind-driven circulation of water in horizontal, spiral bands.

latent heat of evaporation Heat added to a liquid during evaporation (or released from a gas during condensation) that produces a change in state but not a change in temperature. For pure water, 585 calories per gram at 20°C (68°F). Compare *latent heat of vaporization*.

latent heat of fusion Heat removed from a liquid during freezing (or added to a solid during thawing) that produces a change in state but not a change in temperature. For pure water, 80 calories per gram at 0°C (32°F).

latent heat of vaporization Heat added to a liquid during evaporation (or released from a gas during condensation) that produces a change in state but not a change in temperature. For pure water, 540 calories per gram at 100°C (212°F). Compare *latent heat of evaporation*.

lateral-line system A system of sensors and nerves in the head and midbody of fishes and some amphibians that functions to detect low-frequency vibrations in water.

latitude Regularly spaced imaginary lines on Earth's surface running parallel to the equator.

law A large construct explaining events in nature that have been observed to occur with unvarying uniformity under the same conditions.

Law of the Sea Collective term for laws and treaties governing the commercial and practical use of the ocean.

Library of Alexandria The greatest collection of writings in the ancient world, founded

in the third century B.C.E. at the behest of Alexander the Great; could be considered the first university.

light Electromagnetic radiation propagated as small, nearly massless particles that behave like both a wave and a stream of particles.

limiting factor A physical or biological environmental factor whose absence or presence in an inappropriate amount limits the normal actions of an organism.

Linnaeus, Carolus Carl von Linné (1707–1778). Swedish "father" of modern taxonomy.

lithification Conversion of sediment into sedimentary rock by pressure or by the introduction of a mineral cement.

lithosphere The brittle, relatively cool outer layer of Earth, consisting of the oceanic and continental crust and the outermost, rigid layer of mantle.

littoral zone The band of coast alternately covered and uncovered by tidal action; the intertidal zone.

longitude Regularly spaced imaginary lines on Earth's surface running north and south and converging at the poles.

longshore bar A submerged or exposed line of sand lying parallel to shore and accumulated by wave action.

longshore current A current running parallel to shore in the surf zone, caused by the incomplete refraction of waves approaching the beach at an angle.

longshore drift Movement of sediments parallel to shore, driven by wave energy.

longshore trough Submerged excavation parallel to shore adjacent to an exposed sandy beach; caused by the turbulence of water returning to the ocean after each wave.

low-energy coast A coast only rarely exposed to large waves.

lower mantle The rigid portion of Earth's mantle below the asthenosphere.

low tide The low-water position corresponding to a tidal trough.

low-tide terrace The smooth, hard-packed beach seaward of the beach scarp on which waves expend most of their energy. Site of the most vigorous onshore and offshore movement of sand.

lunar tide Tide caused by gravitational and inertial interaction of the moon and Earth.

macroplankton Animal plankters larger than 1 to 2 centimeters (to 1 inch). An example is the jellyfish.

Magellan, Ferdinand (c. 1480–1521) Portuguese navigator in the service of Spain who led the first expedition to circumnavigate Earth, 1519–1522. He was killed in the Philippines.

magma Molten rock capable of fluid flow; called *lava* aboveground.

magnetometer A device that measures the amount and direction of residual magnetism in a rock sample.

Mahan, Alfred Thayer An American naval officer and strategist; the influential author of *The Influence of Sea Power upon History, 1660–1783*.

Mammalia The class of mammals.

mangrove A large flowering shrub or tree that grows in dense thickets or forests along muddy or silty tropical coasts.

mantle The layer of Earth between the crust and the core, composed of silicates of iron and magnesium. The mantle has an average density of about 4.5 g/cm³ and

accounts for about 68% of Earth's mass.

mantle plume Ascending columns of superheated mantle originating at the core–mantle boundary.

map A representation of Earth's surface, usually depicting mostly land areas. See also *chart*.

mariculture The farming of marine organisms, usually in estuaries, bays, or nearshore environments, or in specially designed structures using circulating seawater. Compare *aquaculture*.

marine energy resource Any resource resulting from the direct extraction of energy from the heat or movement of ocean water.

marine pollution The introduction by humans of substances or energy into the ocean that changes the quality of the water or affects the physical and biological environment.

marine science The process (or result) of applying the scientific method to the ocean, its surroundings, and the life-forms within it; also called *oceanography* or *oceanology*.

masking pigment See *accessory pigment*.

mass A measure of the quantity of matter.

mass extinction A catastrophic, global event in which major groups of species perish abruptly.

mathematical model A set of equations that attempts to describe the behavior of a system.

Maury, Matthew (1806–1873) "Father" of physical oceanography. Probably the first person to undertake the systematic study of the ocean as a full-time occupation, and probably the first to understand the global interlocking of currents, wind flow, and weather.

maximum sustainable yield The maximum amount of fish, crustaceans, and mollusks that can be caught without impairing future populations.

mean sea level The height of the ocean surface averaged over a few years' time.

medusa Free-swimming body form of many members of the phylum Cnidaria.

membrane A complex structure of proteins and lipids that forms boundaries around and within the cell. It is usually semipermeable, allowing some kinds of molecules to pass through but not others.

meroplankton The planktonic phase of the life cycle of organisms that spend only part of their life drifting in the plankton.

mesosphere The rigid inner mantle, similar in chemical composition to the asthenosphere.

metabolic rate The rate at which energy-releasing reactions proceed within an organism.

metamerism Segmentation; repeating body parts.

***Meteor* expedition** German Atlantic expedition begun in 1925; the first to use an echo sounder and other modern optical and electronic instrumentation.

meteorological equator The irregular imaginary line of thermal equilibrium between hemispheres. It is situated about 5° north of the geographical equator, and its position changes with the seasons, moving slightly north in northern summer. Also called the *thermal equator*.

meteorological tide A tide influenced by the weather. Arrival of a storm surge will alter the estimate of a tide's height or arrival time, as will

a strong, steady onshore or offshore wind.

metrophagy Tendency for large reptiles to eat entire cities.

microbial loop A cycle of production and consumption of glucose carried on by extremely small organisms.

microtektite Translucent oblong particles of glass, a component of cosmogenous sediment.

Milky Way galaxy The name of our galaxy; sometimes applied to the field of stars in our home spiral arm, which is correctly called the *Orion arm*.

mineral A naturally occurring inorganic crystalline material with a specific chemical composition and structure.

mixed layer See *surface zone*.

mixed tide (or semidiurnal mixed tide) A complex tidal cycle, usually with two high tides and two low tides of unequal height per day.

mixing time The time necessary to mix a substance through the ocean, about 1,600 years.

mixture A close intermingling of different substances that still retain separate identities. The properties of a mixture are heterogeneous; they may vary within the mixture.

molecule A group of atoms held together by chemical bonds. The smallest unit of a compound that retains the characteristics of the compound.

Mollusca The phylum of animals that includes chitons, snails, clams, and octopuses.

molt To shed an external covering.

monsoon A pattern of wind circulation that changes with the season. Also, the rainy season in areas with monsoon wind patterns.

moon tide See *lunar tide*.

moraine Hills or ridges of sediment deposited by glaciers.

motile Able to move about.

multicellular Consisting of more than one cell.

multicellular algae Algae with bodies consisting of more than one cell. Examples are kelp and *Ulva*.

mutation A heritable change in an organism's genes.

mutualism A symbiotic interaction between two species that is beneficial to both.

Mysticeti The suborder of baleen whales.

Nansen bottle A water-sampling instrument perfected early in the 20th century by the Norwegian scientist and explorer Fridtjof Nansen.

National Oceanic and Atmospheric Administration (NOAA) The agency of the U.S. government primarily responsible for oceanic science, service, and stewardship.

natural selection A mechanism of evolution that results in the continuation of only those forms of life best adapted to survive and reproduce in their environment.

natural system of classification A method of classifying an organism based on its ancestry or origin.

nautical chart A chart used for marine navigation.

nautical mile The length of 1 minute of latitude, 6,076 feet, 1.15 statute miles, or 1.85 kilometers. (See Appendix 1.)

neap tide The time of smallest variation between high and low tides occurring when Earth, moon, and sun align at right angles. Neap tides alternate with spring tides, occurring at 2-week intervals.

nebula Diffuse cloud of dust and gas (plural, *nebulae*).

nekton Swimming organisms.

Nematoda The phylum of animals to which round-worms belong.

neritic Of the shore or coast; refers to continental margins and the water covering them, or to nearshore organisms.

neritic sediment Continental shelf sediment consisting primarily of terrigenous material.

neritic zone The zone of open water nearshore, over the continental shelf.

niche Description of an organism's functional role in a habitat; its "job."

nitrifying bacteria Bacteria capable of fixing gaseous nitrogen into nitrite, nitrate, or ammonium ions.

nitrogen cycle The cycle in which nitrogen moves from its largest reservoir (the atmosphere) through the ocean, ocean sediments, and food webs, and then back to the atmosphere.

node The line or point of no wave action in a standing pattern. See also *amphidromic point*.

nodule Solid mass of hydrogenous sediment, most commonly manganese or ferromanganese nodules and phosphorite nodules.

nonconservative constituent An element whose proportion in seawater varies with time and place, depending on biological demand or chemical reactivity. An element with a short residence time; for example, iron, aluminum, silicon, trace nutrients, dissolved oxygen, and carbon dioxide.

nonconservative nutrient A compound or ion that is needed by autotrophs for primary productivity and that changes in concentration with biological activity.

nonextractive resource Any use of the ocean in place, such as transportation of people and commodities by sea, recreation, or waste disposal.

nonrenewable resource Any resource that is present on Earth in fixed amounts and cannot be replenished.

nonvascular Describing photosynthetic autotrophs without vessels for the transport of fluid. Examples are algae.

nor'easter (northeaster) Any energetic extratropical cyclone that sweeps the eastern seaboard of North America in winter.

North Atlantic Deep Water Cold, dense water formed in the Arctic that flows onto the floor of the North Atlantic ocean.

notochord Stiffening structure found at some time in the life cycle of all members of the phylum Chordata.

nuclear energy Energy released when atomic nuclei undergo a nuclear reaction such as the spontaneous emission of radioactivity, nuclear fission, or nuclear fusion. About 17% of the electrical power generated in the United States is provided by the nuclear fission of uranium in civilian power reactors.

nucleus (physics) The small, dense, positively charged center of an atom that contains the protons and neutrons.

nutrient Any needed substance that an organism obtains from its environment except oxygen, carbon dioxide, and water.

ocean (1) The great body of saline water that covers 70.78% of the surface of Earth. (2) One of its primary subdivisions, bounded by continents, the equator, and other imaginary lines.

ocean basin Deep-ocean floor made of basaltic crust. Compare *continental margin*.

oceanic crust The outermost solid surface of Earth beneath ocean floor sediments, composed primarily of basalt.

oceanic ridge Young seabed at the active spreading center of an ocean, often unmasked by sediment, bulging above the abyssal plain. The boundary between diverging plates. Often called a mid-ocean ridge, though less than 60% of the length exists at mid-ocean.

oceanic zone The zone of open water away from shore, past the continental shelf.

oceanography The science of the ocean. See also *marine science*.

oceanus Latin form of *okeanos*, the Greek name for the "ocean river" past Gibraltar.

Odontoceti The suborder of toothed whales.

oolite sand Hydrogenous sediment formed when calcium carbonate precipitates from warmed seawater as pH rises, forming rounded grains around a shell fragment or other particle.

ooze Sediment of at least 30% biological origin.

ophiolite An assemblage of subducting oceanic lithosphere scraped off (obducted) onto the edge of a continent.

Ophiuroidea The class of the phylum Echinodermata to which brittle stars belong.

orbit In ocean waves, the circular pattern of water particle movement at the air–sea interface. Orbital motion contrasts with the side-to-side or back-and-forth motion of pure transverse or longitudinal waves.

orbital inclination The 23°27' "tilt" of Earth's rotational axis relative to the plane of its orbit around the sun.

orbital wave A progressive wave in which particles of the medium move in closed circles.

osmoregulation The ability to adjust internal salt concentration.

osmosis The diffusion of water from a region of high water concentration to a region of lower water concentration through a semipermeable membrane.

Osteichthyes The class of fishes with bony skeletons.

outgassing The volcanic venting of volatile substances.

overfishing Harvesting so many fish that there is not enough breeding stock left to replenish the species.

oxygen minimum zone A zone in which oxygen is depleted by animals and not replaced by phytoplankton.

oxygen revolution The time span, from about 2 billion to 400 million years ago, during which photosynthetic autotrophs changed the composition of Earth's atmosphere to its current oxygen-rich mixture.

ozone O_3, the triatomic form of oxygen. Ozone in the upper atmosphere protects living things from some of the harmful effects of the sun's ultraviolet radiation.

ozone layer A diffuse layer of ozone mixed with other gases surrounding the world at a height of about 20 to 40 kilometers (12–25 miles).

P wave (primary wave) A compressional wave that is associated with an earthquake and that can move through both liquid and rock.

Pacific Ring of Fire The zone of seismic and volcanic activity that encircles the Pacific Ocean.

paleoceanography The study of the ocean's past.

paleomagnetism The "fossil," or remanent, magnetic field of a rock.

Pangaea Name given by Alfred Wegener to the original "protocontinent." The breakup of Pangaea gave rise to the Atlantic Ocean and to the continents we see today.

Panthalassa Name given by Alfred Wegener to the ocean surrounding Pangaea.

parasitism A symbiotic relationship in which one species spends part or all of its life cycle on or within another, using the host species (or food within the host) as a source of nutrients; the most common form of symbiosis.

partially mixed estuary An estuary in which an influx of seawater occurs beneath a surface layer of freshwater flowing seaward. Mixing occurs along the junction.

passive margin The continental margin near an area of lithospheric plate divergence; also called *Atlantic-type margin*.

passive sonar A device that detects the intensity and direction of underwater sounds.

PCBs See *polychlorinated biphenyls*.

pelagic Of the open ocean; refers to the water above the deep-ocean basins, sediments of oceanic origin, or organisms of the open ocean.

pelagic sediment Sediments of the slope, rise, and deep-ocean floor that originate in the ocean.

pelagic zone The realm of open water. See also *benthic zone*.

perigee The point in the orbit of a satellite where it is closest to the main body; opposite of *apogee*.

perihelion The point in the orbit of a satellite where it is closest to the sun; opposite of *aphelion*.

period See *wave period*.

pH scale A measure of the acidity or alkalinity of a solution; numerically, the negative logarithm of the concentration of hydrogen ions in an aqueous solution. A pH of 7 is neutral; lower numbers indicate acidity, and higher numbers indicate alkalinity.

Phaeophyta Brown multicellular algae, including kelps.

photic zone The thin film of lighted water at the top of the world ocean. The photic zone rarely extends deeper than 200 meters (660 feet). Compare *euphotic zone*.

photon The smallest unit of light energy.

photosynthesis The process by which autotrophs bind light energy into the chemical bonds of food with the aid of chlorophyll and other substances. The process uses carbon dioxide and water as raw materials and yields glucose and oxygen.

phycobilin A reddish accessory pigment found in red algae.

phylum One of the major groups of the animal kingdom whose members share a similar body plan, level of complexity, and evolutionary history (see Appendix 6) (plural, *phyla*). (The major groups of the plant kingdom are called *divisions*.)

physical factor An aspect of the physical environment that affects living organisms, such as light, salinity, or temperature.

physical resource Any resource that has resulted from the deposition, precipitation, or accumulation of a useful nonliving substance in the ocean or seabed; also called a *nonliving resource*.

phytoplankton Plantlike, usually single-celled members of the plankton community.

picoplankton Extremely small members of the plankton community, typically 0.2 to 2 micrometers (4–40 millionths of an inch) across.

Pinnipedia The carnivoran suborder that contains the seals, sea lions, and walruses.

piston corer A seabed-sampling device capable of punching through up to 25 meters (80 feet) of sediment and returning an intact plug of material.

planet A smaller, usually nonluminous body orbiting a star.

plankter Informal name for a member of the plankton community.

plankton Drifting or weakly swimming organisms suspended in water. Their horizontal position is to a large extent dependent on the mass flow of water rather than on their own swimming efforts.

plankton bloom A sudden increase in the number of phytoplankton cells in a volume of water.

plankton net Conical net of fine nylon or Dacron fabric used to collect plankton.

Plantae The kingdom to which multicellular vascular autotrophs belong.

plate One of about a dozen rigid segments of Earth's lithosphere that move independently. The plate consists of continental or oceanic crust and the cool, rigid upper mantle directly below the crust.

plate tectonics The theory that Earth's lithosphere is fractured into plates that move relative to each other and are driven by convection currents in the mantle. Most volcanic and seismic activity occurs at plate margins.

Platyhelminthes The phylum of animals to which flatworms belong.

plunging wave A breaking wave in which the upper section topples forward and away from the bottom, forming an air-filled tube.

polar cell The atmospheric circulation cell centered over each pole.

polar front Boundary between the polar cell and the Ferrel cell in each hemisphere.

polar molecule A molecule with unbalanced charge. One end of the molecule has a slight negative charge, and the other end has a slight positive charge.

pollutant A substance that causes damage by interfering directly or indirectly with an organism's biochemical processes.

Polychaeta The largest and most diverse class of phylum Annelida. Nearly all polychaetes are marine.

polychlorinated biphenyls (PCBs) Chlorinated hydrocarbons once widely used to cool and insulate electrical devices and to strengthen wood or concrete. PCBs may be responsible for the changes in and declining fertility of some marine mammals.

Polynesia A large group of Pacific islands lying east of Melanesia and Micronesia, and extending from the Hawai'ian Islands south to New Zealand and east to Easter Island.

Polynesians Inhabitants of the Pacific islands that lie within a triangle formed by Hawai'i, New Zealand, and Easter Island.

polynya A gap in polar pack ice at which liquid water contacts the atmosphere.

polyp One of two body forms of Cnidaria. Polyps are cup-shaped and possess rings of tentacles. Coral animals are polyps.

poorly sorted sediment A sediment in which particles of many sizes are found.

population A group of individuals of the same species occupying the same area.

population density The number of individuals per unit area.

Porifera The phylum of animals to which sponges belong.

potable water Water suitable for drinking.

precipitate (1) A solid substance formed in an aqueous reaction. (2) The process by which a solute forms in and falls from a solution. The falling of water or ice from the atmosphere.

precipitation Liquid or solid water that falls from the air and reaches the surface as rain, hail, or snowfall.

pressure Force per unit area.

prey An organism consumed by a predator.

primary consumer Initial consumer of primary producers. The consumers of autotrophs; the second level in food webs.

primary forces The forces that induce and maintain water flow in ocean current systems: thermal expansion, wind friction, and density differences.

primary producer An organism capable of using energy from light or energy-rich chemicals in the environment to produce energy-rich organic compounds; an autotroph.

primary productivity The synthesis of organic materials from inorganic substances by photosynthesis or chemosynthesis; expressed in grams of carbon bound into carbohydrate per unit area per unit time ($gC/m^2/yr$).

primary wave A compressional wave that is associated with an earthquake and that can move through both liquid and rock.

Prince Henry the Navigator Established a center at Sagres, Portugal, for the study of marine science and navigation in the mid-1450s.

principle of constant proportions The proportions of major conservative elements in seawater remain nearly constant, though total salinity may change with location; also called *Forchhammer's principle*.

progressive wave A wave of moving energy in which the wave form moves in one direction along the surface (or junction) of the transmission medium (or media).

Protista The kingdom of single-celled nucleated organisms to which protozoa, diatoms, and dinoflagellates belong; also called *Protoctista*.

proton A positively charged particle at the center of an atom.

protostar A tightly condensed knot of material that has not yet attained fusion temperature.

pteropod A small planktonic mollusc with a calcareous shell, which contributes to biogenous sediments.

pycnocline The middle zone of the ocean in which density increases rapidly with depth. Temperature falls and salinity rises in this zone.

radial symmetry Body structure in which the body parts radiate from a central axis like spokes from a wheel. An example is a sea star. Compare *bilateral symmetry*.

radioactive decay The disintegration of unstable forms of elements, which releases subatomic particles and heat.

radiolarian One of a group of usually planktonic amoeba-like animals with a siliceous shell, which contributes to biogenous sediments.

radiometric dating The process of determining the age of rocks by observing the ratio of unstable radioactive elements to stable decay products.

random distribution Distribution of organisms within a community whereby the position of one organism is in no way influenced by the positions of other organisms or by physical variations within that community; a rare distribution pattern.

realized niche The niche that an organism actually occupies. It is narrower than an organism's fundamental niche because of interspecific competition.

reef A hazard to navigation; a shoal, a shallow area, or a mass of fish or other marine life.

refraction Bending of light or sound waves as they move at an angle other than 90° between media of different optical or acoustical densities. See also *wave refraction*.

refractive index The degree of refraction from one medium to another expressed as a ratio. The higher the ratio (refractive index), the greater the bending of waves between media.

refractometer A compact optical device that determines the salinity of a water sample by comparing the refractive index of the sample to the refractive index of water of known salinity.

relative dating Determining the age of a geological sample by comparing its position with the positions of other samples.

renewable resource Any resource that is naturally replaced on a seasonal basis by the growth of living organisms or by other natural processes.

Reptilia The class of reptiles, including turtles, crocodiles, iguanas, and snakes.

residence time The average length of time a dissolved substance spends in the ocean.

respiration Release of stored energy from chemical bonds in food; carbon dioxide and water are formed as by-products. (Respiration is a biochemical process and is not the same as the mechanical process of breathing.)

restoring force The dominant force trying to return water to flatness after formation of a wave.

reverse estuary An estuary along a coast in which salinity increases from the ocean to the estuary's upper reaches because of evaporation of seawater and a lack of freshwater input.

Rhodophyta Red, multicellular algae.

Richter scale A logarithmic measure of earthquake magnitude. A great earthquake measures above 8 on the Richter scale.

rift valley A depression that forms between diverging tectonic plates.

rip current A strong, narrow surface current that flows seaward through the surf zone and is caused by the escape of excess water that has piled up in a longshore trough.

rogue wave A single wave crest much higher than usual, caused by constructive interference.

S Dimensionless unit expressing salinity on the Practical Salinity Scale.

S wave Secondary wave; a transverse wave that is associated with an earthquake and that cannot move through liquid.

salinity A measure of the dissolved solids in seawater, usually expressed in grams per kilogram or parts per thousand by weight. Standard seawater has a salinity of 35‰ at 0°C (32°F).

salinometer An electronic device that determines salinity by measuring the electrical conductivity of a seawater sample.

salt gland Specialized tissue responsible for concentration and excretion of excess salt from blood and other body fluids.

salt-wedge estuary An estuary in which rapid river flow and small tidal range cause an inclined wedge of seawater to form at the mouth.

sand Sediment particle between 0.062 and 2 millimeters in diameter.

sandbar A submerged or exposed line of sand accumulated by wave action.

sand spit An accumulation of sand and gravel deposited downcurrent from a headland. Sand spits often curl at their tips.

saturation State of a solution in which no more of the solute will dissolve in the solvent. The rate at which molecules of the solute are being dissolved equals the rate at which they are being precipitated from the solution.

scattering The dispersion (or "bounce") of sound or light waves when they strike particles suspended in water or air. The amount of scatter depends on the number, size, and composition of the particles.

schooling Tendency of small fish of a single species, size, and age to mass in groups. The school moves as a unit, which confuses predators and reduces the effort spent searching for mates.

science A systematic way of asking questions about the natural world and testing the answers to those questions.

scientific method The orderly process by which theories explaining the operation of the natural world are verified or rejected.

scientific name The genus and species name of an organism.

sea Simultaneous wind waves of many wavelengths forming a chaotic ocean surface. Sea is common in an area of wind-wave origin.

sea breeze Onshore movement of air as inland air heats and rises.

sea cave A cave near sea level in a sea cliff cut by processes of marine erosion.

sea cliff A cliff marking the landward limit of marine erosion on an erosional coast.

sea grass Any of several marine angiosperms. Examples are *Zostera* (eelgrass) and *Phyllospadix* (surfgrass). Sea grasses are not seaweeds.

sea ice Ice formed by the freezing of seawater.

sea island An island whose central core was connected to the mainland when sea level was lower. Rising ocean separates these high points from land, and sedimentary processes surround them with beaches. Compare *barrier island*.

sea level The height of the ocean surface. See also *mean sea level*.

sea power The means by which a nation extends its military capacity onto the ocean.

sea state Ocean wave conditions at a specific place and time, usually stated in the Beaufort scale.

seafloor spreading The theory that new ocean crust forms at spreading centers, most of which are on the ocean floor, and pushes the continents aside. Power is thought to be provided by

convection currents in Earth's upper mantle.

seamount A circular or elliptical projection from the seafloor, more than 1 kilometer (0.6 mile) in height, with a relatively steep slope of 20° to 25°.

SEASTAR Satellite capable of measuring the distribution of chlorophyll at the ocean surface, a measure of marine productivity.

seaweed Informal term for large marine multicellular algae.

second law of thermodynamics Disorder (entropy) in a closed system must increase over time. If disorder decreases, it does so at the expense of energy. Because the universe as a whole may be considered a closed system, it follows that an increase in order in one part must result in a decrease in order in another.

secondary consumer Consumer of primary consumers.

sediment Particles of organic or inorganic matter that accumulate in a loose, unconsolidated form.

seiche Pendulum-like rocking of water in an enclosed area; a form of standing wave that can be caused by meteorological or seismic forces, or that may result from normal resonances excited by tides.

seismic Referring to earthquakes and the shock of earthquakes.

seismic sea wave Tsunami caused by displacement of earth along a fault. (Earthquakes and seismic sea waves are caused by the same phenomenon.)

seismic wave A low-frequency wave generated by the forces that cause earthquakes. Some kinds of seismic waves can pass through Earth. See also *P wave; S wave.*

seismograph An instrument that detects and records earth movement associated with earthquakes and other disturbances.

semidiurnal tide A tidal cycle of two high tides and two low tides each lunar day, with the high tides of nearly equal height.

sensible heat Heat whose gain or loss is detectable by a thermometer or other sensor.

sessile Attached; nonmotile; unable to move about.

sewage Waste water with a significant organic content, usually from domestic or industrial sources.

sewage sludge Semisolid mixture of organic matter, microorganisms, toxic metals, and synthetic organic chemicals removed from waste water at a sewage treatment plant.

shadow zone (1) The wide band at Earth's surface 105° to 143° away from an earthquake in which seismic waves are nearly absent. P waves are absent because they are refracted by Earth's liquid outer core; S waves are absent from this band and the zone immediately opposite the earthquake site because they are absorbed by the outer core. (2) In sonar, the volume of ocean from which sound waves diverge and in which a submarine may hide.

shallow-water wave A wave in water shallower than 1/20 wavelength.

shelf break The abrupt increase in slope at the junction between continental shelf and continental slope.

shore The place where ocean meets land. On nautical charts, the limit of high tides.

side-scan sonar A high-resolution sound-imaging system used for geological investigations, archaeological studies, and the location of sunken ships and airplanes.

siliceous ooze Ooze composed mostly of the hard remains of silica-containing organisms.

silicoflagellate A tiny, single-celled phytoplankter with a siliceous skeleton.

silt Sediment particle between 0.004 and 0.062 millimeter in diameter.

Sirenia The order of mammals that includes manatees, dugongs, and the extinct sea cows.

slack water A time of no tide-induced currents that occurs when the current changes direction.

SOFAR *Sound fixing and ranging*. An experimental U.S. Navy technique for locating survivors on life rafts, based on the fact that sound from explosive charges dropped into the layer of minimum sound velocity can be heard for great distances. See also *sofar layer*.

SOFAR layer Layer of minimum sound velocity in which sound transmission is unusually efficient for long distances. Sounds leaving this depth tend to be refracted back into it. The sofar layer usually occurs at mid-latitude depths of around 1,200 meters (4,000 feet).

solar nebula The diffuse cloud of dust and gas from which the solar system originated.

solar system The sun together with the planets and other bodies that revolve around it.

solar tide Tide caused by the gravitational and inertial interaction of the sun and Earth.

solstice One of two times of the year when the overhead position of the sun is farthest from the equator. The time of the solstice is midway between equinoxes.

solute A substance dissolved in a solvent. See also *solution*.

solution A homogeneous substance made of two components, the solvent and the solute.

solvent A substance able to dissolve other substances. See also *solution*.

sonar *Sound navigation and ranging*.

sound A form of energy transmitted by rapid pressure changes in an elastic medium. A relatively narrow passage of water between larger bodies of water or between the mainland and an island.

sounding Measurement of the depth of a body of water.

Southern Oscillation A reversal of airflow between normally low atmospheric pressure over the western Pacific and normally high pressure over the eastern Pacific; the cause of El Niño. See also *El Niño*.

speciation The formation of new species. Charles Darwin suggested that this is accomplished through isolation and natural selection.

species Any group of actually or potentially interbreeding organisms reproductively isolated from all other groups and capable of producing fertile offspring. (Note: The word *species* is both singular and plural.)

species diversity Number of different species in a given area.

species-specific relationship An exclusive relationship between two species. Parasites are usually species-specific; that is, they can usually parasitize only one species of host.

spilling wave A breaking wave whose crest slides down the face of the wave.

spreading center The junction between diverging plates at which new ocean floor is being made; also called *spreading zone*.

spring tide The time of greatest variation between high and low tides occurring when Earth, moon, and sun form a straight line. Spring tides alternate with neap tides throughout the year, occurring at 2-week intervals.

standing wave A wave in which water oscillates without causing progressive wave forward movement. There is no net transmission of energy in a standing wave.

star A massive sphere of incandescent gases powered by the conversion of hydrogen to helium and other heavier elements.

state An expression of the internal form of matter. Water exists in three states: solid, liquid, and gas. A solid has a fixed volume and fixed shape, a liquid has a fixed volume but no fixed shape, and a gas has neither fixed volume nor fixed shape.

stenohaline Describing an organism unable to tolerate a wide range in salinity.

stenothermal Describing an organism unable to tolerate wide variance in temperature.

stipe Multicellular algal equivalent of a vascular plant's stem.

Stokes drift A small net transport of water in the direction a wind wave is moving.

storm Local or regional atmospheric disturbance characterized by strong winds often accompanied by precipitation.

storm surge An unusual rise in sea level as a result of the low atmospheric pressure and strong winds associated with a tropical cyclone. Onrushing seawater precedes landfall of the tropical cyclone and causes most of the damage to life and property.

stratigraphy The branch of geology that deals with the definition and description of natural divisions of rocks; specifically, the analysis of relationships of rock strata.

subduction The downward movement into the asthenosphere of a lithospheric plate.

subduction zone An area at which a lithospheric plate is descending into the asthenosphere. The zone is characterized by linear folds (trenches) in the ocean floor and strong deep-focus earthquakes; also called a *Wadati–Benioff zone*.

sublittoral zone The ocean floor nearshore. The inner sublittoral extends from the littoral (intertidal) zone to the depth at which wind waves have no influence; the outer sublittoral extends to the edge of the continental shelf.

submarine canyon A deep, V-shaped valley running roughly perpendicular to the shoreline and cutting across the edge of the continental shelf and slope.

subsidence Sinking, often of tectonic origin.

Subtropical Convergence Convergence zone marking the boundary between Central Water and either Subarctic or Subantarctic Surface Water. The northern Subtropical Convergence lies at about 45°N in the Pacific and 60°N in the Atlantic; the southern Subtropical Convergence lies at 40° to 50°S.

succession The changes in species composition that lead to a climax community.

sun tide See *solar tide.*

supernova The explosive collapse of a massive star.

superplume A very large mantle plume.

supralittoral zone The splash zone above the highest high tide; not technically part of the ocean bottom.

surf The confused mass of agitated water rushing shoreward during and after a wind wave breaks.

surf beat The pattern of constructive and destructive interference that causes successive breaking waves to grow, shrink, and grow again over a few minutes' time.

surf zone The region between the breaking waves and the shore.

surface current The horizontal flow of water at the ocean's surface.

surface tension A property of water resulting from the cohesion of water molecules by hydrogen bonds.

surface-to-volume ratio A physical constraint on the size of cells. As a cell's linear dimensions grow, its surface area does not increase at the same rate as its volume. As the surface-to-volume ratio decreases, each square unit of outer membrane must serve an increasing interior volume.

surface wave Seismic waves that travel along Earth's surface.

surface zone The upper layer of ocean in which temperature and salinity are relatively constant with depth. Depending on local conditions, the surface zone may reach to 1,000 meters (3,300 feet) or be absent entirely. Also called the *mixed layer.*

surging wave A wave that surges ashore without breaking.

suspension feeder An animal that feeds by straining or otherwise collecting plankton and tiny food particles from the surrounding water.

sverdrup (sv) A unit of volume transport named in honor of oceanographer Harald U. Sverdrup: 1 million cubic meters of water flowing past a fixed point each second.

swash Water from waves washing onto a beach.

swell Mature wind waves of one wavelength that form orderly undulations of the ocean surface.

swim bladder A gas-filled organ that assists in maintaining neutral buoyancy in some bony fishes.

symbiosis The co-occurrence of two species in which the life of one is closely interwoven with the life of the other; mutualism, commensalism, or parasitism.

synoptic sampling Simultaneous sampling at many locations.

$$T = G\left(\frac{m_1 m_2}{r^3}\right)$$ Calculation of the tide-generating force (T).

taxonomy In biology, the laws and principles covering the classification of organisms.

tektite A small, rounded, glassy component of cosmogenous sediments, usually less than 1.5 millimeters (0.06 inches) in length; thought to have formed from the impact of an asteroid or meteor on the crust of Earth or the moon.

Teleostei The osteichthyan order that contains the cod, tuna, halibut, perch, and other species of bony fishes.

telepresence The extension of a person's senses by remote sensors and manipulators.

temperate zone The mid-latitude area between the Tropic of Cancer and the Arctic Circle, and between the Tropic of Capricorn and the Antarctic Circle.

temperature The response of a solid, liquid, or gas to the input or removal of heat energy. A measure of the atomic and molecular vibration in a substance, indicated in degrees.

temperature–salinity (T–S) diagram A graph showing the relationship of temperature and salinity with depth.

terrane An isolated segment of seafloor, island arc, plateau, continental crust, or sediment transported by seafloor spreading to a position adjacent to a larger continental mass; usually different in composition from the larger mass.

terrigenous sediment Sediment derived from the land and transported to the ocean by wind and flowing water.

territorial waters Waters extending 12 miles from shore and in which a nation has the right to jurisdiction.

thallus The body of an alga or other simple plant.

theory A general explanation of a characteristic of nature consistently supported by observation or experiment.

thermal equator See *meteorological equator.*

thermal equilibrium The condition in which the total heat coming into a system (such as a planet) is balanced by the total heat leaving the system.

thermal inertia Tendency of a substance to resist change in temperature with the gain or loss of heat energy.

thermocline The zone of the ocean in which temperature decreases rapidly with depth. See also *pycnocline.*

thermohaline circulation Water circulation produced by differences in temperature and/or salinity (and therefore density).

thermostatic property A property of water that acts to moderate changes in temperature.

tidal bore A high, often breaking wave generated by a tide crest that advances rapidly up an estuary or river.

tidal current Mass flow of water induced by the raising or lowering of sea level caused by passage of tidal crests or troughs. See also *ebb current; flood current.*

tidal datum The reference level (0.0) from which tidal height is measured.

tidal range The difference in height between consecutive high and low tides.

tidal wave The crest of the wave causing tides; another name for a tidal bore; not a tsunami or seismic sea wave.

tide Periodic short-term change in the height of the ocean surface at a particular place, generated by long-wavelength progressive waves that are caused by the interaction of gravitational force and inertia. Movement of Earth beneath tide crests results in the rhythmic rising and falling of sea level.

tombolo Above-water bridge of sand connecting an offshore feature to the mainland.

top consumer An organism at the apex of a trophic pyramid, usually a carnivore.

TOPEX/Poseidon Joint French–U.S. satellite carrying radars that can determine the height of the sea surface with unprecedented accuracy. Other experiments in this 5-year program included sensing water vapor over the ocean, determining the precise location of ocean currents, and determining wind speed and direction.

tornado Localized, narrow, violent funnel of fast-spinning wind, usually generated when two air masses collide; not to be confused with a cyclone. (The tornado's oceanic equivalent is a waterspout.)

trace element A minor constituent of seawater present in amounts of less than 1 part per million.

trade winds Surface winds within the Hadley cells, centered at about 15° latitude, that approach from the northeast in the Northern Hemisphere and from the southeast in the Southern Hemisphere.

transform fault A plane along which rock masses slide horizontally past one another.

transform plate boundary Places where crustal plates shear laterally past one another. Crust is neither produced nor destroyed at this type of junction.

transverse current East-to-west or west-to-east current linking the eastern and western boundary currents. An example is the North Equatorial Current.

trench An arc-shaped depression in the deep-ocean floor with very steep sides and a flat sediment-filled bottom coinciding with a subduction zone. Most trenches occur in the Pacific.

trophic level A feeding step within a trophic pyramid.

trophic pyramid A model of feeding relationships among organisms. Primary producers form the base of the pyramid; consumers eating one another form the higher levels, with the top consumer at the apex.

Tropic of Cancer The imaginary line around Earth parallel to the equator at 23°27'N, marking the point where the sun shines directly overhead at the June solstice.

Tropic of Capricorn The imaginary line around Earth parallel to the equator at 23°27'S, marking the point where the sun shines directly overhead at the December solstice.

tropical cyclone A weather system of low atmospheric pressure around which winds blow counterclockwise in the Northern Hemisphere and clockwise in the Southern Hemisphere. It originates in the tropics within a single air mass but may move into temperate waters if the water temperature is high enough to sustain it. Small tropical cyclones are called *tropical depressions*; larger ones *tropical storms*; and great ones *hurricanes*, *typhoons*, or *willi-willis*, depending on location.

tropics The area between the Tropic of Cancer and the Tropic of Capricorn.

troposphere The lowest layer of Earth's atmosphere.

trough See *wave trough.*

tsunami Long-wavelength, shallow-water wave caused by rapid displacement of water. See also *seismic sea wave.*

tunicate A type of suspension-feeding invertebrate chordate.

turbidite A terrigenous sediment deposited by a turbidity current; typically, coarse-grained layers of nearshore origin interleaved with finer sediments.

turbidity current An underwater "avalanche" of abrasive sediments thought to be responsible for the deep sculpturing of submarine canyons and a means of transport for sediments accumulating on abyssal plains.

turbulence Chaotic fluid flow.

ultraplankton Extremely small plankton, smaller than nanoplankton.

undercurrent A current flowing beneath a surface current, usually in the opposite direction.

unicellular Consisting of a single cell.

unicellular algae Algae with bodies consisting of a single cell. Examples are diatoms and dinoflagellates.

uniform distribution Distribution of organisms within a community characterized by equal space between individuals (the arrangement of trees in an orchard); the rarest natural distribution pattern.

uniformitarianism The theory that all of Earth's geological features and history can be explained by processes occurring today, and that these processes must have been at work for a very long time.

United States Exploring Expedition The first U.S. oceanographic research voyage, launched in 1838.

upwelling Circulation pattern in which deep, cold, usually nutrient-laden water moves toward the surface. Upwelling can be caused by winds blowing parallel to shore or offshore.

U.S. Exclusive Economic Zone The region extending seaward from the coast of the United States for 200 nautical miles, within which the United States claims sovereign rights and jurisdiction over all marine resources.

V velocity

valve In diatoms, each half of the protective silica-rich outer portion of the cell. The complete outer covering is called the *frustule.*

vascular plant Plant having vessels for transport of fluid through leaves, stems, and roots. Examples are sea grasses, mangroves, and maple trees.

velocity Speed in a specified direction.

vertebrate A chordate with a segmented backbone.

Vikings Seafaring Scandinavian raiders who ravaged the coasts of Europe around 780–1070.

viscosity Resistance to fluid flow. A measure of the internal friction in fluids.

voyaging Traveling (usually by sea) with a specific purpose.

Wadati–Benioff zone See *subduction zone.*

water mass A body of water identifiable by its salinity and temperature (and therefore its density) or by its gas content or another indicator.

water vapor The gaseous, invisible form of water.

water-vascular system System of water-filled tubes and canals found in some representatives of the phylum Echinodermata and used for movement, defense, and feeding.

wave Disturbance caused by the movement of energy through a medium.

wave crest Highest part of a progressive wave above average water level.

wave-cut platform The smooth, level terrace sometimes found on erosional coasts that marks the submerged limit of rapid marine erosion.

wave diffraction Bending of waves around obstacles.

wave frequency The number of waves passing a fixed point per second.

wave height Vertical distance between a wave crest and the adjacent wave troughs.

wave period The time it takes for successive wave crests to pass a fixed point.

wave reflection The reflection of progressive waves by a vertical barrier. Reflection occurs with little loss of energy.

wave refraction Slowing and bending of progressive waves in shallow water.

wave shock Physical movement, often sudden, violent, and of great force, caused by the crash of a wave against an organism.

wave steepness Height-to-wavelength ratio of a wave. The theoretical maximum steepness of deep-water waves is 1:7.

wave train A group of waves of similar wavelength and period moving in the same direction across the ocean surface. The group velocity of a wave train is half the velocity of the individual waves.

wave trough The valley between wave crests below the average water level in a progressive wave.

wavelength The horizontal distance between two successive wave crests (or troughs) in a progressive wave.

weather The state of the atmosphere at a specific place and time.

Wegener, Alfred (1880–1930) German scientist who proposed the theory of continental drift in 1912.

well-mixed estuary An estuary in which slow river flow and tidal turbulence mix freshwater and saltwater in a regular pattern through most of its length.

well-sorted sediment A sediment in which particles are of uniform size.

West Wind Drift (Antarctic Circumpolar Current) Current driven by powerful westerly winds north of Antarctica. The largest of all ocean currents, it continues permanently eastward without changing direction. See *Antarctic Circumpolar Current.*

westerlies Surface winds within the Ferrel cells, centered around 45° latitude, that approach from the southwest in the Northern Hemisphere and from the northwest in the Southern Hemisphere.

western boundary current Strong, warm, concentrated, fast-moving current at the western boundary of an ocean (off the east coast of a continent). Examples include the Gulf Stream and the Japan (Kuroshio) Current.

westward intensification The increase in speed of geostrophic currents as they pass along the western boundary of an ocean basin.

Wilson, John Tuzo (1908–1993) Canadian geophysicist who proposed the theory of plate tectonics in 1965.

wind The mass movement of air.

wind duration The length of time the wind blows over the ocean surface, a factor in wind-wave development.

wind-induced vertical circulation Vertical movement in surface water (upwelling or downwelling) caused by wind.

wind strength Average speed of the wind, a factor in wind-wave development.

wind wave Gravity wave formed by transfer of wind energy into water. Wavelengths from 60 to 150 meters (200–500 feet) are most common in the open ocean.

world ocean The great body of saline water that covers 70.78% of Earth's surface.

xanthophyll A yellow or brown accessory pigment that gives some marine autotrophs a yellow or tan appearance.

zone Division or province of the ocean with homogeneous characteristics.

zooplankton Animal members of the plankton community.

zooxanthellae Unicellular dinoflagellates that are symbiotic with coral and that produce the relatively high pH and some of the enzymes essential for rapid calcium-carbonate deposition in coral reefs.

Index

Chthamalus, 458, 462
circumference, 27–28
clams, 424, 425, 450, 474, 475, 508
clamshell sampler, 154
classification schemes, 375–377
clay, 142, 149
climate, 219, 261
climate change, 14, 191–193, 527–536, 539
climate specialists, 4
climax community, 460
CloudSat (satellite), 53
clumped distribution, 459
Cnidaria, 420
cnidoblasts, 420, 421
CO_2. *See* carbon dioxide
coal, 60
coastal cells, 348–349
coastal downwelling, 264
coastal upwelling, 261
coasts
    Atlantic coast, 361–362
    beaches, 344–349
    characteristics of U.S., 361–362
    depositional coasts, 338–352
    Dubai, 366
    effects of biological activity on, 353–355
    effects of land erosion and sea-level
        change, 341
    effects of selective erosion, 341
    effects of volcanism and earth movements,
        343
    erosional coasts, 338–343
    estuaries, 355–360
    fault coasts, 343
    Gulf coast, 362
    high-energy coasts, 338–339
    human interference, 362–365
    investments, 366–367
    low-energy coasts, 339
    mangrove coasts, 355
    marine and terrestrial processes of,
        336–338
    Pacific coast, 361
    reefs, 353–355
    wave action on, 340
cobalt crusts, 490
coccolithophores, 149, 152, 402–404, 508
coccoliths, 147, 401, 402
cod, 435, 496, 508
coelacanths, 435
cold blooded, 384
cold-core eddies, 257
cold front, 235
cold-seep communities, 474–476
cold-seeps, 473–476
colligative properties, 201
Collisella, 458
Colorado Plateau, 148
Colorado River, 148
Columbia University, 50
Columbus, Christopher, 29, 37, 41, 55
Comet Hartley-2, 13
comets, 6, 13
commensalism, 477

commercial development, 360, 524
commercial fishing, 482–483, 493–500
communities. *See* marine communities
compass, 29, 34, 35, 36
compensation depth, 406, 407, 480
compound, 166
condensation theory, 8
conduction, 69
cone-shell toxins, 501
Congo Canyon, 119
Congo River, 119
conical net, 396
conservative constituents, 206
constructive interference, 296
consumers, 378
container cycle, 502–503, 509
container terminals, 502
continental crust, 67, 70
continental drift, 60–62, 73
continental margins
    active margins, 111, 115–116
    deposits within the sediments of, 157
    divisions of, 111
    neritic sediments, 146–148
    passive margins, 111, 114–116
    typical, 114
continental rises, 120
continental shelves
    complex, 117
    continental slopes and, 118–119
    natural resource exploration of, 116
    as seaward extensions of continents,
        111–118
    sediment, 114–115, 145
    submarine canyons, 119
    water depth over, 116
    width of, 115–116
continental slopes, 118–119, 120
continent-continent convergence, 82
continents, 60, 70, 111
convection, 69, 74
convection current, 73, 222, 225
convergent evolution, 373
convergent margins, 82
convergent plate boundaries, 77, 79–82
Cook, James, 25, 39–41, 42, 55
copepods, 389, 428, 430, 452, 467–469
copper, 519
coral animals, 353
coral bleaching, 525
coral reefs
    atoll formation, 43
    autotrophic dinoflagellates and, 407
    coral animals and, 353–355
    effects of environmental change on,
        524–525, 526
    effects of ocean acidification on, 211
    fishes, 539–540
    formation of, 422
    influence on environment, 459
    Kimbe Bay, Papua New Guinea, 454–455
    marine communities in, 466
corals, 421–423, 450
core, 68

core-mantle boundary region, 64
Coriolis effect
    atmospheric circulation cells, 226–227
    deep boundary currents and, 120
    deflection, 223–226, 245
    extratropical cyclone circulation, 234
    influence on tides, 324
    on other planets, 245
    surface currents, 251–253
    westward intensification, 258
    wind waves, 312
Coriolis, Gaspard Gustave de, 223
Corriente del Niño, 266
Coscinodiscus, 401
COSEE. *See* Centers for Ocean Sciences
    Education Excellence
cosmogenous sediments, 145
countercurrents, 258
Cousteau, Jacques, 50
covalent bonds, 166
crab fishers, 495
crabs, 427, 428, 508
Cramer, Alem London, xxi, xxv, 538
Cramer, John Wilson IV, xxv, 456, 502
crayfish, 428
Creation, 72
Cretans, 26
Crete, 25
crocodiles, 439–441
Cromwell Current, 258
Cromwell, Townsend, 258
crude oil, 515
crust, 12, 64, 67, 68, 82, 99, 143
Crustacea, 428
crustaceans, 493–495
crustal rocks, 98, 99
ctenophores, 470
Curie point, 88
Curiosity (rover), 18, 160
currents
    boundary currents, 253–258
    countercurrents, 258
    discovery of patterns of, 44
    eastern boundary currents, 257
    ebb current, 325
    flood current, 325
    Gulf Stream, 43–44
    longshore current, 348
    neritic sediments, 146
    nongeostrophic currents, 280
    plankton and, 396
    rogue waves, 296
    role of wind and gravity, 250
    sediment, 141
    surface currents, 250–253
    tidal currents, 325–327
    transverse currents, 257–258
    turbidity currents, 119, 120, 136, 148, 149,
        150
    undercurrents, 258
    use to generate power, 492
    velocities of, 142
    western boundary currents, 253–257, 258,
        280

echo sounders, 49, 73, 104–105, 471
ecology, 458
economics, 484
Ecteinascidin 743, 501
ectothermic, 384
eddies, 256–257, 569
Ediacara Hills, 419
Edison, Thomas, 104
eels, 436, 437
EEZ. *See* exclusive economic zones
Egypt, 27
Egyptians, 26
Ehrlich, Paul, 540
Ekman spiral, 252
Ekman transport, 252, 264, 407
Elcano, Juan Sebastián, 37
El Chichon, 269
electric eels, 433
electrical power, 280
electricity, 4
electrons, 166
element, 166
elephant seals, 445, 448
Elizabeth I, Queen of England, 55
Ellis, Kalen James, 538
El Niño, 232, 266–271, 365
El Niño/Southern Oscillation (ENSO),
   266–271
Emerita, 328, 464
Emiliania huxleyi, 214, 404
Empedocles, 200
Emperor Seamounts, 93
Enceladus, 17, 18
endothermic, 384
energy, 377–379
ENSO. *See* El Niño/Southern Oscillation
environmental issues
   climate change, 527–536
   habitat destruction, 523–527
   marine protected areas, 527
   pollutants, 512–523
   population growth, 540
   solutions, 536–538
environmental resistance, 458–459
EPA. *See* U.S. Environmental Protection
   Agency
epipelagic zone, 129
EPOXI (spacecraft), 13
equations
   calculation of gravitational attraction,
      316–317
   for expressing primary productivity, 395
   for measuring rate of productivity, 405
   velocity of tsunami, 307–308
   wavelength, 287–289
equator, 30
equatorial upwelling, 261, 263
equilibrium theory, 317
Eratosthenes of Cyrene, 27–28, 42
Eric the Red, 33
Erikson, Leif, 33
Eriocheir, 524
erosion, 71, 119
erosional coasts, 338–343

estuaries
   classification of, 355
   effects of development on, 360, 524
   effects of pollution on, 360, 523–524
   human interference, 360
   influence of water density and flow, 359–
      360
   marine communities in, 360, 464–465
   sediment in, 148
ethane, 19
Ethiopia, 78
Eukarya, 375, 555
eukaryotic food webs, 414
euphotic zone, 383, 390, 407, 452
Eurasian Plates, 82
Europa, 2, 17
Eurypharynx, 472
eustatic change, 337
eutrophication, 519, 520
evaporation, 171–172, 177, 205
evaporites, 153
Everglades National Park, Florida, 411
evolution, 15, 370–373, 390, 418–419, 458
excess volatiles, 201
exclusive economic zones (EEZs), 506–507
exoskeleton, 428
experiments, 4
exploration, 26–35
externalities, 484
extratropical cyclones, 233, 234–235
extremophiles, 413
Exxon Valdez oil spill, 516
Eyjafjallajökull volcano, 58–59, 162

Fahrenheit, 168
fault coasts, 343
faults, 71, 120
favorable traits, 371
femtoplankton, 398
Ferrel cells, 228, 229
Ferrel, William, 228
ferromanganese nodules, 157
fetch, 292–293
fisheries, 495–496, 540
fishes
   adaptations of, 435–439
   aquarium fishes, 539–540
   as biological resources, 493–501
   bony fishes, 435
   cartilaginous fishes, 434–435, 437
   consumption of, 508
   effects of drug trafficking on, 518
   effects of plastic/solid waste on, 521
   feeding and defense, 438–439
   gas exchange, 437–438
   gill membranes, 438
   groups of, 434–435
   jawless fishes, 434
   maintenance of level, 437
   as most abundant vertebrate, 433–435
   movement, shape, and propulsion,
      436–437
   pollutants and, 517, 519
fish farming, 498

Fissipedia, 448
fjord estuaries, 359–360
fjords, 343, 355, 358
flagella, 400, 403
flatworms, 423–424
flood current, 325
flood tide, 325
fog, 173
Folger, Timothy, 44
food, 378
food web, 379, 380, 525
foraminifera, 149, 152
foraminiferan, 147, 470
Forbes, Edward, 46
forced waves, 286, 316
Forchhammer, Georg, 209
Forchhammer's principle, 203, 206
foreshore, 346
formulas
   calculation of gravitational attraction,
      316–317
   for expressing primary productivity, 395
   for measuring rate of productivity, 405
   velocity of tsunami, 307–308
   wavelength, 287–289
fossils, 15, 60, 61, 76, 402, 419
fracking, 486–487
fracture zones, 95, 122, 124
Fram (ship), 47
France, 39, 329
Franklin, Benjamin, 44, 542
free waves, 286, 316
freezing method of desalination, 491
freezing point, 170
Fremont, John C., 43
French Polynesia, 39
freshwater, 491
fringing reefs, 354
frogfish, 416–417
front, 234
frontal storms, 235
frustule, 400
fully developed sea, 292
fundamental niche, 462
Fungi, 555
fusion, 9

Galápagos Islands, 122, 202, 204, 371, 473
galaxy, 7
Galileo (spacecraft), 17
Galveston, Texas, 351, 362
Gama, Vasco da, 29, 36
gamma ray bursts, 21
Ganges–Brahmaputra river system, 120, 352,
   353, 360
Ganges Delta, 325
Ganymede, 17
gas, 169, 170, 207–209
gas exchange, 437–438
Gastropoda, 424
gastropods, 425
genus, 377
geographical equator, 229
geologic time, 546

residence time, 206, 207
restoring force, 286
Reticulofenestra, 401
reverse osmosis desalination, 491
"reverse tribute" system, 35
Rhincodon, 435
Rhizophora, 411
Richter scale, 99
Ricketts, Ed, 479
ridley sea turtles, 440
Riftia, 473, 475
rift valley, 77
Ritter, William, 51
river-dominated deltas, 352
river mouths, 351–352
rivers, 144
River Severn, 325
River Thames, 304
RMS *Titanic* (passenger liner), 104, 110
robots, 108
rock gypsum, 153
rocks, age of, 14, 69, 73
rogue waves, 296, 298, 301
Ross, James Clark, 104
roundworms, 424
ROV *Argo*, 110
ROV *Hercules*, 110
ROVs. *See* remotely operated vehicles
Royal Society, 39, 45
rudder, 34
Russia, 63
R/V *Atlantis* (ship), 125
R/V *Chikyu* (ship), 49, 63, 488
R/V *Robert Gordon Sproul* (ship), 154

S ratio, 205
S waves. *See* secondary waves
Sagan, Carl, 55
Sahara desert, 244
sails, 34–35
Sala, Enric, 467
Salamis, 47
Salazar, Alex, 331
salinity
  average of worldwide ocean surface, 207,
    490
  components of, 201–203
  density stratification and, 182–184,
    271–272
  difference in average, 178
  dissolved inorganic solids and, 201, 203
  global warming and, 177
  influence on cell membrane, 385
  ocean surface conditions and, 177–179
  range in Chesapeake Bay, 360
  sea surface, 180–181
  seawater conductivity and, 209
  temperature and, 178
  water masses and, 184
salinometers, 205
salmon, 508
salps, 432, 470
salt, 490–491
salt glands, 439

salt marshes, 464–465
salt-wedge estuaries, 359
Samoa, 31
sampling bottles, 206
San Andreas Fault, 82, 87, 312, 343
sand, 142, 153, 157, 348–349, 489–490
sand and cobble beach marine communities,
    462–464
sand crab, 464
sand crabs, 328
sand dollars, 431
sand spits, 349–350
sandcastle worms, 501
San Francisco Bay, 355, 359
San Lucas Canyon, 119
Sargasso Sea, 402
satellite altimetry, 136–137
satellites, 52–53, 106–107, 129
saturation, 200
Saturn, 10, 17, 18, 245
scallops, 425
Scandinavia, 61
scattering, 188
science, 4
scientific method, 4–6
scientific names, 377
scientific notation, 543–544
Scotland, 61, 92
Scotoplanes, 473
Scripps Canyon, 136
Scripps, Ellen, 51
Scripps, E. W., 51
Scripps Institution of Oceanography, 50, 51,
    73, 156, 363
sea, 290
sea anemones, 140, 420, 421, 501
sea bed. *See* ocean floor
sea breezes, 230
sea caves, 339–341
sea cliffs, 339
sea cucumbers, 429, 431
sea grasses, 410, 464
sea-ice cover, 176
sea islands, 350–351
sea level
  average sea-surface height, 568
  change in sea-surface height, 568
  changes in, 117, 341–343, 355, 534
  effects on coastlines, 337–338, 350
  influence on continental shelves, 116
  submarine canyons and, 119
sea lions, 443, 445, 448, 479, 521
Sea of Cortez, 325
sea otters, 443, 448
sea power, 47
sea salt, 490–491, 508
sea stars, 429
sea-surface height, 568–569
sea urchin, 470
sea urchins, 429, 431
sea wasp, 422
Sea World, 504
seafloor, 562–567
seafloor spreading, 49, 73

seals, 443, 445–448, 521
seamounts, 125–126, 128, 129
Seasat (satellite), 52
seasons, 221
seawater
  acid-base balance, 209–213, 386
  carbon dioxide in, 208
  characteristics of, 201–206
  conservative constituents, 206
  density of, 173
  drinking, 390
  equivalence in concentration of, 545
  gas concentrations, 209
  gases and, 207–209
  gold in, 214
  latent heat of evaporation, 173
  major constituents of, 202
  minor and trace elements in, 209
  nitrogen in, 208
  nonconservative constituents, 206
  oxygen in, 208
  pressure of, 194
  processes that regulate constituents
    in, 204
  residence time, 206, 207
  resources obtained from, 214
  solubility of gases in, 386
  speed of sound in, 188
  thermal properties, 172–173
seaweeds
  aquarium, 524
  classification of, 410
  as marine resource, 410, 496–497
  as nonvascular organisms, 410
  as primary producers, 408–411
  primary productivity and, 395
SeaWiFS (satellite), 395
secondary consumers, 379
secondary waves (S waves), 63–66, 67,
    312–313
sediment
  accumulation rates, 140
  appearance, 140–141
  authigenic, 145
  average thickness of, 145–146
  beaches, 344–346
  biogenous, 144
  classification of, 142–146
  continental rises, 120
  continental shelves, 114–116
  cosmogenous, 145
  cycle, 143
  dating, 161
  deep-ocean, 120, 153–154
  deep ocean, 501
  deep-sea animals and, 161
  economic importance, 157
  general pattern on ocean floor, 147
  as historical records of ocean processes,
    155–157
  hydrogenous, 145, 152–153
  ice ages and, 148
  lithification, 148
  mangrove coasts, 355, 411

Swedish Vikings, 33
sweet potato, 33
swell, 290, 292
swordfish, 519
Sydney Harbour, Australia, 342
Syene, 27
symbiosis, 476–478
synthetic organic chemicals, 517–519
Systema Naturæ (Linnaeus), 375
Système International (SI), 542

table salt, 490, 508
Tahiti, 39
Taiwan, 33
tankers, 502, 516
tardigrade, 370
TauTona Mine, 62
taxonomy, 375
TEDs. *See* turtle exclusion devices
Teleostei, 435
telepresence, 108
temperature, 168, 544
temperature–salinity (T–S) diagram, 271
terranes, 96–97
Terra (satellite), 53, 223
terrestrial primary productivity, 395, 414
terrigenous sediments, 143, 149
territorial waters, 506
Texas, 485
Thalassiosira, 401
Thalassomedon, 441
thallus, 409, 410
Thames Estuary, 491
Thames tidal barrier, 304, 305
Tharp, Marie, 105
theories, 4, 21
thermal equator, 229
thermal equilibrium, 221
thermal inertia, 174–176, 194
thermocline, 183, 250
thermocline currents, 250
thermohaline circulation, 271–279
thermostatic properties, 174
Thilafushi, 536
Thomson, Charles Wyville, 45
Thomson, William (Lord Kelvin), 68
Thomson, Wyville, 541
tidal bore, 314–315, 325, 331
tidal currents, 325–327
tidal datum, 324
tidal day, 320
tidal friction, 327
tidal power, 329–330
tidal range, 324–325
tidal waves, 301, 303, 325
tide-dominated deltas, 352
tides
    affect on marine organisms, 328–329
    amphidromic points, 322–324
    astronomical tides, 321, 332
    definition of, 316
    disturbing forces for, 285
    diurnal tides, 323

dynamic theory of, 317, 322–327
equilibrium theory of, 317–321
forced waves formed by gravity and inertia,
    317–321
high tides, 319–320
influence of moon, 317–320, 321
influence of sun, 320–321
king tide, 332
lakes, 332
low tides, 319–320, 332
lunar tides, 320
meteorological tide, 322
neap tides, 321
prediction of, 327–328
semidiurnal mixed tides, 323
semidiurnal tides, 323
spring tides, 321, 332
tidal bore, 325, 331
tidal currents, 325–327
tidal datum, 324
tidal friction, 327
tidal power, 329–330, 492–493
tidal range, 324–325
tidal waves, 301, 303, 325–327
tide-related seiches, 305
Tigriopus, 389
Tigris-Euphrates river system, 360
time, 544, 546, 548–549
tin, 519
Titan, 18, 19
Tohoku, Japan earthquake and tsunami, 100
tombolo, 351–352
Tonga, 31, 40
toothed whales, 443–444
top consumers, 379
TOPEX/Poseidon (satellite), 52, 107, 193, 294,
    534
tornado, 233
trace elements, 201
trace metals, 382–383
tractive forces, 317
trade, 26–35
trade winds, 229, 232, 245
Traherne, Thomas, 541
transform faults, 82–83, 87, 95, 120, 124
transform plate boundaries, 77, 82–83
transportation, 501–504
Trash Vortex, 260
trawl net fishing, 497
trenches, 126–127, 130–131, 136
Trichechus, 449
Trieste (bathyscaphe), 49, 132
trilobites, 419
tripod fish, 472, 473
Troll-A (offshore oil drilling rig), 486
trophic pyramid, 379
tropical cyclones, 233, 236–240
tropical depressions, 236
tropical storms, 236
troposphere, 218
Truman, Harry, 505
trunkfish, 437
tsunami, 100, 285, 306–311, 524, 525
tubenoses, 442

tube worms, 473, 475
tuna, 435, 519
Tungurahua, 85
tunicates, 432
turbidites, 149
turbidity currents
    formation of, 150
    sediment transport and, 119, 120, 148
    submarine canyons and, 119
    turbidites, 149
    witnessing, 136
turbines, 329–330
Turner, J. M. W., 541
Turritopsis dohrnii, 451
Tursiops truncatus, 452
turtle exclusion devices (TEDs), 496
turtles, 439–441, 521
Typhoon Haiyan, 245
Typhoon Tip, 245
Typhoon Trami, 327
typhoons, 236

Undaria, 524
undercurrents, 258
uniform distribution, 459
uniformitarianism, 72
United Kingdom, 489
United Nations Conference on Law of the
    Sea, 505
United States Exploring Expedition, 42–43
United States National Research Council,
    533
unity, 370, 380
universe, age of, 14
University of California, 50
University of California, Santa Barbara, 559
University of Rhode Island, 110
University of Toronto, 74
upper mantle, 64
upwelling, 261–264, 267, 381, 405, 406, 407
Uranus, 10
Ursus maritimus, 448
U.S. Coast and Geodetic Survey, 46
U.S. Commission of Fish and Fisheries, 51
U.S. Department of Commerce, 51
U.S. Department of Defense, 53
U.S. Department of Energy, 492
U.S. Environmental Protection Agency
    (EPA), 536
U.S. Exclusive Economic Zone, 507
U.S. Fish and Wildlife Service, 258
U.S. Food and Drug Administration (FDA),
    519
U.S. Global Research Program, 535
U.S. Navy, 49, 51, 106, 137
U.S. Senate, 536
USNS *Bruce C. Heezen* (ship), 49
USS *Albatross* (survey ship), 46
Ussher, James, 72
USS *Nautilus* (submarine), 48
USS *Ramapo* (tanker), 294–295, 296
USS *Stewart* (ship), 104
USS *Thresher* (submarine), 301
USS *Vincennes* (ship), 43